INTERMEDIATE ALGEBRA
Concepts and Graphs
Second Edition

CHARLES P. MCKEAGUE

xyz textbooks

Intermediate Algebra: Concepts and Graphs

Second Edition

Charles P. McKeague

Publisher: XYZ Textbooks

Editorial Team: Anne Scanlan-Rohrer
  and Buddy Galletti

Design & Composition: Tessa Avila

Accuracy Checking: Jana Mooney

Proofreading: Laurie Gibson

Cover Photo © iStockphoto.com/LeeYiuTung

Printed in the United States of America

Full Version ISBN-13: 978-1-63098-207-2 / ISBN-10: 1-63098-207-5

For product information and technology assistance, contact us at
XYZ Textbooks, 1-877-745-3499

For permission to use material from this text or product,
e-mail: **info@mathtv.com**

XYZ Textbooks
1339 Marsh Street
San Luis Obispo, CA 93401
USA

For your course and learning solutions, visit **www.xyztextbooks.com**

Brief Contents

Contents

Preface to the Instructor

Intermediate Algebra: Concepts and Graphs 2/e fits the one-semester intermediate algebra course. The prerequisite is Elementary Algebra, taken in college or in high school.

The common sequence of topics is enriched with applications, study skills, and other features as detailed below. We suggest encouraging your students to make regular use of these features—our objective is to help students build a foundation of successful studying practices that will benefit them in their future courses.

This textbook includes more than just the book itself. While many students still prefer to use the printed book in their studies, the eBook extends the reach of the book, giving students and instructors access to a wide array of supporting tools. Some significant features include:

- The eBook includes free digital access to the Intermediate Algebra text, plus over 20 eBooks covering eight math courses.
- The text and eBook include free access to over 10,000 MathTV videos, with multiple tutorials for every example found in the text.
- Additionally, QR codes printed in each section provide students quick and easy connection to videos for those examples.

NEW TO THIS EDITION

Our main revisions include the following:

- Chapter 1 has been completely reorganized by topics instead of by operations, for a more mainstream organization and a slightly higher level of difficulty.
- A new section on union, intersection, and compound inequalities was added to Chapter 2.
- Material on piecewise functions and horizontal translations was added to Chapter 3.
- New sections on matrices and determinants, Cramer's rule, and using matrices to solve systems of equations were added to Chapter 4.
- Learning Objectives were added to every chapter and keyed to headings in the chapters.
- Applications problems were updated, revised, and added to each chapter.
- Spotlight on Success features added to each chapter offer students a variety of strategies for success.
- Cumulative Reviews were added at the end of chapters.

TEXTBOOK FEATURES

Every section has been written so that it can be discussed in a single class session. A clean layout and conversational style is used by the authors to make it easy for students to read. Important information, such as definitions and properties, are also highlighted so that it can easily be located and referenced by students.

In addition, the following features provide both instructors and students a vast array of resources that can be used to enrich the learning environment and promote student success.

Connecting Print and Digital: QR Codes

We want students to get the most out of their course materials, which is why we think of the textbook as a "toolbox" for students. QR codes are integrated throughout the textbook to connect the printed version to the digital assets easily. As students read their printed book, support material is quickly and conveniently available via one scan of the accompanying QR code. Students don't need to hunt, search, or scroll to find an example video. We find that this direct link to additional support is the appropriate level of technology for all students taking this course; it integrates technology without being overwhelming.

Spotlight on Success

The "Spotlight on Success" feature offers students a variety of strategies from many different sources, and highlights both the challenges and triumphs common to students taking mathematics courses. Students and instructors share their journey and suggestions for success in a math course. These stories highlight the featured individuals' unique and relatable approaches to learning. Many of the spotlights feature the same peer tutors that students will see on the MathTV videos that accompany the text.

SPOTLIGHT ON SUCCESS

Cynthia | student instructor

Each time we face our fear, we gain strength, courage, and confidence in the doing.

—Unknown

I must admit, when it comes to math, it takes me longer to learn the material compared to other students. Because of that, I was afraid to ask questions, especially when it seemed like everyone else understood what was going on. Because I wasn't getting my questions answered, my quiz and exam scores were only getting worse. I realized that I was already paying a lot to go to college and that I couldn't afford to keep doing poorly on my exams. I learned how to overcome my fear of asking questions by studying the material before class, and working on extra problem sets until I was confident enough that at least I understood the main concepts. By preparing myself beforehand, I would often end up answering the question myself. Even when that wasn't the case, the professor knew that I tried to answer the question on my own. If you want to be successful, but you are afraid to ask a question, try putting in a little extra time working on problems before you ask your instructor for help. I think you will find, like I did, that it's not as bad as you imagined it, and you will have overcome an obstacle that was in the way of your success.

Learning Objectives

We have provided a list of the section's important concepts at the beginning of each section, noted with a green badge shape, which is reiterated beside each learning objective header in the section.

OBJECTIVES

A   Factor the greatest common factor from a polynomial.

B   Factor polynomials by grouping.

C   Factor a trinomial whose leading coefficient is 1.

D   Factor a trinomial whose leading coefficient is other than 1 by trial and error.

E   Factor a perfect square trinomial.

F   Factor the difference of two squares.

G   Factor the sum or difference of two cubes.

Example Problems

Example problems, with work, explanations, and solutions are provided for every learning objective.

EXAMPLE 1

Factor the greatest common factor from $8a^3 - 8a^2 - 48a$.

SOLUTION The greatest common factor is $8a$. It is the largest monomial that divides each term of the polynomial. We can write each term in the polynomial as the product of $8a$ and another monomial. Then, we apply the distributive property to factor $8a$ from each term:

$$8a^3 - 8a^2 - 48a = 8a(a^2) - 8a(a) - 8a(6)$$

$$= 8a(a^2 - a - 6)$$

Blueprint for Problem Solving

We provide students with a step-by-step guide for solving application problems in a clear and efficient manner.

BLUEPRINT FOR PROBLEM SOLVING
Using a System of Equations • • • • • • • • • • • • • • • •

Step 1: *Read* the problem, and then mentally *list* the items that are known and the items that are unknown.

Step 2: *Assign variables* to each of the unknown items. That is, let $x =$ one of the unknown items and $y =$ the other unknown item (and $z =$ the third unknown item, if there is a third one). Then *translate* the other *information* in the problem to expressions involving the two (or three) variables.

Step 3: *Reread* the problem, and then *write a system of equations*, using the items and variables listed in steps 1 and 2, that describes the situation.

Getting Ready for Class

Simple writing exercises for students to answer after reading each section increases their understanding of key concepts.

GETTING READY FOR CLASS

After reading through the preceding section, respond in your own words and in complete sentences.

A. Describe the commutative property of multiplication.
B. Give definitions for each of the following:
  1. The opposite of a number
  2. The absolute value of a number
C. Explain why subtraction and division are not commutative operations.
D. Explain why zero does not have a multiplicative inverse.

End-of-Chapter Summaries, Reviews, Cumulative Reviews, and Tests

Problems at the end of each chapter provide students with a comprehensive review of each chapter's concepts.

Order of Operations [1.1]

2. $10 + (2 \cdot 3^2 - 4 \cdot 2)$
$= 10 + (2 \cdot 9 - 4 \cdot 2)$
$= 10 + (18 - 8)$
$= 10 + 10$
$= 20$

When evaluating a mathematical expression, we will perform the operations in the following order, beginning with the expression in the innermost parentheses or brackets and working our way out.

1. Simplify all numbers with exponents, working from left to right if more than one of these numbers is present.
2. Then, do all multiplications and divisions left to right.
3. Finally, perform all additions and subtractions left to right.

Chapter 5 Test

Reduce to lowest terms. [5.1]

1. $\dfrac{x^2 - y^2}{x - y}$ **2.** $\dfrac{2x^2 - 5x + 3}{2x^2 - x - 3}$

Multiply and divide as indicated. [5.2]

3. $\dfrac{a^2 - 16}{5a - 15} \cdot \dfrac{10(a - 3)^2}{a^2 - 7a + 12}$

4. $\dfrac{a^4 - 81}{a^3 + 9} \div \dfrac{a^2 - 8a + 15}{4a - 20}$

5. $\dfrac{x^3 - 8}{2x^2 - 9x + 10} \cdot \dfrac{x^2 + 2x + 4}{x^2 + x - 15}$

Add and subtract as indicated. [5.3]

Solve the following applications. Be sure to show the equation in each case. [5.6]

19. Number Problem What number must be subtracted from the denominator of $\frac{16}{23}$ to make the result $\frac{1}{2}$?

20. Speed of a Boat The current of a river is 2 miles per hour. It takes a motorboat a total of 3 hours to travel 8 miles upstream and return 8 miles downstream. What is the speed of the boat in still water?

21. Filling a Pool An inlet pipe can fill a pool in 10 hours,

Additional textbook features include the following:

Margin Notes   In many sections, you will find pale green notes in the margin that correspond to the section's current discussion. These notes contain important reminders, fun facts, and helpful advice for working through problems.

Colored Boxes   Throughout each section, we have highlighted important definitions, properties, rules, how to's, and strategies in colored boxes for easy reference.

Blueprint for Problem Solving   We provide students with a step-by-step guide for solving application problems in a clear and efficient manner.

Facts from Geometry   Throughout the book, we highlight specific geometric topics and show students how they apply to the concepts they are learning in the course.

Using Technology   Throughout the book, we provide students with step-by-step instructions for how to work some problems on a scientific or graphing calculator.

SUPPLEMENTS FOR THE INSTRUCTOR

Please contact your sales representative or see **xyztextbooks.com/instructors** for more info.

MathTV.com   Every example in every XYZ Textbook is worked on video by multiple student instructors. Students benefit from seeing multiple approaches and gain confidence in learning from their peers. These videos can be used to supplement class time or can be used for alternative instruction such as a flipped classroom approach.

eBook   Through the website, instructors can access the eBook for this course as well as other texts. Using the My Bookshelf feature, you can view the text as well as the resources available to both you and the student.

Complete Solutions Manual   Available online, this manual contains complete solutions to all the exercises in the problem sets.

Printable Test Items   Instructors can choose from a bank of pre-created tests. These are provided in an editable format so that they can be tailored to your teaching style and testing preferences.

Computerized Test Banks   Available for this text and can be found at xyztextbooks. com/instructor.

Matched Problems Worksheets   These worksheets integrate the textbook content and MathTV videos in order to check student progress or provide an alternative to classroom instruction.

PowerPoint Presentations   These presentations can be opened in PowerPoint, Apple Keynote or Google Slides and have been provided as a template for classroom lectures. Instructors can add examples that they want to use in class or use these to supplement their presentation of the material.

SUPPLEMENTS FOR THE STUDENT

MathTV.com   Students have access to math instruction 24 hours a day, seven days a week. Within a few clicks or a scan of the QR code in the textbook, students can access multiple videos that correspond to each example in the text. Many of the video examples are also presented by Spanish-speaking instructors.

XYZ eBook   Students who purchase the textbook have one year of free access to the eBook. Students can also choose to purchase the eBook instead of the printed text. The eBook is tightly integrated with MathTV.com so that students can read the book and watch videos for each example.

QR Codes   QR codes quickly and easily connect the textbook to digital resources. By scanning the QR code located in each section, students will be taken directly to the accompanying MathTV videos for the examples in that section.

Student Solutions Manual   The Student Solutions Manual contains complete solutions to all the odd-numbered exercises in the text. It is available for purchase separately.

ACKNOWLEDGMENTS

We would like to thank Joanna Jacobs of North Platte Community College for updating and revising applications and providing several new applications for each chapter.

We would like to thank the following instructors for providing new applications: Thomas Buss, Community College of Rhode Island; Joseph Crowley, Community College of Rhode Island; Mark Harbison, Sacramento City College.

The following instructors used the first edition and gave us important feedback for the second edition:

Alyson Wilson, *Bowling Green State University— Firelands Campus*

Dan Rothe, *Alpena Community College*

Frank Post, *South Seattle College*

Ignacio Alarcon, *Santa Barbara City College*

Jerry Barkley, *Glen Oaks Community College*

Joanna Jacobs, *North Platte Community College*

Knarik Manukyan, *American River College*

Lisa Rosenbeck, *Alpena Community College*

Mark Harbison, *Sacramento City College*

Tal Avitzur, *Santa Barbara City College*

Matt Register, *American River College*

Meghan Cameron, *Alpena Community College*

Michael Sandelin, *Glen Oaks Community College*

Mohammed Elhakimi, *Santa Barbara City College*

Nicholas Holton, *Alpena Community College*

Ramona Skelly, *Community College of Rhode Island*

Tamara Zilar, *Tulsa Community College— Northeast Campus*

Thomas Buss, *Community College of Rhode Island*

Kenneth Sowden, *West Hills College Coalinga*

A Note to Students

WE WANT YOU TO SUCCEED. ─────────

Welcome to the XYZ Textbooks/MathTV community! We are dedicated to your success in this course. As you will see as you progress through this book, and access the other tools we have for you, we are different than other publishers. Here's how:

Our Authors Are Experienced Teachers   We believe the best textbooks are written by the experienced instructors. These individuals can draw on their background teaching in the classroom and online learning environments and incorporate that expertise into our textbooks. Our objective is to provide the best instruction students can get in written form, produced by award-winning, experienced instructors.

Innovative Products   The foundation of our products is the textbook, which is also a source of tools that you will need to do well in your math course. These include eBooks, videos, worksheets, and a variety of ways to access these resources, from QR codes built into our books, to our MathTV Mobile site.

Peer Tutors   Learning mathematics can be a challenge. Sometimes your class time and textbook are not enough. We understand that, which is why we created MathTV, providing you with a set of instructional videos by students just like you, who have found a way to master the same material you are studying. Since these videos are available on the internet, you can access them whenever and wherever you may be studying. You'll also get to see how your peers solve each problem, sometimes offering a different view from how an instructor solves the problem, and other times, solving the problem in the same way your instructor does, giving you confidence that that is the way the problem should be solved.

Fair Prices   We're small, independently owned, and independently run. Why does that matter to you? Because we do not have the overhead and expenses that the larger publishers have. Yes, we want to be a profitable business, but we believe that we can keep our prices reasonable and still give you everything you need to be successful. Also, we want you to use this book, and the best way to make sure that happens is to make it affordable.

Unlimited Access   When you purchase one of our products, we give you access to all of our products. Why? Because everything you need to know about math is not contained in one book. Suppose you need to review a topic from a math course you completed previously? No problem. Suppose you want to see an alternate approach? It's all yours. As a member of our XYZ Textbooks/MathTV community, you have access to everything we produce, including all our eBooks.

WE KNOW YOU CAN DO IT. ━━━━━━━━━━━━

We believe in you.   We have seen students with all varieties of backgrounds and levels in mathematics do well in the courses we supply books and materials for. In fact, we have never run across a student who could not be successful in algebra. And that carries over to you: We believe in you; we believe you can be successful in whatever math class you are taking. Our job is to supply you with the tools you can use to attain success; you supply the drive and ambition.

We know college can be difficult.   It is not always the material you are studying that makes college difficult. We know that many of you are working, some part time, some full time. We know many of you have families to support, or look after. We understand that your time can be limited. We take all this to heart when we create the materials we think you will need. For example, we make our videos available on your smartphone, tablet, and on the internet. That way, no matter where you are, you will have access to help when you get stuck on a problem.

We believe in what we do.   We are confident that you will see the value in the text and resources we have created. That's why the first chapter in every one of our eBooks is free, and so are all the resources that come with it. We want you to try us out for free. See what you think. We wouldn't do that if we didn't believe in what we do here.

HOW TO BE SUCCESSFUL IN MATHEMATICS ━━━━━━

Mathematics is a challenging subject. Often, it includes abstract concepts that require critical thinking. Concepts also build on one another, requiring that you maintain skills and retain knowledge that you learned in the past. As a result, learning a topic in mathematics isn't always accomplished the first time through the material. If you don't understand a topic the first time you see it, that's perfectly normal. Our advice, though, is to stick with it! Understanding mathematics takes time. You may find that you need to read over new material a number of times before you can begin to work problems. The process of understanding requires reading the book, studying the examples, working problems, and getting your questions answered.

Here are some additional suggestions that will help you succeed in mathematics.

1. If you are in a lecture class, be sure to attend all class sessions on time.   You simply will not know exactly what went on in class unless you were there. Missing class and then expecting to find out what went on from someone else is not a good strategy. Make the time to be there—and to be attentive.

2. Read the book.   It is best to read the section that will be covered in class beforehand. It's OK if you don't fully understand everything you read! Reading in advance at least gives you a sense of what will be discussed, which puts you in a good position when you get to class.

3. Work problems every day and check your answers.   One secret to success in mathematics is working problems. The more problems you work, the better you will perform. It's really that simple. The answers to the odd-numbered problems in each section are given in the back of the book. When you have finished an assignment, be sure to compare your answers with those in the book. If you have made a mistake, find out what it is, and try to correct it.

4. Do it on your own.   Having someone else show you how to work a problem is not the same as working the problem yourself. It is absolutely OK to get help when you are stuck. As a matter of fact, it is a good idea. Just be sure you do the work yourself. After all, when it's test time, it's all you! Get confident in every problem type, and you will do well.

5. Review every day.   After you have finished the problems your instructor has assigned, take another 15 minutes and review a section you have already completed. This simple trick works wonders. Studies have shown, the more you review, the longer you will retain the material you have learned. Since math topics build upon one another, this will help you throughout the term.

6. Don't expect to understand every new topic the first time you see it. Sometimes it will come easy and sometimes it won't. Don't beat yourself up over it—this is perfectly normal. Expecting to understand each new topic the first time you see it can lead to disappointment and frustration. The process of understanding takes time and practice. It requires that you read the book, work problems, and get your questions answered.

7. Spend as much time as it takes for you to master the material.   What's the exact amount of time you need to spend on mathematics to master it? There's no way to know except to do it. You will find out as you go what is or isn't enough time for you. Some sections may take less time, and some may take more. If you end up spending two or more hours on each section, OK. Then that's how much time it takes; trying to get by with less will not work.

8. Know what resources are available to you.   We have provided a multitude of resources within this text to help you succeed in this course. There are probably also resources at your college to help you as well. Does your instructor have time set aside to help students? Does your campus have a tutoring center with tutors? We want to make sure you have all the resources you need to be successful in your math course.

9. Relax.   It's probably not as difficult as you think. You might get stuck at points. That's OK, everyone does. Take a break if you need to. Seek some outside help. Watch a MathTV video of the problem. There is a solution, and you will find it—even if it takes a while.

1

Basic Properties and Definitions

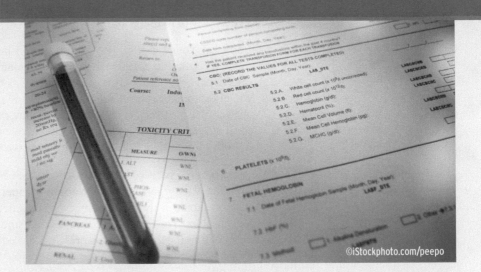

©iStockphoto.com/peepo

The following table and graph show how the concentration of a popular antidepressant changes over time, once the patient stops taking it. In this particular case, the concentration in the patient's system is 80ng/ml (nanograms per milliliter) when the patient stops taking the antidepressant, and the half-life of the antidepressant is 5 days.

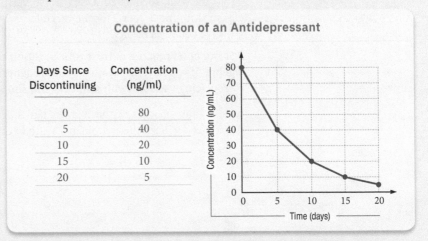

Concentration of an Antidepressant

Days Since Discontinuing	Concentration (ng/ml)
0	80
5	40
10	20
15	10
20	5

The half-life of a medication tells how quickly the medication is eliminated from a person's system: Medications with a long half-life are eliminated slowly, whereas those with a short half-life are more quickly eliminated. Half-life is the key to constructing the preceding table and graph.

In this chapter we will begin working with tables and charts. Many of the charts you will see here are taken directly from the media. As we progress through the book we will learn more about tables and charts and how they are connected.

STUDY SKILLS

The study skills for this chapter cover the way you approach new situations in mathematics. The first study skill is a point of view you hold about your natural instincts for what does and doesn't work in mathematics. The second study skill gives you a way of testing your instincts.

1. **Don't Let Your Intuition Fool You**   As you become more experienced and more successful in mathematics you will be able to trust your mathematical intuition. For now, though, it can get in the way of your success. For example, if you ask some students to "subtract 3 from -5" they will answer -2 or 2. Both answers are incorrect, even though they may seem intuitively true. Likewise, some students will expand $(a + b)^2$ and arrive at $a^2 + b^2$, which is incorrect. In both cases, intuition leads directly to the wrong answer.

2. **Test Properties of Which You Are Unsure**   From time to time, you will be in a situation where you would like to apply a property or rule, but you are not sure it is true. You can always test a property or statement by substituting numbers for variables. For instance, I always have students who rewrite $(x + 3)^2$ as $x^2 + 9$, thinking that the two expressions are equivalent. The fact that the two expressions are not equivalent becomes obvious when we substitute 10 for x in each one.

    When $x = 10$, the expression $(x + 3)^2$ is $(10 + 3)^2 = 13^2 = 169$

    When $x = 10$, the expression $x^2 + 9 = 10^2 + 9 = 100 + 9 = 109$

When you test the equivalence of expressions by substituting numbers for the variable, make it easy on yourself by choosing numbers that are easy to work with, such as 10. Don't try to verify the equivalence of expressions by substituting 0, 1, or 2 for the variable, as using these numbers will occasionally give you false results.

It is not good practice to trust your intuition or instincts in every new situation in algebra. If you have any doubt about the generalizations you are making, test them by replacing variables with numbers and simplifying.

OBJECTIVES

A Translate between phrases written in English and expressions written in symbols.

B Expand and multiply exponents.

C Simplify expressions using the order of operations.

D Find the value of an algebraic expression.

E Solve problems involving sets and subsets.

F Locate numbers on the real number line.

G List the subsets of the real numbers.

H Factor numbers into a product of primes.

©iStockphoto.com/berkay

1.1

The diagram below is called a *bar chart*. This particular chart shows the net price of a popular intermediate algebra textbook. (The net price is the price the bookstore pays for the book.)

From the chart, we can find many relationships between numbers. We may notice that the price of the third edition was less than the price of the fourth edition. In mathematics we use symbols to represent relationships between quantities. If we let $P(n)$ represent the price of the nth edition, then the relationship just mentioned, between the price of the third edition and the price of the fourth edition, can be written this way:

$$P(3) < P(4)$$

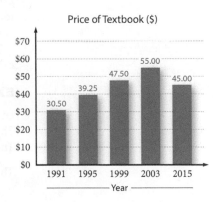

Price of Textbook ($)

30.50 · 39.25 · 47.50 · 55.00 · 45.00

1991 1995 1999 2003 2015

Year

A Phrases into Symbols

This section is, for the most part, simply a list of many of the basic symbols and definitions we will be using throughout the book.

Operation Symbols

Operation	In Symbols	In Words
Addition	$a + b$	The sum of a and b
Subtraction	$a - b$	The difference of a and b
Multiplication	ab, $a \cdot b$, $a(b)$, $(a)b$, $(a)(b)$, or $a \times b$	The product of a and b
Division	$a \div b$, a/b, or $\dfrac{a}{b}$	The quotient of a and b

Comparison Symbols

In Symbols	In Words
$a = b$	a is equal to b
$a \neq b$	a is not equal to b
$a < b$	a is less than b
$a \leq b$	a is less than or equal to b
$a \nless b$	a is not less than b
$a > b$	a is greater than b
$a \geq b$	a is greater than or equal to b
$a \ngtr b$	a is not greater than b
$a \Leftrightarrow b$	a is equivalent to b

The key words are *sum*, *difference*, *product*, and *quotient*. They are used frequently in mathematics. For instance, we may say the product of 3 and 4 is 12. We mean both the statements $3 \cdot 4$ and 12 are called the products of 3 and 4. The important idea here is that the word *product* implies multiplication, regardless of whether it is written $3 \cdot 4$, 12, 3(4), (3)4, or 3×4.

The following example shows how we translate sentences written in English into expressions written in symbols.

EXAMPLE 1

In English	**In Symbols**
The sum of x and 5 is less than 2.	$x + 5 < 2$
The product of 3 and x is 21.	$3x = 21$
The quotient of y and 6 is 4.	$\dfrac{y}{6} = 4$
Twice the difference of b and 7 is greater than 5.	$2(b - 7) > 5$
The difference of twice b and 7 is greater than 5.	$2b - 7 > 5$

B Exponents

Consider the expression 3^4. The 3 is called the *base*, and the 4 is called the *exponent*. The exponent 4 tells us the number of times the base appears in the product. That is,

$$3^4 = 3 \cdot 3 \cdot 3 \cdot 3 = 81$$

The expression 3^4 is said to be in *exponential form*, whereas $3 \cdot 3 \cdot 3 \cdot 3$ is said to be in *expanded form*.

EXAMPLE

Expand and multiply.

2. $5^2 = 5 \cdot 5 = 25$ Base 5, exponent 2

3. $2^5 = 2 \cdot 2 \cdot 2 \cdot 2 \cdot 2 = 32$ Base 2, exponent 5

4. $4^3 = 4 \cdot 4 \cdot 4 = 64$ Base 4, exponent 3

C Order of Operations

It is important when evaluating arithmetic expressions in mathematics that each expression have only one answer in reduced form. Consider the expression

$$3 \cdot 7 + 2$$

If we find the product of 3 and 7 first, then add 2, the answer is 23. On the other hand, if we first combine the 7 and 2, then multiply by 3, we have 27. The problem seems to have two distinct answers depending on whether we multiply first or add first. To avoid this situation, we will decide that multiplication in a situation like this will always be done before addition. In this case, only the first answer, 23, is correct.

Here is the complete set of rules for evaluating expressions.

RULE *Order of Operations*

· ·

When evaluating a mathematical expression, we will perform the operations in the following order, beginning with the expression in the innermost parentheses or brackets and working our way out.
1. Simplify all numbers with exponents, working from left to right if more than one of these expressions is present.
2. Then, do all multiplications and divisions left to right.
3. Perform all additions and subtractions left to right.

EXAMPLE

Simplify each expression using the rule for order of operations.

5. $5 + 3(2 + 4) = 5 + 3(6)$ Simplify inside parentheses.

              $= 5 + 18$ Then, multiply.

              $= 23$ Add.

6. $5 \cdot 2^3 - 4 \cdot 3^2 = 5 \cdot 8 - 4 \cdot 9$ Simplify exponents left to right.

              $= 40 - 36$ Multiply left to right.

              $= 4$ Subtract.

7. $20 - (2 \cdot 5^2 - 30) = 20 - (2 \cdot 25 - 30)$ Simplify inside parentheses,

$ = 20 - (50 - 30)$ exponents first, then multiply,

$ = 20 - (20)$ and finally subtract.

$ = 0$

8. $40 - 20 \div 5 + 8 = 40 - 4 + 8$ Divide first.

$ = 36 + 8$ Then, add and subtract left to right.

$ = 44$

Ⓓ Finding the Value of an Algebraic Expression

An algebraic expression is a combination of numbers, variables, and operation symbols. For example, each of the following is an algebraic expression.

$$5x \qquad x^2 + 5 \qquad 4a^2b^3 \qquad 5t^2 - 6t + 3$$

An expression such as $x^2 + 5$ will take on different values depending on what number we substitute for x. For example:

When	$x = 3$	When	$x = 7$
the expression	$x^2 + 5$	the expression	$x^2 + 5$
becomes	$3^2 + 5 = 9 + 5$	becomes	$7^2 + 5 = 49 + 5$
	$= 14$		$= 54$

As you can see, our expression is 14 when x is 3, and when x is 7, our expression is 54. In each case the value of the expression is found by replacing the variable with a number.

EXAMPLE 9

Evaluate the expression $a^2 + 8a + 16$ when a is 0, 1, 2, 3, and 4.

SOLUTION   We can organize our work efficiently by using a table.

When a is	The expression $a^2 + 8a + 16$ becomes
0	$0^2 + 8 \cdot 0 + 16 = 0 + 0 + 16 = 16$
1	$1^2 + 8 \cdot 1 + 16 = 1 + 8 + 16 = 25$
2	$2^2 + 8 \cdot 2 + 16 = 4 + 16 + 16 = 36$
3	$3^2 + 8 \cdot 3 + 16 = 9 + 24 + 16 = 49$
4	$4^2 + 8 \cdot 4 + 16 = 16 + 32 + 16 = 64$

EXAMPLE 10

Find the value of the expression $3x + 4y + 5$ when x is 6 and y is 7.

SOLUTION We substitute the given values of x and y into the expression and simplify the result:

When $\qquad x = 6$ and $y = 7$

the expression $\qquad 3x + 4y + 5$

becomes $\qquad 3 \cdot 6 + 4 \cdot 7 + 5 = 18 + 28 + 5$

$\qquad\qquad\qquad\qquad\qquad\quad = 51$

The next topic we cover can be considered the starting point for all the branches of mathematics.

Sets and Operations with Sets

> **DEFINITION** *set*
>
> A *set* is a collection of objects or things. The objects in the set are called *elements,* or *members,* of the set.

Sets are usually denoted by capital letters and elements of sets by lowercase letters. We use braces, { }, to enclose the elements of a set.

To show that an element is contained in a set we use the symbol $\in$. That is,

$$x \in A \text{ is read "}x \text{ is an element (member) of set } A\text{"}$$

For example, if A is the set $\{1, 2, 3\}$, then $2 \in A$. On the other hand, $5 \notin A$ means 5 is not an element of set A.

> **DEFINITION** *subset*
>
> Set A is a *subset* of set B, written $A \subset B$ if every element in A is also an element of B. That is,
>
> $$A \subset B \qquad \text{if and only if} \qquad A \text{ is contained in } B$$

> **DEFINITION** *empty/null set*
>
> The set with no members is called the *empty,* or *null, set.* It is denoted by the symbol $\varnothing$. The empty set is considered a subset of every set.

Two basic operations are used to combine sets — union and intersection.

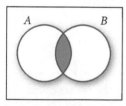

$A \cup B$

FIGURE 1   *The union of two sets*

DEFINITION *union*

The **union** of two sets A and B, written $A \cup B$, is the set of all elements that are either in A or in B, or in both A and B. The key word here is *or*. For an element to be in $A \cup B$, it must be in A or B. In symbols, the definition looks like this:

$$x \in A \cup B \qquad \text{if and only if} \qquad x \in A \text{ or } x \in B$$

$A \cap B$

FIGURE 2   *The intersection of two sets*

DEFINITION *intersection*

The **intersection** of two sets A and B, written $A \cap B$, is the set of elements in both A and B. The key word in this definition is the word *and*. For an element to be in $A \cap B$, it must be in both A and B. In symbols,

$$x \in A \cap B \qquad \text{if and only if} \qquad x \in A \text{ and } x \in B$$

NOTE The diagrams shown here are called *Venn diagrams* after John Venn (1834 – 1923). They can be used to visualize operations with sets. The region inside the circle labeled A is set A; the region inside the circle labeled B is set B.

EXAMPLE

Let $A = \{1, 3, 5\}$, $B = \{0, 2, 4\}$, and $C = \{1, 2, 3, \ldots\}$. Then

11. $A \cup B = \{0, 1, 2, 3, 4, 5\}$

12. $A \cap B = \varnothing$ (*A* and *B* have no elements in common.)

13. $A \cap C = \{1, 3, 5\} = A$

14. $B \cup C = \{0, 1, 2, 3, \ldots\}$

Another notation we can use to describe sets is called *set builder notation*. Here is how we write our definition for the union of two sets A and B using set builder notation:

$$A \cup B = \{x \mid x \in A \quad \text{or} \quad x \in B\}$$

The right side of this statement is read "the set of all x such that x is a member of A or x is a member of B." As you can see, the vertical line after the first x is read "such that."

EXAMPLE 15

If $A = \{1, 2, 3, 4, 5, 6\}$, find $C = \{x \mid x \in A \text{ and } x \geq 4\}$.

SOLUTION We are looking for all the elements of A that are also greater than or equal to 4. They are 4, 5, and 6. Using set notation, we have

$$C = \{4, 5, 6\}$$

The set we will work with most often in this book is the set of real numbers. To develop the real numbers, we start with the real number line.

F The Real Number Line

The real number line is constructed by drawing a straight line and labeling a convenient point with the number 0. Positive numbers are in increasing order to the right of 0; negative numbers are in decreasing order to the left of 0. The point on the line corresponding to 0 is called the *origin*.

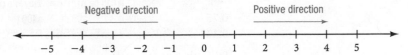

The numbers on the number line increase in size as we move to the right. When we compare the size of two numbers on the number line, the one on the left is always the smaller number.

The numbers associated with the points on the line are called *coordinates* of those points. Every point on the line has a number associated with it. The set of all these numbers makes up the set of real numbers.

> **NOTE** In this book, we will refer to real numbers as being on the real number line. Actually, real numbers are *not* on the line; only the points representing them are on the line. We can save some writing, however, if we simply refer to real numbers as being on the number line.

> **DEFINITION** *real number*
>
> A *real number* is any number that is the coordinate of a point on the real number line.

EXAMPLE 16

Locate the numbers -4.5, -0.75, $\frac{1}{2}$, $\sqrt{2}$, π, and 4.1 on the real number line.

SOLUTION

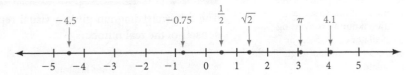

G Subsets of the Real Numbers

Next, we consider some of the more important subsets of the real numbers. Each set listed here is a subset of the real numbers:

$$\textit{Counting (or natural) numbers} = 1, 2, 3, \ldots$$

$$\textit{Whole numbers} = 0, 1, 2, 3, \ldots$$

$$\textit{Integers} = \ldots, -3, -2, -1, 0, 1, 2, 3, \ldots$$

$$\textit{Rational numbers} = \left\{ \frac{a}{b} \,\middle|\, a \text{ and } b \text{ are integers}, b \neq 0 \right\}$$

Remember, the notation used to write the rational numbers is read "the set of numbers a/b, such that a and b are integers and b is not equal to 0." Any number that can be written in the form

$$\frac{\text{Integer}}{\text{Integer}}$$

is a rational number. Rational numbers are numbers that can be written as the ratio of two integers. Each of the following is a rational number:

$\frac{3}{4}$ Because it is the ratio of the integers 3 and 4.

-8 Because it can be written as the ratio of -8 to 1.

0.75 Because it is the ratio of 75 to 100 (or 3 to 4 if reduced to lowest terms).

$0.333\ldots$ Because it can be written as the ratio of 1 to 3.

Still other numbers on the number line are not members of the subsets we have listed so far. They are real numbers, but they cannot be written as the ratio of two integers. That is, they are not rational numbers. For that reason, we call them irrational numbers.

$$\textbf{\textit{Irrational numbers}} = \{x \mid x \text{ is real, but not rational}\}$$

The following are irrational numbers:

$$\sqrt{2}, \qquad -\sqrt{3}, \qquad 4 + 2\sqrt{3}, \qquad \pi, \qquad \frac{\sqrt{3}}{2}$$

EXAMPLE 17

For the set $\{-5, -3.5, 0, \frac{3}{4}, \sqrt{3}, \sqrt{5}, 9\}$, list the numbers that are

a. whole numbers   **d.** irrational numbers

b. integers   **e.** real numbers

c. rational numbers

SOLUTION

a. Whole numbers: $0, 9$ **d.** Irrational numbers: $\sqrt{3}, \sqrt{5}$

b. Integers: $-5, 0, 9$ **e.** They are all real numbers.

c. Rational numbers: $-5, -3.5, 0, \frac{3}{4}, 9$

NOTE We can find decimal approximations to some irrational numbers by using a calculator or a table. For example, on an eight-digit calculator

$$\sqrt{2} = 1.4142136$$

This is not exactly $\sqrt{2}$ but simply an approximation of it. There is no decimal that gives $\sqrt{2}$ exactly.

The following diagram gives a visual representation of the relationships among subsets of the real numbers.

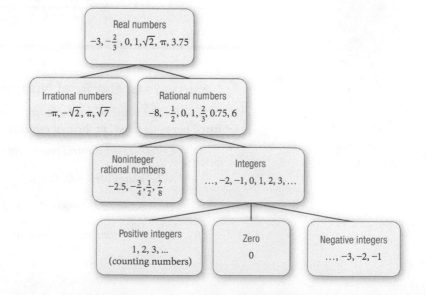

🔢 Prime Numbers and Factoring

The following diagram shows the relationship between multiplication and factoring:

$$\text{Multiplication}$$

$$\text{Factors} \longrightarrow 3 \cdot 4 = 12 \longleftarrow \text{Product}$$

$$\text{Factoring}$$

When we read the problem from left to right, we say the product of 3 and 4 is 12. Or we multiply 3 and 4 to get 12. When we read the problem in the other direction, from right to left, we say we have *factored* 12 into 3 times 4, or 3 and 4 are *factors* of 12.

The number 12 can be factored still further:

$$12 = 4 \cdot 3$$
$$= 2 \cdot 2 \cdot 3$$
$$= 2^2 \cdot 3$$

The numbers 2 and 3 are called *prime* factors of 12 because neither can be factored any further.

DEFINITION *factor*

If a and b represent integers, then a is said to be a ***factor*** (or divisor) of b if a divides b evenly — that is, if a divides b with no remainder.

DEFINITION *prime number*

A ***prime number*** is any positive integer larger than 1 whose only positive factors (divisors) are itself and 1. An integer greater than 1 that is not prime is said to be ***composite***.

Here is a list of the first few prime numbers:

Prime numbers = {2, 3, 5, 7, 11, 13, 17, 19, 23, 29, 31, 37, 41, . . .}

When a number is not prime, we can factor it into the product of prime numbers. To factor a number into the product of primes, we simply factor it until it cannot be factored further.

EXAMPLE 18

Factor 525 into the product of primes.

SOLUTION Because 525 ends in 5, it is divisible by 5:

$$525 = 5 \cdot 105$$
$$= 5 \cdot 5 \cdot 21$$
$$= 5 \cdot 5 \cdot 3 \cdot 7$$
$$= 3 \cdot 5^2 \cdot 7$$

EXAMPLE 19

Reduce $\dfrac{210}{231}$ to lowest terms.

SOLUTION   First we factor 210 and 231 into the product of prime factors. Then we reduce to lowest terms by dividing the numerator and denominator by any factors they have in common.

$$\frac{210}{231} = \frac{2 \cdot 3 \cdot 5 \cdot 7}{3 \cdot 7 \cdot 11}$$ Factor the numerator and denominator completely.

$$= \frac{2 \cdot \cancel{3} \cdot 5 \cdot \cancel{7}}{\cancel{3} \cdot \cancel{7} \cdot 11}$$ Divide the numerator and denominator by 3 and 7.

$$= \frac{2 \cdot 5}{11}$$

$$= \frac{10}{11}$$

The small lines we have drawn through the factors that are common to the numerator and denominator are used to indicate that we have divided the numerator and denominator by those factors.

GETTING READY FOR CLASS

Each section of the book will end with some problems and questions like the ones that follow. They are for you to answer after you have read through the section, but before you go to class. All of them require that you give written responses, in complete sentences. Writing about mathematics is a valuable exercise. If you write with the intention of explaining and communicating what you know to someone else, you will find that you understand the topic you are writing about even better than you did before you started writing. As with all problems in this course, you want to approach these writing exercises with a positive point of view. You will get better at giving written responses to questions as you progress through the course. Even if you are never comfortable writing about mathematics, just the process of attempting to do so will increase your understanding and ability in mathematics.

   After reading through the preceding section, respond in your own words and in complete sentences.

A. What is the meaning of the expression 2^3?
B. Why do we have a rule for the order of operations?
C. What is an irrational number?
D. Explain the operations that are associated with the words *sum*, *difference*, *product*, and *quotient*.

Problem Set 1.1

Translate each of the following statements into symbols.

1. The sum of x and 5 is 2.
2. The sum of y and -3 is 9.
3. The difference of 6 and x is y.
4. The difference of x and 6 is $-y$.
5. The product of 2 and t is less than y.
6. The product of $5x$ and y is equal to z.
7. The sum of x and y is less than the difference of x and y.
8. Twice the sum of a and b is 15.
9. Half the difference of t and s is less than twice the sum of t and s.
10. Half the product of r and s is less than twice the difference of r and s.

The problems that follow are intended to give you practice using the rule for order of operations. Some of them are arranged so they model some of the properties of real numbers. Simplify each expression.

11. **a.** $3 \cdot 5 + 4$ **b.** $3 \cdot (5 + 4)$ **c.** $3 \cdot 5 + 3 \cdot 4$
12. **a.** $3 \cdot 7 - 6$ **b.** $3(7 - 6)$ **c.** $3 \cdot 7 - 3 \cdot 6$
13. **a.** $6 + 3 \cdot 4 - 2$ **b.** $6 + 3(4 - 2)$ **c.** $(6 + 3)(4 - 2)$
14. **a.** $8 + 2 \cdot 7 - 3$ **b.** $8 + 2(7 - 3)$ **c.** $(8 + 2)(7 - 3)$
15. **a.** $(7 - 4)(7 + 4)$ **b.** $7^2 - 4^2$
16. **a.** $(8 - 5)(8 + 5)$ **b.** $8^2 - 5^2$
17. **a.** $(5 + 7)^2$ **b.** $5^2 + 7^2$ **c.** $5^2 + 2 \cdot 5 \cdot 7 + 7^2$
18. **a.** $(8 - 3)^2$ **b.** $8^2 - 3^2$ **c.** $8^2 - 2 \cdot 8 \cdot 3 + 3^2$
19. **a.** $2 + 3 \cdot 2^2 + 3^2$ **b.** $2 + 3(2^2 + 3^2)$ **c.** $(2 + 3)(2^2 + 3^2)$
20. **a.** $3 + 4 \cdot 4^2 + 5^2$ **b.** $3 + 4(4^2 + 5^2)$ **c.** $(3 + 4)(4^2 + 5^2)$
21. **a.** $40 - 10 \div 5 + 1$ **b.** $(40 - 10) \div 5 + 1$ **c.** $(40 - 10) \div (5 + 1)$
22. **a.** $20 - 10 \div 2 + 3$ **b.** $(20 - 10) \div 2 + 3$ **c.** $(20 - 10) \div (2 + 3)$
23. **a.** $40 - [10 - (4 - 2)]$ **b.** $40 - 10 - 4 - 2$
24. **a.** $50 - [17 - (8 - 3)]$ **b.** $50 - 17 - 8 - 3$

The problems below will make certain you are using the rule for order of operations correctly.

25. $25 - 17 + 3$
26. $38 - 19 + 1$
27. $109 - 36 + 14$
28. $200 - 150 + 20$
29. $20 - 13 - 3$
30. $57 - 18 - 8$
31. $63 - 37 - 4$
32. $71 - 11 - 1$
33. $36 \div 9 \cdot 4$
34. $28 \div 7 \cdot 2$
35. $75 \div 3 \cdot 25$
36. $48 \div 3 \cdot 2$
37. $64 \div 16 \div 4$
38. $125 \div 25 \div 5$
39. $75 \div 25 \div 5$
40. $36 \div 12 \div 4$

The problems below are problems you will see later in the book. Work these problems without using a calculator.

41. $18,000 - 9,300$
42. $18,000 - 4,500$
43. $3.45 + 2.6 - 1.004$
44. $24.3 + 6(8.1)$
45. $275 \div 55$
46. $4.8 \div 2.4$
47. $4(2)(4)5^2$
48. $230(5) - 20(5)^2$
49. $250(5) - 25(5)^2$

50. $3(3)^2 + 2(3) - 1$ **51.** $5 \cdot 2^3 - 3 \cdot 2^2 + 4 \cdot 2 - 5$

52. $125 \cdot 2^2$ **53.** $125 \cdot 2^3$ **54.** $7.5(10)$ **55.** $500(1.5)$

56. $39.3 \cdot 60$ **57.** $5(0.10)$ **58.** $2(0.25)$ **59.** $0.20(8)$

60. $0.30(12)$ **61.** $0.08(4,000)$ **62.** $0.09(6,000)$

We are assuming that you know how to use a calculator to do simple arithmetic problems. Use a calculator to simplify each expression. If rounding is necessary, round your answers to the nearest ten-thousandth (4 places past the decimal point). You will see these problems again later in the book.

Simplify

63. $0.6931 + 1.0986$ **64.** $1.6094 + 1.9459$ **65.** $3(0.6931)$

66. $2(1.9459)$ **67.** $250(3.14)$ **68.** $165(3.14)$

69. $4,628 \div 25$ **70.** $7,546 \div 35$ **71.** $65,000 \div 5,280$

72. $2,358 \div 5,280$ **73.** $0.8413 - 1$ **74.** $1.2052 - 1$

75. $16(3.5)^2$ **76.** $4(3.14)3^2$ **77.** $11.5(130) - 0.05(130)^2$

78. $10(130) - 0.04(130)^2$

79. Find the value of each expression when x is 5.
   a. $x + 2$ **b.** $2x$ **c.** x^2 **d.** 2^x

80. Find the value of each expression when x is 3.
   a. $x + 5$ **b.** $5x$ **c.** x^5 **d.** 5^x

81. Find the value of each expression when x is 10.
   a. $x^2 + 2x + 1$ **b.** $(x + 1)^2$ **c.** $x^2 + 1$ **d.** $(x - 1)^2$

82. Find the value of each expression when x is 8.
   a. $x^2 - 6x + 9$ **b.** $(x - 3)^2$ **c.** $x^2 - 3$ **d.** $x^2 - 9$

83. Find the value of $b^2 - 4ac$ if
   a. $a = 2$, $b = 5$, and $c = 3$ **b.** $a = 10$, $b = 60$, and $c = 30$
   c. $a = 0.4$, $b = 1$, and $c = 0.3$

84. Find the value of $6x + 5y + 4$ if
   a. $x = 3$ and $y = 2$ **b.** $x = 2$ and $y = 3$ **c.** $x = 0$ and $y = 0$

Let $A = \{0, 2, 4, 6\}$, $B = \{1, 2, 3, 4, 5\}$, and $C = \{1, 3, 5, 7\}$. List the elements of each of the following sets.

85. $A \cup B$ **86.** $A \cup C$ **87.** $A \cap B$ **88.** $A \cap C$

89. $B \cap C$ **90.** $B \cup C$ **91.** $A \cup (B \cap C)$ **92.** $C \cup (A \cap B)$

93. $\{x \mid x \in A \text{ and } x < 4\}$ **94.** $\{x \mid x \in B \text{ and } x > 3\}$ **95.** $\{x \mid x \in A \text{ and } x \notin B\}$

96. $\{x \mid x \in B \text{ and } x \notin C\}$ **97.** $\{x \mid x \in A \text{ or } x \in C\}$ **98.** $\{x \mid x \in A \text{ or } x \in B\}$

For the set $\{-6, -5.2, -\sqrt{7}, -\pi, 0, 1, 2, 2.3, \frac{9}{2}, \sqrt{17}\}$, list all the elements named in each of the following problems.

99. Counting numbers   **100.** Whole numbers   **101.** Rational numbers

102. Integers   **103.** Irrational numbers   **104.** Real numbers

Label the following true or false.

105. The number 39 is a prime number.

106. Some irrational numbers are also rational numbers.

107. All whole numbers are integers.

108. Every real number is a rational number.

109. All integers are rational numbers.

110. Zero is not considered a real number.

111. An irrational number is also a real number.

112. The number 0.25 is a rational number.

Factor each number into the product of prime factors.

113. 266   **114.** 385   **115.** 111   **116.** 735

117. 369   **118.** 1,155

Reduce each fraction to lowest terms.

119. $\dfrac{165}{385}$ **120.** $\dfrac{550}{735}$ **121.** $\dfrac{385}{735}$ **122.** $\dfrac{266}{285}$

123. $\dfrac{111}{185}$ **124.** $\dfrac{279}{310}$ **125.** $\dfrac{75}{135}$ **126.** $\dfrac{38}{30}$

127. $\dfrac{6}{8}$ **128.** $\dfrac{10}{25}$ **129.** $\dfrac{200}{5}$ **130.** $\dfrac{240}{6}$

131. $\dfrac{10}{22}$ **132.** $\dfrac{39}{13}$

133. **Minutes and Seconds** The chart shows the five fastest winning times for the Fifth Avenue Mile run in New York City. The times are given in minutes and seconds, to the nearest hundredth of a second.

Fastest on Fifth	Continental Airlines Fifth Avenue Mile
Craig Mottram, AUS	3:49.90
Alan Webb, USA	3:51.40
Elkanah Angwenyi, KEN	3:54.30
Anthony Famiglietti, USA	3:57.10
Rui Silva, POR	3:57.40

Source: www.coolrunning.com

a. How much faster was Alan Webb's time than Elkanah Angwenyi's time?

b. How much faster was Craig Mottram's time than Rui Silva's time?

134. Solar Power The chart shows the annual sums of solar thermal collector shipments in the U.S.

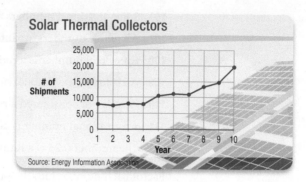

Solar Thermal Collectors

Source: Energy Information Association

a. For which year was the sum of solar thermal collector shipments closest to 12,500 units?

b. Estimate the increase in solar thermal collector shipments from Year 9 to Year 10.

135. Fermat's Last Theorem The postage stamp shown here was issued by France in 2001 to commemorate the 400th anniversary of the birth of the French mathematician Pierre de Fermat. The stamp shows Fermat's last theorem, which states that if n is an integer greater than 2, then there are no positive integers x, y, and z that will make the formula $x^n + y^n = z^n$ true.

However, there are many ways to make the formula $x^n + y^n = z^n$ true when n is 1 or 2. Show that this formula is true for each case below.

a. $n = 1$, $x = 5$, $y = 8$, and $z = 13$

b. $n = 1$, $x = 2$, $y = 3$, and $z = 5$

c. $n = 2$, $x = 3$, $y = 4$, and $z = 5$

d. $n = 2$, $x = 7$, $y = 24$, and $z = 25$

EXTENDING THE CONCEPTS

Many of the problem sets in this book end with a few problems like the ones that follow. These problems challenge you to extend your knowledge of the material in the problem set. In most cases, there are no examples in the text similar to these problems. You should approach them with a positive point of view, because even though you may not complete them correctly, just the process of attempting them will increase your knowledge and ability in algebra.

The expression $n!$ is read "n factorial" and is the product of all the consecutive integers from n down to 1. For example,

$$1! = 1$$

$$2! = 2 \cdot 1 = 2$$

$$3! = 3 \cdot 2 \cdot 1 = 6$$

$$4! = 4 \cdot 3 \cdot 2 \cdot 1 = 24$$

136. Calculate 5!

137. Calculate 6!

138. Show that this statement is true: $6! = 6 \cdot 5!$

139. Show that this statement is false: $(2 + 3)! = 2! + 3!$

PROPERTIES OF REAL NUMBERS

OBJECTIVES

A Identify opposites.

B Multiply fractions and give reciprocals of numbers.

C Simplify expressions containing absolute value.

D Simplify expressions using properties of the real numbers.

E Combine similar terms.

F Find LCDs and solve problems involving fractions with different denominators.

G Simplify expressions using the order of operations.

© 2010 Katherine Heistand Shields

1.2

In this section, we will list all the facts (properties) that you know from past experience are true about numbers in general. We will give each property a name so we can refer to it later in this book. Mathematics is very much like a game. The game involves numbers. The rules of the game are the properties we are developing in this chapter. The goal of the game is to extend the basic rules to as many situations as possible.

You know from past experience with numbers that it makes no difference in which order you add two numbers; that is, $3 + 5$ is the same as $5 + 3$. Imagine a surfer catching some waves on a beautiful sunny day. He is able to stand up on the first 3 waves but then immediately wipes out into the water. He has successful rides on the next 5 waves. Using our knowledge of addition, we know that the surfer attempted a total of 8 waves. Even if the surfer had successful rides on the first 5 waves and wiped out on the next 3, the total waves he attempted would still be 8. This is an example of the *commutative property of addition*. We say addition is a commutative operation. Changing the order of the numbers does not change the answer. This fact about numbers is one of the properties of real numbers we will cover in this section. Before we arrive at the distributive property, we need to review some basic definitions and vocabulary.

A Opposites

> **DEFINITION** *opposites*
>
> Any two real numbers the same distance from 0, but in opposite directions from 0 on the number line, are called *opposites*, or *additive inverses*.

1.2 VIDEOS

EXAMPLE 1

The numbers -3 and 3 are opposites, so are π and $-\pi$, $\frac{3}{4}$ and $-\frac{3}{4}$, and $\sqrt{2}$ and $-\sqrt{2}$.

18

The negative sign in front of a number can be read in a few different ways. It can be read as "negative" or "the opposite of." We say -4 is the opposite of 4, or negative 4. The term we use will depend on the situation. For instance, the expression $-(-3)$ is best read, "the opposite of negative 3." Because the opposite of -3 is 3, we have $-(-3) = 3$. In general, if a is any positive real number, then

$$-(-a) = a \quad \text{(The opposite of a negative is a positive)}$$

B Review of Multiplication with Fractions

Before we go further with our study of the number line, we need to review multiplication with fractions. Recall that for the fraction $\frac{a}{b}$, a is called the numerator and b is called the denominator. To multiply two fractions, we simply multiply the numerators and multiply the denominators.

EXAMPLE

Multiply.

2. $\frac{3}{5} \cdot \frac{7}{8} = \frac{3 \cdot 7}{5 \cdot 8} = \frac{21}{40}$

3. $8 \cdot \frac{1}{5} = \frac{8}{1} \cdot \frac{1}{5} = \frac{8 \cdot 1}{1 \cdot 5} = \frac{8}{5}$

4. $\left(\frac{2}{3}\right)^4 = \frac{2}{3} \cdot \frac{2}{3} \cdot \frac{2}{3} \cdot \frac{2}{3} = \frac{16}{81}$

NOTE In past math classes you may have written fractions like $\frac{8}{5}$ (improper fractions) as mixed numbers, such as $1\frac{3}{5}$. In algebra, it is usually better to leave them as improper fractions.

The idea of multiplication of fractions is useful in understanding the concept of the reciprocal of a number. Here is the definition.

DEFINITION   *reciprocals*

Any two real numbers whose product is 1 are called *reciprocals*, or *multiplicative inverses*.

EXAMPLE

Give the reciprocal of each number.

	Number	Reciprocal	
5.	3	$\frac{1}{3}$	Because $3 \cdot \frac{1}{3} = \frac{3}{1} \cdot \frac{1}{3} = \frac{3}{3} = 1$
6.	$\frac{1}{6}$	6	Because $\frac{1}{6} \cdot 6 = \frac{1}{6} \cdot \frac{6}{1} = \frac{6}{6} = 1$
7.	$\frac{4}{5}$	$\frac{5}{4}$	Because $\frac{4}{5} \cdot \frac{5}{4} = \frac{20}{20} = 1$
8.	a	$\frac{1}{a}$	Because $a \cdot \frac{1}{a} = \frac{a}{1} \cdot \frac{1}{a} = \frac{a}{a} = 1$ $(a \neq 0)$

Although we will not address multiplication with negative numbers until later in this chapter, you should know that the reciprocal of a negative number is also a negative number. For example, the reciprocal of -5 is $-\frac{1}{5}$.

C The Absolute Value of a Real Number

DEFINITION *absolute value*

The *absolute value* of a number (also called its *magnitude*) is the distance the number is from 0 on the number line. If x represents a real number, then the absolute value of x is written $|x|$.

This definition of absolute value is geometric in form because it defines absolute value in terms of the number line. Here is an alternative definition of absolute value that is algebraic in form because it only involves symbols.

NOTE It is important to recognize that if x is a real number, $-x$ is not necessarily negative. For example, if x is 5, then $-x$ is -5. On the other hand, if x were -5, then $-x$ would be $-(-5)$, which is 5.

ALTERNATIVE DEFINITION *absolute value*

If x represents a real number, then the *absolute value* of x is written $|x|$, and is given by

$$|x| = \left\{ \begin{array}{ll} x & \text{if } x \geq 0 \\ -x & \text{if } x < 0 \end{array} \right\}$$

If the original number is positive or 0, then its absolute value is the number itself. If the number is negative, its absolute value is its opposite (which must be positive).

EXAMPLE

Write each expression without absolute value symbols.

9. $|5| = 5$

10. $|-2| = 2$

11. $\left| -\dfrac{1}{2} \right| = \dfrac{1}{2}$

12. $-|-3| = -3$

13. $-|5| = -5$

14. $-|-\sqrt{2}| = -\sqrt{2}$

D Properties of Real Numbers

We know that adding 3 and 7 gives the same answer as adding 7 and 3. The order of two numbers in an addition problem can be changed without changing the result. This fact about numbers and addition is called the *commutative property of addition*.

For all the properties listed in this section, a, b, and c represent real numbers.

PROPERTY *Commutative Property of Addition*

In symbols: $a + b = b + a$
In words: The *order* of the numbers in a sum does not affect the result.

> **PROPERTY**   *Commutative Property of Multiplication*
> . . . .
> *In symbols:* $a \cdot b = b \cdot a$
> *In words:*   The *order* of the numbers in a product does not affect the result.

EXAMPLE

15. The statement $3 + 7 = 7 + 3$ is an example of the commutative property of addition.

16. The statement $3 \cdot x = x \cdot 3$ is an example of the commutative property of multiplication.

The other two basic operations (subtraction and division) are not commutative. If we change the order in which we are subtracting or dividing two numbers, we change the result.

Another property of numbers you have used many times has to do with grouping. When adding $3 + 5 + 7$, we can add the 3 and 5 first and then the 7, or we can add the 5 and 7 first and then the 3. Mathematically, it looks like this: $(3 + 5) + 7 = 3 + (5 + 7)$. Operations that behave in this manner are called *associative* operations.

> **PROPERTY**   *Associative Property of Addition*
> . . . .
> *In symbols:* $a + (b + c) = (a + b) + c$
> *In words:*   The *grouping* of the numbers in a sum does not affect the result.

> **PROPERTY**   *Associative Property of Multiplication*
> . . . .
> *In symbols:* $a(bc) = (ab)c$
> *In words:*   The *grouping* of the numbers in a product does not affect the result.

The following examples illustrate how the associative properties can be used to simplify expressions that involve both numbers and variables.

EXAMPLE

Simplify by using the associative property.

17. $2 + (3 + y) = (2 + 3) + y$ Associative property

$\qquad\qquad\quad = 5 + y$ Addition

18. $5(4x) = (5 \cdot 4)x$ Associative property

$\qquad\quad = 20x$ Multiplication

19. $\dfrac{1}{4}(4a) = \left(\dfrac{1}{4} \cdot 4\right)a$ Associative property

$\qquad\quad = 1a$ Multiplication

$\qquad\quad = a$

20. $2\left(\dfrac{1}{2}x\right) = \left(2 \cdot \dfrac{1}{2}\right)x$ Associative property

$\qquad\qquad = 1x$ Multiplication

$\qquad\qquad = x$

21. $6\left(\dfrac{1}{3}x\right) = \left(6 \cdot \dfrac{1}{3}\right)x$ Associative property

$\qquad\qquad = 2x$ Multiplication

Our next property involves both addition and multiplication. It is called the *distributive property* and is stated as follows.

> **PROPERTY**   *Distributive Property*
>
> *In symbols:* $a(b + c) = ab + ac$
> *In words:*   Multiplication *distributes* over addition.

NOTE Although the properties we are listing are stated for only two or three real numbers, they hold for as many numbers as needed. For example, the distributive property holds for expressions like $3(x + y + z + 2)$. That is,

$3(x + y + z + 2) = 3x + 3y + 3z + 6$

You will see as we progress through the book that the distributive property is used very frequently in algebra. To see that the distributive property works, compare the following:

$$3(4 + 5) = 3(9) \qquad\qquad 3(4) + 3(5) = 12 + 15$$

$$= 27 \qquad\qquad\qquad\qquad = 27$$

In both cases the result is 27. Because the results are the same, the original two expressions must be equal, or $3(4 + 5) = 3(4) + 3(5)$.

EXAMPLE ——————

Apply the distributive property to each expression and then simplify the result.

22. $5(4x + 3) = 5(4x) + 5(3)$ Distributive property

$\qquad\qquad = 20x + 15$ Multiplication

23. $6(3x + 2y) = 6(3x) + 6(2y)$ Distributive property

$\qquad\qquad = 18x + 12y$ Multiplication

24. $\dfrac{1}{2}(3x + 6) = \dfrac{1}{2}(3x) + \dfrac{1}{2}(6)$ Distributive property

$\qquad\qquad = \dfrac{3}{2}x + 3$ Multiplication

25. $2(3y + 4) + 2 = 2(3y) + 2(4) + 2$ Distributive property

$\qquad\qquad = 6y + 8 + 2$ Multiplication

$\qquad\qquad = 6y + 10$ Addition

We can combine our knowledge of the distributive property with multiplication of fractions to manipulate expressions involving fractions. Here are some examples that show how we do this.

EXAMPLE

Apply the distributive property to each expression and then simplify the result.

26. $a\left(1 + \dfrac{1}{a}\right) = a \cdot 1 + a \cdot \dfrac{1}{a} = a + 1$

27. $3\left(\dfrac{1}{3}x + 5\right) = 3 \cdot \dfrac{1}{3}x + 3 \cdot 5 = x + 15$

28. $6\left(\dfrac{1}{3}x + \dfrac{1}{2}y\right) = 6 \cdot \dfrac{1}{3}x + 6 \cdot \dfrac{1}{2}y = 2x + 3y$

E Combining Similar Terms

The distributive property can also be used to combine similar terms. (For now, a term is the product of a number with one or more variables.) Similar terms are terms with the same variable part. The terms $3x$ and $5x$ are similar, as are $2y$, $7y$, and $-3y$, because the variable parts are the same.

EXAMPLE

Use the distributive property to combine similar terms.

29. $3x + 5x = (3 + 5)x$ Distributive property

            $= 8x$ Addition

30. $3y + y = (3 + 1)y$ Distributive property

            $= 4y$ Addition

F Review of Addition with Fractions

To add fractions, each fraction must have the same denominator.

> **DEFINITION** *least common denominator*
>
> The *least common denominator* (LCD) for a set of denominators is the smallest number divisible by *all* the denominators.

The first step in adding fractions is to find a common denominator for all the denominators. We then rewrite each fraction (if necessary) as an equivalent fraction with the common denominator. Finally, we add the numerators and reduce to lowest terms if necessary.

EXAMPLE 31

Add $\dfrac{5}{12} + \dfrac{7}{18}$.

SOLUTION The least common denominator for the denominators 12 and 18 must be the smallest number divisible by both 12 and 18. We can factor 12 and 18 completely and then build the LCD from these factors.

$$12 = 2 \cdot 2 \cdot 3 \atop 18 = 2 \cdot 3 \cdot 3 \Big\} \quad \text{LCD} = 2 \cdot 2 \cdot 3 \cdot 3 = 36$$

12 divides the LCD

18 divides the LCD

Next, we rewrite our original fractions as equivalent fractions with denominators of 36. To do so, we multiply each original fraction by an appropriate form of the number 1:

$$\frac{5}{12} + \frac{7}{18} = \frac{5}{12} \cdot \frac{3}{3} + \frac{7}{18} \cdot \frac{2}{2}$$

$$= \frac{15}{36} + \frac{14}{36}$$

Finally, we add numerators and place the result over the common denominator, 36. (Remember, this is an application of the distributive property.)

$$\frac{15}{36} + \frac{14}{36} = \frac{15 + 14}{36} = \frac{29}{36}$$

G Simplifying Expressions

We can use the commutative, associative, and distributive properties together to simplify expressions.

EXAMPLE 32

Simplify $7x + 4 + 6x + 3$.

SOLUTION We begin by applying the commutative and associative properties to group similar terms:

$$7x + 4 + 6x + 3 = (7x + 6x) + (4 + 3) \qquad \text{Commutative and associative properties}$$

$$= (7 + 6)x + (4 + 3) \qquad \text{Distributive property}$$

$$= 13x + 7 \qquad \text{Addition}$$

EXAMPLE 33

Simplify $4 + 3(2y + 5) + 8y$.

SOLUTION Because our rule for order of operations indicates that we are to multiply before adding, we must distribute the 3 across $2y + 5$ first:

$$4 + 3(2y + 5) + 8y = 4 + 6y + 15 + 8y \qquad \text{Distributive property}$$

$$= (6y + 8y) + (4 + 15) \qquad \text{Commutative and associative properties}$$

$$= (6 + 8)y + (4 + 15) \qquad \text{Distributive property}$$

$$= 14y + 19 \qquad \text{Addition}$$

The remaining properties of real numbers have to do with the numbers 0 and 1.

NOTE 0 and 1 are called the *additive identity* and *multiplicative identity*, respectively. Combining 0 with a number, under addition, does not change the identity of the number. Likewise, combining 1 with a number, under multiplication, does not alter the identity of the number. We see that 0 is to addition what 1 is to multiplication.

PROPERTY *Additive Identity Property*

There exists a unique number 0 such that
In symbols: $a + 0 = a$ and $0 + a = a$
In words: Adding zero to any number is the number.

PROPERTY *Multiplicative Identity Property*

There exists a unique number 1 such that
In symbols: $a \cdot 1 = a$ and $1 \cdot a = a$
In words: Any number multiplied by 1 is the number.

PROPERTY *Additive Inverse Property*

For each real number a, there exists a unique real number $-a$ such that
In symbols: $a + (-a) = 0$
In words: Opposites add to 0.

PROPERTY *Multiplicative Inverse Property*

For every real number a, except 0, there exists a unique real number $\frac{1}{a}$ such that
In symbols: $a\left(\dfrac{1}{a}\right) = 1$ *In words:*   Reciprocals multiply to 1.

EXAMPLE 34

a. $7(1) = 7$ Multiplicative identity property

b. $4 + (-4) = 0$ Additive inverse property

c. $6\left(\dfrac{1}{6}\right) = 1$ Multiplicative inverse property

d. $(5 + 0) + 2 = 5 + 2$ Additive identity property

GETTING READY FOR CLASS

After reading through the preceding section, respond in your own words and in complete sentences.

A. Describe the commutative property of multiplication.
B. Give definitions for each of the following:
   1. The opposite of a number
   2. The absolute value of a number
C. Explain why subtraction and division are not commutative operations.
D. Explain why zero does not have a multiplicative inverse.

Problem Set 1.2

Complete the following table.

	Number	Opposite	Reciprocal
1.	4		
2.	-3		
3.	$-\frac{1}{2}$		
4.	$\frac{5}{6}$		
5.		-5	
6.		7	
7.		$-\frac{3}{8}$	
8.		$\frac{1}{2}$	
9.			-6
10.			-3
11.			$\frac{1}{3}$
12.			$-\frac{1}{4}$

13. Name two numbers that are their own reciprocals.

14. Give the number that has no reciprocal.

15. Name the number that is its own opposite.

16. The reciprocal of a negative number is negative — true or false?

Write each of the following without absolute value symbols.

17. $|-2|$ **18.** $|-7|$ **19.** $\left|-\frac{3}{4}\right|$ **20.** $\left|\frac{5}{6}\right|$ **21.** $|\pi|$ **22.** $|-\sqrt{2}|$

23. $-|4|$ **24.** $-|5|$ **25.** $-|-2|$ **26.** $-|-10|$ **27.** $-\left|-\frac{3}{4}\right|$ **28.** $-\left|\frac{7}{8}\right|$

Multiply the following.

29. $\frac{3}{5} \cdot \frac{7}{8}$ **30.** $\frac{6}{7} \cdot \frac{9}{5}$ **31.** $\frac{1}{3} \cdot 6$ **32.** $\frac{1}{4} \cdot 8$

33. $\left(\frac{2}{3}\right)^3$ **34.** $\left(\frac{4}{5}\right)^2$ **35.** $\left(\frac{1}{10}\right)^4$ **36.** $\left(\frac{1}{2}\right)^5$

37. $\frac{3}{5} \cdot \frac{4}{7} \cdot \frac{6}{11}$ **38.** $\frac{4}{5} \cdot \frac{6}{7} \cdot \frac{3}{11}$ **39.** $\frac{4}{3} \cdot \frac{3}{4}$ **40.** $\frac{5}{8} \cdot \frac{8}{5}$

Use the associative property to rewrite each of the following expressions and then simplify the result.

41. $4 + (2 + x)$ **42.** $6 + (5 + 3x)$ **43.** $(a + 3) + 5$ **44.** $(4a + 5) + 7$

45. $5(3y)$ **46.** $7(4y)$ **47.** $\frac{1}{3}(3x)$ **48.** $\frac{1}{5}(5x)$

49. $4\left(\frac{1}{4}a\right)$ **50.** $7\left(\frac{1}{7}a\right)$ **51.** $\frac{2}{3}\left(\frac{3}{2}x\right)$ **52.** $\frac{4}{3}\left(\frac{3}{4}x\right)$

Apply the distributive property to each expression. Simplify when possible.

53. $3(x + 6)$ **54.** $5(x + 9)$ **55.** $2(6x + 4)$ **56.** $3(7x + 8)$

57. $5(3a + 2b)$ **58.** $7(2a + 3b)$ **59.** $\frac{1}{3}(4x + 6)$ **60.** $\frac{1}{2}(3x + 8)$

61. $\frac{1}{5}(10 + 5y)$ **62.** $\frac{1}{6}(12 + 6y)$ **63.** $(5t + 1)8$ **64.** $(3t + 2)5$

The problems below are problems you will see later in the book. Apply the distributive property, then simplify if possible.

65. $3(3x + y - 2z)$ **66.** $2(2x - y + z)$ **67.** $10(0.3x + 0.7y)$

68. $10(0.2x + 0.5y)$ **69.** $100(0.06x + 0.07y)$ **70.** $100(0.09x + 0.08y)$

71. $3\left(x + \dfrac{1}{3}\right)$ **72.** $5\left(x - \dfrac{1}{5}\right)$ **73.** $2\left(x - \dfrac{1}{2}\right)$

74. $7\left(x + \dfrac{1}{7}\right)$ **75.** $x\left(1 + \dfrac{2}{x}\right)$ **76.** $x\left(1 - \dfrac{1}{x}\right)$

77. $a\left(1 - \dfrac{3}{a}\right)$ **78.** $a\left(1 + \dfrac{1}{a}\right)$ **79.** $8\left(\dfrac{1}{8}x + 3\right)$

80. $4\left(\dfrac{1}{4}x - 9\right)$ **81.** $6\left(\dfrac{1}{2}x - \dfrac{1}{3}y\right)$ **82.** $12\left(\dfrac{1}{4}x - \dfrac{1}{6}y\right)$

83. $12\left(\dfrac{1}{4}x + \dfrac{2}{3}y\right)$ **84.** $12\left(\dfrac{2}{3}x - \dfrac{1}{4}y\right)$ **85.** $20\left(\dfrac{2}{5}x + \dfrac{1}{4}y\right)$

86. $15\left(\dfrac{2}{3}x + \dfrac{2}{5}y\right)$

Apply the distributive property to each expression. Simplify when possible.

87. $3(5x + 2) + 4$ **88.** $4(3x + 2) + 5$ **89.** $4(2y + 6) + 8$ **90.** $6(2y + 3) + 2$

91. $5(1 + 3t) + 4$ **92.** $2(1 + 5t) + 6$ **93.** $3 + (2 + 7x)4$ **94.** $4 + (1 + 3x)5$

Add the following fractions.

95. $\dfrac{2}{5} + \dfrac{1}{15}$ **96.** $\dfrac{5}{8} + \dfrac{1}{4}$ **97.** $\dfrac{17}{30} + \dfrac{11}{42}$ **98.** $\dfrac{19}{42} + \dfrac{13}{70}$

99. $\dfrac{9}{48} + \dfrac{3}{54}$ **100.** $\dfrac{6}{28} + \dfrac{5}{42}$ **101.** $\dfrac{25}{84} + \dfrac{41}{90}$ **102.** $\dfrac{23}{70} + \dfrac{29}{84}$

The problems below are problems you will see later in the book. Simplify each expression.

103. $\left(\dfrac{3}{14} + \dfrac{7}{30}\right)$ **104.** $\left(\dfrac{3}{10} + \dfrac{11}{42}\right)$

105. $32\left(\dfrac{3}{4}\right) - 16\left(\dfrac{3}{4}\right)^2$ **106.** $32\left(\dfrac{3}{2}\right) - 16\left(\dfrac{3}{2}\right)^2$

Use the commutative, associative, and distributive properties to simplify the following.

107. $5a + 7 + 8a + a$ **108.** $3y + y + 5 + 2y + 1$

109. $2(5x + 1) + 2x$ **110.** $7 + 2(4y + 2)$

111. $3 + 4(5a + 3) + 4a$ **112.** $5x + 2(3x + 8) + 4$

113. $5x + 3(x + 2) + 7$ **114.** $2a + 4(2a + 6) + 3$

115. $5(x + 2y) + 4(3x + y)$

116. $3x + 4(2x + 3y) + 7y$

117. $5b + 3(4b + a) + 6a$

118. $4 + 3(2x + 3y) + 6(x + 4)$

Identify the property of real numbers that justifies each of the following.

119. $3 + 2 = 2 + 3$

120. $3(ab) = (3a)b$

121. $5x = x5$

122. $2 + 0 = 2$

123. $4 + (-4) = 0$

124. $1(6) = 6$

125. $x + (y + 2) = (y + 2) + x$

126. $(a + 3) + 4 = a + (3 + 4)$

127. $4(5 \cdot 7) = 5(4 \cdot 7)$

128. $6(xy) = (xy)6$

129. $4 + (x + y) = (4 + y) + x$

130. $(r + 7) + s = (r + s) + 7$

131. $3(4x + 2) = 12x + 6$

132. $5\left(\dfrac{1}{5}\right) = 1$

133. TV Viewing Use the chart to answer the questions below.

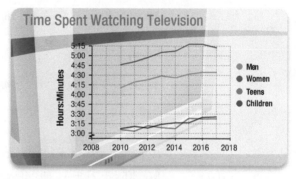

a. For the years 2016 through 2017, which group of people does not follow the growth trend found in the other three groups?

b. Is the rate of increase for men from 2012 to 2013 a negative or positive number?

c. What year did the amount of television that men watched decrease?

134. Drinking and Driving The chart shows that if the risk of getting in an accident with no alcohol in your system is 1, then the risk of an accident with a blood-alcohol level of 0.24 is 147 times as high.

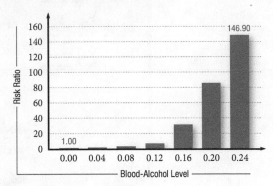

a. A person driving with a blood-alcohol level of 0.20 is how many times more likely to get in an accident than if she was driving with a blood-alcohol level of 0?

b. If the probability of getting in an accident while driving for an hour on surface streets in a certain city is 0.02%, what is the probability of getting in an accident in the same circumstances with a blood-alcohol level of 0.20?

EXTENDING THE CONCEPTS

Simplify.

135. $x(x + 8)$

136. $2a(a + 4)$

137. True or False: $-x$ is always a negative number.

138. True or False: $|x| = x$

139. True or False: $|-x| = x$

140. True or False: $|-x| = |x|$

OBJECTIVES

A   Add any combination of positive and negative numbers.

B   Subtract any combination of positive and negative numbers.

C   Multiply any combination of positive and negative numbers.

D   Divide any combination of positive and negative numbers.

E   Divide fractions and reduce to lowest terms.

F   Evaluate expressions using a table.

1.3

©iStockphoto.com/aapsky

The temperature at the airport is 70°F. A plane takes off and reaches its cruising altitude of 28,000 feet, where the temperature is −40°F. Find the difference in the temperatures at takeoff and at cruising altitude.

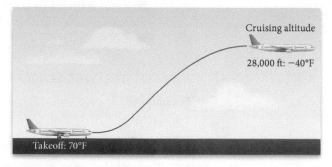

Cruising altitude

28,000 ft: −40°F

Takeoff: 70°F

We know intuitively that the difference in temperature is 110°F. If we write this problem using symbols, we have

$$70 - (-40) = 110$$

In this section we review the rules for arithmetic with real numbers, which will include problems such as this one.

Adding Real Numbers

This section reviews the rules for arithmetic with real numbers and the justification for those rules. We can justify the rules for addition of real numbers geometrically by use of the real number line. Consider the sum of −5 and 3:

$$-5 + 3$$

We can interpret this expression as meaning "start at the origin and move 5 units in the negative direction and then 3 units in the positive direction." With the aid of a number line, we can visualize the process.

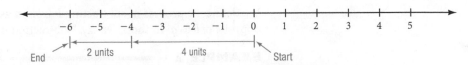

Because the process ends at -2, we say the sum of -5 and 3 is -2:

$$-5 + 3 = -2$$

We can use the real number line in this way to add any combination of positive and negative numbers.

The sum of -4 and -2, $-4 + (-2)$, can be interpreted as starting at the origin, moving 4 units in the negative direction, and then 2 more units in the negative direction:

Because the process ends at -6, we say the sum of -4 and -2 is -6:

$$-4 + (-2) = -6$$

We can eliminate actually drawing a number line by simply visualizing it mentally. The following example gives the results of all possible sums of positive and negative 5 and 7.

EXAMPLE 1

Add all combinations of positive and negative 5 and 7.

SOLUTION

$$5 + 7 = 12$$
$$-5 + 7 = 2$$
$$5 + (-7) = -2$$
$$-5 + (-7) = -12$$

Looking closely at the relationships in Example 1 (and trying other similar examples, if necessary), we can arrive at the following rule for adding two real numbers.

HOW TO   Add two real numbers

With the *same* sign:

Step 1:   Add their absolute values.

Step 2:   Attach their common sign. If both numbers are positive, their sum is positive; if both numbers are negative, their sum is negative.

With *opposite* signs:

Step 1:   Subtract the smaller absolute value from the larger.

Step 2:   Attach the sign of the number whose absolute value is larger.

ⓑ Subtracting Real Numbers

To have as few rules as possible, we will not attempt to list new rules for the *difference* of two real numbers. We will instead define it in terms of addition and apply the rule for addition.

DEFINITION

If a and b are any two real numbers, then the **difference** of a and b, written $a - b$, is given by

$$a - b \qquad = \qquad a + (-b)$$

To **subtract** b,   add the opposite of b.

We define the process of subtracting b from a as being equivalent to adding the opposite of b to a. In short, we say, "subtraction is addition of the opposite."

EXAMPLE 2

Subtract.

a. $5 - 3 = 5 + (-3)$ Subtracting 3 is equivalent to adding -3.

       $= 2$

b. $-7 - 6 = -7 + (-6)$ Subtracting 6 is equivalent to adding -6.

       $= -13$

c. $9 - (-2) = 9 + 2$ Subtracting -2 is equivalent to adding 2.

       $= 11$

d. $-6 - (-5) = -6 + 5$ Subtracting -5 is equivalent to adding 5.

       $= -1$

EXAMPLE 3

Subtract -3 from -9.

SOLUTION   Because subtraction is not commutative, we must be sure to write the numbers in the correct order. Because we are subtracting -3, the problem looks like this when translated into symbols:

$$-9 - (-3) = -9 + 3 \qquad \text{Change to addition of the opposite.}$$
$$= -6 \qquad \text{Add.}$$

EXAMPLE 4

Add -4 to the difference of -2 and 5.

SOLUTION   The difference of -2 and 5 is written $-2 - 5$. Adding -4 to that difference gives us

$$(-2 - 5) + (-4) = -7 + (-4) \qquad \text{Simplify inside parentheses.}$$
$$= -11 \qquad \text{Add.}$$

🄲 Multiplying Real Numbers

Multiplication with whole numbers is simply a shorthand way of writing repeated addition.

For example, $3(-2)$ can be evaluated as follows:

$$3(-2) = -2 + (-2) + (-2) = -6$$

We can evaluate the product $-3(2)$ in a similar manner if we first apply the commutative property of multiplication:

$$-3(2) = 2(-3) = -3 + (-3)$$

From these results, it seems reasonable to say that the product of a positive and a negative is a negative number.

The last case we must consider is the product of two negative numbers, such as $-3(-2)$. To evaluate this product we will look at the expression $-3[2 + (-2)]$ in two different ways. First, since $2 + (-2) = 0$, we know the expression $-3[2 + (-2)]$ is equal to 0. On the other hand, we can apply the distributive property to get

$$-3[2 + (-2)] = -3(2) + (-3)(-2) = -6 + ?$$

Because we know the expression is equal to 0, it must be true that our ? is 6; 6 is the only number we can add to -6 to get 0. Therefore, we have

$$-3(-2) = 6$$

Here is a summary of what we have so far:

Original Numbers Have		Answer Is
The same sign	$3(2) = 6$	Positive
Different signs	$3(-2) = -6$	Negative
Different signs	$-3(2) = -6$	Negative
The same sign	$-3(-2) = 6$	Positive

HOW TO *Multiply two real numbers*

To multiply two real numbers, simply

Step 1: Multiply their absolute values.

Step 2: If the two numbers have the *same* sign, the product is positive. If the two numbers have *opposite* signs, the product is negative.

EXAMPLE 5

Multiply all combinations of positive and negative 7 and 3.

SOLUTION

$$7(3) = 21$$

$$7(-3) = -21$$

$$-7(3) = -21$$

$$-7(-3) = 21$$

⬡ Dividing Real Numbers

DEFINITION

If a and b are any two real numbers, where $b \neq 0$, then the quotient of a and b, written $\frac{a}{b}$, is given by

$$\frac{a}{b} = a \cdot \frac{1}{b}$$

Dividing a by b is equivalent to multiplying a by the reciprocal of b. In short, we say, "division is multiplication by the reciprocal."

Because division is defined in terms of multiplication, the same rules hold for assigning the correct sign to a quotient as held for assigning the correct sign to a product. That is, *the quotient of two numbers with like signs is positive, while the quotient of two numbers with unlike signs is negative.*

EXAMPLE 6

Divide.

a. $\dfrac{6}{3} = 6 \cdot \left(\dfrac{1}{3}\right) = 2$

Notice these examples indicate that if a and b are positive real numbers, then

b. $\dfrac{6}{-3} = 6 \cdot \left(-\dfrac{1}{3}\right) = -2$

$$\frac{-a}{b} = \frac{a}{-b} = -\frac{a}{b}$$

c. $\dfrac{-6}{3} = -6 \cdot \left(\dfrac{1}{3}\right) = -2$

and

d. $\dfrac{-6}{-3} = -6 \cdot \left(-\dfrac{1}{3}\right) = 2$

$$\frac{-a}{-b} = \frac{a}{b}$$

The second step in the preceding examples is written only to show that each quotient can be written as a product. It is not actually necessary to show this step when working problems.

In the examples that follow, we find a combination of operations. In each case, we use the rule for order of operations.

EXAMPLE

Simplify each expression as much as possible.

7. $(-2 - 3)(5 - 9) = (-5)(-4)$ Simplify inside parentheses.

$\qquad\qquad\qquad\quad = 20$ Multiply.

8. $2 - 5(7 - 4) - 6 = 2 - 5(3) - 6$ Simplify inside parentheses.

$\qquad\qquad\qquad\quad\; = 2 - 15 - 6$ Then, multiply.

$\qquad\qquad\qquad\quad\; = -19$ Finally, subtract, left to right.

9. $2(4 - 7)^3 + 3(-2 - 3)^2 = 2(-3)^3 + 3(-5)^2$ Simplify inside parentheses.

$\qquad\qquad\qquad\qquad\quad = 2(-27) + 3(25)$ Evaluate numbers with exponents.

$\qquad\qquad\qquad\qquad\quad = -54 + 75$ Multiply.

$\qquad\qquad\qquad\qquad\quad = 21$ Add.

10. $6\left(\dfrac{t}{3}\right) = 6\left(\dfrac{1}{3}t\right)$ Dividing by 3 is the same as multiplying by $\frac{1}{3}$.

$\qquad\qquad = \left(6 \cdot \dfrac{1}{3}\right)t$ Associative property

$\qquad\qquad = 2t$ Multiplication

11. $3\left(\dfrac{t}{3} - 2\right) = 3 \cdot \dfrac{t}{3} - 3 \cdot 2$ Distributive property

$\qquad\qquad = t - 6$ Multiplication

Our next examples involve more complicated fractions. The fraction bar works like parentheses to separate the numerator from the denominator. Although we don't write expressions this way, here is one way to think of the fraction bar:

$$\frac{-8 - 8}{-5 - 3} = (-8 - 8) \div (-5 - 3)$$

As you can see, if we apply the rule for order of operations to the expression on the right, we would work inside each set of parentheses first, then divide. Applying this to the expression on the left, we work on the numerator and denominator separately, then we divide, or reduce the resulting fraction to the lowest terms.

EXAMPLE

12. $\dfrac{-8 - 8}{-5 - 3} = \dfrac{-16}{-8}$ Simplify numerator and denominator separately.

$\qquad\qquad = 2$ Divide.

13. $\dfrac{-5(-4) + 2(-3)}{2(-1) - 5} = \dfrac{20 - 6}{-2 - 5}$ Simplify numerator and denominator separately.

$\qquad\qquad\qquad = \dfrac{14}{-7}$ Divide.

$\qquad\qquad\qquad = -2$

14. $\dfrac{2^3 + 3^3}{2^2 - 3^2} = \dfrac{8 + 27}{4 - 9}$ Simplify numerator and denominator separately.

$\qquad\qquad = \dfrac{35}{-5}$ Divide.

$\qquad\qquad = -7$

Remember, since subtraction is defined in terms of addition, we can restate the distributive property in terms of subtraction. That is, if a, b, and c are real numbers, then $a(b - c) = ab - ac$.

EXAMPLE 15

Simplify $3(2y - 1) + y$.

SOLUTION We begin by multiplying the 3 and $2y - 1$. Then, we combine similar terms:

$$3(2y - 1) + y = 6y - 3 + y \qquad \text{Distributive property}$$

$$= 7y - 3 \qquad \text{Combine similar terms.}$$

EXAMPLE 16

Simplify $8 - 3(4x - 2) + 5x$.

SOLUTION First, we distribute the -3 across the $4x - 2$.

$$8 - 3(4x - 2) + 5x = 8 - 12x + 6 + 5x$$

$$= -7x + 14$$

EXAMPLE 17

Simplify $5(2a + 3) - (6a - 4)$.

SOLUTION We begin by applying the distributive property to remove the parentheses. The expression $-(6a - 4)$ can be thought of as $-1(6a - 4)$. Thinking of it in this way allows us to apply the distributive property.

$$-1(6a - 4) = -1(6a) - (-1)(4) = -6a + 4$$

Here is the complete problem:

$$5(2a + 3) - (6a - 4) = 10a + 15 - 6a + 4 \qquad \text{Distributive property}$$

$$= 4a + 19 \qquad \text{Combine similar terms.}$$

We end this section by reviewing division with fractions and division with the number 0.

E Dividing Fractions

EXAMPLE

Divide and reduce to lowest terms.

18. $\dfrac{3}{4} \div \dfrac{6}{11} = \dfrac{3}{4} \cdot \dfrac{11}{6}$ Definition of division

$$= \dfrac{33}{24} \qquad \text{Multiply numerators; multiply denominators.}$$

$$= \dfrac{11}{8} \qquad \text{Divide numerator and denominator by 3.}$$

19. $10 \div \dfrac{5}{6} = \dfrac{10}{1} \cdot \dfrac{6}{5}$ Definition of division

$$= \dfrac{60}{5} \qquad \text{Multiply numerators; multiply denominators.}$$

$$= 12 \qquad \text{Divide.}$$

20. $-\dfrac{3}{8} \div 6 = -\dfrac{3}{8} \cdot \dfrac{1}{6}$ Definition of division

$\qquad\qquad\;\; = -\dfrac{3}{48}$ Multiply numerators; multiply denominators.

$\qquad\qquad\;\; = -\dfrac{1}{16}$ Divide numerator and denominator by 3.

Division with the Number 0

For every division problem an associated multiplication problem involving the same numbers exists. For example, the following two problems say the same thing about the numbers 2, 3, and 6:

Division   *Multiplication*

$\dfrac{6}{3} = 2 \qquad\qquad 6 = 2(3)$

We can use this relationship between division and multiplication to clarify division involving the number 0.

First, dividing 0 by a number other than 0 is allowed and always results in 0. To see this, consider dividing 0 by 5. We know the answer is 0 because of the relationship between multiplication and division. This is how we write it:

$$\frac{0}{5} = 0 \qquad \text{because} \qquad 0 = 0(5)$$

On the other hand, dividing a nonzero number by 0 is not allowed in the real numbers. Suppose we were attempting to divide 5 by 0. We don't know whether there is an answer to this problem, but if there is, let's say the answer is a number that we can represent with the letter n. If 5 divided by 0 is a number n, then

$$\frac{5}{0} = n \qquad \text{and} \qquad 5 = n(0)$$

But this is impossible, because no matter what number n is, when we multiply it by 0 the answer must be 0. It can never be 5. In algebra, we say expressions like $\frac{5}{0}$ are undefined, because there is no answer to them. That is, division by 0 is not allowed in the real numbers.

Ⓕ Finding the Value of an Algebraic Expression

As we mentioned earlier in this chapter, an algebraic expression is a combination of numbers, variables, and operation symbols. Each of the following is an algebraic expression:

$7a \qquad x^2 - y^2 \qquad 2(3t - 4) \qquad \dfrac{2x - 5}{6}$

An expression such as $2(3t - 4)$ will take on different values depending on what number we substitute for t. For example, if we substitute -8 for t then the expression $2(3t - 4)$ becomes $2[3(-8) - 4]$, which simplifies to -56. If we apply the distributive property to $2(3t - 4)$ we have

$$2(3t - 4) = 6t - 8$$

Substituting -8 for t in the simplified expression gives us $6(-8) - 8 = -56$, which is the same result we obtained previously. As you would expect, substituting the same number into an expression, and any simplified form of that expression, will yield the same result.

EXAMPLE 21

Evaluate the expressions $(a + 4)^2$, $a^2 + 16$, and $a^2 + 8a + 16$ when a is $-2, 0$, and 3.

SOLUTION   Organizing our work with a table, we have

a	$(a + 4)^2$	$a^2 + 16$	$a^2 + 8a + 16$
-2	$(-2 + 4)^2 = 4$	$(-2)^2 + 16 = 20$	$(-2)^2 + 8(-2) + 16 = 4$
0	$(0 + 4)^2 = 16$	$0^2 + 16 = 16$	$0^2 + 8(0) + 16 = 16$
3	$(3 + 4)^2 = 49$	$3^2 + 16 = 25$	$3^2 + 8(3) + 16 = 49$

When we study polynomials later in the book, you will see that the expressions $(a + 4)^2$ and $a^2 + 8a + 16$ are equivalent, and that neither one is equivalent to $a^2 + 16$.

GETTING READY FOR CLASS

After reading through the preceding section, respond in your own words and in complete sentences.

A. Explain the steps you use to add two real numbers.

B. Use symbols to define the word *difference*.

C. Why do we define division as multiplication by the reciprocal?

D. Find the value of the expression $3(4t - 2)$ if $t = 2$.

Problem Set 1.3

Find each of the following sums.

1. $6 + (-2)$ **2.** $11 + (-5)$ **3.** $-6 + 2$ **4.** $-11 + 5$

Find each of the following differences.

5. $-7 - 3$ **6.** $-6 - 9$ **7.** $-7 - (-3)$ **8.** $-6 - (-9)$

9. $\frac{3}{4} - \left(-\frac{5}{6}\right)$ **10.** $\frac{2}{3} - \left(-\frac{7}{5}\right)$ **11.** $\frac{11}{42} - \frac{17}{30}$ **12.** $\frac{13}{70} - \frac{19}{42}$

13. Subtract 5 from -3. **14.** Subtract -3 from 5.

15. Find the difference of -4 and 8. **16.** Find the difference of 8 and -4.

17. Subtract $4x$ from $-3x$. **18.** Subtract $-5x$ from $7x$.

19. What number do you subtract from 5 to get -8?

20. What number do you subtract from -3 to get 9?

21. Add -7 to the difference of 2 and 9.

22. Add -3 to the difference of 9 and 2.

23. Subtract $3a$ from the sum of $8a$ and a.

24. Subtract $-3a$ from the sum of $3a$ and $5a$.

Find the following products.

25. $3(-5)$ **26.** $-3(5)$ **27.** $-3(-5)$ **28.** $4(-6)$

29. $2(-3)(4)$ **30.** $-2(3)(-4)$ **31.** $-2(5x)$ **32.** $-5(4x)$

33. $-\frac{1}{3}(-3x)$ **34.** $-\frac{1}{6}(-6x)$ **35.** $-\frac{2}{3}\left(-\frac{3}{2}y\right)$ **36.** $-\frac{2}{5}\left(-\frac{5}{2}y\right)$

37. $-2(4x - 3)$ **38.** $-2(-5t + 6)$ **39.** $-\frac{1}{2}(6a - 8)$ **40.** $-\frac{1}{3}(6a - 9)$

Simplify each expression as much as possible. You will see each of these problems later in the book.

41. $1(-2) - 2(-16) + 1(9)$ **42.** $6(1) - 1(-5) + 1(2)$

43. $1(1) - 3(-2) + (-2)(-2)$ **44.** $-2(-14) + 3(-4) - 1(-10)$

45. $-4(0)(-2) - (-1)(1)(1) - 1(2)(3)$ **46.** $1(0)(1) + 3(1)(4) + (-2)(2)(-1)$

47. $1[0 - (-1)] - 3(2 - 4) + (-2)(-2 - 0)$

48. $-3(-1 - 1) + 4(-2 + 2) - 5[2 - (-2)]$

49. $3(-2)^2 + 2(-2) - 1$ **50.** $4(-1)^2 + 3(-1) - 2$

51. $2(-2)^3 - 3(-2)^2 + 4(-2) - 8$ **52.** $5 \cdot 2^3 - 3 \cdot 2^2 + 4 \cdot 2 - 5$

53. $\frac{0 - 4}{0 - 2}$ **54.** $\frac{0 + 6}{0 - 3}$ **55.** $\frac{-4 - 4}{-4 - 2}$ **56.** $\frac{6 + 6}{6 - 3}$

57. $\frac{-6 + 6}{-6 - 3}$ **58.** $\frac{4 - 4}{4 - 2}$ **59.** $\frac{-2 - 4}{2 - 2}$ **60.** $\frac{3 + 6}{3 - 3}$

61. $\frac{3 - (-1)}{-3 - 3}$ **62.** $\frac{-1 - 3}{3 - (-3)}$

The problems below are problems you will see later in the book. Apply the distributive property, then simplify if possible.

63. $-1(5 - x)$

64. $-1(a - b)$

65. $-1(7 - x)$

66. $-1(6 - y)$

67. $-3(2x - 3y)$

68. $-1(x - 2z)$

69. $6\left(\dfrac{x}{2} - 3\right)$

70. $6\left(\dfrac{x}{3} + 1\right)$

71. $12\left(\dfrac{a}{4} + \dfrac{1}{2}\right)$

72. $15\left(\dfrac{a}{3} + 2\right)$

73. $15\left(\dfrac{x}{5} + 4\right)$

74. $10\left(\dfrac{x}{2} - 9\right)$

75. $12\left(\dfrac{y}{2} + \dfrac{y}{4} + \dfrac{y}{6}\right)$

76. $12\left(\dfrac{y}{3} - \dfrac{y}{6} + \dfrac{y}{2}\right)$

Simplify each expression.

77. $3(5x + 4) - x$

78. $4(7x + 3) - x$

79. $6 - 7(3 - m)$

80. $3 - 5(5 - m)$

81. $7 - 2(3x - 1) + 4x$

82. $8 - 5(2x - 3) + 4x$

83. $5(3y + 1) - (8y - 5)$

84. $4(6y + 3) - (6y - 6)$

85. $4(2 - 6x) - (3 - 4x)$

86. $7(1 - 2x) - (4 - 10x)$

87. $10 - 4(2x + 1) - (3x - 4)$

88. $7 - 2(3x + 5) - (2x - 3)$

89. $3x - 5(x - 3) - 2(1 - 3x)$

90. $4x - 7(x - 3) + 2(4 - 5x)$

Use the definition of division to write each division problem as a multiplication problem, then simplify.

91. $\dfrac{4}{0}$

92. $\dfrac{-7}{0}$

93. $\dfrac{0}{-3}$

94. $\dfrac{0}{5}$

95. $-\dfrac{3}{4} \div \dfrac{9}{8}$

96. $-\dfrac{2}{3} \div \dfrac{4}{9}$

97. $-8 \div \left(-\dfrac{1}{4}\right)$

98. $-12 \div \left(-\dfrac{2}{3}\right)$

99. $-40 \div \left(-\dfrac{5}{8}\right)$

100. $-30 \div \left(-\dfrac{5}{6}\right)$

101. $\dfrac{4}{9} \div (-8)$

102. $\dfrac{3}{7} \div (-6)$

Simplify as much as possible.

103. $\dfrac{3(-1) - 4(-2)}{8 - 5}$

104. $\dfrac{6(-4) - 5(-2)}{7 - 6}$

105. $8 - (-6)\left[\dfrac{2(-3) - 5(4)}{-8(6) - 4}\right]$

106. $-9 - 5\left[\dfrac{11(-1) - 9}{4(-3) + 2(5)}\right]$

107. $6 - (-3)\left[\dfrac{2 - 4(3 - 8)}{1 - 5(1 - 3)}\right]$

108. $8 - (-7)\left[\dfrac{6 - 1(6 - 10)}{4 - 3(5 - 7)}\right]$

109.

		Sum	Difference	Product	Quotient
a	b	$a + b$	$a - b$	ab	$\dfrac{a}{b}$
3	12				
-3	12				
3	-12				
-3	-12				

110.

a	b	Sum $a + b$	Difference $a - b$	Product ab	Quotient $\frac{a}{b}$
8	2				
−8	2				
8	−2				
−8	−2				

111.

x	$3(5x - 2)$	$15x - 6$	$15x - 2$
−2			
−1			
0			
1			
2			

112.

x	$(x + 1)^2$	$x^2 + 1$	$x^2 + 2x + 1$
−2			
−1			
0			
1			
2			

113. Find the value of $-\frac{b}{2a}$ when

    a. $a = 3, b = -6$ **c.** $a = -1, b = -2$

    b. $a = -2, b = 6$ **d.** $a = -0.1, b = 27$

114. Find the value of $b^2 - 4ac$ when

    a. $a = 3, b = -2$, and $c = 4$

    b. $a = 1, b = -3$, and $c = -28$

    c. $a = 1, b = -6$, and $c = 9$

    d. $a = 0.1, b = -27$, and $c = 1700$

Use a calculator to simplify each expression. If rounding is necessary, round your answers to the nearest ten-thousandth (4 places past the decimal point). You will see these problems later in the book.

115. $\dfrac{1.380}{0.903}$ **116.** $\dfrac{1.0792}{0.6690}$ **117.** $\dfrac{1}{2}(-0.1587)$ **118.** $\dfrac{1}{2}(-0.7948)$

119. $\dfrac{1}{2}\left(\dfrac{1.2}{1.4} - 1\right)$ **120.** $\dfrac{1}{2}\left(\dfrac{1.3}{1.1} - 1\right)$ **121.** $\dfrac{(6.8)(3.9)}{7.8}$ **122.** $\dfrac{(2.4)(1.8)}{1.2}$

123. $\dfrac{0.0005(200)}{(0.25)^2}$ **124.** $\dfrac{0.0006(400)}{(0.25)^2}$

125. $-500 + 27(100) - 0.1(100)^2$ **126.** $-500 + 27(170) - 0.1(170)^2$

127. $-0.05(130)^2 + 9.5(130) - 200$ **128.** $-0.04(130)^2 + 8.5(130) - 210$

129. Time Zones The continental United States is divided into four time zones.

When it is 4:00 in the Pacific zone, it is 5:00 in the Mountain zone, 6:00 in the Central zone, and 7:00 in the Eastern zone. You board a plane at 6:55 p.m. in California (Pacific zone), and take 2 hours and 15 minutes to arrive in Santa Fe, New Mexico (Mountain zone), and another 3 hours and 20 minutes to arrive at your final destination, Detroit, Michigan (Eastern zone). At what local times did you arrive at Santa Fe and Detroit?

130. Ironman Triathlon In our number system, everything is in terms of powers of 10. With minutes and seconds, we think in terms of 60's. The format for the times in the chart is:

hours:minutes:seconds

Find the difference between each of the following Triathlon times:

a. Daniela Ryf and Mirinda Carfrae

b. Heather Jackson and Ania Beranek

c. Daniela Ryf and Kaisa Lehtonen

131. **Ironman Triathlon** The Ironman Triathlon World Championship, held each October in Kona on the island of Hawaii, consists of three parts: a 2.4-mile ocean swim, a 112-mile bike race, and a 26.2-mile marathon. The table shows results from the 2017 event.

Triathlete	Swim	Bike	Run	Total
Patrick Lange	0:48:45	4:28:53	2:39:59	
Lucy Charles	0:48:48	4:58:19	3:08:09	

a. Fill in the total time column.

b. How much faster was Patrick's total time than Lucy's?

c. How much faster was Patrick than Lucy in the bike race?

d. How much faster was Patrick than Lucy in the run?

132. **Ironman Triathlon** The Ironman Triathlon World Championship, held each October in Kona on the island of Hawaii, consists of three parts: a 2.4-mile ocean swim, a 112-mile bike race, and a 26.2-mile marathon. The table shows results from the 2017 event.

Triathlete	Swim	Bike	Run	Total
Heather Jackson	0:57:58	4:53:54	3:06:59	
Liz Lyles	1:00:08	5:04:10	3:11:21	

a. Fill in the total time column.

b. How much faster was Heather's total time than Liz's?

c. How much faster was Heather than Liz in the bike race?

d. How much faster was Heather than Liz in the run?

OBJECTIVES

A Simplify expressions using properties of exponents.

B Write numbers in scientific notation and in expanded form.

©iStockphoto.com/VLPA

1.4

The following figure shows a square and a cube, each with a side of length 1.5 centimeters. To find the area of the square, we raise 1.5 to the second power: 1.5^2. To find the volume of the cube, we raise 1.5 to the third power: 1.5^3.

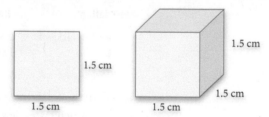

1.5 cm

1.5 cm

1.5 cm

1.5 cm

1.5 cm

Because the area of the square is 1.5^2, we say second powers are *squares*; that is, x^2 is read "x squared." Likewise, because the volume of the cube is 1.5^3, we say third powers are *cubes*; that is, x^3 is read "x cubed." Exponents and the vocabulary associated with them are topics we will study in this section.

Ⓐ Properties of Exponents

In this section, we will be concerned with the simplification of expressions that involve exponents. We begin by making some generalizations about exponents, based on specific examples.

EXAMPLE 1

Write the product $x^3 \cdot x^4$ with a single exponent.

SOLUTION

$$x^3 \cdot x^4 = (x \cdot x \cdot x)(x \cdot x \cdot x \cdot x)$$
$$= (x \cdot x \cdot x \cdot x \cdot x \cdot x \cdot x)$$
$$= x^7 \quad \textit{Notice: } 3 + 4 = 7$$

1.4 VIDEOS

We can generalize this result into the first property of exponents.

> **PROPERTY** *Property 1 for Exponents*
> ●
> If a is a real number and r and s are integers, then
> $$a^r \cdot a^s = a^{r+s}$$

EXAMPLE 2

Write $(5^3)^2$ with a single exponent.

SOLUTION

$$(5^3)^2 = 5^3 \cdot 5^3$$
$$= 5^6 \qquad \textit{Notice: } 3 \cdot 2 = 6.$$

Generalizing this result, we have a second property of exponents.

> **PROPERTY** *Property 2 for Exponents*
> ●
> If a is a real number and r and s are integers, then
> $$(a^r)^s = a^{r \cdot s}$$

A third property of exponents arises when we have the product of two or more numbers raised to an integer power.

EXAMPLE 3

Expand $(3x)^4$ and then multiply.

SOLUTION

$$(3x)^4 = (3x)(3x)(3x)(3x)$$
$$= (3 \cdot 3 \cdot 3 \cdot 3)(x \cdot x \cdot x \cdot x)$$
$$= 3^4 \cdot x^4 \qquad \textit{Notice: } \text{The exponent 4 distributes over the product } 3x.$$
$$= 81x^4$$

Generalizing Example 3, we have Property 3 for exponents.

> **PROPERTY** *Property 3 for Exponents*
> ●
> If a and b are any two real numbers and r is an integer, then
> $$(ab)^r = a^r \cdot b^r$$

Here are some examples that use combinations of the first three properties of exponents to simplify expressions involving exponents.

EXAMPLE

Simplify each expression using the properties of exponents.

4. $(-3x^2)(5x^4) = -3(5)(x^2 \cdot x^4)$ Commutative and associative properties

$\qquad\qquad\quad = -15x^6$ Property 1 for exponents

5. $(-2x^2)^3(4x^5) = (-2)^3(x^2)^3(4x^5)$ Property 3

$\qquad\qquad\quad = -8x^6 \cdot (4x^5)$ Property 2

$\qquad\qquad\quad = (-8 \cdot 4)(x^6 \cdot x^5)$ Commutative and associative properties

$\qquad\qquad\quad = -32x^{11}$ Property 1

6. $(x^2)^4(x^2y^3)^2(y^4)^3 = x^8 \cdot x^4 \cdot y^6 \cdot y^{12}$ Properties 2 and 3

$\qquad\qquad\qquad\quad = x^{12}y^{18}$ Property 1

The next property of exponents deals with negative integer exponents.

NOTE This property is actually a definition. That is, we are defining negative integer exponents as indicating reciprocals. Doing so gives us a way to write an expression with a negative exponent as an equivalent expression with a positive exponent.

PROPERTY   *Property 4 for Exponents*

• •

If a is any nonzero real number and r is a positive integer, then

$$a^{-r} = \frac{1}{a^r}$$

EXAMPLE

Write with positive exponents, and then simplify.

7. $5^{-2} = \dfrac{1}{5^2} = \dfrac{1}{25}$

8. $(-2)^{-3} = \dfrac{1}{(-2)^3} = \dfrac{1}{-8} = -\dfrac{1}{8}$

9. $\left(\dfrac{3}{4}\right)^{-2} = \dfrac{1}{\left(\frac{3}{4}\right)^2} = \dfrac{1}{\frac{9}{16}} = \dfrac{16}{9}$

If we generalize the result in Example 9, we have the following extension of Property 4,

$$\left(\frac{a}{b}\right)^{-r} = \left(\frac{b}{a}\right)^{r}$$

which indicates that raising a fraction to a negative power is equivalent to raising the reciprocal of the fraction to the positive power.

Property 3 indicated that exponents distribute over products. Because division is defined in terms of multiplication, we can expect that exponents will distribute over quotients as well. Property 5 is the formal statement of this fact.

> **PROPERTY** *Property 5 for Exponents*
>
> If a and b are any two real numbers with $b \neq 0$, and r is an integer, then
>
> $$\left(\frac{a}{b}\right)^r = \frac{a^r}{b^r}$$

Proof of Property 5

$$\left(\frac{a}{b}\right)^r = \underbrace{\left(\frac{a}{b}\right)\left(\frac{a}{b}\right)\left(\frac{a}{b}\right) \cdots \left(\frac{a}{b}\right)}_{r \text{ factors}}$$

$$= \frac{a \cdot a \cdot a \cdots a}{b \cdot b \cdot b \cdots b} \quad \begin{matrix} \leftarrow r \text{ factors} \\ \leftarrow r \text{ factors} \end{matrix}$$

$$= \frac{a^r}{b^r}$$

Because multiplication with the same base resulted in addition of exponents, it seems reasonable to expect division with the same base to result in subtraction of exponents.

> **PROPERTY** *Property 6 for Exponents*
>
> If a is any nonzero real number, and r and s are any two integers, then
>
> $$\frac{a^r}{a^s} = a^{r-s}$$

Notice again that we have specified r and s to be any integers. Our definition of negative exponents is such that the properties of exponents hold for all integer exponents, whether positive or negative integers. Here is a proof of Property 6.

Proof of Property 6
Our proof is centered on the fact that division by a number is equivalent to multiplication by the reciprocal of the number.

$$\frac{a^r}{a^s} = a^r \cdot \frac{1}{a^s} \qquad \text{Dividing by } a^s \text{ is equivalent to multiplying by } \frac{1}{a^s}.$$

$$= a^r a^{-s} \qquad \text{Property 4}$$

$$= a^{r+(-s)} \qquad \text{Property 1}$$

$$= a^{r-s} \qquad \text{Definition of subtraction}$$

EXAMPLE 10

Apply Property 6 to each expression, and then simplify the result. All answers that contain exponents should contain positive exponents only.

a. $\dfrac{2^8}{2^3} = 2^{8-3} = 2^5 = 32$

b. $\dfrac{x^2}{x^{18}} = x^{2-18} = x^{-16} = \dfrac{1}{x^{16}}$

c. $\dfrac{a^6}{a^{-8}} = a^{6-(-8)} = a^{14}$

d. $\dfrac{m^{-5}}{m^{-7}} = m^{-5-(-7)} = m^2$

Let's complete our list of properties by looking at how the numbers 0 and 1 behave when used as exponents.

We can use the original definition for exponents when the number 1 is used as an exponent.

$$a^1 = a$$
$$\uparrow$$
1 factor

For 0 as an exponent, consider the expression $\dfrac{3^4}{3^4}$. Because $3^4 = 81$, we have

$$\frac{3^4}{3^4} = \frac{81}{81} = 1$$

However, because we have the quotient of two expressions with the same base, we can subtract exponents.

$$\frac{3^4}{3^4} = 3^{4-4} = 3^0$$

Hence, 3^0 must be the same as 1.

Summarizing these results, we have our last property for exponents.

PROPERTY   *Property 7 for Exponents*
· ·

If a is any real number, then

$$a^1 = a$$

and

$$a^0 = 1 \quad \text{(as long as } a \neq 0\text{)}$$

EXAMPLE 11

Simplify.

a. $(2x^2y^4)^0 = 1$

b. $(2x^2y^4)^1 = 2x^2y^4$

Here are some examples that use many of the properties of exponents. There are a number of ways to proceed on problems like these. You should use the method that works best for you.

EXAMPLE

Simplify.

12. $\dfrac{(x^3)^{-2}(x^4)^5}{(x^{-2})^7} = \dfrac{x^{-6}x^{20}}{x^{-14}}$ Property 2

$\qquad\qquad\qquad = \dfrac{x^{14}}{x^{-14}}$ Property 1

$\qquad\qquad\qquad = x^{28}$ Property 6: $x^{14-(-14)} = x^{28}$

13. $\dfrac{6a^5b^{-6}}{12a^3b^{-9}} = \dfrac{6}{12} \cdot \dfrac{a^5}{a^3} \cdot \dfrac{b^{-6}}{b^{-9}}$ Write as separate fractions.

$\qquad\qquad\quad = \dfrac{1}{2}a^2b^3$ Property 6

14. $\dfrac{(4x^{-5}y^3)^2}{(x^4y^{-6})^{-3}} = \dfrac{16x^{-10}y^6}{x^{-12}y^{18}}$ Properties 2 and 3

$\qquad\qquad\qquad = 16x^2y^{-12}$ Property 6

$\qquad\qquad\qquad = 16x^2 \cdot \dfrac{1}{y^{12}}$ Property 4

$\qquad\qquad\qquad = \dfrac{16x^2}{y^{12}}$ Multiplication

> **NOTE** Example 13 can also be written as $\dfrac{a^2b^3}{2}$. Either answer is correct.

B   Scientific Notation

Scientific notation is a method for writing very large or very small numbers in a more manageable form. Here is the definition.

> **DEFINITION** *scientific notation*
>
> A number is written in *scientific notation* if it is written as the product of a number between 1 and 10 and an integer power of 10. A number written in scientific notation has the form
>
> $$n \times 10^r$$
>
> where $1 \le n < 10$ and $r =$ an integer.

EXAMPLE 15

Write 376,000 in scientific notation.

SOLUTION We must rewrite 376,000 as the product of a number between 1 and 10 and a power of 10. To do so, we move the decimal point five places to the left so that it appears between the 3 and the 7. Then we multiply this number by 10^5. The number that results has the same value as our original number and is written in scientific notation.

$$376,000 = 3.76 \times 10^5$$

Move decimal
five places

Decimal point
originally here

Keep track of the
five places we moved
the decimal point

If a number written in expanded form is greater than or equal to 10, then when the number is written in scientific notation the exponent on 10 will be positive. A number that is less than 1 and greater than 0 will have a negative exponent when written in scientific notation.

EXAMPLE 16

Write 4.52×10^3 in expanded form.

SOLUTION   Because 10^3 is 1,000, we can think of this as simply a multiplication problem. That is,

$$4.52 \times 10^3 = 4.52 \times 1,000 = 4,520$$

On the other hand, we can think of the exponent 3 as indicating the number of places we need to move the decimal point to write our number in expanded form. Because our exponent is positive 3, we move the decimal point three places to the right.

$$4.52 \times 10^3 = 4,520$$

The following table lists some additional examples of numbers written in expanded form and in scientific notation. In each case, note the relationship between the number of places the decimal point is moved and the exponent on 10.

Number Written in Expanded Form		Number Written in Scientific Notation
376,000	=	3.76×10^5
49,500	=	4.95×10^4
3,200	=	3.2×10^3
591	=	5.91×10^2
46	=	4.6×10^1
8	=	8×10^0
0.47	=	4.7×10^{-1}
0.093	=	9.3×10^{-2}
0.00688	=	6.88×10^{-3}
0.0002	=	2×10^{-4}
0.000098	=	9.8×10^{-5}

Calculator Note   Some scientific calculators have a key that allows you to enter numbers in scientific notation. The key is labeled

$$\boxed{\text{EXP}} \quad \text{or} \quad \boxed{\text{EE}} \quad \text{or} \quad \boxed{\text{SCI}}$$

To enter the number 3.45×10^6, you would enter the decimal number, press the scientific notation key, and then enter the exponent:

$$3.45 \ \boxed{\text{EXP}} \ 6$$

To enter 6.2×10^{-27}, you would use the following sequence:

$$6.2 \ \boxed{\text{EXP}} \ 27 \ \boxed{+/-}$$

We can use the properties of exponents to do arithmetic with numbers written in scientific notation. Here are some examples.

EXAMPLE 17

Simplify each expression and write all answers in scientific notation.

a. $(2 \times 10^8)(3 \times 10^{-3}) = (2)(3) \times (10^8)(10^{-3})$
$$= 6 \times 10^5$$

b. $\dfrac{4.8 \times 10^9}{2.4 \times 10^{-3}} = \dfrac{4.8}{2.4} \times \dfrac{10^9}{10^{-3}}$
$$= 2 \times 10^{9-(-3)}$$
$$= 2 \times 10^{12}$$

c. $\dfrac{(6.8 \times 10^5)(3.9 \times 10^{-7})}{7.8 \times 10^{-4}} = \dfrac{(6.8)(3.9)}{7.8} \times \dfrac{(10^5)(10^{-7})}{10^{-4}}$
$$= 3.4 \times 10^2$$

GETTING READY FOR CLASS

After reading through the preceding section, respond in your own words and in complete sentences.

A. Explain the difference between -2^4 and $(-2)^4$.

B. Explain the difference between 2^5 and 2^{-5}.

C. If a positive base is raised to a negative exponent, can the result be a negative number?

D. State Property 1 for exponents in your own words.

Problem Set 1.4

Evaluate each of the following.

1. 4^2 **2.** $(-4)^2$ **3.** -4^2 **4.** $-(-4)^2$

5. -0.3^3 **6.** $(-0.3)^3$ **7.** 2^5 **8.** 2^4

9. $\left(\dfrac{1}{2}\right)^3$ **10.** $\left(\dfrac{3}{4}\right)^2$ **11.** $\left(-\dfrac{5}{6}\right)^2$ **12.** $\left(-\dfrac{7}{8}\right)^2$

Use the properties of exponents to simplify each of the following as much as possible.

13. $x^5 \cdot x^4$ **14.** $x^6 \cdot x^3$ **15.** $(2^3)^2$ **16.** $(3^2)^2$

17. $-3a^2(2a^4)$ **18.** $5a^7(-4a^6)$

Write each of the following with positive exponents. Then simplify as much as possible.

19. 3^{-2} **20.** $(-5)^{-2}$ **21.** $(-2)^{-5}$

22. 2^{-5} **23.** $\left(\dfrac{3}{4}\right)^{-2}$ **24.** $\left(\dfrac{3}{5}\right)^{-2}$

25. $\left(\dfrac{1}{3}\right)^{-2} + \left(\dfrac{1}{2}\right)^{-3}$ **26.** $\left(\dfrac{1}{2}\right)^{-2} + \left(\dfrac{1}{3}\right)^{-3}$

Problems 27–44 are problems you will see later in the book. Multiply.

27. $8x^3 \cdot 10y^6$ **28.** $5y^2 \cdot 4x^2$ **29.** $8x^3 \cdot 9y^3$ **30.** $4y^3 \cdot 3x^2$

31. $3x \cdot 5y$ **32.** $3xy \cdot 5z$ **33.** $4x^6y^6 \cdot 3x$ **34.** $16x^4y^4 \cdot 3y$

35. $27a^6c^3 \cdot 2b^2c$ **36.** $8a^3b^3 \cdot 5a^2b$

Divide. (Assume all variables are nonzero.)

37. $\dfrac{10x^5}{5x^2}$ **38.** $\dfrac{-15x^4}{5x^2}$ **39.** $\dfrac{20x^3}{5x^2}$ **40.** $\dfrac{25x^7}{-5x^2}$

41. $\dfrac{8x^3y^5}{-2x^2y}$ **42.** $\dfrac{-16x^2y^2}{-2x^2y}$ **43.** $\dfrac{4x^4y^3}{-2x^2y}$ **44.** $\dfrac{10a^4b^2}{4a^2b^2}$

Use the properties of exponents to simplify each expression. Write all answers with positive exponents only. (Assume all variables are nonzero.)

45. $\dfrac{x^{-1}}{x^9}$ **46.** $\dfrac{x^{-3}}{x^5}$ **47.** $\dfrac{a^4}{a^{-6}}$ **48.** $\dfrac{a^5}{a^{-2}}$

49. $\dfrac{t^{-10}}{t^{-4}}$ **50.** $\dfrac{t^{-8}}{t^{-5}}$ **51.** $\left(\dfrac{x^5}{x^3}\right)^6$ **52.** $\left(\dfrac{x^7}{x^4}\right)^5$

53. $\dfrac{(x^5)^6}{(x^3)^4}$ **54.** $\dfrac{(x^7)^3}{(x^4)^5}$ **55.** $\dfrac{(x^{-2})^3(x^3)^{-2}}{x^{10}}$ **56.** $\dfrac{(x^{-4})^3(x^3)^{-4}}{x^{10}}$

57. $\dfrac{5a^8b^3}{20a^5b^{-4}}$ **58.** $\dfrac{7a^6b^{-2}}{21a^2b^{-5}}$ **59.** $\dfrac{(3x^{-2}y^8)^4}{(9x^4y^{-3})^2}$ **60.** $\dfrac{(6x^{-3}y^{-5})^2}{(3x^{-4}y^{-3})^4}$

61. $\left(\dfrac{8x^2y}{4x^4y^{-3}}\right)^4$ **62.** $\left(\dfrac{5x^4y^5}{10xy^{-2}}\right)^3$ **63.** $\left(\dfrac{x^{-5}y^2}{x^{-3}y^5}\right)^{-2}$ **64.** $\left(\dfrac{x^{-8}y^{-3}}{x^{-5}y^6}\right)^{-1}$

65. $\left(\dfrac{ab^{-3}c^{-2}}{a^{-3}b^0c^{-5}}\right)^0$ **66.** $\left(\dfrac{a^3b^2c^1}{a^{-1}b^{-2}c^{-3}}\right)^0$

Write each number in scientific notation.

67. 378,000 **68.** 3,780,000 **69.** 4,900 **70.** 490

71. 0.00037 **72.** 0.000037 **73.** 0.00495 **74.** 0.0495

Write each number in expanded form.

75. 5.34×10^3 **76.** 5.34×10^2 **77.** 7.8×10^6 **78.** 7.8×10^4

79. 3.44×10^{-3} **80.** 3.44×10^{-5} **81.** 4.9×10^{-1} **82.** 4.9×10^{-2}

Use the properties of exponents to simplify each of the following expressions. Write all answers in scientific notation.

83. $(4 \times 10^{10})(2 \times 10^{-6})$

84. $(3 \times 10^{-12})(3 \times 10^4)$

85. $\dfrac{8 \times 10^{14}}{4 \times 10^5}$

86. $\dfrac{6 \times 10^8}{2 \times 10^3}$

87. $\dfrac{(5 \times 10^6)(4 \times 10^{-8})}{8 \times 10^4}$

88. $\dfrac{(6 \times 10^{-7})(3 \times 10^9)}{5 \times 10^6}$

89. $\dfrac{(2.4 \times 10^{-3})(3.6 \times 10^{-7})}{(4.8 \times 10^6)(1 \times 10^{-9})}$

90. $\dfrac{(7.5 \times 10^{-6})(1.5 \times 10^9)}{(1.8 \times 10^4)(2.5 \times 10^{-2})}$

Simplify. Write answers without using scientific notation, rounding to the nearest whole number

91. $\dfrac{2.00 \times 10^8}{3.98 \times 10^6}$

92. $\dfrac{2.00 \times 10^8}{3.16 \times 10^5}$

Use a calculator to find each of the following. Write your answer in scientific notation with the first number in each answer rounded to the nearest tenth.

93. Complete the following table.

Expanded Form	Scientific Notation $n \times 10^r$
0.000357	3.57×10^{-4}
0.00357	
0.0357	
0.357	
3.57	
35.7	
357	
3,570	
35,700	

94. Complete the following table.

Expanded Form	Scientific Notation $n \times 10^r$
0.000123	1.23×10^{-4}
	1.23×10^{-3}
	1.23×10^{-2}
	1.23×10^{-1}
	1.23×10^0
	1.23×10^1
	1.23×10^2
	1.23×10^3
	1.23×10^4

NASA

Galilean Moons The planet Jupiter has about 60 known moons. In the year 1610 Galileo first discovered the four largest moons of Jupiter: Io, Europa, Ganymede, and Callisto. These moons are known as the Galilean moons. Each moon has a unique period, or the time it takes to make a trip around Jupiter. Fill in the tables below.

95.

Jupiter's Moon	Period (seconds)
Io	153,000
Europa	3.07×10^5
Ganymede	618,000
Callisto	1.44×10^6

96.

Jupiter's Moon	Distance from Jupiter (Kilometers)
Io	422,000
Europa	6.17×10^5
Ganymede	1,070,000
Callisto	1.88×10^6

97. Large Numbers If you are 20 years old, you have been alive for more than 630,000,000 seconds. Write this last number in scientific notation.

98. Fingerprints The FBI has been collecting fingerprint cards since 1924. Their collection has grown to over 200 million cards. They are digitizing the fingerprints. Each fingerprint card turns into about 10 MB of data. (A megabyte [MB] is $2^{20} \approx 1$ million bytes.)

  a. How many bytes of storage will they need?

  b. A compression routine called the WSQ method will compress the bytes by ratio of 12.9 to 1. Approximately how many bytes of storage will the FBI need for the compressed data? (Hint: Divide by 12.9.)

99. Ad Networks The chart shows the number of ads served from different ad networks.

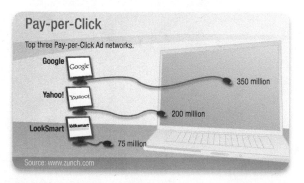

For each of the following search engines, write the number of ads served in scientific notation.

  a. Google   **b.** Yahoo   **c.** LookSmart

100. Our Galaxy The galaxy in which Earth resides is called the Milky Way galaxy. It is a spiral galaxy that contains approximately 200,000,000,000 stars (our Sun is one of them). Write this number in words and in scientific notation.

101. Credit Card Debt Outstanding credit card debt in the United States is over $422 billion. Write the number 422 billion in scientific notation.

102. Computer Science We all use the language of computers to indicate how much memory our computers hold or how much information we can put on a storage device such as a keychain drive. Scientific notation gives us a way to compare the actual numbers associated with the words we use to describe data storage in computers. The smallest amount of data that a computer can hold is measured in bits. A byte is the next largest unit and is equal to 8, or 2^3, bits. Fill in the table below.

NASA/JPL-CalTech

Number of Bytes		
Unit	Exponential Form	Scientific Notation
Kilobyte	$2^{10} = 1{,}024$	
Megabyte	$2^{20} \approx 1{,}048{,}000$	
Gigabyte	$2^{30} \approx 1{,}074{,}000{,}000$	
Terabyte	$2^{40} \approx 1{,}099{,}500{,}000{,}000$	

EXTENDING THE CONCEPTS

Assume all variable exponents represent positive integers, and simplify each expression.

103. $x^{m+2} \cdot x^{-2m} \cdot x^{m-5}$

104. $x^{m-4}x^{m+9}x^{-2m}$

105. $(y^m)^2(y^{-3})^m(y^{m+3})$

106. $(y^m)^{-4}(y^3)^m(y^{m+6})$

107. $\dfrac{x^{n+2}}{x^{n-3}}$

108. $\dfrac{x^{n-3}}{x^{n-7}}$

109. $x^{a+3}x^{2a-5}x^{7-9a}$

110. $x^{2b-3}x^{5-8b}x^{-4-3b}$

111. $(x^{m+3})^2(x^{2m-5})^3$

112. $(x^{4m-1})^3(x^{m+2})^4$

113. $\dfrac{x^{a+5}}{x^{3a-7}}$

114. $\dfrac{x^{2b-6}}{x^{5b+2}}$

115. $\dfrac{(x^{m+3})^2}{(x^{2m-1})^3}$

116. $\dfrac{(x^{4m-3})^4}{(x^{2m+1})^3}$

117. $\left(\dfrac{x^{4a-3}}{x^{-2a+1}}\right)^2$

118. $\left(\dfrac{x^{-3a+2}}{x^{-5a-4}}\right)^3$

 OBJECTIVES

A   Identify monomials and polynomials.

B   Add polynomials.

C   Subtract polynomials.

D   Multiply polynomials.

E   Multiply binomials using the FOIL method.

F   Use formulas to expand binomial squares.

G   Find products using the difference of two squares.

1.5

©iStockphoto.com/PierreOlivierClementMantion

The following chart is from a company that duplicates DVDs. It shows the revenue and cost to duplicate a 30-minute DVD. From the chart, you can see that 300 copies will bring in $900 in revenue, with a cost of $600. The profit is the difference between revenue and cost, or $300.

Revenue and Cost to Duplicate a 30-Minute DVD

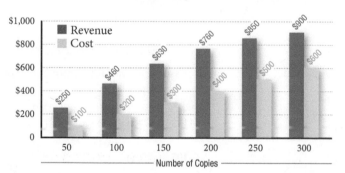

The relationship between profit, revenue, and cost is one application of the polynomials we will study in this section. Let's begin with a definition that we will use to build polynomials.

Polynomials in General

> **DEFINITION**   *term/monomial*
>
> A *term,* or *monomial,* is a constant or the product of a constant and one or more variables raised to whole-number exponents.

The following are monomials, or terms:

$$-16 \qquad 3x^2y \qquad -\frac{2}{5}a^3b^2c \qquad xy^2z$$

The numerical part of each monomial is called the *numerical coefficient,* or just *coefficient* for short. For the preceding terms, the coefficients are -16, 3, $-\frac{2}{5}$, and 1. Notice that the coefficient for xy^2z is understood to be 1.

> **DEFINITION** *polynomial*
>
> A *polynomial* is any finite sum of terms. Because subtraction can be written in terms of addition, finite differences are also included in this definition.

The following are polynomials:

$$2x^2 - 6x + 3 \qquad -5x^2y + 2xy^2 \qquad 4a - 5b + 6c + 7d$$

Polynomials can be classified further according to the number of terms present. If a polynomial consists of two terms, it is said to be a *binomial.* If it has three terms, it is called a *trinomial.* And, as stated, a polynomial with only one term is said to be a *monomial.*

> **DEFINITION** *degree*
>
> The *degree* of a polynomial with one variable is the highest power to which the variable is raised in any one term.

EXAMPLE

1. $6x^2 + 2x - 1$ A trinomial of degree 2

2. $5x - 3$ A binomial of degree 1

3. $7x^6 - 5x^3 + 2x - 4$ A polynomial of degree 6

4. $-7x^4$ A monomial of degree 4

5. 15 A monomial of degree 0

Polynomials with one variable are usually written in decreasing powers of the variable. When this is the case, the coefficient of the first term is called the *leading coefficient.* In Example 1, the leading coefficient is 6. In Example 2, it is 5. The leading coefficient in Example 3 is 7.

> **DEFINITION** *similar/like terms*
>
> Two or more terms that differ only in their numerical coefficients are called *similar,* or *like,* terms. Because similar terms differ only in their coefficients, they have identical variable parts.

B Adding Polynomials

To add two polynomials, we simply apply the commutative and associative properties to group similar terms together and then use the distributive property as we have in the following example.

EXAMPLE 6

Add $5x^2 - 4x + 2$ and $3x^2 + 9x - 6$.

SOLUTION

$(5x^2 - 4x + 2) + (3x^2 + 9x - 6)$

$\qquad = (5x^2 + 3x^2) + (-4x + 9x) + (2 - 6)$ Commutative and associative properties

$\qquad = (5 + 3)x^2 + (-4 + 9)x + (2 - 6)$ Distributive property

$\qquad = 8x^2 + 5x + (-4)$

$\qquad = 8x^2 + 5x - 4$

EXAMPLE 7

Find the sum of $-8x^3 + 7x^2 - 6x + 5$ and $10x^3 + 3x^2 - 2x - 6$.

SOLUTION   We can add the two polynomials using the method of Example 6, or we can arrange similar terms in columns and add vertically. Using the column method, we have

$$\begin{array}{r} -8x^3 + 7x^2 - 6x + 5 \\ \underline{10x^3 + 3x^2 - 2x - 6} \\ 2x^3 + 10x^2 - 8x - 1 \end{array}$$

C Subtracting Polynomials

To find the difference of two polynomials, we need to use the fact that the opposite of a sum is the sum of the opposites. That is,

$$-(a + b) = -a + (-b)$$

One way to remember this is to observe that $-(a + b)$ is equivalent to $-1(a + b) = (-1)a + (-1)b = -a + (-b)$.

If there is a negative sign directly preceding the parentheses surrounding a polynomial, we may remove the parentheses and preceding negative sign by changing the sign of each term within the parentheses. For example:

$$-(3x + 4) = -3x + (-4) = -3x - 4$$

$$-(5x^2 - 6x + 9) = -5x^2 + 6x - 9$$

$$-(-x^2 + 7x - 3) = x^2 - 7x + 3$$

EXAMPLE 8

Subtract $4x^2 + 2x - 3$ from $9x^2 - 3x + 5$.

SOLUTION First, we write the problem in terms of subtraction. Then, we subtract by adding the opposite of each term in the polynomial that follows the subtraction sign:

$(9x^2 - 3x + 5) - (4x^2 + 2x - 3)$

$= 9x^2 - 3x + 5 + (-4x^2) + (-2x) + 3$ The opposite of a sum is the sum of the opposites.

$= (9x^2 - 4x^2) + (-3x - 2x) + (5 + 3)$ Commutative and associative properties

$= 5x^2 - 5x + 8$ Combine similar terms.

When one set of grouping symbols is contained within another, it is best to begin the process of simplification within the innermost grouping symbols and work outward from there.

EXAMPLE 9

Simplify: $4x - 3[2 - (3x + 4)]$.

SOLUTION Removing the innermost parentheses first, we have

$4x - 3[2 - (3x + 4)]$

$= 4x - 3(2 - 3x - 4)$

$= 4x - 3(-3x - 2)$

$= 4x + 9x + 6$

$= 13x + 6$

In the example that follows, we will find the value of a polynomial for a given value of the variable.

EXAMPLE 10

Find the value of $5x^3 - 3x^2 + 4x - 5$ when x is 2.

SOLUTION We begin by substituting 2 for x in the original polynomial:

When $\qquad\qquad\qquad x = 2$

the polynomial $\qquad\quad 5x^3 - 3x^2 + 4x - 5$

becomes $\qquad\qquad 5 \cdot 2^3 - 3 \cdot 2^2 + 4 \cdot 2 - 5 = 5 \cdot 8 - 3 \cdot 4 + 4 \cdot 2 - 5$

$= 40 - 12 + 8 - 5$

$= 31$

Three quantities that occur very frequently in business and economics are profit, revenue, and cost. If a company manufactures and sells x items, then the revenue R is the total amount of money obtained by selling all x items. The cost C is the total amount of money it costs the company to manufacture the x items. The profit P obtained by selling all x items is the difference between the revenue and the cost and is given by the equation

$$P = R - C$$

EXAMPLE 11

A company produces and sells copies of an accounting program for home computers. The total weekly cost (in dollars) for the company to produce x copies of the program is $C = 8x + 500$. Find the weekly profit if the total revenue obtained from selling all x programs is $R = 35x - 0.1x^2$. How much profit will the company make if they produce and sell 100 copies of the program a week?

SOLUTION   Using the equation $P = R - C$ and the information given in the problem, we have

$$P = R - C$$
$$= 35x - 0.1x^2 - (8x + 500)$$
$$= 35x - 0.1x^2 - 8x - 500$$
$$= -500 + 27x - 0.1x^2 \qquad \text{Weekly profit}$$

If the company produces and sells 100 copies of the program, the weekly profit will be

$$P = -500 + 27(100) - 0.1(100)^2$$
$$= -500 + 27(100) - 0.1(10{,}000)$$
$$= -500 + 2{,}700 - 1{,}000$$
$$= 1{,}200$$

The weekly profit is $1,200 when 100 copies are sold.

Ⓓ Multiplying Polynomials

The distributive property is the key to multiplying polynomials. The simplest type of multiplication occurs when we multiply a polynomial by a monomial.

EXAMPLE 12

Find the product of $4x^3$ and $5x^2 - 3x + 1$.

SOLUTION   To multiply, we apply the distributive property:

$$4x^3(5x^2 - 3x + 1) = 4x^3(5x^2) + 4x^3(-3x) + 4x^3(1) \quad \text{Distributive property}$$
$$= 20x^5 - 12x^4 + 4x^3$$

NOTE We multiply coefficients and add exponents.

The distributive property can also be applied to multiply a polynomial by a polynomial. Let's consider the case where both polynomials have two terms.

EXAMPLE 13

Multiply $2x - 3$ and $x + 5$.

SOLUTION   Distributing the $2x - 3$ across the sum $x + 5$ gives us

$$(2x - 3)(x + 5) = (2x - 3)x + (2x - 3)5 \qquad \text{Distributive property}$$
$$= 2x(x) + (-3)x + 2x(5) + (-3)5 \qquad \text{Distributive property}$$
$$= 2x^2 - 3x + 10x - 15$$
$$= 2x^2 + 7x - 15 \qquad \text{Combine like terms.}$$

Notice the third line in Example 13. It consists of all possible products of terms in the first binomial and those of the second binomial. We can generalize this into a rule for multiplying two polynomials.

> **RULE** *Multiplying Two Polynomials*
>
> Multiply each term in the first polynomial by each term in the second polynomial.

Multiplying polynomials can be accomplished by a method that looks very similar to long multiplication with whole numbers. Reworking Example 13, we have

$$
\begin{array}{ll}
2x - 3 & \\
\underline{x + 5} & \\
2x^2 - 3x & \text{Multiply } x \text{ by } 2x - 3. \\
\underline{\quad\quad 10x - 15} & \text{Multiply } + 5 \text{ by } 2x - 3. \\
2x^2 + 7x - 15 & \text{Add in columns.}
\end{array}
$$

EXAMPLE 14

Multiply $2x - 3y$ and $3x^2 - xy + 4y^2$ vertically.

SOLUTION

$$
\begin{array}{ll}
3x^2 - \quad xy + 4y^2 & \\
\underline{2x - 3y} & \\
6x^3 - \quad 2x^2y + \quad 8xy^2 & \text{Multiply } 3x^2 - xy + 4y^2 \text{ by } 2x. \\
\underline{\quad - \quad 9x^2y + \quad 3xy^2 - 12y^3} & \text{Multiply } 3x^2 - xy + 4y^2 \text{ by } -3y. \\
6x^3 - 11x^2y + 11xy^2 - 12y^3 & \text{Add similar terms.}
\end{array}
$$

E Multiplying Binomials—The FOIL Method

The product of two binomials occurs very frequently in algebra. Because this type of product is so common, we have a special method of multiplication that applies only to products of binomials.

Consider the product of $2x - 5$ and $3x - 2$. Distributing $3x - 2$ over $2x$ and -5, we have

$$(2x - 5)(3x - 2) = (2x)(3x - 2) + (-5)(3x - 2)$$

$$= (2x)(3x) + (2x)(-2) + (-5)(3x) + (-5)(-2)$$

$$= 6x^2 - 4x - 15x + 10$$

$$= 6x^2 - 19x + 10$$

Looking closely at the second and third lines, we notice the following relationships:

a. $6x^2$ comes from multiplying the *first* terms in each binomial:

$$(2x - 5)(3x - 2) \qquad 2x(3x) = 6x^2 \qquad \textit{First terms}$$

b. $-4x$ comes from multiplying the *outside* terms in the product:

$$(2x - 5)(3x - 2) \qquad 2x(-2) = -4x \qquad \textit{Outside} \text{ terms}$$

c. $-15x$ comes from multiplying the *inside* terms in the product:

$$(2x - 5)(3x - 2) \qquad -5(3x) = -15x \quad \textit{Inside} \text{ terms}$$

d. 10 comes from multiplying the *last* two terms in the product:

$$(2x - 5)(3x - 2) \qquad -5(-2) = 10 \qquad \textit{Last} \text{ terms}$$

Once we know where the terms in the answer come from, we can reduce the number of steps used in finding the product:

$$(2x - 5)(3x - 2) = \underset{\text{First}}{6x^2} - \underset{\text{Outside}}{4x} - \underset{\text{Inside}}{15x} + \underset{\text{Last}}{10}$$

$$= 6x^2 - 19x + 10$$

This is called the *FOIL method*.

EXAMPLE

Multiply using the FOIL method.

15. $(4a - 5b)(3a + 2b) = \underset{F}{12a^2} + \underset{O}{8ab} - \underset{I}{15ab} - \underset{L}{10b^2}$

$$= 12a^2 - 7ab - 10b^2$$

16. $(3 - 2t)(4 + 7t) = \underset{F}{12} + \underset{O}{21t} - \underset{I}{8t} - \underset{L}{14t^2}$

$$= 12 + 13t - 14t^2$$

17. $\left(2x + \dfrac{1}{2}\right)\left(4x - \dfrac{1}{2}\right) = \underset{F}{8x^2} - \underset{O}{x} + \underset{I}{2x} - \underset{L}{\dfrac{1}{4}}$

$$= 8x^2 + x - \dfrac{1}{4}$$

18. $(a^5 + 3)(a^5 - 7) = \underset{F}{a^{10}} - \underset{O}{7a^5} + \underset{I}{3a^5} - \underset{L}{21}$

$$= a^{10} - 4a^5 - 21$$

19. $(2x + 3)(5y - 4) = \underset{F}{10xy} - \underset{O}{8x} + \underset{I}{15y} - \underset{L}{12}$

F The Square of a Binomial

EXAMPLE 20

Find $(4x - 6)^2$.

SOLUTION Applying the definition of exponents and then the FOIL method, we have

$$(4x - 6)^2 = (4x - 6)(4x - 6)$$

$$= \underset{\underset{\text{F}\quad\text{O}\quad\text{I}\quad\text{L}}{}}{16x^2 - 24x - 24x + 36}$$

$$= 16x^2 - 48x + 36$$

The preceding example is the square of a binomial. This type of product occurs frequently enough in algebra that we have special formulas for squares of binomials:

$$(a + b)^2 = (a + b)(a + b) = a^2 + ab + ab + b^2 = a^2 + 2ab + b^2$$
$$(a - b)^2 = (a - b)(a - b) = a^2 - ab - ab + b^2 = a^2 - 2ab + b^2$$

Observing the results in both cases, we have the following rule.

The Square of a Binomial

The square of a binomial is the sum of the square of the first term, twice the product of the two terms, and the square of the last term. That is:

$$(a + b)^2 = \quad a^2 \quad + \quad 2ab \quad + \quad b^2$$

| Square of first term | Twice the product of the two terms | Square of last term |

$$(a - b)^2 = \quad a^2 \quad - \quad 2ab \quad + \quad b^2$$

EXAMPLE

Use the preceding formulas to expand each binomial square.

21. $(x + 7)^2 = x^2 + 2(x)(7) + 7^2 = x^2 + 14x + 49$

22. $(3t - 5)^2 = (3t)^2 - 2(3t)(5) + 5^2 = 9t^2 - 30t + 25$

23. $(4x + 2y)^2 = (4x)^2 + 2(4x)(2y) + (2y)^2 = 16x^2 + 16xy + 4y^2$

24. $(5 - a^3)^2 = 5^2 - 2(5)(a^3) + (a^3)^2 = 25 - 10a^3 + a^6$

25. $\left(x + \dfrac{1}{3}\right)^2 = x^2 + 2(x)\left(\dfrac{1}{3}\right) + \left(\dfrac{1}{3}\right)^2 = x^2 + \dfrac{2}{3}x + \dfrac{1}{9}$

G Products That Are the Difference of Two Squares

Another frequently occurring kind of product is found when multiplying two binomials that differ only in the sign between their terms.

EXAMPLE 26

Multiply $(3x - 5)$ and $(3x + 5)$.

SOLUTION   Applying the FOIL method, we have

$$(3x - 5)(3x + 5) = 9x^2 + 15x - 15x - 25 \qquad \text{The middle terms add to 0.}$$
$$ \text{F} \quad \text{O} \quad \text{I} \quad \text{L}$$
$$= 9x^2 - 25$$

The outside and inside products in Example 26 are opposites and therefore add to 0. Here is a generalization of Example 26:

$$(a - b)(a + b) = a^2 + ab - ab - b^2 \qquad \text{The middle terms add to 0.}$$
$$= a^2 - b^2$$

The Product of the Sum and Difference of Two Terms

To multiply two binomials that differ only in the sign between their two terms, simply subtract the square of the second term from the square of the first term:

$$(a - b)(a + b) = a^2 - b^2$$

The expression $a^2 - b^2$ is called the *difference of two squares*.

Once we memorize and understand this rule, we can multiply binomials of this form with a minimum amount of work.

EXAMPLE

Find the following products.

27. $(x - 5)(x + 5) = x^2 - 25$

28. $(2a - 3)(2a + 3) = 4a^2 - 9$

29. $(x^2 + 4)(x^2 - 4) = x^4 - 16$

30. $(x^3 - 2a)(x^3 + 2a) = x^6 - 4a^2$

GETTING READY FOR CLASS

After reading through the preceding section, respond in your own words and in complete sentences.

A. What are similar terms?
B. Explain how you subtract one polynomial from another.
C. Explain why $(x + 3)^2 \neq x^2 + 9$.
D. When will the product of two binomials result in a binomial?

Problem Set 1.5

Identify which of the following are monomials, binomials, or trinomials. Give the degree of each, and name the leading coefficient.

1. $5x^2 - 3x + 2$

2. $2x^2 + 4x - 1$

3. $3x - 5$

4. $5y + 3$

5. $8a^2 + 3a - 5$

6. $9a^2 - 8a - 4$

7. $4x^3 - 6x^2 + 5x - 3$

8. $9x^4 + 4x^3 - 2x^2 + x$

9. $-\dfrac{3}{4}$

10. -16

11. $4x - 5 + 6x^3$

12. $9x + 2 + 3x^3$

Simplify each of the following by combining similar terms.

13. $2x^2 - 3x + 10x - 15$

14. $12a^2 + 8ab - 15ab - 10b^2$

15. $(3x^2 + 4x - 5) - (2x^2 - 5x - 9)$

16. $(-4x^2 + 5x - 3) - (2x^2 + x - 7)$

17. $(-2a^2 + 4a - 5) - (-3a^2 - a - 9)$

18. $(-6a^2 + a - 4) - (-2a^2 - 3a - 4)$

19. $(5x^2 - 6x + 1) - (4x^2 + 7x - 2)$

20. $(11x^2 - 8x) - (4x^2 - 2x - 7)$

21. Subtract $2x^2 - 4x$ from $2x^2 - 7x$.

22. Subtract $-3x + 6$ from $-3x + 9$.

23. Subtract $x^2 - 2x + 1$ from $2x^2 - 8x - 4$.

24. Subtract $-5x^2 + x - 4$ from $-2x^2 - 3x + 1$.

Simplify.

25. $-5[-(x - 3) - (x + 2)]$

26. $-6[(2x - 5) - 3(8x - 2)]$

27. $4x - 5[3 - (x - 4)]$

28. $x - 7[3x - (2 - x)]$

29. $3x - 2[4(x + 3) - 5(2x - 4)]$

30. $-5x - 3[-2(2x - 1) - 3(4x + 2)]$

31. Find the value of $2x^2 - 3x - 4$ when x is 2.

32. Find the value of $4x^2 + 3x - 2$ when x is -1.

33. Find the value of $\dfrac{3}{2}x^2 + \dfrac{3}{4}x + 1$ when x is 12.

34. Find the value of $\dfrac{2}{5}x^2 - \dfrac{1}{10}x + 2$ when x is 10.

35. Find the value of $x^3 - x^2 + x - 1$ when x is -2.

36. Find the value of $x^3 + x^2 + x + 1$ when x is -2.

Multiply.

37. $2x(6x^2 - 5x + 4)$

38. $-3x(5x^2 - 6x - 4)$

39. $(a - b)(a^2 + ab + b^2)$

40. $(a + b)(a^2 - ab + b^2)$

41. $(2x + y)(4x^2 - 2xy + y^2)$

42. $(x - 3y)(x^2 + 3xy + 9y^2)$

43. $(2a + 3)(3a + 2)$

44. $(5a - 4)(2a + 1)$

45. $(5 - 3t)(4 + 2t)$

46. $(7 - t)(6 - 3t)$

47. $(x^3 + 3)(x^3 - 5)$ **48.** $(x^3 + 4)(x^3 - 7)$

49. $\left(3t + \dfrac{1}{3}\right)\left(6t - \dfrac{2}{3}\right)$ **50.** $\left(5t - \dfrac{1}{5}\right)\left(10t + \dfrac{3}{5}\right)$

Find the following special products.

51. $(2a - 3)^2$ **52.** $(3a + 2)^2$ **53.** $(5x + 2y)^2$ **54.** $(3x - 4y)^2$

55. $(5 - 3t^3)^2$ **56.** $(7 - 2t^4)^2$

57. $(2a + 3b)(2a - 3b)$ **58.** $(6a - 1)(6a + 1)$

59. $(3r^2 + 7s)(3r^2 - 7s)$ **60.** $(5r^2 - 2s)(5r^2 + 2s)$

Expand and simplify.

61. $(x - 2)^3$ **62.** $(x + 4)^3$ **63.** $(2x - 1)^3$ **64.** $(4x + 1)^3$

65. $(b^2 + 8)(a^2 + 1)$ **66.** $(b^2 + 1)(a^4 - 5)$

67. $(x - 2)(3y^2 + 4)$ **68.** $(x - 4)(2y^3 + 1)$

69. $(x + 1)^2 + (x + 2)^2 + (x + 3)^2$ **70.** $(x - 1)^2 + (x - 2)^2 + (x - 3)^2$

71. $(2x + 3)^2 - (2x - 3)^2$ **72.** $(5x - 4)^2 - (5x + 4)^2$

Here are a variety of problems that you will see again as we progress through the book.

Simplify.

73. $-4x + 9x$ **74.** $-6x - 2x$ **75.** $5x^2 + 3x^2$

76. $7x^2 + 3x^2$ **77.** $-8x^3 + 10x^3$ **78.** $4x^3 - 7x^3$

79. $2x + 3 - 2x - 8$ **80.** $9x - 4 - 9x - 10$ **81.** $-1(2x - 3)$

82. $-1(-3x + 1)$ **83.** $-3(-3x - 2)$ **84.** $-4(-5x + 3)$

85. a. $(3x - 5) - (3a - 5)$ **b.** $(2x + 3) - (2a + 3)$

86. a. $(x^2 - 4) - (a^2 - 4)$ **b.** $(x^2 - 1) - (a^2 - 1)$

87. $2x^3 - 3x + 10x - 15$ **88.** $12a^2 + 8ab - 15ab - 10b^2$

89. $(6x^3 - 2x^2y + 8xy^2) + (-9x^2y + 3xy^2 - 12y^3)$

90. $(3x^3 - 15x^2 + 18x) + (2x^2 - 10x + 12)$

91. $4x^3(-3x)$ **92.** $5x^2(-4x)$ **93.** $4x^3(5x^2)$ **94.** $5x^2(3x^2)$

95. $(a^3)^2$ **96.** $(a^4)^2$ **97.** $-1(5 - x)$ **98.** $-1(a - b)$

99. $-1(7 - x)$ **100.** $-1(6 - y)$ **101.** $5\left(x - \dfrac{1}{5}\right)$ **102.** $7\left(x + \dfrac{1}{7}\right)$

103. $x\left(1 - \dfrac{1}{x}\right)$ **104.** $a\left(1 + \dfrac{1}{a}\right)$ **105.** $12\left(\dfrac{1}{4}x + \dfrac{2}{3}y\right)$ **106.** $20\left(\dfrac{2}{5}x + \dfrac{1}{4}y\right)$

107. $(x + 3)^2 - 2(x + 3) - 8$ **108.** $(x - 2)^2 - 3(x - 2) - 10$

109. $(2a - 3)^2 - 9(2a - 3) + 20$ **110.** $(3a - 2)^2 + 2(3a - 2) - 3$

111. $2(4a + 2)^2 - 3(4a + 2) - 20$ **112.** $6(2a + 4)^2 - (2a + 4) - 2$

113. Height of an Object The formula for the height of an object that has been thrown straight up with a velocity of 64 feet/second is

$$h = -16t^2 + 64t$$

Find the height after 1 second and after 3 seconds.

114. Profits Suppose it costs a company selling patterns $C = 800 + 6.5x$ dollars to produce and sell x patterns a month. If the revenue obtained by selling x patterns is $R = 10x - 0.002x^2$, what is the profit equation? How much profit will be made if 1,000 patterns are produced and sold in May?

115. Interest If you deposit P dollars in an account with an annual interest rate r that is compounded twice a year, then at the end of a year the amount of money in that account is given by the formula

$$A = P\left(1 + \frac{r}{2}\right)^2$$

Expand the right side of this formula.

EXTENDING THE CONCEPTS

Assume n is a positive integer and multiply.

116. $(x^n - 2)(x^n - 3)$

117. $(x^{2n} + 3)(x^{2n} - 3)$

118. $(2x^n + 3)(5x^n - 1)$

119. $(4x^n - 3)(7x^n + 2)$

120. $(x^n + 5)^2$

121. $(x^n + 2)^3$

122. $(x^n + 1)(x^{2n} + x^n - 1)$

123. $(x^n + 5)(2x^{2n} - 3x^n - 4)$

FACTORING

©iStockphoto.com/rihardzz

A Factor the greatest common factor from a polynomial.

B Factor polynomials by grouping.

C Factor a trinomial whose leading coefficient is 1.

D Factor a trinomial whose leading coefficient is other than 1 by trial and error.

E Factor a perfect square trinomial.

F Factor the difference of two squares.

G Factor the sum or difference of two cubes.

In flight archery, a person uses a bow to shoot an arrow into the air with a goal of achieving the greatest distance, as opposed to aiming for a specific target. The curved path that the arrow takes from the bow to the ground creates the shape of a parabola, which we will discuss further in Chapter 10. For now, let's say the arrow's path can be represented by the polynomial $-16x^2 + 62x + 8$. To begin factoring this polynomial, we need to find the greatest common factor, which is the primary focus of this section.

A Factoring the Greatest Common Factor

As we mentioned before, factoring is the reverse of multiplication. The following diagram illustrates the relationship between factoring and multiplication:

Multiplication

Factors → $3 \cdot 7 = 21$ ← Product

Factoring

Reading from left to right, we say the product of 3 and 7 is 21. Reading in the other direction, from right to left, we say 21 factors into 3 times 7, or 3 and 7 are factors of 21.

> **DEFINITION** *greatest common factor*
> . . . .
> The *greatest common factor* for a polynomial is the largest monomial that divides (is a factor of) each term of the polynomial.

The greatest common factor for the polynomial $25x^5 + 20x^4 - 30x^3$ is $5x^3$ because it is the largest monomial that is a factor of each term. We can apply the distributive property and write

$$25x^5 + 20x^4 - 30x^3 = 5x^3(5x^2) + 5x^3(4x) - 5x^3(6)$$

$$= 5x^3(5x^2 + 4x - 6)$$

The last line is written in factored form.

EXAMPLE 1

Factor the greatest common factor from $8a^3 - 8a^2 - 48a$.

SOLUTION The greatest common factor is $8a$. It is the largest monomial that divides each term of the polynomial. We can write each term in the polynomial as the product of $8a$ and another monomial. Then, we apply the distributive property to factor $8a$ from each term:

$$8a^3 - 8a^2 - 48a = 8a(a^2) - 8a(a) - 8a(6)$$

$$= 8a(a^2 - a - 6)$$

EXAMPLE 2

Factor the greatest common factor from $5x^2(a + b) - 6x(a + b) - 7(a + b)$.

SOLUTION The greatest common factor is $a + b$. Factoring it from each term, we have

$$5x^2(a + b) - 6x(a + b) - 7(a + b) = (a + b)(5x^2 - 6x - 7)$$

NOTE The phrase *largest monomial,* as used in Example 1, refers to the monomial with the largest integer exponent and whose coefficient has the greatest absolute value. We could have factored the polynomial in Example 1 correctly by taking out $-8a$, instead of $8a$. However, when factoring like this, it is usually better to have a positive coefficient on the greatest common factor, even though it is not incorrect to have a negative coefficient.

B Factoring by Grouping

Many polynomials have no greatest common factor other than the number 1. Some of these can be factored using the distributive property if those terms with a common factor are grouped together.

For example, the polynomial $5x + 5y + x^2 + xy$ can be factored by noticing that the first two terms have a 5 in common, whereas the last two have an x in common. Applying the distributive property, we have

$$5x + 5y + x^2 + xy = 5(x + y) + x(x + y)$$

This last expression can be thought of as having two terms, $5(x + y)$ and $x(x + y)$, each of which has a common factor $(x + y)$. We apply the distributive property again to factor $(x + y)$ from each term:

$$5(x + y) + x(x + y)$$

$$= (x + y)(5 + x)$$

EXAMPLE 3

Factor $a^2b^2 + b^2 + 8a^2 + 8$.

SOLUTION The first two terms have b^2 in common; the last two have 8 in common:

$$a^2b^2 + b^2 + 8a^2 + 8 = b^2(a^2 + 1) + 8(a^2 + 1)$$

$$= (a^2 + 1)(b^2 + 8)$$

C Factoring Trinomials with Leading Coefficient 1

Previously we multiplied binomials as follows:

$$(x - 2)(x + 3) = x^2 + x - 6$$
$$(x + 5)(x + 2) = x^2 + 7x + 10$$

In each case, the product of two binomials is a trinomial. The first term in the resulting trinomial is obtained by multiplying the first term in each binomial. The middle term comes from adding the products of the two inside terms and the two outside terms. The last term is the product of the last term in each binomial.

In general,

$$(x + a)(x + b) = x^2 + ax + bx + ab$$
$$= x^2 + (a + b)x + ab$$

Writing this as a factoring problem, we have

$$x^2 + (a + b)x + ab = (x + a)(x + b)$$

To factor a trinomial with a leading coefficient of 1, we simply find the two numbers a and b whose sum is the coefficient of the middle term and whose product is the constant term.

EXAMPLE 4

Factor $x^2 + 2x - 15$.

SOLUTION   The leading coefficient is 1. We need two integers whose product is -15 and whose sum is $+2$. The integers are $+5$ and -3:

$$x^2 + 2x - 15 = (x + 5)(x - 3)$$

If a trinomial is not factorable, we say that it is a *prime polynomial*.

EXAMPLE 5

Factor $x^2 - xy - 12y^2$.

SOLUTION   We need two terms whose product is $-12y$ and whose sum is $-y$. The terms are $-4y$ and $3y$:

$$x^2 - xy - 12y^2 = (x - 4y)(x + 3y)$$

Check   Checking this result gives

$$(x - 4y)(x + 3y) = x^2 + 3xy - 4xy - 12y^2$$
$$= x^2 - xy - 12y^2$$

EXAMPLE 6

Factor $x^2 - 8x + 6$.

SOLUTION   Because there is no pair of integers whose product is 6 and whose sum is -8, the trinomial $x^2 - 8x + 6$ is not factorable. It is a prime polynomial.

D Factoring Other Trinomials by Trial and Error

We turn our attention now to trinomials with leading coefficients other than 1 and with no greatest common factor other than 1.

Suppose we want to factor $3x^2 - x - 2$. The factors will be a pair of binomials. The product of the first terms will be $3x^2$ and the product of the last terms will be -2. We can list all the possible factors along with their products as follows:

Possible Factors	First Term	Middle Term	Last Term
$(x + 2)(3x - 1)$	$3x^2$	$+5x$	-2
$(x - 2)(3x + 1)$	$3x^2$	$-5x$	-2
$(x + 1)(3x - 2)$	$3x^2$	$+x$	-2
$(x - 1)(3x + 2)$	$3x^2$	$-x$	-2

From the last line, we see that the factors of $3x^2 - x - 2$ are $(x - 1)(3x + 2)$. That is,

$$3x^2 - x - 2 = (x - 1)(3x + 2)$$

To factor trinomials with leading coefficients other than 1, when the greatest common factor is 1, we must use trial and error or list all the possible factors. In either case, the idea is this: Look only at pairs of binomials whose products give the correct first and last terms, then look for a combination that will give the correct middle term.

EXAMPLE 7

Factor $2x^2 + 13xy + 15y^2$.

SOLUTION Listing all possible factors that give $2x^2$ as the product of the first terms and $+15y^2$ as the product of the last terms yields the table below.

The last line has the correct middle term:

$$2x^2 + 13xy + 15y^2 = (2x + 3y)(x + 5y)$$

Possible Factors	Middle Term of Product
$(2x + y)(x + 15y)$	$31xy$
$(2x - y)(x - 15y)$	$-31xy$
$(2x + 15y)(x + y)$	$+17xy$
$(2x - 15y)(x - y)$	$-17xy$
$(2x - 5y)(x - 3y)$	$-11xy$
$(2x + 5y)(x + 3y)$	$+11xy$
$(2x - 3y)(x - 5y)$	$-13xy$
$(2x + 3y)(x + 5y)$	$+13xy$

Actually, we did not need to check four of the pairs of possible factors in the preceding list. All the signs in the trinomial $2x^2 + 13xy + 15y^2$ are positive. The binomial factors must then be of the form $(ax + b)(cx + d)$, where a, b, c, and d are all positive.

There are other ways to reduce the number of possible factors under consideration. For example, if we were to factor the trinomial $2x^2 - 11x + 12$, we would not have to consider the pair of possible factors $(2x - 4)(x - 3)$. The reason is, if the original trinomial has no greatest common factor other than 1, then neither of its binomial factors will either. The trinomial $2x^2 - 11x + 12$ has a greatest common factor of 1, but the possible factor $2x - 4$ has a greatest common factor of 2: $2x - 4 = 2(x - 2)$. Therefore, we do not need to consider $2x - 4$ as a possible factor.

EXAMPLE 8

Factor $18x^3y + 3x^2y^2 - 36xy^3$.

SOLUTION First, factor out the greatest common factor $3xy$. Then, factor the remaining trinomial:

$$18x^3y + 3x^2y^2 - 36xy^3 = 3xy(6x^2 + xy - 12y^2)$$
$$= 3xy(3x - 4y)(2x + 3y)$$

EXAMPLE 9

Factor $12x^4 + 17x^2 + 6$.

SOLUTION This is a trinomial in x^2:

$$12x^4 + 17x^2 + 6 = (4x^2 + 3)(3x^2 + 2)$$

EXAMPLE 10

Factor $2x^2(x - 3) - 5x(x - 3) - 3(x - 3)$.

SOLUTION We begin by factoring out the greatest common factor $(x - 3)$. Then, we factor the trinomial that remains:

$$2x^2(x - 3) - 5x(x - 3) - 3(x - 3) = (x - 3)(2x^2 - 5x - 3)$$
$$= (x - 3)(2x + 1)(x - 3)$$
$$= (x - 3)^2(2x + 1)$$

E Perfect Square Trinomials

We previously listed some special products found in multiplying polynomials. Two of the formulas looked like this:

$$(a + b)^2 = a^2 + 2ab + b^2$$
$$(a - b)^2 = a^2 - 2ab + b^2$$

If we exchange the left and right sides of each formula, we have two special formulas for factoring:

$$a^2 + 2ab + b^2 = (a + b)^2$$
$$a^2 - 2ab + b^2 = (a - b)^2$$

The left side of each factoring formula is called a *perfect square trinomial*. The right sides are binomial squares. Perfect square trinomials can always be factored using the usual methods for factoring trinomials. However, if we notice that the first and last terms of a trinomial are perfect squares, it is wise to see whether the trinomial factors as a binomial square before attempting to factor by the usual method.

EXAMPLE 11

Factor $x^2 - 6x + 9$.

SOLUTION Because the first and last terms are perfect squares, we attempt to factor according to the preceding formulas:

$$x^2 - 6x + 9 = (x - 3)^2$$

If we expand $(x - 3)^2$, we have $x^2 - 6x + 9$, indicating that we have factored correctly.

EXAMPLE

Factor each of the following perfect square trinomials.

12. $16a^2 + 40ab + 25b^2 = (4a + 5b)^2$

13. $49 - 14t + t^2 = (7 - t)^2$

14. $9x^4 - 12x^2 + 4 = (3x^2 - 2)^2$

15. $(y + 3)^2 + 10(y + 3) + 25 = [(y + 3) + 5]^2 = (y + 8)^2$

EXAMPLE 16

Factor $8x^2 - 24xy + 18y^2$.

SOLUTION We begin by factoring the greatest common factor 2 from each term:

$$8x^2 - 24xy + 18y^2 = 2(4x^2 - 12xy + 9y^2)$$
$$= 2(2x - 3y)^2$$

F The Difference of Two Squares

Recall the formula that results in the difference of two squares:

$$(a - b)(a + b) = a^2 - b^2$$

Writing this as a factoring formula, we have

$$a^2 - b^2 = (a - b)(a + b)$$

EXAMPLE

Each of the following is the difference of two squares. Use the formula $a^2 - b^2 = (a - b)(a + b)$ to factor each one.

17. $x^2 - 25 = x^2 - 5^2 = (x - 5)(x + 5)$

18. $49 - t^2 = 7^2 - t^2 = (7 - t)(7 + t)$

19. $81a^2 - 25b^2 = (9a)^2 - (5b)^2 = (9a - 5b)(9a + 5b)$

20. $4x^6 - 1 = (2x^3)^2 - 1^2 = (2x^3 - 1)(2x^3 + 1)$

21. $x^2 - \dfrac{4}{9} = x^2 - \left(\dfrac{2}{3}\right)^2 = \left(x - \dfrac{2}{3}\right)\left(x + \dfrac{2}{3}\right)$

As our next example shows, the difference of two fourth powers can be factored as the difference of two squares.

EXAMPLE 22

Factor $16x^4 - 81y^4$.

SOLUTION The first and last terms are perfect squares. We factor according to the preceding formula:

$$16x^4 - 81y^4 = (4x^2)^2 - (9y^2)^2$$
$$= (4x^2 - 9y^2)(4x^2 + 9y^2)$$

Notice that the first factor is also the difference of two squares. Factoring completely, we have

$$16x^4 - 81y^4 = (2x - 3y)(2x + 3y)(4x^2 + 9y^2)$$

Here is another example of the difference of two squares.

EXAMPLE 23

Factor: $(x - 3)^2 - 25$.

SOLUTION This example has the form $a^2 - b^2$, where a is $x - 3$ and b is 5. We factor it according to the formula for the difference of two squares:

$$(x - 3)^2 - 25 = (x - 3)^2 - 5^2 \qquad \text{Write 25 as } 5^2.$$
$$= [(x - 3) - 5][(x - 3) + 5] \qquad \text{Factor.}$$
$$= (x - 8)(x + 2) \qquad \text{Simplify.}$$

Notice in this example that we could have expanded $(x - 3)^2$, subtracted 25, and then factored to obtain the same result:

$$(x - 3)^2 - 25 = x^2 - 6x + 9 - 25 \qquad \text{Expand } (x - 3)^2.$$
$$= x^2 - 6x - 16 \qquad \text{Simplify.}$$
$$= (x - 8)(x + 2) \qquad \text{Factor.}$$

EXAMPLE 24

Factor $x^2 - 10x + 25 - y^2$.

SOLUTION Notice that the first three terms form a perfect square trinomial. That is, $x^2 - 10x + 25 = (x - 5)^2$. If we replace the first three terms with $(x - 5)^2$, the expression that results has the form $a^2 - b^2$. We can then factor as we did in Example 23:

$$x^2 - 10x + 25 - y^2 = (x^2 - 10x + 25) - y^2 \qquad \text{Group first three terms together.}$$
$$= (x - 5)^2 - y^2 \qquad \text{This has the form } a^2 - b^2.$$
$$= [(x - 5) - y][(x - 5) + y] \qquad \text{Factor according to the formula } a^2 - b^2 = (a - b)(a + b).$$
$$= (x - 5 - y)(x - 5 + y) \qquad \text{Simplify.}$$

We could check this result by multiplying the two factors together. (You may want to do that to convince yourself that we have the correct result.)

EXAMPLE 25

Factor completely $x^3 + 2x^2 - 9x - 18$.

SOLUTION   We use factoring by grouping to begin and then factor the difference of two squares:

$$x^3 + 2x^2 - 9x - 18 = x^2(x + 2) - 9(x + 2)$$
$$= (x + 2)(x^2 - 9)$$
$$= (x + 2)(x - 3)(x + 3)$$

Ⓖ The Sum and Difference of Two Cubes

Here are the formulas for factoring the sum and difference of two cubes:

$$a^3 + b^3 = (a + b)(a^2 - ab + b^2)$$
$$a^3 - b^3 = (a - b)(a^2 + ab + b^2)$$

EXAMPLE 26

Verify the two formulas.

SOLUTION   We verify the formulas by multiplying the right sides and comparing the results with the left sides:

$$
\begin{array}{r}
a^2 - ab\ + b^2 \\
a\ + b \\
\hline
a^3 - a^2b + ab^2 \\
a^2b - ab^2 + b^3 \\
\hline
a^3 \qquad\qquad + b^3
\end{array}
\qquad\qquad
\begin{array}{r}
a^2 + ab \qquad\quad + b^2 \\
a\ - b \\
\hline
a^3 + a^2b\ + ab^2 \\
- a^2b - ab^2\ - b^3 \\
\hline
a^3 \qquad\qquad - b^3
\end{array}
$$

Here are some examples using the formulas for factoring the sum and difference of two cubes.

EXAMPLE 27

Factor $64 + t^3$.

SOLUTION   The first term is the cube of 4 and the second term is the cube of t. Therefore,

$$64 + t^3 = 4^3 + t^3$$

$$= (4 + t)(16 - 4t + t^2)$$

EXAMPLE 28

Factor $27x^3 + 125y^3$.

SOLUTION   Writing both terms as perfect cubes, we have

$$27x^3 + 125y^3 = (3x)^3 + (5y)^3$$

$$= (3x + 5y)(9x^2 - 15xy + 25y^2)$$

EXAMPLE 29

Factor $a^3 - \frac{1}{8}$.

SOLUTION   The first term is the cube of a, while the second term is the cube of $\frac{1}{2}$:

$$a^3 - \frac{1}{8} = a^3 - \left(\frac{1}{2}\right)^3$$

$$= \left(a - \frac{1}{2}\right)\left(a^2 + \frac{1}{2}a + \frac{1}{4}\right)$$

EXAMPLE 30

Factor $x^6 - y^6$.

SOLUTION   We have a choice of how we initially want to write the two terms. We can write the expression as the difference of two squares, $(x^3)^2 - (y^3)^2$, or as the difference of two cubes, $(x^2)^3 - (y^2)^3$. It is better to start with the difference of two squares if we have a choice:

$$x^6 - y^6 = (x^3)^2 - (y^3)^2$$

Factor completely, using the sum and difference of two cubes.

$$= (x^3 - y^3)(x^3 + y^3)$$

$$= (x - y)(x^2 + xy + y^2)(x + y)(x^2 - xy + y^2)$$

Try this example again, writing the first line as the difference of two cubes rather than the difference of two squares. It will become apparent why it is better to use the difference of two squares.

GETTING READY FOR CLASS

After reading through the preceding section, respond in your own words and in complete sentences.

A. In what cases can you factor a binomial?

B. What is a perfect square trinomial?

C. How do you know when you have factored completely?

D. If a polynomial has four terms, what method of factoring should you try?

Problem Set 1.6

Factor the greatest common factor from each of the following. (The answers in the back of the book all show greatest common factors whose coefficients are positive.)

1. $10x^3 - 15x^2$ **2.** $12x^5 + 18x^7$

3. $9y^6 + 18y^3$ **4.** $24y^4 - 8y^2$

5. $9a^2b - 6ab^2$ **6.** $30a^3b^4 + 20a^4b^3$

7. $21xy^4 + 7x^2y^2$ **8.** $14x^6y^3 - 6x^2y^4$

9. $3a^2 - 21a + 30$ **10.** $3a^2 - 3a - 6$

11. $4x^3 - 16x^2 - 20x$ **12.** $2x^3 - 14x^2 + 20x$

13. $5x(a - 2b) - 3y(a - 2b)$ **14.** $3a(x - y) - 7b(x - y)$

15. $3x^2(x + y)^2 - 6y^2(x + y)^2$ **16.** $10x^3(2x - 3y) - 15x^2(2x - 3y)$

17. $2x^2(x + 5) + 7x(x + 5) + 6(x + 5)$

18. $2x^2(x + 2) + 13x(x + 2) + 15(x + 2)$

Factor each of the following by grouping.

19. $3xy + 3y + 2ax + 2a$ **20.** $5xy^2 + 5y^2 + 3ax + 3a$

21. $x^2y + x + 3xy + 3$ **22.** $x^3y^3 + 2x^3 + 5x^2y^3 + 10x^2$

23. $x^2 - ax - bx + ab$ **24.** $ax - x^2 - bx + ab$

25. $ab + 5a - b - 5$ **26.** $x^2 - xy - ax + ay$

Factor each of the following trinomials.

27. $x^2 + 7x + 12$ **28.** $x^2 - 7x + 12$ **29.** $x^2 - x - 12$

30. $x^2 + x - 12$ **31.** $y^2 + y - 6$ **32.** $y^2 - y - 6$

33. $16 - 6x - x^2$ **34.** $3 + 2x - x^2$ **35.** $12 + 8x + x^2$

36. $15 - 2x - x^2$ **37.** $x^2 + 3xy + 2y^2$ **38.** $x^2 - 5xy - 24y^2$

39. $a^2 + 3ab - 18b^2$ **40.** $a^2 - 8ab - 9b^2$ **41.** $x^2 - 2xa - 48a^2$

42. $x^2 + 14xa + 48a^2$ **43.** $x^2 - 12xb + 36b^2$ **44.** $x^2 + 10xb + 25b^2$

Factor completely, if possible. Be sure to factor out the greatest common factor first if it is other than 1.

45. $2x^2 + 7x - 15$ **46.** $2x^2 - 7x - 15$ **47.** $2x^2 + x - 15$

48. $2x^2 - x - 15$ **49.** $2x^2 - 13x + 15$ **50.** $2x^2 + 13x + 15$

51. $2x^2 - 11x + 15$ **52.** $2x^2 + 11x + 15$ **53.** $2x^2 + 7x + 15$

54. $2x^2 - 7x + 15$ **55.** $2 + 7a + 6a^2$ **56.** $2 - 7a + 6a^2$

57. $60y^2 - 15y - 45$ **58.** $72y^2 + 60y - 72$ **59.** $6x^4 - x^3 - 2x^2$

60. $3x^4 + 2x^3 - 5x^2$ **61.** $40r^3 - 120r^2 + 90r$ **62.** $40r^3 + 200r^2 + 250r$

63. $4x^2 - 11xy - 3y^2$ **64.** $3x^2 + 19xy - 14y^2$ **65.** $10x^2 - 3xa - 18a^2$

66. $9x^2 + 9xa - 10a^2$ **67.** $18a^2 + 3ab - 28b^2$ **68.** $6a^2 - 7ab - 5b^2$

69. $600 + 800t - 800t^2$ **70.** $200 - 600t - 350t^2$ **71.** $9y^4 + 9y^3 - 10y^2$

72. $4y^5 + 7y^4 - 2y^3$ **73.** $24a^2 - 2a^3 - 12a^4$ **74.** $60a^2 + 65a^3 - 20a^4$

75. $8x^4y^2 - 2x^3y^3 - 6x^2y^4$ **76.** $8x^4y^2 - 47x^3y^3 - 6x^2y^4$

77. $300x^4 + 1,000x^2 + 300$ **78.** $600x^4 - 100x^2 - 200$

Factor each perfect square trinomial.

79. $x^2 - 6x + 9$ **80.** $x^2 + 10x + 25$ **81.** $a^2 - 12a + 36$

82. $36 - 12a + a^2$ **83.** $25 - 10t + t^2$ **84.** $64 - 16t + t^2$

85. $\frac{1}{4}x^2 - 3x + 9$ **86.** $\frac{1}{4}x^2 - \frac{1}{3}x + \frac{1}{9}$ **87.** $16a^2 + 40ab + 25b^2$

88. $25a^2 - 40ab + 16b^2$ **89.** $\frac{1}{25} + \frac{1}{10}t^2 + \frac{1}{16}t^4$ **90.** $\frac{1}{9} - \frac{1}{3}t^3 + \frac{1}{4}t^6$

91. $(x + 2)^2 + 6(x + 2) + 9$ **92.** $(x + 5)^2 + 4(x + 5) + 4$

Factor completely.

93. $49x^2 - 64y^2$ **94.** $81x^2 - 49y^2$ **95.** $4a^2 - \frac{1}{4}$ **96.** $25a^2 - \frac{1}{25}$

97. $x^2 - \frac{9}{25}$ **98.** $x^2 - \frac{25}{36}$ **99.** $25 - \frac{1}{100}t^2$ **100.** $64 - \frac{1}{16}t^2$

101. $16a^4 - 81$ **102.** $81a^4 - 16b^4$

103. $x^2 - 10x + 25 - y^2$ **104.** $x^2 - 6x + 9 - y^2$

105. $a^2 + 8a + 16 - 25b^2$ **106.** $a^2 + 12a + 36 - 100b^2$

107. $x^3 + 2x^2 - 25x - 50$ **108.** $x^3 + 4x^2 - 9x - 36$

109. $2x^3 + 3x^2 - 8x - 12$ **110.** $3x^3 + 2x^2 - 27x - 18$

111. $4x^3 + 12x^2 - 9x - 27$ **112.** $9x^3 + 18x^2 - 4x - 8$

Factor each of the following as the sum or difference of two cubes.

113. $x^3 - y^3$ **114.** $x^3 + y^3$ **115.** $a^3 + 8$ **116.** $a^3 - 8$

117. $\frac{1}{8}y^3 - 64$ **118.** $64y^3 - \frac{1}{8}$ **119.** $10r^3 - 1,250$ **120.** $10r^3 + 1,250$

121. $64 + 27a^3$ **122.** $27 - 64a^3$ **123.** $125t^3 + \frac{1}{27}$ **124.** $27t^3 - \frac{1}{27}$

125. Height of a Bullet A bullet is fired into the air with an initial upward velocity of 80 feet per second from the top of a building 96 feet high. The equation that gives the height of the bullet at any time t is

$$h(t) = 96 + 80t - 16t^2$$

Factor the right side of this equation, and then find h when t is 6 seconds and when t is 3 seconds. [Find $h(6)$ and $h(3)$.]

126. Height of an Arrow An arrow is shot into the air with an upward velocity of 16 feet per second from a hill 32 feet high. The equation that gives the height of the arrow at any time t is

$$h(t) = 32 + 16t - 16t^2$$

Factor the right side of this equation, and then find h when t is 2 seconds and when t is 1 second. [Find $h(2)$ and $h(1)$.]

127. Compound Interest If P dollars are placed in a savings account in which the rate of interest r is compounded yearly, then

at the end of 1 year the amount of money in the account can be written as $P + Pr$. At the end of 2 years, the amount of money in the account is

$$P + Pr + (P + Pr)r$$

Use factoring by grouping to show that this last expression can be written as $P(1 + r)^2$.

128. Compound Interest At the end of 3 years, the amount of money in the savings account in Problem 127 will be

$$P(1 + r)^2 + P(1 + r)^2 r$$

Use factoring to show that this last expression can be written as $P(1 + r)^3$.

EXTENDING THE CONCEPTS

Factor completely.

129. $8x^6 + 26x^3 y^2 + 15y^4$

130. $24x^4 + 6x^2 y^3 - 45y^6$

131. $3x^2 + 295x - 500$

132. $3x^2 + 594x - 1{,}200$

133. $\dfrac{1}{8}x^2 + x + 2$

134. $\dfrac{1}{9}x^2 + x + 2$

135. $2x^2 + 1.5x + 0.25$

136. $6x^2 + 2x + 0.16$

OBJECTIVE

A Factor a variety of polynomials.

©iStockphoto.com/skynesher

1.7

If you watch professional football on television, you will hear the announcers refer to "hang time" when the punter punts the ball. Hang time is the amount of time the ball is in the air, and it depends on only one thing—the initial vertical velocity imparted to the ball by the kicker's foot. Table 1 shows the equations to solve for hang time, given various initial vertical velocities. Figure 1 is a visual representation of some equations associated with those in Table 1.

TABLE 1
Hang Time for a Football

Initial Vertical Velocity	Equation in factored form	Hang Time
16	$16t(1 - t) = 0$	1
32	$16t(2 - t) = 0$	2
48	$16t(3 - t) = 0$	3
64	$16t(4 - t) = 0$	4
80	$16t(5 - t) = 0$	5

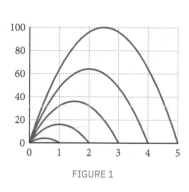

FIGURE 1

The equations in the second column of the table are in what is called "factored form." Once the equation is in factored form, hang time can be read from the second factor.

A Factoring Polynomials of Any Type

In this section we review the different methods of factoring that we have presented previously in this book. This section is important because it will give you an opportunity to factor a variety of polynomials.

HOW TO *Factor a Polynomial*

Step 1: If the polynomial has a greatest common factor other than 1, then factor out the greatest common factor.

Step 2: If the polynomial has two terms (it is a binomial), then see if it is the difference of two squares or the sum or difference of two cubes, and then factor accordingly. Remember, if it is the sum of two squares it will not factor.

Step 3: If the polynomial has three terms (a trinomial), then it is either a perfect square trinomial, which will factor into the square of a binomial, or it is not a perfect square trinomial, in which case we try to write it as the product of two binomials. Remember, not all trinomials are factorable.

Step 4: If the polynomial has more than three terms, then try to factor it by grouping.

Step 5: As a final check, see if any of the factors you have written can be factored further. If you have overlooked a common factor, you can catch it here.

Here are some examples illustrating how we use the steps on our list.

EXAMPLE 1

Factor $2x^5 - 8x^3$.

SOLUTION First we check to see if the greatest common factor is other than 1. Since the greatest common factor is $2x^3$, we begin by factoring it out. Once we have done so, we notice that the binomial that remains is the difference of two squares, which we factor according to the formula $a^2 - b^2 = (a + b)(a - b)$.

$$2x^5 - 8x^3 = 2x^3(x^2 - 4) \qquad \text{Factor out the greatest common factor, } 2x^3.$$

$$= 2x^3(x + 2)(x - 2) \qquad \text{Factor the difference of two squares.}$$

EXAMPLE 2

Factor $2x^5 + 8x^3$.

SOLUTION This problem is very similar to the problem in Example 1. We factor out the greatest common factor, $2x^3$, first.

$$2x^5 + 8x^3 = 2x^3(x^2 + 4) \qquad \text{Factor out } 2x^3.$$

In this case we cannot factor further because the sum of two squares, $x^2 + 4$, does not factor.

EXAMPLE 3

Factor $3x^4 - 18x^3 + 27x^2$.

SOLUTION Step 1 is to factor out the greatest common factor $3x^2$. After we have done so, we notice that the trinomial remains a perfect square trinomial, which will factor as the square of a binomial.

$$3x^4 - 18x^3 + 27x^2 = 3x^2(x^2 - 6x + 9) \qquad \text{Factor out } 3x^2.$$

$$= 3x^2(x - 3)^2 \qquad x^2 - 6x + 9 \text{ is the square of } x - 3.$$

EXAMPLE 4

Factor $6a^2 - 11a + 4$.

SOLUTION Here we have a trinomial that does not have a greatest common factor other than 1. Since it is not a perfect square trinomial, we factor it by trial and error. Without showing all the different possibilities, here is the answer.

$$6a^2 - 11a + 4 = (3a - 4)(2a - 1)$$

EXAMPLE 5

Factor $6a^2 - 11a - 4$.

SOLUTION This trinomial is very similar to the trinomial in Example 4. However, this trinomial cannot be factored further. It is a prime polynomial. Convince yourself that this trinomial is prime.

> **NOTE** When we say a trinomial cannot be factored, we mean it cannot be written as the product of two binomials with coefficients that are whole numbers.

EXAMPLE 6

Factor $2x^4 + 16x$.

SOLUTION This binomial has a greatest common factor of $2x$. The binomial that remains after the $2x$ has been factored from each term is the sum of two cubes, which we factor according to the formula $a^3 + b^3 = (a + b)(a^2 - ab + b^2)$.

$$2x^4 + 16x = 2x(x^3 + 8) \qquad \text{Factor } 2x \text{ from each term.}$$

$$= 2x(x + 2)(x^2 - 2x + 4) \qquad \text{The sum of two cubes}$$

EXAMPLE 7

Factor $2ab^5 + 8ab^4 + 2ab^3$.

SOLUTION The greatest common factor is $2ab^3$. We begin by factoring it from each term. After that we find that the trinomial that remains cannot be factored further.

$$2ab^5 + 8ab^4 + 2ab^3 = 2ab^3(b^2 + 4b + 1)$$

EXAMPLE 8

Factor $4x^2 - 6x + 2ax - 3a$.

SOLUTION Our polynomial has four terms, so we factor by grouping.

$$4x^2 - 6x + 2ax - 3a = 2x(2x - 3) + a(2x - 3)$$

$$= (2x - 3)(2x + a)$$

EXAMPLE 9

Factor $4x^4 - 35x^2 - 9$.

SOLUTION This is a trinomial in x^2. We factor it into the product of two binomials, and then we continue to factor.

$$4x^4 - 35x^2 - 9 = (x^2 - 9)(4x^2 + 1)$$

$$= (x + 3)(x - 3)(4x^2 + 1)$$

Notice that $4x^2 + 1$ does not factor further because it is the sum of two squares. You must convince yourself that this is true.

EXAMPLE 10

Factor $2x^2(x - 3) - 5x(x - 3) - 3(x - 3)$.

SOLUTION We begin by factoring out the greatest common factor $(x - 3)$. Then we factor the trinomial that remains.

$$2x^2(x - 3) - 5x(x - 3) - 3(x - 3) = (x - 3)(2x^2 - 5x - 3)$$

$$= (x - 3)(2x + 1)(x - 3)$$

GETTING READY FOR CLASS

After reading through the preceding section, respond in your own words and in complete sentences.

A. How do you know when you've factored completely?
B. If a polynomial has four terms, what method of factoring should you try?
C. What is the first step in factoring a polynomial?
D. What do we call a polynomial that does not factor?

Problem Set 1.7

The problems below are representative of the type of factoring problems you will see throughout the rest of this book. To make a successful start with intermediate algebra, be sure you are comfortable with all the factoring problems shown here.

Factor the following trinomials.

1. $x^2 - 2x - 24$ **2.** $x^2 + 2x - 24$ **3.** $x^2 - 5x - 6$

4. $x^2 + 5x - 6$ **5.** $x^2 - 5x + 6$ **6.** $x^2 - x - 6$

7. $x^2 - 10x + 25$ **8.** $4x^2 + 4x + 1$ **9.** $2x^2 - 5x - 3$

10. $20x^2 - 93x + 34$ **11.** $21x^2 - 23x + 6$ **12.** $42x^2 + 23x - 10$

Factor the following binomials as the difference of two squares.

13. $x^2 - 16$ **14.** $x^2 - 25$ **15.** $a^2 - 1$ **16.** $a^2 - 9$

17. $a^2 - 16b^2$ **18.** $a^2 - 25b^2$ **19.** $9x^2 - 49$ **20.** $49x^2 - 144$

21. $16x^4 - 49$ **22.** $4x^4 - 25$ **23.** $t^4 - 81$ **24.** $t^4 - 16$

Factor the following binomials as the sum or difference of two cubes.

25. $a^3 + b^3$ **26.** $a^3 - b^3$ **27.** $x^3 - 8$ **28.** $x^3 + 125$

29. $x^3 + 1$ **30.** $x^3 - 27$ **31.** $8x^3 + 1$ **32.** $27x^3 + 64$

Factor each of the following by first factoring out the greatest common factor and then factoring the polynomial that remains, if possible.

33. $60x^2 - 130x + 60$ **34.** $90x^2 + 60x - 80$ **35.** $x^3 + 5x^2 + 6x$

36. $x^3 - 5x^2 - 6x$ **37.** $2x^3 - 5x^2 - 3x$ **38.** $6x^3 - 5x^2 - x$

39. $x^3 - 2x^2 - 24x$ **40.** $x^3 + 2x^2 - 24x$ **41.** $6x + 24$

42. $3x^2 - 3xy$ **43.** $100x^2 - 300x$ **44.** $10x^2 + 100x$

45. $20a^2 - 45$ **46.** $50a - 2ax^2$ **47.** $9a^3 - 16a$

48. $16a^3 - 25a$ **49.** $12y - 2xy - 2x^2y$ **50.** $6y - 4xy - 2x^2y$

Use factoring by grouping to factor each of the following.

51. $ax + 2x + 3a + 6$ **52.** $ay + 2y - 4a - 8$ **53.** $x^2 - 3ax - 2x + 6a$

54. $x^2 - 3ax + 2x - 6a$ **55.** $x^3 + 2x^2 - 9x - 18$ **56.** $x^3 + 5x^2 - 4x - 20$

57. $x^3 + 3x^2 - 4x - 12$ **58.** $x^3 - 3x^2 - 4x + 12$ **59.** $4x^3 + 12x^2 - 9x - 27$

60. $9x^3 + 18x^2 - 4x - 8$ **61.** $2x^3 + x^2 - 18x - 9$ **62.** $3x^3 - x^2 - 12x + 4$

Factor completely, if possible.

63. $4x^2 - 31x - 8$ **64.** $6x^2 - 55xy + 9y^2$

65. $x^2 + 49$ **66.** $25 + a^2$

67. $150x^3 + 65x^2 - 280x$ **68.** $360x^3 - 490x$

69. $24x^2 + 2x - 5$ **70.** $12x^2 - 49x + 4$

71. $x^6 - 1$ **72.** $x^6 - 64$

73. $x^2(a + 5) + 6x(a + 5) + 9(a + 5)$ **74.** $a^2(x - 2) + 4a(x - 2) + 4(x - 2)$

75. $12a^2(x - 7) - 75(x - 7)$ **76.** $18a^2(2x + 3) - 50(2x + 3)$

77. $r^2 - \dfrac{1}{9}$

78. $4a^2 + 2a + \dfrac{1}{4}$

79. $t^2 + 4t + 4 - y^2$

80. $x^2 - 10x + 25 - b^2$

81. $125t^3 + \dfrac{1}{27}$

82. $\dfrac{1}{25} + \dfrac{1}{10}t^2 + \dfrac{1}{16}t^4$

83. $15t^2 + t - 16$

84. $48x^2 - 74x + 3$

85. $100x^2 - 100x - 600$

86. $100x^2 - 100x - 1200$

87. $4x^3 + 16xy^2$

88. $50 - 2a^2$

89. $25 - 10t + t^2 - x^2$

90. $25 - 20t + 4t^2 - y^2$

91. $30x^2 + 97x + 77$

92. $96a^2 + 44a - 35$

MODELING PRACTICE

Next are a few problems where an equation or expression is used to model something in the world around us. In each case, you will be asked to factor an expression. As we progress through the book, you will become more familiar with the equations used here. For now, we want to see how factoring helps us solve a few modeling problems.

93. Investing If you invest P dollars in an account with an annual interest rate of 8% compounded annually, then the amount of money in that account after one year is given by the formula:

$$A = P + 0.08P$$

Rewrite this formula with the right side in factored form, and then find the amount of money in the account at the end of one year if $500 was the initial investment.

94. Biological Growth If there are B E. coli bacteria present initially in a petri dish and their growth rate is r (a percent expressed as a decimal) per hour, then after one hour there will be $A = B + Br$ bacteria present.

a. Factor the right side of this equation.

b. The following bar graph shows the number of E. coli bacteria present initially and the number of bacteria present hours later. Use the bar chart to find B and r in the preceding equation.

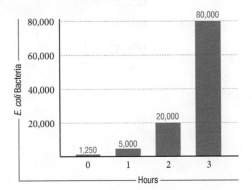

95. Archery Margaret shoots an arrow into the air. The equation for the height (in feet) of the tip of the arrow is:

$$h = 8 + 62t - 16t^2$$

Factor the right side of this equation. Then fill in the table for various heights of the arrow, using the factored form of the equation.

Time t (seconds)	0	1	2	3	4
Height h (feet)					

96. Constructing a Box Yesterday I was experimenting with how to cut and fold a certain piece of cardboard to make a box with different volumes. Unfortunately, today I have lost both the cardboard and most of my notes. I remember that I made the box by cutting equal squares from the corners then folding up the side flaps:

I don't remember how big the cardboard was, and I can only find the last page of notes, which says that if x is the length of a side of a small square (in inches), then the volume is $99x - 40x^2 + 4x^3$.

a. Factor the right side of this expression completely.

b. What were the dimensions of the original piece of cardboard?

EXTENDING THE CONCEPTS

Assume m and n are positive integers and factor completely.

97. $x^{4n} - 1$ **98.** $x^{2n} - 25$

99. $x^{3n} - 27$ **100.** $8x^{3n} + 1$

101. $x^{4n} - y^{4m}$ **102.** $x^{6n} - 1$

103. $x^{9n} - 1$ **104.** $x^{8m} - 1$

105. $x^{6n} + 1$ **106.** $x^{6m} - y^{6n}$

SPOTLIGHT ON SUCCESS

Cynthia | student instructor

Each time we face our fear, we gain strength,
courage, and confidence in the doing.

—Unknown

I must admit, when it comes to math, it takes me longer to learn the material compared to other students. Because of that, I was afraid to ask questions, especially when it seemed like everyone else understood what was going on. Because I wasn't getting my questions answered, my quiz and exam scores were only getting worse. I realized that I was already paying a lot to go to college and that I couldn't afford to keep doing poorly on my exams. I learned how to overcome my fear of asking questions by studying the material before class, and working on extra problem sets until I was confident enough that at least I understood the main concepts. By preparing myself beforehand, I would often end up answering the question myself. Even when that wasn't the case, the professor knew that I tried to answer the question on my own. If you want to be successful, but you are afraid to ask a question, try putting in a little extra time working on problems before you ask your instructor for help. I think you will find, like I did, that it's not as bad as you imagined it, and you will have overcome an obstacle that was in the way of your success.

OBJECTIVE

A Solve problems using unit analysis and conversion factors.

©iStockphoto.com/rancho_runner

1.8

In the 1950s, the United States had a spy plane, the U-2, that could fly at an altitude of 65,000 feet. Do you know how many miles are in 65,000 feet?

A Solving Problems with Unit Analysis

We can solve problems like this by using a method called *unit analysis*. With unit analysis, we analyze the units we are given and the units for which we are asked, and then multiply by the appropriate *conversion factor*. Because 1 mile is 5,280 feet, the conversion factor we use is

$$\frac{1 \text{ mile}}{5{,}280 \text{ feet}}$$

which is the number 1. Multiplying 65,000 feet by this conversion factor we have the following:

$$65{,}000 \text{ feet} = \frac{65{,}000 \text{ feet}}{1} \cdot \frac{1 \text{ mile}}{5{,}280 \text{ feet}}$$

We treat the units common to the numerator and denominator in the same way we treat factors common to the numerator and denominator: We divide out common units, just as we divide out common factors. In the preceding expression, we have feet common to the numerator and denominator. Dividing them out leaves us with miles only. Here is the complete problem.

$$65{,}000 \text{ feet} = \frac{65{,}000 \, \cancel{\text{feet}}}{1} \cdot \frac{1 \text{ mile}}{5{,}280 \, \cancel{\text{feet}}}$$

$$= \frac{65{,}000}{5{,}280} \text{ mile}$$

$$= 12.3 \text{ miles, to the nearest tenth of a mile}$$

The key to changing from feet to miles lies in choosing the appropriate conversion factor. The fact that 1 mile = 5,280 feet yields two conversion factors, each of which is equal to the number 1. They are

$$\frac{1 \text{ mile}}{5{,}280 \text{ feet}} \quad \text{and} \quad \frac{5{,}280 \text{ feet}}{1 \text{ mile}}$$

The conversion factor we choose depends on the units we are given and the units with which we want to end up. Multiplying any expression by either of the two conversion factors leaves the value of the original expression unchanged because each of the conversion factors is simply the number 1.

EXAMPLE 1

The London Eye Ferris wheel travels at approximately 52.8 feet per minute. Convert 52.8 feet per minute to miles per hour.

SOLUTION We know that 5,280 feet = 1 mile and 60 minutes = 1 hour. Therefore, we have the following conversion factors, each of which is equal to 1.

$$\frac{5{,}280 \text{ feet}}{1 \text{ mile}} \qquad \frac{1 \text{ mile}}{5{,}280 \text{ feet}} \qquad \frac{60 \text{ minutes}}{1 \text{ hour}} \qquad \frac{1 \text{ hour}}{60 \text{ minutes}}$$

The conversion factors we choose to multiply by are the ones that will allow us to divide out the units we are converting from and leave us with the units we are converting to. Specifically, we want to get rid of feet and be left with miles. Likewise, we want to get rid of minutes and be left with hours. Here is the conversion process that will accomplish these goals:

$$52.8 \text{ feet per minute} = \frac{52.8 \text{ feet}}{1 \text{ minute}} \cdot \frac{1 \text{ mile}}{5{,}280 \text{ feet}} \cdot \frac{60 \text{ minutes}}{1 \text{ hour}}$$

$$= \frac{52.8 \cdot 60 \text{ miles}}{5{,}280 \text{ hours}}$$

$$= 0.6 \text{ miles per hour, to the nearest hundredth}$$

EXAMPLE 2

A ski resort in Vermont advertised their new high-speed chair lift as "the world's fastest chair lift, with a speed of 1,100 feet per second." Show why the speed cannot be correct.

SOLUTION To solve this problem, we can convert feet per second into miles per hour, a unit of measure we are more familiar with on an intuitive level.

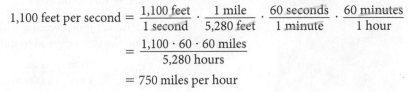

$$1{,}100 \text{ feet per second} = \frac{1{,}100 \text{ feet}}{1 \text{ second}} \cdot \frac{1 \text{ mile}}{5{,}280 \text{ feet}} \cdot \frac{60 \text{ seconds}}{1 \text{ minute}} \cdot \frac{60 \text{ minutes}}{1 \text{ hour}}$$

$$= \frac{1{,}100 \cdot 60 \cdot 60 \text{ miles}}{5{,}280 \text{ hours}}$$

$$= 750 \text{ miles per hour}$$

iStockPhoto.com/©Paul Laliberte

Obviously, there is a mistake in the advertisement.

EXAMPLE 3

Population change information given by the United States Census Bureau states that there is one birth every 8 seconds, one death every 12 seconds, and one international migrant every 29 seconds. Use this information to find the average number of births in one week.

SOLUTION Along with the rates given, we need the following relationships.

1 minute = 60 seconds   1 hour = 60 minutes   1 day = 24 hours   1 week = 7 days

Writing each of these relationships as a conversion factor equal to 1, we have

$$\frac{60 \text{ sec}}{1 \text{ min}} \qquad \frac{60 \text{ min}}{1 \text{ hr}} \qquad \frac{24 \text{ hr}}{1 \text{ day}} \qquad \frac{7 \text{ days}}{1 \text{ week}}$$

To find the average number of births in one week, we convert from births per second to births per week:

$$\frac{1 \text{ birth}}{8 \text{ seconds}} = \frac{1 \text{ birth}}{8 \text{ sec}} \cdot \frac{60 \text{ sec}}{1 \text{ min}} \cdot \frac{60 \text{ min}}{1 \text{ hr}} \cdot \frac{24 \text{ hr}}{1 \text{ day}} \cdot \frac{7 \text{ days}}{1 \text{ week}}$$

$$= 75,600 \text{ births/week}$$

EXAMPLE 4

The volcanic rock on the island of Kauai is approximately 5 million years old. Kauai is 519 kilometers from Kilauea, the volcano on the island of Hawaii. This means the Pacific plate is moving 519 km every 5 million years. What is the rate of movement in cm/year?

SOLUTION   We divide 519 kilometers by 5,000,000 years, then multiply by the appropriate conversion factors to arrive at our answer.

$$\frac{519 \text{ km}}{5,000,000 \text{ yr}} = \frac{519 \text{ km}}{5,000,000 \text{ yr}} \cdot \frac{1,000 \text{ m}}{1 \text{ km}} \cdot \frac{100 \text{ cm}}{1 \text{ m}}$$

$$= 10.38 \text{ cm/yr}$$

This is approximately the same rate at which your fingernails grow.

iStockPhoto.com/©Ashok Rodrigues

EXAMPLE 5

Pyroclastic flows are high-speed avalanches of volcanic gases and ash that accompany some volcanic eruptions. They travel at speeds of 10 meters/second to 100 meters/second. Could you outrun a pyroclastic flow on foot, on a bicycle, or in a car? Use the conversion factors

$$1 \text{ mile} = 1.61 \text{ kilometers} \qquad 1 \text{ kilometer} = 1,000 \text{ meters}$$

SOLUTION   Since we are most familiar with miles per hour as a unit of speed, let's start by converting the slower speed to miles per hour.

$$10 \text{ m/sec} = \frac{10 \text{ m}}{1 \text{ sec}} \cdot \frac{1 \text{ km}}{1,000 \text{ m}} \cdot \frac{1 \text{ mi}}{1.61 \text{ km}} \cdot \frac{60 \text{ sec}}{1 \text{ min}} \cdot \frac{60 \text{ min}}{1 \text{ hr}}$$

$$= \frac{36,000 \text{ mi}}{1,610 \text{ hr}}$$

$$\approx 22.4 \text{ mi/hr}$$

USGS/Photo by Peter W. Lipman

We could not outrun this on foot. Some people could outrun it on a bicycle, or a car.

To convert 100 m/sec to miles per hour, we simply multiply our previous result by 10:

$$100 \text{ m/sec} \approx 224 \text{ mi/hr}$$

At this speed, the pyroclastic flow cannot be outrun. You can see why many people have lost their lives in pyroclastic flows that accompany some volcanic eruptions.

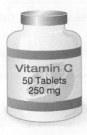

EXAMPLE 6

A bottle of vitamin C contains 50 tablets. Each tablet contains 250 milligrams of vitamin C. What is the total number of grams of vitamin C in the bottle?

SOLUTION   We begin by finding the total number of milligrams of vitamin C in the bottle. Since there are 50 tablets, and each contains 250 mg of vitamin C, we can multiply 50 by 250 to get the total number of milligrams of vitamin C:

$$\text{Milligrams of vitamin C} = 50 \times 250 \text{ mg}$$
$$= 12,500 \text{ mg}$$

Next we convert 12,500 mg to grams:

$$12,500 \text{ mg} = 12,500 \text{ mg} \times \frac{1 \text{ g}}{1,000 \text{ mg}}$$
$$= \frac{12,500}{1,000} \text{ g}$$
$$= 12.5 \text{ g}$$

The bottle contains 12.5 g of vitamin C.

EXAMPLE 7

The engine in a car has a 2-liter displacement. What is the displacement in cubic inches? Use the conversion factor $1 \text{ in}^3 = 16.39 \text{ mL}$.

SOLUTION   We convert liters to milliliters and then milliliters to cubic inches:

$$2 \text{ liters} = 2 \text{ liters} \times \frac{1,000 \text{ mL}}{1 \text{ liter}} \times \frac{1 \text{ in}^3}{16.39 \text{ mL}}$$
$$= \frac{2 \times 1,000}{16.39} \text{ in}^3 \qquad \text{This calculation should be done on a calculator.}$$
$$= 122 \text{ in}^3 \qquad\qquad \text{To the nearest cubic inch}$$

EXAMPLE 8

The Cone Nebula, as photographed by the Hubble telescope, is shown here. The distance across the photograph is about 2.5 light-years, where a light-year is the distance light travels in a year. If we assume light travels 186,000 miles in one second, find the distance across the photograph in miles.

NASA

SOLUTION   We can find the number of miles in 1 light-year by converting 186,000 miles/second to miles/year.

$$186,000 \text{ miles/second} = \frac{186,000 \text{ miles}}{1 \text{ sec}} \cdot \frac{60 \text{ sec}}{1 \text{ min}} \cdot \frac{60 \text{ min}}{1 \text{ hr}} \cdot \frac{24 \text{ hr}}{1 \text{ day}} \cdot \frac{365 \text{ days}}{1 \text{ year}}$$
$$= 5,865,696,000,000 \text{ miles/year}$$
$$\approx 5.87 \times 10^{12} \text{ miles/year}$$

Multiplying this number by 2.5 will give us the distance across the photograph in miles.

$$5.87 \times 10^{12} \text{ miles/year} \times 2.5 \text{ years} \approx 1.47 \times 10^{13} \text{ miles}$$

GETTING READY FOR CLASS

After reading through the preceding section, respond in your own words and in complete sentences.

A. How do you divide two expressions containing exponents when they each have the same base?

B. Explain the difference between 3^2 and 3^{-2}.

C. If a positive base is raised to a negative exponent, can the result be a negative number?

D. Explain what happens when we use 0 as an exponent.

Problem Set 1.8

1. **Mount Whitney** The top of Mount Whitney, the highest point in California, is 14,494 feet above sea level. Give this height in miles to the nearest tenth of a mile.

2. **Motor Displacement** The relationship between liters and cubic inches, both of which are measures of volume, is 0.0164 liters = 1 cubic inch. If a Ford Mustang has a motor with a displacement of 4.9 liters, what is the displacement in cubic inches? Round your answer to the nearest cubic inch.

3. **Speed of Sound** The speed of sound is 1,088 feet per second. Convert the speed of sound to miles per hour. Round your answer to the nearest whole number.

4. **Average Speed** A car travels 122 miles in 3 hours. Find the average speed of the car in feet per second. Round to the nearest whole number.

5. **Ferris Wheel** The first Ferris wheel was built in 1893. It was a large wheel with a circumference of 785 feet. If one trip around the circumference of the wheel took 20 minutes, find the average speed of a rider in miles per hour. Round to the nearest hundredth.

6. **Ferris Wheel** A Ferris wheel called *Colossus* has a circumference of 518 feet. If a trip around the circumference of *Colossus* takes 40 seconds, find the average speed of a rider in miles per hour. Round to the nearest tenth.

7. **Track and Field** A person who runs the 100-yard dash in 10.5 seconds has an average speed of 9.52 yards/second. If 1 yard = 3 feet, convert 9.52 yards/second to miles/hour. (Round to the nearest tenth.)

8. **Track and Field** A person who runs a mile in 8 minutes has an average speed of 0.125 miles/minute. Convert 0.125 miles/minute to miles/hour.

9. **Fitness Walking** Guidelines for fitness indicate that a person who walks 10,000 steps daily is physically fit. According to www.thewalkingsite.com, "The average person's stride length is approximately 2.5 feet long. That means it takes just over 2,000 steps to walk one mile, and 10,000 steps is close to 5 miles." Use your knowledge of unit analysis to determine if these facts are correct.

10. **Football** The diagrams below show the dimensions of playing fields for Arena Football, the National Football League (NFL), and the Canadian Football League. Find the area of each field and then convert each area to acres if 1 acre = 43,560 square feet and 9 square feet = 1 square yard. Round answers to the nearest hundredth.

iStockPhoto.com/©Holger Mette

iStockPhoto.com/©Phillip Dyer

Football Fields

Arena

 $28\frac{1}{3}$ yd

50 yd

NFL

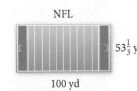 $53\frac{1}{3}$ yd

100 yd

Canadian

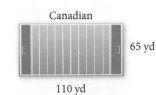 65 yd

110 yd

11. **Soccer** The rules for soccer state that the playing field must be from 100 to 120 yards long and 55 to 75 yards wide. The 1999 Women's World Cup was played at the Rose Bowl on a playing field 116 yards long and 72 yards wide. The diagram below shows the smallest possible soccer field, the largest possible soccer field, and the soccer field at the Rose Bowl. Find the area of each one and then convert the area of each to acres. Round to the nearest tenth, if necessary.

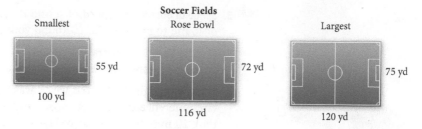

Soccer Fields

Smallest   Rose Bowl   Largest

55 yd   72 yd   75 yd

100 yd   116 yd   120 yd

12. **Fish Oil** A bottle of fish oil contains 50 soft gels, each containing 300 mg of the omega-6 fatty acid. How many total grams of the omega-6 fatty acid are in this bottle?

13. **B-Complex** A certain B-complex vitamin supplement contains 30 mg of thiamine, or vitamin B1. A bottle contains 80 vitamins. How many total grams of thiamine are in this bottle?

14. **Aspirin** A bottle of maximum-strength aspirin contains 90 tablets. Each tablet contains 500 mg of aspirin. How many total grams of aspirin are in this bottle?

15. **Vitamin C** A certain brand of vitamin C contains 500 mg per tablet. A bottle contains 240 vitamins. How many total grams of vitamin C are in this bottle?

16. **Vitamin C** A certain brand of vitamin C contains 600 mg per tablet. A bottle contains 150 vitamins. How many total grams of vitamin C are in this bottle?

17. **Light-Year** A light-year, the distance light travels in 1 year, is approximately 5.9×10^{12} miles. The Andromeda galaxy is approximately 1.7×10^6 light-years from our galaxy. Find the distance in miles between our galaxy and the Andromeda galaxy.

18. **Distance to the Sun** The distance from Earth to the Sun is approximately 9.3×10^7 miles. If light travels 1.2×10^7 miles in 1 minute, how many minutes does it take the light from the Sun to reach Earth?

The numbers in brackets refer to the section(s) in which the topic can be found.

Exponents [1.1]

EXAMPLES

The margins of the chapter summaries will be used for brief examples of the topics being reviewed, whenever it is convenient.

1. $2^5 = 2 \cdot 2 \cdot 2 \cdot 2 \cdot 2 = 32$
$5^2 = 5 \cdot 5 = 25$
$10^3 = 10 \cdot 10 \cdot 10 = 1,000$
$1^4 = 1 \cdot 1 \cdot 1 \cdot 1 = 1$

Exponents represent notation used to indicate repeated multiplication. In the expression 3^4, 3 is the *base* and 4 is the *exponent*.

$$3^4 = 3 \cdot 3 \cdot 3 \cdot 3 = 81$$

The expression 3^4 is said to be in *exponential form*, whereas the expression $3 \cdot 3 \cdot 3 \cdot 3$ is in *expanded form*.

Order of Operations [1.1]

2. $10 + (2 \cdot 3^2 - 4 \cdot 2)$
$= 10 + (2 \cdot 9 - 4 \cdot 2)$
$= 10 + (18 - 8)$
$= 10 + 10$
$= 20$

When evaluating a mathematical expression, we will perform the operations in the following order, beginning with the expression in the innermost parentheses or brackets and working our way out.

1. Simplify all numbers with exponents, working from left to right if more than one of these numbers is present.

2. Then, do all multiplications and divisions left to right.

3. Finally, perform all additions and subtractions left to right.

Sets [1.1]

3. If $A = \{0, 1, 2\}$ and $B = \{2, 3\}$, then $A \cup B = \{0, 1, 2, 3\}$ and $A \cap B = \{2\}$.

A *set* is a collection of objects or things.

The *union* of two sets A and B, written $A \cup B$, is all the elements that are in A *or* are in B.

The *intersection* of two sets A and B, written $A \cap B$, is the set consisting of all elements common to both A *and* B.

Set A is a *subset* of set B, written $A \subset B$, if all elements in set A are also in set B.

Special Sets [1.1]

4. 5 is a counting number, a whole number, an integer, a rational number, and a real number.

$\frac{3}{4}$ is a rational number and a real number.

$\sqrt{2}$ is an irrational number and a real number.

Counting numbers $= \{1, 2, 3, \dots\}$
Whole numbers $= \{0, 1, 2, 3, \dots\}$
Integers $= \{\dots, -3, -2, -1, 0, 1, 2, 3, \dots\}$
Rational numbers $= \left\{ \dfrac{a}{b} \mid a \text{ and } b \text{ are integers, } b \neq 0 \right\}$
Irrational numbers $= \{x \mid x \text{ is real, but not rational}\}$
Real numbers $= \{x \mid x \text{ is rational or } x \text{ is irrational}\}$
Prime numbers $= \{2, 3, 5, 7, 11, \dots\} = \{x \mid x \text{ is a positive integer greater than 1}$
$\text{whose only positive divisors are itself and 1}\}$

Opposites [1.2, 1.3]

5. The numbers 5 and -5 are opposites; their sum is 0.

$$5 + (-5) = 0$$

Any two real numbers the same distance from 0 on the number line, but in opposite directions from 0, are called *opposites*, or *additive inverses*. Opposites always add to 0.

Reciprocals [1.2, 1.3]

6. The numbers 3 and $\frac{1}{3}$ are reciprocals; their product is 1.

$$3\left(\frac{1}{3}\right) = 1$$

Any two real numbers whose product is 1 are called *reciprocals*. Every real number has a reciprocal except 0.

Absolute Value [1.2]

7. $|5| = 5$
$|-5| = 5$

The *absolute value* of a real number is its distance from 0 on the number line. If $|x|$ represents the absolute value of x, then

$$|x| = \begin{cases} x & \text{if} & x \geq 0 \\ -x & \text{if} & x < 0 \end{cases}$$

The absolute value of a real number is never negative.

Properties of Real Numbers [1.2]

	For Addition	*For Multiplication*
Commutative	$a + b = b + a$	$ab = ba$
Associative	$a + (b + c) = (a + b) + c$	$a(bc) = (ab)c$
Identity	$a + 0 = a$	$a \cdot 1 = a$
Inverse	$a + (-a) = 0$	$a\left(\dfrac{1}{a}\right) = 1$
Distributive	$a(b + c) = ab + ac$	

Addition [1.3]

8. $\quad 5 + 3 = 8$
$\quad 5 + (-3) = 2$
$\quad -5 + 3 = -2$
$\quad -5 + (-3) = -8$

To add two real numbers with

1. *The same sign:* Simply add absolute values and use the common sign.

2. *Different signs:* Subtract the smaller absolute value from the larger absolute value. The answer has the same sign as the number with the larger absolute value.

Subtraction [1.3]

9. $\quad 6 - 2 = 6 + (-2) = 4$
$\quad 6 - (-2) = 6 + 2 = 8$

If a and b are real numbers,

$$a - b = a + (-b)$$

To subtract b, add the opposite of b.

Multiplication [1.3]

10.
$5(4) = 20$
$5(-4) = -20$
$-5(4) = -20$
$-5(-4) = 20$

To multiply two real numbers, simply multiply their absolute values. Like signs give a positive answer. Unlike signs give a negative answer.

Division [1.3]

11. $\frac{12}{-3} = -4$

$\frac{-12}{-3} = 4$

If a and b are real numbers and $b \neq 0$, then

$$\frac{a}{b} = a \cdot \left(\frac{1}{b}\right)$$

To divide by b, multiply by the reciprocal of b.

Properties of Exponents [1.4]

12. These expressions illustrate the properties of exponents.

a. $x^2 \cdot x^3 = x^{2+3} = x^5$

b. $(x^2)^3 = x^{2 \cdot 3} = x^6$

c. $(3x)^2 = 3^2 \cdot x^2 = 9x^2$

d. $2^{-3} = \frac{1}{2^3} = \frac{1}{8}$

e. $\left(\frac{x}{5}\right)^2 = \frac{x^2}{5^2} = \frac{x^2}{25}$

f. $\frac{x^7}{x^5} = x^{7-5} = x^2$

g. $3^1 = 3$
$3^0 = 1$

If a and b represent real numbers and r and s represent integers, then

1. $a^r \cdot a^s = a^{r+s}$

2. $(a^r)^s = a^{r \cdot s}$

3. $(ab)^r = a^r \cdot b^r$

4. $a^{-r} = \frac{1}{a^r}$ $(a \neq 0)$

5. $\left(\frac{a}{b}\right)^r = \frac{a^r}{b^r}$ $(b \neq 0)$

6. $\frac{a^r}{a^s} = a^{r-s}$ $(a \neq 0)$

7. $a^1 = a$
$a^0 = 1$ $(a \neq 0)$

Scientific Notation [1.4]

13. $49{,}800{,}000 = 4.98 \times 10^7$
$0.00462 = 4.62 \times 10^{-3}$

A number is written in scientific notation when it is written as the product of a number between 1 and 10 and an integer power of 10; that is, when it has the form

$$n \times 10^r$$

where $1 \leq n < 10$ and $r =$ an integer.

Operations with Polynomials [1.5]

14. $(3x^2 + 2x - 5) + (4x^3 - 7x + 2)$
$= 7x^2 - 5x - 3$

15. $(3x - 5)(x + 2)$
$3x^2 + 6x - 5x - 10$
$3x^2 + x - 10$

To add two polynomials, combine the coefficients of similar terms.

To multiply two polynomials, multiply each term in the first polynomial by each term in the second polynomial.

Special Products [1.3]

16. The following are examples of the three special products:
$(x + 3)^2 = x^2 + 6x + 9$
$(5 - x)^2 = 25 - 10x + x^2$
$(x + 7)(x - 7) = x^2 - 49$

$$(a + b)^2 = a^2 + 2ab + b^2$$
$$(a - b)^2 = a^2 - 2ab + b^2$$
$$(a + b)(a - b) = a^2 - b^2$$

Greatest Common Factor [1.4]

17. The greatest common factor of $10x^5 - 15x^4 + 30x^3$ is $5x^3$. Factoring it out of each term, we have

$$5x^3(2x^2 - 3x + 6)$$

The greatest common factor of a polynomial is the largest monomial (the monomial with the largest coefficient and highest exponent) that divides each term of the polynomial. The first step in factoring a polynomial is to factor the greatest common factor (if it is other than 1) out of each term.

Factoring Trinomials [1.5]

18. $x^2 + 5x + 6 = (x + 2)(x + 3)$
$x^2 - 5x + 6 = (x - 2)(x - 3)$
$x^2 + x - 6 = (x - 2)(x + 3)$
$x^2 - x - 6 = (x + 2)(x - 3)$

We factor a trinomial by writing it as the product of two binomials. (This refers to trinomials whose greatest common factor is 1.) Each factorable trinomial has a unique set of factors. Finding the factors is sometimes a matter of trial and error.

Special Factoring [1.6]

19. Here are some binomials that have been factored this way.
$x^2 + 6x + 9 = (x + 3)^2$
$x^2 - 6x + 9 = (x - 3)^2$
$x^2 - 9 = (x + 3)(x - 3)$
$x^3 - 27 = (x - 3)(x^2 + 3x + 9)$
$x^3 + 27 = (x + 3)(x^2 - 3x + 9)$

$$a^2 + 2ab + b^2 = (a + b)^2$$ Perfect square trinomials
$$a^2 - 2ab + b^2 = (a - b)^2$$
$$a^2 - b^2 = (a - b)(a + b)$$ Difference of two squares
$$a^3 - b^3 = (a - b)(a^2 + ab + b^2)$$ Difference of two cubes
$$a^3 + b^3 = (a + b)(a^2 - ab + b^2)$$ Sum of two cubes

Unit Analysis [1.8]

20. Convert 52.8 feet/minute into miles/hour.

$52.8 \text{ ft/min} = \dfrac{52.8 \text{ ft}}{1 \text{ min}} \cdot \dfrac{1 \text{ mi}}{5,280 \text{ ft}} \cdot \dfrac{60 \text{ min}}{1 \text{ hr}}$

$= \dfrac{52.8 \cdot 60 \text{ mi}}{5,280 \text{ hr}}$

$= 0.6 \text{ mph, to the nearest hundredth}$

Unit analysis is a process that we use to convert one set of units to another set of units. To do so, we analyze the units we are given and the units we are asked for, and then multiply by the appropriate conversion factor or factors.

COMMON MISTAKES *Order of Operations*

1. Interpreting absolute value as changing the sign of the number inside the absolute value symbols. That is, $|-5| = +5, |+5| = -5$. To avoid this mistake, remember: absolute value is defined as a distance and distance is always measured in non-negative units.

2. Confusing $-(-5)$ with $-|-5|$. The first answer is $+5$, but the second answer is -5.

3. When we subtract one polynomial from another, it is common to forget to add the opposite of each term in the second polynomial. For example:

$$(6x - 5) - (3x + 4) = 6x - 5 - 3x + 4 \quad \text{Mistake}$$
$$= 3x - 1$$

This mistake occurs if the negative sign outside the second set of parentheses is not distributed over all terms inside the parentheses. To avoid this mistake, remember: The opposite of a sum is the sum of the opposites, or

$$-(3x + 4) = -3x + (-4)$$

4. Interpreting the square of a sum to be the sum of the squares. That is,

$$(x + y)^2 = x^2 + y^2 \quad \text{Mistake}$$

This can easily be shown as false by trying a couple of numbers for x and y. If $x = 4$ and $y = 3$, we have

$$(4 + 3)^2 = 4^2 + 3^2 \quad \text{Mistake}$$
$$7^2 = 16 + 9$$
$$49 = 25$$

There has obviously been a mistake. The correct formula for $(a + b)^2$ is

$$(a + b)^2 = a^2 + 2ab + b^2$$

Chapter 1 Review

At the end of each chapter, we will provide a review of problems similar to those you have seen in the chapter's previous sections. These problems are meant to supplement, not replace, a thorough review of the concepts, examples, and exercise sets presented in the chapter.

Write each fraction as an equivalent fraction with denominator 36. [1.1]

1. $\dfrac{2}{3}$

2. $\dfrac{7}{9}$

Write each fraction as an equivalent fraction with denominator 60. [1.1]

3. $\dfrac{9}{10}$

4. $-\dfrac{7}{12}$

For each of the following numbers, give the opposite, the reciprocal, and the absolute value. (Assume all variables are nonzero.) [1.1]

5. $\dfrac{16}{3}$

6. -5

7. $-\dfrac{3}{8}$

8. a

Write each of the following without absolute value symbols. [1.2]

9. $|-7|$

10. $\left|\dfrac{5}{6}\right|$

11. $\left|-\sqrt{2}\right|$

12. $-|5|$

Insert a $<$ or $>$ to make the statement true. [1.1]

13. $-4 \quad 9$

14. $-6 \quad -3$

15. $5 \quad 3$

16. $-\dfrac{2}{3} \quad -\dfrac{1}{6}$

Simplify each expression. [1.1, 1.2]

17. $|7 - 12 \div 3|$

18. $|9 - 12| - |3 - 11|$

19. $13 - |7(2 - 5)|$

20. $7 + |5 \cdot 3 - 7| - 6$

21. $12 - |9 - 3(7 - 5)|$

22. $25 - |9 - 3(4 \cdot 5 - 18)| - 20$

Factor each number into the product of prime factors. [1.1]

23. 385

24. 735

Reduce to lowest terms. [1.1]

25. $\dfrac{252}{468}$

26. $\dfrac{208}{496}$

27. $\dfrac{38}{30}$

28. $\dfrac{240}{6}$

Find the perimeter and area of each figure. [1.4]

29.

30.

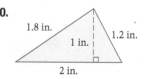

Identify the property of real numbers that justifies each of the following. [1.2]

31. $6(xy) = (xy)6$

32. $(r + 7) + s = (r + s) + 7$

33. $5\left(\dfrac{1}{5}\right) = 1$

34. $3(2x - 6) = 6x - 18$

Simplify. [1.3]

35. $3(2x + 5) - 4x$

36. $5x - 5 + 2x + 7$

37. $3x + 2(3x + 2y) + 4y$

38. $4a + 3(2b + 4) - 8$

39. $4y + 2(2x - 4) - 2y$

40. $16\left(\dfrac{5}{8}x + \dfrac{1}{4}\right) - 2$

Add the following fractions. [1.2]

41. $\dfrac{9}{14} + \dfrac{7}{24}$

42. $\dfrac{9}{16} + \dfrac{12}{36}$

Find each of the following differences. [1.3]

43. $\dfrac{2}{3} - \left(-\dfrac{7}{5}\right)$

44. $\dfrac{13}{70} - \dfrac{19}{42}$

45. Add -3 to the difference of 9 and 2. [1.1}

46. Subtract $-3a$ from the sum of $3a$ and $5a$. [1.1]

Simplify each expression. [1.3]

47. $3 - (5 - x) + 3x$

48. $(6 - 3x) - (2x - 4)$

49. $3x - 4(2x - 3) - 12$

50. $\dfrac{3(-2) - 4(4)}{6 - 4}$

51. $\dfrac{3 + 6}{3 - 3}$

52. $\dfrac{-2^2 + 4}{-5 - 5}$

53. $-9 - 5\left[\dfrac{11(-1) - 9}{4(-3) + 2(5)}\right]$

54. $8 - (-7)\left[\dfrac{6 - 1(6 - 10)}{4 - 3(5 - 7)}\right]$

55. Complete the following table. [1.3]

x	$(x + 1)^2$	$x^2 + 1$	$x^2 + 2x + 1$
-2			
-1			
0			
1			
2			

56. Find the value of $b^2 - 4ac$ when [1.3]

  a. $a = 3$, $b = -2$, and $c = 4$

  b. $a = 1$, $b = -3$, and $c = -28$

  c. $a = 1$, $b = -6$, and $c = 9$

  d. $a = 0.1$, $b = -27$, and $c = 1700$

Evaluate each of the following. [1.4]

57. $(-4)^2$

58. $-(-4)^2$

59. $\left(\dfrac{3}{4}\right)^2$

60. $\left(-\dfrac{4}{5}\right)^3$

Simplify as much as possible. Write all answers with positive exponents. [1.4]

61. $(-2x^2)^{-2}$

62. $(3x^2y^2)^3$

63. $\left(\dfrac{2}{3}\right)^{-3}$

64. $2x^4y \cdot -2xy^4$

65. $-3x^2y \cdot 7x^5y^2$

66. $\dfrac{36a^5b}{6a^2b^4}$

67. $\dfrac{24a^6b^3}{-30ab^4}$

68. $\left(\dfrac{7a^4b}{14a^6b^5}\right)^{-2}$

Simplify the following expressions. [1.5]

69. $\left(\frac{6}{5}x^3 - 2x - \frac{3}{5}\right) - \left(\frac{6}{5}x^2 - \frac{2}{5}x + \frac{3}{5}\right)$

70. $5 - 7[9(2x + 1) - 16x]$

Multiply. [1.5]

71. $(x + 7)(-5x + 4)$

72. $(3x - 2)(2x^2 + 6x - 5)$

73. $(3a^4 - 7)^2$

74. $(2x + 3)(2x - 3)$

75. $x(x - 7)(3x + 4)$

76. $\left(2x - \frac{1}{7}\right)\left(7x + \frac{1}{2}\right)$

Factor the following expressions. [1.5, 1.6]

77. $x^2 - 6x + 5$

78. $15x^4 + 33x^2 - 36$

79. $81x^4 - 16y^4$

80. $6ax - ay + 18b^2x - 3b^2y$

81. $y^3 - \frac{1}{27}$

82. $3x^4y^4 + 15x^3y^5 - 72x^2y^6$

Write in scientific notation. [1.4]

83. 482,000

84. 7,280,000

85. 0.00421

86. 0.0526

Write in expanded form. [1.4]

87. 6.29×10^{-3}

88. 3.29×10^3

89. 6.31×10^{-2}

90. 4.82×10^7

Simplify. Write in scientific notation. [1.4]

91. $\dfrac{(3.6 \times 10^7)(2 \times 10^{-4})}{2.4 \times 10^6}$

92. $\dfrac{(4 \times 10^{-6})(1.6 \times 10^4)}{(3.2 \times 10^9)(1 \times 10^{-7})}$

Chapter 1 Test

1. Write the following in symbols. [1.1]

   The difference of $2a$ and $3b$ is less than their sum.

2. If $A = \{1, 2, 3, 4\}$ and $B = \{2, 4, 6\}$, find $A \cup B$. [1.1]

3. Factor 770 into the product of prime factors. [1.1]

State the property or properties that justify each of the following. [1.2]

4. $4 + x = x + 4$

5. $5(1) = 5$

Simplify each of the following as much as possible. [1.3]

6. $5(-4) + 1$

7. $3(2 - 4)^3 - 5(2 - 7)^2$

8. $\dfrac{-4(-1) - (-10)}{5 - (-2)}$

9. $-\dfrac{3}{8} + \dfrac{5}{12} - \left(-\dfrac{7}{9}\right)$

10. $-\dfrac{1}{2}(8x)$

11. $-4(3x + 2) + 7x$

12. Subtract $\dfrac{3}{4}$ from the product of -4 and $\dfrac{7}{16}$ and simplify. [1.3]

Simplify. (Assume all variables are nonzero.) [1.4]

13. $x^4 \cdot x^7 \cdot x^{-3}$

14. 2^{-5}

15. $\dfrac{a^{-5}}{a^{-7}}$

16. $\dfrac{(2ab^3)^{-2}(a^4b^{-3})}{(a^{-4}b^3)^4(2a^{-2}b^2)^{-3}}$

17. Write 6,530,000 in scientific notation. [1.4]

18. Perform the indicated operations and write your answer in scientific notation. [1.4]

$$\frac{(6 \times 10^{-4})(4 \times 10^9)}{8 \times 10^{-3}}$$

Simplify. [1.5]

19. $\left(\dfrac{3}{4}x^3 - x^2 - \dfrac{3}{2}\right) - \left(\dfrac{1}{4}x^2 + 2x - \dfrac{1}{2}\right)$

20. $3 - 4[2x - 3(x + 6)]$

21. $(3y - 7)(2y + 5)$

22. $(2x - 5)(x^2 + 4x - 3)$

23. $(8 - 3t^3)^2$

24. $(1 - 6y)(1 + 6y)$

Factor completely. [1.6]

25. $12x^4 + 26x^2 - 10$

26. $16a^4 - 81y^4$

27. $7ax^2 - 14ay - b^2x^2 + 2b^2y$

28. $t^3 + \dfrac{1}{8}$

29. $x^2 - 10x + 25 - b^2$

30. $81 - x^4$

31. The speed of sound is 1,088 feet per second. Convert the speed of sound to miles per hour. Round your answer to the nearest whole number. [1.8]

32. A bullet fired from a gun travels a distance of 4,750 feet in 3.2 seconds. Find the average speed of the bullet in miles per hour. Round to the nearest whole number. [1.8]

2

Equations and Inequalities in One Variable

Chapter Outline

©iStockphoto.com/Rawpixel

A recent newspaper article gave the following guideline for college students taking out loans to finance their education: The maximum monthly payment on the amount borrowed should not exceed 8% of their monthly starting salary. In this situation, the maximum monthly payment can be described mathematically with the formula

$$y = 0.08x$$

Using this formula, we can construct a table and a line graph that show the maximum student loan payments that can be made for a variety of starting salaries.

Maximum Student Loan Payments

Monthly Starting Salary	Maximum Loan Payment
$2,000	$160
$2,500	$200
$3,000	$240
$3,500	$280
$4,000	$320
$4,500	$360
$5,000	$400

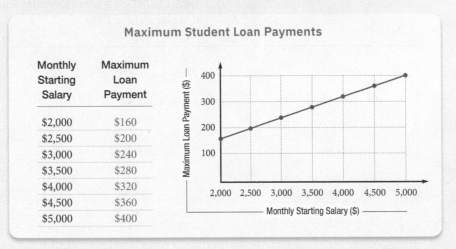

In this chapter, we begin our work with connecting tables and line graphs with algebraic formulas.

STUDY SKILLS

If you have successfully completed Chapter 1, then you have made a good start at developing the study skills necessary to succeed in all math classes. Some of the study skills for this chapter are a continuation of the skills from Chapter 1, while others are new to this chapter.

1. **Continue to Set and Keep a Schedule**   Sometimes I find students do well in Chapter 1 and then become overconfident. They will begin to put in less time with their homework. Don't do it. Keep to the same schedule.

2. **Increase Effectiveness**   You want to become more and more effective with the time you spend on your homework. Increase those activities that are the most beneficial and decrease those that have not given you the results you want.

3. **List Difficult Problems**   Begin to make lists of problems that give you the most difficulty. These are the problems in which you are repeatedly making mistakes.

4. **Begin to Develop Confidence with Word Problems**   It seems that the main difference between people who are good at working word problems and those who are not is confidence. People with confidence know that no matter how long it takes them, they will eventually be able to solve the problem. Those without confidence begin by saying to themselves, "I'll never be able to work this problem." If you are in this second category, then instead of telling yourself that you can't do word problems, decide to do whatever it takes to master them. The more word problems you work, the better you will become at them.

Many of my students keep a notebook that contains everything that they need for the course: class notes, homework, quizzes, tests, and research projects. A three-ring binder with tabs is ideal. Organize your notebook so that you can easily get to any item you want to look at.

OBJECTIVES

A Solve equations using the addition and multiplication properties of equality.

B Solve a quadratic equation by writing it in standard form and then factoring.

C Determine if an equation is an identity or has no solution.

©iStockphoto.com/Milkos

According to the introduction to this chapter, the formula

$$y = 0.08x$$

gives the monthly starting salary of x dollars per month a student should earn after they graduate, if they are to pay back a student loan at y dollars per month. From this formula, we find that the solution to the equation

$$300 = 0.08x$$

is the starting salary needed to pay back a student loan at $300 per month. Solving this type of equation is one of the things we will do in this section.

A *linear equation in one variable* is any equation that can be put in the form

$$ax + b = c$$

where a, b, and c are constants and $a \neq 0$. For example, each of the equations

$$5x + 3 = 2 \qquad 2x = 7 \qquad 2x + 5 = 0$$

are linear because they can be put in the form $ax + b = c$. In the first equation, $5x$, 3, and 2 are called *terms* of the equation: $5x$ is a variable term; 3 and 2 are constant terms.

DEFINITION *solution set*

The *solution set* for an equation is the set of all numbers that, when used in place of the variable, make the equation a true statement.

DEFINITION *equivalent equations*

Two or more equations with the same solution set are called *equivalent equations.*

The equations $2x - 5 = 9$, $x - 1 = 6$, and $x = 7$ are all equivalent equations because the solution set for each is {7}.

Properties of Equality

The first property of equality states that adding the same quantity to both sides of an equation preserves equality. Or, more importantly, adding the same amount to both sides of an equation *never changes* the solution set. This property is called the *addition property of equality* and is stated in symbols as follows:

> **PROPERTY** *Addition Property of Equality*
> ....
>
> For any three algebraic expressions A, B, and C,
>
> $$\text{if} \qquad A = B$$
> $$\text{then} \qquad A + C = B + C$$
>
> *In words:* Adding the same quantity to both sides of an equation will not change the solution set.

Our second property is called the *multiplication property of equality* and is stated as follows:

NOTE Because subtraction is defined in terms of addition and division is defined in terms of multiplication, we do not need to introduce separate properties for subtraction and division. The solution set for an equation will never be changed by subtracting the same amount from both sides or by dividing both sides by the same nonzero quantity.

> **PROPERTY** *Multiplication Property of Equality*
> ....
>
> For any three algebraic expressions A, B, and C, where $C \neq 0$,
>
> $$\text{if} \qquad A = B$$
> $$\text{then} \qquad AC = BC$$
>
> *In words:* Multiplying both sides of an equation by the same nonzero quantity will not change the solution set.

EXAMPLE 1

Find the solution set for $3a - 5 = -6a + 1$.

SOLUTION   To solve for a, we must isolate it on one side of the equation. Let's decide to isolate a on the left side. We start by adding $6a$ to both sides of the equation.

$$3a - 5 = -6a + 1$$
$$3a + 6a - 5 = -6a + 6a + 1 \qquad \text{Add } 6a \text{ to both sides.}$$
$$9a - 5 = 1$$
$$9a - 5 + 5 = 1 + 5 \qquad \text{Add 5 to both sides.}$$
$$9a = 6$$
$$\frac{1}{9}(9a) = \frac{1}{9}(6) \qquad \text{Multiply both sides by } \frac{1}{9}.$$
$$a = \frac{2}{3} \qquad \frac{1}{9}(6) = \frac{6}{9} = \frac{2}{3}.$$

the solution set is $\left\{ \dfrac{2}{3} \right\}$.

NOTE We know that multiplication by a number and division by its reciprocal always produce the same result. Because of this fact, instead of multiplying each side of our equation by $\frac{1}{9}$, we could just as easily divide each side by 9. If we did so, the last two lines in our solution would look like this:

$$\frac{9a}{9} = \frac{6}{9}$$
$$a = \frac{2}{3}$$

The next example involves fractions. The least common denominator, which is the smallest expression that is divisible by each of the denominators, can be used with the multiplication property of equality to simplify equations containing fractions.

EXAMPLE 2

Solve $\frac{2}{3}x + \frac{1}{2} = -\frac{3}{8}$.

SOLUTION   We can solve this equation by applying our properties and working with fractions, or we can begin by eliminating the fractions. Let's work the problem using both methods.

Method 1: *Working with the fractions*

$$\frac{2}{3}x + \frac{1}{2} + \left(-\frac{1}{2}\right) = -\frac{3}{8} + \left(-\frac{1}{2}\right) \qquad \text{Add } -\tfrac{1}{2} \text{ to each side.}$$

$$\frac{2}{3}x = -\frac{7}{8} \qquad\qquad -\tfrac{3}{8} + \left(-\tfrac{1}{2}\right) = -\tfrac{3}{8} + \left(-\tfrac{4}{8}\right).$$

$$\frac{3}{2}\left(\frac{2}{3}x\right) = \frac{3}{2}\left(-\frac{7}{8}\right) \qquad \text{Multiply each side by } \tfrac{3}{2}.$$

$$x = -\frac{21}{16}$$

Method 2: *Eliminating the fractions in the beginning*

Our original equation has denominators of 3, 2, and 8. The least common denominator, abbreviated LCD, for these three denominators is 24, and it has the property that all three denominators will divide it evenly. Therefore, if we multiply both sides of our equation by 24, each denominator will divide into 24, and we will be left with an equation that does not contain any denominators other than 1.

$$24\left(\frac{2}{3}x + \frac{1}{2}\right) = 24\left(-\frac{3}{8}\right) \qquad \text{Multiply each side by the LCD 24.}$$

$$24\left(\frac{2}{3}x\right) + 24\left(\frac{1}{2}\right) = 24\left(-\frac{3}{8}\right) \qquad \text{Distributive property on the left side}$$

$$16x + 12 = -9 \qquad \text{Multiply.}$$

$$16x = -21 \qquad \text{Add } -12 \text{ to each side.}$$

$$x = -\frac{21}{16} \qquad \text{Multiply each side by } \tfrac{1}{16}.$$

As the third line above indicates, multiplying each side of the equation by the LCD eliminates all the fractions from the equation. Both methods yield the same solution.

EXAMPLE 3

Solve the equation $0.06x + 0.05(10{,}000 - x) = 560$.

SOLUTION   We can solve the equation in its original form by working with the decimals, or we can eliminate the decimals first by using the multiplication property of equality and solve the resulting equation. Here are both methods.

Method 1: *Working with the decimals*

$$0.06x + 0.05(10{,}000 - x) = 560 \qquad\qquad \text{Original equation}$$

$$0.06x + 0.05(10{,}000) - 0.05x = 560 \qquad\qquad \text{Distributive property}$$

$$0.01x + 500 = 560 \qquad\qquad \text{Simplify the left side.}$$

$$0.01x + 500 + (-500) = 560 + (-500) \qquad\qquad \text{Add } -500 \text{ to each side.}$$

$$0.01x = 60$$

$$\frac{0.01x}{0.01} = \frac{60}{0.01} \qquad \text{Divide each side by 0.01.}$$

$$x = 6{,}000$$

Method 2: *Eliminating the decimals in the beginning.*

To move the decimal point two places to the right in $0.06x$ and 0.05, we multiply each side of the equation by 100.

$0.06x + 0.05(10{,}000 - x) = 560$	Original equation
$0.06x + 500 - 0.05x = 560$	Distributive property
$100(0.06x) + 100(500) - 100(0.05x) = 100(560)$	Multiply each side by 100.
$6x + 50{,}000 - 5x = 56{,}000$	Multiply.
$x + 50{,}000 = 56{,}000$	Simplify the left side.
$x = 6{,}000$	Add $-50{,}000$ to each side.

Using either method, the solution to our equation is 6,000. We check our work (to be sure we have not made a mistake in applying the properties or an arithmetic mistake) by substituting 6,000 into our original equation and simplifying each side of the result separately.

NOTE We are placing question marks over the equal signs because we don't know yet if the expressions on the left will be equal to the expressions on the right.

Check: Substituting 6,000 for x in the original equation, we have

$$0.06(6{,}000) + 0.05(10{,}000 - 6{,}000) \stackrel{?}{=} 560$$

$$0.06(6{,}000) + 0.05(4{,}000) \stackrel{?}{=} 560$$

$$360 + 200 \stackrel{?}{=} 560$$

$$560 = 560 \qquad \text{A true statement}$$

Here is a list of steps to use as a guideline for solving linear equations in one variable.

HOW TO *Solve Linear Equations in One Variable*

Step 1a: Use the distributive property to separate terms, if necessary.

1b: If fractions are present, consider multiplying both sides by the LCD to eliminate the fractions. If decimals are present, consider multiplying both sides by a power of 10 to clear the equation of decimals.

1c: Combine similar terms on each side of the equation.

Step 2: Use the addition property of equality to get all variable terms on one side of the equation and all constant terms on the other side. A variable term is a term that contains the variable. A constant term is a term that does not contain the variable (the number 3, for example).

Step 3: Use the multiplication property of equality to get the variable by itself on one side of the equation.

Step 4: Check your solution in the original equation to be sure that you have not made a mistake in the solution process.

As you work through the problems in the problem set, you will see that it is not always necessary to use all four steps when solving equations. The number of steps used depends on the equation. In Example 4, there are no fractions or decimals in the original equation, so step 1b will not be used.

EXAMPLE 4

Solve the equation $8 - 3(4x - 2) + 5x = 35$.

SOLUTION   We must begin by distributing the -3 across the quantity $4x - 2$. (It would be a mistake to subtract 3 from 8 first, because the rule for order of operations indicates we are to do multiplication before subtraction.) After we have simplified the left side of our equation, we apply the addition property and the multiplication property. In this example, we will show only the result:

	$8 - 3(4x - 2) + 5x = 35$	Original equation
Step 1a:	$8 - 12x + 6 + 5x = 35$	Distributive property
Step 1c:	$-7x + 14 = 35$	Simplify.
Step 2:	$-7x = 21$	Add -14 to each side.
Step 3:	$x = -3$	Multiply by $-\frac{1}{7}$.

Step 4: When x is replaced by -3 in the original equation, a true statement results. Therefore, -3 is the solution to our equation.

B Solving Equations by Factoring

Next we will use our knowledge of factoring to solve equations. Most of the equations we will see are *quadratic equations*.

> ### DEFINITION   *quadratic equations*
> ·
> Any equation that can be written in the form
>
> $$ax^2 + bx + c = 0$$
>
> where a, b, and c are constants and a is not 0 ($a \neq 0$), is called a ***quadratic equation***. The form $ax^2 + bx + c = 0$ is called ***standard form*** for quadratic equations.

NOTE For a quadratic equation written in standard form, the first term ax^2 is called the *quadratic term*, the second term bx is the *linear term,* and the last term c is called the *constant term.*

Each of the following is a quadratic equation:

$$2x^2 = 5x + 3 \qquad 5x^2 = 75 \qquad 4x^2 - 3x + 2 = 0$$

The number 0 is a special number, and is the key to solving quadratic equations. If we multiply two expressions and get 0, then one, or both, of the expressions must have been 0. In other words, the only way to multiply and get 0 for an answer is to multiply by 0. This fact allows us to solve certain quadratic equations. We state this fact as follows:

> **PROPERTY** *Zero-Factor Property*
> ·
> For all real numbers r and s,
>
> $$r \cdot s = 0 \qquad \text{if and only if} \qquad r = 0 \qquad \text{or} \qquad s = 0 \qquad \text{(or both)}$$

EXAMPLE 5

Solve $x^2 - 2x - 24 = 0$.

SOLUTION   We begin by factoring the left side as $(x - 6)(x + 4)$ and get

$$(x - 6)(x + 4) = 0$$

Now both $(x - 6)$ and $(x + 4)$ represent real numbers. We notice that their product is 0. By the zero-factor property, one or both of them must be 0:

$$x - 6 = 0 \qquad \text{or} \qquad x + 4 = 0$$

We have used factoring and the zero-factor property to rewrite our original second-degree equation as two first-degree equations connected by the word *or*. Completing the solution, we solve the two first-degree equations:

$$x - 6 = 0 \qquad \text{or} \qquad x + 4 = 0$$
$$x = 6 \qquad\qquad\qquad x = -4$$

We check our solutions in the original equation as follows:

Check $x = 6$	Check $x = -4$
$6^2 - 2(6) - 24 \stackrel{?}{=} 0$	$(-4)^2 - 2(-4) - 24 \stackrel{?}{=} 0$
$36 - 12 - 24 \stackrel{?}{=} 0$	$16 + 8 - 24 \stackrel{?}{=} 0$
$0 = 0$	$0 = 0$

In both cases the result is a true statement, which means that both 6 and -4 are solutions to the original equation.

To generalize, here are the steps used in solving a quadratic equation by factoring.

> **HOW TO**   *Solve an Equation by Factoring*
> ·
> **Step 1:** Write the equation in standard form.
> **Step 2:** Factor the left side.
> **Step 3:** Use the zero-factor property to set each factor equal to 0.
> **Step 4:** Solve the resulting linear equations.
> **Step 5:** Check the solutions in the original equation.

EXAMPLE 6

Solve $100x^2 = 300x$.

SOLUTION   We begin by writing the equation in standard form and factoring:

$$100x^2 = 300x$$
$$100x^2 - 300x = 0 \qquad \text{Standard form}$$
$$100x(x - 3) = 0 \qquad \text{Factor.}$$

Using the zero-factor property to set each factor to 0, we have:

$$100x = 0 \quad \text{or} \quad x - 3 = 0$$
$$x = 0 \qquad\qquad x = 3$$

The two solutions are 0 and 3.

EXAMPLE 7

Solve $(x - 2)(x + 1) = 4$.

SOLUTION   We begin by multiplying the two factors on the left side. (Notice that it would be incorrect to set each of the factors on the left side equal to 4. The fact that the product is 4 does not imply that either of the factors must be 4.)

$$(x - 2)(x + 1) = 4$$

$$x^2 - x - 2 = 4 \qquad\qquad \text{Multiply the left side.}$$

$$x^2 - x - 6 = 0 \qquad\qquad \text{Standard form}$$

$$(x - 3)(x + 2) = 0 \qquad\qquad \text{Factor.}$$

$$x - 3 = 0 \quad \text{or} \quad x + 2 = 0 \qquad \text{Zero-factor property}$$

$$x = 3 \qquad\qquad x = -2$$

EXAMPLE 8

Solve $2x^3 = 5x^2 + 3x$.

SOLUTION   First we add $-5x^2$ and $-3x$ to each side so the right side will become 0.

$$2x^3 - 5x^2 - 3x = 0 \qquad\qquad \text{Standard form}$$

We factor the left side and then use the zero-factor property to set each factor to 0.

$$x(2x^2 - 5x - 3) = 0 \qquad \text{Factor out the greatest} \\ \text{common factor.}$$

$$x(2x + 1)(x - 3) = 0 \qquad \text{Continue factoring.}$$

$$x = 0 \quad \text{or} \quad 2x + 1 = 0 \quad \text{or} \quad x - 3 = 0 \qquad \text{Zero-factor property}$$

Solving each of the resulting equations, we have

$$x = 0 \quad \text{or} \quad x = -\frac{1}{2} \quad \text{or} \quad x = 3$$

EXAMPLE 9

Solve for x: $x^3 + 2x^2 - 9x - 18 = 0$

SOLUTION   We start with factoring by grouping.

$$x^3 + 2x^2 - 9x - 18 = 0$$

$$x^2(x + 2) - 9(x + 2) = 0$$

$$(x + 2)(x^2 - 9) = 0$$

$$(x + 2)(x - 3)(x + 3) = 0 \qquad\qquad \text{The difference of two squares}$$

$$x + 2 = 0 \quad \text{or} \quad x - 3 = 0 \quad \text{or} \quad x + 3 = 0$$

$$x = -2 \qquad\qquad x = 3 \qquad\qquad x = -3$$

We have three solutions: -2, 3, and -3.

Identities and Equations with No Solution

There are two special cases associated with solving linear equations in one variable, which are illustrated in the following examples.

EXAMPLE 10

Solve for x: $2(3x - 4) = 3 + 6x$

SOLUTION   Applying the distributive property to the left side gives us

$$6x - 8 = 3 + 6x \qquad \text{Distributive property}$$

Now, if we add $-6x$ to each side, we are left with

$$-8 = 3$$

which is a false statement. This means that there is no solution to our equation. Any number we substitute for x in the original equation will lead to a similar false statement.

EXAMPLE 11

Solve for x: $-15 + 3x = 3(x - 5)$

SOLUTION   We start by applying the distributive property to the right side.

$$-15 + 3x = 3x - 15 \qquad \text{Distributive property}$$

If we add $-3x$ to each side, we are left with the true statement

$$-15 = -15$$

In this case, our result tells us that any number we use in place of x in the original equation will lead to a true statement. Therefore, all real numbers are solutions to our equation. We say the original equation is an *identity* because the left side is always identically equal to the right side.

GETTING READY FOR CLASS

After reading through the preceding section, respond in your own words and in complete sentences.

A. Name the constant terms in the equation $5x + 3 = 2$.

B. In your own words, explain how you would use the multiplication property of equality.

C. What is the first step in solving the equation $100x^2 = 300x$?

D. How do you use the zero-factor property to help solve a quadratic equation by factoring?

Problem Set 2.1

Each odd/even pair of problems below is matched to an example in the text. If you have any trouble with any of these problems, go to the example that is matched with that problem.

Solve each of the following equations.

1. $7y - 4 - 2y + 11$

2. $5 - 2x = 3x + 1$

3. $-\dfrac{2}{5}x + \dfrac{2}{15} = \dfrac{2}{3}$

4. $\dfrac{1}{2}x + \dfrac{1}{4} = \dfrac{1}{3}x + \dfrac{5}{4}$

5. $0.14x + 0.08(10{,}000 - x) = 1220$

6. $-0.3y + 0.1 = 0.5$

7. $5(y + 2) - 4(y + 1) = 3$

8. $6(y - 3) - 5(y + 2) = 8$

9. $x^2 - 5x - 6 = 0$

10. $x^2 - x - 12 = 0$

11. $9a^3 = 16a$

12. $-100x = 10x^2$

13. $(x + 6)(x - 2) = -7$

14. $(x - 7)(x + 5) = -20$

15. $2y^3 - 9y = -3y^2$

16. $3y^2 + 10y = 17y^2$

17. $4x^3 + 12x^2 - 9x - 27 = 0$

18. $2x^3 + x^2 - 18x - 9 = 0$

PAYING ATTENTION TO INSTRUCTIONS

The next four problems are intended to give you practice reading, and paying attention to, the instructions that accompany the problems you are working. Working these problems is an excellent way to get ready for a test or a quiz.

19. Work each problem according to the instructions given.

  a. Solve: $8x - 5 = 0$

  b. Add: $(8x - 5) + (2x - 3)$

  c. Multiply: $(8x - 5)(2x - 3)$

  d. Solve: $16x^2 - 34x + 15 = 0$

20. Work each problem according to the instructions given.

  a. Subtract: $(3x + 5) - (7x - 4)$

  b. Solve: $3x + 5 = 7x - 4$

  c. Multiply: $(3x + 5)(7x - 4)$

  d. Solve: $21x^2 + 23x - 20 = 0$

21. Solve each equation.

  a. $9x - 25 = 0$

  b. $9x^2 - 25 = 0$

  c. $9x^2 - 25 = 56$

  d. $9x^2 - 25 = 30x - 50$

22. Solve each equation.

  a. $5x - 6 = 0$

  b. $(5x - 6)^2 = 0$

  c. $25x^2 - 36 = 0$

  d. $25x^2 - 36 = 28$

Now that you have practiced solving a variety of equations, we can turn our attention to the type of equation you will see as you progress through the book. Each equation appears later in the book exactly as you see it below.

Solve each equation.

23. $-3 - 4x = 15$

24. $-\dfrac{3}{5}a + 2 = 8$

25. $x^3 - 5x^2 + 6x = 0$

26. $x^3 + 3x^2 - 4x - 12 = 0$

27. $0 = 6400a + 70$

28. $.07x = 1.4$

29. $5(2x + 1) = 12$

30. $50 = \dfrac{K}{48}$

31. $100P = 2{,}400$

32. $2x - 3(3x - 5) = -6$

33. $5\left(-\dfrac{19}{15}\right) + 5y = 9$

34. $2\left(-\dfrac{29}{22}\right) - 3y = 4$

35. $3x^2 + x = 10$

36. $12(x + 3) + 12(x - 3) = 3(x^2 - 9)$

37. $(y + 3)^2 + y^2 = 9$

38. $3x + (x - 2) \cdot 2 = 6$

39. $15 - 3(x - 1) = x - 2$

40. $2(2x - 3) + 2x = 45$

41. $2(20 + x) = 3(20 - x)$

42. $2x + 1.5(75 - x) = 127.5$

43. $0.08x + 0.09(9{,}000 - x) = 750$

44. $0.12x + 0.10(15{,}000 - x) = 1{,}600$

45. $(x + 3)^2 + 1^2 = 2$

46. $(x + 2)(x) = 2^3$

Solve each equation, if possible.

47. $3x - 6 = 3(x + 4)$

48. $4y + 2 - 3y + 5 = 3 + y + 4$

49. $2(4t - 1) + 3 = 5t + 4 + 3t$

50. $7x - 3(x - 2) = -4(5 - x)$

51. $7(x + 2) - 4(2x - 1) = 18 - x$

52. $2x^2 + x - 1 = (2x + 3)(x - 1)$

53. Temperature and Altitude As an airplane gains altitude, the temperature outside the plane decreases. The relationship between temperature T and altitude A can be described with the formula

$$T = -0.0035A + 70$$

when the temperature on the ground is 70°F. Solve the equation below to find the altitude at which the temperature outside the plane is -35°F.

$$-35 = -0.0035A + 70$$

54. Revenue A company manufactures and sells headphones. The revenue obtained by selling x sets of headphones is given by the formula

$$R = 11.5x - 0.05x^2$$

Solve the equation below to find the number of headphones they must sell to receive \$650 in revenue.

$$650 = 11.5x - 0.05x^2$$

55. Fuel Economy The combined city and highway fuel economy, in miles per gallon (mpg), for certain automobiles is related to the automobile's curb weight by the formula

$$F = -0.004w + 40$$

where F is the fuel economy and w is the curb weight in pounds. Solve to find the weight of a car that has a combined fuel economy of 25 mpg.

56. Fuel Economy The combined city and highway fuel economy, in miles per gallon (mpg), for certain automobiles is related to the automobile's curb weight by the formula

$$F = -0.004w + 40$$

where F is the fuel economy and w is the curb weight in pounds. Solve to find the weight of a car that has a combined fuel economy of 30 mpg.

GETTING READY FOR THE NEXT SECTION

Problems under this heading, "Getting Ready for the Next Section," are problems that you must be able to work in order to understand the material in the next section. In this case, the problems below are variations on the type of problems you have already worked in this problem set. They are exactly the type of problems you will see in the explanations and examples in the next section.

Solve each equation.

57. $x \cdot 42 = 21$

58. $x \cdot 84 = 21$

59. $25 = 0.4x$

60. $35 = 0.4x$

61. $12 - 4y = 12$

62. $-6 - 3y = 6$

63. $525 = 900 - 300p$

64. $375 = 900 - 300p$

65. $48 = 64t - 16t^2$

66. $4{,}000 = (1{,}300 - 100p)p$

67. $486.7 = 78.5 + 31.4h$

68. $486.23 = 113.0 + 37.7h$

EXTENDING THE CONCEPTS

Solve each of the following problems.

69. $\dfrac{x+4}{5} - \dfrac{x+3}{3} = -\dfrac{7}{15}$

70. $\dfrac{x+1}{7} - \dfrac{x-2}{2} = \dfrac{1}{14}$

71. $\dfrac{1}{x} - \dfrac{2}{3} = \dfrac{2}{x}$

72. $\dfrac{1}{x} - \dfrac{3}{5} = \dfrac{2}{x}$

73. $\dfrac{x+3}{2} - \dfrac{x-4}{4} = -\dfrac{1}{8}$

74. $\dfrac{x-3}{5} - \dfrac{x+1}{10} = -\dfrac{1}{10}$

75. $\dfrac{x-1}{2} - \dfrac{x+2}{3} = \dfrac{x+3}{6}$

76. $\dfrac{x+2}{4} - \dfrac{x-1}{3} = \dfrac{x+2}{6}$

77. $x^4 - 13x^2 + 36 = 0$

78. $x^4 - 26x^2 + 25 = 0$

FORMULAS

OBJECTIVES

A Solve a formula for one of its variables.

B Solve revenue problems using basic formulas.

C Solve problems involving rate equations.

©iStockphoto.com/shansekala

The study of design, architecture, or construction is not complete without an introduction to the ideas presented in the book *A Pattern Language*, published by the Center for Environmental Structure. The book presents a language for design and construction of houses, neighborhoods, and cities, based on what is best for the people who will live there. Here is a passage from that book concerning the size of public places.

> For public squares, courts, pedestrian streets, any place where crowds are drawn together, estimate the mean number of people in the place at any given moment (P), and make the area of the place between $150P$ and $300P$ square feet.

This passage gives us a formula for the area A of a public square that averages P people at any given time. In the language of algebra, the minimum size of such a public square should be

$$A = 150P$$

In this section we review formulas, such as the one shown here. We want to evaluate formulas given specific values for some of its variables, and second, we must learn how to change the form of a formula without changing its meaning.

A Solving Formulas

A formula in mathematics is an equation that contains more than one variable. Some formulas are probably already familiar to you. For example, the formula for the area A of a rectangle with length l and width w is $A = lw$.

To begin our work with formulas, we will consider some examples in which we are given numerical replacements for all but one of the variables.

EXAMPLE 1

Find y when x is 4 in the formula $3x - 4y = 2$.

SOLUTION   We substitute 4 for x in the formula and then solve for y:

When	$x = 4$	
the formula	$3x - 4y - 2$	
becomes	$3(4) - 4y = 2$	
	$12 - 4y = 2$	Multiply 3 and 4.
	$-4y = -10$	Add -12 to each side.
	$y = \dfrac{5}{2}$	Divide each side by -4.

Note that, in the last line of Example 1, we divided each side of the equation by -4. Remember that this is equivalent to multiplying each side of the equation by $-\frac{1}{4}$. For the rest of the examples in this section, it will be more convenient to think in terms of division rather than multiplication.

EXAMPLE 2

A store selling art supplies finds that they can sell x sketch pads each week at a price of p dollars each, according to the formula $x = 900 - 300p$. What price should they charge for each sketch pad if they want to sell 525 pads each week?

SOLUTION   Here we are given a formula, $x = 900 - 300p$, and asked to find the value of p if x is 525. To do so, we simply substitute 525 for x and solve for p:

When	$x = 525$	
the formula	$x = 900 - 300p$	
becomes	$525 = 900 - 300p$	
	$-375 = -300p$	Add -900 to each side.
	$1.25 = p$	Divide each side by -300.

To sell 525 sketch pads, the store should charge \$1.25 for each pad.

EXAMPLE 3

A boat is traveling upstream against a current. If the speed of the boat in still water is r and the speed of the current is c, then the formula for the distance traveled by the boat is $d = (r - c) \cdot t$, where t is the length of time. Find c if $d = 52$ miles, $r = 16$ miles per hour, and $t = 4$ hours.

SOLUTION   Substituting 52 for d, 16 for r, and 4 for t into the formula, we have

$52 = (16 - c) \cdot 4$	
$13 = 16 - c$	Divide each side by 4.
$-3 = -c$	Add -16 to each side.
$3 = c$	Divide each side by -1.

The speed of the current is 3 miles per hour.

EXAMPLE 4

If an object is projected into the air with an initial vertical velocity v (in feet/second), its height h (in feet) after t seconds will be given by

$$h = vt - 16t^2$$

Find t if $v = 64$ feet/second and $h = 48$ feet.

SOLUTION   Substituting $v = 64$ and $h = 48$ into the formula, we have

$$48 = 64t - 16t^2$$

which is a quadratic equation. We write it in standard form and solve by factoring:

$$16t^2 - 64t + 48 = 0$$
$$t^2 - 4t + 3 = 0 \qquad \text{Divide each side by 16.}$$
$$(t - 1)(t - 3) = 0 \qquad \text{Factor.}$$
$$t - 1 = 0 \quad \text{or} \quad t - 3 = 0$$
$$t = 1 \qquad\qquad t = 3$$

Here is how we interpret our results: If an object is projected upward with an initial vertical velocity of 64 feet/second, it will be 48 feet above the ground after 1 second and after 3 seconds. That is, it passes 48 feet going up and also coming down.

B Revenue

The most basic business formula is the formula for profit:

$$\text{Profit} = \text{Revenue} - \text{Cost}$$

where revenue is the total amount of money a company brings in by selling their product and cost is the total cost to produce the product they sell. As we mentioned in Chapter 1, revenue itself can be broken down further by another formula common in the business world. The revenue obtained from selling x items is the product of the number of items sold and the price per item. That is,

$$\text{Revenue} = (\text{Number of items sold})(\text{Price of each item})$$

For example, if 100 items are sold for \$9 each, the revenue is $100(9) = \$900$. Likewise, if 500 items are sold for \$11 each, then the revenue is $500(11) = \$5,500$. In general, if x is the number of items sold and p is the selling price of each item, then we can write

$$R = xp$$

EXAMPLE 5

A manufacturer of small calculators knows that the number of calculators she can sell each week is related to the price of the calculators by the equation $x = 1{,}300 - 100p$, where x is the number of calculators and p is the price per calculator. What price should she charge for each calculator if she wants the weekly revenue to be \$4,000?

SOLUTION   The formula for total revenue is $R = xp$. Because we want R in terms of p, we substitute $1,300 - 100p$ for x in the equation $R = xp$:

$$\text{If} \qquad R = xp$$

$$\text{and} \qquad x = 1,300 - 100p$$

$$\text{then} \qquad R = (1,300 - 100p)p$$

We want to find p when R is 4,000. Substituting for R in the formula gives us

$$4,000 = (1,300 - 100p)p$$

$$4,000 = 1,300p - 100p^2$$

This is a quadratic equation. To write it in standard form, we add $100p^2$ and $-1,300p$ to each side, giving us

$$100p^2 - 1,300p + 4,000 = 0$$

$$p^2 - 13p + 40 = 0 \qquad \text{Divide each side by 100.}$$

$$(p - 5)(p - 8) = 0$$

$$p - 5 = 0 \qquad \text{or} \qquad p - 8 = 0$$

$$p = 5 \qquad\qquad\qquad p = 8$$

If she sells the calculators for \$5 each or for \$8 each, she will have a weekly revenue of \$4,000.

FACTS FROM GEOMETRY   *Formulas for Area and Perimeter*

To review, here are the formulas for the area and perimeter of some common geometric objects.

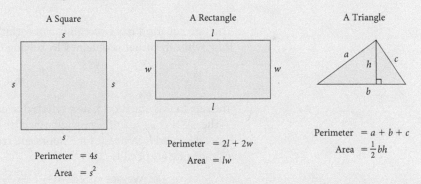

A Square

Perimeter $= 4s$

Area $= s^2$

A Rectangle

Perimeter $= 2l + 2w$

Area $= lw$

A Triangle

Perimeter $= a + b + c$

Area $= \frac{1}{2}bh$

The formula for perimeter gives us the distance around the outside of the object along its sides, while the formula for area gives us a measure of the amount of surface the object covers.

EXAMPLE 6

Given the formula $P = 2w + 2l$, solve for w.

SOLUTION   To solve for w, we must isolate it on one side of the equation. We can accomplish this if we delete the $2l$ term and the coefficient 2 from the right side of the equation.

To begin, we add $-2l$ to both sides:

$$P + (-2l) = 2w + 2l + (-2l)$$

$$P - 2l = 2w$$

To delete the 2 from the right side, we can multiply both sides by $\frac{1}{2}$:

$$\frac{1}{2}(P - 2l) = \frac{1}{2}(2w)$$

$$\frac{P - 2l}{2} = w$$

The two formulas

$$P = 2w + 2l \qquad \text{and} \qquad w = \frac{P - 2l}{2}$$

give the relationship between P, l, and w. They look different, but they both say the same thing about P, l, and w. The first formula gives P in terms of l and w, and the second formula gives w in terms of P and l.

C Rate Equation and Average Speed

Now we will look at some problems that use what is called the *rate equation*. You use this equation on an intuitive level when you are estimating how long it will take you to drive long distances. For example, if you drive at 50 miles per hour for 2 hours, you will travel 100 miles. Here is the rate equation:

$$\text{Distance} = \text{rate} \cdot \text{time, or } d = r \cdot t$$

The rate equation has two equivalent forms, one of which is obtained by solving for r, while the other is obtained by solving for t. Here they are:

$$r = \frac{d}{t} \quad \text{and} \quad t = \frac{d}{r}$$

The rate in this equation is also referred to as average speed.

The *average speed* of a moving object is defined to be the ratio of distance to time. If you drive your car for 5 hours and travel a distance of 200 miles, then your average rate of speed is

$$\text{Average speed} = \frac{200 \text{ miles}}{5 \text{ hours}} = 40 \text{ miles per hour}$$

Our next example involves both the formula for the circumference of a circle and the rate equation.

EXAMPLE 7

The first Ferris wheel was designed and built by George Ferris in 1893. The diameter of the wheel was 250 feet. It had 36 carriages, equally spaced around the wheel, each of which held a maximum of 40 people. One trip around the wheel took 20 minutes. Find the average speed of a rider on the first Ferris wheel. (Use 3.14 as an approximation for π.)

SOLUTION   The distance traveled is the circumference of the wheel, which is

$$C = \pi d = 250\pi = 250(3.14) = 785 \text{ feet}$$

To find the average speed, we divide the distance traveled by the amount of time it took to go once around the wheel.

$$r = \frac{d}{t} = \frac{785 \text{ feet}}{20 \text{ minutes}} = 39.3 \text{ feet per minute (to the nearest tenth)}$$

EXAMPLE 8

Solve for x: $ax - 3 = bx + 5$.

SOLUTION   In this example, we must begin by collecting all the variable terms on the left side of the equation and all the constant terms on the other side (just like we did when we were solving linear equations in Section 2.1):

$$ax - 3 = bx + 5$$

$$ax - bx - 3 = 5 \qquad \text{Add } -bx \text{ to each side.}$$

$$ax - bx = 8 \qquad \text{Add 3 to each side.}$$

At this point, we need to apply the distributive property to write the left side as $(a - b)x$. After that, we divide each side by $a - b$:

$$(a - b)x = 8 \qquad \text{Distributive property}$$

$$x = \frac{8}{a - b} \qquad \text{Divide each side by } a - b.$$

EXAMPLE 9

Solve for y: $\dfrac{y - b}{x - 0} = m$.

SOLUTION   Although we will do more extensive work with formulas of this form later in the book, we need to know how to solve this particular formula for y in order to understand some things in the next chapter. We begin by simplifying the denominator on the left side and then multiplying each side of the formula by x. Doing so makes the rest of the solution process simple.

$$\frac{y - b}{x - 0} = m \qquad \text{Original formula}$$

$$\frac{y - b}{x} = m \qquad x - 0 = x.$$

$$x \cdot \frac{y - b}{x} = m \cdot x \qquad \text{Multiply each side by } x.$$

$$y - b = mx \qquad \text{Simplify each side.}$$

$$y = mx + b \qquad \text{Add } b \text{ to each side.}$$

This is our solution. If we look back to the first step, we can justify our result on the left side of the equation this way: Dividing by x is equivalent to multiplying by its reciprocal $\frac{1}{x}$. Here is what it looks like when written out completely:

$$x \cdot \frac{y - b}{x} = x \cdot \frac{1}{x} \cdot (y - b) = 1(y - b) = y - b$$

EXAMPLE 10

Solve for y: $\dfrac{y-4}{x-5} = 3$.

SOLUTION We proceed as we did in the previous example, but this time we clear the formula of fractions by multiplying each side of the formula by $x - 5$.

$$\frac{y-4}{x-5} = 3 \qquad \text{Original formula}$$

$$(x-5) \cdot \frac{y-4}{x-5} = 3 \cdot (x-5) \qquad \text{Multiply each side by } (x-5).$$

$$y - 4 = 3x - 15 \qquad \text{Simplify each side.}$$

$$y = 3x - 11 \qquad \text{Add 4 to each side.}$$

We have solved for y. We can justify our result on the left side of the equation this way: Dividing by $x - 5$ is equivalent to multiplying by its reciprocal $\frac{1}{x-5}$. Here are the details:

$$(x-5) \cdot \frac{y-4}{x-5} = (x-5) \cdot \frac{1}{x-5} \cdot (y-4) = 1(y-4) = y-4$$

EXAMPLE 11

Solve the formula $S = 2\pi rh + 2\pi r^2$ for h.

SOLUTION This is the formula for the surface area of a right circular cylinder, with radius r and height h, that is closed at both ends. To isolate h, we first add $-2\pi r^2$ to both sides:

$$S + (-2\pi r^2) = 2\pi rh + 2\pi r^2 + (-2\pi r^2)$$

$$S - 2\pi r^2 = 2\pi rh$$

Next, we divide each side by $2\pi r$.

$$\frac{S - 2\pi r^2}{2\pi r} = h$$

Exchanging the two sides of our equation, we have

$$h = \frac{S - 2\pi r^2}{2\pi r}$$

GETTING READY FOR CLASS

After reading through the preceding section, respond in your own words and in complete sentences.

A. What is a formula in mathematics?

B. Give two equivalent forms of the rate equation $d = rt$.

C. What are the formulas for the area and perimeter of a triangle?

D. Explain in words the formula for the perimeter of a rectangle.

Use the formula $3x - 4y = 12$ to find y if

1. x is 0   **2.** x is -2 **3.** x is 4   **4.** x is -4

Use the formula $y = 2x - 3$ to find x when

5. y is 0   **6.** y is -3 **7.** y is 5   **8.** y is -5

Problems 9 through 24 are problems that you will see later in the text.

9. If $x - 2y = 4$ and $y = -\dfrac{6}{5}$, find x

10. If $x - 2y = 4$ and $x = \dfrac{8}{5}$, find y

11. Let $x = 160$ and $y = 0$ in $y = a(x - 80)^2 + 70$ and solve for a.

12. Let $x = 0$ and $y = 0$ in $y = a(x - 80)^2 + 70$ and solve for a.

13. Find R if $p = 1.5$ and $R = (900 - 300p)p$

14. Find R if $p = 2.5$ and $R = (900 - 300p)p$

15. Find P if $P = -0.1x^2 + 27x + 1{,}700$ and
     a. $x = 100$ **b.** $x = 170$

16. Find P if $P = -0.1x^2 + 27x + 1{,}820$ and
     a. $x = 130$ **b.** $x = 140$

17. Find h if $h = 16 + 32t - 16t^2$ and
     a. $t = \dfrac{1}{4}$ **b.** $t = \dfrac{7}{4}$

18. Find h if $h = 64t - 16t^2$ and
     a. $t = 1$ **b.** $t = 3$

19. Use the formula $d = (r - c)t$ to find c if $d = 30$, $r = 12$, and $t = 3$.

20. Use the formula $d = (r - c)t$ to find r if $d = 49$, $c = 4$, and $t = 3.5$.

21. If $y = Kx$, find K if $x = 5$ and $y = 15$

22. If $d = Kt^2$, find K if $t = 2$ and $d = 64$

23. If $V = \dfrac{K}{P}$, find K if $P = 48$ and $V = 50$

24. If $y = Kxz^2$, find K if $x = 5$, $z = 3$, and $y = 180$

Use the formula $5x - 3y = -15$ to find y if

25. $x = 2$ **26.** $x = -3$ **27.** $x = -\dfrac{1}{5}$ **28.** $x = 3$

Solve each of the following formulas for the indicated variable.

29. $d = rt$ for r **30.** $d = rt$ for t

31. $d = (r + c)t$ for t **32.** $d = (r + c)t$ for r

33. $A = lw$ for l **34.** $A = \dfrac{1}{2}bh$ for b

35. $I = prt$ for t **36.** $I = prt$ for r

37. $PV = nRT$ for T **38.** $PV = nRT$ for R

39. $y = mx + b$ for x **40.** $A = P + Prt$ for t

41. $C = \dfrac{5}{9}(F - 32)$ for F **42.** $F = \dfrac{9}{5}C + 32$ for C

43. $h = vt + 16t^2$ for v **44.** $h = vt - 16t^2$ for v

45. $A = a + (n - 1)d$ for d **46.** $A = a + (n - 1)d$ for n

47. $2x + 3y = 6$ for y **48.** $2x - 3y = 6$ for y

49. $-3x + 5y = 15$ for y **50.** $-2x - 7y = 14$ for y

51. $2x - 6y + 12 = 0$ for y **52.** $7x - 2y - 6 = 0$ for y

53. $ax + 4 = bx + 9$ for x **54.** $ax - 5 = cx - 2$ for x

55. $S = \pi r^2 + 2\pi rh$ for h **56.** $A = P + Prt$ for P

57. $-3x + 4y = 12$ for x **58.** $-3x + 4y = 12$ for y

59. $ax + 3 = cx - 7$ for x **60.** $by - 9 = dy + 3$ for y

Problems 61 through 68 are problems that you will see later in the text. Solve each formula for y.

61. $x = 2y - 3$ **62.** $x = 4y + 1$

63. $y - 3 = -2(x + 4)$ **64.** $y - 1 = \dfrac{1}{4}(x - 3)$

65. $y - 3 = -\dfrac{2}{3}(x + 3)$ **66.** $y + 1 = -\dfrac{2}{3}(x - 3)$

67. $y - 4 = -\dfrac{1}{2}(x + 1)$ **68.** $y - 2 = \dfrac{1}{3}(x - 1)$

69. Solve for y.

     a. $\dfrac{y + 1}{x - 0} = 4$ **b.** $\dfrac{y + 2}{x - 4} = -\dfrac{1}{2}$ **c.** $\dfrac{y + 3}{x - 7} = 0$

70. Solve for y.

     a. $\dfrac{y - 1}{x - 0} = -3$ **b.** $\dfrac{y - 2}{x - 6} = \dfrac{2}{3}$ **c.** $\dfrac{y - 3}{x - 1} = 0$

Solve for y.

71. $\dfrac{x}{8} + \dfrac{y}{2} = 1$ **72.** $\dfrac{x}{7} + \dfrac{y}{9} = 1$ **73.** $\dfrac{x}{5} + \dfrac{y}{-3} = 1$ **74.** $\dfrac{x}{16} + \dfrac{y}{-2} = 1$

PAYING ATTENTION TO INSTRUCTIONS

The next two problems are intended to give you practice reading, and paying attention to, the instructions that accompany the problems you are working. As we have mentioned previously, working these problems is an excellent way to get ready for a test or a quiz.

75. Work each problem according to the instructions given.

     a. Solve: $-4x + 5 = 20$ **b.** Find the value of $-4x + 5$ when x is 3

     c. Solve for y: $-4x + 5y = 20$ **d.** Solve for x: $-4x + 5y = 20$

76. Work each problem according to the instructions given.

     a. Solve: $2x + 1 = -4$ **b.** Find the value of $2x + 1$ when x is 8

     c. Solve for y: $2x + y = 20$ **d.** Solve for x: $2x + y = 20$

XYZ Textbooks/©Matthew Hoy

Estimating Vehicle Weight   If you can measure the area where the tires on your car contact the ground and know the air pressure in the tires, then you can estimate the weight of your car with the following formula:

$$W = \frac{APN}{2,000}$$

where W is the vehicle's weight in tons, A is the average tire contact area with a hard surface in square inches, P is the air pressure in the tires in pounds per square inch (psi, or lb/in²), and N is the number of tires.

77. What is the approximate weight of a car if the average tire contact area is a rectangle 6 inches by 5 inches and if the air pressure in the tires is 30 psi?

78. What is the approximate weight of a car if the average tire contact area is a rectangle 5 inches by 4 inches and the tire pressure is 30 psi?

79. **Height of a Bullet**   A bullet is fired into the air with an initial upward velocity of 80 feet per second from the top of a building 96 feet high. The equation that gives the height of the bullet at any time t is $h = 96 + 80t - 16t^2$. At what times will the bullet be 192 feet in the air?

80. **Height of an Arrow**   An arrow is shot into the air with an upward velocity of 48 feet per second from a hill 32 feet high. The equation that gives the height of the arrow at any time t is $h = 32 + 48t - 16t^2$. Find the times at which the arrow will be 64 feet above the ground.

81. **Current**   It takes a boat 2 hours to travel 18 miles upstream against the current. If the speed of the boat in still water is 15 miles per hour, what is the speed of the current?

82. **Current**   It takes a boat 6.5 hours to travel 117 miles upstream against the current. If the speed of the current is 5 miles per hour, what is the speed of the boat in still water?

83. **Wind**   An airplane takes 4 hours to travel 864 miles while flying against the wind. If the speed of the airplane on a windless day is 258 miles per hour, what is the speed of the wind?

84. **Wind**   A cyclist takes 3 hours to travel 39 miles while pedaling against the wind. If the speed of the wind is 4 miles per hour, how fast would the cyclist be able to travel on a windless day?

For problems 85 and 86, use 3.14 as an approximation for π. Round answers to the nearest tenth.

85. **Average Speed**   A person riding a Ferris wheel with a diameter of 65 feet travels once around the wheel in 30 seconds. What is the average speed of the rider in feet per second?

86. **Average Speed**   A person riding a Ferris wheel with a diameter of 102 feet travels once around the wheel in 3.5 minutes. What is the average speed of the rider in feet per minute?

iStockPhoto.com/©Nikada

87. **Raindrop Volume**   A small raindrop will take on the shape of a sphere. The volume V (in cubic millimeters) of a raindrop, then, follows $V = \frac{4}{3}\pi r^3$, where r is the radius of the raindrop (in mm). If a small raindrop has a radius of 0.9 mm, what is its volume? Use 3.14 as an approximation for π. Round answer to the nearest tenth.

88. **Raindrop Volume**   A small raindrop will take on the shape of a sphere. The volume V (in cubic millimeters) of a raindrop, then, follows $V = \frac{4}{3}\pi r^3$, where r is the radius of the raindrop (in mm). If a small raindrop has a radius of 0.7 mm, what is its volume? Use 3.14 as an approximation for π. Round answer to the nearest tenth.

89. **Price and Revenue**   The relationship between the number of calculators x a company sells per day and the price of each calculator p is given by the equation $x = 1{,}700 - 100p$. At what price should the calculators be sold if the daily revenue is to be $7,000?

90. **Price and Revenue**   The relationship between the number of pencil sharpeners x a company can sell each week and the price of each sharpener p is given by the equation $x = 1{,}800 - 100p$. At what price should the sharpeners be sold if the weekly revenue is to be $7,200?

Digital Video   The *Star Wars: The Phantom Menace* movie trailer was the largest file download to that point and changed the way movies were marketed.

© Lucasfilm/The Kobal Collection/Hamshere, Keith

The video was compressed so it would be small enough for people to download over the internet. A formula for estimating the size, in kilobytes, of a compressed video is

$$S = \frac{height \cdot width \cdot fps \cdot time}{35{,}000}$$

where *height* and *width* are in pixels, *fps* is the number of frames per second the video is to play (television plays at 30 fps), and time is given in seconds.

91. Estimate the size in kilobytes of the *Star Wars* trailer that has a height of 480 pixels, a width of 216 pixels, plays at 30 fps, and runs for 150 seconds.

92. Estimate the size in kilobytes of the *Star Wars* trailer that has a height of 320 pixels, a width of 144 pixels, plays at 15 fps, and runs for 150 seconds.

Fermat's Last Theorem   This postage stamp shows Fermat's last theorem, which states that if n is an integer greater than 2, then there are no positive integers x, y, and z that will make the formula $x^n + y^n = z^n$ true. Use the formula $x^n + y^n = z^n$ to

93. Find x if $n = 1$, $y = 7$, and $z = 15$.

94. Find y if $n = 1$, $x = 23$, and $z = 37$.

Exercise Physiology In exercise physiology, a person's maximum heart rate, in beats per minute, is found by subtracting his age, in years, from 220. So, if A represents your age in years, then your maximum heart rate is

$$M = 220 - A$$

A person's training heart rate, in beats per minute, is her resting heart rate plus 60% of the difference between her maximum heart rate and her resting heart rate. If resting heart rate is R and maximum heart rate is M, then the formula that gives training heart rate is

$$T = R + 0.6(M - R)$$

95. Training Heart Rate Shar is 46 years old. Her daughter, Sara, is 26 years old. If they both have a resting heart rate of 60 beats per minute, find the training heart rate for each.

96. Training Heart Rate Shane is 30 years old and has a resting heart rate of 68 beats per minute. Her mother, Carol, is 52 years old and has the same resting heart rate. Find the training heart rate for Shane and for Carol.

GETTING READY FOR THE NEXT SECTION

Translate into symbols.

97. Three less than twice a number

98. Ten less than four times a number

99. The sum of x and y is 180

100. The sum of a and b is 90

Solve each equation.

101. $x + 2x = 90$

102. $x + 5x = 180$

103. $2(2x - 3) + 2x = 45$

104. $2(4x - 10) + 2x = 12.5$

105. $0.06x + 0.05(10,000 - x) = 560$

106. $x + 0.0725x = 17,481.75$

EXTENDING THE CONCEPTS

107. Solve for x: $\dfrac{x}{a} + \dfrac{y}{b} = 1$

108. Solve for y: $\dfrac{x}{a} + \dfrac{y}{b} = 1$

109. Solve for a: $\dfrac{1}{a} + \dfrac{1}{b} = \dfrac{1}{c}$

110. Solve for b: $\dfrac{1}{a} + \dfrac{1}{b} = \dfrac{1}{c}$

OBJECTIVES

A Use the Blueprint for Problem Solving to solve application problems.

B Construct tables and graphs using basic formulas.

©2010 Katherine Heistand Shields

2.3

The air-powered Stomp Rocket can be propelled over 200 feet using a blast of air. The harder you stomp on the Launch Pad, the farther the rocket flies.

If the rocket is launched straight up into the air with a velocity of 112 feet per second, then the formula

$$h = -16t^2 + 112t$$

gives the height h of the rocket t seconds after it is launched. We can use this formula to find the height of the rocket 3.5 seconds after launch by substituting $t = 3.5$:

$$h = -16(3.5)^2 + 112(3.5) = 196$$

At 3.5 seconds, the rocket reaches a height of 196 feet.

 Problem Solving

In this section we begin our work with application problems and use the skills we have developed for solving equations to solve problems written in words. You may find that some of the examples and problems are more realistic than others. Because we are just beginning our work with application problems, even the ones that seem unrealistic are good practice. What is important in this section is the *method* we use to solve application problems, not the applications themselves. The method, or strategy, that we use to solve application problems is called the *Blueprint for Problem Solving*. It is an outline that will overlay the solution process we use on all application problems.

2.3 VIDEOS

> ### BLUEPRINT FOR PROBLEM SOLVING
> #### Using a System of Equations • • • • • • • • • • • • • • • • • • •
>
> **Step 1:** *Read* the problem, and then mentally *list* the items that are known and the items that are unknown.
>
> **Step 2:** *Assign a variable* to one of the unknown items. (In most cases, this will amount to letting $x =$ the item that is asked for in the problem.) Then *translate* the other *information* in the problem to expressions involving the variable.
>
> **Step 3:** *Reread* the problem, and then *write an equation,* using the items and variable listed in steps 1 and 2, that describes the situation.
>
> **Step 4:** *Solve the equation* found in step 3.
>
> **Step 5:** *Write your answer* using a complete sentence.
>
> **Step 6:** *Reread* the problem, and *check* your solution with the original words in the problem.

A number of substeps occur within each of the steps in our blueprint. For instance, with steps 1 and 2 it is always a good idea to draw a diagram or picture if it helps you visualize the relationship among the items in the problem.

EXAMPLE 1

The length of a rectangle is 3 inches less than twice the width. The perimeter is 45 inches. Find the length and width.

SOLUTION   When working problems that involve geometric figures, a sketch of the figure helps organize and visualize the problem.

Step 1: *Read and list.*
Known items: The figure is a rectangle. The length is 3 inches less than twice the width. The perimeter is 45 inches.
Unknown items: The length and the width

Step 2: *Assign a variable and translate information.*
Because the length is given in terms of the width (the length is 3 less than twice the width), we let $x =$ the width of the rectangle. The length is 3 less than twice the width, so it must be $2x - 3$. The diagram in Figure 1 is a visual description of the relationships we have listed so far.

Step 3: *Reread and write an equation.*
The equation that describes the situation is

Twice the length $+$ twice the width $=$ perimeter

$2(2x - 3)\quad + \qquad 2x \qquad = \qquad 45$

Step 4: *Solve the equation.*

$$2(2x - 3) + 2x = 45$$
$$4x - 6 + 2x = 45$$
$$6x - 6 = 45$$
$$6x = 51$$
$$x = 8.5$$

Step 5: *Write the answer.*
The width is 8.5 inches. The length is $2x - 3 = 2(8.5) - 3 = 14$ inches.

x

$2x - 3$

FIGURE 1

Step 6: **Reread and check.**
If the length is 14 inches and the width is 8.5 inches, then the perimeter must be 2(14) + 2(8.5) = 28 + 17 = 45 inches. Also, the length, 14, is 3 less than twice the width.

Remember as you read through the steps in the solutions to the examples in this section that step 1 is done mentally. Read the problem and then *mentally* list the items that you know and the items that you don't know. The purpose of step 1 is to give you direction as you begin to work application problems. Finding the solution to an application problem is a process; it doesn't happen all at once. The first step is to read the problem with a purpose in mind. That purpose is to mentally note the items that are known and the items that are unknown.

EXAMPLE 2

In April, Pat bought a Ford Mustang with a 5.0-liter engine. The total price, which includes the price of the used car plus sales tax, was $17,481.75. If the sales tax rate was 7.25%, what was the price of the car?
SOLUTION

PRE-OWNED
CONVERTIBLE

$16,300
plus tax

Step 1: **Read and list.**
Known items: The total price is $17,481.75. The sales tax rate is 7.25%, which is 0.0725 in decimal form.
Unknown item: The price of the car

Step 2: **Assign a variable and translate information.**
If we let x = the price of the car, then to calculate the sales tax we multiply the price of the car x by the sales tax rate:

Sales tax = (sales tax rate)(price of the car)
= $0.0725x$

Step 3: **Reread and write an equation.**

$$\text{Car price} + \text{sales tax} = \text{total price}$$
$$x \qquad + 0.0725x = 17,481.75$$

Step 4: **Solve the equation.**

$$x + 0.0725x = 17,481.75$$
$$1.0725x = 17,481.75$$
$$x = \frac{17,481.75}{1.0725}$$
$$= 16,300.00$$

Step 5: **Write the answer.**
The price of the car is $16,300.00.

Step 6: **Reread and check.**
The price of the car is $16,300.00. The tax is
$$0.0725(16,300) = \$1,181.75.$$
Adding the retail price and the sales tax we have a total bill of $17,481.75.

FACTS FROM GEOMETRY   *Angles*

An angle is formed by two rays with the same endpoint. The common endpoint is called the *vertex* of the angle, and the rays are called the *sides* of the angle.

In Figure 2, angle θ (theta) is formed by the two rays *OA* and *OB*. The vertex of θ is *O*. Angle θ is also denoted as angle *AOB*, where the letter associated with the vertex is always the middle letter in the three letters used to denote the angle.

Degree Measure

The angle formed by rotating a ray through one complete revolution around its endpoint (Figure 3) has a measure of 360 degrees, which we write as 360°.

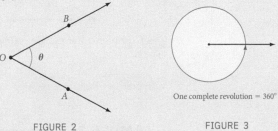

FIGURE 2   FIGURE 3

One degree of angle measure, written 1°, is $\frac{1}{360}$ of a complete rotation of a ray around its endpoint; there are 360° in one full rotation. (The number 360 was decided upon by early civilizations because it was believed that Earth was at the center of the universe and the sun would rotate once around Earth every 360 days.) Similarly, 180° is half of a complete rotation, and 90° is a quarter of a full rotation. Angles that measure 90° are called *right angles*, and angles that measure 180° are called *straight angles*. If an angle measures between 0° and 90° it is called an *acute angle*, and an angle that measures between 90° and 180° is an *obtuse angle*. Figure 4 illustrates further.

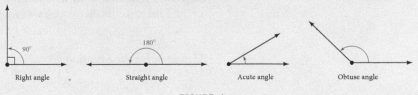

FIGURE 4

FACTS FROM GEOMETRY *Complementary Angles and Supplementary Angles*

· ·

If two angles add up to 90°, we call them *complementary angles*, and each is called the *complement* of the other. If two angles have a sum of 180°, we call them *supplementary angles*, and each is called the *supplement* of the other. Figure 5 illustrates the relationship between angles that are complementary and angles that are supplementary.

  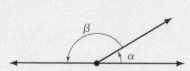

Complementary angles: $\alpha + \beta = 90°$ Supplementary angles: $\alpha + \beta = 180°$

FIGURE 5

EXAMPLE 3

Two complementary angles are such that one is twice as large as the other. Find the two angles.

SOLUTION   Applying the Blueprint for Problem Solving, we have:

Step 1: **Read and list.**
Known items: Two complementary angles. One is twice as large as the other.
Unknown items: The size of the angles

Step 2: **Assign a variable and translate information.**
Let x = the smaller angle. The larger angle is twice the smaller, so we represent the larger angle with $2x$.

Step 3: **Reread and write an equation.**
Because the two angles are complementary, their sum is 90. Therefore,

$x + 2x = 90$

Step 4: **Solve the equation.**
$$x + 2x = 90$$
$$3x = 90$$
$$x = 30$$

Step 5: **Write the answer.**
The smaller angle is 30°, and the larger angle is $2 \cdot 30 = 60°$.

Step 6: **Reread and check.**
The larger angle is twice the smaller angle, and their sum is 90°.

Suppose we know that the sum of two numbers is 50. If we let x represent one of the two numbers, how can we represent the other? Let's suppose for a moment that x turns out to be 30. Then the other number will be 20, because their sum is 50.

That is, if two numbers add up to 50, and one of them is 30, then the other must be $50 - 30 = 20$. Generalizing this to any number x, we see that if two numbers have a sum of 50, and one of the numbers is x, then the other must be $50 - x$. The following table shows some additional examples:

If Two Numbers Have a Sum of	And One of Them Is	Then the Other Must Be
50	x	$50 - x$
10	y	$10 - y$
12	n	$12 - n$

EXAMPLE 4

Suppose a person invests a total of $10,000 in two accounts. One account earns 5% annually, and the other earns 6% annually. If the total interest earned from both accounts in a year is $560, how much is invested in each account?

SOLUTION

Step 1: *Read and list.*

Known items: Two accounts. One pays interest of 5%, and the other pays 6%. The total invested is $10,000.

Unknown items: The number of dollars invested in each individual account

Step 2: *Assign a variable and translate information.*

If we let $x =$ the amount invested at 6%, then $10,000 - x$ is the amount invested at 5%. The total interest earned from both accounts is $560. The amount of interest earned on x dollars at 6% is $.06x$, whereas the amount of interest earned on $10,000 - x$ dollars at 5% is $0.05(10,000 - x)$.

	Dollars at 6%	Dollars at 5%	Total
Number of	x	$10,000 - x$	10,000
Interest on	$0.06x$	$0.05(10,000) - x$	560

Step 3: *Reread and write an equation.*

The last line gives us the equation we are after:

$$0.06x + 0.05(10,000 - x) = 560$$

Step 4: *Solve the equation.*

To make the equation a little easier to solve, we begin by multiplying both sides by 100 to move the decimal point two places to the right.

$$6x + 5(10,000 - x) = 56,000$$
$$6x + 50,000 - 5x = 56,000$$
$$x + 50,000 = 56,000$$
$$x = 6,000$$

Step 5: *Write the answer.*

The amount of money invested at 6% is $6,000. The amount of money invested at 5% is $10,000 − $6,000 = $4,000.

Step 6: *Reread and check.*

To check our results, we find the total interest from the two accounts:

The interest earned on $6,000 at 6% is 0.06(6,000) =   360
The interest earned on $4,000 at 5% is 0.05(4,000) =   200

The total interest = $560

FACTS FROM GEOMETRY *Triangles*

Special Triangles

It is not unusual to have the terms we use in mathematics show up in the descriptions of things we find in the world around us. The flag of Puerto Rico shown here is described on the government website as "Five equal horizontal bands of red (top and bottom) alternating with  white; a blue isosceles triangle based on the hoist side bears a large white five-pointed star in the center." An *isosceles triangle* as shown here and in Figure 6, is a triangle with two sides of equal length.

Angles A and B in the isosceles triangle in Figure 6 are called the *base angles*: they are the angles opposite the two equal sides. In every isosceles triangle, the base angles are equal.

An equilateral triangle (Figure 7) is a triangle with three sides of equal

NOTE As you can see from Figures 6 and 7, one way to label the important parts of a triangle is to label the vertices with capital letters and the sides with small letters: side a is opposite vertex A, side b is opposite vertex B, and side c is opposite vertex C.

Also, because each vertex is the vertex of one of the angles of the triangle, we refer to the three interior angles as A, B, and C.

Finally, in any triangle, the sum of the interior angles is 180°. For the triangles shown in Figures 6 and 7, the relationship is written

$$A + B + C = 180°$$

Isosceles Triangle

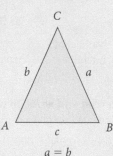

$a = b$

FIGURE 6

Equilateral Triangle

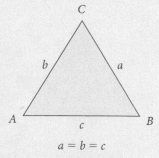

$a = b = c$

FIGURE 7

length. If all three sides in a triangle have the same length, then the three interior angles in the triangle also must be equal. Because the sum of the interior angles in a triangle is always 180°, the three interior angles in any equilateral triangle must be 60°.

A *right triangle* (Figure 8) is a triangle with an interior angle of 90°. The side opposite the right angle is called the *hypotenuse*. The other two sides are called *legs*. The *Pythagorean theorem* states a special relationship between the hypotenuse and legs of a right triangle.

THEOREM *Pythagorean Theorem*

In any right triangle, the square of the longest side (hypotenuse) is equal to the sum of the squares of the other two sides (legs).

$$c^2 = a^2 + b^2$$

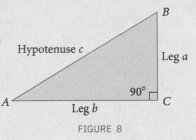

FIGURE 8

EXAMPLE 5

The lengths of the three sides of a right triangle are given by three consecutive integers. Find the lengths of the three sides.

SOLUTION

Step 1: *Read and list.*

Known items: A right triangle. The three sides are three consecutive integers.

Unknown items: The three sides.

Step 2: *Assign a variable and translate information.*

Let x = First integer (shortest side)

Then $x + 1$ = Next consecutive integer

$x + 2$ = Last consecutive integer (longest side)

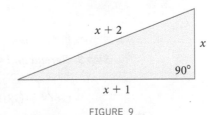

FIGURE 9

Step 3: *Reread and write an equation.* By the Pythagorean theorem, we have

$$(x + 2)^2 = (x + 1)^2 + x^2$$

Step 4: *Solve the equation.*

$$x^2 + 4x + 4 = x^2 + 2x + 1 + x^2$$

$$x^2 - 2x - 3 = 0$$

$$(x - 3)(x + 1) = 0$$

$$x = 3 \quad \text{or} \quad x = -1$$

Step 5: *Write the answer.* Because x is the length of a side in a triangle, it must be a positive number. Therefore, $x = -1$ cannot be used. The shortest side is 3. The other two sides are 4 and 5.

Step 6: *Reread and check.* The three sides are given by consecutive integers. The square of the longest side is equal to the sum of the squares of the two shorter sides ($5^2 = 3^2 + 4^2$).

EXAMPLE 6

Two boats leave from an island port at the same time. One travels due north at a speed of 12 miles per hour, and the other travels due west at a speed of 16 miles per hour. How long until the distance between the two boats is 60 miles?

SOLUTION

Step 1: *Read and list.*

Known items: The speed and direction of both boats. The distance between the boats.

Unknown items: The distance traveled by each boat, and the time.

Step 2: *Assign a variable and translate information.*

Let $\quad t =$ the time.

Then $\quad 12t =$ the distance traveled by boat going north.

$\qquad 16t =$ the distance traveled by boat going west.

If we draw a diagram for the problem (Figure 10), we see that the distances traveled by the two boats form the legs of a right triangle. The hypotenuse of the triangle will be the distance between the boats, which is 60 miles.

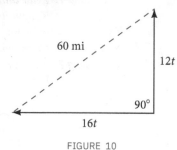

FIGURE 10

Step 3: *Reread and write an equation.* By the Pythagorean theorem, we have

$$(16t)^2 + (12t)^2 = 60^2$$

Step 4: *Solve the equation.*

$$256t^2 + 144t^2 = 3{,}600$$
$$400t^2 = 3{,}600$$
$$400t^2 - 3{,}600 = 0$$
$$t^2 - 9 = 0$$
$$(t + 3)(t - 3) = 0$$
$$t = -3 \quad \text{or} \quad t = 3$$

Step 5: *Write the answer.* Because t is measuring time, it must be a positive number. Therefore, $t = -3$ cannot be used. The two boats will be 60 miles apart after 3 hours.

Step 6: *Reread and check.* The boat going north will travel $12 \cdot 3 = 36$ miles in 3 hours, and the boat going west will travel $16 \cdot 3 = 48$ miles. The distance between them after 3 hours will be 60 miles ($48^2 + 36^2 = 60^2$).

Table Building

We can use our knowledge of formulas from Section 2.2 to build tables of paired data. As you will see, equations or formulas that contain exactly two variables produce pairs of numbers that can be used to construct tables.

EXAMPLE 7

A piece of string 12 inches long is to be formed into a rectangle. Build a table that gives the length of the rectangle if the width is 1, 2, 3, 4, or 5 inches. Then find the area of each of the rectangles formed.

SOLUTION   Because the formula for the perimeter of a rectangle is $P = 2l + 2w$, and our piece of string is 12 inches long, the formula we will use to find the lengths for the given widths is $12 = 2l + 2w$. To solve this formula for l, we divide each side by 2 and then subtract w. The result is $l = 6 - w$. Table 1 organizes our work so that the formula we use to find l for a given value of w is shown, and we have added a last column to give us the areas of the rectangles formed. The units for the first three columns are inches, and the units for the numbers in the last column are square inches.

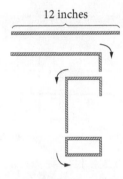
12 inches

TABLE 1 Length, Width, and Area			
Width (in.)	Length (in.)		Area (in.²)
w	$l = 6 - w$	l	$A = lw$
1	$l = 6 - 1$	5	5
2	$l = 6 - 2$	4	8
3	$l = 6 - 3$	3	9
4	$l = 6 - 4$	2	8
5	$l = 6 - 5$	1	5

Figures 11 and 12 show two *bar charts* constructed from the information in Table 1.

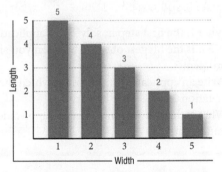

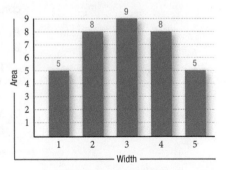

FIGURE 11 *Length and width of rectangles with perimeters fixed at 12 inches*

FIGURE 12 *Area and width of rectangles with perimeters fixed at 12 inches*

USING TECHNOLOGY   *Graphing Calculators*

A number of graphing calculators have table-building capabilities. We can let the calculator variable X represent the widths of the rectangles in Example 7. To find the lengths, we set variable Y_1 equal to $6 - X$. The area of each rectangle can be found by setting variable Y_2 equal to $X * Y_1$. To have the calculator produce the table automatically, we use a table minimum of 0 and a table increment of 1. Here is a summary of how the graphing calculator is set up:

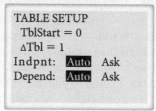

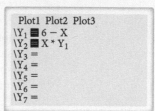

The table will look like this:

X	Y_1	Y_2
0	6	0
1	5	5
2	4	8
3	3	9
4	2	8
5	1	5
6	0	0

GETTING READY FOR CLASS

After reading through the preceding section, respond in your own words and in complete sentences.

A. What is the first step in solving an application problem?

B. What is the biggest obstacle between you and success in solving application problems?

C. Write an application problem for which the solution depends on solving the equation $2x + 2 \cdot 3 = 18$.

D. What is the last step in solving an application problem? Why is this step important?

Problem Set 2.3

Solve each application problem. Be sure to follow the steps in the Blueprint for Problem Solving.

GEOMETRY PROBLEMS

1. **Rectangle** A rectangle is twice as long as it is wide. The perimeter is 60 feet. Find the dimensions.

2. **Rectangle** The length of a rectangle is 5 times the width. The perimeter is 48 inches. Find the dimensions.

3. **Square** A square has a perimeter of 28 feet. Find the length of each side.

4. **Square** A square has a perimeter of 36 centimeters. Find the length of each side.

5. **Triangle** A triangle has a perimeter of 23 inches. The medium side is 3 inches more than the shortest side, and the longest side is twice the shortest side. Find the shortest side.

6. **Triangle** The longest side of a triangle is two times the shortest side, while the medium side is 3 meters more than the shortest side. The perimeter is 27 meters. Find the dimensions.

7. **Rectangle** The length of a rectangle is 3 meters less than twice the width. The perimeter is 18 meters. Find the width.

8. **Rectangle** The length of a rectangle is 1 foot more than twice the width. The perimeter is 20 feet. Find the dimensions.

9. **Livestock Pen** A livestock pen is built in the shape of a rectangle that is twice as long as it is wide. The perimeter is 48 feet. If the material used to build the pen is $1.75 per foot for the longer sides and $2.25 per foot for the shorter sides (the shorter sides have gates, which increase the cost per foot), find the cost to build the pen.

10. **Garden** A garden is in the shape of a square with a perimeter of 42 feet. The garden is surrounded by two fences. One fence is around the perimeter of the garden, whereas the second fence is 3 feet from the first fence, as Figure 12 indicates. If the material used to build the two fences is $1.28 per foot, what was the total cost of the fences?

$p = 48$ ft.

3 feet

FIGURE 12

PERCENT PROBLEMS

11. **Money** Shane returned from a trip to Las Vegas with $300.00, which was 50% more money than he had at the beginning of the trip. How much money did Shane have at the beginning of his trip?

12. **Items Sold** Every item in the Just a Dollar store is priced at $1.00. When Mary Jo opens the store, there is $125.50 in the cash register. When she counts the money in the cash register at the end of the day, the total is $1,058.60. If the sales tax rate is 8.5%, how many items were sold that day?

13. **Textbook Price** Suppose a college bookstore buys a textbook from a publishing company and then marks up the price they paid for the book 33% and sells it to a student at the marked-up price. If the student pays $115.00 for the textbook, what did the bookstore pay for it? Round your answer to the nearest cent.

14. Hourly Wage A sheet metal worker earns $26.80 per hour after receiving a 4.5% raise. What was the sheet metal worker's hourly pay before the raise? Round your answer to the nearest cent.

15. Movies *Star Wars: The Force Awakens* grossed $248 million on its opening weekend and had one of the most successful movie launches in history. If *Star Wars: The Force Awakens* accounted for approximately 53% of all box office receipts that weekend, what were the total box office receipts?

16. Fat Content in Milk I was reading the information on the milk carton in Figure 13 at breakfast one morning when I was working on this book. According to the carton, this milk contains 70% less fat than whole milk. The nutrition label on the other side of the carton states that one serving of this milk contains 2.5 grams of fat. How many grams of fat are in an equivalent serving of whole milk?

FIGURE 13

MORE GEOMETRY PROBLEMS

17. Angles Two supplementary angles are such that one is eight times larger than the other. Find the two angles.

18. Angles Two complementary angles are such that one is five times larger than the other. Find the two angles.

19. Angles One angle is 12° less than four times another. Find the measure of each angle if

  a. they are complements of each other.

  b. they are supplements of each other.

20. Angles One angle is 4° more than three times another. Find the measure of each angle if

  a. they are complements of each other.

  b. they are supplements of each other.

21. Triangles A triangle is such that the largest angle is three times the smallest angle. The third angle is 9° less than the largest angle. Find the measure of each angle.

22. Triangles The smallest angle in a triangle is half of the largest angle. The third angle is 15° less than the largest angle. Find the measure of all three angles.

23. Triangles The smallest angle in a triangle is one-third of the largest angle. The third angle is 10° more than the smallest angle. Find the measure of all three angles.

24. Triangles The third angle in an isosceles triangle is half as large as each of the two base angles. Find the measure of each angle.

25. Isosceles Triangles The third angle in an isosceles triangle is 8° more than twice as large as each of the two base angles. Find the measure of each angle.

26. Isosceles Triangles The third angle in an isosceles triangle is 4° more than one-fifth of each of the two base angles. Find the measure of each angle.

7 ft

27. **Right Triangle** A 25-foot ladder is leaning against a building. The base of the ladder is 7 feet from the side of the building. How high does the ladder reach along the side of the building?

28. **Right Triangle** Noreen wants to place a 13-foot ramp against the side of her house so the top of the ramp rests on a ledge that is 5 feet above the ground. How far will the base of the ramp be from the house?

29. **Right Triangle** The lengths of the three sides of a right triangle are given by three consecutive even integers. Find the lengths of the three sides.

30. **Right Triangle** The longest side of a right triangle is 3 less than twice the shortest side. The third side measures 12 inches. Find the length of the shortest side.

31. **Rectangle** The length of a rectangle is 2 feet more than three times the width. If the area is 16 square feet, find the width and the length.

32. **Rectangle** The length of a rectangle is 4 yards more than twice the width. If the area is 70 square yards, find the width and the length.

33. **Triangle** The base of a triangle is 2 inches more than four times the height. If the area is 36 square inches, find the base and the height.

34. **Triangle** The height of a triangle is 4 feet less than twice the base. If the area is 48 square feet, find the base and the height.

INTEREST PROBLEMS

35. **Investing** A woman has a total of $9,000 to invest. She invests part of the money in an account that pays 8% per year and the rest in an account that pays 9% per year. If the interest earned in the first year is $750, how much did she invest in each account?

	Dollars at 8%	Dollars at 9%	Total
Number of			
Interest on			

36. **Investing** A man invests $12,000 in two accounts. If one account pays 10% per year and the other pays 7% per year, how much was invested in each account if the total interest earned in the first year was $960?

	Dollars at 10%	Dollars at 7%	Total
Number of			
Interest on			

37. **Investing** A total of $15,000 is invested in two accounts. One of the accounts earns 12% per year, and the other earns 10% per year. If the total interest earned in the first year is $1,600, how much was invested in each account?

38. **Investing** A total of $11,000 is invested in two accounts. One of the two accounts pays 9% per year, and the other account pays 11% per year. If the total interest paid in the first year is $1,150, how much was invested in each account?

39. Investing Stacey has a total of $6,000 in two accounts. The total amount of interest she earns from both accounts in the first year is $500. If one of the accounts earns 8% interest per year and the other earns 9% interest per year, how much did she invest in each account?

40. Investing Travis has a total of $6,000 invested in two accounts. The total amount of interest he earns from the accounts in the first year is $410. If one account pays 6% per year and the other pays 8% per year, how much did he invest in each account?

DISTANCE, RATE, AND TIME PROBLEMS

41. Distance Two cyclists leave from an intersection at the same time. One travels due north at a speed of 15 miles per hour, and the other travels due east at a speed of 20 miles per hour. How long until the distance between the two cyclists is 75 miles?

42. Distance Two airplanes leave from an airport at the same time. One travels due south at a speed of 480 miles per hour, and the other travels due west at a speed of 360 miles per hour. How long until the distance between the two airplanes is 2,400 miles?

43. Distance A search is being conducted for someone guilty of a hit-and-run felony. In order to set up roadblocks at appropriate points, the police must determine how far the guilty party might have traveled during the past half-hour. Use the formula $d = rt$ with $t = 0.5$ hour to complete the following table.

Speed (miles per hour)	Distance (miles)
20	
30	
40	
50	
60	
70	

44. Speed To determine the average speed of a bullet when fired from a rifle, the time is measured from when the gun is fired until the bullet hits a target that is 1,000 feet away. Use the formula $r = \frac{d}{t}$ with $d = 1,000$ feet to complete the following table.

Time (seconds)	Rate (feet per second)
1.00	
0.80	
0.64	
0.50	
0.40	
0.32	

45. Current A boat that can travel 10 miles per hour in still water is traveling along a stream with a current of 4 miles per hour. The distance the boat will travel upstream is given by the formula $d = (r - c) \cdot t$, and the distance it will travel downstream is given by the formula $d = (r + c) \cdot t$. Use these formulas with $r = 10$ and $c = 4$ to complete the following table.

Time (hours)	Distance Upstream (miles)	Distance Downstream (miles)
1		
2		
3		
4		
5		
6		

46. Wind A plane that can travel 300 miles per hour in still air is traveling in a wind stream with a speed of 20 miles per hour. The distance the plane will travel against the wind is given by the formula $d = (r - w) \cdot t$, and the distance it will travel with the wind is given by the formula $d = (r + w) \cdot t$. Use these formulas with $r = 300$ and $w = 20$ to complete the following table.

Time (hours)	Distance against the wind (miles)	Distance with the wind (miles)
0.50		
1.00		
1.50		
2.00		
2.50		
3.00		

MISCELLANEOUS PROBLEMS

47. Tickets Tickets for the father-and-son breakfast were $2.00 for fathers and $1.50 for sons. If a total of 75 tickets were sold for $127.50, how many fathers and how many sons attended the breakfast?

48. Tickets A Girl Scout troop sells 62 tickets to their mother-and-daughter dinner, for a total of $216. If the tickets cost $4.00 for mothers and $3.00 for daughters, how many of each ticket did they sell?

49. Sales Tax A woman owns a small, cash-only business in a state that requires her to charge 6% sales tax on each item she sells. At the beginning of the day, she has $250 in the cash register. At the end of the day, she has $1,204 in the register. How much money should she send to the state government for the sales tax she collected?

50. **Sales Tax** A store is located in a state that requires 6% tax on all items sold. If the store brings in $3,392 in one day, how much of that total was sales tax?

51. **Long-Distance Phone Calls** Patrick goes away to college. The first week he is away from home he calls his girlfriend, using his parents' telephone credit card, and talks for a long time. The telephone company charges 40 cents for the first minute and 30 cents for each additional minute, and then adds on a 50-cent service charge for using the credit card. If his parents receive a bill for $13.80 for Patrick's call, how long did he talk?

52. **Long-Distance Phone Calls** A person makes a long-distance person-to-person call to Santa Barbara, California. The telephone company charges 41 cents for the first minute and 32 cents for each additional minute. Because the call is person-to-person, there is also a service charge of $3.00. If the cost of the call is $6.29, how many minutes did the person talk?

Problems 53–62 may be solved using a graphing calculator.

TABLE BUILDING

53. Use $h = 32t - 16t^2$ to complete the table.

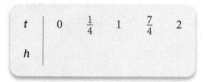

t	0	$\frac{1}{4}$	1	$\frac{7}{4}$	2
h					

54. Use $h = \dfrac{60}{t}$ to complete the table.

t	4	6	8	10
h				

Coffee Sales The chart below appeared in the news recently. Use the information to complete the following tables. Round to the nearest tenth of a billion dollars.

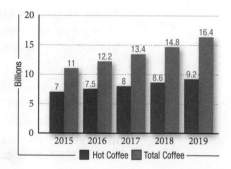

55.

Hot Coffee Sales

Year	Sales (billions of dollars)
2015	
2016	
2017	
2018	
2019	

56.

Total Coffee Sales

Year	Sales (billions of dollars)
2015	
2016	
2017	
2018	
2019	

57. Livestock Pen A farmer buys 48 feet of fencing material to build a rectangular livestock pen. Fill in the second column of the table to find the length of the pen if the width is 2, 4, 6, 8, 10, or 12 feet. Then find the area of each of the pens formed.

w (ft)	l (ft)	A (ft²)
2		
4		
6		
8		
10		
12		

58. Model Rocket A small rocket is projected straight up into the air with a velocity of 128 feet per second. The formula that gives the height h of the rocket t seconds after it is launched is

$$h = -16t^2 + 128t$$

Use this formula to find the height of the rocket after 1, 2, 3, 4, 5, and 6 seconds.

Time (seconds)	Height (feet)
1	
2	
3	
4	
5	
6	

Maximum Heart Rate In exercise physiology, a person's maximum heart rate, in beats per minute, is found by subtracting his age in years from 220. So, if A represents your age in years, then your maximum heart rate is

$$M = 220 - A$$

Use this formula to complete the following tables.

59.

Age (years)	Maximum Heart Rate (beats per minute)
18	
19	
20	
21	
22	
23	

60.

Age (years)	Maximum Heart Rate (beats per minute)
15	
20	
25	
30	
35	
40	

Training Heart Rate A person's training heart rate, in beats per minute, is his resting heart rate plus 60% of the difference between his maximum heart rate and his resting heart rate. If resting heart rate is R and maximum heart rate is M, then the formula that gives training heart rate is

$$T = R + 0.6(M - R)$$

Use this formula along with the results of Problems 59 and 60 to fill in the following two tables.

61. For a 20-year-old person

Resting Heart Rate (beats per minute)	Training Heart Rate (beats per minute)
60	
62	
64	
68	
70	
72	

62. For a 40-year-old person

Resting Heart Rate (beats per minute)	Training Heart Rate (beats per minute)
60	
62	
64	
68	
70	
72	

GETTING READY FOR THE NEXT SECTION

To understand all of the explanations and examples in the next section you must be able to work the problems below. Solve each equation.

63. $-2x - 3 = 7$

64. $3x + 3 = 2x - 1$

65. $3(2x - 4) - 7x = -3x$

66. $3(2x + 5) = -3x$

OBJECTIVES

A Solve linear inequalities using the addition property for inequalities.

B Solve inequalities using the multiplication property for inequalities.

©iStockphoto.com/kellington1

2.4

The ʻApapane, a native Hawaiian bird, feeds mainly on the nectar of the ʻohiʻa lehua blossom, although the adult diet also includes insects and spiders. The ʻApapane live in high altitude regions where the ʻohiʻa blossoms are found and where the birds are protected from mosquitoes which transmit avian malaria and avian pox. Predators of the ʻApapane include the rat, feral cat, mongoose, and owl. According to the U.S. Geological Survey:

Annual survival rates based on 1,584 recaptures of 429 banded individuals: 0.72 ± 0.11 for adults and 0.13 ± 0.07 for juveniles.

We can write the survival rate for the adults as an inequality:

$$0.61 \leq r \leq 0.83$$

Inequalities are what we will study in this section.

A Addition Property for Inequalities

A linear inequality in one variable is any inequality that can be put in the form

$$ax + b < c \qquad (a, b, \text{ and } c \text{ constants}, a \neq 0)$$

where the inequality symbol ($<$) can be replaced with any of the other three inequality symbols ($\leq$, $>$, or $\geq$).

Some examples of *linear inequalities* are

$$3x - 2 \geq 7 \qquad -5y < 25 \qquad 3(x - 4) > 2x$$

Our first property for inequalities is similar to the addition property we used when solving equations.

NOTE Because subtraction is defined as addition of the opposite, our new property holds for subtraction as well as addition. That is, we can subtract the same quantity from each side of an inequality and always be sure that we have not changed the solution.

2.4 VIDEOS

PROPERTY *Addition Property for Inequalities*

For any algebraic expressions, A, B, and C,

$$\text{if} \qquad A < B$$

$$\text{then} \qquad A + C < B + C$$

In words: Adding the same quantity to both sides of an inequality will not change the solution set.

149

EXAMPLE 1

Solve $3x + 3 < 2x - 1$, and graph the solution.

SOLUTION   We use the addition property for inequalities to write all the variable terms on one side and all constant terms on the other side:

$$3x + 3 < 2x - 1$$
$$3x + (-2x) + 3 < 2x + (-2x) - 1 \qquad \text{Add } -2x \text{ to each side.}$$
$$x + 3 < -1$$
$$x + 3 + (-3) < -1 + (-3) \qquad \text{Add } -3 \text{ to each side.}$$
$$x < -4$$

The solution set is all real numbers that are less than -4. To show this we can use *set notation* and write

$$\{x \mid x < -4\}$$

We can graph the solution set on the number line using an open circle at -4 to show that -4 is not part of the solution set. This is the format you used when graphing inequalities in beginning algebra.

Here is an equivalent graph that uses a parenthesis opening left, instead of an open circle, to represent the endpoint of the graph.

This graph gives rise to the following notation, called *interval notation*, that is an alternative way to write the solution set.

$$(-\infty, -4)$$

The preceding expression indicates that the solution set is all real numbers from negative infinity up to, but not including, -4.

We have three equivalent representations for the solution set to our original inequality. Here are all three together.

Set Notation	Line Graph	Interval Notation

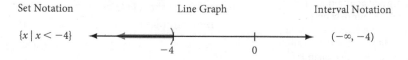

NOTE   The English mathematician John Wallis (1616–1703) was the first person to use the ∞ symbol to represent infinity. When we encounter the interval $(3, \infty)$, we read it as "the interval from 3 to infinity," and we mean the set of real numbers that are greater than 3. Likewise, the interval $(-\infty, -4)$ is read "the interval from negative infinity to -4," which is all real numbers less than -4.

B Multiplication Property for Inequalities

The following table shows the connection between set notation, interval notation, and number line graphs. We have included the graphs with open and closed circles for those of you who have used this type of graph previously. In this book, we will continue to show our graphs using the parentheses/brackets method.

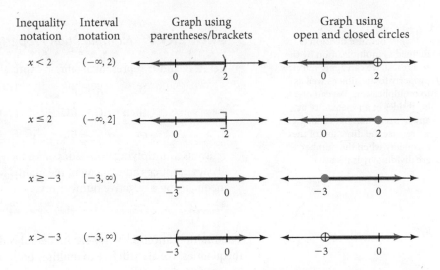

Inequality notation	Interval notation	Graph using parentheses/brackets	Graph using open and closed circles
$x < 2$	$(-\infty, 2)$		
$x \le 2$	$(-\infty, 2]$		
$x \ge -3$	$[-3, \infty)$		
$x > -3$	$(-3, \infty)$		

Before we state the multiplication property for inequalities, we will take a look at what happens to an inequality statement when we multiply both sides by a positive number and what happens when we multiply by a negative number.

We begin by writing three true inequality statements:

$$3 < 5 \qquad -3 < 5 \qquad -5 < -3$$

We multiply both sides of each inequality by a positive number, say, 4:

$$4(3) < 4(5) \qquad 4(-3) < 4(5) \qquad 4(-5) < 4(-3)$$

$$12 < 20 \qquad -12 < 20 \qquad -20 < -12$$

Notice in each case that the resulting inequality symbol points in the same direction as the original inequality symbol. Multiplying both sides of an inequality by a positive number preserves the *sense* of the inequality.

Let's take the same three original inequalities and multiply both sides by -4:

$$3 < 5 \qquad\qquad -3 < 5 \qquad\qquad -5 < -3$$

$$-4(3) > -4(5) \qquad -4(-3) > -4(5) \qquad -4(-5) > -4(-3)$$

$$-12 > -20 \qquad\qquad 12 > -20 \qquad\qquad 20 > 12$$

Notice in this case that the resulting inequality symbol always points in the opposite direction from the original one. Multiplying both sides of an inequality by a negative number *reverses* the sense of the inequality. Keeping this in mind, we will now state the multiplication property for inequalities.

NOTE Because division is defined as multiplication by the reciprocal, we can apply our new property to division as well as to multiplication. We can divide both sides of an inequality by any nonzero number as long as we reverse the direction of the inequality when the number we are dividing by is negative.

PROPERTY *Multiplication Property for Inequalities*

Let A, B, and C represent algebraic expressions.

$$\text{If} \quad A < B$$

$$\text{then} \quad AC < BC \quad \text{if} \quad C \text{ is positive } (C > 0)$$

$$\text{or} \quad AC > BC \quad \text{if} \quad C \text{ is negative } (C < 0)$$

In words: Multiplying both sides of an inequality by a positive number always produces an equivalent inequality. Multiplying both sides of an inequality by a negative number reverses the sense of the inequality.

The multiplication property for inequalities does not limit what we can do with inequalities. We are still free to multiply both sides of an inequality by any nonzero number we choose. If the number we multiply by happens to be *negative*, then we *must also reverse* the direction of the inequality.

EXAMPLE 2

Solve $3a < 15$ and graph the solution.

SOLUTION We begin by multiplying each side by $\frac{1}{3}$. Because $\frac{1}{3}$ is a positive number, we do not reverse the direction of the inequality symbol:

$$3a < 15$$

$$\frac{1}{3}(3a) < \frac{1}{3}(15) \qquad \text{Multiply each side by } \frac{1}{3}.$$

$$a < 5$$

Set Notation	Line Graph	Interval Notation
$\{x \mid x < 5\}$		$(-\infty, 5)$

EXAMPLE 3

Solve $-3a \le 18$, and graph the solution.

SOLUTION We begin by multiplying both sides by $-\frac{1}{3}$. Because $-\frac{1}{3}$ is a negative number, we must reverse the direction of the inequality symbol at the same time that we multiply by $-\frac{1}{3}$.

$$-3a \le 18$$

$$-\frac{1}{3}(-3a) \ge -\frac{1}{3}(18) \qquad \text{Multiply both sides by } -\frac{1}{3} \text{ and reverse the direction of the inequality symbol.}$$

$$a \ge -6$$

Set Notation	Line Graph	Interval Notation
$\{x \mid x \ge -6\}$		$[-6, \infty)$

EXAMPLE 4

Solve $-\dfrac{x}{4} > 2$ and graph the solution.

SOLUTION To isolate x, we multiply each side by -4. Because -4 is a negative number, we also must reverse the direction of the inequality symbol:

$$-\frac{x}{4} > 2$$

$$-4\left(-\frac{x}{4}\right) < -4(2) \qquad \text{Multiply each side by } -4, \text{ and reverse the}$$
$$\text{direction of the inequality symbol.}$$

$$x < -8$$

Set Notation	Line Graph	Interval Notation
$\{x \mid x < -8\}$	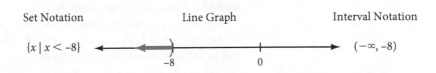	$(-\infty, -8)$

To solve more complicated inequalities, we use the following steps.

> **How to Solve Linear Inequalities in One Variable**
>
> **Step 1a:** Use the distributive property to separate terms, if necessary.
>
> **1b:** If fractions are present, consider multiplying both sides by the LCD to eliminate the fractions. If decimals are present, consider multiplying both sides by a power of 10 to clear the inequality of decimals.
>
> **1c:** Combine similar terms on each side of the inequality.
>
> **Step 2:** Use the addition property for inequalities to get all variable terms on one side of the inequality and all constant terms on the other side.
>
> **Step 3:** Use the multiplication property for inequalities to get x by itself on one side of the inequality.
>
> **Step 4:** Graph the solution set.

EXAMPLE 5

Solve $2.5x - 3.48 < -4.9x + 2.07$.

SOLUTION We have two methods we can use to solve this inequality. We can simply apply our properties to the inequality the way it is currently written and work with the decimal numbers, or we can eliminate the decimals to begin with and solve the resulting inequality.

Method 1 Working with the decimals.

$$2.5x - 3.48 < -4.9x + 2.07$$ Original inequality

$$2.5x + 4.9x - 3.48 < -4.9x + 4.9x + 2.07$$ Add 4.9x to each side.

$$7.4x - 3.48 < 2.07$$

$$7.4x - 3.48 + 3.48 < 2.07 + 3.48$$ Add 3.48 to each side.

$$7.4x < 5.55$$

$$\frac{7.4x}{7.4} < \frac{5.55}{7.4}$$ Divide each side by 7.4.

$$x < 0.75$$

Method 2 Eliminating the decimals in the beginning.

Because the greatest number of places to the right of the decimal point in any of the numbers is 2, we can multiply each side of the inequality by 100 and we will be left with an equivalent inequality that contains only whole numbers.

$$2.5x - 3.48 < -4.9x + 2.07$$ Original inequality

$$100(2.5x - 3.48) < 100(-4.9x + 2.07)$$ Multiply each side by 100.

$$100(2.5x) - 100(3.48) < 100(-4.9x) + 100(2.07)$$ Distributive property

$$250x - 348 < -490x + 207$$ Multiplication

$$740x - 348 < 207$$ Add 490x to each side.

$$740x < 555$$ Add 348 to each side.

$$\frac{740x}{740} < \frac{555}{740}$$ Divide each side by 740.

$$x < 0.75$$

The solution by either method is $x < 0.75$. Here is the graph:

EXAMPLE 6

Find the solution set for $-2y - 3 \le 7$.

SOLUTION   We begin by adding 3 to each side of the inequality:

$$-2y - 3 \le 7$$

$$-2y \le 10$$ Add 3 to both sides.

$$-\frac{1}{2}(-2y) \ge -\frac{1}{2}(10)$$ Multiply by $-\frac{1}{2}$ and reverse the direction of the inequality symbol.

$$y \ge -5$$

The solution set is all real numbers that are greater than or equal to -5. The following are three equivalent ways to represent this solution set.

Set Notation	Line Graph	Interval Notation
$\{y \mid y \ge -5\}$	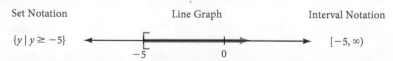	$[-5, \infty)$

EXAMPLE 7

Solve $3(2x - 4) - 7x \le -3x$.

SOLUTION   We begin by using the distributive property to separate terms. Next, simplify both sides.

$3(2x - 4) - 7x \le -3x$	Original inequality
$6x - 12 - 7x \le -3x$	Distributive property
$-x - 12 \le -3x$	$6x - 7x = (6 - 7)x = -x.$
$-12 \le -2x$	Add x to both sides.

$$-\frac{1}{2}(-12) \ge -\frac{1}{2}(-2x)$$

Multiply both sides by $-\frac{1}{2}$ and reverse the direction of the inequality symbol.

$$6 \ge x$$

This last line is equivalent to $x \le 6$. The solution set can be represented with any of the three following items.

Set Notation	Line Graph	Interval Notation
$\{x \mid x \le 6\}$		$(-\infty, 6]$

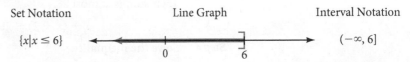

EXAMPLE 8

Solve $2x - 3y < 6$ for y.

SOLUTION   We can solve this formula for y by first adding $-2x$ to each side and then multiplying each side by $-\frac{1}{3}$. When we multiply by $-\frac{1}{3}$ we must reverse the direction of the inequality symbol. Because this is a formula, we will not graph the solution.

$2x - 3y < 6$	Original formula
$2x + (-2x) - 3y < (-2x) + 6$	Add $-2x$ to each side.
$-3y < -2x + 6$	
$-\frac{1}{3}(-3y) > -\frac{1}{3}(-2x + 6)$	Multiply each side by $-\frac{1}{3}$.
$y > \frac{2}{3}x - 2$	Distributive property

When working application problems that involve inequalities, the phrases "at least" and "at most" translate as follows:

In Words	In Symbols
x is at least 30	$x \ge 30$
x is at most 20	$x \le 20$

Our next example is similar to an example done earlier in this chapter. This time it involves an inequality instead of an equation.

> **NOTE** In Examples 6 and 7, notice that each time we multiplied both sides of the inequality by a negative number we also reversed the direction of the inequality symbol. Failure to do so would cause our graph to lie on the wrong side of the endpoint.

We can modify our Blueprint for Problem Solving to solve application problems whose solutions depend on writing and then solving inequalities.

EXAMPLE 9

The sum of two consecutive odd integers is at most 28. What are the possibilities for the first of the two integers?

SOLUTION   When we use the phrase "their sum is at most 28," we mean that their sum is less than or equal to 28.

Step 1:   Read and list.
Known items: Two consecutive odd integers. Their sum is less than or equal to 28.
Unknown items: The numbers in question.

Step 2:   **Assign a variable, and translate information.**
If we let x = the first of the two consecutive odd integers, then $x + 2$ is the next consecutive one.

Step 3:   **Reread, and write an inequality.**
Their sum is at most 28.

$$x + (x + 2) \le 28$$

Step 4:   **Solve the inequality.**

$2x + 2 \le 28$	Simplify the left side.
$2x \le 26$	Add -2 to each side.
$x \le 13$	Multiply each side by $\frac{1}{2}$.

Step 5:   **Write the answer.**
The first of the two integers must be an odd integer that is less than or equal to 13. The second of the two integers will be 2 more than whatever the first one is.

Step 6:   **Reread, and check.**
Suppose the first integer is 13. The next consecutive odd integer is 15. The sum of 15 and 13 is 28. If the first odd integer is less than 13, the sum of it and the next consecutive odd integer will be less than 28.

GETTING READY FOR CLASS

After reading through the preceding section, respond in your own words and in complete sentences.

A. State the addition property for inequalities.
B. How is the multiplication property for inequalities different from the multiplication property of equality?
C. When do we reverse the direction of an inequality symbol?
D. Under what conditions do we not change the direction of the inequality symbol when we multiply both sides of an inequality by a number?

Problem Set 2.4

Solve each of the following inequalities, and graph each solution.

1. $2x \le 3$ **2.** $5x \ge -115$ **3.** $\frac{1}{2}x > 2$ **4.** $\frac{1}{3}x > 4$

5. $-5x \le 25$ **6.** $-7x \ge 35$ **7.** $-\frac{3}{2}x > -6$ **8.** $-\frac{2}{3}x < -8$

9. $-12 \le 2x$ **10.** $-20 \ge 4x$ **11.** $-1 \ge -\frac{1}{4}x$ **12.** $-1 \le -\frac{1}{5}x$

Solve each of the following inequalities, and graph each solution.

13. $-3x + 1 > 10$ **14.** $-2x - 5 \le 15$

15. $\frac{1}{2} - \frac{m}{12} \le \frac{7}{12}$ **16.** $\frac{1}{2} - \frac{m}{10} > -\frac{1}{5}$

17. $\frac{1}{2} \ge -\frac{1}{6} - \frac{2}{9}x$ **18.** $\frac{9}{5} > -\frac{1}{5} - \frac{1}{2}x$

19. $-40 \le 30 - 20y$ **20.** $-20 > 50 - 30y$

21. $\frac{2}{3}x - 3 < 1$ **22.** $\frac{3}{4}x - 2 > 7$

23. $10 - \frac{1}{2}y \le 36$ **24.** $8 - \frac{1}{3}y \ge 20$

25. $4 - \frac{1}{2}x < \frac{2}{3}x - 5$ **26.** $5 - \frac{1}{3}x > \frac{1}{4}x + 2$

27. $0.03x - 0.4 \le 0.08x + 1.2$ **28.** $2.0 - 0.7x < 1.3 - 0.3x$

29. $3 - \frac{x}{5} < 5 - \frac{x}{4}$ **30.** $-2 + \frac{x}{3} \ge \frac{x}{2} - 5$

Simplify each side first, then solve the following inequalities. Write your answers with interval notation.

31. $2(3y + 1) \le -10$ **32.** $3(2y - 4) > 0$

33. $-(a + 1) - 4a \le 2a - 8$ **34.** $-(a - 2) - 5a \le 3a + 7$

35. $\frac{1}{3}t - \frac{1}{2}(5 - t) < 0$ **36.** $\frac{1}{4}t - \frac{1}{3}(2t - 5) < 0$

37. $-2 \le 5 - 7(2a + 3)$ **38.** $1 < 3 - 4(3a - 1)$

39. $-\frac{1}{3}(x + 5) \le -\frac{2}{9}(x - 1)$ **40.** $-\frac{1}{2}(2x + 1) \le -\frac{3}{8}(x + 2)$

41. $5(x - 2) - 7(x + 1) \le -4x + 3$ **42.** $-3(1 - 2x) - 3(x - 4) < -3 - 4x$

43. $\frac{2}{3}x - \frac{1}{3}(4x - 5) < 1$ **44.** $\frac{1}{4}x - \frac{1}{2}(3x + 1) \ge 2$

45. $20x + 9,300 > 18,000$ **46.** $20x + 4,800 > 18,000$

Solve each of the following formulas for y.

47. $3x + 2y < 6$ **48.** $-3x + 2y < 6$ **49.** $2x - 5y > 10$

50. $-2x - 5y > 5$ **51.** $-3x + 7y \le 21$ **52.** $-7x + 3y \le 21$

53. $2x - 4y \ge -4$ **54.** $4x - 2y \ge -8$

PAYING ATTENTION TO INSTRUCTIONS

The next two problems are intended to give you practice reading, and paying attention to, the instructions that accompany the problems you are working.

55. Work each problem according to the instructions given.

    a. Evaluate when $x = 0$: $-5x + 3$ **b.** Solve: $-5x + 3 = -7$

    c. Is 0 a solution to $-5x + 3 < -7$?   **d.** Solve: $-5x + 3 < -7$

56. Work each problem according to the instructions given.

    a. Evaluate when $x = 0$: $-2x - 5$ **b.** Solve: $-2x - 5 = 1$

    c. Is 0 a solution to $-2x - 5 > 1$?   **d.** Solve: $-2x - 5 > 1$

APPLYING THE CONCEPTS

57. Consecutive Integers The sum of two consecutive integers is at least 583. What are the possibilities for the first of the two integers?

58. Consecutive Integers The sum of two consecutive integers is at most 583. What are the possibilities for the first of the two integers?

59. Number Problems The sum of twice a number and 6 is less than 10. Find all solutions.

60. Number Problems Twice the difference of a number and 3 is greater than or equal to the number increased by 5. Find all solutions.

61. Number Problems The product of a number and 4 is greater than the number minus 8. Find the solution set.

62. Number Problems The quotient of a number and 5 is less than the sum of 7 and 2. Find the solution set.

63. Geometry Problems The length of a rectangle is 3 times the width. If the perimeter is to be at least 48 meters, what are the possible values for the width? (If the perimeter is at least 48 meters, then it is greater than or equal to 48 meters.)

64. Geometry Problems The length of a rectangle is 3 more than twice the width. If the perimeter is to be at least 51 meters, what are the possible values for the width? (If the perimeter is at least 51 meters, then it is greater than or equal to 51 meters.)

65. Geometry Problems The numerical values of the three sides of a triangle are given by three consecutive even integers. If the perimeter is greater than 24 inches, what are the possibilities for the shortest side?

66. Geometry Problems The numerical values of the three sides of a triangle are given by three consecutive odd integers. If the perimeter is greater than 27 inches, what are the possibilities for the shortest side?

GETTING READY FOR THE NEXT SECTION

Graph each inequality.

67. $x < 2$ **68.** $x \leq 2$ **69.** $x \geq -3$ **70.** $x > -3$

EXTENDING THE CONCEPTS

Assume that a, b, and c are positive, and solve each formula for x.

71. $ax + b < c$

72. $\dfrac{x}{a} + \dfrac{y}{b} < 1$

73. $-c < ax + b < c$

74. $-1 < \dfrac{ax + b}{c} < 1$

Assume that $0 < a < b < c < d$ and solve each formula for x.

75. $ax + b < cx + d$

76. $cx + b < ax + d$

77. $-d < \dfrac{ax + b}{c} < d$

78. $\dfrac{x}{a} + \dfrac{1}{b} < \dfrac{x}{c}$

79. $\dfrac{x}{a} + \dfrac{1}{b} < \dfrac{x}{c} + \dfrac{1}{d}$

80. $\dfrac{3}{a} - \dfrac{x}{b} \leq \dfrac{5}{c} - \dfrac{x}{d}$

iStockphoto.com/nathanaparise

2.5

OBJECTIVES

A Graph compound inequalities.

B Solve and graph compound inequalities.

Before we solve more inequalities, let's review some of the details involved with graphing more complicated inequalities.

The *union* of two sets A and B is said to be the set of all elements that are in either A or B. The word *or* is the key word in the definition. The *intersection* of two sets A and B is the set of all elements contained in both A and B, the key word here being *and*. We can use the words *and* and *or*, together with our methods of graphing inequalities, to graph some *compound inequalities*.

 Compound Inequalities

EXAMPLE 1

Graph: $\{x \mid x \leq -2 \text{ or } x > 3\}$.

SOLUTION The two inequalities connected by the word *or* are referred to as a *compound inequality*. We begin by graphing each inequality separately:

$$x \leq -2$$

$$\text{or}$$

$$x > 3$$

NOTE The square bracket indicates -2 is included in the solution set, and the parenthesis indicates 3 is not included.

Because the two inequalities are connected by the word *or*, we graph their union. That is, we graph all points on either graph:

To represent this set of numbers with interval notation, we use two intervals connected with the symbol for the union of two sets. Here is the equivalent set of numbers described with interval notation:

$$(-\infty, -2] \cup (3, \infty)$$

Next we will show two examples similar to Example 1, but with small changes to the inequalities. As you will see, these changes to the inequalities make big changes to the solutions.

EXAMPLE 2

Graph: $\{x \mid x \geq -2 \text{ or } x > 3\}$.

SOLUTION   We begin by graphing each inequality separately:

$x \geq -2$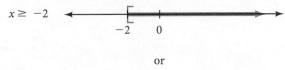

or

$x > 3$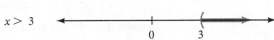

Because the two inequalities are connected by the word *or*, we graph their union. That is, we graph all points on either graph. Any point that is on either graph is on the final graph.

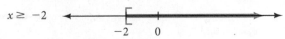

Here is the equivalent set of numbers described with interval notation:

$$[-2, \infty)$$

EXAMPLE 3

Graph: $\{x \mid x \geq -2 \text{ or } x < 3\}$.

SOLUTION   We begin by graphing each inequality separately:

$x \geq -2$

or

$x < 3$

Again the two inequalities are connected by the word *or*, so we graph their union: all points that are on either graph. All points on the number line are on one graph or the other, so the solution set is all real numbers.

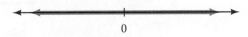

Here is the solution described with interval notation:

$$(-\infty, \infty)$$

EXAMPLE 4

Graph: $\{x \mid x > -1 \text{ and } x < 2\}$.

SOLUTION   We first graph each inequality separately:

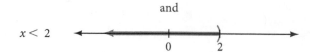

$x > -1$

and

$x < 2$

Because the two inequalities are connected by the word *and*, we graph their intersection—the part they have in common:

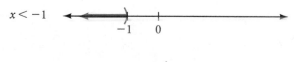

This graph corresponds to the interval $(-1, 2)$, which is called an *open interval* because neither endpoint is included in the interval.

Next we make some small changes to the inequalities in Example 4, to see how those changes affect the solution set.

EXAMPLE 5

Graph: $\{x \mid x < -1 \text{ and } x < 2\}$.

SOLUTION   We first graph each inequality separately:

$x < -1$

and

$x < 2$

Because the two inequalities are connected by the word *and*, we graph their intersection—the part they have in common, which is where they overlap.

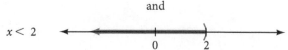

This graph corresponds to the interval $(-\infty, -1)$, which is called an *open interval* because neither endpoint is included in the interval.

EXAMPLE 6

Graph: $\{x \mid x < -1 \text{ and } x > 2\}$.

SOLUTION   We first graph each inequality separately:

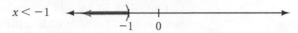

$x < -1$

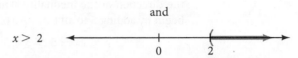

and

$x > 2$

Again, the two inequalities are connected by the word *and*, so we graph their intersection—the part they have in common. But in this case, there are no points that lie on both graphs. Therefore, there is no solution. The solution set is the empty set, $\varnothing$.

Notation   Sometimes compound inequalities that use the word *and* as the connecting word can be written in a shorter form. For example, the compound inequality $-3 \le x$ and $x \le 4$ can be written $-3 \le x \le 4$. The word *and* does not appear when an inequality is written in this form. It is implied. Inequalities of the form $-3 \le x \le 4$ are called *continued inequalities*. This new notation is useful because writing it takes fewer symbols. The graph of $-3 \le x \le 4$ is

The corresponding interval is $[-3, 4]$, which is called a *closed interval* because both endpoints are included in the interval.

ⓑ Solving Inequalities and Graphing

The table below shows the connection between set notation, interval notation, and number line graphs for a variety of continued inequalities. Again, we have included the graphs with open and closed circles for those of you who have used this type of graph previously. Remember, however, that in this book we will be using the parentheses/brackets method of graphing.

Inequality notation	Interval notation	Graph using parentheses/brackets	Graph using open and closed circles
$-3 < x < 2$	$(-3, 2)$		
$-3 \le x \le 2$	$[-3, 2]$		
$-3 \le x < 2$	$[-3, 2)$		
$-3 < x \le 2$	$(-3, 2]$		

EXAMPLE 7

Solve and graph $-3 \leq 2x - 5 \leq 3$.

SOLUTION   We can extend our properties for addition and multiplication to cover this situation. If we add a number to the middle expression, we must add the same number to the outside expressions. If we multiply the center expression by a number, we must do the same to the outside expressions, remembering to reverse the direction of the inequality symbols if we multiply by a negative number. We begin by adding 5 to all three parts of the inequality:

$$-3 \leq 2x - 5 \leq 3$$

$$2 \leq \quad 2x \quad \leq 8 \qquad \text{Add 5 to all three members.}$$

$$1 \leq \quad x \quad \leq 4 \qquad \text{Multiply through by } \tfrac{1}{2}.$$

Here are three ways to write this solution set:

Set Notation	Line Graph	Interval Notation
$\{x \mid 1 \leq x \leq 4\}$	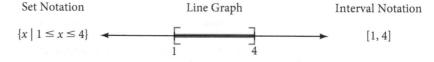	$[1, 4]$

EXAMPLE 8

Solve the compound inequality.

$$3t + 7 \leq -4 \qquad \text{or} \qquad 3t + 7 \geq 4$$

SOLUTION   We solve each half of the compound inequality separately, then we graph the solution set:

$$3t + 7 \leq -4 \qquad \text{or} \qquad 3t + 7 \geq 4$$

$$3t \leq -11 \qquad \text{or} \qquad 3t \geq -3 \qquad \text{Add } -7.$$

$$t \leq -\frac{11}{3} \qquad \text{or} \qquad t \geq -1 \qquad \text{Multiply by } \tfrac{1}{3}.$$

The solution set can be written in any of the following ways:

Set Notation	Line Graph	Interval Notation
$\{t \mid t \leq -\frac{11}{3} \text{ or } t \geq -1\}$		$(-\infty, -\frac{11}{3}] \cup [-1, \infty)$

EXAMPLE 9

A company that manufactures ink cartridges for printers finds that they can sell x cartridges each week at a price of p dollars each, according to the formula $x = 1{,}300 - 100p$. What price should they charge for each cartridge if they want to sell at least 300 cartridges a week?

SOLUTION   Because x is the number of cartridges they sell each week, an inequality that corresponds to selling at least 300 cartridges a week is

$$x \geq 300$$

Substituting $1{,}300 - 100p$ for x gives us an inequality in the variable p.

$$1{,}300 - 100p \geq 300$$

$$-100p \geq -1{,}000 \qquad \text{Add } -1{,}300 \text{ to each side.}$$

$$p \leq 10 \qquad \text{Divide each side by } -100, \text{ and reverse the direction of the inequality symbol.}$$

To sell at least 300 cartridges each week, the price per cartridge should be no more than \$10. That is, selling the cartridges for \$10 or less will produce weekly sales of 300 or more cartridges.

EXAMPLE 10

The formula $F = \frac{9}{5}C + 32$ gives the relationship between the Celsius and Fahrenheit temperature scales. If the temperature range on a certain day is 86° to 104° Fahrenheit, what is the temperature range in degrees Celsius?

SOLUTION From the given information, we can write $86 \leq F \leq 104$. However, because F is equal to $\frac{9}{5}C + 32$, we can also write:

$$86 \leq \frac{9}{5}C + 32 \leq 104$$

$$54 \leq \frac{9}{5}C \qquad \leq 72 \qquad \text{Add } -32 \text{ to each number.}$$

$$\frac{5}{9}(54) \leq \frac{5}{9}\left(\frac{9}{5}C\right) \leq \frac{5}{9}(72) \qquad \text{Multiply each number by } \frac{5}{9}.$$

$$30 \leq \quad C \quad \leq 40$$

A temperature range of 86° to 104° Fahrenheit corresponds to a temperature range of 30° to 40° Celsius.

GETTING READY FOR CLASS

After reading through the preceding section, respond in your own words and in complete sentences.

A. What is the addition property for inequalities?

B. When we use interval notation to denote a section of the real number line, when do we use parentheses ( ) and when do we use brackets [ ]?

C. Explain the difference between the multiplication property of equality and the multiplication property for inequalities.

D. When solving an inequality, when do we change the direction of the inequality symbol?

Problem Set 2.5

Graph the solution to each compound inequality and write your answer using interval notation.

1. **a.** $\{x \mid x < -3 \text{ or } x > 1\}$
   b. $\{x \mid x > -3 \text{ or } x > 1\}$
   c. $\{x \mid x > -3 \text{ or } x < 1\}$

2. **a.** $\{x \mid x \leq -1 \text{ or } x \geq 3\}$
   b. $\{x \mid x \leq -1 \text{ or } x \leq 3\}$
   c. $\{x \mid x \geq -1 \text{ or } x \leq 3\}$

3. **a.** $\{x \mid x < 1 \text{ or } x \geq 4\}$
   b. $\{x \mid x < 1 \text{ or } x \leq 4\}$
   c. $\{x \mid x > 1 \text{ or } x \leq 4\}$

4. **a.** $\{x \mid x \leq -5 \text{ or } x > -2\}$
   b. $\{x \mid x \geq -5 \text{ or } x > -2\}$
   c. $\{x \mid x \geq -5 \text{ or } x < -2\}$

5. **a.** $\{x \mid x > -6 \text{ and } x \leq 2\}$
   b. $\{x \mid x < -6 \text{ and } x \leq 2\}$
   c. $\{x \mid x > -6 \text{ and } x \geq 2\}$
   d. $\{x \mid x < -6 \text{ and } x \geq 2\}$

6. **a.** $\{x \mid x \geq 1 \text{ and } x < 3\}$
   b. $\{x \mid x \geq 1 \text{ and } x > 3\}$
   c. $\{x \mid x \leq 1 \text{ and } x < 3\}$
   d. $\{x \mid x \leq 1 \text{ and } x > 3\}$

Solve the following continued inequalities. Use both a line graph and interval notation to write each solution set.

7. $-2 \leq m - 5 \leq 2$

8. $-3 \leq m + 1 \leq 3$

9. $-60 < 20a + 20 < 60$

10. $-60 < 50a - 40 < 60$

11. $0.5 \leq 0.3a - 0.7 \leq 1.1$

12. $0.1 \leq 0.4a + 0.1 \leq 0.3$

13. $3 < \frac{1}{2}x + 5 < 6$

14. $5 < \frac{1}{4}x + 1 < 9$

15. $4 < 6 + \frac{2}{3}x < 8$

16. $3 < 7 + \frac{4}{5}x < 15$

Graph the solution sets for the following compound inequalities. Then write each solution set using interval notation.

17. $x + 5 \leq -2$ or $x + 5 \geq 2$

18. $3x + 2 < -3$ or $3x + 2 > 3$

19. $5y + 1 \leq -4$ or $5y + 1 \geq 4$

20. $7y - 5 \leq -2$ or $7y - 5 \geq 2$

21. $2x + 5 < 3x - 1$ or $x - 4 < 2x + 6$

22. $3x - 1 > 2x + 4$ or $5x - 2 > 3x + 4$

23. $3x + 1 < -8$ or $-2x + 1 \leq -3$

24. $2x - 5 \leq -1$ or $-3x - 6 < -15$

PAYING ATTENTION TO INSTRUCTIONS

The next two problems are intended to give you practice reading, and paying attention to, the instructions that accompany the problems you are working.

25. Work each problem according to the instructions given.

   a. Evaluate when $x = 0$: $-\frac{1}{2}x + 1$ **b.** Solve: $-\frac{1}{2}x + 1 = -7$

   c. Is 0 a solution to $-\frac{1}{2}x + 1 < -7$?   **d.** Solve: $-\frac{1}{2}x + 1 < -7$

26. Work each problem according to the instructions given.

   a. Evaluate when $x = 0$: $-\frac{2}{3}x - 5$ **b.** Solve: $-\frac{2}{3}x - 5 = 1$

   c. Is 0 a solution to $-\frac{2}{3}x - 5 > 1$?   **d.** Solve: $-\frac{2}{3}x - 5 > 1$

Translate each of the following phrases into an equivalent inequality statement.

27. x is greater than -2 and at most 4   **28.** x is less than 9 and at least -3

29. x is less than -4 or at least 1   **30.** x is at most 1 or more than 6

31. Art Supplies A store selling art supplies finds that they can sell x sketch pads each week at a price of p dollars each, according to the formula $x = 900 - 300p$. What price should they charge if they want to sell

  a. At least 300 pads each week?   **b.** More than 600 pads each week?

  c. Fewer than 525 pads each week?   **d.** At most 375 pads each week?

32. Temperature Range Each of the following temperature ranges is in degrees Fahrenheit. Use the formula $F = \frac{9}{5}C + 32$ to find the corresponding temperature range in degrees Celsius.

  a. 95° to 113°   **b.** 68° to 86°   **c.** $-13°$ to 14°   **d.** $-4°$ to 23°

33. Fuel Efficiency The fuel efficiency (mpg rating) for cars has been increasing steadily since 1980. The formula for a car's fuel efficiency for a given year between 1980 and 2016 is

$$E = 0.36x + 15.9$$

where E is miles per gallon and x is the number of years after 1980 (U.S. Federal Highway Administration, *Highway Statistics,* annual).

  a. In what years was the average fuel efficiency for cars less than 17 mpg?

  b. In what years was the average fuel efficiency for cars more than 20 mpg?

34. Organic Groceries The map shows a range of how many organic grocery stores are found in each state.

  a. Find the difference, d, in the number of organic stores between California and the state of Washington.

  b. If Oregon had 24 organic grocery stores, what is the difference, d, between the number of stores in Oregon and the number of stores found in Nevada?

35. Student Loan When considering how much debt to incur in student loans, you learn that it is wise to keep your student loan payment to 8% or less of your starting monthly income. Suppose you anticipate a starting annual salary of $24,000. Set up and solve an inequality that represents the amount of monthly debt for student loans that would be considered manageable.

36. Here is what the United States Geological Survey has to say about the survival rates of the ʻApapane, one of the endemic birds of Hawaii.

> *Annual survival rates based on 1,584 recaptures of 429 banded individuals 0.72 ± 0.11 for adults and 0.13 ± 0.07 for juveniles.*

Write the survival rates using inequalities. Then give the survival rates in terms of percent.

37. Survival Rates for Sea Gulls Here is part of a report concerning the survival rates of Western Gulls that appeared on the website of Cornell University.

> Survival of eggs to hatching is 70%–80%; of hatched chicks to fledglings 50%–70%; of fledglings to age of first breeding <50%.

Write the survival rates using inequalities without percent.

38. Wedding Costs For a wedding reception, you wish to rent a hall for 150 guests. Your budget, including hall rental fees and food charges, cannot exceed $6,500. Set up and solve an inequality to find the highest cost per person for food to budget for at a venue if typical hall rental costs are $2,000.

39. Wedding Costs For a wedding reception, you wish to rent a hall for 100 guests. Your budget, including hall rental fees and food charges, cannot exceed $7,000. Set up and solve an inequality to find the highest cost per person for food to budget for at a venue if typical hall rental costs are $3,000.

GETTING READY FOR THE NEXT SECTION ━━━━

Solve each equation.

40. $2a - 1 = -7$ **41.** $3x - 6 = 9$ **42.** $\dfrac{2}{3}x - 3 = 7$

43. $\dfrac{2}{3}x - 3 = -7$ **44.** $x - 5 = x - 7$ **45.** $x + 3 = x + 8$

46. $x - 5 = -x - 7$ **47.** $x + 3 = -x + 8$

2.6

 OBJECTIVE

A   Solve equations with absolute value

Amtrak's annual passenger revenue for the years 1985–1995 is modeled approximately by the formula

$$R = -60|x - 11| + 962$$

where R is the annual revenue in millions of dollars and x is the number of years after 1980 (Association of American Railroads, Washington, DC, *Railroad Facts, Statistics of Railroads of Class 1*, annual). In what year was the passenger revenue $722 million?

You may recall that the *absolute value* of x, $|x|$, is the distance between x and 0 on the number line. The absolute value of a number measures its distance from 0.

Solving Equations with Absolute Value

EXAMPLE 1

Solve for x: $|x| = 5$.

SOLUTION   Using the definition of absolute value, we can read the equation as, "The distance between x and 0 on the number line is 5." If x is 5 units from 0, then x can be 5 or -5:

If $|x| = 5$ then $x = 5$ or $x = -5$

In general, then, we can see that any equation of the form $|a| = b$ is equivalent to the equations $a = b$ or $a = -b$, as long as $b > 0$.

EXAMPLE 2

Solve $|2a - 1| = 7$.

SOLUTION   We can read this question as "$2a - 1$ is 7 units from 0 on the number line." The quantity $2a - 1$ must be equal to 7 or -7:

$$|2a - 1| = 7$$

$$2a - 1 = 7 \quad \text{or} \quad 2a - 1 = -7$$

We have transformed our absolute value equation into two equations that do not involve absolute value. We can solve each equation using the method in Section 2.1:

$$2a - 1 = 7 \quad \text{or} \quad 2a - 1 = -7$$

$$2a = 8 \qquad\qquad 2a = -6 \qquad \text{Add 1 to both sides.}$$

$$a = 4 \qquad\qquad a = -3 \qquad \text{Multiply by } \tfrac{1}{2}.$$

Our solution set is {4, −3}.

To check our solutions, we put them into the original absolute value equation:

When $a = 4$ When $a = -3$

the equation $|2a - 1| = 7$ the equation $|2a - 1| = 7$

becomes $|2(4) - 1| = 7$ becomes $|2(-3) - 1| = 7$

$$|7| = 7 \qquad\qquad |-7| = 7$$

$$7 = 7 \qquad\qquad 7 = 7$$

EXAMPLE 3

Solve $\left|\dfrac{2}{3}x - 3\right| + 5 = 12$.

SOLUTION   To use the definition of absolute value to solve this equation, we must isolate the absolute value on the left side of the equal sign. To do so, we add -5 to both sides of the equation to obtain

$$\left|\frac{2}{3}x - 3\right| = 7$$

Now that the equation is in the correct form, we can write

$$\frac{2}{3}x - 3 = 7 \quad \text{or} \quad \frac{2}{3}x - 3 = -7$$

$$\frac{2}{3}x = 10 \qquad\qquad \frac{2}{3}x = -4 \qquad \text{Add 3 to both sides.}$$

$$x = 15 \qquad\qquad x = -6 \qquad \text{Multiply by } \tfrac{3}{2}.$$

The solution set is {15, −6}.

EXAMPLE 4

Solve $|3a - 6| = -4$.

SOLUTION   The solution set is $\varnothing$ because the right side is negative but the left side cannot be negative. No matter what we try to substitute for the variable a, the quantity $|3a - 6|$ will always be positive or zero. It can never be -4.

NOTE Recall that $\varnothing$ is the symbol we use to denote the empty set. When we use it to indicate the solutions to an equation, then we are saying the equation has no solution.

NOTE $\Leftrightarrow$ means "if and only if" and "is equivalent to"

Consider the statement $|a| = |b|$. What can we say about a and b? We know they are equal in absolute value. By the definition of absolute value, they are the same distance from 0 on the number line. They must be equal to each other or opposites of each other. In symbols, we write

$$|a| = |b| \qquad \Leftrightarrow \qquad a = b \quad \text{or} \quad a = -b$$

    ↑   ↑   ↑

Equal in   Equals   or   Opposites

absolute value

EXAMPLE 5

Solve $|3a + 2| = |2a + 3|$.

SOLUTION   The quantities $3a + 2$ and $2a + 3$ have equal absolute values. They are, therefore, the same distance from 0 on the number line. They must be equals or opposites:

$$|3a + 2| = |2a + 3|$$

Equals		Opposites
$3a + 2 = 2a + 3$	or	$3a + 2 = -(2a + 3)$
$a + 2 = 3$		$3a + 2 = -2a - 3$
$a = 1$		$5a + 2 = -3$
		$5a = -5$
		$a = -1$

The solution set is $\{1, -1\}$.

   It makes no difference in the outcome of the problem if we take the opposite of the first or second expression. It is very important, once we have decided which one to take the opposite of, that we take the opposite of both its terms and not just the first term. That is, the opposite of $2a + 3$ is $-(2a + 3)$, which we can think of as $-1(2a + 3)$. Distributing the -1 across *both* terms, we have

$$-1(2a + 3) = -2a - 3$$

EXAMPLE 6

Solve $|x - 5| = |x - 7|$.

SOLUTION   As was the case in Example 5, the quantities $x - 5$ and $x - 7$ must be equal or they must be opposites, because their absolute values are equal:

Equals		Opposites
$x - 5 = x - 7$	or	$x - 5 = -(x - 7)$
$-5 = -7$		$x - 5 = -x + 7$
No solution here		$2x - 5 = 7$
		$2x = 12$
		$x = 6$

Because the first equation leads to a false statement, it will not give us a solution. (If either of the two equations were to reduce to a true statement, it would mean all real numbers would satisfy the original equation.) In this case, our only solution is $x = 6$.

GETTING READY FOR CLASS

After reading through the preceding section, respond in your own words and in complete sentences.

A. Why do some of the equations in this section have two solutions instead of one?
B. Translate $|x| = 6$ into words using the definition of absolute value.
C. Explain in words what the equation $|x - 3| = 4$ means with respect to distance on the number line.
D. When is the statement $|x| = x$ true?

Problem Set 2.6

Use the definition of absolute value to solve each of the following problems.

1. $|x| = 4$

2. $|x| = 7$

3. $2 = |a|$

4. $5 = |a|$

5. $|x| = -3$

6. $|x| = -4$

7. $|a| + 2 = 3$

8. $|a| - 5 = 2$

9. $|y| + 4 = 3$

10. $|y| + 3 = 1$

11. $|a - 4| = \dfrac{5}{3}$

12. $|a + 2| = \dfrac{7}{5}$

13. $\left|\dfrac{3}{5}a + \dfrac{1}{2}\right| = 1$

14. $\left|\dfrac{2}{7}a + \dfrac{3}{4}\right| = 1$

15. $60 = |20x - 40|$

16. $800 = |400x - 200|$

17. $|2x + 1| = -3$

18. $|2x - 5| = -7$

19. $\left|\dfrac{3}{4}x - 6\right| = 9$

20. $\left|\dfrac{4}{5}x - 5\right| = -15$

21. $\left|1 - \dfrac{1}{2}a\right| = 3$

22. $\left|2 - \dfrac{1}{3}a\right| = 10$

23. $|2x - 5| = 3$

24. $|3x + 1| = 4$

25. $|4 - 7x| = 5$

26. $|9 - 4x| = 1$

27. $\left|3 - \dfrac{2}{3}y\right| = 5$

28. $\left|-2 - \dfrac{3}{4}y\right| = 6$

Solve each equation.

29. $|3x + 4| + 1 = 7$

30. $|5x - 3| + 4 = 3$

31. $|3 - 2y| + 4 = 3$

32. $|8 - 7y| + 9 = 1$

33. $3 + |4t - 1| = 8$

34. $2 + |2t - 6| = 10$

35. $\left|9 - \dfrac{3}{5}x\right| + 6 = 12$

36. $\left|4 - \dfrac{2}{7}x\right| + 2 = 14$

37. $5 = \left|\dfrac{2}{7}x + \dfrac{4}{7}\right| - 3$

38. $7 = \left|\dfrac{3}{5}x + \dfrac{1}{5}\right| + 2$

39. $2 = -8 + \left|4 - \dfrac{1}{2}y\right|$

40. $1 = -3 + \left|2 - \dfrac{1}{4}y\right|$

41. $|3(x + 1)| - 4 = -1$

42. $|2(2x + 3)| - 5 = -1$

43. $|1 + 3(2x - 1)| = 5$

44. $|3 + 4(3x + 1)| = 7$

45. $3 = -2 + \left|5 - \dfrac{2}{3}a\right|$

46. $4 = -1 + \left|6 - \dfrac{4}{5}a\right|$

47. $6 = |7(k + 3) - 4|$

48. $5 = |6(k - 2) + 1|$

Solve the following equations.

49. $|3a + 1| = |2a - 4|$

50. $|5a + 2| = |4a + 7|$

51. $\left|x - \dfrac{1}{3}\right| = \left|\dfrac{1}{2}x + \dfrac{1}{6}\right|$

52. $\left|\dfrac{1}{10}x - \dfrac{1}{2}\right| = \left|\dfrac{1}{5}x + \dfrac{1}{10}\right|$

53. $|y - 2| = |y + 3|$

54. $|y - 5| = |y - 4|$

55. $|3x - 1| = |3x + 1|$

56. $|5x - 8| = |5x + 8|$

57. $|0.03 - 0.01x| = |0.04 + 0.05x|$

58. $|0.07 - 0.01x| = |0.08 - 0.02x|$

59. $|x - 2| = |2 - x|$

60. $|x - 4| = |4 - x|$

61. $\left|\dfrac{x}{5} - 1\right| = \left|1 - \dfrac{x}{5}\right|$

62. $\left|\dfrac{x}{3} - 1\right| = \left|1 - \dfrac{x}{3}\right|$

63. $\left|\dfrac{2}{3}b - \dfrac{1}{4}\right| = \left|\dfrac{1}{6}b + \dfrac{1}{2}\right|$ **64.** $\left|-\dfrac{1}{4}x + 1\right| = \left|\dfrac{1}{2}x - \dfrac{1}{3}\right|$

65. $|0.1a - 0.04| = |0.3a + 0.08|$ **66.** $|-0.4a + 0.6| = |1.3 - 0.2a|$

67. Work each problem according to the instructions given.

    a. Solve: $4x - 5 = 0$ **b.** Solve: $|4x - 5| = 0$

    c. Solve: $4x - 5 = 3$ **d.** Solve: $|4x - 5| = 3$

    e. Solve: $|4x - 5| = |2x + 3|$

68. Work each problem according to the instructions given.

    a. Solve: $3x + 6 = 0$ **b.** Solve: $|3x + 6| = 0$

    c. Solve: $3x + 6 = 4$ **d.** Solve: $|3x + 6| = 4$

    e. Solve: $|3x + 6| = |7x + 4|$

APPLYING THE CONCEPTS

69. Amtrak Amtrak's annual passenger revenue for the years 1985–1995 is modeled approximately by the formula

$$R = -60|x - 11| + 962$$

where R is the annual revenue in millions of dollars and x is the number of years after 1980 (Association of American Railroads, Washington, DC, *Railroad Facts, Statistics of Railroads of Class 1*, annual). In what year was the passenger revenue $722 million?

70. Corporate Profits The corporate profits for various U.S. industries vary from year to year. An approximate model for profits of U.S. "communications companies" during a given year between 1990 and 1997 is given by

$$P = -3400|x - 5.5| + 36000$$

where P is the annual profits (in millions of dollars) and x is the number of years after 1990 (U.S. Bureau of Economic Analysis, Income and Product Accounts of the U.S. (1929–1994), *Survey of Current Business*, September 1998). Use the model to determine the years in which profits of "communication companies" were $31.5 billion ($31,500 million).

GETTING READY FOR THE NEXT SECTION

Solve each inequality. Do not graph the solution set.

71. $2x - 5 < 3$ **72.** $-3 < 2x - 5$ **73.** $-4 \le 3a + 7$

74. $3a + 2 \le 4$ **75.** $4t - 3 \le -9$ **76.** $4t - 3 \ge 9$

EXTENDING THE CONCEPTS

Solve each formula for x. (Assume a, b, and c are positive.)

77. $|x - a| = b$ **78.** $|x + a| - b = 0$

79. $|ax + b| = c$ **80.** $|ax - b| - c = 0$

81. $\left|\dfrac{x}{a} + \dfrac{y}{b}\right| = 1$ **82.** $\left|\dfrac{x}{a} + \dfrac{y}{b}\right| = c$

OBJECTIVE

A Solve inequalities with absolute value and graph the solution.

©iStockphoto.com/ArisSu

2.7

In a student survey conducted by the University of Minnesota, it was found that 30% of students were solely responsible for their finances. The survey was reported to have a margin of error plus or minus 3.74%. This means that the difference between the sample estimate of 30% and the actual percent of students who are responsible for their own finances is most likely less than 3.74%. We can write this as an inequality:

$$|x - 0.30| \le 0.0374$$

where x represents the true percent of students who are responsible for their own finances.

A Solving Inequalities with Absolute Value

In this section, we will apply the definition of absolute value to solve inequalities involving absolute value. Again, the absolute value of x, which is denoted x represents the distance that x is from 0 on the number line. We will begin by considering three absolute value expressions and their verbal translations:

Expression	In Words
$|x| = 7$	x is exactly 7 units from 0 on the number line.
$|a| < 5$	a is less than 5 units from 0 on the number line.
$|y| \ge 4$	y is greater than or equal to 4 units from 0 on the number line.

Once we have translated the expression into words, we can use the translation to graph the original equation or inequality. The graph is then used to write a final equation or inequality that does not involve absolute value.

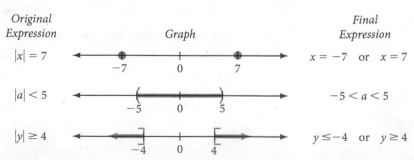

Original Expression	Graph	Final Expression
$|x| = 7$		$x = -7$ or $x = 7$
$|a| < 5$		$-5 < a < 5$
$|y| \ge 4$		$y \le -4$ or $y \ge 4$

Although we will not always write out the verbal translation of an absolute value inequality, it is important that we understand the translation. Our second expression, $|a| < 5$, means a is within 5 units of 0 on the number line. The graph of this relationship is

which can be written with the following continued inequality:

$$-5 < a < 5$$

We can follow this same kind of reasoning to solve more complicated absolute value inequalities.

EXAMPLE 1

Graph the solution set: $|2x - 5| < 3$

SOLUTION   The absolute value of $2x - 5$ is the distance that $2x - 5$ is from 0 on the number line. We can translate the inequality as "$2x - 5$ is less than 3 units from 0 on the number line." That is, $2x - 5$ must appear between -3 and 3 on the number line.

A picture of this relationship is

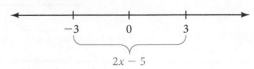

Using the picture, we can write an inequality without absolute value that describes the situation:

$$-3 < 2x - 5 < 3$$

Next, we solve the continued inequality by first adding 5 to all three members and then multiplying all three by $\frac{1}{2}$.

$$-3 < 2x - 5 < 3$$
$$\quad 2 < 2x \quad\;\; < 8 \qquad \text{Add 5 to all three expressions.}$$
$$\quad 1 < x \quad\;\;\; < 4 \qquad \text{Multiply each expression by } \frac{1}{2}.$$

The graph of the solution set is

We can see from the solution that for the absolute value of $2x - 5$ to be within 3 units of 0 on the number line, x must be between 1 and 4.

EXAMPLE 2

Solve and graph $|3a + 7| \leq 4$.

SOLUTION   We can read the inequality as "The distance between $3a + 7$ and 0 is less than or equal to 4." Or, "$3a + 7$ is within 4 units of 0 on the number line." This relationship can be written without absolute value as

$$-4 \leq 3a + 7 \leq 4$$

Solving as usual, we have

$$-4 \leq 3a + 7 \leq 4$$

$$-11 \leq 3a \quad \leq -3 \qquad \text{Add } -7 \text{ to all three members.}$$

$$-\frac{11}{3} \leq \quad a \quad \leq -1 \qquad \text{Multiply each expression by } \tfrac{1}{3}.$$

We can see from Examples 1 and 2 that to solve an inequality involving absolute value, we must be able to write an equivalent expression that does not involve absolute value.

EXAMPLE 3

Solve $|x - 3| > 5$, and graph the solution.

SOLUTION   We interpret the absolute value inequality to mean that $x - 3$ is more than 5 units from 0 on the number line. The quantity $x - 3$ must be either above $+5$ or below -5. Here is a picture of the relationship:

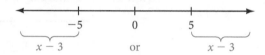

An inequality without absolute value that also describes this situation is

$$x - 3 < -5 \quad \text{or} \quad x - 3 > 5$$

Adding 3 to both sides of each inequality, we have

$$x < -2 \quad \text{or} \quad x > 8$$

Here are three ways to write our result.

Set Notation

$\{x \mid x < -2 \text{ or } x > 8\}$

Interval Notation

$(-\infty, -2) \cup (8, \infty)$

EXAMPLE 4

Graph the solution set: $|4t - 3| \geq 9$.

SOLUTION   The quantity $4t - 3$ is greater than or equal to 9 units from 0. It must be either above $+9$ or below -9.

$$4t - 3 \leq -9 \qquad \text{or} \qquad 4t - 3 \geq 9$$

$$4t \leq -6 \qquad\qquad 4t \geq 12 \qquad \text{Add 3.}$$

$$t \leq -\frac{6}{4} \qquad\qquad t \geq \frac{12}{4} \qquad \text{Multiply by } \tfrac{1}{4}.$$

$$t \leq -\frac{3}{2} \qquad\qquad t \geq 3$$

We can use the results of our first few examples and the material in the previous section to summarize the information we have related to absolute value equations and inequalities.

Rewriting Absolute Value Equations and Inequalities

If c is a positive real number, then each of the following statements on the left is equivalent to the corresponding statement on the right.

With Absolute Value	Without Absolute Value
$\lvert x \rvert = c$	$x = -c \quad \text{or} \quad x = c$
$\lvert x \rvert < c$	$-c < x < c$
$\lvert x \rvert > c$	$x < -c \quad \text{or} \quad x > c$
$\lvert ax + b \rvert = c$	$ax + b = -c \quad \text{or} \quad ax + b = c$
$\lvert ax + b \rvert < c$	$-c < ax + b < c$
$\lvert ax + b \rvert > c$	$ax + b < -c \quad \text{or} \quad ax + b > c$

EXAMPLE 5

Solve and graph $\lvert 2x + 3 \rvert + 4 < 9$.

SOLUTION Before we can apply the method of solution we used in the previous examples, we must isolate the absolute value on one side of the inequality. To do so, we add -4 to each side.

$$\lvert 2x + 3 \rvert + 4 < 9$$
$$\lvert 2x + 3 \rvert + 4 + (-4) < 9 + (-4)$$
$$\lvert 2x + 3 \rvert < 5$$

From this last line, we know that $2x + 3$ must be between -5 and $+5$.

$$-5 < 2x + 3 < 5$$
$$-8 < \quad 2x \quad < 2 \qquad \text{Add } -3x \text{ to each expression.}$$
$$-4 < \quad x \quad < 1 \qquad \text{Multiply each expression by } \tfrac{1}{2}.$$

Here are three equivalent ways to write our solution.

Set Notation $\qquad$ Interval Notation
$\{x \mid -4 < x < 1\}$ $\qquad\qquad\qquad\qquad$ $(-4, 1)$

EXAMPLE 6

Solve and graph $\lvert 4 - 2t \rvert > 2$.

SOLUTION The inequality indicates that $4 - 2t$ is less than -2 or greater than $+2$. Writing this without absolute value symbols, we have

$$4 - 2t < -2 \quad \text{or} \quad 4 - 2t > 2$$

To solve these inequalities, we begin by adding -4 to each side.

$$4 + (-4) - 2t < -2 + (-4) \qquad \text{or} \qquad 4 + (-4) - 2t > 2 + (-4)$$
$$-2t < -6 \qquad\qquad\qquad\qquad\qquad -2t > -2$$

Next we must multiply both sides of each inequality by $-\frac{1}{2}$. When we do so, we must also reverse the direction of each inequality symbol.

$$-2t < -6 \qquad \text{or} \qquad -2t > -2$$

$$-\frac{1}{2}(-2t) > -\frac{1}{2}(-6) \qquad \text{or} \qquad -\frac{1}{2}(-2t) < -\frac{1}{2}(-2)$$

$$t > 3 \qquad\qquad\qquad t < 1$$

Although in situations like this we are used to seeing the "less than" symbol written first, the meaning of the solution is clear. We want to graph all real numbers that are either greater than 3 or less than 1. Here is the graph.

Because absolute value always results in a nonnegative quantity, we sometimes come across special solution sets when a negative number appears on the right side of an absolute value inequality.

EXAMPLE 7

Solve $|7y - 1| < -2$.

SOLUTION   The *left* side is never negative because it is an absolute value. The *right* side is negative. We have a positive quantity less than a negative quantity, which is impossible. The solution set is the empty set, $\varnothing$. There is no real number to substitute for y to make this inequality a true statement.

EXAMPLE 8

Solve $|6x + 2| > -5$.

SOLUTION   This is the opposite case from that in Example 7. No matter what real number we use for x on the *left* side, the result will always be positive, or zero. The *right* side is negative. We have a positive quantity (or zero) greater than a negative quantity. Every real number we choose for x gives us a true statement. The solution set is the set of all real numbers.

GETTING READY FOR CLASS

After reading through the preceding section, respond in your own words and in complete sentences.

A. Write an inequality containing absolute value, the solution to which is all the numbers between -5 and 5 on the number line.

B. Translate $x \geq |3|$ into words using the definition of absolute value.

C. Explain in words what the inequality $|x - 5| < 2$ means with respect to distance on the number line.

D. Why is there no solution to the inequality $|2x - 3| < 0$?

Problem Set 2.7

Solve each of the following inequalities using the definition of absolute value. Graph the solution set in each case.

1. $|x| < 3$ **2.** $|x| \leq 7$ **3.** $|x| \geq 2$ **4.** $|x| > 4$

5. $|x| + 2 < 5$ **6.** $|x| - 3 < -1$ **7.** $|t| - 3 > 4$ **8.** $|t| + 5 > 8$

9. $|y| < -5$ **10.** $|y| > -3$ **11.** $|x| \geq -2$ **12.** $|x| \leq -4$

13. $|x - 3| < 7$ **14.** $|x + 4| < 2$ **15.** $|a + 5| \geq 4$ **16.** $|a - 6| \geq 3$

Solve each inequality and graph the solution set.

17. $|a - 1| < -3$ **18.** $|a + 2| \geq -5$ **19.** $|2x - 4| < 6$

20. $|2x + 6| < 2$ **21.** $|3y + 9| \geq 6$ **22.** $|5y - 1| \geq 4$

23. $|2k + 3| \geq 7$ **24.** $|2k - 5| \geq 3$ **25.** $|x - 3| + 2 < 6$

26. $|x + 4| - 3 < -1$ **27.** $|2a + 1| + 4 \geq 7$ **28.** $|2a - 6| - 1 \geq 2$

29. $|3x + 5| - 8 < 5$ **30.** $|6x - 1| - 4 \leq 2$

Solve each inequality and write your answer using interval notation. Keep in mind that if you multiply or divide both sides of an inequality by a negative number, you must reverse the sense of the inequality.

31. $|x - 3| \leq 5$ **32.** $|a + 4| < 6$ **33.** $|3y + 1| < 5$

34. $|2x - 5| \leq 3$ **35.** $|a + 4| \geq 1$ **36.** $|y - 3| > 6$

37. $|2x + 5| > 2$ **38.** $|-3x + 1| \geq 7$ **39.** $|-5x + 3| \leq 8$

40. $|-3x + 4| \leq 7$ **41.** $|-3x + 7| < 2$ **42.** $|-4x + 2| < 6$

Solve each inequality and graph the solution set.

43. $|5 - x| > 3$ **44.** $|7 - x| > 2$ **45.** $\left|3 - \dfrac{2}{3}x\right| \geq 5$

46. $\left|3 - \dfrac{3}{4}x\right| \geq 9$ **47.** $\left|2 - \dfrac{1}{2}x\right| > 1$ **48.** $\left|3 - \dfrac{1}{3}x\right| > 1$

Solve each inequality.

49. $|x - 1| < 0.01$ **50.** $|x + 1| < 0.01$ **51.** $|2x + 1| \geq \dfrac{1}{5}$

52. $|2x - 1| \geq \dfrac{1}{8}$ **53.** $|3x - 2| \leq \dfrac{1}{3}$ **54.** $|2x + 5| < \dfrac{1}{2}$

55. $\left|\dfrac{3x + 1}{2}\right| > \dfrac{1}{2}$ **56.** $\left|\dfrac{2x - 5}{3}\right| \geq \dfrac{1}{6}$ **57.** $\left|\dfrac{4 - 3x}{2}\right| \geq 1$

58. $\left|\dfrac{2x - 3}{4}\right| < 0.35$ **59.** $\left|\dfrac{3x - 2}{5}\right| \leq \dfrac{1}{2}$ **60.** $\left|\dfrac{4x - 3}{2}\right| \leq \dfrac{1}{3}$

61. $\left|2x - \dfrac{1}{5}\right| < 0.3$ **62.** $\left|3x - \dfrac{3}{5}\right| < 0.2$

63. Write the continued inequality $-4 \leq x \leq 4$ as a single inequality involving absolute value.

64. Write the continued inequality $-8 \leq x \leq 8$ as a single inequality involving absolute value.

65. Write $-1 \leq x - 5 \leq 1$ as a single inequality involving absolute value.

66. Write $-3 \leq x + 2 \leq 3$ as a single inequality involving absolute value.

67. Work each problem according to the instructions given.

    a. Evaluate when $x = 0$: $|5x + 3|$ **b.** Solve: $|5x + 3| = 7$

    c. Is 0 a solution to $|5x + 3| > 7$?   **d.** Solve: $|5x + 3| > 7$

68. Work each problem according to the instructions given.

    a. Evaluate when $x = 0$: $|-2x - 5|$ **b.** Solve: $|-2x - 5| = 1$

    c. Is 0 a solution to $|-2x - 5| > 1$?   **d.** Solve: $|-2x - 5| > 1$

69. Speed Limits The interstate speed limit for cars is 75 miles per hour in Nebraska, Nevada, New Mexico, Oklahoma, South Dakota, Utah, and Wyoming and is the highest in the United States. To discourage passing, minimum speeds are also posted, so that the difference between the fastest and slowest moving traffic is no more than 20 miles per hour. Write an absolute value inequality that describes the relationship between the minimum allowable speed and a maximum speed of 75 miles per hour.

70. Wavelengths of Light When white light from the Sun passes through a prism, it is broken down into bands of light that form colors. The wavelength, v, (in nanometers) of some common colors are:

Blue: $424 < v < 491$
Green: $491 < v < 575$
Yellow: $575 < v < 585$
Orange: $585 < v < 647$
Red: $647 < v < 700$

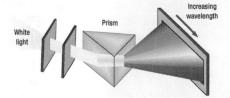

When a fireworks display made of copper is burned, it lets out light with wavelengths, v, that satisfy the relationship $|v - 455| < 23$. Write this inequality without absolute values, find the range of possible values for v, and then using the preceding list of wavelengths, determine the color of that copper fireworks display.

71. Pressure of a Football The official pressure p, in pounds per square inch (psi), of an NFL football must conform to a specific range of values according to the relationship $|p - 13| \le 0.5$. Determine the range of possible values for p.

MAINTAINING YOUR SKILLS

Simplify each expression. Assume all variables represent nonzero real numbers, and write your answer with positive exponents only.

72. 3^{-2} **73.** $\dfrac{x^6}{x^{-4}}$ **74.** $\dfrac{15x^3y^8}{5xy^{10}}$

75. $(2a^{-3}b^4)^2$ **76.** $\dfrac{(3x^{-3}y^5)^{-2}}{(9xy^{-2})^{-1}}$ **77.** $(3x^4y)^2(5x^3y^4)^3$

Write each number in scientific notation.

78. 54,000   **79.** 0.0359

Write each number in expanded form.

80. 6.44×10^3 **81.** 2.5×10^{-2}

Simplify each expression as much as possible. Write all answers in scientific notation.

82. $(3 \times 10^8)(4 \times 10^{-5})$

83. $\dfrac{8 \times 10^5}{2 \times 10^{-8}}$

EXTENDING THE CONCEPTS

Assume a, b, and c are positive, and solve each formula for x.

84. $|x - a| < b$

85. $|x + a| > b$

86. $|ax - b| > c$

87. $|ax - b| < c$

88. $|ax + b| \le c$

89. $|ax + b| \ge c$

90. $\left| \dfrac{x}{a} + \dfrac{1}{b} \right| > c$

91. $\left| \dfrac{x}{a} + \dfrac{1}{b} \right| < c$

92. $|(x + a)^2 - x^2| < 3a^2$

93. $|(x - a)^2 - x^2| < 2a^2$

SPOTLIGHT ON SUCCESS

Lauren | student instructor

There are a lot of word problems in algebra and many of them involve topics that I don't know much about. I am better off solving these problems if I know something about the subject. So, I try to find something I can relate to. For instance, an example may involve the amount of fuel used by a pilot in a jet airplane engine. In my mind, I'd change the subject to something more familiar, like the mileage I'd be getting in my car and the amount spent on fuel, driving from my hometown to my college. Changing these problems to more familiar topics makes math much more interesting and gives me a better chance of getting the problem right. It also helps me to understand how greatly math affects and influences me in my everyday life. We really do use math more than we would like to admit—budgeting our income, purchasing gasoline, planning a day of shopping with friends—almost everything we do is related to math. So the best advice I can give with word problems is to learn how to associate the problem with something familiar to you.

You should know that I have always enjoyed math. I like working out problems and love the challenges of solving equations like individual puzzles. Although there are more interesting subjects to me, and I don't plan on pursuing a career in math or teaching, I do think it's an important subject that will help you in any profession.

EXAMPLES

1. We can solve

$$x + 3 = 5$$

by adding -3 to both sides:

$$x + 3 + (-3) = 5 + (-3)$$
$$x = 2$$

Addition Property of Equality [2.1]

For algebraic expressions A, B, and C,

if $A = B$

then $A + C = B + C$

This property states that we can add the same quantity to both sides of an equation without changing the solution set.

Multiplication Property of Equality [2.1]

2. We can solve $3x = 12$ by multiplying both sides by $\frac{1}{3}$.

$$3x = 12$$
$$\frac{1}{3}(3x) = \frac{1}{3}(12)$$
$$x = 4$$

For algebraic expressions A, B, and C,

if $A = B$

then $AC = BC \, (C \neq 0)$

Multiplying both sides of an equation by the same nonzero quantity never changes the solution set.

Strategy for Solving Linear Equations in One Variable [2.1]

3. Solve: $3(2x - 1) = 9$.

$$3(2x - 1) = 9$$
$$6x - 3 = 9$$
$$6x - 3 + 3 = 9 + 3$$
$$6x = 12$$
$$\frac{1}{6}(6x) = \frac{1}{6}(12)$$
$$x = 2$$

Step 1: **a.** Use the distributive property to separate terms, if necessary.

b. If fractions are present, consider multiplying both sides by the LCD to eliminate the fractions. If decimals are present, consider multiplying both sides by a power of 10 to clear the equation of decimals.

c. Combine similar terms on each side of the equation.

Step 2: Use the addition property of equality to get all variable terms on one side of the equation and all constant terms on the other side. A variable term is a term that contains the variable (for example, $5x$). A constant term is a term that does not contain the variable (the number 3, for example).

Step 3: Use the multiplication property of equality to get the variable by itself on one side of the equation.

Step 4: Check your solution in the original equation to be sure that you have not made a mistake in the solution process.

Formulas [2.2]

4. Solve for w:

$$P = 2l + 2w$$
$$P - 2l = 2w$$
$$\frac{P - 2l}{2} = w$$

A *formula* in algebra is an equation involving more than one variable. To solve a formula for one of its variables, simply isolate that variable on one side of the equation.

Blueprint for Problem Solving [2.3]

5. The perimeter of a rectangle is 32 inches. If the length is 3 times the width, find the dimensions.

    Step 1: This step is done mentally.

    Step 2: Let x = the width. Then the length is $3x$.

    Step 3: The perimeter is 32; therefore

$$2x + 2(3x) = 32$$

    Step 4:
$$8x = 32$$
$$x = 4$$

    Step 5: The width is 4 inches. The length is $3(4) = 12$ inches.

    Step 6: The perimeter is $2(4) + 2(12)$, which is 32. The length is 3 times the width.

Step 1:   *Read* the problem, and then mentally *list* the items that are known and the items that are unknown.

Step 2:   *Assign a variable* to one of the unknown items. (In most cases this will amount to letting x = the item that is asked for in the problem.) Then *translate* the other *information* in the problem to expressions involving the variable.

Step 3:   *Reread* the problem, and then *write an equation,* using the items and variables listed in steps 1 and 2, that describes the situation.

Step 4:   *Solve the equation* found in step 3.

Step 5:   *Write your answer* using a complete sentence.

Step 6:   *Reread* the problem, and *check* your solution with the original words in the problem.

Addition Property for Inequalities [2.4]

6. Adding 5 to both sides of the inequality $x - 5 < -2$ gives

$$x - 5 + 5 < -2 + 5$$
$$x < 3$$

For expressions A, B, and C,

    if $A < B$

    then $A + C < B + C$

Adding the same quantity to both sides of an inequality never changes the solution set.

Multiplication Property for Inequalities [2.4]

7. Multiplying both sides of $-2x \geq 6$ by $-\frac{1}{2}$ gives

$$-2x \geq 6$$
$$-\frac{1}{2}(-2x) \leq -\frac{1}{2}(6)$$
$$x \leq -3$$

For expressions A, B, and C,

    if $A < B$

    then $AC < BC$ if $C > 0$ (C is positive)

    or $AC > BC$ if $C < 0$ (C is negative)

We can multiply both sides of an inequality by the same nonzero number without changing the solution set as long as each time we multiply by a negative number we also reverse the direction of the inequality symbol.

Strategy for Solving Linear Inequalities in One Variable [2.4]

8. Solve $3(x - 4) \geq -2$.

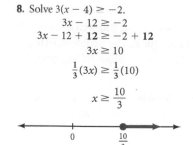

Step 1a: Use the distributive property to separate terms, if necessary.

Step 1b: If fractions are present, consider multiplying both sides by the LCD to eliminate the fractions. If decimals are present, consider multiplying both sides by a power of 10 to clear the inequality of decimals.

Step 1c: Combine similar terms on each side of the inequality.

Step 2: Use the addition property for inequalities to get all variable terms on one side of the inequality and all constant terms on the other side.

Step 3: Use the multiplication property for inequalities to get x by itself on one side of the inequality.

Step 4: Graph the solution set.

Compound Inequalities [2.5]

9. Graph $x < -3$ or $x > 1$

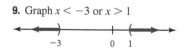

Graph $-2 \leq x \leq 3$

Two inequalities connected by the word *and* or *or* form a compound inequality. If the connecting word is *or*, we graph all points that are on either graph. If the connecting word is *and*, we graph only those points that are common to both graphs. The inequality $-2 \leq x \leq 3$ is equivalent to the compound inequality $-2 \leq x$ and $x \leq 3$.

Absolute Value Equations [2.6]

10. To solve

$$|2x - 1| + 2 = 7$$

we first isolate the absolute value on the left side by adding -2 to each side to obtain

$$|2x - 1| = 5$$
$$2x - 1 = 5 \quad \text{or} \quad 2x - 1 = -5$$
$$2x = 6 \quad \text{or} \quad 2x = -4$$
$$x = 3 \quad \text{or} \quad x = -2$$

To solve an equation that involves absolute value, we isolate the absolute value on one side of the equation and then rewrite the absolute value equation as two separate equations that do not involve absolute value. In general, if b is a positive real number, then

$$|a| = b \quad \text{is equivalent to} \quad a = b \quad \text{or} \quad a = -b$$

Absolute Value Inequalities [2.7]

11. To solve

$$|x - 3| + 2 < 6$$

we first add -2 to both sides to obtain

$$|x - 3| < 4$$

which is equivalent to

$$-4 < x - 3 < 4$$
$$-1 < x < 7$$

To solve an inequality that involves absolute value, we first isolate the absolute value on the left side of the inequality symbol. Then we rewrite the absolute value inequality as an equivalent continued or compound inequality that does not contain absolute value symbols. In general, if b is a positive real number, then

$$|a| < b \quad \text{is equivalent to} \quad -b < a < b$$

and

$$|a| > b \quad \text{is equivalent to} \quad a < -b \quad \text{or} \quad a > b$$

COMMON MISTAKES

A very common mistake in solving inequalities is to forget to reverse the direction of the inequality symbol when multiplying both sides by a negative number. When this mistake occurs, the graph of the solution set is always drawn on the wrong side of the endpoint.

Chapter 2 Review

Determine whether each number is a solution to the given equation. [2.1]

1. $5x + 1 = -6; -1$

2. $2x + 1 = 3x - 4; 5$

3. $\dfrac{x}{3} - 1 = 1; -6$

4. $-4(x + 3) + 2 = 5(x - 1) + 4; -1$

Solve the following equations. [2.1]

5. $x - 4 = 7$

6. $3y = -5$

7. $5 - 2x = 3x + 1$

8. $\dfrac{1}{2}x + \dfrac{1}{4} = \dfrac{1}{3}x + \dfrac{5}{4}$

9. $a^2 - a - 6 = 0$

10. $2x^2 - 5x = 12$

11. $10y^2 = 3y + 4$

12. $9x^2 - 25 = 0$

13. $81a^2 = 1$

14. $0.08x + 0.07(900 - x) = 67$

15. $(x - 2)(x - 3) = 2$

16. $x^3 + 4x^2 - 9x - 36 = 0$

17. $9x^3 + 18x^2 - 4x - 8 = 0$

18. $2(x + 2) - 4(2x - 3) = 4(2x - 1)$

19. $-0.05x - 0.08 = 0.03 - 0.04(2x - 1)$

Use the formula $6x - 9y = 15$ to find y when [2.2]

20. $x = 3$

21. $x = -2$

Use the formula $-5x + 7y = 35$ to find x when [2.2]

22. $y = 0$

23. $y = -5$

Solve for the indicated variable. [2.2]

24. $5x - 10y = 20$ for y

25. $ax + by = cx + dy$ for x

26. $d = (r + c)t$ for r

27. $A = \dfrac{1}{2}bh$ for b

28. $F = \dfrac{9}{5}C + 32$ for C

29. $h = vt - 16t^2$ for v

30. $-2x - 7y = 14$ for y

31. $A = P + Prt$ for P

Solve for y. [2.2]

32. $5x - y = 5$

33. $6x - 3y = 15$

34. $y - 2 = \dfrac{2}{3}(x + 6)$

35. $y - 5 = \dfrac{1}{4}(12x + 8)$

Solve each geometry problem. [2.3]

36. Find the dimensions of the rectangle below if its length is twice its width and the perimeter is 102 inches.

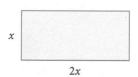

37. The perimeter of the triangle below is 24 meters. Find the value of x.

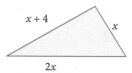

38. Find the measures of the two supplementary angles.

39. Find the measures of the two acute angles in the given right triangle.

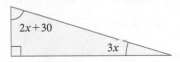

Solve each of the following. [2.3]

40. Garden A garden is in the shape of a square with a perimeter of 42 feet. The garden is surrounded by two fences. One fence is around the perimeter of the garden, whereas the second fence is 3 feet out from the first fence, as Figure 10 indicates. If the material used to build the two fences is $1.28 per foot, what was the total cost of the fences?

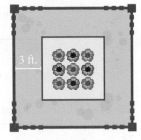

FIGURE 10

41. Fat Content in Milk According to the carton, this milk contains 75% less fat than whole milk. The nutrition label on the other side of the carton states that one serving of this milk contains 3.5 grams of fat. How many grams of fat are in an equivalent serving of whole milk?

42. Sales Tax At the beginning of the day, the cash register at a coffee shop contains $75. At the end of the day, it contains $1,021.40. If the sales tax rate is 8.2%, how much of the total is sales tax?

43. Investing Suppose you invest a certain amount of money in an account that pays 9% interest annually, and $2,000 more than that in an account that pays 11% annually. How much money do you have in each account if the total interest for a year is $320?

Graph the solution set for the following inequalities. [2.4]

44. $5x \geq -115$

45. $-\frac{2}{3}x < -8$

46. $\frac{1}{2} - \frac{m}{10} > -\frac{1}{5}$

47. $\frac{9}{5} > -\frac{1}{5} - \frac{1}{2}x$

48. $2.0 - 0.7x < 1.3 - 0.3x$

49. $-2 + \frac{x}{3} \geq \frac{x}{2} - 5$

Simplify each side first, then solve the following inequalities. Write your answers with interval notation. [2.4]

50. $-(a - 2) - 5a \leq 3a + 7$ **51.** $\frac{1}{4}t - \frac{1}{3}(2t - 5) < 0$

52. $-3(1 - 2x) - 3(x - 4) < -3 - 4x$

53. $0.08x + 0.05(900 - x) \geq 60$

Solve the following inequalities. Write the solution set using interval notation, then graph the solution set. [2.4]

54. $-3x < 12$ **55.** $-6 - \frac{4}{3}x \geq 2$

56. $1.5x - 2.6 \geq 0.5x + 1.4$ **57.** $3(5x - 6) < 6(3x + 5)$

If $A = \{1, 2, 3, 4\}$, $B = \{-2, 0, 2, 4\}$, and $C = \{0, 3, 5\}$, find each of the following. [2.5]

58. $A \cap C$

59. $\{x \mid x \in A$ and $x > 2\}$

For the following problems, let $A = \{0, 2, 4, 6\}$, $B = \{1, 2, 3, 4, 5\}$, and $C = \{1, 3, 5, 7\}$.

60. $C \cup (A \cap B)$

61. $\{x \mid x \in C$ and $x \neq 5\}$

62. $A \cup C$

63. $\{x \mid x \in B$ and $x > 3\}$

Draw a Venn diagram for the following sets. [2.5]

64. $A = \{2, 3, 4\}$
$B = \{4, 5, 6\}$

Use interval notation to describe the graph. [2.5]

65.

$$\longleftrightarrow \quad \overset{0}{\mid} \qquad \overset{7}{\mid} \quad \longrightarrow$$

Solve the following compound inequalities. [2.5]

66. $-2 \le \dfrac{2}{3}x - 4 \le 8$

67. $x - 5 > 3$ or $2x - 7 < 3$

Solve the following continued inequalities. Use both a line graph and interval notation to write each solution set. [2.5]

68. $0.1 \le 0.4a + 0.1 \le 0.3$

69. $-1.25 \le \dfrac{2x + 3}{4} \le .75$

Graph the solution sets for the following compound inequalities. Then write each solution set using interval notation. [2.5]

70. $3x - 1 > 2x + 4$ or $5x - 2 < 3x + 4$

$$\longleftarrow \hspace{5cm} \longrightarrow$$

71. $2x - 5 \le -1$ or $-3x - 6 < -15$

$$\longleftarrow \hspace{5cm} \longrightarrow$$

Solve the following equations. [2.6]

72. $\left| \dfrac{1}{8}x - 4 \right| = \dfrac{1}{2}$

73. $\left| \dfrac{3}{2}a + 7 \right| = 4$

74. $|4y - 5| - 3 = 10$

75. $\left| \dfrac{5}{6} - \dfrac{1}{9}x \right| + 5 = 2$

Solve the following inequalities and graph the solutions. [2.7]

76. $|3y - 4| \ge 2$

77. $|5x + 3| - 4 \le 9$

78. $|4x - 6| > -8$

79. $|2 - 5t| < -3$

At the end of each chapter from here on, we will also provide a review that consists of problems similar to those found in the current and previous chapters. Use these reviews as a study tool for exams, and as a way to check your progress, attending to any difficulties you come across. Remember that math concepts build on each other. It is important to master the skills that come before in order to find success as you proceed through the book.

Simplify each of the following.

1. 6^2

2. $|-5|$

3. $4 \cdot (-3) + 7$

4. $3(4 - 5)$

5. $9 - 2 \cdot 4 + 1$

6. $(7 - 1)(5 + 4)$

7. $8 + 2(3 + 2)$

8. $45 - 24 \div 8 + 3$

9. $-(4 + 1) - (8 - 11)$

10. $3(-4)^2 + 2(-3)^3$

11. $\dfrac{-4 - 4}{-1 - 3}$

12. $\dfrac{7(-2) + 4}{9(-2) + 6(2)}$

Reduce the following fractions to lowest terms.

13. $\dfrac{57}{76}$

14. $\dfrac{129}{387}$

Add the following fractions.

15. $\dfrac{7}{28} + \dfrac{5}{12}$

16. $\dfrac{45}{56} + \dfrac{17}{21}$

17. Find the difference of 4 and -5.

18. Subtract -3 from the product of 4 and -7.

Let $A = \{1, 3, 5\}$; $B = \{3, 6, 9, 12\}$; $C = \{5, 6, 7, 8\}$. Find the following.

19. $A \cup C$

20. $A \cap B$

21. $\{x \mid x \in B \text{ and } x > 6\}$

22. $\{x \mid x \in C \text{ and } x \le 7\}$

Simplify each of the following expressions.

23. $14\left(\dfrac{x}{7} + 1\right)$

24. $4(3x - 2) + 7$

25. $x\left(1 + \dfrac{3}{x}\right)\left(1 + \dfrac{3}{x}\right)$

26. $1{,}000(0.03x + 0.01y)$

Simplify each of the following expressions.

27. $(4x + 1) - (5x - 2)$

28. $(4x^3 + 8) - (5x^2 - 8) + (3x^3 - 2) + (6x^2 + 1)$

Multiply or divide as indicated.

29. $(2x^2y)^5$

30. $\dfrac{48x^2y}{16xy^9}$

31. $(4x + 5)(2x - 1)$

32. $\left(x - \dfrac{3}{4}\right)^2$

33. $(8 \times 10^4)(2 \times 10^3)$

Factor each of the following equations.

34. $x^2 + ax - bx - ab$

35. $(x - 1)^2 - 16$

Solve each of the following equations.

36. $3 - 6y = -9$

37. $3(2 - x) = 4(1 + 3x)$

38. $6 - \dfrac{4}{3}a = -2$

39. $300x = 100x + 50$

40. $|x| + 3 = 5$

41. $|5y + 3| = 18$

Substitute the given values in each formula, and then solve for the variable that does not have a numerical replacement.

42. $A = P + Prt$: $A = 1{,}550$, $P = 50$, $r = 0.3$

43. $A = a + (n - 1)d$: $A = 15$, $a = 3$, $d = 6$

Solve each formula for the indicated variable.

44. $y = mx + b$ for m

45. $S = \dfrac{a}{1 - r}$ for r

46. $C = \dfrac{5}{9}(F - 32)$ for F

47. $ax - 2 = bx + 3$ for x

Solve each application. In each case, be sure to show the equation that describes the situation.

48. Angles Two angles are complementary. If the larger angle is 5° more than four times the smaller angle, find the measure of each angle.

49. Rectangle A rectangle is three times as long as it is wide. The perimeter is 64 feet. Find the dimensions.

Solve each inequality. Write you answer using interval notation.

50. $500 + 200x > 300$

51. $-\dfrac{1}{8} \le \dfrac{1}{16}x \le \dfrac{1}{4}$

Solve each inequality and graph the solution set.

52. $|3x - 1| > 7$

53. $|4t - 3| + 2 < 4$

The chart shows the average hours per day spent commuting in different cities in the United States. Use the information to answer the following questions.

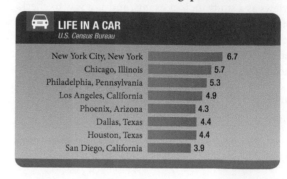

54. What is the difference between New York City and San Diego in daily commuting hours?

55. What is the difference between Philadelphia and Houston in daily commuting hours?

Solve the following equations. [2.1]

1. $5 - \dfrac{4}{7}a = -11$

2. $\dfrac{1}{5}x - \dfrac{1}{2} - \dfrac{1}{10}x + \dfrac{2}{5} = \dfrac{3}{10}x + \dfrac{1}{2}$

3. $3x^2 = 5x + 2$

4. $100x^3 = 500x^2$

5. $5(x - 1) - 2(2x + 3) = 5x - 4$

6. $0.07 - 0.02(3x + 1) = -0.04x + 0.01$

7. $(x + 1)(x + 2) = 12$

8. $x^3 + 2x^2 - 16x - 32 = 0$

Solve for the indicated variable. [2.2]

9. $P = 2l + 2w$; for w

10. $A = \dfrac{1}{2}h(b + B)$; for B

Solve for y. [2.2]

11. $5x - 2y = 10$

12. $\dfrac{y - 5}{x - 4} = 3$

Solve each of the following. [2.3]

13. Geometry A rectangle is twice as long as it is wide. The perimeter is 36 inches. Find the dimensions.

14. Geometry Two angles are supplementary. If the larger angle is 15° more than twice the smaller angle, find the measure of each angle.

15. Right Triangle The longest side of a right triangle is 4 inches more than the shortest side. The third side is 2 inches more than the shortest side. Find the length of each side.

16. Velocity and Height If an object is thrown straight up into the air with an initial velocity of 32 feet per second, then its height h (in feet) above the ground at any time t (in seconds) is given by the formula $h = 32t - 16t^2$. Find the times at which the object is on the ground by letting $h = 0$ in the equation and solving for t.

Solve each inequality, and graph the solution. [2.4]

17. $\dfrac{1}{2}x - 2 > 3$

18. $-6y \leq 24$

19. $0.3 - 0.2x < 1.1$

20. $3 - 2(n - 1) \geq 9$

Solve the following inequalities. Write the solution set using interval notation, then graph the solution set. [2.4]

21. $-5t \leq 30$

22. $5 - \dfrac{3}{2}x > -1$

23. $1.6x - 2 < 0.8x + 2.8$

24. $3(2y + 4) \geq 5(y - 8)$

Solve the following equations. [2.6]

25. $\left| \dfrac{1}{4}x - 1 \right| = \dfrac{1}{2}$

26. $|3 - 2x| + 5 = 2$

Solve the following inequalities and graph the solutions. [2.7]

27. $|6x - 1| > 7$

28. $|3x - 5| - 4 \leq 3$

29. $|5 - 4x| \geq -7$

30. $|4t - 1| < -3$

3

Equations and Inequalities in Two Variables

©iStockphoto.com/mrloz

A student is heating water in a chemistry lab. As the water heats, she records the temperature readings from two thermometers, one giving temperature in degrees Fahrenheit and the other in degrees Celsius. The table below shows some of the data she collects. The scatter diagram gives a visual representation of the data in the table.

Corresponding Temperatures

Degrees Fahrenheit	Degrees Celsius
77	25
95	35
167	75
212	100

The exact relationship between the Fahrenheit and Celsius temperature scales is given by the formula

$$C = \frac{5}{9}(F - 32)$$

We have three ways to describe the relationship between the two temperature scales: a table, a graph, and an equation. But, most important to us, we don't need to accept this formula on faith. Later, you will derive the formula from the data in the table above.

STUDY SKILLS

The study skills for this chapter are about attitude. They are perspectives that point toward success.

1. **Be Focused, Not Distracted**   I have students who begin their assignments by asking themselves, "Why am I taking this class?" If you are asking yourself similar questions, you are distracting yourself from doing the things that will produce the results you want in this course. Don't dwell on questions and evaluations of the class that can be used as excuses for not doing well. If you want to succeed in this course, focus your energy and efforts toward success, rather than distracting yourself from your goals.

2. **Be Resilient**   Don't let setbacks keep you from your goals. You want to put yourself on the road to becoming a person who can succeed in this class, or any class in college. Failing a test or quiz, or having a difficult time on some topics, is normal. No one goes through college without some setbacks. Don't let a temporary disappointment keep you from succeeding in this course. A low grade on a test or quiz is simply a signal that you need to reevaluate your study habits.

3. **Intend to Succeed**   I have a few students who simply go through the motions of studying without intending to master the material. It is more important to them to look like they are studying than to actually study. You need to study with the intention of being successful in the course. Intend to master the material, no matter what it takes.

PAIRED DATA AND THE
RECTANGULAR COORDINATE SYSTEM

©iStockphoto.com/scanrail

OBJECTIVES

A Graph ordered pairs on a rectangular coordinate system.

B Graph a linear equation in two variables.

C Graph lines through the origin.

D Graph equations of vertical translations.

E Find intercepts and graph their equations.

F Graph horizontal and vertical lines.

In this section we place our work with charts and graphs in a more formal setting. Our foundation will be the *rectangular coordinate system*, because it gives us a link between algebra and geometry. With it we notice relationships between certain equations and different lines and curves.

Table 1 gives the net price of a popular intermediate algebra text at the beginning of each year in which a new edition was published. (The net price is the price the bookstore pays for the book, not the price you pay for it.)

TABLE 1	Price of a Textbook	
		Net Price
Edition	Year Published	($)
First	1995	39.25
Second	1999	47.50
Third	2003	55.00
Fourth	2007	65.75
Fifth	2017	71.00

The information in Table 1 is represented visually in Figures 1 and 2. The diagram in Figure 1 is called a *bar chart*. The diagram in Figure 2 is called a *line graph*. The data in Table 1 is called *paired data* because each number in the year column is paired with a specific number in the price column.

3.1 VIDEOS

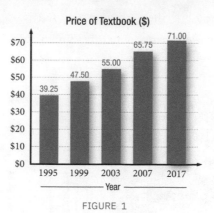

FIGURE 1

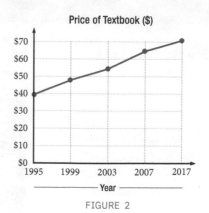

FIGURE 2

A Ordered Pairs and the Rectangular Coordinate System

Paired data play an important role in equations that contain two variables. Working with these equations is easier if we standardize the terminology and notation associated with paired data. So here is a definition that will do just that.

> **DEFINITION**
> •
>
> A pair of numbers enclosed in parentheses and separated by a comma, such as $(-2, 1)$, is called an *ordered pair* of numbers. The first number in the pair is called the *x-coordinate* of the ordered pair; the second number is called the *y-coordinate*. For the ordered pair $(-2, 1)$, the x-coordinate is -2 and the y-coordinate is 1.

A *rectangular coordinate system* is made by drawing two real number lines at right angles to each other. The two number lines, called *axes*, cross each other at 0. This point is called the origin. Positive directions are to the right and up. Negative directions are to the left and down. The rectangular coordinate system is shown in Figure 3.

NOTE A rectangular coordinate system allows us to connect algebra and geometry by associating geometric shapes (the curves shown in the diagrams) with algebraic equations. The French philosopher and mathematician René Descartes (1596 – 1650) is usually credited with the invention of the rectangular coordinate system, which is often referred to as the Cartesian coordinate system in his honor. As a philosopher, Descartes is responsible for the statement, "I think, therefore, I am." Until Descartes invented his coordinate system in 1637, algebra and geometry were treated as separate subjects.

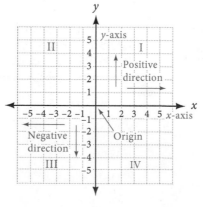

FIGURE 3

The horizontal number line is called the *x-axis*, and the vertical number line is called the *y-axis*. The two number lines divide the coordinate system into four quadrants, which we number I through IV in a counterclockwise direction. Points on the axes are not considered as being in any quadrant.

To graph the ordered pair (a, b) on a rectangular coordinate system, we start at the origin and move a units right or left (right if a is positive, left if a is negative). Then we move b units up or down (up if b is positive, down if b is negative). The point where we end up is the graph of the ordered pair (a, b).

EXAMPLE 1

Plot (graph) the ordered pairs $(2, 5)$, $(-2, 5)$, $(-2, -5)$, and $(2, -5)$.

NOTE From Example 1, we see that any point in quadrant I has both its *x*- and *y*-coordinates positive $(+, +)$. Points in quadrant II have negative *x*-coordinates and positive *y*-coordinates $(-, +)$. In quadrant III, both coordinates are negative $(-, -)$. In quadrant IV, the form is $(+, -)$.

SOLUTION   To graph the ordered pair $(2, 5)$, we start at the origin and move 2 units to the right, then 5 units up. We are now at the point whose coordinates are $(2, 5)$. We graph the other three ordered pairs in a similar manner (see Figure 4).

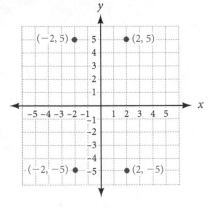

FIGURE 4

EXAMPLE 2

Graph the ordered pairs $(1, -3)$, $\left(\dfrac{1}{2}, 2\right)$, $(3, 0)$, $(0, -2)$, $(-1, 0)$, and $(0, 5)$.

SOLUTION

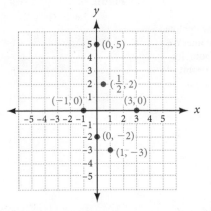

FIGURE 5

From Figure 5, we see that any point on the *x*-axis has a *y*-coordinate of 0 (it has no vertical displacement), and any point on the *y*-axis has an *x*-coordinate of 0 (no horizontal displacement).

🅑 Graphing Equations

We can plot a single point from an ordered pair, but to draw a line or a curve, we need more points.

To graph an equation in two variables, we simply graph its solution set. That is, we draw a line or smooth curve through all the points whose coordinates satisfy the equation.

EXAMPLE 3

Graph the equation $y = -\dfrac{1}{3}x$.

SOLUTION   The graph of this equation will be a straight line. We need to find three ordered pairs that satisfy the equation. To do so, we can let x equal any numbers we choose and find corresponding values of y. However, because every value of x we substitute into the equation is going to be multiplied by $-\frac{1}{3}$, let's use numbers for x that are divisible by 3, like -3, 0, and 3. That way, when we multiply them by $-\frac{1}{3}$, the result will be an integer.

$$\text{Let } x = -3; \qquad y = -\frac{1}{3}(-3) = 1$$

The ordered pair $(-3, 1)$ is one solution.

$$\text{Let } x = 0; \qquad y = -\frac{1}{3}(0) = 0$$

The ordered pair $(0, 0)$ is a second solution.

$$\text{Let } x = 3; \qquad y = -\frac{1}{3}(3) = -1$$

The ordered pair $(3, -1)$ is a third solution.

In table form

x	y
-3	1
0	0
3	-1

Plotting the ordered pairs $(-3, 1)$, $(0, 0)$, and $(3, -1)$ and drawing a straight line through their graphs, we have the graph of the equation $y = -\frac{1}{3}x$ as shown in Figure 6.

NOTE   It takes only two points to determine a straight line. We have included a third point for "insurance." If all three points do not line up in a straight line, we have made a mistake.

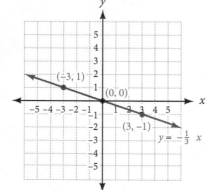

FIGURE 6

Example 3 illustrates again the connection between algebra and geometry that we mentioned earlier in this section. Descartes's rectangular coordinate system allows us to associate the equation $y = -\frac{1}{3}x$ (an algebraic concept) with a specific straight line (a geometric concept). The study of the relationship between equations in algebra and their associated geometric figures is called *analytic geometry*.

C Lines Through the Origin

As you can see from Figure 6, the graph of the equation $y = -\frac{1}{3}x$ is a straight line that passes through the origin. The same will be true of the graph of any equation that has the same form as $y = -\frac{1}{3}x$. Here are three more equations that have that form, along with their graphs.

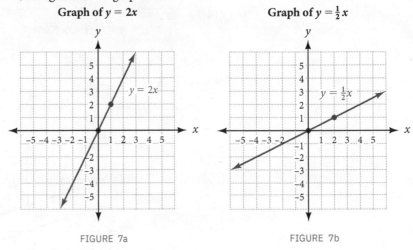

Graph of $y = 2x$

FIGURE 7a

Graph of $y = \frac{1}{2}x$

FIGURE 7b

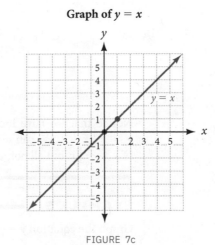

Graph of $y = x$

FIGURE 7c

Here is a summary of this discussion.

> ### Lines Through the Origin
>
> The graph of any equation of the form
> $$y = mx$$
> will be a straight line through the origin.

EXAMPLE 4

Graph the equation $y = -\frac{1}{3}x + 2$.

SOLUTION   Again, we need ordered pairs that are solutions to our equation. Noticing the similarity of this equation to our previous equation, we choose the same values of x for our inputs.

Input x	Calculate Using the Equation	Output y	Form Ordered Pairs
-3	$y = -\frac{1}{3}(-3) + 2 = 1 + 2 =$	3	$(-3, 3)$
0	$y = -\frac{1}{3}(0) + 2 = 0 + 2 =$	2	$(0, 2)$
3	$y = -\frac{1}{3}(3) + 2 = -1 + 2 =$	1	$(3, 1)$

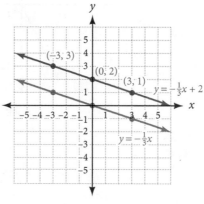

FIGURE 8

Notice that we have included the graph of $y = -\frac{1}{3}x$ along with the graph of $y = -\frac{1}{3}x + 2$. We can see that the graph of $y = -\frac{1}{3}x + 2$ looks just like the graph of $y = -\frac{1}{3}x$, but all points are moved up vertically 2 units.

EXAMPLE 5

Graph the equation $y = -\frac{1}{3}x - 4$.

SOLUTION   We create a table from ordered pairs, then we graph the information in the table. However, even before we start, we are expecting the graph of $y = -\frac{1}{3}x - 4$ to be 4 units below the graph of $y = -\frac{1}{3}x$

Input x	Calculate Using the Equation	Output y	Form Ordered Pairs
-3	$y = -\frac{1}{3}(-3) - 4 = 1 - 4 =$	-3	$(-3, -3)$
0	$y = -\frac{1}{3}(0) - 4 = 0 - 4 =$	-4	$(0, -4)$
3	$y = -\frac{1}{3}(3) - 4 = -1 - 4 =$	-5	$(3, -5)$

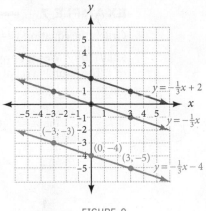

FIGURE 9

D Vertical Translations

We know that the graph of $y = mx$ is a line that passes through the origin. From our previous two examples we can generalize as follows.

If K is a positive number, then:	
The Graph of	Is the Graph of $y = mx$ translated
$y = mx + K$	K units up
$y = mx - K$	K units down

EXAMPLE 6

Find an equation for the blue line.

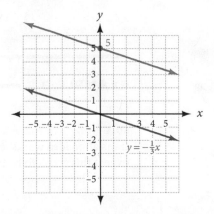

FIGURE 10

SOLUTION   The blue line is parallel to the line $y = -\frac{1}{3}x$, but translated up 5 units from $y = -\frac{1}{3}x$. According to what we have done up to this point, the equation for the blue line is

$$y = -\frac{1}{3}x + 5$$

EXAMPLE 7

Graph the equation $y = x^2$.

SOLUTION We input values of x then calculate using the equation to output values of y. The result is a set of ordered pairs that we plot and then connect with a smooth curve.

Input x	Calculate Using the Equation	Output y	Form Ordered Pairs
-3	$y = (-3)^2 =$	9	$(-3, 9)$
-2	$y = (-2)^2 =$	4	$(-2, 4)$
-1	$y = (-1)^2 =$	1	$(-1, 1)$
0	$y = (0)^2 =$	0	$(0, 0)$
1	$y = (1)^2 =$	1	$(1, 1)$
2	$y = (2)^2 =$	4	$(2, 4)$
3	$y = (3)^2 =$	9	$(3, 9)$

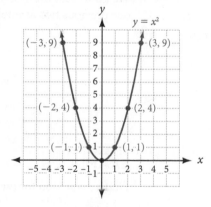

FIGURE 11

EXAMPLE 8

Graph $y = x^2 - 4$.

SOLUTION We make a table as we did in the previous example. If the vertical translation idea works for this type of equation as it did with our straight lines, we expect this graph to be the graph in Example 7 translated down 4 units.

Input x	Calculate Using the Equation	Output y	Form Ordered Pairs
-3	$y = (-3)^2 - 4 =$	5	$(-3, 5)$
-2	$y = (-2)^2 - 4 =$	0	$(-2, 0)$
-1	$y = (-1)^2 - 4 =$	-3	$(-1, -3)$
0	$y = (0)^2 - 4 =$	-4	$(0, -4)$
1	$y = (1)^2 - 4 =$	-3	$(1, -3)$
2	$y = (2)^2 - 4 =$	0	$(2, 0)$
3	$y = (3)^2 - 4 =$	5	$(3, 5)$

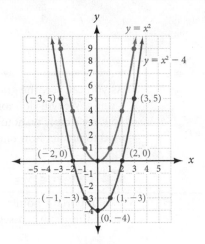

FIGURE 12

As you can see, the graph of $y = x^2 - 4$ is the graph of $y = x^2$ translated down 4 units. We generalize this as follows.

If K is a positive number, then:	
The Graph of	**Is the Graph of $y = x^2$ translated**
$y = x^2 + K$	K units up
$y = x^2 - K$	K units down

EXAMPLE 9

Graph $y = |x|$ and $y = |x| - 4$

SOLUTION   We let x take on values of -4, -3, -2, -1, 0, 1, 2, 3, and 4. The corresponding values of y are shown in the tables.

y = |x|	
Input x	Output y
−4	4
−3	3
−2	2
−1	1
0	0
1	1
2	2
3	3
4	4

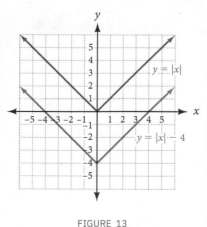

FIGURE 13

y = |x| − 4	
Input x	Output y
−4	0
−3	−1
−2	−2
−1	−3
0	−4
1	−3
2	−2
3	−1
4	0

As you can see, the graph of $y = |x| - 4$ is the graph of $y = |x|$ translated down 4 units. We will do more with these vertical translations later in the chapter. For now, we simply want to notice how the relationship between the equations can be used to predict how one graph will look when the other graph is given.

E Intercepts

Two important points on the graph of a straight line, if they exist, are the points where the graph crosses the axes.

> **DEFINITION** *intercepts*
> •
> An *x-intercept* of the graph of an equation is the *x*-coordinate of a point where the graph crosses the *x*-axis. The *y-intercept* is defined similarly.

Because any point on the *x*-axis has a *y*-coordinate of 0, we can find the *x*-intercept by letting $y = 0$ and solving the equation for *x*. We find the *y*-intercept by letting $x = 0$ and solving for *y*.

EXAMPLE 10

Find the *x*- and *y*-intercepts for $2x + 3y = 6$; then graph the equation.

SOLUTION   To find the *y*-intercept, we let $x = 0$.

$$\text{When} \qquad x = 0$$
$$\text{we have} \qquad 2(0) + 3y = 6$$
$$3y = 6$$
$$y = 2$$

The *y*-intercept is 2 so the graph crosses the *y*-axis at the point (0, 2). To find the *x*-intercept, we let $y = 0$.

$$\text{When} \qquad\qquad y = 0$$

$$\text{we have} \qquad 2x + 3(0) = 6$$

$$2x = 6$$

$$x = 3$$

The x-intercept is 3, so the graph crosses the x-axis at the point $(3, 0)$. We use these results to graph the solution set for $2x + 3y = 6$. The graph is shown in Figure 14.

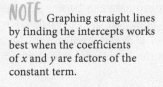

NOTE Graphing straight lines by finding the intercepts works best when the coefficients of x and y are factors of the constant term.

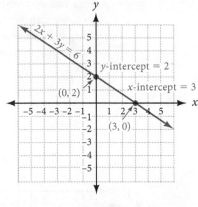

FIGURE 14

EXAMPLE 11

Find the intercepts for $y = x^2 - 4$.

SOLUTION We drew the graph of this equation in Example 8. Looking back to that graph we see that the x-intercepts are -2 and 2, and the y-intercept is -4. Let's see if we obtain the same results using the algebraic method shown in Example 10.

x-intercept

When $y = 0$, we have

$$0 = x^2 - 4$$

$$0 = (x + 2)(x - 2)$$

$$x + 2 = 0 \quad \text{or} \quad x - 2 = 0$$

$$x = -2 \qquad\qquad x = 2$$

The x-intercepts are -2 and 2

y-intercept

When $x = 0$, we have

$$y = 0^2 - 4 = -4$$

The y-intercept is -4

F Horizontal and Vertical Lines

EXAMPLE 12

Graph each of the following lines.

a. $x = 3$ **b.** $y = -2$

SOLUTION

a. The line $x = 3$ is the set of all points whose x-coordinate is 3. The variable y does not appear in the equation, so the y-coordinate can be any number. Note that we can write our equation as a linear equation in two variables by writing it as $x + 0y = 3$. Because the product of 0 and y will always be 0, y can be any number. The graph of $x = 3$ is the vertical line shown in Figure 15a.

b. The line $y = -2$ is the set of all points whose y-coordinate is -2. The variable x does not appear in the equation, so the x-coordinate can be any number. Again, we can write our equation as a linear equation in two variables by writing it as $0x + y = -2$. Because the product of 0 and x will always be 0, x can be any number. The graph of $y = -2$ is the horizontal line shown in Figure 15b.

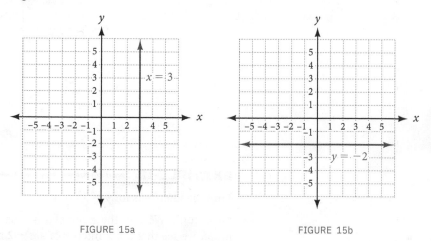

FIGURE 15a   FIGURE 15b

FACTS FROM GEOMETRY *Special Equations and Their Graphs*

For the equations below, m, a, and b are real numbers.

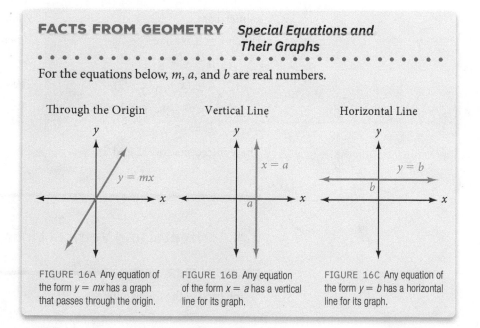

Through the Origin

Vertical Line

Horizontal Line

FIGURE 16A Any equation of the form $y = mx$ has a graph that passes through the origin.

FIGURE 16B Any equation of the form $x = a$ has a vertical line for its graph.

FIGURE 16C Any equation of the form $y = b$ has a horizontal line for its graph.

USING TECHNOLOGY   *Graphing Calculators and Computer Graphing Programs*

Graphing with Trace and Zoom

All graphing calculators have the ability to graph a function and then trace over the points on the graph, giving their coordinates. Furthermore, all graphing calculators can zoom in and out on a graph that has been drawn. To graph an equation on a graphing calculator, we first set the graph window. The counterpart values of y are Ymin and Ymax. We will use the notation

Window: $-5 \le x \le 4$ and $-3 \le y \le 2$

to stand for a window in which Xmin $= -5$, Xmax $= 4$, Ymin $= -3$, and Ymax $= 2$

Set your calculator to the following window:

Window: $-10 \le x \le 10$ and $-10 \le y \le 10$

Graph the equation $Y = -X + 8$ and compare your results with this graph:

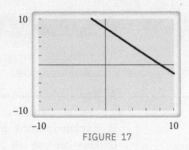

FIGURE 17

Use the Trace feature of your calculator to name three points on the graph. Next, use the Zoom feature of your calculator to zoom out so your window is twice as large.

Solving for y First

To graph the equation from Example 10, $2x + 3y = 6$, on a graphing calculator, you must first solve it for y. When you do so, you will get $y = -\frac{2}{3} x + 2$, which results with the graph in Figure 14. Use this window:

Window: $-6 \le x \le 6$ and $-6 \le y \le 6$

Hint on Tracing

If you are going to use the Trace feature and want the x-coordinates to be exact numbers, set your window so the range of X inputs is a multiple of the number of horizontal pixels on your calculator screen. On the TI-82/83, the screen is 94 pixels wide. Here are a few convenient trace windows:

To trace with x to the nearest tenth, use $-4.7 \le x \le 4.7$ or $0 \le x \le 9.4$

To trace with x to the nearest integer, use $-47 \le x \le 47$ or $0 \le x \le 94$

USING TECHNOLOGY *Graphing Calculators and Computer Graphing Programs*

· ·

continued

Graph each equation using the indicated window.

1. $y = \frac{1}{2}x - 3$ $-10 \leq x \leq 10$ and $-10 \leq y \leq 10$

2. $y = \frac{1}{2}x^2 - 3$ $-10 \leq x \leq 10$ and $-10 \leq y \leq 10$

3. $y = \frac{1}{2}x^2 - 3$ $-4.7 \leq x \leq 4.7$ and $-10 \leq y \leq 10$

4. $y = x^3$ $-10 \leq x \leq 10$ and $-10 \leq y \leq 10$

5. $y = x^3 - 5$ $-4.7 \leq x \leq 4.7$ and $-10 \leq y \leq 10$

GETTING READY FOR CLASS

After reading through the preceding section, respond in your own words and in complete sentences.

A. Explain how you would construct a rectangular coordinate system from two real number lines.

B. Explain in words how you would graph the ordered pair $(2, -3)$.

C. How can you tell if an ordered pair is a solution to the equation $y = 2x - 5$?

D. If you were looking for solutions to the equation $y = \frac{1}{3}x + 5$, why would it be easier to substitute 6 for x than to substitute 5 for x?

Problem Set 3.1

Graph each of the following ordered pairs on a rectangular coordinate system.

1. **a.** $(-1, 2)$ **b.** $(-1, -2)$ **c.** $(5, 0)$ **d.** $(0, 2)$ **e.** $(-5, -5)$ **f.** $\left(\frac{1}{2}, 2\right)$

2. **a.** $(1, 2)$ **b.** $(1, -2)$ **c.** $(0, -3)$ **d.** $(4, 0)$ **e.** $(-4, -1)$ **f.** $\left(3, \frac{1}{4}\right)$

Give the coordinates of each point.

3.

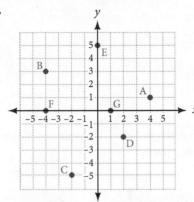

4.

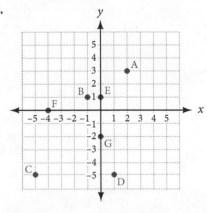

5. Which of the following tables could be produced from the equation $y = 2x - 6$?

   a.

x	y
0	6
1	4
2	2
3	0

   b.

x	y
0	-6
1	-4
2	-2
3	0

   c.

x	y
0	-6
1	-5
2	-4
3	-3

6. Which of the following tables could be produced from the equation $3x - 5y = 15$?

   a.

x	y
0	5
-3	0
10	3

   b.

x	y
0	-3
5	0
10	3

   c.

x	y
0	-3
-5	0
10	-3

7. The graph shown here is the graph of which of the following equations?

   a. $y = \frac{3}{2}x - 3$

   b. $y = \frac{2}{3}x - 2$

   c. $y = -\frac{2}{3}x + 2$

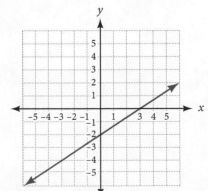

8. The graph shown here is the graph of which of the following equations?

    a. $3x - 2y = 8$

    b. $2x - 3y = 8$

    c. $2x + 3y = 8$

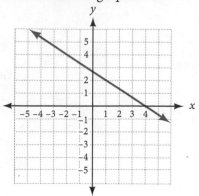

For each problem below, the equation of the red graph is given. Find the equation for the blue graph.

9.

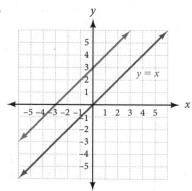

10.

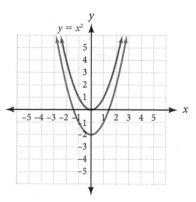

11.

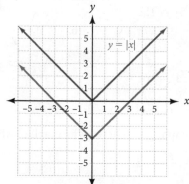

12.

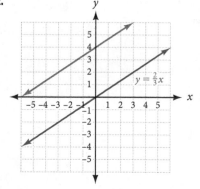

Graph each of the following. Use one coordinate system for each problem.

13. a. $y = 2x$ **b.** $y = 2x + 3$ **c.** $y = 2x - 5$

14. a. $y = \frac{1}{3}x$ **b.** $y = \frac{1}{3}x + 1$ **c.** $y = \frac{1}{3}x - 3$

15. a. $y = \frac{1}{2}x^2$ **b.** $y = \frac{1}{2}x^2 - 2$ **c.** $y = \frac{1}{2}x^2 + 2$

16. a. $y = 2x^2$ **b.** $y = 2x^2 - 8$ **c.** $y = 2x^2 + 1$

17. Graph the straight line $0.02x + 0.03y = 0.06$.

18. Graph the straight line $0.05x - 0.03y = 0.15$.

19. Graph each of the following lines.

    a. $y = 2x$ **b.** $x = -3$ **c.** $y = 2$

20. Graph each of the following lines.

    a. $y = 3x$ **b.** $x = -2$ **c.** $y = 4$

21. Graph each of the following lines.

    a. $y = -\frac{1}{2}x$ **b.** $x = 4$ **c.** $y = -3$

22. Graph each of the following lines.

    a. $y = -\frac{1}{3}x$ **b.** $x = 1$ **c.** $y = -5$

Find the intercepts for each graph. Then use them to help sketch the graph.

23. $y = x^2 - 9$ **24.** $y = x^2$ **25.** $y = 2x - 4$ **26.** $y = 4x - 2$

27. $y = \frac{1}{2}x + 1$ **28.** $y = -\frac{1}{2}x + 1$ **29.** $y = 3x$ **30.** $y = -\frac{1}{3}x$

31. $y = x^2 - x$ **32.** $y = x^2 + 3$ **33.** $y = x - 3$ **34.** $y = x + 2$

PAYING ATTENTION TO INSTRUCTIONS

35. Work each problem according to the instructions given:

    a. Solve: $4x + 12 = -16$

    b. Find x when y is 0: $4x + 12y = -16$

    c. Find y when x is 0: $4x + 12y = -16$

    d. Graph: $4x + 12y = -16$

    e. Solve for y: $4x + 12y = -16$

36. Work each problem according to the instructions given:

    a. Solve: $3x - 8 = -12$

    b. Find x when y is 0: $3x - 8y = -12$

    c. Find y when x is 0: $3x - 8y = -12$

    d. Graph: $3x - 8y = -12$

    e. Solve for y: $3x - 8y = -12$

APPLYING THE CONCEPTS

37. **Solar Energy** The graph shows the rise in shipments of solar thermal collectors over a 10-year period. Use the chart to answer the following questions.

    a. Does the graph contain the point (4, 7,500)?

    b. Does the graph contain the point (8, 15,000)?

    c. Does the graph contain the point (9, 15,000)?

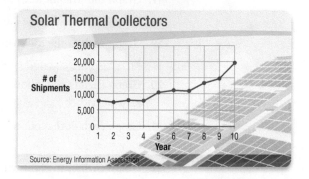

38. **Health Care Costs** The graph shows the projected rise in the cost of health care from 2005 to 2019. Using years as x and billions of dollars as y, write five ordered pairs that describe the information in the graph.

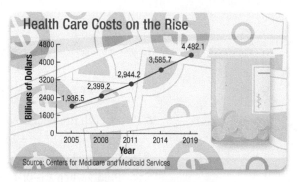

39. **Hourly Wages** Suppose you have a job that pays $7.50 per hour, and you work anywhere from 0 to 40 hours per week. Table 2 gives the amount of money you will earn in one week for working various hours. Construct a line graph from the information in Table 2.

TABLE 2 Weekly Wages	
Hours Worked	Pay ($)
0	0
10	75
20	150
30	225
40	300

40. **Softball Toss** Chaudra is tossing a softball into the air with an underhand motion. It takes exactly 2 seconds for the ball to come back to her. Table 3 shows the distance the ball is above her hand at quarter-second intervals. Construct a line graph from the information in the table.

TABLE 3	Tossing a softball into the air
Time (sec)	Distance (ft)
0	0
0.25	7
0.5	12
0.75	15
1	16
1.25	15
1.5	12
1.75	7
2	0

41. **Intensity of Light** Table 4 gives the intensity of light that falls on a surface at various distances from a 100-watt light bulb. Construct a bar chart from the information in Table 4.

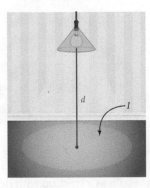

TABLE 4	Light intensity from a 100-watt light bulb
Distance Above Surface (ft)	Intensity (lumens/sq ft)
1	120.0
2	30.0
3	13.3
4	7.5
5	4.8
6	3.3

42. **Value of a Painting** A piece of abstract art was purchased in 1995 for $125. Table 5 shows the value of the painting at various times, assuming that it doubles in value every five years. Construct a bar chart from the information in the table.

TABLE 5	Value of a Painting
Year	Value ($)
1995	125
2000	250
2005	500
2010	1,000
2015	2,000

43. Reading Graphs This graph shows the number of people in line at a theater box office to buy tickets for a movie that starts at 7:30. The box office opens at 6:45.

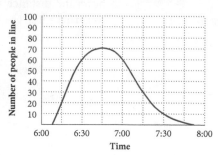

a. How many people are in line at 6:30?

b. How many people are in line when the box office opens?

c. How many people are in line when the movie starts?

d. At what times are there 60 people in line?

e. How long after the movie starts is there no one left in line?

44. Kentucky Derby The graph gives the monetary bets placed at the Kentucky Derby for specific years. If x represents the year in question and y represents the total wagering for that year, write five ordered pairs that describe the information in the graph.

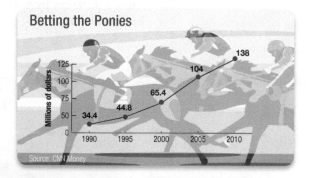

GETTING READY FOR THE NEXT SECTION

45. Write -0.06 as a fraction with denominator 100.

46. Write -0.07 as a fraction with denominator 100.

47. If $y = 2x - 3$, find y when $x = 2$

48. If $y = 2x - 3$, find x when $y = 5$

Simplify.

49. $\dfrac{1 - (-3)}{-5 - (-2)}$ **50.** $\dfrac{-3 - 1}{-2 - (-5)}$ **51.** $\dfrac{-1 - 4}{3 - 3}$ **52.** $\dfrac{-3 - (-3)}{2 - (-1)}$

53. The product of $\dfrac{2}{3}$ and what number will result in

    a. 1?   **b.** -1?

54. The product of 3 and what number will result in

    a. 1?   **b.** -1?

EXTENDING THE CONCEPTS

Each of the following equations is a graph that is not a straight line. Set up a table of values, then sketch the graph of the equation.

55. $y = x^2$

56. $y = x^2 - 3$

57. $y = |x|$

58. $y = |x + 2|$

59. $y = x^3$

60. $y = x^3 + 2$

Find the x- and y-intercepts for each equation. Your answers will contain the constants a, b, and c.

61. $ax + by = c$

62. $ax - by = c$

63. $\dfrac{x}{a} + \dfrac{y}{b} = 1$

64. $y = ax + b$

3.2

©iStockphoto.com/Jorge Villalba

OBJECTIVES

A Find the slope of a line.

B Find the slope representing the rate of change.

C Use the slope of a line to find average speed.

A highway sign tells us we are approaching a 6% downgrade. As we drive down this hill, each 100 feet we travel horizontally is accompanied by a 6-foot drop in elevation.

In mathematics we say the slope of the highway is $-0.06 = -\frac{6}{100} = -\frac{3}{50}$. The slope is the ratio of the vertical change to the accompanying horizontal change.

 Finding the Slope

In defining the slope of a straight line, we want to associate a number with the line. This does two things. First, we want the slope of a line to measure the "steepness" of the line. That is, in comparing two lines, the slope of the steeper line should have the larger numerical value. Second, we want a line that rises going from left to right to have a *positive* slope. We want a line that falls going from left to right to have a *negative* slope. (A line that neither rises nor falls going from left to right must, therefore, have 0 slope.)

3.2 VIDEOS

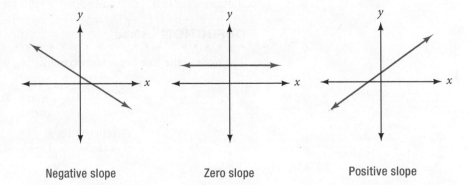

Negative slope   Zero slope   Positive slope

Geometrically, we can define the *slope* of a line as the ratio of the vertical change to the horizontal change encountered when moving from one point to another on the line. The vertical change is sometimes called the *rise*. The horizontal change is called the *run*.

EXAMPLE 1

Find the slope of the line $y = 2x - 3$.

SOLUTION   To use our geometric definition, we first graph $y = 2x - 3$ (Figure 1). We then pick any two convenient points and find the ratio of rise to run. By convenient points we mean points with integer coordinates. If we let $x = 2$ in the equation, then $y = 1$. Likewise, if we let $x = 4$, then y is 5.

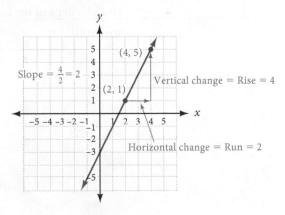

FIGURE 1

The ratio of vertical change to horizontal change is 4 to 2, giving us a slope of $\frac{4}{2} = 2$. Our line has a slope of 2.

Notice that we can measure the vertical change (rise) by subtracting the y-coordinates of the two points shown in Figure 1: $5 - 1 = 4$. The horizontal change (run) is the difference of the x-coordinates: $4 - 2 = 2$. This gives us a second way of defining the slope of a line.

DEFINITION *slope*

The *slope* of the line between two points (x_1, y_1) and (x_2, y_2) is given by

$$\text{Slope} = m = \frac{\text{Rise}}{\text{Run}} = \frac{y_2 - y_1}{x_2 - x_1}$$

Geometric Form   Algebraic Form

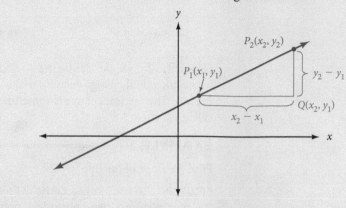

EXAMPLE 2

Find the slope of the line through $(-2, -3)$ and $(-5, 1)$.

SOLUTION

$$m = \frac{y_2 - y_1}{x_2 - x_1} = \frac{1 - (-3)}{-5 - (-2)} = \frac{4}{-3} = -\frac{4}{3}$$

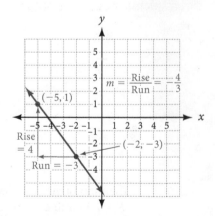

FIGURE 2

Looking at the graph of the line between the two points (Figure 2), we can see our geometric approach does not conflict with our algebraic approach.

We should note here that it does not matter which ordered pair we call (x_1, y_1) and which we call (x_2, y_2). If we were to reverse the order of subtraction of both the x- and y-coordinates in the precceding example, we would have

$$m = \frac{-3 - 1}{-2 - (-5)} = \frac{-4}{3} = -\frac{4}{3}$$

which is the same as our previous result.

EXAMPLE 3

Find the slope of the line containing $(3, -1)$ and $(3, 4)$.

SOLUTION Using the definition for slope, we have

$$m = \frac{-1 - 4}{3 - 3} = \frac{-5}{0}$$

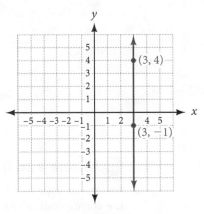

FIGURE 3

The expression $\frac{-5}{0}$ is undefined. That is, there is no real number to associate with it. In this case, we say the line has an *undefined slope*.

The graph of our line is shown in Figure 3. Our line with an undefined slope is a vertical line. All vertical lines have an undefined slope. (All horizontal lines, as we mentioned earlier, have 0 slope.)

Slopes of Parallel and Perpendicular Lines

In geometry, we call lines in the same plane that never intersect "parallel." For two lines to be nonintersecting, they must rise or fall at the same rate. In other words, two lines are *parallel* if and only if they have the *same slope*.

Although it is not as obvious, it is also true that two nonvertical lines are *perpendicular* if and only if the *product of their slopes is* -1. This is the same as saying their slopes are negative reciprocals.

We can state these facts with symbols as follows: If line l_1 has slope m_1 and line l_2 has slope m_2, then

$$l_1 \text{ is parallel to } l_2 \Leftrightarrow m_1 = m_2$$
$$\text{and}$$
$$l_1 \text{ is perpendicular to } l_2 \Leftrightarrow m_1 \cdot m_2 = -1 \text{ or } \left(m_1 = \frac{-1}{m_2} \right)$$

For example, if a line has a slope of $\frac{2}{3}$, then any line parallel to it has a slope of $\frac{2}{3}$. Any line perpendicular to it has a slope of $-\frac{3}{2}$ (the negative reciprocal of $\frac{2}{3}$).

Although we cannot give a formal proof of the relationship between the slopes of perpendicular lines at this level of mathematics, we can offer some justification for the relationship. Figure 4 shows the graphs of two lines. One of the lines has a slope of $\frac{2}{3}$; the other has a slope of $-\frac{3}{2}$. As you can see, the lines are perpendicular.

NOTE The two most common mistakes students make when first working with the formula for the slope of a line are

1. Putting the difference of the x-coordinates over the difference of the y-coordinates.

2. Subtracting in one order in the numerator and then subtracting in the opposite order in the denominator. You would make this mistake in Example 2 if you wrote $1 - (-3)$ in the numerator and then $-2 - (-5)$ in the denominator.

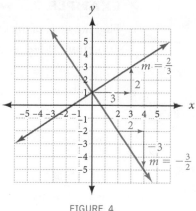

FIGURE 4

ⓑ Slope and Rate of Change

So far, the slopes we have worked with represent the ratio of the change in *y* to the corresponding change in *x*, or, on the graph of the line, the slope is the ratio of vertical change to horizontal change in moving from one point on the line to another. However, when our variables represent quantities from the world around us, slope can have additional interpretations.

EXAMPLE 4

On the chart below, find the slope of the line connecting the first point (1955, 0.29) with the last point (2015, 2.40). Explain the significance of the result.

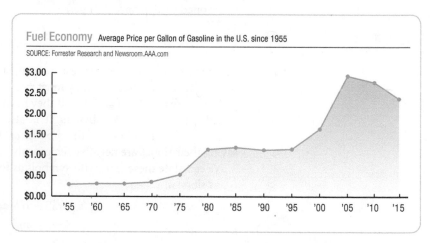

SOLUTION The slope of the line connecting the first point (1955, 0.29) with the last point (2015, 2.40), is

$$m = \frac{2.40 - 0.29}{2015 - 1955} = \frac{2.11}{60} = 0.0352$$

The units are dollars/year. If we write this in terms of cents, we have

$$m = 3.52 \text{ cents/year}$$

which is the average change in the price of a gallon of gasoline over a 60-year period of time.

Likewise, if we connect the points (2005, 2.93) and (2015, 2.40), the line that results has a slope of

$$m = \frac{2.40 - 2.93}{2015 - 2005} = \frac{-0.53}{10} = -0.053 \text{ dollars/year} = -5.3 \text{ cents/year}$$

which is the average change in the price of a gallon of gasoline over a 10-year period. As you can imagine by looking at the chart, the line connecting the first and last point is not as steep as the line connecting the points from 2005 and 2015, and this is what we are seeing numerically with our slope calculations. If we were summarizing this information for an article in the newspaper, we could say, "Although the price of a gallon of gasoline has increased 3.52 cents per year over the last 60 years, in the last 10 years the average annual rate has actually decreased, to 5.3 cents per year."

C Slope and Average Speed

Previously we introduced the rate equation $d = rt$. Suppose that a boat is traveling at a constant speed of 15 miles per hour in still water. The following table shows the distance the boat will have traveled in the specified number of hours. The graph of this data shown in Figure 5. Notice that the points all lie along a line.

t (Hours)	d (Miles)
0	0
1	15
2	30
3	45
4	60
5	75

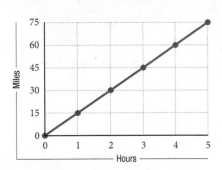

FIGURE 5

We can calculate the slope of this line using any two points from the table. Notice we have graphed the data with t on the horizontal axis and d on the vertical axis. Using the points (2, 30) and (3, 45), the slope will be

$$m = \frac{\text{rise}}{\text{run}} = \frac{45 - 30}{3 - 2} = \frac{15}{1} = 15$$

The units of the rise are miles and the units of the run are hours, so the slope will be in units of miles per hour. We see that the slope is simply the change in distance divided by the change in time, which is how we compute the average speed. Since the speed is constant, the slope of the line represents the speed of 15 miles per hour.

EXAMPLE 5

A car is traveling at a constant speed. A graph (Figure 6) of the distance the car has traveled over time is shown below. Use the graph to find the speed of the car.

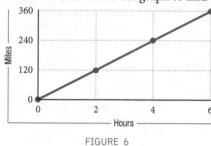

FIGURE 6

SOLUTION   Using the second and third points, we see the rise is $240 - 120 = 120$ miles, and the run is $4 - 2 = 2$ hours. The speed is given by the slope, which is

$$m = \frac{\text{rise}}{\text{run}}$$

$$= \frac{120 \text{ miles}}{2 \text{ hours}}$$

$$= 60 \text{ miles per hour}$$

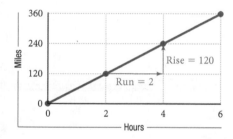

GETTING READY FOR CLASS

After reading through the preceding section, respond in your own words and in complete sentences.

A. If you were looking at a graph that described the performance of a stock you had purchased, why would it be better if the slope of the line were positive, rather than negative?

B. Describe the behavior of a line with a negative slope.

C. Would you rather climb a hill with a slope of $\frac{1}{2}$ or a slope of 3? Explain why.

D. Describe how to obtain the slope of a line if you know the coordinates of two points on the line.

USING TECHNOLOGY   *Families of Curves*

We can use a graphing calculator to investigate the effects of the numbers a and b on the graph of $y = ax + b$. To see how the number b affects the graph, we can hold a constant and let b vary. Doing so will give us a *family of curves*. Suppose we set $a = 1$ and then let b take on integer values from -3 to 3.

We will give three methods of graphing this set of equations on a graphing calculator.

Method 1: Y-Variables List

To use the Y-variables list, enter each equation at one of the Y variables, set the graph window, then graph. The calculator will graph the equations in order, starting with Y_1 and ending with Y_7. Following is the Y-variables list, an appropriate window, and a sample of the type of graph obtained (Figure 7).

$$Y_1 = X - 3$$
$$Y_2 = X - 2$$
$$Y_3 = X - 1$$
$$Y_4 = X$$
$$Y_5 = X + 1$$
$$Y_6 = X + 2$$
$$Y_7 = X + 3$$

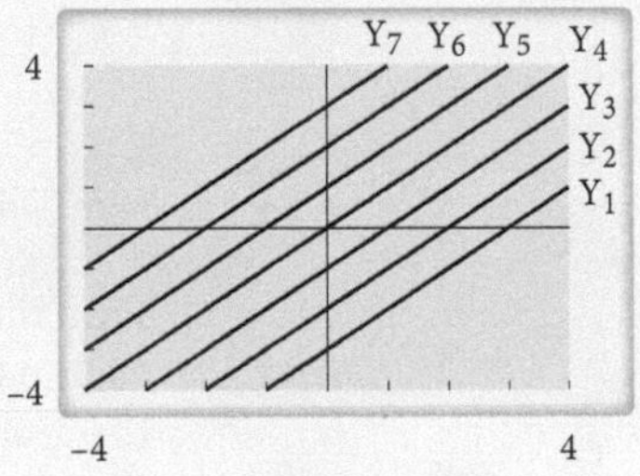

FIGURE 7

Window: X from -4 to 4, Y from -4 to 4

Method 2: Programming

The same result can be obtained by programming your calculator to graph $y = x + b$ for $b = -3, -2, -1, 0, 1, 2$, and 3. Here is an outline of a program that will do this. Check the manual that came with your calculator to find the commands for your calculator.

Step 1:   Clear screen
Step 2:   Set window for X from -4 to 4 and Y from -4 to 4
Step 3: $-3 \rightarrow B$
Step 4:   Label 1
Step 5:   Graph Y = X + B
Step 6: $B + 1 \rightarrow B$
Step 7:   If $B < 4$, Go to 1
Step 8:   End

Method 3: Using Lists

On the TI-82/83 you can set Y_1 as follows

$$Y_1 = X + \{-3, -2, -1, 0, 1, 2, 3\}$$

When you press $\boxed{\text{GRAPH}}$, the calculator will graph each line from $y = x + (-3)$ to $y = x + 3$.

Each of the three methods will produce graphs similar to those in Figure 7.

Problem Set 3.2

Find the slope of each of the following lines from the given graph.

1.
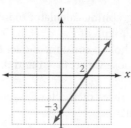

2.
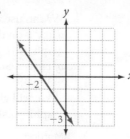

3.
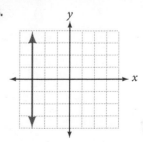

4.
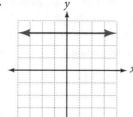

5.
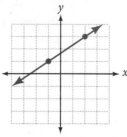

6.
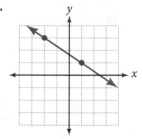

Find the slope of the line through each of the following pairs of points. Then, plot each pair of points, draw a line through them, and indicate the rise and run in the graph in the manner shown in Example 2.

7. $(2, 1), (4, 4)$ **8.** $(3, 1), (5, 4)$ **9.** $(1, 4), (5, 2)$

10. $(1, 3), (5, 2)$ **11.** $(1, -3), (4, 2)$ **12.** $(2, -3), (5, 2)$

13. $(2, -4)$ and $(5, -9)$ **14.** $(-3, 2)$ and $(-1, 6)$ **15.** $(-3, 5)$ and $(1, -1)$

16. $(-2, -1)$ and $(3, -5)$ **17.** $(-4, 6)$ and $(2, 6)$ **18.** $(2, -3)$ and $(2, 7)$

19. $(a, -3)$ and $(a, 5)$ **20.** $(x, 2y)$ and $(4x, 8y)$

Solve for the indicated variable if the line through the two given points has the given slope.

21. $(a, 3)$ and $(2, 6)$, $m = -1$ **22.** $(a, -2)$ and $(4, -6)$, $m = -3$

23. $(2, b)$ and $(-1, 4b)$, $m = -2$ **24.** $(-4, y)$ and $(-1, 6y)$, $m = 2$

25. $(2, 4)$ and (x, x^2), $m = 5$ **26.** $(3, 9)$ and (x, x^2), $m = -2$

27. $(1, 3)$ and $(x, 2x^2 + 1)$, $m = -6$ **28.** $(3, 7)$ and $(x, x^2 - 2)$, $m = -4$

For each of the equations in Problems 29–32, complete the table, and then use the results to find the slope of the graph of the equation.

29. $2x + 3y = 6$ **30.** $3x - 2y = 6$ **31.** $y = \dfrac{2}{3}x - 5$ **32.** $y = -\dfrac{3}{4}x + 2$

x	y
0	
	0

x	y
0	
	0

x	y
0	
3	

x	y
0	
4	

33. **Finding Slope from Intercepts** Graph the line that has an x-intercept of 3 and a y-intercept of -2. What is the slope of this line?

34. **Finding Slope from Intercepts** Graph the line with x-intercept -4 and y-intercept -2. What is the slope of this line?

35. **Parallel Lines** Find the slope of any line parallel to the line through $(2, 3)$ and $(-8, 1)$.

36. **Parallel Lines** Find the slope of any line parallel to the line through $(2, 5)$ and $(5, -3)$.

37. **Perpendicular Lines** Line l contains the points $(5, -6)$ and $(5, 2)$. Give the slope of any line perpendicular to l.

38. **Perpendicular Lines** Line l contains the points $(3, 4)$ and $(-3, 1)$. Give the slope of any line perpendicular to l.

39. **Parallel Lines** Line l contains the points $(-2, 1)$ and $(4, -5)$. Find the slope of any line parallel to l.

40. **Parallel Lines** Line l contains the points $(3, -4)$ and $(-2, -6)$. Find the slope of any line parallel to l.

41. **Perpendicular Lines** Line l contains the points $(-2, -5)$ and $(1, -3)$. Find the slope of any line perpendicular to l.

42. **Perpendicular Lines** Line l contains the points $(6, -3)$ and $(-2, 7)$. Find the slope of any line perpendicular to l.

43. Determine if each of the following tables could represent ordered pairs from an equation of a line.

a.

x	y
0	5
1	7
2	9
3	11

b.

x	y
-2	-5
0	-2
2	0
4	1

44. The following lines have slope 2, $\frac{1}{2}$, 0, and -1. Match each line to its slope value.

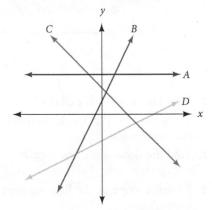

APPLYING THE CONCEPTS

An object is traveling at a constant speed. The distance and time data are shown on the given graph. Use the graph to find the speed of the object.

45.

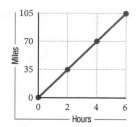

46.

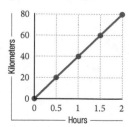

47.

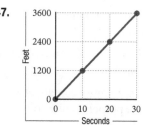

48.

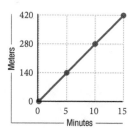

49. Heating a Block of Ice A block of ice with an initial temperature of $-20°C$ is heated at a steady rate. The graph shows how the temperature changes as the ice melts to become water and the water boils to become water and steam.

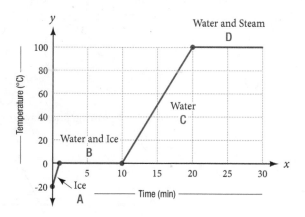

a. How long does it take all the ice to melt?

b. From the time the heat is applied to the block of ice, how long is it before the water boils?

c. Find the slope of the line segment labeled A. What units would you attach to this number?

d. Find the slope of the line segment labeled C. Be sure to attach units to your answer.

e. Is the temperature changing faster during the 1st minute or the 16th minute?

50. Slope of a Highway A sign at the top of the Cuesta Grade, outside of San Luis Obispo, reads "7% downgrade next 3 miles." The following diagram is a model of the Cuesta Grade that takes into account the information on that sign.

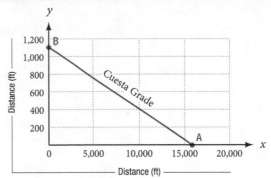

a. At point *B*, the graph crosses the *y*-axis at 1,106 feet. How far is it from the origin to point *A*?

b. What is the slope of the Cuesta Grade?

51. Solar Energy The graph below shows the annual shipments of solar thermal collectors in the United States. Using the graph below, find the slope of the line connecting the first (1, 8,000) and last (10, 20,000) endpoints and then explain in words what the slope represents.

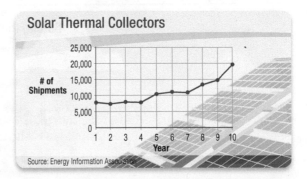

52. Age of New Mothers The graph shows the increase in average age of first-time mothers in the U.S. since 1970. Find the slope of the line that connects the points (1975, 21.75) to (1990, 24.25). Round to the nearest hundredth. Explain in words what the slope represents.

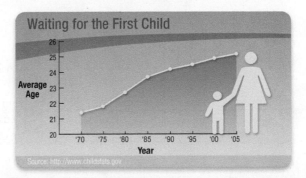

53. **Light Bulbs** The chart below shows a comparison of power usage between incandescent and energy-efficient light bulbs. Use the chart to work the following problems involving slope.

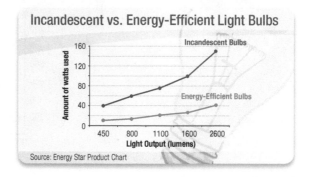

a. Find the slope of the line for the incandescent bulb from the two endpoints and then explain in words what the slope represents.

b. Find the slope of the line for the energy efficient bulb from the two endpoints and then explain in words what the slope represents.

c. Which light bulb is better? Why?

54. **Horse Racing** The graph shows the amount of money bet on horse racing from 1990 to 2015. Use the chart to work the following problems involving slope.

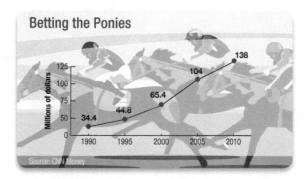

a. Find the slope of the line from 1990 to 1995, and then explain in words what the slope represents.

b. Find the slope of the line from 2005 to 2015, and then explain in words what the slope represents.

GETTING READY FOR THE NEXT SECTION

Simplify.

55. $2\left(-\dfrac{1}{2}\right)$ **56.** $\dfrac{3-(-1)}{-3-3}$ **57.** $-\dfrac{5-(-3)}{2-6}$ **58.** $3\left(-\dfrac{2}{3}x+1\right)$

Solve for y.

59. $\dfrac{y-b}{x-0}=m$

60. $2x+3y=6$

61. $y-3=-2(x+4)$

62. $y+1=-\dfrac{2}{3}(x-3)$

63. If $y=-\dfrac{4}{3}x+5$, find y when x is 0   **64.** If $y=-\dfrac{4}{3}x+5$, find y when x is 3

EXTENDING THE CONCEPTS

65. Find the slope of the line passing through the two points $(2, 4)$ and (x, x^2).

66. Find the slope of the line passing through the two points (x, x^2) and $(x+2, (x+2)^2)$.

67. Find a point P on the line $y=3x-2$ such that the slope of the line through $(1, 2)$ and P is equal to -1. (Hint: Any point on $y=3x-2$ has coordinates $(x, 3x-2)$.

68. Find a point P on the line $y=-2x+3$ such that the slope of the line through $(2, 3)$ and P is equal to 2.

69. Find two points P on the curve $y=x^2+2$ such that the slope of the line through $(1, 4)$ and P is equal to 2.

70. Find two points P on the curve $y=x^2-3$ such that the slope of the line through $(2, 3)$ and P is equal to 3.

71. Use your Y-variables list or write a program to graph the family of curves $Y=2X+B$ for $B=-3, -2, -1, 0, 1, 2,$ and 3.

72. Use your Y-variables list or write a program to graph the family of curves $Y=-2X+B$ for $B=-3, -2, -1, 0, 1, 2,$ and 3.

73. Use your Y-variables list or write a program to graph the family of curves $Y=AX$ for $A=-3, -2, -1, 0, 1, 2,$ and 3.

74. Use your Y-variables list or write a program to graph the family of curves $Y=AX$ for $A=\frac{1}{4}, \frac{1}{3}, \frac{1}{2}, 1, 2,$ and 3.

75. Use your Y-variables list or write a program to graph the family of curves $Y=AX+2$ for $A=-3, -2, -1, 0, 1, 2,$ and 3.

76. Use your Y-variables list or write a program to graph the family of curves $Y=AX-2$ for $A=\frac{1}{4}, \frac{1}{3}, \frac{1}{2}, 1, 2,$ and 3.

THE EQUATION OF A LINE

3.3

OBJECTIVES

A Find the equation of a line given the slope and *y*-intercept of a line.

B Find the equation of a line given a point on the line and the slope of the line.

C Write the equation of the line in standard form.

©iStockphoto.com/MarianVejcik

The table and illustrations below show some corresponding temperatures on the Fahrenheit and Celsius temperature scales. For example, water freezes at 32°F and 0°C, and boils at 212°F and 100°C.

°F °C
32°– – – – – – –0°
Ice

Degrees Celsius	Degrees Fahrenheit
0	32
25	77
50	122
75	167
100	212

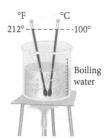

°F °C
212°– – – – – – –100°
Boiling water

If we plot all the points in the table using the *x*-axis for temperatures on the Celsius scale and the *y*-axis for temperatures on the Fahrenheit scale, we see that they line up in a straight line (Figure 1).

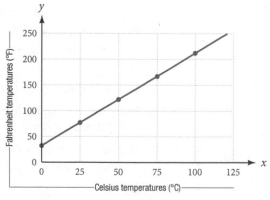

FIGURE 1

This means that a linear equation in two variables will give a perfect description of the relationship between the two scales. That equation is

$$F = \frac{9}{5}C + 32$$

The techniques we use to find the equation of a line from a set of points is what this section is all about.

Slope-Intercept Form

Suppose line l has slope m and y-intercept b. What is the equation of l? Because the y-intercept is b, we know the point $(0, b)$ is on the line. If (x, y) is any other point on l, then using the definition for slope, we have

$$\frac{y - b}{x - 0} = m \qquad \text{Definition of slope}$$

$$y - b = mx \qquad \text{Multiply both sides by } x.$$

$$y = mx + b \qquad \text{Add } b \text{ to both sides.}$$

This last equation is known as the *slope-intercept form* of the equation of a straight line.

PROPERTY *Slope-Intercept Form of the Equation of a Line*

The equation of any line with slope m and y-intercept b is given by

$$y = mx + b$$

$\nearrow$ $\uparrow$

Slope y-intercept

When the equation is in this form, the *slope* of the line is always the *coefficient* of x and the y-intercept is always the *constant term*.

EXAMPLE 1

Find the equation of the line with slope $-\frac{4}{3}$ and y-intercept 5. Then graph the line.

SOLUTION Substituting $m = -\frac{4}{3}$ and $b = 5$ into the equation $y = mx + b$, we have

$$y = -\frac{4}{3}x + 5$$

Finding the equation from the slope and y-intercept is just that easy. If the slope is m and the y-intercept is b, then the equation is always $y = mx + b$. Now, let's graph the line.

Because the y-intercept is 5, the graph goes through the point $(0, 5)$. To find a second point on the graph, we start at $(0, 5)$ and move 4 units down (that's a rise of -4) and 3 units to the right (a run of 3). The point we end up at is $(3, 1)$. Drawing a line that passes through $(0, 5)$ and $(3, 1)$, we have the graph of our equation. (Note that we could also let the rise $= 4$ and the run $= -3$ and obtain the same graph.)

The graph is shown in Figure 2.

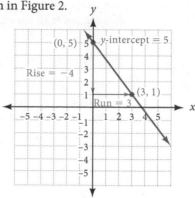

FIGURE 2

EXAMPLE 2

Give the slope and y-intercept for the line $2x - 3y = 5$.

SOLUTION To use the slope-intercept form, we must solve the equation for y in terms of x:

$$2x - 3y = 5$$

$$-3y = -2x + 5 \qquad \text{Add } -2x \text{ to both sides.}$$

$$y = \frac{2}{3}x - \frac{5}{3} \qquad \text{Divide by } -3.$$

The last equation has the form $y = mx + b$. The slope must be $m = \frac{2}{3}$ and the y-intercept is $b = -\frac{5}{3}$.

EXAMPLE 3

Graph the linear function $y = -\frac{2}{3}x + 2$ using the slope and y-intercept.

SOLUTION The slope is $m = -\frac{2}{3}$ and the y-intercept is $b = 2$. Therefore, the point $(0, 2)$ is on the graph, and the ratio of rise to run going from $(0, 2)$ to any other point on the line is $-\frac{2}{3}$. If we start at $(0, 2)$ and move 2 units up (that's a rise of 2) and 3 units to the left (a run of -3), we will be at another point on the graph. (We could also go down 2 units and right 3 units and still be assured of ending up at another point on the line because $\frac{2}{-3}$ is the same as $\frac{-2}{3}$.)

> **NOTE** As we mentioned earlier in this chapter, the rectangular coordinate system is the tool we use to connect algebra and geometry. Example 3 illustrates this connection, as do the many other examples in this chapter. In Example 3, Descartes's rectangular coordinate system allows us to associate the equation $y = -\frac{2}{3}x + 2$ (an algebraic concept) with the straight line (a geometric concept) shown in Figure 3.

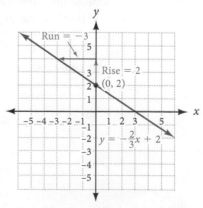

FIGURE 3

B Point-Slope Form

A second useful form of the equation of a straight line is the point-slope form.

Let line l contain the point (x_1, y_1) and have slope m. If (x, y) is any other point on l, then by the definition of slope we have

$$\frac{y - y_1}{x - x_1} = m$$

Multiplying both sides by $(x - x_1)$ gives us

$$(x - x_1) \cdot \frac{y - y_1}{x - x_1} = m(x - x_1)$$

$$y - y_1 = m(x - x_1)$$

This last equation is known as the *point-slope form* of the equation of a straight line.

PROPERTY *Point-Slope Form of the Equation of a Line*

The equation of the line through (x_1, y_1) with slope m is given by

$$y - y_1 = m(x - x_1)$$

This form of the equation of a straight line is used to find the equation of a line, either given one point on the line and the slope, or given two points on the line.

EXAMPLE 4

Find the equation of the line with slope -2 that contains the point $(-4, 3)$. Write the answer in slope-intercept form.

SOLUTION

Using $(x_1, y_1) = (-4, 3)$ and $m = -2$

in $y - y_1 = m(x - x_1)$ Point-slope form

gives us $y - 3 = -2(x + 4)$ *Note:* $x - (-4) = x + 4$.

$y - 3 = -2x - 8$ Multiply out right side.

$y = -2x - 5$ Add 3 to each side.

Figure 4 is the graph of the line that contains $(-4, 3)$ and has a slope of -2. Notice that the y-intercept on the graph matches that of the equation we found.

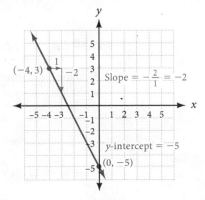

FIGURE 4

EXAMPLE 5

Find the equation of the line that passes through the points $(-3, 3)$ and $(3, -1)$.

SOLUTION   We begin by finding the slope of the line:

$$m = \frac{3 - (-1)}{-3 - 3} = \frac{4}{-6} = -\frac{2}{3}$$

Using $(x_1, y_1) = (3, -1)$ and $m = -\frac{2}{3}$ in $y - y_1 = m(x - x_1)$ yields

$$y + 1 = -\frac{2}{3}(x - 3)$$

$$y + 1 = -\frac{2}{3}x + 2 \qquad \text{Multiply out right side.}$$

$$y = -\frac{2}{3}x + 1 \qquad \text{Add } -1 \text{ to each side.}$$

Figure 5 shows the graph of the line that passes through the points $(-3, 3)$ and $(3, -1)$. As you can see, the slope and y-intercept are $-\frac{2}{3}$ and 1, respectively.

> **NOTE**   We could have used the point $(-3, 3)$ instead of $(3, -1)$ and obtained the same equation. That is, using $(x_1, y_1) = (-3, 3)$ and $m = -\frac{2}{3}$ in
>
> $y - y_1 = m(x - x_1)$ gives us
>
> $$y - 3 = -\frac{2}{3}(x + 3)$$
>
> $$y - 3 = -\frac{2}{3}x - 2$$
>
> $$y = -\frac{2}{3}x + 1$$
>
> which is the same result we obtained using $(3, -1)$.

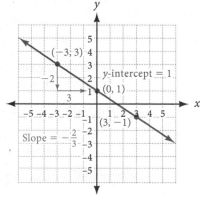

FIGURE 5

C Standard Form

The last form of the equation of a line that we will consider in this section is called the *standard form*. It is used mainly to write equations in a form that is free of fractions and is easy to compare with other equations.

> **PROPERTY   Standard Form for the Equation of a Line**
>
> If a, b, and c are integers, then the equation of a line is in standard form when it has the form
>
> $$ax + by = c$$

If we were to write the equation

$$y = -\frac{2}{3}x + 1$$

in standard form, we would first multiply both sides by 3 to obtain

$$3y = -2x + 3$$

Then we would add $2x$ to each side, yielding

$$2x + 3y = 3$$

which is a linear equation in standard form.

EXAMPLE 6

Give the equation of the line through $(-1, 4)$ whose graph is perpendicular to the graph of $2x - y = -3$. Write the answer in standard form.

SOLUTION To find the slope of $2x - y = -3$, we solve for y:

$$2x - y = -3$$
$$y = 2x + 3$$

The slope of this line is 2. The line we are interested in is perpendicular to the line with slope 2 and must, therefore, have a slope of $-\frac{1}{2}$.

Using $(x_1, y_1) = (-1, 4)$ and $m = -\frac{1}{2}$, we have

$$y - y_1 = m(x - x_1)$$
$$y - 4 = -\frac{1}{2}(x + 1)$$

Because we want our answer in standard form, we multiply each side by 2.

$$2y - 8 = -1(x + 1)$$
$$2y - 8 = -x - 1$$
$$x + 2y - 8 = -1$$
$$x + 2y = 7$$

The last equation is in standard form.

As a final note, the following summary reminds us that all horizontal lines have equations of the form $y = b$, and slopes of 0. Since they cross the y-axis at b, the y-intercept is b; there is no x-intercept. Vertical lines have an undefined slope, and equations of the form $x = a$. Each will have an x-intercept at a, and no y-intercept. Finally, equations of the form $y = mx$ have graphs that pass through the origin. The slope is always m and both the x-intercept and the y-intercept are 0.

GETTING READY FOR CLASS

After reading through the preceding section respond in your own words and in complete sentences.

A. How would you graph the line $y = \frac{1}{2}x + 3$?

B. What is the slope-intercept form of the equation of a line?

C. Describe how you would find the equation of a line if you knew the slope and the y-intercept of the line.

D. If you had the graph of a line, how would you use it to find the equation of the line?

FACTS FROM GEOMETRY *Special Equations: Their Graphs, Slopes, and Intercepts*

For the equations below, m, a, and b are real numbers.

Through the Origin

Equation: $y = mx$
Slope $= m$
x-intercept $= 0$
y-intercept $= 0$

Vertical Line

Equation: $x = a$
Undefined
x-intercept $= a$
No y-intercept

Horizontal Line

Equation: $y = b$
Slope $= 0$
No x-intercept
y-intercept $= b$

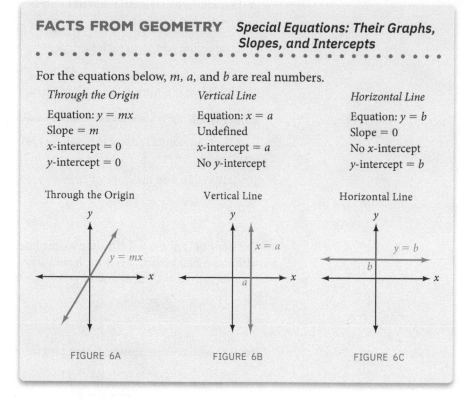

Through the Origin

Vertical Line

Horizontal Line

FIGURE 6A

FIGURE 6B

FIGURE 6C

USING TECHNOLOGY *Graphing Calculators*

One advantage of using a graphing calculator to graph lines is that a calculator does not care whether the equation has been simplified or not. To illustrate, in Example 5 we found that the equation of the line with slope $-\frac{2}{3}$ that passes through the point $(3, -1)$ is

$$y + 1 = -\frac{2}{3}(x - 3)$$

Normally, to graph this equation we would simplify it first. With a graphing calculator, we add -1 to each side and enter the equation this way:

$$Y_1 = -(2/3)(X - 3) - 1$$

No simplification is necessary. We can graph the equation in this form, and the graph will be the same as the simplified form of the equation, which is $y = -\frac{2}{3}x + 1$. To convince yourself that this is true, graph both the simplified form for the equation and the unsimplified form in the same window. As you will see, the two graphs coincide.

Problem Set 3.3

Give the equation of the line with the following slope and y-intercept.

1. $m = -4, b = -3$ **2.** $m = -6, b = \frac{4}{3}$ **3.** $m = -\frac{2}{3}, b = 0$

4. $m = 0, b = \frac{3}{4}$ **5.** $m = -\frac{2}{3}, b = \frac{1}{4}$ **6.** $m = \frac{5}{12}, b = -\frac{3}{2}$

Find the slope of a line **a.** parallel and **b.** perpendicular to the given line.

7. $y = 3x - 4$ **8.** $y = -4x + 1$ **9.** $3x + y = -2$

10. $2x - y = -4$ **11.** $2x + 5y = -11$ **12.** $3x - 5y = -4$

Give the slope and y-intercept for each of the following equations. Sketch the graph using the slope and y-intercept. Give the slope of any line perpendicular to the given line.

13. $y = 3x - 2$ **14.** $y = 2x + 3$ **15.** $2x - 3y = 12$

16. $3x - 2y = 12$ **17.** $4x + 5y = 20$ **18.** $5x - 4y = 20$

For each of the following lines, name the slope and y-intercept. Then write the equation of the line in slope-intercept form.

19.
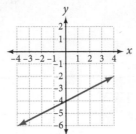

20.
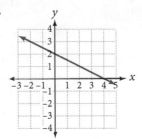

21.
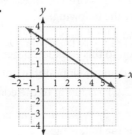

22.
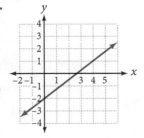

For each of the following problems, the slope and one point on the line are given. In each case, find the equation of that line. (Write the equation for each line in slope-intercept form.)

23. $(-2, -5); m = 2$ **24.** $(-1, -5); m = 2$ **25.** $(-4, 1); m = -\frac{1}{2}$

26. $(-2, 1); m = -\frac{1}{2}$ **27.** $\left(-\frac{1}{3}, 2\right); m = -3$ **28.** $\left(-\frac{2}{3}, 5\right); m = -3$

29. $(-4, 2), m = \frac{2}{3}$ **30.** $(3, -4), m = -\frac{1}{3}$ **31.** $(-5, -2), m = -\frac{1}{4}$

32. $(-4, -3), m = \frac{1}{6}$

Find the equation of the line that passes through each pair of points. Write your answers in standard form.

33. $(3, -2), (-2, 1)$ **34.** $(-4, 1), (-2, -5)$ **35.** $\left(-2, \frac{1}{2}\right), \left(-4, \frac{1}{3}\right)$

36. $(-6, -2), (-3, -6)$ **37.** $\left(\frac{1}{3}, -\frac{1}{5}\right), \left(-\frac{1}{3}, -1\right)$ **38.** $\left(-\frac{1}{2}, -\frac{1}{2}\right), \left(\frac{1}{2}, \frac{1}{10}\right)$

For each of the following lines, name the coordinates of any two points on the line. Then use those two points to find the equation of the line.

39.

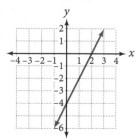

40.

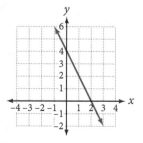

41.

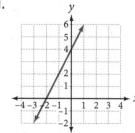

42.

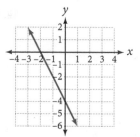

43. The equation $3x - 2y = 10$ is a linear equation in standard form. From this equation, answer the following:

  a. Find the x and y intercepts

  b. Find a solution to this equation other than the intercepts in part **a**.

  c. Write this equation in slope-intercept form

  d. Is the point $(2, 2)$ a solution to the equation?

44. The equation $4x + 3y = 8$ is a linear equation in standard form. From this equation, answer the following:

  a. Find the x and y intercepts

  b. Find a solution to this equation other than the intercepts in part **a**.

  c. Write this equation in slope-intercept form

  d. Is the point $(-3, 2)$ a solution to the equation?

PAYING ATTENTION TO INSTRUCTIONS

The next two problems are intended to give you practice reading, and paying attention to the instructions that accompany the problems you are working. Working these problems is an excellent way to get ready for a test or a quiz.

45. Work each problem according to the instructions given:

    a. Solve: $-2x + 1 = -3$ **b.** Find x when y is 0: $-2x + y = -3$

    c. Find y when x is 0: $-2x + y = -3$ **d.** Graph: $-2x + y = -3$

    e. Solve for y: $-2x + y = -3$

46. Work each problem according to the instructions given:

    a. Solve: $\dfrac{x}{3} + \dfrac{1}{4} = 1$ **b.** Find x when y is 0: $\dfrac{x}{3} + \dfrac{y}{4} = 1$

    c. Find y when x is 0: $\dfrac{x}{3} + \dfrac{y}{4} = 1$ **d.** Graph: $\dfrac{x}{3} + \dfrac{y}{4} = 1$

    e. Solve for y: $\dfrac{x}{3} + \dfrac{y}{4} = 1$

47. Graph each of the following lines. In each case, name the slope, the x-intercept, and the y-intercept.

    a. $y = \dfrac{1}{2}x$ **b.** $x = 3$ **c.** $y = -2$

48. Graph each of the following lines. In each case, name the slope, the x-intercept, and the y-intercept.

    a. $y = -2x$ **b.** $x = 2$ **c.** $y = -4$

49. Find the equation of the line parallel to the graph of $3x - y = 5$ that contains the point $(-1, 4)$.

50. Find the equation of the line parallel to the graph of $2x - 4y = 5$ that contains the point $(0, 3)$.

51. Line l is perpendicular to the graph of the equation $2x - 5y = 10$ and contains the point $(-4, -3)$. Find the equation for l.

52. Line l is perpendicular to the graph of the equation $-3x - 5y = 2$ and contains the point $(2, -6)$. Find the equation for l.

53. Give the equation of the line perpendicular to the graph of $y = -4x + 2$ that has an x-intercept of -1.

54. Write the equation of the line parallel to the graph of $7x - 2y = 14$ that has an x-intercept of 5.

55. Find the equation of the line with x-intercept 3 and y-intercept 2.

56. Find the equation of the line with x-intercept 2 and y-intercept 3.

APPLYING THE CONCEPTS

57. Deriving the Temperature Equation The table below resembles the table from the introduction to this section. The rows of the table give us ordered pairs (C, F).

Degrees Celsius	Degrees Fahrenheit
C	F
0	32
25	77
50	122
75	167
100	212

a. Use any two of the ordered pairs from the table to derive the equation $F = \frac{9}{5}C + 32$.

b. Use the equation from part **a.** to find the Fahrenheit temperature that corresponds to a Celsius temperature of 30°.

58. Maximum Heart Rate The table below gives the maximum heart rate for adults 30, 40, 50, and 60 years old. Each row of the table gives us an ordered pair (A, M).

Age (years)	Maximum Heart Rate (beats per minute)
A	M
30	190
40	180
50	170
60	160

a. Use any two of the ordered pairs from the table to derive the equation $M = 220 - A$, which gives the maximum heart rate M for an adult whose age is A.

b. Use the equation from part **a.** to find the maximum heart rate for a 25-year-old adult.

59. Textbook Cost To produce this textbook, suppose the publisher spent $125,000 for typesetting and $6.50 per book for printing and binding. The total cost to produce and print n books can be written as

$$C = 125{,}000 + 6.5n$$

a. Suppose the number of books printed in the first printing is 10,000. What is the total cost?

b. If the average cost is the total cost divided by the number of books printed, find the average cost of producing 10,000 textbooks.

c. Find the cost to produce one more textbook when you have already produced 10,000 textbooks.

60. Exercise Heart Rate In an aerobics class, the instructor indicates that her students' exercise heart rate is 60% of their maximum heart rate, where maximum heart rate is 220 minus their age.

   a. Determine the equation that gives exercise heart rate E in terms of age A.

   b. Use the equation to find the exercise heart rate of a 22-year-old student.

   c. Sketch the graph of the equation for students from 18 to 80 years of age.

61. Horse Racing The graph shows the total amount of money wagered on the Kentucky Derby. Find an equation for the line segment between 2005 and 2015. Write the equation in slope-intercept form. Round the slope and intercept to the nearest tenth.

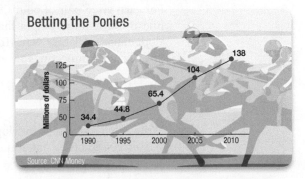

62. Solar Energy The graph shows the annual number of solar thermal collector shipments in the United States. Find an equation for the line segment that connects the points (8, 13,750) and (9, 15,000). Write your answer in slope-intercept form.

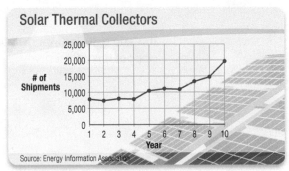

GETTING READY FOR THE NEXT SECTION

63. Which of the following are solutions to $x + y \le 4$?

$$(0, 0) \quad (4, 0) \quad (2, 3)$$

64. Which of the following are solutions to $y < 2x - 3$?

$$(0, 0) \quad (3, -2) \quad (-3, 2)$$

65. Which of the following are solutions to $y \le \frac{1}{2}x$?

$$(0, 0) \quad (2, 0) \quad (-2, 0)$$

66. Which of the following are solutions to $y > -2x$?

$$(0, 0) \quad (2, 0) \quad (-2, 0)$$

EXTENDING THE CONCEPTS

The midpoint M of two points (x_1, y_1) and (x_2, y_2) is defined to be the average of each of their coordinates, so

$$M = \left(\frac{x_1 + x_2}{2}, \frac{y_1 + y_2}{2} \right)$$

For example, the midpoint of $(-2, 3)$ and $(6, 8)$ is given by

$$\left(\frac{-2 + 6}{2}, \frac{3 + 8}{2} \right) = \left(2, \frac{11}{2} \right)$$

For each given pair of points, find the equation of the line perpendicular to the line through these points that passes through their midpoint. Answer using slope-intercept form. **Note:** This line is called the perpendicular bisector of the line segment connecting the two points.

67. $(1, 4)$ and $(7, 8)$

68. $(-2, 1)$ and $(6, 7)$

69. $(-5, 1)$ and $(-1, 4)$

70. $(-6, -2)$ and $(2, 1)$

Write each equation in slope-intercept form. Then name the slope, the y-intercept, and the x-intercept.

71. $\dfrac{x}{2} + \dfrac{y}{3} = 1$

72. $\dfrac{x}{5} + \dfrac{y}{4} = 1$

73. $\dfrac{x}{-2} + \dfrac{y}{3} = 1$

74. $\dfrac{x}{2} + \dfrac{y}{-3} = 1$

75. When a linear equation is written in the form

$$\frac{x}{a} + \frac{y}{b} = 1$$

it is said to be in *two-intercept form*. Find the x-intercept, the y-intercept, and the slope of this line.

OBJECTIVE

A Graph the solution set for a linear equality in two variables.

© Tessa McGlasson Avila, 2018

3.4

A small movie theater holds 100 people. The owner charges more for adults than for children, so it is important to know the different combinations of adults and children that can be seated at one time. The shaded region in Figure 1 contains all the seating combinations. The line $x + y = 100$ shows the combinations for a full theater: The y-intercept corresponds to a theater full of adults, and the x-intercept corresponds to a theater full of children. In the shaded region below the line $x + y = 100$ are the combinations that occur if the theater is not full.

Shaded regions like the one shown in Figure 1 are produced by linear inequalities in two variables, which is the topic of this section.

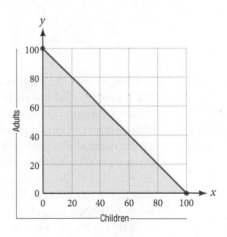

FIGURE 1

A Solution Set of a Linear Inequality

A *linear inequality in two variables* is any expression that can be put in the form

$$ax + by < c$$

where a, b, and c are real numbers (a and b not both 0). The inequality symbol can be any one of the following four: $<, \leq, >, \geq$.

Some examples of linear inequalities are

$$2x + 3y < 6 \qquad y \geq 2x + 1 \qquad x - y \leq 0$$

Although not all of these examples have the form $ax + by < c$, each one can be put in that form.

The solution set for a linear inequality is a *section of the coordinate plane*. The *boundary* for the section is found by replacing the inequality symbol with an equal sign and graphing the resulting equation. The boundary is included in the solution set (and is represented with a *solid line*) if the inequality symbol used originally is $\leq$ or $\geq$. The boundary is not included (and is represented with a *broken line*) if the original symbol is $<$ or $>$.

EXAMPLE 1

Graph the solution set for $x + y \leq 4$.

SOLUTION The boundary for the graph is the graph of $x + y = 4$. The boundary is included in the solution set because the inequality symbol is $\leq$.
Figure 2 is the graph of the boundary:

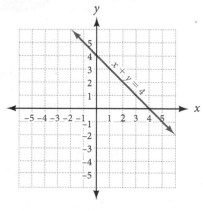

FIGURE 2

The boundary separates the coordinate plane into two regions: the region above the boundary and the region below it. The solution set for $x + y \leq 4$ is one of these two regions along with the boundary. To find the correct region, we simply choose any convenient point that is *not* on the boundary. We then substitute the coordinates of the point into the original inequality $x + y \leq 4$. If the point we choose satisfies the inequality, then it is a member of the solution set, and we can assume that all points on the same side of the boundary as the chosen point are also in the solution set. If the coordinates of our point do not satisfy the original inequality, then the solution set lies on the other side of the boundary.

In this example, a convenient point that is not on the boundary is the origin.

Substituting	$(0, 0)$
into	$x + y \leq 4$
gives us	$0 + 0 \leq 4$
	$0 \leq 4$ A true statement

Because the origin is a solution to the inequality $x + y \leq 4$ and the origin is below the boundary, all other points below the boundary are also solutions.

Figure 3 is the graph of $x + y \leq 4$.

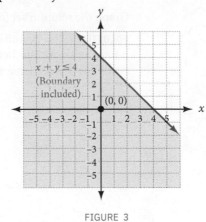

FIGURE 3

The region above the boundary is described by the inequality $x + y > 4$.

Here is a list of steps to follow when graphing the solution sets for linear inequalities in two variables.

> ### HOW TO   *Graph a Linear Inequality in Two Variables*
>
> **Step 1:** Replace the inequality symbol with an equal sign. The resulting equation represents the boundary for the solution set.
>
> **Step 2:** Graph the boundary found in step 1. Use a *solid line* if the boundary is included in the solution set (i.e., if the original inequality symbol was either $\leq$ or $\geq$). Use a *broken line* to graph the boundary if it is *not* included in the solution set. (It is not included if the original inequality was either $<$ or $>$.)
>
> **Step 3:** Choose any convenient point not on the boundary and substitute the coordinates into the *original* inequality. If the resulting statement is *true*, the solution set lies on the *same* side of the boundary as the chosen point. If the resulting statement is *false,* the solution set lies on the *opposite* side of the boundary.

EXAMPLE 2

Graph the solution set for $y < 2x - 3$.

SOLUTION   The boundary is the graph of $y = 2x - 3$, a line with slope 2 and y-intercept -3. The boundary is not included because the original inequality symbol is $<$. We therefore use a broken line to represent the boundary in Figure 4.

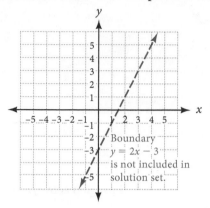

FIGURE 4

A convenient test point is again the origin:

Using	$(0, 0)$
in	$y < 2x - 3$
we have	$0 < 2(0) - 3$
	$0 < -3$ A false statement

Because our test point gives us a false statement and it lies above the boundary, the solution set must lie on the other side of the boundary (Figure 5).

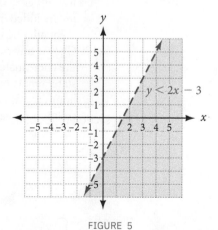

FIGURE 5

USING TECHNOLOGY   *Graphing Calculators*

Most graphing calculators have a Shade command that allows a portion of a graphing screen to be shaded. With this command we can visualize the solution sets to linear inequalities in two variables. Because most graphing calculators cannot draw a dotted line, however, we are not actually "graphing" the solution set, only visualizing it.

Strategy for Visualizing a Linear Inequality in Two Variables on a Graphing Calculator

Step 1: Solve the inequality for y.

Step 2: Replace the inequality symbol with an equal sign. The resulting equation represents the boundary for the solution set.

Step 3: Graph the equation in an appropriate viewing window.

Step 4: Use the Shade command to indicate the solution set:
For inequalities having the $<$ or $\leq$ sign, use Shade (Xmin, Y_1).
For inequalities having the $>$ or $\geq$ sign, use Shade (Y_1, Xmax).
Figures 6 and 7 show the graphing calculator screens that help us visualize the solution set to the inequality $y < 2x - 3$ that we graphed in Example 2.

Windows: X from -5 to 5, Y from -5 to 5

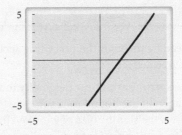

FIGURE 6 $Y_1 = 2x - 3$

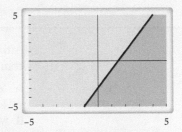

FIGURE 7 Shade (Xmin, Y_1)

EXAMPLE 3

Graph the solution set for $x \le 5$.

SOLUTION   The boundary is $x = 5$, which is a vertical line. All points in Figure 8 to the left have x-coordinates less than 5 and all points to the right have x-coordinates greater than 5.

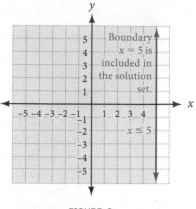

FIGURE 8

GETTING READY FOR CLASS

After reading through the preceding section, respond in your own words and in complete sentences.

A. When graphing a linear inequality in two variables, how do you find the equation of the boundary line?

B. What is the significance of a broken line in the graph of an inequality?

C. When graphing a linear inequality in two variables, how do you know which side of the boundary line to shade?

D. Describe the set of ordered pairs that are solutions to $x + y < 6$.

Problem Set 3.4

Graph the solution set for each of the following.

1. $x + y < 5$ **2.** $x + y \leq 5$ **3.** $x - y \geq -3$

4. $x - y > -3$ **5.** $2x + 3y < 6$ **6.** $2x - 3y > -6$

7. $-x + 2y > -4$ **8.** $-x - 2y < 4$ **9.** $2x + y < 5$

10. $2x + y < -5$ **11.** $y < 2x - 1$ **12.** $y \leq 2x - 1$

13. $3x - 4y < 12$ **14.** $-2x + 3y < 6$ **15.** $-5x + 2y \leq 10$

16. $4x - 2y \leq 8$

For each graph shown here, name the linear inequality in two variables that is represented by the shaded region.

17.

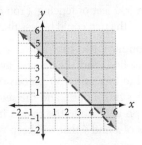

18.

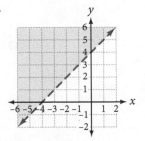

19.

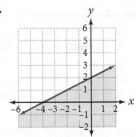

20.

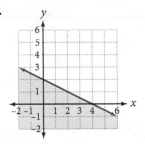

Graph each inequality.

21. $x \geq 3$ **22.** $x > -2$ **23.** $y \leq 4$

24. $y > -5$ **25.** $y < 2x$ **26.** $y > -3x$

27. $y \geq \dfrac{1}{2}x$ **28.** $y \leq \dfrac{1}{3}x$ **29.** $y \geq \dfrac{3}{4}x - 2$

30. $y > -\dfrac{2}{3}x + 3$ **31.** $\dfrac{x}{3} + \dfrac{y}{2} > 1$ **32.** $\dfrac{x}{5} + \dfrac{y}{4} < 1$

33. $\dfrac{x}{3} - \dfrac{y}{2} > 1$ **34.** $-\dfrac{x}{4} - \dfrac{y}{3} > 1$ **35.** $y \leq -\dfrac{2}{3}x$

36. $y \geq \dfrac{1}{4}x - 1$ **37.** $5x - 3y < 0$ **38.** $2x + 3y > 0$

39. $\dfrac{x}{4} + \dfrac{y}{5} \leq 1$ **40.** $\dfrac{x}{2} + \dfrac{y}{3} < 1$

APPLYING THE CONCEPTS

41. **Number of People in a Dance Club** A dance club holds a maximum of 200 people. The club charges one price for students and a higher price for nonstudents. If the number of students in the club at any time is x and the number of nonstudents is y, sketch the graph and shade the region in the first quadrant that contains all combinations of students and nonstudents who are in the club at any time.

42. **Many Perimeters** Suppose you have 500 feet of fencing that you will use to build a rectangular livestock pen. Let x represent the length of the pen and y represent the width. Sketch the graph and shade the region in the first quadrant that contains all possible values of x and y that will give you a rectangle from 500 feet of fencing. (You don't have to use all of the fencing, so the perimeter of the pen could be less than 500 feet.)

43. **Gas Mileage** You have two cars. The first car travels an average of 12 miles on a gallon of gasoline, and the second averages 22 miles per gallon. Suppose you can afford to buy up to 30 gallons of gasoline this month. If the first car is driven x miles this month, and the second car is driven y miles this month, sketch the graph and shade the region in the first quadrant that gives all the possible values of x and y that will keep you from buying more than 30 gallons of gasoline this month.

44. **Student Loan Payments** When considering how much debt to incur in student loans, it is advisable to keep your student loan payment after graduation to 8% or less of your starting monthly income. Let x represent your starting monthly salary and let y represent your monthly student loan payment. Write an inequality that describes this situation. Sketch the graph and shade the region in the first quadrant that is a solution to your inequality.

GETTING READY FOR THE NEXT SECTION

Complete each table using the given equation.

45. $y = 7.5x$

x	y
0	
10	
20	

46. $h = 32t - 16t^2$

t	h
0	
1	
-1	

47. $y = 7.5x$

x	y
0	
$\frac{1}{2}$	
1	

48. $h = 32t - 16t^2$

t	h
-3	
0	
3	

EXTENDING THE CONCEPTS

Graph each inequality.

49. $y < |x + 2|$

50. $y > |x - 2|$

51. $y > |x - 3|$

52. $y < |x + 3|$

53. $y \leq |x - 1|$

54. $y \geq |x + 1|$

55. $y < x^2$

56. $y > x^2 - 2$

57. The Associated Students organization holds a *Night at the Movies* fundraiser. Student tickets are $1.00 each and nonstudent tickets are $2.00 each. The theater holds a maximum of 200 people. The club needs to collect at least $100 to make money. Shade the region in the first quadrant that contains all combinations of students and nonstudents who could attend the movie night so that the club makes money.

3.5

OBJECTIVES

A Construct a table and graph from a function.

B Identify the domain and range of a function.

C Determine whether a relation is also a function.

D Determine whether a graph is a function using the vertical line test.

The ad shown here is from the Help Wanted section of the local newspaper. If you held the job described in the ad, you would earn $10.00 for every hour you worked. The amount of money you make in one week depends on the number of hours you work that week. In mathematics, we say that your weekly earnings are a *function* of the number of hours you work.

A An Informal Look at Functions

Suppose you have a job that pays $10.00 per hour and that you work anywhere from 0 to 40 hours per week. If we let the variable x represent hours and the variable y represent the money you make, then the relationship between x and y can be written as

$$y = 10x \quad \text{for} \quad 0 \le x \le 40$$

EXAMPLE 1

Construct a table and graph for the function

$$y = 10x \quad \text{for} \quad 0 \le x \le 40$$

SOLUTION Table 1 gives some of the paired data that satisfy the equation $y = 10x$. Figure 1 is the graph of the equation with the restriction $0 \le x \le 40$.

TABLE 1 Weekly Wages		
Hours Worked	Rule	Pay
x	$y = 10x$	y
0	$y = 10(0)$	0
10	$y = 10(10)$	100
20	$y = 10(20)$	200
30	$y = 10(30)$	300
40	$y = 10(40)$	400

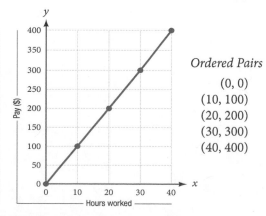

Ordered Pairs

(0, 0)
(10, 100)
(20, 200)
(30, 300)
(40, 400)

FIGURE 1 *Weekly wages at $10.00 per hour*

3.5 VIDEOS

The equation $y = 10x$ with the restriction $0 \leq x \leq 40$, Table 1, and Figure 1 are three ways to describe the same relationship between the number of hours you work in one week and your gross pay for that week. In all three, we *input* values of x, and then use the function rule to *output* values of y.

Domain and Range of a Function

We began this discussion by saying that the number of hours worked during the week was from 0 to 40, so these are the values that x can assume. From the line graph in Figure 1, we see that the values of y range from 0 to 400. We call the complete set of values that x can assume the *domain* of the function. The values that are assigned to y are called the *range* of the function.

EXAMPLE 2

State the domain and range for the function

$$y = 7.5x, \quad 0 \leq x \leq 40$$

SOLUTION   From the previous discussion, we have

Domain $= \{x \mid 0 \leq x \leq 40\}$

Range $= \{y \mid 0 \leq y \leq 400\}$

Function Maps

Another way to visualize the relationship between x and y is with the diagram in Figure 2, which we call a *function map*.

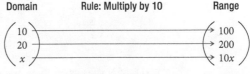

FIGURE 2 *A function map*

Although the diagram in Figure 2 does not show all the values that x and y can assume, it does give us a visual description of how x and y are related. It shows that values of y in the range come from values of x in the domain according to a specific rule (multiply by 10 each time).

C A Formal Look at Functions

We are now ready for the formal definition of a function.

> **DEFINITION** *function*
>
> A *function* is a rule that pairs each element in one set, called the **domain**, with exactly one element from a second set, called the **range**.

In other words, a function is a rule for which each input is paired with exactly one output.

EXAMPLE 3

Kendra tosses a softball into the air with an underhand motion. The distance of the ball above her hand is given by the function

$$h = 32t - 16t^2 \quad \text{for} \quad 0 \le t \le 2$$

where h is the height of the ball in feet and t is the time in seconds. Construct a table that gives the height of the ball at quarter-second intervals, starting with $t = 0$ and ending with $t = 2$, then graph the function.

SOLUTION We construct Table 2 using the following values of t: $0, \frac{1}{4}, \frac{1}{2}, \frac{3}{4}, 1, \frac{5}{4}, \frac{3}{2}, \frac{7}{4}, 2$. Then we construct the graph in Figure 3 from the table. The graph appears only in the first quadrant because neither t nor h can be negative.

TABLE 2	Tossing a Softball into the Air	
Input		Output
Time (sec) t	Function Rule $h = 32t - 16t^2$	Distance (ft) h
0	$h = 32(0) - 16(0)^2 = 0 - 0 = 0$	0
$\frac{1}{4}$	$h = 32\left(\frac{1}{4}\right) - 16\left(\frac{1}{4}\right)^2 = 8 - 1 = 7$	7
$\frac{1}{2}$	$h = 32\left(\frac{1}{2}\right) - 16\left(\frac{1}{2}\right)^2 = 16 - 4 = 12$	12
$\frac{3}{4}$	$h = 32\left(\frac{3}{4}\right) - 16\left(\frac{3}{4}\right)^2 = 24 - 9 = 15$	15
1	$h = 32(1) - 16(1)^2 = 32 - 16 = 16$	16
$\frac{5}{4}$	$h = 32\left(\frac{5}{4}\right) - 16\left(\frac{5}{4}\right)^2 = 40 - 25 = 15$	15
$\frac{3}{2}$	$h = 32\left(\frac{3}{2}\right) - 16\left(\frac{3}{2}\right)^2 = 48 - 36 = 12$	12
$\frac{7}{4}$	$h = 32\left(\frac{7}{4}\right) - 16\left(\frac{7}{4}\right)^2 = 56 - 49 = 7$	7
2	$h = 32(2) - 16(2)^2 = 64 - 64 = 0$	0

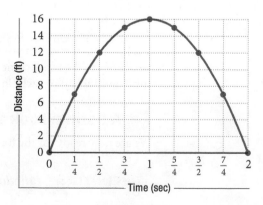

FIGURE 3

Here is a summary of what we know about functions as it applies to this example: We input values of t and output values of h according to the function rule

$$h = 32t - 16t^2 \qquad \text{for} \qquad 0 \le t \le 2$$

The domain is given by the inequality that follows the equation; it is

$$\text{Domain} = \{t \,|\, 0 \le t \le 2\}$$

The range is the set of all outputs that are possible by substituting the values of t from the domain into the equation. From our table and graph, it seems that the range is

$$\text{Range} = \{h \,|\, 0 \le h \le 16\}$$

USING TECHNOLOGY   *More About Example 3*

Most graphing calculators can easily produce the information in Table 2. Simply set Y_1 equal to $32X - 16X^2$. Then set up the table so it starts at 0 and increases by an increment of 0.25 each time. (On a TI-82/83, use the $\boxed{\text{TBLSET}}$ key to set up the table.)

The table will look like this:

X	Y_1	
0	0	
.25	7	
.5	12	
.75	15	
1	16	
1.25	15	
1.5	12	

Graph each equation and build a table as indicated.

1. $y = 64t - 16t^2$ TblStart $= 0$ ΔTbl $= 1$

2. $y = \dfrac{1}{2}x - 4$ TblStart $= -5$ ΔTbl $= 1$

3. $y = \dfrac{12}{x}$ TblStart $= 0.5$ ΔTbl $= 0.5$

Functions as Ordered Pairs

As you can see from the examples we have presented to this point, the function rule produces ordered pairs of numbers. We use this result to write an alternative definition for a function.

ALTERNATE DEFINITION *function*

A *function* is a set of ordered pairs in which no two different ordered pairs have the same first coordinate. The set of all first coordinates is called the *domain* of the function. The set of all second coordinates is called the *range* of the function.

The restriction on first coordinates in the alternative definition keeps us from assigning a number in the domain to more than one number in the range.

A Relationship That Is Not a Function

You may be wondering if any sets of paired data fail to qualify as functions. The answer is yes, as the next example reveals.

EXAMPLE 4

Table 3 shows the prices of used Ford Mustangs that were listed in the local newspaper. The diagram in Figure 4 is called a *scatter diagram*. It gives a visual representation of the data in Table 3. Why is this data not a function?

TABLE 3	Used Mustang Prices
Year *x*	Price ($) *y*
1997	13,925
1997	11,850
1997	9,995
1996	10,200
1996	9,600
1995	9,525
1994	8,675
1994	7,900
1993	6,975

Used Mustang Prices

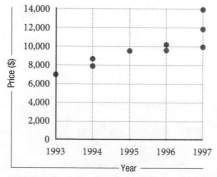

Ordered Pairs

(1997, 13,925)
(1997, 11,850)
(1997, 9,995)
(1996, 10,200)
(1996, 9,600)
(1995, 9,525)
(1994, 8,675)
(1994, 7,900)
(1993, 6,975)

FIGURE 4   *Scatter diagram of data in Table 3*

SOLUTION   In Table 3, the year 1997 is paired with three different prices: $13,925, $11,850, and $9,995. That is enough to disqualify the data from belonging to a function. For a set of paired data to be considered a function, each number in the domain must be paired with exactly one number in the range.

Still, there is a relationship between the first coordinates and second coordinates in the used car data. It is not a function relationship, but it is a relationship. To classify all relationships specified by ordered pairs, whether they are functions or not, we include the following two definitions.

> **DEFINITION**   *relation*
>
> A *relation* is a rule that pairs each element in one set, called the domain, with *one or more elements* from a second set, called the *range*.

> **ALTERNATE DEFINITION**   *relation*
>
> A *relation* is a set of ordered pairs. The set of all first coordinates is the *domain* of the relation. The set of all second coordinates is the *range* of the relation.

Here are some facts that will help clarify the distinction between relations and functions:

1. Any rule that assigns numbers from one set to numbers in another set is a relation. If that rule makes the assignment so no input has more than one output, then it is also a function.
2. Any set of ordered pairs is a relation. If none of the first coordinates of those ordered pairs is repeated, the set of ordered pairs is also a function.
3. Every function is a relation.
4. Not every relation is a function.

EXAMPLE 5

Sketch the graph of $x = y^2$.

SOLUTION   Without going into much detail, we graph the equation $x = y^2$ by finding a number of ordered pairs that satisfy the equation, plotting these points, then drawing a smooth curve that connects them. A table of values for x and y that satisfy the equation follows, along with the graph of $x = y^2$ shown in Figure 5.

x	y
0	0
1	1
1	−1
4	2
4	−2
9	3
9	−3

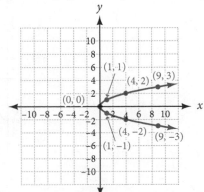

FIGURE 5

As you can see from looking at the table and the graph in Figure 5, several ordered pairs whose graphs lie on the curve have repeated first coordinates, for instance (1, 1) and (1, −1), (4, 2) and (4, −2), as well as (9, 3) and (9, −3). The graph is therefore not the graph of a function.

D Vertical Line Test

Look back at the scatter diagram for used Mustang prices shown in Figure 4. Notice that some of the points on the diagram lie above and below each other along vertical lines. This is an indication that the data do not constitute a function. Two data points that lie on the same vertical line must have come from two ordered pairs with the same first coordinates.

Now, look at the graph shown in Figure 5. The reason this graph is the graph of a relation, but not of a function, is that some points on the graph have the same first coordinates, for example, the points (4, 2) and (4, −2). Furthermore, any time two points on a graph have the same first coordinates, those points must lie on a vertical line. [To convince yourself, connect the points (4, 2) and (4, −2) with a straight line. You will see that it must be a vertical line.] This allows us to write the following test that uses the graph to determine whether a relation is also a function.

> ### RULE *Vertical Line Test*
> If a vertical line crosses the graph of a relation in more than one place, the relation cannot be a function. If no vertical line can be found that crosses a graph in more than one place, then the graph represents a function.

If we look back to the graph of $h = 32t − 16t^2$ as shown in Figure 3, we see that no vertical line can be found that crosses this graph in more than one place. The graph shown in Figure 3 is therefore the graph of a function.

EXAMPLE 6

Match each relation with its graph, then indicate which relations are functions.
 a. $y = |x| − 4$ **b.** $y = x^2 − 4$ **c.** $y = 2x + 2$

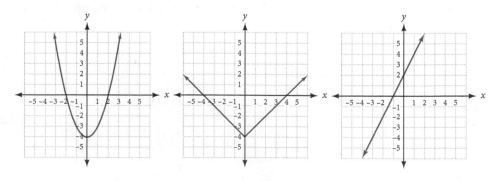

FIGURE 6   FIGURE 7   FIGURE 8

SOLUTION   Using the basic graphs for a guide along with our knowledge of translations, we have the following:

a. Figure 7   **b.** Figure 6   **c.** Figure 8

And, since all graphs pass the vertical line test, all are functions.

GETTING READY FOR CLASS

After reading through the preceding section, respond in your own words and in complete sentences.

A. What is a function?

B. What is the vertical line test?

C. Is every line the graph of a function? Explain.

D. Which variable is usually associated with the domain of a function?

Problem Set 3.5

For each of the following relations, give the domain and range, and indicate which are also functions.

1. (1, 2), (3, 4), (5, 6), (7, 8)
2. (2, 1), (4, 3), (6, 5), (8, 7)

3. (2, 5), (3, 4), (1, 4), (0, 6)
4. (0, 4), (1, 6), (2, 4), (1, 5)

5. $(a, 3), (b, 4), (c, 3), (d, 5)$
6. $(a, 5), (b, 5), (c, 4), (d, 5)$

7. $(a, 1), (a, 2), (a, 3), (a, 4)$
8. $(a, 1), (b, 1), (c, 1), (d, 1)$

State whether each of the following graphs represents a function.

9.
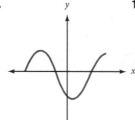

10.
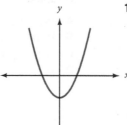

11.
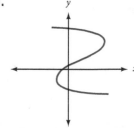

12.
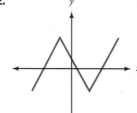

13.
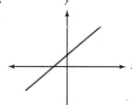

14.
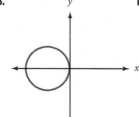

15.
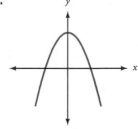

16.
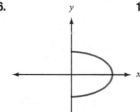

17.

18.
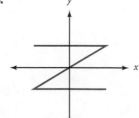

Determine the domain and range of the following functions. Assume the *entire* function is shown.

19.

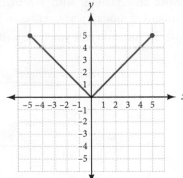

20.

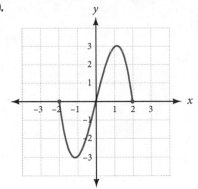

21.

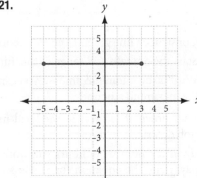

22.

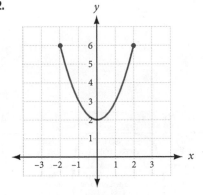

Graph each of the following relations. In each case, use the graph to find the domain and range, and indicate whether the graph is the graph of a function.

23. $y = x^2 - 1$ **24.** $y = x^2 + 1$ **25.** $y = x^2 + 4$ **26.** $y = x^2 - 9$

27. $x = y^2 - 1$ **28.** $x = y^2 + 1$ **29.** $y = (x + 2)^2$ **30.** $y = (x - 3)^2$

31. $x = (y + 1)^2$ **32.** $x = 3 - y^2$

33. Suppose you have a job that pays \$8.50 per hour and you work anywhere from 10 to 40 hours per week.

   a. Write an equation, with a restriction on the variable x, that gives the amount of money, y, you will earn for working x hours in one week.

   b. Use the function rule you have written in part **a.** to complete Table 4.

TABLE 4 Weekly Wages

Hours Worked	Function Rule	Gross Pay (\$)
x		y
10		
20		
30		
40		

   c. Construct a line graph from the information in Table 4.

   d. State the domain and range of this function.

   e. What is the minimum amount you can earn in a week with this job? What is the maximum amount?

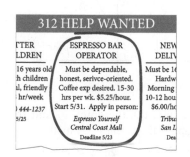

34. The ad shown at left was in the local newspaper. Suppose you are hired for the job described in the ad.

   a. If x is the number of hours you work per week and y is your weekly gross pay, write the equation for y. (Be sure to include any restrictions on the variable x that are given in the ad.)

   b. Use the function rule you have written in part **a.** to complete Table 5.

TABLE 5 Weekly Wages

Hours Worked	Function Rule	Gross Pay (\$)
x		y
15		
20		
25		
30		

   c. Construct a line graph from the information in Table 5.

   d. State the domain and range of this function.

   e. What is the minimum amount you can earn in a week with this job? What is the maximum amount?

35. Camera Phones The chart shows the estimated number of smartphones and digital cameras sold from 2010 to 2015. Using the chart, list all the values in the domain and range for the total phones sales.

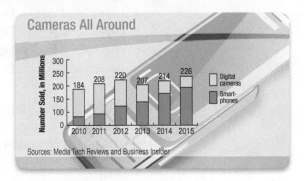

36. Light Bulbs The chart shows a comparison of power usage between incandescent and energy-efficient light bulbs. Use the chart to state the domain and range of the function for an energy-efficient bulb.

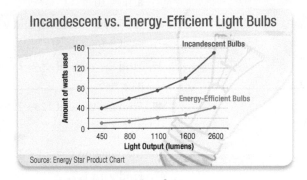

37. Profits Match each of the following statements to the appropriate graph indicated by labels I–IV.

   a. Sarah works 25 hours to earn $250.

   b. Justin works 35 hours to earn $560.

   c. Rosemary works 30 hours to earn $360.

   d. Marcus works 40 hours to earn $320.

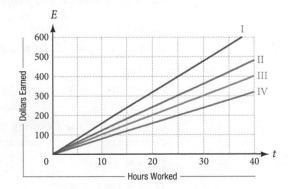

38. Find an equation for each of the functions shown in the graph. Show dollars earned, E, as a function of hours worked, t. Then, indicate the domain and range of each function.

a. Graph I: $E =$ ___ Domain $= \{t| \quad \}$ Range $= \{E| \quad \}$
b. Graph II: $E =$ ___ Domain $= \{t| \quad \}$ Range $= \{E| \quad \}$
c. Graph III: $E =$ ___ Domain $= \{t| \quad \}$ Range $= \{E| \quad \}$
d. Graph IV: $E =$ ___ Domain $= \{t| \quad \}$ Range $= \{E| \quad \}$

GETTING READY FOR THE NEXT SECTION

Simplify. Round to the nearest whole number if necessary.

39. $4(3.14)(9)$

40. $\frac{4}{3}(3.14) \cdot 3^3$

41. $4(-2) - 1$

42. $3(3)^2 + 2(3) - 1$

43. If $s = \dfrac{60}{t}$, find s when

    a. $t = 10$ **b.** $t = 8$

44. If $y = 3x^2 + 2x - 1$, find y when

    a. $x = 0$ **b.** $x = -2$

45. Find the value of $x^2 + 2$ for

    a. $x = 5$ **b.** $x = -2$

46. Find the value of $125 \cdot 2^t$ for

    a. $t = 0$ **b.** $t = 1$

For the equation $y = x^2 - 3$:

47. Find y if x is 2.

48. Find y if x is -2.

49. Find y if x is 0.

50. Find y if x is -4.

The problems that follow review some of the more important skills you have learned in previous sections and chapters.

51. If $x - 2y = 4$, and $x = \dfrac{8}{5}$ find y.

52. If $\dfrac{x^2}{25} + \dfrac{y^2}{9} = 1$, find y when x is -4.

53. Let $x = 0$ and $y = 0$ in $y = a(x - 8)^2 + 70$ and solve for a.

54. Find R if $p = 2.5$ and $R = (900 - 300p)p$.

EXTENDING THE CONCEPTS

Graph each of the following relations. In each case, use the graph to find the domain and range, and indicate whether the graph is the graph of a function.

55. $y = 5 - |x|$

56. $y = |x| - 3$

57. $x = |y| + 3$

58. $x = 2 - |y|$

59. $|x| + |y| = 4$

60. $2|x| + |y| = 6$

FUNCTION NOTATION

©iStockphoto.com/todaydesign

A Find the value of a function from a given value of the variable using function notation.

B Visualize the relationship between a function and a variable using a graph.

C Solve problems with formulas using function notation.

Let's return to the discussion that introduced us to functions. If a job pays $10.00 per hour for working from 0 to 40 hours a week, then the amount of money y earned in one week is a function of the number of hours worked x. The exact relationship between x and y is written

$$y = 10x \quad \text{for} \quad 0 \leq x \leq 40$$

Because the amount of money earned y depends on the number of hours worked x, we call y the *dependent variable* and x the *independent variable*. Furthermore, if we let f represent all the ordered pairs produced by the equation, then we can write

$$f = \{(x, y) \mid y = 10x \quad \text{and} \quad 0 \leq x \leq 40\}$$

Once we have named a function with a letter, we can use an alternative notation to represent the dependent variable y. The alternative notation for y is $f(x)$. It is read "f of x" and can be used instead of the variable y when working with functions. The notation y and the notation $f(x)$ are equivalent. That is,

$$y = 10x \Leftrightarrow f(x) = 10x$$

A Notation

When we use the notation $f(x)$ we are using *function notation*. The benefit of using function notation is that we can write more information with fewer symbols than we can by using just the variable y. For example, asking how much money a person will make for working 20 hours is simply a matter of asking for $f(20)$. Without function notation, we would have to say, "Find the value of y that corresponds to a value of $x = 20$." To illustrate further, using the variable y, we can say "y is 150 when x is 20." Using the notation $f(x)$, we simply say "$f(20) = 150$." Each expression indicates that you will earn $150 for working 20 hours.

EXAMPLE 1

If $f(x) = 10x$, find $f(0)$, $f(10)$, and $f(20)$.

SOLUTION To find $f(0)$, we substitute 0 for x in the expression $10x$ and simplify. We find $f(10)$ and $f(20)$ in a similar manner — by substitution.

$$\begin{aligned} \text{If} \qquad & f(x) = 10x \\ \text{then} \qquad & f(0) = 10(0) = 0 \\ & f(10) = 10(10) = 75 \\ & f(20) = 10(20) = 150 \end{aligned}$$

NOTE Some students like to think of functions as machines. Values of x are put into the machine, which transforms them into values of $f(x)$, which are then output by the machine.

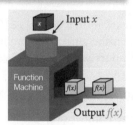

Input x

Function Machine

$f(x)$ $f(x)$

Output $f(x)$

NOTE The symbol $\in$ means "is a member of".

If we changed the example in the discussion that opened this section so the hourly wage was \$8.50 per hour, we would have a new equation to work with, namely,

$$y = 8.5x \qquad \text{for} \qquad 0 \le x \le 40$$

Suppose we name this new function with the letter g. Then

$$g = \{(x, y) \mid y = 8.5x \quad \text{and} \quad 0 \le x \le 40\}$$

and

$$g(x) = 8.5x$$

If we want to talk about both functions in the same discussion, having two different letters, f and g, makes it easy to distinguish between them. For example, since $f(x) = 10x$ and $g(x) = 8.5x$, asking how much money a person makes for working 20 hours is simply a matter of asking for $f(20)$ or $g(20)$, avoiding any confusion over which hourly wage we are talking about.

The diagrams shown in Figure 1 further illustrate the similarities and differences between the two functions we have been discussing.

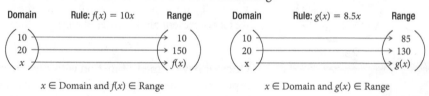

FIGURE 1 *Function maps*

Ⓑ Function Notation and Graphs

We can visualize the relationship between x and $f(x)$ on the graph of the function. Figure 2 shows the graph of $f(x) = 10x$ along with two additional line segments. The horizontal line segment corresponds to $x = 20$, and the vertical line segment corresponds to $f(20)$. (Note that the domain is restricted to $0 \le x \le 40$.)

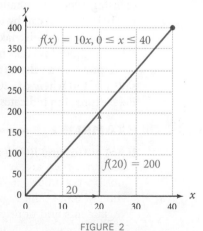

FIGURE 2

We can use functions and function notation to talk about numbers in the chart of gasoline prices. Let's let *x* represent one of the years in the chart.

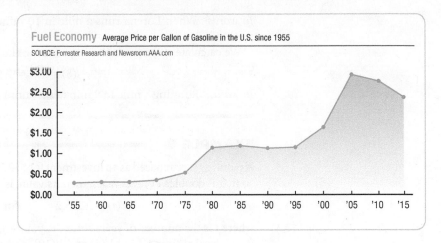

Fuel Economy Average Price per Gallon of Gasoline in the U.S. since 1955

SOURCE: Forrester Research and Newsroom.AAA.com

If the function *f* pairs each year in the chart with the average price of regular gasoline for that year, then each statement below is true:

$$f(1955) = \$0.29$$

The domain of *f* =
{1955, 1960, 1965, 1970, 1975, 1980, 1985, 1990, 1995, 2000, 2005, 2010, 2015}

In general, when we refer to the function *f* we are referring to the domain, the range, and the rule that takes elements in the domain and outputs elements in the range. When we talk about $f(x)$ we are talking about the rule itself, or an element in the range, or the variable *y*.

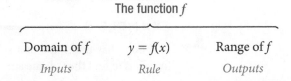

The function *f*

Domain of *f*	$y = f(x)$	Range of *f*
Inputs	*Rule*	*Outputs*

🅲 Using Function Notation

The remaining examples in this section show a variety of ways to use and interpret function notation.

EXAMPLE 2

If it takes Lorena *t* minutes to run a mile, then her average speed *s*, in miles per hour, is given by the formula

$$s(t) = \frac{60}{t} \qquad \text{for} \qquad t > 0$$

Find $s(10)$ and $s(8)$, and then explain what they mean.

SOLUTION   To find $s(10)$, we substitute 10 for t in the equation and simplify:

$$s(10) = \frac{60}{10} = 6$$

In words: When Lorena runs a mile in 10 minutes, her average speed is 6 miles per hour.

We calculate $s(8)$ by substituting 8 for t in the equation. Doing so gives us

$$s(8) = \frac{60}{8} = 7.5$$

In words: Running a mile in 8 minutes is running at a rate of 7.5 miles per hour.

EXAMPLE 3

A painting is purchased as an investment for $125. If its value increases continuously so that it doubles every 5 years, then its value is given by the function

$$V(t) = 125 \cdot 2^{t/5} \qquad \text{for} \qquad t \geq 0$$

where t is the number of years since the painting was purchased, and V is its value (in dollars) at time t. Find $V(5)$ and $V(10)$, and explain what they mean.

SOLUTION   The expression $V(5)$ is the value of the painting when $t = 5$ (5 years after it is purchased). We calculate $V(5)$ by substituting 5 for t in the equation $V(t) = 125 \cdot 2^{t/5}$. Here is our work:

$$V(5) = 125 \cdot 2^{5/5} = 125 \cdot 2^1 = 125 \cdot 2 = 250$$

In words: After 5 years, the painting is worth $250.

The expression $V(10)$ is the value of the painting after 10 years. To find this number, we substitute 10 for t in the equation:

$$V(10) = 125 \cdot 2^{10/5} = 125 \cdot 2^2 = 125 \cdot 4 = 500$$

In words: The value of the painting 10 years after it is purchased is $500.

EXAMPLE 4

A balloon has the shape of a sphere with a radius of 3 inches. Use the following formulas to find the volume and surface area of the balloon.

$$V(r) = \frac{4}{3}\pi r^3 \qquad S(r) = 4\pi r^2$$

SOLUTION   As you can see, we have used function notation to write the formulas for volume and surface area, because each quantity is a function of the radius. To find these quantities when the radius is 3 inches, we evaluate $V(3)$ and $S(3)$:

$$V(3) = \frac{4}{3}\pi \cdot 3^3 = \frac{4}{3}\pi \cdot 27$$

$$= 36\pi \text{ cubic inches, or } 113 \text{ cubic inches}$$
(to the nearest whole number)

$$S(3) = 4\pi \cdot 3^2$$

$$= 36\pi \text{ square inches, or } 113 \text{ square inches}$$
(to the nearest whole number)

The fact that $V(3) = 36\pi$ means that the ordered pair $(3, 36\pi)$ belongs to the function V. Likewise, the fact that $S(3) = 36\pi$ tells us that the ordered pair $(3, 36\pi)$ is a member of function S.

We can generalize the discussion at the end of Example 4 this way:

$$(a, b) \in f \quad \text{if and only if} \quad f(a) = b$$

USING TECHNOLOGY *More About Example 4*

· ·

If we look at Example 4, we see that when the radius of a sphere is 3, the numerical values of the volume and surface area are equal. How unusual is this? Are there other values of r for which $V(r)$ and $S(r)$ are equal? We can answer this question by looking at the graphs of both V and S.

To graph the function $V(r) = \frac{4}{3}\pi r^3$, set $Y_1 = 4\pi X^3/3$. To graph $S(r) = 4\pi r^2$, set $Y_2 = 4\pi X^2$. Graph the two functions in each of the following windows:

Window 1:   X from -4 to 4, Y from -2 to 10

Window 2:   X from 0 to 4, Y from 0 to 50

Window 3:   X from 0 to 4, Y from 0 to 150

Then use the Trace and Zoom features of your calculator to locate the point in the first quadrant where the two graphs intersect. How do the coordinates of this point compare with the results in Example 4?

EXAMPLE 5

If $f(x) = 3x^2 + 2x - 1$, find $f(0)$, $f(3)$, and $f(-2)$.

SOLUTION   Since $f(x) = 3x^2 + 2x - 1$, we have

$$f(0) = 3(0)^2 + 2(0) - 1 = 0 - 1 = -1$$
$$f(3) = 3(3)^2 + 2(3) - 1 = 27 + 6 - 1 = 32$$
$$f(-2) = 3(-2)^2 + 2(-2) - 1 = 12 - 4 - 1 = 7$$

In Example 5, the function f is defined by the equation $f(x) = 3x^2 + 2x - 1$. We could just as easily have said $y = 3x^2 + 2x - 1$. That is, $y = f(x)$. Saying $f(-2) = 7$ is exactly the same as saying y is 7 when x is -2.

EXAMPLE 6

If $f(x) = 4x - 1$ and $g(x) = x^2 + 2$, then

$$f(5) = 4(5) - 1 = 19 \qquad \text{and} \qquad g(5) = 5^2 + 2 = 27$$
$$f(-2) = 4(-2) - 1 = -9 \qquad \text{and} \qquad g(-2) = (-2)^2 + 2 = 6$$
$$f(0) = 4(0) - 1 = -1 \qquad \text{and} \qquad g(0) = 0^2 + 2 = 2$$
$$f(z) = 4z - 1 \qquad \text{and} \qquad g(z) = z^2 + 2$$
$$f(a) = 4a - 1 \qquad \text{and} \qquad g(a) = a^2 + 2$$

$$f(a + 3) = 4(a + 3) - 1 \qquad\qquad g(a + 3) = (a + 3)^2 + 2$$
$$= 4a + 12 - 1 \qquad\qquad\qquad = (a^2 + 6a + 9) + 2$$
$$= 4a + 11 \qquad\qquad\qquad\qquad = a^2 + 6a + 11$$

USING TECHNOLOGY   *More About Example 6*

Most graphing calculators can use tables to evaluate functions. To work Example 6 using a graphing calculator table, set Y_1 equal to $4X - 1$ and Y_2 equal to $X^2 + 2$. Then set the independent variable in the table to Ask instead of Auto. Go to your table and input 5, -2, and 0. Under Y_1 in the table, you will find $f(5)$, $f(-2)$, and $f(0)$. Under Y_2, you will find $g(5)$, $g(-2)$, and $g(0)$. The table will look like this:

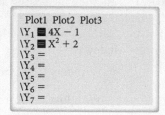

Plot1 Plot2 Plot3
\Y₁ ■ 4X − 1
\Y₂ ■ X² + 2
\Y₃ =
\Y₄ =
\Y₅ =
\Y₆ =
\Y₇ =

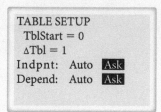

TABLE SETUP
TblStart = 0
ΔTbl = 1
Indpnt: Auto Ask
Depend: Auto Ask

Although the calculator asks us for a table increment, the increment doesn't matter because we are inputting the X values ourselves.

X	Y_1	Y_2
5	19	27
-2	-9	6
0	-1	2

EXAMPLE 7

If the function f is given by

$$f = \{(-2, 0), (3, -1), (2, 4), (7, 5)\}$$

then $f(-2) = 0$, $f(3) = -1$, $f(2) = 4$, and $f(7) = 5$.

EXAMPLE 8

If $f(x) = 2x^2$ and $g(x) = 3x - 1$, find

a. $f[g(2)]$ **b.** $g[f(2)]$

SOLUTION   The expression $f[g(2)]$ is read "f of g of 2."

a. Because $g(2) = 3(2) - 1 = 5$, $f[g(2)] = f(5) = 2(5)^2 = 50$
b. Because $f(2) = 2(2)^2 = 8$, $g[f(2)] = g(8) = 3(8) - 1 = 23$

GETTING READY FOR CLASS

After reading through the preceding section, respond in your own words and in complete sentences.

A. Explain what you are calculating when you find $f(2)$ for a given function f.

B. If $s(t) = \frac{60}{t}$ how do you find $s(10)$?

C. If $f(2) = 3$ for a function f, what is the relationship between the numbers 2 and 3 and the graph of f?

D. If $f(6) = 0$ for a particular function f, then you can immediately graph one of the intercepts. Explain.

Let $f(x) = 2x - 5$ and $g(x) = x^2 + 3x + 4$. Evaluate the following.

1. $f(2)$ **2.** $f(3)$ **3.** $f(-3)$ **4.** $g(-2)$

5. $g(-1)$ **6.** $f(-4)$ **7.** $g(-3)$ **8.** $g(2)$

9. $g(a)$ **10.** $f(a)$ **11.** $f(a + 6)$ **12.** $g(a + 6)$

Let $f(x) = 3x^2 - 4x + 1$ and $g(x) = 2x - 1$. Evaluate the following.

13. $f(0)$ **14.** $g(0)$ **15.** $g(-4)$ **16.** $f(1)$

17. $f(-1)$ **18.** $g(-1)$ **19.** $g\left(\dfrac{1}{2}\right)$ **20.** $g\left(\dfrac{1}{4}\right)$

21. $f(a)$ **22.** $g(a)$ **23.** $f(a + 2)$ **24.** $g(a + 2)$

If $f = \{(1, 4), (-2, 0), \left(3, \frac{1}{2}\right), (\pi, 0)\}$ and $g = \{(1, 1), (-2, 2), \left(\frac{1}{2}, 0\right)\}$, find each of the following values of f and g.

25. $f(1)$ **26.** $g(1)$ **27.** $g\left(\dfrac{1}{2}\right)$

28. $f(3)$ **29.** $g(-2)$ **30.** $f(\pi)$

Let $f(x) = x^2 - 2x$ and $g(x) = 5x - 4$. Evaluate the following.

31. $f(-4)$ **32.** $g(-3)$ **33.** $f(-2) + g(-1)$

34. $f(-1) + g(-2)$ **35.** $2f(x) - 3g(x)$ **36.** $f(x) - g(x^2)$

37. $f[g(3)]$ **38.** $g[f(3)]$

Let $f(x) = \dfrac{1}{x + 3}$ and $g(x) = \dfrac{1}{x} + 1$. Evaluate the following.

39. $f\left(\dfrac{1}{3}\right)$ **40.** $g\left(\dfrac{1}{3}\right)$ **41.** $f\left(-\dfrac{1}{2}\right)$

42. $g\left(-\dfrac{1}{2}\right)$ **43.** $f(-3)$ **44.** $g(0)$

45. For the function $f(x) = x^2 - 4$, evaluate each of the following expressions.

     a. $f(a) - 3$ **b.** $f(a - 3)$ **c.** $f(x) + 2$

     d. $f(x + 2)$ **e.** $f(a + b)$ **f.** $f(x + h)$

46. For the function $f(x) = 3x^2$, evaluate each of the following expressions.

     a. $f(a) - 2$ **b.** $f(a - 2)$ **c.** $f(x) + 5$

     d. $f(x + 5)$ **e.** $f(a + b)$ **f.** $f(x + h)$

47. Graph the function $f(x) = \frac{1}{2}x + 2$. Then draw and label the line segments that represent $x = 4$ and $f(4)$.

48. Graph the function $f(x) = -\frac{1}{2}x + 6$. Then draw and label the line segments that represent $x = 4$ and $f(4)$.

49. For the function $f(x) = \frac{1}{2}x + 2$, find the value of x for which $f(x) = x$.

50. For the function $f(x) = -\frac{1}{2}x + 6$, find the value of x for which $f(x) = x$.

51. Graph the function $f(x) = x^2$. Then draw and label the line segments that represent $x = 1$ and $f(1)$, $x = 2$ and $f(2)$ and, finally, $x = 3$ and $f(3)$.

52. Graph the function $f(x) = x^2 - 2$. Then draw and label the line segments that represent $x = 2$ and $f(2)$ and the line segments corresponding to $x = 3$ and $f(3)$.

APPLYING THE CONCEPTS

53. Investing in Art A painting is purchased as an investment for $150. If its value increases continuously so that it doubles every 3 years, then its value is given by the function

$$V(t) = 150 \cdot 2^{t/3} \qquad \text{for} \qquad t \geq 0$$

where t is the number of years since the painting was purchased, and $V(t)$ is its value (in dollars) at time t. Find $V(3)$ and $V(6)$, and then explain what they mean.

54. Average Speed If it takes Minke t minutes to run a mile, then her average speed $s(t)$, in miles per hour, is given by the formula

$$s(t) = \frac{60}{t} \qquad \text{for} \qquad t > 0$$

Find $s(4)$ and $s(5)$, and then explain what they mean.

55. Antidepressant Sales Suppose x represents one of the years in the chart. Suppose further that we have three functions f, g, and h that do the following:

f pairs each year with the total sales of Zoloft in billions of dollars for that year.
g pairs each year with the total sales of Effexor in billions of dollars for that year.
h pairs each year with the total sales of Wellbutrin in billions of dollars for that year.

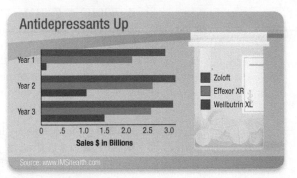

For each statement below, indicate whether the statement is true or false.

a. The domain of g is $\{1, 2, 3\}$
b. The domain of g is $\{x \mid 1 \leq x \leq 3\}$
c. $f(2) > g(2)$
d. $h(3) > 1.5$
e. $h(3) > h(2) > h(1)$

56. Mobile Phone Sales Suppose x represents one of the years in the chart. Suppose further that we have three functions f, g, and h that do the following:

f pairs each year with the number of smartphones sold that year.
g pairs each year with the number of digital cameras sold that year.
h is such that $h(x) = f(x) + g(x)$.

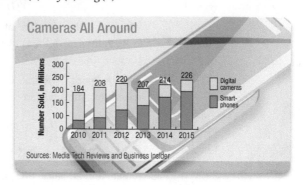

Sources: Media Tech Reviews and Business Insider

For each statement below, indicate whether the statement is true or false.

 a. The domain of f is {2010, 2011, 2012, 2013, 2014, 2015}
 b. $h(2011) = 188,000,000$
 c. $f(2014) > g(2014)$
 d. $f(2010) < f(2011)$
 e. $h(2015) > h(2013) > h(2010)$

Straight-Line Depreciation Straight-line depreciation is an accounting method used to help spread the cost of new equipment over a number of years. It takes into account both the cost when new and the salvage value, which is the value of the equipment at the time it gets replaced.

57. Value of a Copy Machine The function $V(t) = -3,300t + 18,000$, where V is value and t is time in years, can be used to find the value of a large copy machine during the first 5 years of use.

 a. What is the value of the copier after 3 years and 9 months?
 b. What is the salvage value of this copier if it is replaced after 5 years?
 c. State the domain of this function.
 d. Sketch the graph of this function.
 e. What is the range of this function?
 f. After how many years will the copier be worth only $10,000?

58. Step Function Figure 3 shows the graph of the step function C that is used to calculate the first-class postage on a letter weighing x ounces in 2018. Use this graph to answer questions **a.** through **d.**

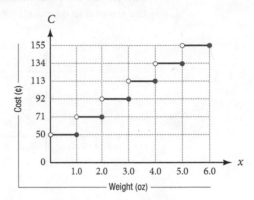

FIGURE 3 *The graph of C(x)*

a. Fill in the following table:

Weight (ounces)	0.6	1.0	1.1	2.5	3.0	4.8	5.0	5.3
Cost (cents)								

b. If a letter cost 92 cents to mail, how much does it weigh? State your answer in words. State your answer as an inequality.

c. If the entire function is shown in Figure 3, state the domain.

d. State the range of the function shown in Figure 3.

GETTING READY FOR THE NEXT SECTION

Simplify.

59. $16(3.5)^2$

60. $\dfrac{2{,}400}{100}$

61. $\dfrac{180}{45}$

62. $4(2)(4)^2$

63. $\dfrac{0.0005(200)}{(0.25)^2}$

64. $\dfrac{0.2(0.5)^2}{100}$

65. If $y = Kx$, find K if $x = 5$ and $y = 15$.

66. If $d = Kt^2$, find K if $t = 2$ and $d = 64$.

67. If $P = \dfrac{K}{V}$, find K if $P = 48$ and $V = 50$.

68. If $y = Kxz^2$, find K if $x = 5$, $z = 3$, and $y = 180$.

EXTENDING THE CONCEPTS

The graphs of two functions are shown in Figures 4 and 5. Use the graphs to find the answers to questions 69 and 70.

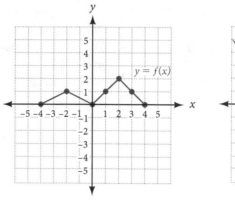

Figure 4   Figure 5

69. a. $f(2)$ **b.** $f(-4)$ **c.** $g(0)$ **d.** $g(3)$

70. a. $g(2) - f(2)$ **b.** $f(1) + g(1)$ **c.** $f[g(3)]$ **d.** $g[f(3)]$

The fixed points of a function f are the values of x such that $f(x) = x$. Find the fixed points of each function.

71. $f(x) = x^2$

72. $f(x) = 3x - 5$

73. $f(x) = \frac{1}{2}x - 3$

74. $f(x) = x^2 - 6$

The average rate of change of a function f on the interval $a \leq x \leq b$ is defined by the ratio $[f(b) - f(a)]/(b - a)$. Find the average rate of change of each function on the interval $1 \leq x \leq 3$.

75. $f(x) = 2x - 3$

76. $f(x) = x^2 + 3$

77. $f(x) = \frac{1}{x}$

78. $f(x) = \frac{1}{x^2}$

The difference quotient of a function f at $x = a$ is defined to be the ratio $[f(a + h) - f(a)]/h$. Find the difference quotient of each function at $a = 2$.

79. $f(x) = x^2$

80. $f(x) = \frac{1}{2}x - 3$

81. $f(x) = \frac{1}{x}$

82. $f(x) = x^2 - 3x$

3.7

OBJECTIVES

A Solve problems using direction variation.

B Solve problems using inverse variation.

C Solve problems using joint variation.

If you are a runner and you average t minutes for every mile you run during one of your workouts, then your speed s in miles per hour is given by the equation and graph shown here. The graph (Figure 1) is shown in the first quadrant only because both t and s are positive.

$$s = \frac{60}{t}$$

Input	Output
t	s
4	15
6	10
8	7.5
10	6
12	5
14	4.3

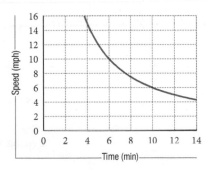

FIGURE 1

You know intuitively that as your average time per mile t increases, your speed s decreases. Likewise, lowering your time per mile will increase your speed. The equation and Figure 1 also show this to be true: Increasing t decreases s, and decreasing t increases s. Quantities that are connected in this way are said to *vary inversely* with each other. Inverse variation is one of the topics we will study in this section.

There are two main types of variation: *direct variation* and *inverse variation*. Variation problems are most common in the sciences, particularly in chemistry and physics.

3.7 VIDEOS

A Direct Variation

When we say the variable y *varies directly* with the variable x, we mean that the relationship can be written in symbols as $y = Kx$, where K is a nonzero constant called the *constant of variation* (or *proportionality constant*).

Another way of saying y varies directly with x is to say y is *directly proportional* to x.

Study the following list. It gives the mathematical equivalent of some direct variation statements.

Verbal Phrase	Algebraic Equation
y varies directly with x.	$y = Kx$
s varies directly with the square of t.	$s = Kt^2$
y is directly proportional to the cube of z.	$y = Kz^3$
u is directly proportional to the square root of v.	$u = K\sqrt{v}$

EXAMPLE 1

y varies directly with x. If y is 15 when x is 5, find y when x is 7.

SOLUTION The first sentence gives us the general relationship between x and y. The equation equivalent to the statement "y varies directly with x" is

$$y = Kx$$

The first part of the second sentence in our example gives us the information necessary to evaluate the constant K:

When	$y = 15$
and	$x = 5$
the equation	$y = Kx$
becomes	$15 = K \cdot 5$
or	$K = 3$

The equation can now be written specifically as

$$y = 3x$$

Letting $x = 7$, we have

$$y = 3 \cdot 7$$
$$y = 21$$

EXAMPLE 2

A skydiver jumps from a plane. Like any object that falls toward Earth, the distance the skydiver falls is directly proportional to the square of the time he has been falling, until he reaches his terminal velocity. If the skydiver falls 64 feet in the first 2 seconds of the jump, then

a. How far will he have fallen after 3.5 seconds?
b. Graph the relationship between distance and time.
c. How long will it take him to fall 256 feet?

SOLUTION We let t represent the time the skydiver has been falling, then we can let $d(t)$ represent the distance he has fallen.

a. Since $d(t)$ is directly proportional to the square of t, we have the general function that describes this situation:

$$d(t) = Kt^2$$

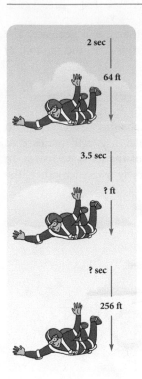

2 sec

64 ft

3.5 sec

? ft

? sec

256 ft

Next, we use the fact that $d(2) = 64$ to find K.

$$64 = K(2)^2$$

$$K = 16$$

The specific equation that describes this situation is

$$d(t) = 16t^2$$

To find how far a skydiver will fall after 3.5 seconds, we find $d(3.5)$,

$$d(3.5) = 16(3.5)^2$$

$$d(3.5) = 196$$

A skydiver will fall 196 feet after 3.5 seconds.

b. To graph this equation, we use a table:

Input	Output
t	$d(t)$
0	0
1	16
2	64
3	144
4	256
5	400

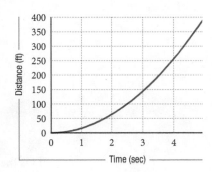

FIGURE 2

c. From the table or the graph (Figure 2), we see that it will take 4 seconds for the skydiver to fall 256 feet.

B Inverse Variation

Running

From the introduction to this section, we know that the relationship between the number of minutes t it takes a person to run a mile and his or her average speed in miles per hour s can be described with the following equation and table, and with Figure 3.

$$s = \frac{60}{t}$$

Input	Output
t	s
4	15
6	10
8	7.5
10	6
12	5
14	4.3

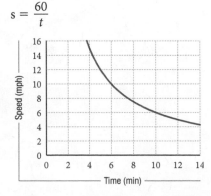

FIGURE 3

If t decreases, then s will increase, and if t increases, then s will decrease. The variable s is *inversely proportional* to the variable t. In this case, the *constant of proportionality* is 60.

Photography

If you are familiar with the terminology and mechanics associated with photography, you know that the f-stop for a particular lens will increase as the aperture (the maximum diameter of the opening of the lens) decreases. In mathematics, we say that f-stop and aperture vary inversely with each other. The following diagram illustrates this relationship.

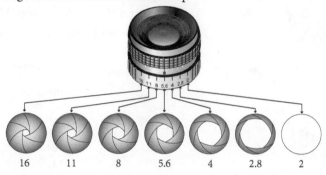

If f is the f-stop and d is the aperture, then their relationship can be written

$$f = \frac{K}{d}$$

In this case, K is the constant of proportionality. (Those of you familiar with photography know that K is also the focal length of the camera lens.)

In General

We generalize this discussion of inverse variation as follows: If y varies inversely with x, then

$$y = K\frac{1}{x} \qquad \text{or} \qquad y = \frac{K}{x}$$

We can also say y is inversely proportional to x. The constant K is again called the constant of variation or proportionality constant.

Verbal Phrase	Algebraic Equation
y is inversely proportional to x.	$y = \dfrac{K}{x}$
s varies inversely with the square of t.	$s = \dfrac{K}{t^2}$
y is inversely proportional to x^4.	$y = \dfrac{K}{x^4}$
z varies inversely with the cube root of t.	$z = \dfrac{K}{\sqrt[3]{t}}$

EXAMPLE 3

The volume of a gas is inversely proportional to the pressure of the gas on its container. If a pressure of 48 pounds per square inch corresponds to a volume of 50 cubic feet, what pressure is needed to produce a volume of 100 cubic feet?

SOLUTION We can represent volume with V and pressure with P:

$$V = \frac{K}{P}$$

Using $P = 48$ and $V = 50$, we have

$$50 = \frac{K}{48}$$

$$K = 50(48)$$

$$K = 2{,}400$$

The equation that describes the relationship between P and V is

$$V = \frac{2{,}400}{P}$$

Here is a graph of this relationship.

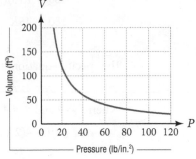

FIGURE 4

NOTE The relationship between pressure and volume as given in this example is known as Boyle's law and applies to situations such as those encountered in a piston-cylinder arrangement. It was Robert Boyle (1627–1691) who, in 1662, published the results of some of his experiments that showed, among other things, that the volume of a gas decreases as the pressure increases. This is an example of inverse variation.

Substituting $V = 100$ into our last equation, we get

$$100 = \frac{2{,}400}{P}$$

$$100P = 2{,}400$$

$$P = \frac{2{,}400}{100}$$

$$P = 24$$

A volume of 100 cubic feet is produced by a pressure of 24 pounds per square inch.

🄲 Joint and Other Variation Combinations

Many times relationships among different quantities are described in terms of more than two variables. If the variable y varies directly with *two* other variables, say x and z, then we say y varies *jointly* with x and z. In addition to *joint variation*, there are many other combinations of direct and inverse variation involving more than two variables. The following table is a list of some variation statements and their equivalent mathematical forms.

Verbal Phrase	Algebraic Equation
y varies jointly with x and z.	$y = Kxz$
z varies jointly with r and the square of s.	$z = Krs^2$
V is directly proportional to T and inversely proportional to P.	$V = \dfrac{KT}{P}$
F varies jointly with m_1 and m_2 and inversely with the square of r.	$F = \dfrac{Km_1 m_2}{r^2}$

EXAMPLE 4

y varies jointly with x and the square of z. When x is 5 and z is 3, y is 180. Find y when x is 2 and z is 4.

SOLUTION   The general equation is given by

$$y = Kxz^2$$

Substituting $x = 5$, $z = 3$, and $y = 180$, we have

$$180 = K(5)(3)^2$$
$$180 = 45K$$
$$K = 4$$

The specific equation is

$$y = 4xz^2$$

When $x = 2$ and $z = 4$, the last equation becomes

$$y = 4(2)(4)^2$$
$$y = 128$$

EXAMPLE 5

In electricity, the resistance of a cable is directly proportional to its length and inversely proportional to the square of the diameter. If a 100-foot cable 0.5 inch in diameter has a resistance of 0.2 ohm, what will be the resistance of a cable made from the same material if it is 200 feet long with a diameter of 0.25 inch?

SOLUTION   Let R = resistance, l = length, and d = diameter. The equation is

$$R = \frac{Kl}{d^2}$$

When $R = 0.2$, $l = 100$, and $d = 0.5$, the equation becomes

$$0.2 = \frac{K(100)}{(0.5)^2}$$

or

$$K = 0.0005$$

Using this value of K in our original equation, the result is

$$R = \frac{0.0005l}{d^2}$$

When $l = 200$ and $d = 0.25$, the equation becomes

$$R = \frac{0.0005(200)}{(0.25)^2}$$

$$R = 1.6 \text{ ohms}$$

GETTING READY FOR CLASS

After reading through the preceding section, respond in your own words and in complete sentences.

A. Give an example of a direct variation statement, and then translate it into symbols.

B. Translate the equation $y = \frac{K}{X}$ into words.

C. For the inverse variation equation $y = \frac{3}{x}$ what happens to the values of y as x gets larger?

D. How are direct variation statements and linear equations in two variables related?

Problem Set 3.7

For the following problems, y varies directly with x.

1. If y is 10 when x is 2, find y when x is 6.

2. If y is -32 when x is 4, find x when y is -40.

For the following problems, r is inversely proportional to s.

3. If r is -3 when s is 4, find r when s is 2.

4. If r is 8 when s is 3, find s when r is 48.

For the following problems, d varies directly with the square of r.

5. If $d = 10$ when $r = 5$, find d when $r = 10$.

6. If $d = 12$ when $r = 6$, find d when $r = 9$.

For the following problems, y varies inversely with the square of x.

7. If $y = 45$ when $x = 3$, find y when x is 5.

8. If $y = 12$ when $x = 2$, find y when x is 6.

For the following problems, z varies jointly with x and the square of y.

9. If z is 54 when x and y are 3, find z when $x = 2$ and $y = 4$.

10. If z is 27 when $x = 6$ and $y = 3$, find x when $z = 50$ and $y = 4$.

For the following problems, I varies inversely with the cube of w.

11. If $I = 32$ when $w = \dfrac{1}{2}$, find I when $w = \dfrac{1}{3}$.

12. If $I = \dfrac{1}{25}$ when $w = 5$, find I when $w = 10$.

For the following problems, z varies jointly with y and the square of x.

13. If $z = 72$ when $x = 3$ and $y = 2$, find z when $x = 5$ and $y = 3$.

14. If $z = 240$ when $x = 4$ and $y = 5$, find z when $x = 6$ and $y = 3$.

15. If $x = 1$ when $z = 25$ and $y = 5$, find x when $z = 160$ and $y = 8$.

16. If $x = 4$ when $z = 96$ and $y = 2$, find x when $z = 108$ and $y = 1$.

For the following problems, F varies directly with m and inversely with the square of d.

17. If $F = 150$ when $m = 240$ and $d = 8$, find F when $m = 360$ and $d = 3$.

18. If $F = 72$ when $m = 50$ and $d = 5$, find F when $m = 80$ and $d = 6$.

19. If $d = 5$ when $F = 24$ and $m = 20$, find d when $F = 18.75$ and $m = 40$.

20. If $d = 4$ when $F = 75$ and $m = 20$, find d when $F = 200$ and $m = 120$.

APPLYING THE CONCEPTS

21. **Length of a Spring** The length a spring stretches is directly proportional to the force applied. If a force of 5 pounds stretches a spring 3 inches, how much force is necessary to stretch the same spring 10 inches?

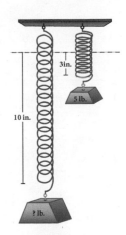

22. **Weight and Surface Area** The weight of a certain material varies directly with the surface area of that material. If 8 square feet weighs half a pound, how much will 10 square feet weigh?

23. **Pressure and Temperature** The temperature of a gas varies directly with its pressure. A temperature of 200 *K* produces a pressure of 50 pounds per square inch.
    a. Find the equation that relates pressure and temperature.
    b. Graph the equation from part **a.** in the first quadrant only.
    c. What pressure will the gas have at 280 *K*?

24. **Circumference and Diameter** The circumference of a wheel is directly proportional to its diameter. A wheel has a circumference of 8.5 feet and a diameter of 2.7 feet.
    a. Find the equation that relates circumference and diameter.
    b. Graph the equation from part **a.** in the first quadrant only.
    c. What is the circumference of a wheel that has a diameter of 11.3 feet?

25. **Volume and Pressure** The volume of a gas is inversely proportional to the pressure. If a pressure of 36 pounds per square inch corresponds to a volume of 25 cubic feet, what pressure is needed to produce a volume of 75 cubic feet?

26. **Wave Frequency** The frequency of an electromagnetic wave varies inversely with the wavelength. If a wavelength of 200 meters has a frequency of 800 kilocycles per second, what frequency will be associated with a wavelength of 500 meters?

27. **f-Stop and Aperture Diameter** The relative aperture, or *f*-stop, for a camera lens is inversely proportional to the diameter of the aperture. An *f*-stop of 2 corresponds to an aperture diameter of 40 millimeters for the lens on an automatic camera.
    a. Find the equation that relates *f*-stop and diameter.
    b. Graph the equation from part **a.** in the first quadrant only.
    c. What is the *f*-stop of this camera when the aperture diameter is 10 millimeters?

28. **f-Stop and Aperture Diameter** The relative aperture, or f-stop, for a camera lens is inversely proportional to the diameter of the aperture. An f-stop of 2.8 corresponds to an aperture diameter of 75 millimeters for a particular telephoto lens.

   a. Find the equation that relates f-stop and diameter.

   b. Graph the equation from part **a.** in the first quadrant only.

   c. What aperture diameter corresponds to an f-stop of 5.6?

29. **Surface Area of a Cylinder** The surface area of a hollow cylinder varies jointly with the height and radius of the cylinder. If a cylinder with radius 3 inches and height 5 inches has a surface area of 94 square inches, what is the surface area of a cylinder with radius 2 inches and height 8 inches?

30. **Capacity of a Cylinder** The capacity of a cylinder varies jointly with its height and the square of its radius. If a cylinder with a radius of 3 centimeters and a height of 6 centimeters has a capacity of 169.56 cubic centimeters, what will be the capacity of a cylinder with radius 4 centimeters and height 9 centimeters?

31. **Electrical Resistance** The resistance of a wire varies directly with its length and inversely with the square of its diameter. If 100 feet of wire with diameter 0.01 inch has a resistance of 10 ohms, what is the resistance of 60 feet of the same type of wire if its diameter is 0.02 inch?

32. **Volume and Temperature** The volume of a gas varies directly with its temperature and inversely with the pressure. If the volume of a certain gas is 30 cubic feet at a temperature of 300 K and a pressure of 20 pounds per square inch, what is the volume of the same gas at 340 K when the pressure is 30 pounds per square inch?

33. **Period of a Pendulum** The time it takes for a pendulum to complete one period varies directly with the square root of the length of the pendulum. A 100-centimeter pendulum takes 2.1 seconds to complete one period.

   a. Find the equation that relates period and pendulum length.

   b. Graph the equation from part **a.** in quadrant I only.

   c. How long does it take to complete one period if the pendulum hangs 225 centimeters?

GETTING READY FOR THE NEXT SECTION

Multiply.

34. $x(35 - 0.1x)$

35. $0.6(M - 70)$

36. $(4x - 3)(x - 1)$

37. $(4x - 3)(4x^2 - 7x + 3)$

Simplify.

38. $(35x - 0.1x^2) - (8x + 500)$

39. $(4x - 3) + (4x^2 - 7x + 3)$

40. $(4x^2 + 3x + 2) - (2x^2 - 5x - 6)$

41. $(4x^2 + 3x + 2) + (2x^2 - 5x - 6)$

42. $4(2)^2 - 3(2)$

43. $4(-1)^2 - 7(-1)$

ALGEBRA AND COMPOSITION WITH FUNCTIONS, AND PIECEWISE FUNCTIONS

 OBJECTIVES

A Combine functions using algebra.

B Create new functions using composition of functions.

C Graph a piecewise function.

©iStockphoto.com/Natee Meepian

A company produces and sells copies of an accounting program for home computers. The price they charge for the program is related to the number of copies sold by the demand function

$$p(x) = 35 - 0.1x$$

We find the revenue for this business by multiplying the number of items sold by the price per item. When we do so, we are forming a new function by combining two existing functions. That is, if $n(x) = x$ is the number of items sold and $p(x) = 35 - 0.1x$ is the price per item, then revenue is

$$R(x) = n(x) \cdot p(x) = x(35 - 0.1x) = 35x - 0.1x^2$$

In this case, the revenue function is the product of two functions. When we combine functions in this manner, we are applying our rules for algebra to functions.

To carry this situation further, we know the profit function is the difference between two functions. If the cost function for producing x copies of the accounting program is $C(x) = 8x + 500$, then the profit function is

$$P(x) = R(x) - C(x) = (35x - 0.1x^2) - (8x + 500) = -500 + 27x - 0.1x^2$$

The relationship between these last three functions is represented visually in Figure 1.

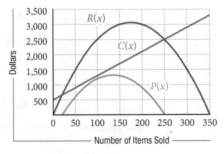

FIGURE 1

287

Algebra with Functions

Again, when we combine functions in the manner shown, we are applying our rules for algebra to functions. To begin this section, we take a formal look at addition, subtraction, multiplication, and division with functions.

If we are given two functions f and g with a common domain, we can define four other functions as follows.

DEFINITION

$(f + g)(x) = f(x) + g(x)$ The function $f + g$ is the sum of the functions f and g.

$(f - g)(x) = f(x) - g(x)$ The function $f - g$ is the difference of the functions f and g.

$(fg)(x) = f(x)g(x)$ The function fg is the product of the functions f and g.

$\left(\dfrac{f}{g}\right)(x) = \dfrac{f(x)}{g(x)}$ The function $\dfrac{f}{g}$ is the quotient of the functions f and g, where $g(x) \neq 0$.

EXAMPLE 1

If $f(x) = 4x^2 + 3x + 2$ and $g(x) = 2x^2 - 5x - 6$, write the formulas for the functions $f + g$, $f - g$, fg, and $\dfrac{f}{g}$.

SOLUTION   The function $f + g$ is defined by

$$(f + g)(x) = f(x) + g(x)$$
$$= (4x^2 + 3x + 2) + (2x^2 - 5x - 6)$$
$$= 6x^2 - 2x - 4$$

The function $f - g$ is defined by

$$(f - g)(x) = f(x) - g(x)$$
$$= (4x^2 + 3x + 2) - (2x^2 - 5x - 6)$$
$$= 4x^2 + 3x + 2 - 2x^2 + 5x + 6$$
$$= 2x^2 + 8x + 8$$

The function fg is defined by

$$(fg)(x) = f(x)g(x)$$
$$= (4x^2 + 3x + 2)(2x^2 - 5x - 6)$$
$$= 8x^4 - 20x^3 - 24x^2 + 6x^3 - 15x^2 - 18x + 4x^2 - 10x - 12$$
$$= 8x^4 - 14x^3 - 35x^2 - 28x - 12$$

The function $\dfrac{f}{g}$ is defined by

$$\left(\dfrac{f}{g}\right)(x) = \dfrac{f(x)}{g(x)}$$
$$= \dfrac{4x^2 + 3x + 2}{2x^2 - 5x - 6}$$

EXAMPLE 2

Let $f(x) = 4x - 3$, $g(x) = 4x^2 - 7x + 3$, and $h(x) = x - 1$. Find $f + g$, fh, fg and $\dfrac{g}{f}$.

SOLUTION   The function $f + g$, the sum of functions f and g, is defined by

$$(f + g)(x) = f(x) + g(x)$$
$$= (4x - 3) + (4x^2 - 7x + 3)$$
$$= 4x^2 - 3x$$

The function fh, the product of functions f and h, is defined by

$$(fh)(x) = f(x)h(x)$$
$$= (4x - 3)(x - 1)$$
$$= 4x^2 - 7x + 3$$
$$= g(x)$$

The function fg, the product of the functions f and g, is defined by

$$(fg)(x) = f(x)g(x)$$
$$= (4x - 3)(4x^2 - 7x + 3)$$
$$= 16x^3 - 28x^2 + 12x - 12x^2 + 21x - 9$$
$$= 16x^3 - 40x^2 + 33x - 9$$

The function $\dfrac{g}{f}$, the quotient of the functions g and f, is defined by

$$\left(\frac{g}{f}\right)(x) = \frac{g(x)}{f(x)}$$
$$= \frac{4x^2 - 7x + 3}{4x - 3}$$

Factoring the numerator, we can reduce to lowest terms:

$$\left(\frac{g}{f}\right)(x) = \frac{(4x - 3)(x - 1)}{4x - 3}$$
$$= x - 1$$
$$= h(x)$$

EXAMPLE 3

If f, g, and h are the same functions defined in Example 2, evaluate $(f + g)(2)$, (fh) (-1), $(fg)(0)$, and $\left(\dfrac{g}{f}\right)(5)$.

SOLUTION We use the formulas for $f + g, fh, fg$ and $\dfrac{g}{f}$ found in Example 2:

$$(f + g)(2) = 4(2)^2 - 3(2)$$
$$= 16 - 6$$
$$= 10$$
$$(fh)(-1) = 4(-1)^2 - 7(-1) + 3$$
$$= 4 + 7 + 3$$
$$= 14$$
$$(fg)(0) = 16(0)^3 - 40(0)^2 + 33(0) - 9$$
$$= 0 - 0 + 0 - 9$$
$$= -9$$
$$\left(\frac{g}{f}\right)(5) = 5 - 1$$
$$= 4$$

B Composition of Functions

In addition to the four operations used to combine functions shown so far in this section, there is a fifth way to combine two functions to obtain a new function. It is called *composition of functions*. To illustrate the concept, recall from Chapter 2 the definition of training heart rate: training heart rate, in beats per minute, is resting heart rate plus 60% of the difference between maximum heart rate and resting heart rate. If your resting heart rate is 70 beats per minute, then your training heart rate is a function of your maximum heart rate M.

$$T(M) = 70 + 0.6(M - 70) = 70 + 0.6M - 42 = 28 + 0.6M$$

But your maximum heart rate is found by subtracting your age in years from 220. So, if x represents your age in years, then your maximum heart rate is

$$M(x) = 220 - x$$

Therefore, if your resting heart rate is 70 beats per minute and your age in years is x, then your training heart rate can be written as a function of x.

$$T(x) = 28 + 0.6(220 - x)$$

This last line is the composition of functions T and M. We input x into function M, which outputs $M(x)$. Then, we input $M(x)$ into function T, which outputs $T(M(x))$, which is the training heart rate as a function of age x. Here is a diagram, called a function map, of the situation:

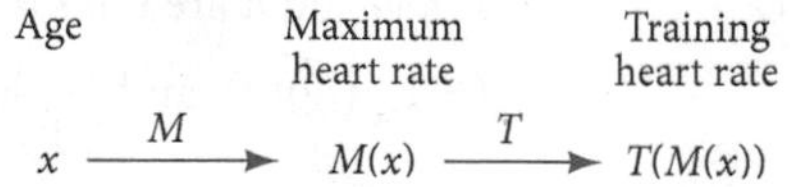

Age	Maximum heart rate	Training heart rate

$$x \xrightarrow{\ \ M\ \ } M(x) \xrightarrow{\ \ T\ \ } T(M(x))$$

FIGURE 2

Now let's generalize the preceding ideas into a formal development of composition of functions. To find the composition of two functions f and g, we first require that the range of g have numbers in common with the domain of f. Then the composition of f with g is defined this way:

$$(f \circ g)(x) = f(g(x))$$

To understand this new function, we begin with a number x, and we operate on it with g, giving us $g(x)$. Then we take $g(x)$ and operate on it with f, giving us $f(g(x))$. The only numbers we can use for the domain of the composition of f with g are numbers x in the domain of g, for which $g(x)$ is in the domain of f. The diagrams in Figure 3 illustrate the composition of f with g.

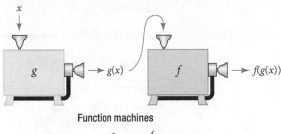

Function machines

$$x \xrightarrow{g} g(x) \xrightarrow{f} f(g(x))$$

FIGURE 3

Composition of functions is not commutative. The composition of f with g, $f \circ g$, may therefore be different from the composition of g with f, $g \circ f$.

$$(g \circ f)(x) = g(f(x))$$

Again, the only numbers we can use for the domain of the composition of g with f are numbers in the domain of f, for which $f(x)$ is in the domain of g. The diagrams in Figure 4 illustrate the composition of g with f.

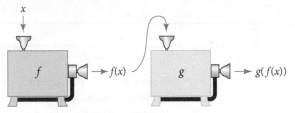

Function machines

$$x \xrightarrow{f} f(x) \xrightarrow{g} g(f(x))$$

FIGURE 4

EXAMPLE 4

If $f(x) = x + 5$ and $g(x) = x^2 - 2x$, find $(f \circ g)(x)$ and $(g \circ f)(x)$.

SOLUTION   The composition of f with g is

$$(f \circ g)(x) = f(g(x))$$
$$= f(x^2 - 2x)$$
$$= (x^2 - 2x) + 5$$
$$= x^2 - 2x + 5$$

The composition of g with f is

$$(g \circ f)(x) = g(f(x))$$
$$= g(x + 5)$$
$$= (x + 5)^2 - 2(x + 5)$$
$$= (x^2 + 10x + 25) - 2x - 10$$
$$= x^2 + 8x + 15$$

⚙ Piecewise Functions

A function may also be defined by two or more different equations pieced together, called a *piecewise function*.

EXAMPLE 5

Graph the function $f(x) = \begin{cases} -1 & \text{if } x \le 0 \\ 1 & \text{if } x > 0 \end{cases}$.

SOLUTION   The graph of this function is the same as the graph of $y = -1$ for all x values less than or equal to 0, then it becomes $y = 1$ for all x values greater than 0. The graph looks like this:

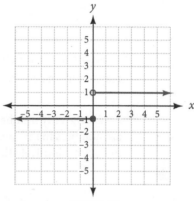

FIGURE 10

Notice at $x = 0$, the function value is $f(0) = -1$. We indicate this by a closed circle showing that the point $(0, -1)$ is included in the solution set. For all values of x greater than 0, we have $f(x) = 1$. To show this we use an open circle at 1 on the y-axis.

EXAMPLE 6

Graph the function $f(x) = \begin{cases} x - 1 & \text{if } x < -1 \\ 2x + 3 & \text{if } x \ge -1 \end{cases}$. Find the domain and range.

SOLUTION   For all x values less than -1, the function is defined by $y = x - 1$, then it switches to $y = 2x + 3$ for values of x greater than -1. Because $f(-1) = 2(-1) + 3 = 1$, we have a closed circle at $(-1, 1)$. We use an open circle at $(-1, -2)$ to show that the left side of the graph starts there, but does not include that point.

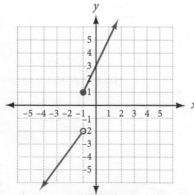

FIGURE 11

The domain of this function is still all real numbers, $\mathbb{R}$. Because there is a gap between $y = -2$ and $y = 1$ for the values y can take on, we write this as a union, $y \in (-\infty, -2) \cup [1, \infty]$. The range is $\{y \mid y < -2 \text{ or } y \geq 1\}$.

EXAMPLE 7

Graph $f(x) = \begin{cases} -2 & \text{if } x < -2 \\ x & \text{if } -2 \leq x \leq 2 \\ 2 & \text{if } x > 2 \end{cases}$ and find the domain and range.

SOLUTION   This piecewise function has 3 parts. Sometimes it is helpful to draw the vertical lines where the function changes before we graph.

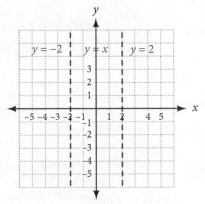

FIGURE 12

Now place each equation in the corresponding area.

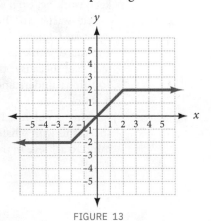

FIGURE 13

The domain of this function is all real numbers and the range is $\{y \mid -2 \leq y \leq 2\}$.

GETTING READY FOR CLASS

Respond in your own words and in complete sentences.

A. How are profit, revenue, and cost related?

B. How do you find maximum heart rate?

C. For functions f and g, how do you find the composition of f with g?

D. For functions f and g, how do you find the composition of g with f?

SPOTLIGHT ON SUCCESS

Julieta | student instructor

Success is no accident. It is hard work, perseverance, learning, studying, sacrifice, and most of all, love of what you are doing or learning to do.

—Pelé

Success really is no accident, nor is it something that happens overnight. Sure, you may be sitting there wondering why you don't understand a certain lesson or topic, but you are not alone. There are many others who are sitting in your exact position. Throughout my first year in college (and more specifically in Calculus I), I learned that it is normal for any student to feel stumped every now and then. The students who do well are the ones who keep working, even when they are confused.

Pelé wasn't just born with all that legendary talent. It took dedication and hard work as well. Don't ever feel bad because there's something you don't understand—it's not worth it. Stick with it 100% and just keep working problems; I'm sure you'll be successful with whatever you set your mind to achieve in this course.

Problem Set 3.8

Let $f(x) = 4x - 3$ and $g(x) = 2x + 5$. Write a formula for each of the following functions.

1. $f + g$ **2.** $f - g$ **3.** $g - f$ **4.** $g + f$

5. fg **6.** $\dfrac{f}{g}$ **7.** $\dfrac{g}{f}$ **8.** ff

If the functions f, g, and h are defined by $f(x) = 3x - 5$, $g(x) = x - 2$ and $h(x) = 3x^2 - 11x + 10$, write a formula for each of the following functions.

9. $g + f$ **10.** $f + h$ **11.** $g + h$ **12.** $f - g$

13. $g - f$ **14.** $h - g$ **15.** fg **16.** gf

17. fh **18.** gh **19.** $\dfrac{h}{f}$ **20.** $\dfrac{h}{g}$

21. $\dfrac{f}{h}$ **22.** $\dfrac{g}{h}$ **23.** $f + g + h$ **24.** $h - g + f$

25. $h + fg$ **26.** $h - fg$

Let $f(x) = 2x + 1$, $g(x) = 4x + 2$, and $h(x) = 4x^2 + 4x + 1$, and find the following.

27. $(f + g)(2)$ **28.** $(f - g)(-1)$ **29.** $(fg)(3)$ **30.** $\left(\dfrac{f}{g}\right)(-3)$

31. $\left(\dfrac{h}{g}\right)(1)$ **32.** $(hg)(1)$ **33.** $(fh)(0)$ **34.** $(h - g)(-4)$

35. $(f + g + h)(2)$ **36.** $(h - f + g)(0)$ **37.** $(h + fg)(3)$ **38.** $(h - fg)(5)$

39. Let $f(x) = x^2$ and $g(x) = x + 4$, and find

     a. $(f \circ g)(5)$ **b.** $(g \circ f)(5)$ **c.** $(f \circ g)(x)$ **d.** $(g \circ f)(x)$

40. Let $f(x) = 3 - x$ and $g(x) = x^3 - 1$, and find

     a. $(f \circ g)(0)$ **b.** $(g \circ f)(0)$ **c.** $(f \circ g)(x)$ **d.** $(g \circ f)(x)$

41. Let $f(x) = x^2 + 3x$ and $g(x) = 4x - 1$, and find

     a. $(f \circ g)(0)$ **b.** $(g \circ f)(0)$ **c.** $(f \circ g)(x)$ **d.** $(g \circ f)(x)$

42. Let $f(x) = (x - 2)^2$ and $g(x) = x + 1$, and find the following

     a. $(f \circ g)(-1)$ **b.** $(g \circ f)(-1)$ **c.** $(f \circ g)(x)$ **d.** $(g \circ f)(x)$

For each of the following pairs of functions f and g, show that
$(f \circ g)(x) = (g \circ f)(x) = x$.

43. $f(x) = 5x - 4$ and $g(x) = \dfrac{x + 4}{5}$ **44.** $f(x) = \dfrac{x}{6} - 2$ and $g(x) = 6x + 12$

Graph the following piecewise functions.

45. $f(x) = \begin{cases} -2 & \text{if } x \le -3 \\ 1 & \text{if } x > -3 \end{cases}$

46. $f(x) = \begin{cases} -4 & \text{if } x \le -1 \\ 2x & \text{if } x > -1 \end{cases}$

47. $f(x) = \begin{cases} -3x + 6 & \text{if } x < 2 \\ \frac{1}{2}x + 5 & \text{if } x \ge 2 \end{cases}$

48. $f(x) = \begin{cases} \frac{1}{4}x - 4 & \text{if } x \le 4 \\ -3x + 9 & \text{if } x > 4 \end{cases}$

49. $f(x) = \begin{cases} -3 & \text{if } x < -3 \\ x & \text{if } -3 \le x \le 3 \\ 3 & \text{if } x > 3 \end{cases}$

50. $f(x) = \begin{cases} x + 1 & \text{if } x < -2 \\ 1 & \text{if } -2 \le x \le 2 \\ x + 3 & \text{if } x > 2 \end{cases}$

APPLYING THE CONCEPTS

51. Profit, Revenue, and Cost A company manufactures and sells wireless speakers. Here are the equations they use in connection with their business.

Number of wireless speakers sold each day: $n(x) = x$
Selling price for each speaker: $p(x) = 11.5 - 0.05x$
Daily fixed costs: $f(x) = 200$
Daily variable costs: $v(x) = 2x$
Find the following functions.

a. Revenue $= R(x) =$ the product of the number of wireless speakers sold each day and the selling price of each speaker.

b. Cost $= C(x) =$ the sum of the fixed costs and the variable costs.

c. Profit $= P(x) =$ the difference between revenue and cost.

d. Average cost $= \overline{C}(x) =$ the quotient of cost and the number of speakers sold each day.

52. Profit, Revenue, and Cost A company manufactures and sells headphones. Here are the equations they use in connection with their business.

Number of headphones sold each day: $n(x) = x$

Selling price for each set of headphones: $p(x) = 3 - \frac{1}{300}x$
Daily fixed costs: $f(x) = 200$
Daily variable costs: $v(x) = 2x$
Find the following functions.

a. Revenue $= R(x) =$ the product of the number of headphones sold each day and the selling price of each set.

b. Cost $= C(x) =$ the sum of the fixed costs and the variable costs.

c. Profit $= P(x) =$ the difference between revenue and cost.

d. Average cost $= \overline{C}(x) =$ the quotient of cost and the number of sets of headphones sold each day.

53. **Training Heart Rate** Find the training heart rate function, $T(M)$ for a person with a resting heart rate of 62 beats per minute, then find the following.

   a. Find the maximum heart rate function, $M(x)$, for a person x years of age.

   b. What is the maximum heart rate for a 24-year-old person?

   c. What is the training heart rate for a 24-year-old person with a resting heart rate of 62 beats per minute?

   d. What is the training heart rate for a 36-year-old person with a resting heart rate of 62 beats per minute?

   e. What is the training heart rate for a 48-year-old person with a resting heart rate of 62 beats per minute?

54. **Training Heart Rate** Find the training heart rate function, $T(M)$ for a person with a resting heart rate of 72 beats per minute, then find the following to the nearest whole number.

   a. Find the maximum heart rate function, $M(x)$, for a person x years of age.

   b. What is the maximum heart rate for a 20-year-old person?

   c. What is the training heart rate for a 20-year-old person with a resting heart rate of 72 beats per minute?

   d. What is the training heart rate for a 30-year-old person with a resting heart rate of 72 beats per minute?

   e. What is the training heart rate for a 40-year-old person with a resting heart rate of 72 beats per minute?

EXTENDING THE CONCEPTS

For the following problems, y varies inversely with the square of x.

55. What is the effect on y when x is tripled?

56. What is the effect on y when x is cut in half?

For the following problems, z varies jointly with x and the square of y.

57. What is the effect on z when x is doubled and y is tripled?

58. What is the effect on z when x is tripled and y is doubled?

OBJECTIVES

A Graph the four basic graphs.

B Graph using vertical translations.

C Graph using horizontal translations.

D Graph using scale factors.

©iStockphoto.com/Minerva Studio

3.9

The waitstaff at a restaurant are paid an hourly wage of $15 per hour. The amount of money they make from working 0 to 40 hours a week from their hourly wages is given by the function $f(x) = 15x$. The graph is shown below.

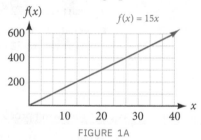

FIGURE 1A

At this particular restaurant, the owner puts all the tips earned by the waitstaff into a pool and then divides it evenly among them, adding it to their hourly wages the next week. If the previous week's pool gives each server an additional $100, then the total amount of money that each one will make this week is $g(x) = 15x + 100$; their hourly earnings plus their share of the tips from the previous week. The graph of this function is shown below.

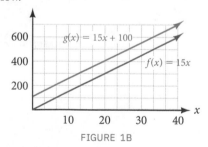

FIGURE 1B

As you can see, the graph of $g(x)$ is the graph of $f(x)$ moved up 100. We say the graph of $g(x)$ is the graph of $f(x)$, translated vertically 100 units. We start this section with a look at vertical translations of graphs.

Why Learn Graphing Techniques?

With all the graphing software available to us, you may be wondering why it is necessary to learn any graphing techniques at all. The answer is that we want to

3.9 VIDEOS

obtain a basic, intuitive, sense of what our graphs will look like before we do any graphing. It is also important to know how different numbers in the expression of a function affect the graph of that function. For example, you know intuitively what the graph of $y = x^2$ looks like. Do you also know how the number 2, in the functions below, affects the graph of $y = x^2$?

$$y = 2x^2 \qquad\qquad y = x^2 + 2 \qquad\qquad y = (x + 2)^2$$

If not, you will after you study this section. Knowing how these numbers affect the graph of the underlying function means that you have a more complete understanding of the relationship between a function and its graph.

We are not going to go into extreme detail with our techniques of graphing. We are going to do just enough to give you some good graphing tools that you can use in the next math class you take. To begin, let's list the four basic graphs that you should have memorized by this point in the course.

A Four Basic Graphs

$y = f(x) = x,$ **linear**

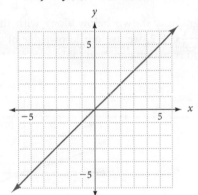

FIGURE 2A

Domain: All real numbers
Range: All real numbers

$f(x) = x^2,$ **parabolic**

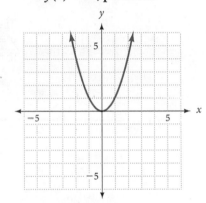

FIGURE 2B

Domain: All real numbers
Range: $\{y \mid y \geq 0\}$

$f(x) = \dfrac{1}{x},$ **rational**

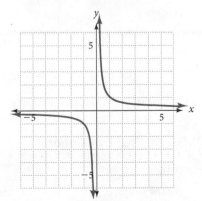

FIGURE 2C

Domain: All reals except $x = 0$
Range: All reals except $y = 0$

$f(x) = |x|,$ **absolute value**

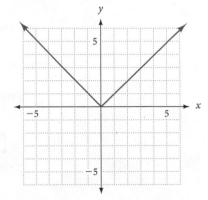

FIGURE 2D

Domain: All real numbers
Range: $\{y \mid y \geq 0\}$

ⓑ Review of Vertical Translations

EXAMPLE 1

Graph the equations $y = -\frac{1}{3}x - 4$, $y = -\frac{1}{3}x$, and $y = -\frac{1}{3}x + 2$.

SOLUTION For our first equation, we create a table from ordered pairs, then we graph the information in the table. However, even before we start, we are expecting the graph of $y = -\frac{1}{3}x - 4$ to be 4 units below the graph of $y = -\frac{1}{3}x$.

Input x	Calculate Using the Equation	Output y	Form Ordered Pairs
-3	$y = -\frac{1}{3}(-3) - 4 = 1 - 4 =$	-3	$(-3, -3)$
0	$y = -\frac{1}{3}(0) - 4 = 0 - 4 =$	-4	$(0, -4)$
3	$y = -\frac{1}{3}(3) - 4 = -1 - 4 =$	-5	$(3, -5)$

The graph is shown below along with the other two graphs. For the graph of $y = -\frac{1}{3}x + 2$, we use our knowledge of vertical translations to draw the graph, instead of making another table.

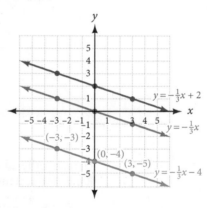

FIGURE 3

Vertical Translations

From our previous two examples we can generalize as follows.

> **RULE** *Vertical Translations*
> •
> If $k > 0$, then
>
> | The Graph of | Is the Graph of $y = f(x)$ translated |
> |---|---|
> | $y = f(x) + k$ | k units up |
> | $y = f(x) - k$ | k units down |

EXAMPLE 2

Match each relation with its graph, then indicate which relations are functions.

a. $y = |x| - 4$ **b.** $y = x^2 - 4$ **c.** $y = 2x + 2$

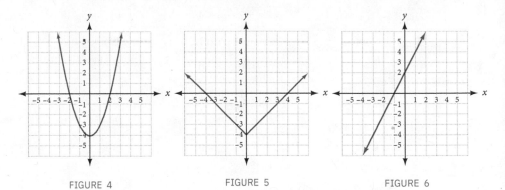

FIGURE 4   FIGURE 5   FIGURE 6

SOLUTION   Using the basic graphs for a guide along with our knowledge of translations, we have the following:

a. Figure 5   **b.** Figure 4   **c.** Figure 6

And, since all of these graphs pass the vertical line test, all are functions.

🅒 Horizontal Translations

Now consider the graphs of $f(x) = (x + 2)^2$ and $g(x) = (x - 2)^2$ shown in Figure 7 on the next page. Compared with the graph of the basic function $y = x^2$, the graph of $f(x) = (x + 2)^2$ is shifted two units to the *left*, as shown by the arrows. Similarly, the graph of $g(x) = (x - 2)^2$ is shifted two units to the *right* compared to the graph of $y = x^2$. In general, we have the following principle:

RULE   *Horizontal Translations*

If $h > 0$, then

The Graph of	Is the Graph of $y = f(x)$ translated
$y = f(x + h)$	h units to the left
$y = f(x - h)$	h units to the right

x	$y = x^2$	$f(x) = (x + 2)^2$
-3	9	1
-2	4	0
-1	1	1
0	0	4
1	1	9
2	4	16
3	9	25

x	$y = x^2$	$g(x) = (x - 2)^2$
-3	9	25
-2	4	16
-1	1	9
0	0	4
1	1	1
2	4	0
3	9	1

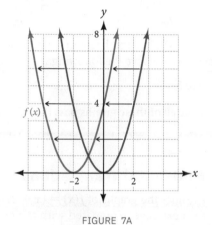

FIGURE 7A

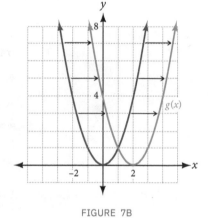

FIGURE 7B

EXAMPLE 3

Graph $y = |x + 3|$.

SOLUTION The graph of this equation will be a transformation of the basic graph $f(x) = |x|$. Since $y = f(x + 3)$, we know its graph is the horizontal translation of $|x|$ by 3 units to the left.

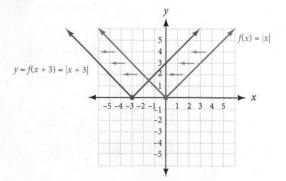

FIGURE 8

Our next example involves both a horizontal translation and a vertical translation of a graph.

EXAMPLE 4

Graph $y = |x - 3| + 4$.

SOLUTION   Our graph will be a transformation of the basic graph $y = |x|$. We proceed according to the rule for order of operations, which tells us to go within the first grouping symbol first and then work our way out. Here is a diagram showing how we approach this, along with the graphs that result:

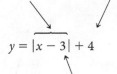

Step 1:   The basic graph is $y = |x|$ **Step 3:**   Translate 4 units up

$$y = |x - 3| + 4$$

Step 2:   Translate 3 units right

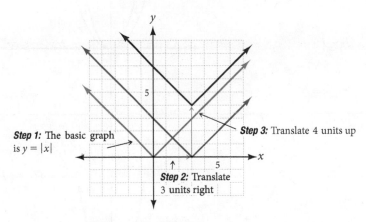

Step 1: The basic graph is $y = |x|$

Step 3: Translate 4 units up

Step 2: Translate 3 units right

FIGURE 9

🅓 Scale Factors

We have seen that adding a constant to the expression defining a function results in a translation of its graph. What happens if we multiply the expression by a constant? Consider the graphs of the functions

$$f(x) = 2x^2, g(x) = \frac{1}{2}x^2, \text{ and } h(x) = -x^2$$

shown in Figure 10A, B, and C on the next page, and compare each to the graph of $y = x^2$.

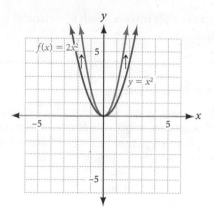

FIGURE 10A

x	$y = x^2$	$f(x) = 2x^2$
−2	4	8
−1	1	2
0	0	0
1	1	2
2	4	8

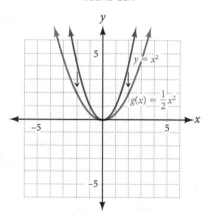

FIGURE 10B

x	$y = x^2$	$g(x) = \frac{1}{2}x^2$
−2	4	2
−1	1	$\frac{1}{2}$
0	0	0
1	1	$\frac{1}{2}$
2	4	2

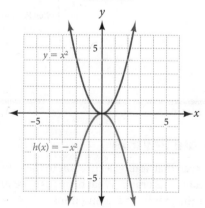

FIGURE 10C

x	$y = x^2$	$h(x) = -x^2$
−2	4	−4
−1	1	−1
0	0	0
1	1	−1
2	4	−4

Compared to the graph of $y = x^2$, the graph of $f(x) = 2x^2$ is expanded, or stretched, vertically by a factor of 2. The y-coordinate of each point on the graph has been doubled, as you can see in the table of values, so each point on the graph of f is twice as far from the x-axis as its counterpart on the basic graph $y = x^2$. The graph of $g(x) = \frac{1}{2}x^2$ is compressed vertically by a factor of $\frac{1}{2}$; each point is half as far from the x-axis as its counterpart on the graph of $y = x^2$. The graph of $h(x) = -x^2$ is flipped, or reflected, about the x-axis; the y-coordinate of each point on the graph of $y = x^2$ is replaced by its opposite.

In general, we have the following:

RULE *Scale Factors and Reflections*
• •

Compared with the graph of $y = f(x)$, the graph of $y = af(x)$, where $a \neq 0$, is

1. stretched vertically by a factor of $|a|$ if $|a| > 1$,

2. compressed vertically by a factor of $|a|$ if $0 < |a| < 1$, and

3. reflected about the x-axis if $a < 0$.

The constant a is called the *scale factor* for the graph.

EXAMPLE 5

Graph $f(x) = \dfrac{1}{4}(x + 4)^2 + 2$.

SOLUTION   We identify the basic graph from the structure of the formula for $f(x)$. In this case, the basic graph is $y = x^2$. We will perform the translations separately, following the order of operations. Here is a diagram showing how we interpret the information in the equation, followed by Figure 11, which shows how we construct the graph.

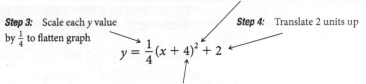

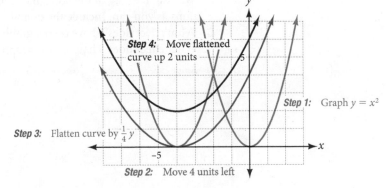

FIGURE 11

EXAMPLE 6

Graph $f(x) = 5 - 3(x - 1)^2$, and state the vertex and range.

SOLUTION   Writing the function in the form $f(x) = -3(x - 1)^2 + 5$, we see that there is a vertical expansion by a factor of 3 and reflection about the x-axis, a horizontal translation of 1 unit to the right, and a vertical translation of 5 units upward.

The process is shown in Figure 12. We begin by graphing the basic parabola $y = x^2$. Next, we multiply each y-coordinate from the basic parabola by -3 to obtain the graph of $y = -3x^2$. Finally, we shift each point on this new graph 1 unit to the right and 5 units upward.

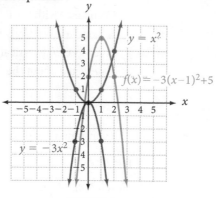

FIGURE 12

From the graph, we can see that the vertex is (1, 5) and the range of the function is $\{y \mid y \leq 5\}$, or $(-\infty, 5]$ using interval notation. The negative sign in front of the 3 has reflected the parabola so it opens downward.

GETTING READY FOR CLASS

Respond in your own words and in complete sentences.

A. Why do we learn graphing techniques when there are so many graphing programs available to us?

B. Without looking in the book, name and graph as many of the four basic functions as you can. Include the domain and range for each one.

C. What effect does 5 have on the graph of $y = 5x^2$?

D. What effect does 5 have on the graph of $y = x^2 - 5$?

Problem Set 3.9

For each problem below, the equation of the red curve is given. Find the equation for the blue graph.

1.

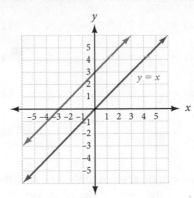

2.

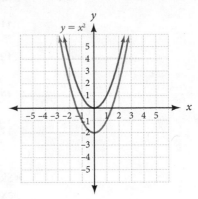

3.

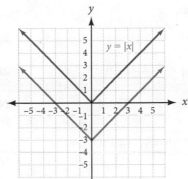

4.

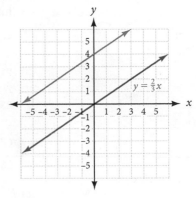

Graph each of the following. Use one coordinate system for each problem.

5.   **a.** $y = 2x$ **6.**   **a.** $y = \dfrac{1}{3}x$ **7.**   **a.** $y = \dfrac{1}{2}x^2$ **8.**   **a.** $y = 2x^2$

     b. $y = 2x + 3$ **b.** $y = \dfrac{1}{3}x + 1$ **b.** $y = \dfrac{1}{2}x^2 - 2$ **b.** $y = 2x^2 - 8$

     c. $y = 2x - 5$ **c.** $y = \dfrac{1}{3}x - 3$ **c.** $y = \dfrac{1}{2}x^2 + 2$ **c.** $y = 2x^2 + 1$

Identify each graph as a translation of a basic function, and write a formula for the graph.

9.

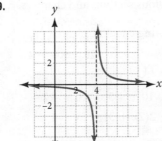

10.

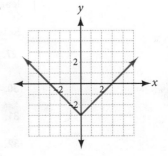

307

Sketch the basic graph and the graph of the given function on the same axes. Label the coordinates of two or three points on the graph of the given function.

11. $f(x) = |x| - 2$ **12.** $f(s) = s^2 + 3$

13. $F(r) = \dfrac{1}{r - 4}$ **14.** $h(v) = \dfrac{1}{v + 6}$

15. $f(x) = \dfrac{1}{3}|x|$ **16.** $H(x) = -3|x|$

17. $g(z) = \dfrac{2}{z}$ **18.** $H(x) = \dfrac{1}{3x}$

Sketch the basic graph and the graph of the given function by hand on the same axes. Label the coordinates of two or three points on the graph of the given function.

19. $f(x) = 2 + (x - 3)^2$ **20.** $f(x) = (x + 4)^2 + 1$

21. $g(z) = \dfrac{1}{z + 2} - 3$ **22.** $g(z) = \dfrac{1}{z - 1} + 1$

23. $G(t) = 2|t - 5| - 1$ **24.** $G(t) = 2 - |t + 4|$

Give an equation for each graph.

25.

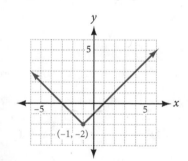

26.

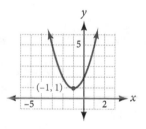

MAINTAINING YOUR SKILLS

The problems that follow review some of the more important skills you have learned in previous sections and chapters.

Solve the following equations.

27. $x - 5 = 7$ **28.** $3y = -4$ **29.** $5 - \dfrac{4}{7}a = -11$

30. $\dfrac{1}{5}x - \dfrac{1}{2} - \dfrac{1}{10}x + \dfrac{2}{5} = \dfrac{3}{10}x + \dfrac{1}{2}$

31. $5(x - 1) - 2(2x + 3) = 5x - 4$

32. $0.07 - 0.02(3x + 1) = -0.04x + 0.01$

Solve for the indicated variable.

33. $P = 2l + 2w$ for w

34. $A = \dfrac{1}{2}h(b + B)$ for B

Solve the following inequalities. Write the solution set using interval notation, then graph the solution set.

35. $-5t \le 30$

36. $5 - \dfrac{3}{2}x > -1$

37. $1.6x - 2 < 0.8x + 2.8$

38. $3(2y + 4) \ge 5(y - 8)$

Solve the following equations.

39. $\left|\dfrac{1}{4}x - 1\right| = \dfrac{1}{2}$

40. $\left|\dfrac{2}{3}a + 4\right| = 6$

41. $|3 - 2x| + 5 = 2$

42. $5 = |3y + 6| - 4$

EXAMPLES

Linear Equations in Two Variables [3.1, 3.3]

1. The equation $3x + 2y = 6$ is an example of a linear equation in two variables.

A *linear equation in two variables* is any equation that can be put in *standard form* $ax + by = c$. The graph of every linear equation is a straight line.

Intercepts [3.1]

2. To find the x-intercept for $3x + 2y = 6$, we let $y = 0$ and get

$$3x = 6$$
$$x = 2$$

In this case the x-intercept is 2, and the graph crosses the x-axis at $(2, 0)$.

The *x-intercept* of an equation is the *x-coordinate* of the point where the graph crosses the *x-axis*. The *y-intercept* is the *y-coordinate* of the point where the graph crosses the *y-axis*. We find the y-intercept by substituting $x = 0$ into the equation and solving for y. The x-intercept is found by letting $y = 0$ and solving for x.

The Slope of a Line [3.2]

3. The slope of the line through $(6, 9)$ and $(1, -1)$ is

$$m = \frac{9 - (-1)}{6 - 1} = \frac{10}{5} = 2$$

The *slope* of the line containing points (x_1, y_1) and (x_2, y_2) is given by

$$\text{Slope} = m = \frac{\text{Rise}}{\text{Run}} = \frac{y_2 - y_1}{x_2 - x_1}$$

Horizontal lines have 0 slope, and vertical lines have an undefined slope. Parallel lines have equal slopes, and perpendicular lines have slopes that are negative reciprocals.

The Slope-Intercept Form of a Line [3.3]

4. The equation of the line with slope 5 and y-intercept 3 is

$$y = 5x + 3$$

The equation of a line with slope m and y-intercept b is given by

$$y = mx + b$$

The Point-Slope Form of a Line [3.3]

5. The equation of the line through $(3, 2)$ with slope -4 is

$$y - 2 = -4(x - 3)$$

which can be simplified to

$$y = -4x + 14$$

The equation of the line through (x_1, y_1) that has slope m can be written as

$$y - y_1 = m(x - x_1)$$

Linear Inequalities in Two Variables [3.4]

6. The graph of

$$x - y \leq 3$$

is

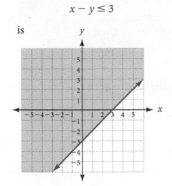

An inequality of the form $ax + by < c$ is a *linear inequality in two variables*. The equation for the boundary of the solution set is given by $ax + by = c$. (This equation is found by simply replacing the inequality symbol with an equal sign.)

To graph a linear inequality, first graph the boundary, using a solid line if the boundary is included in the solution set and a broken line if the boundary is not included in the solution set. Next, choose any point not on the boundary and substitute its coordinates into the original inequality. If the resulting statement is true, the graph lies on the same side of the boundary as the test point. A false statement indicates that the solution set lies on the other side of the boundary.

Relations and Functions [3.5]

7. The relation

$$\{(8, 1), (6, 1), (-3, 0)\}$$

is also a function because no ordered pairs have the same first coordinates. The domain is $\{8, 6, -3\}$ and the range is $\{1, 0\}$.

A *function* is a rule that pairs each element in one set, called the *domain*, with exactly one element from a second set, called the *range*.

A *relation* is any set of ordered pairs. The set of all first coordinates is called the *domain* of the relation, and the set of all second coordinates is the *range* of the relation. A function is a relation in which no two different ordered pairs have the same first coordinates.

Vertical Line Test [3.5]

8. The graph of $x = y^2$ shown in Figure 5 in Section 3.5 fails the vertical line test. It is not the graph of a function.

If a vertical line crosses the graph of a relation in more than one place, the relation cannot be a function. If no vertical line can be found that crosses the graph in more than one place, the relation must be a function.

Function Notation [3.6]

9. If $f(x) = 5x - 3$ then
$$f(0) = 5(0) - 3$$
$$= -3$$
$$f(1) = 5(1) - 3$$
$$= 2$$
$$f(-2) = 5(-2) - 3$$
$$= -13$$
$$f(a) = 5a - 3$$

The alternative notation for y is $f(x)$. It is read "f of x" and can be used instead of the variable y when working with functions. The notation y and the notation $f(x)$ are equivalent; that is, $y = f(x)$.

Variation [3.7]

10. If y varies directly with x, then

$$y = Kx$$

Then if y is 18 when x is 6,

$$18 = K \cdot 6$$

or

$$K = 3$$

So the equation can be written more specifically as

$$y = 3x$$

If we want to know what y is when x is 4, we simply substitute:

$$y = 3 \cdot 4$$
$$y = 12$$

If y *varies directly* with x (y is directly proportional to x), then

$$y = Kx$$

If y *varies inversely* with x (y is inversely proportional to x), then

$$y = \frac{K}{x}$$

If z *varies jointly* with x and y (z is directly proportional to both x and y), then

$$z = Kxy$$

In each case, K is called the *constant of variation*.

Algebra with Functions [3.8]

If f and g are any two functions with a common domain, then:

$(f + g)(x) = f(x) + g(x)$	The function $f + g$ is the sum of the functions f and g.
$(f - g)(x) = f(x) - g(x)$	The function $f - g$ is the difference of the functions f and g.
$(fg)(x) = f(x)g(x)$	The function fg is the product of the functions f and g.
$\dfrac{f}{g}(x) = \dfrac{f(x)}{g(x)}$	The function $\frac{f}{g}$ is the quotient of the functions f and g, where $g(x) \neq 0$.

Composition of Functions [3.8]

If f and g are two functions for which the range of each has numbers in common with the domain of the other, then we have the following definitions:

$$\text{The composition of } f \text{ with } g : (f \circ g)(x) = f[g(x)]$$

$$\text{The composition of } g \text{ with } f : (g \circ f)(x) = g[f(x)]$$

Piecewise Functions [3.8]

11. To graph the function

$$f(x) = \begin{cases} -2x - 5 & \text{if } x \geq 0 \\ 3 & \text{if } x \geq 0 \end{cases},$$

you would graph $y = 2x - 5$ for all $x < 0$ and then $y = 3$ for all $x \geq 0$. For $f(0) = 3$ and the domain is all real numbers. The range is $y \in (5,)$.

A function is called a *piecewise function* when it is defined as two or more functions pieced together.

Vertical and Horizontal Translations [3.9]

12. The graph of $f(x) = (x + 1)^2 - 4$ translated 1 unit to the left and 4 units down as compared to the basic graph $y = x^2$. The vertex is $(-1, -4)$.

If $f(x) = (x - h)^2 + k$, and h and k are real numbers, then h represents a *horizontal translation* and k represents a *vertical translation*.

The graph of	is the graph of $y = x^2$ translated
$f(x) = (x - h)^2$	h units to the right
$f(x) = (x + h)^2$	h units to the left
$f(x) = x^2 + k$	k units upward
$f(x) = x^2 - k$	k units downward

COMMON MISTAKES

1. When graphing ordered pairs, the most common mistake is to associate the first coordinate with the y-axis and the second with the x-axis. If you make this mistake you would graph (3, 1) by going up 3 and to the right 1, which is just the reverse of what you should do. Remember, the first coordinate is always associated with the horizontal axis, and the second coordinate is always associated with the vertical axis.

2. The two most common mistakes students make when first working with the formula for the slope of a line are the following:
   a. Putting the difference of the x-coordinates over the difference of the y-coordinates.
   b. Subtracting in one order in the numerator and then subtracting in the opposite order in the denominator.

3. When graphing linear inequalities in two variables, remember to graph the boundary with a broken line when the inequality symbol is $<$ or $>$. The only time you use a solid line for the boundary is when the inequality symbol is $\leq$ or $\geq$.

Graph the following. [3.1]

1. $4x - 8y = 16$

2. $y = -\dfrac{5}{3}x + 4$

3. $y = 5$

4. $y = 2x^2 - 5$

Find the intercepts for each graph. Then use them to help sketch the graph. [3.1]

5. $y = -\dfrac{1}{2}x + 1$

6. $y = x^2 + 3$

7. $y = x + 2$

State the domain and range of each relation, and then indicate which relations are also functions. [3.5]

8. $\{(3, 0), (2, -4), (2, 1)\}$

9. $\{(3, 0), (4, 2), (5, -3)\}$

10. $(a, 5), (b, 5), (c, 4), (d, 5)$

11. $(a, 1), (b, 1), (c, 1), (d, 1)$

Graph each of the following relations. In each case, use the graph to find the domain and range, and indicate whether the graph is the graph of a function. [3.5]

12. $y = x^2 - 9$

13. $x = y^2 + 1$

14. $x = 3 - y^2$

15. $x = |y| - 3$

If $f = \{(2, 3), (-4, 4), (5, 0), (-3, -4)\}$ and $g = \{(\pi, 3), (-2, 0), (0, -4)\}$, find the following. [3.3]

16. $f(2)$

17. $f(5) + g(-2)$

Let $f(x) = 2x - 5$ and $g(x) = x^2 + 3x + 4$. Evaluate the following. [3.3]

18. $f(-4)$

19. $g(2)$

20. $f(a)$

21. $g(a + 6)$

If $f = \{(1, 4), (-2, 0), \left(3, \frac{1}{2}\right), (\pi, 0)\}$ and $g = \{(1, 1), (-2, 2), \left(\frac{1}{2}, 0\right)\}$, find each of the following values of f and g.

22. $f(3)$

23. $f(-2) - g(-2)$

Let $f(x) = 3x^2 - 4x + 1$ and $g(x) = 4x - 3$, and evaluate each of the following. [3.4]

24. $f(0)$

25. $g(3)$

26. $f(g(1))$

27. $f(g(0))$

Let $f(x) = x^2 - 2x$ and $g(x) = 5x - 4$. Evaluate the following.

28. $f(x) - g(x^2)$

29. $g(f(3))$

Let $f(x) = \dfrac{1}{x + 3}$ and $g(x) = \dfrac{1}{x} + 1$. Evaluate the following.

30. $g(0)$

31. $f(-1)$

Let $f(x) = 4x - 3$, $g(x) = x^2 + 4$, and $h(x) = x + 5$, and find the following. [3.4]

32. $(g \circ h)(-2)$

33. $(f + g)(-4)$

34. $\left(\dfrac{g}{f}\right)(2)$

35. $(g - fh)(-5)$

Find the slope of the line through the following pairs of points. [3.5]

36. $(6, 3), (-4, 8)$

37. $(-3, 2), (-5, 7)$

Find the variable if the line through the two given points has the given slope. [3.5]

38. $(5, y), (4, 4); m = -1$

39. $(-3, 7), (x, 9); m = \dfrac{1}{4}$

40. Find the slope of any line parallel to the line through $(5, -3)$ and $(-7, 4)$. [3.5]

41. The line through $(4, 2y)$ and $(6, 3y)$ is parallel to a line with slope $-\frac{1}{2}$. What is the value of y? [3.5]

42. Perpendicular Lines Line l contains the points $(3, 4)$ and $(-3, 1)$. Give the slope of any line perpendicular to l. [3.5]

43. Perpendicular Lines Line l contains the points $(6, -3)$ and $(-2, 7)$. Find the slope of any line perpendicular to l. [3.5]

Find the average rate of change of each function on the given interval. [3.5]

44. $f(x) = -4x + 7, [-2, 5]$ **45.** $f(x) = 3x^2 + 7, [0, 2]$

Give the equation of the line with the following slope and y-intercept. [3.6]

46. $m = -5, b = 4$ **47.** $m = -7, b = -3$

Give the slope and y-intercept of each equation. [3.6]

48. $4x - y = 3$ **49.** $4x - 5y = 10$

Find the equation of the line that contains the given point and has the given slope. [3.6]

50. $(3, 5), m = 3$ **51.** $(-2, 5), m = \dfrac{3}{2}$

For each of the following problems, the slope and one point on the line are given. In each case, find the equation of the linear function. [3.6]

52. $(3, -4), m = -\dfrac{1}{3}$ **53.** $(6, 2), m = 0$

Find the equation of the line that contains the given pair of points. [3.6]

54. $(3, 0), (-2, 4)$ **55.** $(-4, 2), (3, -5)$

56. $\left(-\dfrac{1}{2}, -\dfrac{1}{2}\right), \left(\dfrac{1}{2}, \dfrac{1}{10}\right)$ **57.** $\left(-\dfrac{2}{3}, \dfrac{2}{3}\right), \left(\dfrac{1}{6}, -\dfrac{1}{6}\right)$

58. $(-2, 3), (4, 1)$

59. Find the equation of the line that is parallel to $3x + y = 5$ and contains the point $(0, 4)$. [3.6]

60. Find the equation of the line perpendicular to $y = -\frac{2}{3}x + 2$ that has an x-intercept of -4. [3.6]

Graph each linear inequality. [3.7]

61. $x + y \leq 5$ **62.** $x - y > -3$

63. $4y \leq 2x + 8$ **64.** $-2y < 6$

65. $x > -2$ **66.** $6y > 4y - 10$

67. $y \geq \dfrac{1}{4}x - 1$ **68.** $\dfrac{x}{2} + \dfrac{y}{3} < 1$

69. Piecewise Functions Graph the piecewise function. [3.8]

$$f(x) = \begin{cases} 0 & \text{if } x < -2 \\ \dfrac{1}{2}x - 1 & \text{if } -2 \leq x \leq 2 \\ 2 & \text{if } x > 2 \end{cases}$$

70. Translations Sketch the basic graph and the graph of the given function by hand on the same axes. Label the coordinates of two or three points on the graph of the given function. [3.9]

$$g(z) = \dfrac{1}{z - 1} + 1$$

71. Inverse Variation y varies inversely with the square of x. If $y = 6$ when $x = 3$, find y when $x = 6$. [1.8]

72. Joint Variation z varies jointly with x and the cube of y. If $z = 216$ when $x = 4$ and $y = 3$, find z when $x = 7$ and $y = 2$. [1.8]

Simplify each of the following.

1. -6^2

2. $-|-4|$

3. $3^3 + 4^2$

4. $(4 + 9)^2$

5. $36 \div 12 \cdot 6$

6. $108 \div 6 \cdot 3$

7. $36 - 24 \div 6 + 2$

8. $36 - 24 \div (6 + 2)$

Find the value of each expression when x is 5.

9. $x^2 + 3x - 7$

10. $x^2 - 36$

Reduce the following fractions to lowest terms.

11. $\dfrac{108}{126}$

12. $\dfrac{252}{468}$

Perform the indicated operation.

13. $\dfrac{7}{18} + \dfrac{7}{24}$

14. $\dfrac{7}{32} - \dfrac{5}{24}$

15. Add $-\dfrac{5}{4}$ to the product of 7 and $\dfrac{1}{4}$.

16. Subtract $\dfrac{7}{9}$ from the product of 8 and $\dfrac{5}{6}$.

Simplify each of the following expressions.

17. $12\left(\dfrac{2}{3}x - \dfrac{5}{6}y\right)$

18. $6\left(\dfrac{1}{3}x - \dfrac{3}{2}y\right)$

Simplify each of the following expressions.

19. $5a - 3[4 + (a - 2)]$

20. $(4x^3 + 8) - (5x^2 - 8) + (3x^3 - 2) + (6x^2 + 1)$

Multiply or divide as indicated.

21. $(-5ab^3)(4a^2b^2)$

22. $\dfrac{(5a^2b)^3}{(a^4b)^2}$

23. $2x(9x^2 + 4x + 7)$

24. $\left(x + \dfrac{3}{4}\right)^3$

25. $\dfrac{9 \times 10^8}{3 \times 10^2}$

Factor each of the following expressions.

26. $8a^2 + 10a - 7$

27. $\dfrac{1}{27} + t^3$.

Solve the following equations.

28. $(x + 5)^2 = 2x + 9$

Solve the following equations.

29. $\dfrac{2}{3}x + 3 = -9$

30. $\dfrac{1}{2}(4x - 7) + \dfrac{1}{6} = -1$

31. $-3 + 4(2x - 5) = 9$

32. $4x - (2x - 7) = -9$

33. $|3x - 5| = 4$

34. $|2x - 5| = 2x + 7$

Solve the following expressions for y and simplify.

35. $y - 3 = -4(x - 2)$

36. $y + 4 = \dfrac{3}{2}(4x - 6)$

Solve each inequality and graph the solution set.

37. $|2x - 5| \le 9$

38. $|3x + 5| - 3 > 7$

Graph on a rectangular coordinate system.

39. $4x - 3y = -12$

40. $y = -3$

41. $y = \dfrac{2}{3}x - 2$

42. $x = 2$

43. Find the slope of the line through $(-6, 4)$ and $(5, -7)$.

44. Find y if the line through $(-6, 2)$ and $(3, y)$ has a slope of $\frac{2}{3}$.

45. Find the slope of the line $x = -3$.

46. Give the equation of a line with slope $-\frac{3}{5}$ and y-intercept $= 7$.

47. Find the slope and y-intercept of $3x - 6y = 12$.

48. Find the equation of the line with x-intercept -3 and y-intercept 6.

49. Find the equation of the line that is parallel to $3x - 6y = 4$ and contains the point $(6, 7)$.

50. Find the equation of a line with slope $\frac{5}{7}$ that contains the point $(14, -3)$. Write the answer in slope-intercept form.

If $f(x) = x^2 + 5x$, $g(x) = 2x + 3$ and $h(x) = 3x - 1$, find the following.

51. $g(0)$

52. $f(-1) + h(2)$

53. $(f \circ h)(3)$

54. $(f - g)(x)$

55. Specify the domain and range for the relation $\{(-2, 3), (-5, 3), (1, 4)\}$. Is the relation also a function?

56. Sarah left a $2.25 tip for a $15 dinner. What percent of the cost of dinner was the tip?

57. If the sales tax rate is 8.2% of the purchase price, how much sales tax will Randy pay on a $200 iPod?

Use the information in the illustration to work Problems 58–59.

CONVENIENCE STORES AROUND THE U.S.
U.S. Census Bureau, 2002

Texas	9,143
California	5,768
Florida	5,676
Ohio	3,537
Illinois	3,280
Michigan	3,402
Pennsylvania	2,996
Georgia	4,050
North Carolina	4,081

*Convenience Store Establishments per State

58. How many more convenience stores are there in Texas as compared to California? Express your answer as a percent rounded to the nearest tenth.

59. How many more convenience stores are there in Michigan as compared to Pennsylvania? Express your answer as a percent rounded to the nearest tenth.

For each of the following straight lines, identify the x-intercept, y-intercept, and slope, and sketch the graph. [3.1–3.3]

1. $2x + y = 6$

2. $y = -2x - 3$

3. $y = \dfrac{3}{2}x + 4$

4. $x = -2$

Find the equation for each line. [3.3]

5. Give the equation of the line through $(-1, 3)$ that has slope $m = 2$.

6. Give the equation of the line through $(-3, 2)$ and $(4, -1)$.

7. Line l contains the point $(5, -3)$ and has a graph parallel to the graph of $2x - 5y = 10$. Find the equation for l.

8. Line l contains the point $(-1, -2)$ and has a graph perpendicular to the graph of $y = 3x - 1$. Find the equation for l.

9. Give the equation of the vertical line through $(4, -7)$.

Graph the following linear inequalities. [3.4]

10. $3x - 4y < 12$ **11.** $y \le -x + 2$

State the domain and range for the following relations, and indicate which relations are also functions. [3.5]

12. $\{(-2, 0), (-3, 0), (-2, 1)\}$

13. $y = x^2 - 9$

Let $f(x) = x - 2$, $g(x) = 3x + 4$ and $h(x) = 3x^2 - 2x - 8$, and find the following. [3.6, 3.7]

14. $f(3) + g(2)$ **15.** $h(0) + g(0)$

16. $(f \circ g)(2)$ **17.** $(g \circ f)(2)$

Solve the following variation problems. [3.8]

18. Direct Variation Quantity y varies directly with the square of x. If y is 50 when x is 5, find y when x is 3.

19. Joint Variation Quantity z varies jointly with x and the cube of y. If z is 15 when x is 5 and y is 2, find z when x is 2 and y is 3.

20. Maximum Load The maximum load (L) a horizontal beam can safely hold varies jointly with the width (w) and the square of the depth (d) and inversely with the length (l). If a 10-foot beam with width 3 feet and depth 4 feet will safely hold up to 800 pounds, how many pounds will a 12-foot beam with width 3 feet and depth 4 feet hold?

21. Piecewise Functions Graph the piecewise function. [3.8]

$$f(x) = \begin{cases} -2x - 5 & \text{if } x < -2 \\ -1 & \text{if } -2 \le x \le 2. \\ 2x - 5 & \text{if } x > 2 \end{cases}$$

22. Translations Sketch the basic graph and the graph of the given function by hand on the same axes. Label the coordinates of two or three points on the graph of the given function. [3.9]

$$g(z) = \frac{1}{z - 1} + 1$$

4

Systems of Equations

© iStockphoto/ViewApart

Suppose you decide to buy a cellular phone and are trying to decide between two rate plans. Plan A is $18.95 per month plus $.48 for each minute, or fraction of a minute, that you use the phone. Plan B is $34.95 per month plus $.36 for each minute, or fraction of a minute. The monthly cost $C(x)$ for each plan can be represented with a linear equation in two variables:

$$\text{Plan A:} \quad C(x) = 0.48x + 18.95$$

$$\text{Plan B:} \quad C(x) = 0.36x + 34.95$$

To compare the two plans, we use the table and graph shown below.

Monthly Cellular Phone Charges

Number of Minutes x	Monthly Cost ($) Plan A	Plan B
0	18.95	34.95
40	38.15	49.35
80	57.35	63.75
120	76.55	78.15
160	95.75	92.55
200	114.95	106.95
240	134.15	121.35

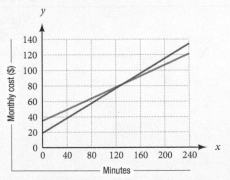

The point of intersection of the two lines in the graph is the point at which the monthly costs of the two plans are equal. In this chapter, we will develop methods of finding that point of intersection.

STUDY SKILLS

The study skills for this chapter are concerned with getting ready to take an exam.

1. **Preparing for an Exam** Try to arrange your daily study habits so you have little studying to do the night before your next exam. The next two goals will help you achieve goal number 1.

2. **Review with the Exam in Mind** You should review material that will be covered on the next exam every day. Your review should consist of working problems. Preferably, the problems you work should be problems from your list of difficult problems.

3. **Continue to List Difficult Problems** You should continue to list and rework the problems that give you the most difficulty. It is this list that you will use to study for the next exam. Your goal is to go into the next exam knowing you can successfully work any problem from your list of hard problems.

4. **Pay Attention to Instructions** Taking a test is different from doing homework. When you take a test, the problems will usually be in random order. When you do your homework, you usually work a number of similar problems. Sometimes students who do well on their homework become confused when they see the same problems on a test, because they have not paid attention to the instructions on their homework. For example, suppose you see the equation $y = 3x - 2$ on your next test. By itself, the equation is simply a statement. There isn't anything to do unless the equation is accompanied by instructions. Each of the following is a valid instruction with respect to the equation $y = 3x - 2$ and the result of applying the instructions will be different in each case:

   Find x when y is 10.
   Solve for x.
   Graph the equation.
   Find the intercepts.
   Find the slope.

There are many things to do with the equation. If you train yourself to pay attention to the instructions that accompany a problem as you work through the assigned problems, you will not find yourself confused about what to do with a problem when you see it on a test.

4.1

OBJECTIVES

A Solve a system of linear equations in two variables using the addition method.

B Solve a system of linear equations in two variables using the substitution method.

© iStockphoto/rappensuncle

Suppose you need to rent a moving truck for two days. If U-Haul charges $50 per day and $0.50 per mile, and Budget Truck Rental charges $45 per day and $0.75 per mile, which company offers the lowest price? After working through this section, you will be able to set up a *linear system of equations* for this problem and use an algebraic method, called the substitution method, to solve the system and answer the question. But first, let's examine how graphing is a useful method for finding the solution set to a system of equations.

Previously, we found the graph of an equation of the form $ax + by = c$ to be a straight line. Because the graph is a straight line, the equation is said to be a linear equation. Two linear equations considered together form a *linear system* of equations. For example,

$$3x - 2y = 6$$

$$2x + 4y = 20$$

is a linear system. The solution set to the system is the set of all ordered pairs that satisfy both equations. If we graph each equation on the same set of axes, we can see the solution set (see Figure 1).

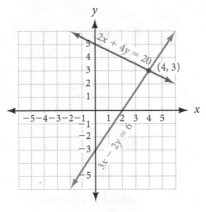

FIGURE 1

The point (4, 3) lies on both lines and therefore must satisfy both equations. It is obvious from the graph that it is the only point that does so. The solution set for the system is {(4, 3)}.

More generally, if $a_1x + b_1y = c_1$ and $a_2x + b_2y = c_2$ are linear equations, then the solution set for the system

$$a_1x + b_1y = c_1$$

$$a_2x + b_2y = c_2$$

can be illustrated through one of the graphs in Figure 2.

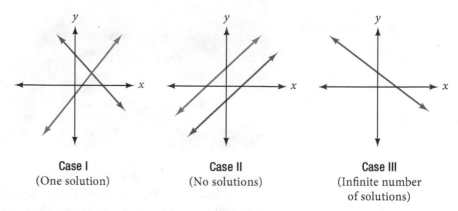

| Case I | Case II | Case III |
| (One solution) | (No solutions) | (Infinite number of solutions) |

FIGURE 2

Case I   The two lines intersect at one and only one point. The coordinates of the point give the solution to the system. This is what usually happens.

Case II   The lines are parallel and therefore have no points in common. The solution set to the system is the empty set, $\varnothing$. In this case, we say the equations are *inconsistent*.

Case III   The lines coincide. That is, their graphs represent the same line. The solution set consists of all ordered pairs that satisfy either equation. In this case, the equations are said to be *dependent*.

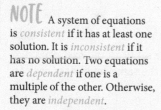 A system of equations is *consistent* if it has at least one solution. It is *inconsistent* if it has no solution. Two equations are *dependent* if one is a multiple of the other. Otherwise, they are *independent*.

In the beginning of this section, we found the solution set for the system

$$3x - 2y = 6$$

$$2x + 4y = 20$$

by graphing each equation and then reading the solution set from the graph. Solving a system of linear equations by graphing is the least accurate method. If the coordinates of the point of intersection are not integers, it can be difficult to read the solution set from the graph. There is another method of solving a linear system that does not depend on the graph. It is called the *addition method*.

A   The Addition Method

EXAMPLE 1

Solve the system.

$$4x + 3y = 10$$

$$2x + y = 4$$

SOLUTION   If we multiply the bottom equation by -3, the coefficients of y in the resulting equation and the top equation will be opposites:

$$4x + 3y = 10 \xrightarrow{\text{No Change}} 4x + 3y = 10$$

$$2x + y = 4 \xrightarrow[\text{Multiply by }-3]{} -6x - 3y = -12$$

Adding the left and right sides of the resulting equations, we have

$$
\begin{array}{r}
4x + 3y = 10 \\
-6x - 3y = -12 \\
\hline
-2x = -2
\end{array}
$$

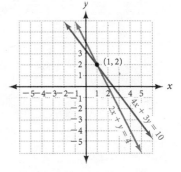

FIGURE 3 *A visual representation of the solution to the system in Example 1*

The result is a linear equation in one variable. We have eliminated the variable y from the equations by addition. (It is for this reason we call this method of solving a linear system the *addition method*.) Solving $-2x = -2$ for x, we have

$$x = 1$$

> **NOTE** If we had put $x = 1$ into the first equation in our system, we would have obtained $y = 2$ also:
>
> $$4(1) + 3y = 10$$
> $$3y = 6$$
> $$y = 2$$

This is the x-coordinate of the solution to our system. To find the y-coordinate, we substitute $x = 1$ into any of the equations containing both the variables x and y. Let's try the second equation in our original system:

$$
\begin{array}{r}
2(1) + y = 4 \\
2 + y = 4 \\
\hline
y = 2
\end{array}
$$

This is the y-coordinate of the solution to our system. The ordered pair $(1, 2)$ is the solution to the system.

Checking Solutions

We can check our solution by substituting it into both of our equations.

Substituting $x = 1$ and $y = 2$ into $4x + 3y = 10$, we have

$$4(1) + 3(2) \overset{?}{=} 10$$
$$4 + 6 \overset{?}{=} 10$$
$$10 = 10 \quad \text{A true statement}$$

Substituting $x = 1$ and $y = 2$ into $2x + y = 4$, we have

$$2(1) + 2 \overset{?}{=} 4$$
$$2 + 2 \overset{?}{=} 4$$
$$4 = 4 \quad \text{A true statement}$$

Our solution satisfies both equations; therefore, it is a solution to our system of equations.

EXAMPLE 2

Solve the system.

$$3x - 5y = -2$$

$$2x - 3y = 1$$

SOLUTION We can eliminate either variable. Let's decide to eliminate the variable x. We can do so by multiplying the top equation by 2 and the bottom equation by -3, and then adding the left and right sides of the resulting equations:

$$3x - 5y = -2 \xrightarrow{\text{Multiply by 2}} 6x - 10y = -4$$
$$2x - 3y = 1 \xrightarrow[\text{Multiply by } -3]{} -6x + 9y = -3$$
$$-y = -7$$
$$y = 7$$

The y-coordinate of the solution to the system is 7. Substituting this value of y into any of the equations with both x- and y-variables gives $x = 11$. The solution to the system is $(11, 7)$. It is the only ordered pair that satisfies both equations.

Checking Solutions

Checking $(11, 7)$ in each equation looks like this:

Substituting $x = 11$ and $y = 7$ into $3x - 5y = -2$, we have

$$3(11) - 5(7) \overset{?}{=} -2$$

$$33 - 35 \overset{?}{=} -2$$

$$-2 = -2 \qquad \text{A true statement}$$

Substituting $x = 11$ and $y = 7$ into $2x - 3y = 1$, we have

$$2(11) - 3(7) \overset{?}{=} 1$$

$$22 - 21 \overset{?}{=} 1$$

$$1 = 1 \qquad \text{A true statement}$$

Our solution satisfies both equations; therefore, $(11, 7)$ is a solution to our system.

EXAMPLE 3

Solve the system.

$$2x - 3y = 4$$
$$4x + 5y = 3$$

SOLUTION We can eliminate x by multiplying the top equation by -2 and adding it to the bottom equation:

$$2x - 3y = 4 \xrightarrow{\text{Multiply by } -2} -4x + 6y = -8$$
$$4x + 5y = 3 \xrightarrow[\text{No Change}]{} 4x + 5y = 3$$
$$11y = -5$$
$$y = -\frac{5}{11}$$

The y-coordinate of our solution is $-\frac{5}{11}$. If we were to substitute this value of y back into either of our original equations, we would find the arithmetic necessary to solve for x cumbersome. For this reason, it is probably best to go back to the original system and solve it a second time—for x instead of y. Here is how we do that:

$$2x - 3y = 4 \xrightarrow{\text{Multiply by 5}} 10x - 15y = 20$$
$$4x + 5y = 3 \xrightarrow[\text{Multiply by 3}]{} 12x + 15y = 9$$
$$22x = 29$$
$$x = \frac{29}{22}$$

The solution to our system is $\left(\dfrac{29}{22}, -\dfrac{5}{11}\right)$.

The main idea in solving a system of linear equations by the addition method is to use the multiplication property of equality on one or both of the original equations, if necessary, to make the coefficients of either variable opposites. The following box shows some steps to follow when solving a system of linear equations by the addition method.

HOW TO   *Solve a System of Linear Equations by the Addition Method*

· ·

Step 1: Decide which variable to eliminate. (In some cases, one variable will be easier to eliminate than the other. With practice, you will notice which one it is.)

Step 2: Use the multiplication property of equality on each equation separately to make the coefficients of the variable that is to be eliminated opposites.

Step 3: Add together the respective left and right sides of the system.

Step 4: Solve for the remaining variable.

Step 5: Substitute the value of the variable from step 4 into an equation containing both variables and solve for the other variable. (Or repeat steps 2–4 to eliminate the other variable.)

Step 6: Check your solution in both equations, if necessary.

EXAMPLE 4

Solve the system.

$$5x - 2y = 5$$
$$-10x + 4y = 15$$

SOLUTION We can eliminate y by multiplying the first equation by 2 and adding the result to the second equation:

$$5x - 2y = 5 \xrightarrow{\text{Multiply by 2}} 10x - 4y = 10$$
$$-10x + 4y = 15 \xrightarrow{\text{No change}} \underline{-10x + 4y = 15}$$
$$0 = 25$$

The result is the false statement $0 = 25$, which indicates there is no solution to the system. If we were to graph the two lines, we would find that they are parallel. In a case like this, we say the system is *inconsistent*. Whenever both variables have been eliminated and the resulting statement is false, the solution set for the system will be the empty set, $\varnothing$.

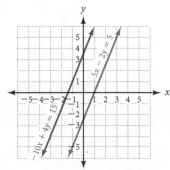

FIGURE 4 *A visual representation of the situation in Example 4 — the two lines are parallel*

EXAMPLE 5

Solve the system.

$$4x + 3y = 2$$
$$8x + 6y = 4$$

SOLUTION Multiplying the top equation by -2 and adding, we can eliminate the variable x:

$$4x + 3y = 2 \xrightarrow{\text{Multiply by } -2} -8x - 6y = -4$$
$$8x + 6y = 4 \xrightarrow{\text{No change}} \underline{8x + 6y = 4}$$
$$0 = 0$$

Both variables have been eliminated and the resulting statement $0 = 0$ is true. In this case, the lines coincide and the system is said to be *dependent*. The solution set consists of all ordered pairs that satisfy either equation. We can write the solution set as $\{(x, y) | 4x + 3y = 2\}$ or $\{(x, y) | 8x + 6y = 4\}$.

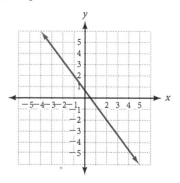

FIGURE 5 *A visual representation of the situation in Example 5 — both equations produce the same graph*

Special Cases

The previous two examples illustrate the two special cases in which the graphs of the equations in the system either coincide or are parallel. In both cases the left-hand sides of the equations were multiples of each other. In the case of the dependent equations, the right-hand sides were also multiples. We can generalize these observations for the system

$$a_1x + b_1y = c_1$$
$$a_2x + b_2y = c_2$$

Inconsistent System

What happens	*Geometric Intrepretation*	*Algebraic Intrepretation*
Both variables are eliminated, and the resulting statement is false.	The lines are parallel, and there is no solution to the system.	$\dfrac{a_1}{a_2} = \dfrac{b_1}{b_2} \neq \dfrac{c_1}{c_2}$

Dependent Equations

What happens	*Geometric Intrepretation*	*Algebraic Intrepretation*
Both variables are eliminated, and the resulting statement is true.	The lines coincide, and there are an infinite number of solutions to the system.	$\dfrac{a_1}{a_2} = \dfrac{b_1}{b_2} = \dfrac{c_1}{c_2}$

EXAMPLE 6

Solve the system.

$$\frac{1}{2}x - \frac{1}{3}y = 2$$

$$\frac{1}{4}x + \frac{2}{3}y = 6$$

SOLUTION   Although we could solve this system without clearing the equations of fractions, there is probably less chance for error if we have only integer coefficients to work with. So let's begin by multiplying both sides of the top equation by 6, and both sides of the bottom equation by 12, to clear each equation of fractions:

$$\frac{1}{2}x - \frac{1}{3}y = 2 \xrightarrow{\text{Multiply by 6}} 3x - 2y = 12$$

$$\frac{1}{4}x + \frac{2}{3}y = 6 \xrightarrow[\text{Multiply by 12}]{} 3x + 8y = 72$$

Now we can eliminate x by multiplying the top equation by -1 and leaving the bottom equation unchanged:

$$3x - 2y = 12 \xrightarrow{\text{Multiply by } -1} -3x + 2y = -12$$

$$3x + 8y = 72 \xrightarrow[\text{No change}]{} \underline{3x + 8y = 72}$$

$$10y = 60$$

$$y = 6$$

We can substitute $y = 6$ into any equation that contains both x and y. Let's use $3x - 2y = 12$.

$$3x - 2(6) = 12$$

$$3x - 12 = 12$$

$$3x = 24$$

$$x = 8$$

The solution to the system is $(8, 6)$.

B The Substitution Method

We end this section by considering another method of solving a linear system. The method is called the *substitution method* and is shown in the following examples.

EXAMPLE 7

Solve the system.

$$2x - 3y = -6$$

$$y = 3x - 5$$

SOLUTION   The second equation tells us y is $3x - 5$. Substituting the expression $3x - 5$ for y in the first equation, we have

$$2x - 3(3x - 5) = -6$$

The result of the substitution is the elimination of the variable y. Solving the resulting linear equation in x as usual, we have

$$2x - 9x + 15 = -6$$
$$-7x + 15 = -6$$
$$-7x = -21$$
$$x = 3$$

Putting $x = 3$ into the second equation in the original system, we have

$$y = 3(3) - 5$$
$$= 9 - 5$$
$$= 4$$

The solution to the system is $(3, 4)$.

Checking Solutions

Checking $(3, 4)$ in each equation looks like this:

Substituting $x = 3$ and $y = 4$ into $2x - 3y = -6$, we have

$$2(3) - 3(4) \overset{?}{=} -6$$
$$6 - 12 \overset{?}{=} -6$$
$$-6 = -6 \qquad \text{A true statement}$$

Substituting $x = 3$ and $y = 4$ into $y = 3x - 5$, we have

$$4 \overset{?}{=} 3(3) - 5$$
$$4 \overset{?}{=} 9 - 5$$
$$4 = 4 \qquad \text{A true statement}$$

Our solution satisfies both equations; therefore, $(3, 4)$ is a solution to our system.

Here are the steps to use in solving a system of equations by the substitution method.

> **HOW TO** *Solve a System of Equations by the Substitution Method*
>
> **Step 1:** Solve either one of the equations for x or y. (This step is not necessary if one of the equations is already in the correct form, as in Example 7.)
>
> **Step 2:** Substitute the expression for the variable obtained in step 1 into the other equation and solve it.
>
> **Step 3:** Substitute the solution from step 2 into any equation in the system that contains both variables and solve it.
>
> **Step 4:** Check your results, if necessary.

EXAMPLE 8

Solve by substitution

$$2x + 3y = 5$$
$$x - 2y = 6$$

NOTE Both the substitution method and the addition method can be used to solve any system of linear equations in two variables. Systems like the one in Example 7, however, are easier to solve using the substitution method, because one of the variables is already written in terms of the other. A system like the one in Example 6 is easier to solve using the addition method, because solving for one of the variables would lead to an expression involving fractions. The system in Example 8 could be solved easily by either method, because solving the second equation for x is a one-step process.

SOLUTION   To use the substitution method, we must solve one of the two equations for x or y. We can solve for x in the second equation by adding $2y$ to both sides:

$$x - 2y = 6$$

$$x = 2y + 6 \qquad \text{Add } 2y \text{ to both sides.}$$

Substituting the expression $2y + 6$ for x in the first equation of our system, we have

$$2(2y + 6) + 3y = 5$$

$$4y + 12 + 3y = 5$$

$$7y + 12 = 5$$

$$7y = -7$$

$$y = -1$$

Using $y = -1$ in either equation in the original system, we find $x = 4$. The solution is $(4, -1)$.

GETTING READY FOR CLASS

After reading through the preceding section, respond in your own words and in complete sentences.

A. Two linear equations, each with the same two variables, form a system of equations. How do we define a solution to this system? That is, what form will a solution have, and what properties does a solution possess?

B. When would substitution be more efficient than the addition method in solving two linear equations?

C. Explain what an inconsistent system of linear equations looks like graphically and what would result algebraically when attempting to solve the system.

D. When might the graphing method of solving a system of equations be more desirable than the other techniques, and when might it be less desirable?

USING TECHNOLOGY   *Graphing Calculators*

Solving Systems that Intersect in Exactly One Point

A graphing calculator can be used to solve a system of equations in two variables if the equations intersect in exactly one point. To solve the system shown in Example 3, we first solve each equation for *y*. Here is the result:

$$2x - 3y = 4 \quad \text{becomes} \quad y = \frac{4 - 2x}{-3}$$

$$4x + 5y = 3 \quad \text{becomes} \quad y = \frac{3 - 4x}{5}$$

Graphing these two functions on the calculator gives a diagram similar to the one in Figure 6.

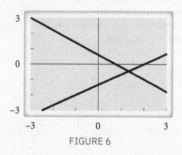

FIGURE 6

Using the Trace and Zoom features, we find that the two lines intersect at $x = 1.32$ and $y = -0.45$, which are the decimal equivalents (accurate to the nearest hundredth) of the fractions found in Example 3.

Special Cases

We cannot assume that two lines that look parallel in a calculator widow are in fact parallel. If you graph the functions $y = x - 5$ and $y = 0.99x + 2$ in a window where x and y range from -10 to 10, the lines look parallel. We know this is not the case, however, since their slopes are different. As we zoom out repeatedly, the lines begin to look as if they coincide. We know this is not the case, because the two lines have different y-intercepts. To summarize: If we graph two lines on a calculator and the graphs look as if they are parallel or coincide, we should use algebraic methods, not the calculator, to determine the solution to the system.

Problem Set 4.1

Solve each system by graphing both equations on the same set of axes and then reading the solution from the graph.

1. $3x - 2y = 6$

$x - y = 1$

2. $5x - 2y = 10$

$x - y = -1$

3. $y = \dfrac{3}{5}x - 3$

$2x - y = -4$

4. $y = \dfrac{1}{2}x - 2$

$2x - y = -1$

5. $y = \dfrac{1}{2}x$

$y = -\dfrac{3}{4}x + 5$

6. $y = \dfrac{2}{3}x$

$y = -\dfrac{1}{3}x + 6$

7. $3x + 3y = -2$

$y = -x + 4$

8. $2x - y = 5$

$y = 2x - 5$

Solve each of the following systems by the addition method.

9. $3x + y = 5$
$3x - y = 3$

10. $-x - y = 4$
$-x + 2y = -3$

11. $x + 2y = 0$
$2x - 6y = 5$

12. $x + 3y = 3$
$2x - 9y = 1$

13. $2x - 5y = 16$
$4x - 3y = 11$

14. $5x - 3y = -11$
$7x + 6y = -12$

15. $6x + 3y = -1$
$9x + 5y = 1$

16. $5x + 4y = -1$
$7x + 6y = -2$

17. $4x + 3y = 14$
$9x - 2y = 14$

18. $7x - 6y = 13$
$6x - 5y = 11$

19. $2x - 5y = 3$
$-4x + 10y = 3$

20. $-3x - 2y = -1$
$-6x + 4y = -2$

21. $\dfrac{1}{2}x + \dfrac{1}{3}y = 13$

$\dfrac{2}{5}x + \dfrac{1}{4}y = 10$

22. $\dfrac{1}{2}x + \dfrac{1}{3}y = \dfrac{2}{3}$

$\dfrac{2}{3}x + \dfrac{2}{5}y = \dfrac{14}{15}$

23. $\dfrac{2}{3}x + \dfrac{2}{5}y = -4$

$\dfrac{1}{3}x - \dfrac{1}{2}y = -\dfrac{1}{3}$

24. $\dfrac{1}{2}x - \dfrac{1}{3}y = \dfrac{5}{6}$

$-\dfrac{2}{5}x + \dfrac{1}{2}y = -\dfrac{9}{10}$

Solve each of the following systems by the substitution method.

25. $7x - y = 24$

$x = 2y + 9$

26. $3x - y = -8$

$y = 6x + 3$

27. $6x - y = 10$

$y = -\dfrac{3}{4}x - 1$

28. $2x - y = 6$

$y = -\dfrac{4}{3}x + 1$

29. $y = 3x - 2$

$y = 4x - 4$

30. $y = 5x - 2$

$y = -2x + 5$

31. $2x - y = 5$

$4x - 2y = 10$

32. $-10x + 8y = -6$

$y = \dfrac{5}{4}x$

33. $\dfrac{1}{3}x - \dfrac{1}{2}y = 0$

$x = \dfrac{3}{2}y$

34. $\frac{2}{5}x - \frac{2}{3}y = 0$

$y = \frac{3}{5}x$

You may want to read Example 3 again before solving the systems that follow.

35. $4x - 7y = 3$
$5x + 2y = -3$

36. $3x - 4y = 7$
$6x - 3y = 5$

37. $9x - 8y = 4$
$2x + 3y = 6$

38. $4x - 7y = 10$
$-3x + 2y = -9$

39. $3x - 5y = 2$
$7x + 2y = 1$

40. $4x - 3y = -1$
$5x + 8y = 2$

Solve each of the following systems by using either the addition or substitution method. Choose the method that is most appropriate for the problem.

41. $x - 3y = 7$

$2x + y = -6$

42. $2x - y = 9$

$x + 2y = -11$

43. $y = \frac{1}{2}x + \frac{1}{3}$

$y = -\frac{1}{3}x + 2$

44. $y = \frac{3}{4}x - \frac{4}{5}$

$y = \frac{1}{2}x - \frac{1}{2}$

45. $3x - 4y = 12$

$x = \frac{2}{3}y - 4$

46. $-5x + 3y = -15$

$x = \frac{4}{5}y - 2$

47. $4x - 3y = -7$

$-8x + 6y = -11$

48. $3x - 4y = 8$

$y = \frac{3}{4}x - 2$

49. $3y + z = 17$

$5y + 20z = 65$

50. $x + y = 850$

$1.5x + y = 1,100$

51. $\frac{3}{4}x - \frac{1}{3}y = 1$

$y = \frac{1}{4}x$

52. $-\frac{2}{3}x + \frac{1}{2}y = -1$

$y = -\frac{1}{3}x$

53. $\frac{1}{4}x - \frac{1}{2}y = \frac{1}{3}$

$\frac{1}{3}x - \frac{1}{4}y = -\frac{2}{3}$

54. $\frac{1}{5}x - \frac{1}{10}y = -\frac{1}{5}$

$\frac{2}{3}x - \frac{1}{2}y = -\frac{1}{6}$

PAYING ATTENTION TO INSTRUCTIONS ━━━━━━━━

55. Work each problem according to the instructions given.
  a. Simplify: $(3x - 4y) - 3(x - y)$
  b. Find y when x is 0 in $3x - 4y = 8$
  c. Find the y-intercept: $3x - 4y = 8$
  d. Graph: $3x - 4y = 8$
  e. Find the point where the graphs of $3x - 4y = 8$ and $x - y = 2$ cross

56. Work each problem according to the instructions given.

    a. Solve: $4x - 5 = 20$

    b. Solve for y: $4x - 5y = 20$

    c. Solve for x: $x - y = 5$

    d. Solve the system:

$$4x - 5 = 20$$
$$x - y = 5$$

57. Multiply both sides of the second equation in the following system by 100, and then solve as usual.

$$x + y = 10{,}000$$
$$0.06x + 0.05y = 560$$

58. What value of c will make the following system a dependent system (one in which the lines coincide)?

$$6x - 9y = 3$$
$$4x - 6y = c$$

59. Where do the graphs of the lines $x + y = 4$ and $x - 2y = 4$ intersect?

60. Where do the graphs of the line $x = -1$ and $x - 2y = 4$ intersect?

GETTING READY FOR THE NEXT SECTION

Simplify.

61. $2 - 2(6)$

62. $2(1) - 2 + 3$

63. $(x + 3y) - 1(x - 2z)$

64. $(x + y + z) + (2x - y + z)$

Solve.

65. $-9y = -9$

66. $30x = 38$

67. $3(1) + 2z = 9$

68. $4\left(\dfrac{19}{15}\right) - 2y = 4$

Apply the distributive property, then simplify if possible.

69. $2(5x - z)$

70. $-1(x - 2z)$

71. $3(3x + y - 2z)$

72. $2(2x - y + z)$

EXTENDING THE CONCEPTS

73. The height of an object thrown in the air is given by the function $h(t) = at^2 + bt$, where t is measured in seconds and $h(t)$ is measured in feet. After 3 seconds, the object reaches a height of 24 feet; after 6 seconds, it hits the ground. Find the values of a and b.

74. The height of an object thrown in the air is given by the function $h(t) = at^2 + bt$, where t is measured in seconds and $h(t)$ is measured in feet. After 4 seconds, the object reaches a height of 40 feet; after 6 seconds, it reaches a height of 54 feet. How long will it take to reach the ground?

75. The lines $y = 0$, $y = 4x$, and $y = 12 - 2x$ form a triangle in the first quadrant. Find its area. *Hint:* Draw a picture of the situation first.

76. The lines $y = 2$, $y = x$, and $y = 10 - x$ form a triangle in the first quadrant. Find its area. *Hint:* Draw a picture of the situation first.

Solve the following systems.

77. $\dfrac{1}{x} + \dfrac{2}{y} = 6$

$\dfrac{3}{x} - \dfrac{1}{y} = 4$

(*Hint:* Let $s = \dfrac{1}{x}$ and $t = \dfrac{1}{y}$.)

78. $\dfrac{4}{x} - \dfrac{3}{y} = 12$

$\dfrac{3}{x} + \dfrac{2}{y} = -6$

OBJECTIVE

A Solve systems of linear equations in three variables.

© iStockphoto/AlexBrylov

4.2

A solution to an equation in three variables such as

$$2x + y - 3z = 6$$

is an ordered triple of numbers (x, y, z). For example, the ordered triples $(0, 0, -2)$, $(2, 2, 0)$, and $(0, 9, 1)$ are solutions to the equation $2x + y - 3z = 6$, because they produce a true statement when their coordinates are substituted for x, y, and z in the equation.

A Solving Systems in Three Variables

> **DEFINITION** *solution set*
>
> The *solution set* for a system of three linear equations in three variables is the set of ordered triples that satisfies all three equations.

EXAMPLE 1

Solve the system.

$$\begin{array}{rcrcrcrr} x & + & y & + & z & = & 6 & (1) \\ 2x & - & y & + & z & = & 3 & (2) \\ x & + & 2y & - & 3z & = & -4 & (3) \end{array}$$

SOLUTION We want to find the ordered triple (x, y, z) that satisfies all three equations. We have numbered the equations so it will be easier to keep track of where they are and what we are doing.

There are many ways to proceed. The main idea is to take two different pairs of equations and eliminate the same variable from each pair. We begin by adding equations (1) and (2) to eliminate the y-variable. The resulting equation is numbered (4):

$$\begin{array}{rll} x + y + z = 6 & \quad (1) \\ 2x - y + z = 3 & \quad (2) \\ \hline 3x + 2z = 9 & \quad (4) \end{array}$$

Adding twice equation (2) to equation (3) will also eliminate the variable y. The resulting equation is numbered (5):

$$
\begin{array}{ll}
4x - 2y + 2z = 6 & \text{Twice (2)} \\
\underline{x + 2y - 3z = -4} & \text{(3)} \\
5x - z = 2 & \text{(5)}
\end{array}
$$

Equations (4) and (5) form a linear system in two variables. By multiplying equation (5) by 2 and adding the result to equation (4), we succeed in eliminating the variable z from the new pair of equations:

$$
\begin{array}{ll}
3x + 2z = 9 & \text{(4)} \\
\underline{10x - 2z = 4} & \text{Twice (5)} \\
13x = 13 & \\
x = 1 &
\end{array}
$$

Substituting $x = 1$ into equation (4), we have

$$
\begin{aligned}
3(1) + 2z &= 9 \\
2z &= 6 \\
z &= 3
\end{aligned}
$$

Using $x = 1$ and $z = 3$ in equation (1) gives us

$$
\begin{aligned}
1 + y + 3 &= 6 \\
y + 4 &= 6 \\
y &= 2
\end{aligned}
$$

The solution is the ordered triple $(1, 2, 3)$.

EXAMPLE 2

Solve the system.

$$
\begin{array}{ll}
2x + y - z = 3 & \text{(1)} \\
3x + 4y + z = 6 & \text{(2)} \\
2x - 3y + z = 1 & \text{(3)}
\end{array}
$$

SOLUTION   It is easiest to eliminate z from the equations. The equation produced by adding (1) and (2) is

$$5x + 5y = 9 \qquad \text{(4)}$$

The equation that results from adding (1) and (3) is

$$4x - 2y = 4 \qquad \text{(5)}$$

Equations (4) and (5) form a linear system in two variables. We can eliminate the variable y from this system as follows:

$$
\begin{array}{l}
5x + 5y = 9 \xrightarrow{\text{Multiply by 2}} 10x + 10y = 18 \\
4x - 2y = 4 \xrightarrow[\text{Multiply by 5}]{} \underline{20x - 10y = 20} \\
\phantom{4x - 2y = 4 \xrightarrow[\text{Multiply by 5}]{} 20x - 1} 30x = 38 \\
\phantom{4x - 2y = 4 \xrightarrow[\text{Multiply by 5}]{} 20x - 10y 00} x = \dfrac{38}{30} \\
\phantom{4x - 2y = 4 \xrightarrow[\text{Multiply by 5}]{} 20x - 10y 0000} = \dfrac{19}{15}
\end{array}
$$

Substituting $x = \frac{19}{15}$ into equation (5) or equation (4) and solving for y gives

$$y = \frac{8}{15}$$

Using $x = \frac{19}{15}$ and $y = \frac{8}{15}$ in equation (1), (2), or (3) and solving for z results in

$$z = \frac{1}{15}$$

The ordered triple that satisfies all three equations is $\left(\frac{19}{15}, \frac{8}{15}, \frac{1}{15} \right)$.

EXAMPLE 3

Solve the system.

$$\begin{array}{ll} 2x + 3y - z = 5 & (1) \\ 4x + 6y - 2z = 10 & (2) \\ x - 4y + 3z = 5 & (3) \end{array}$$

SOLUTION Multiplying equation (1) by -2 and adding the result to equation (2) looks like this:

$$\begin{array}{ll} -4x - 6y + 2z = -10 & -2 \text{ times (1)} \\ \underline{4x + 6y - 2z = 10} & (2) \\ 0 = 0 & \end{array}$$

All three variables have been eliminated, and we are left with a true statement. This implies that the two equations are dependent. With a system of three equations in three variables, however, a dependent system can have no solution or an infinite number of solutions. After we have concluded the examples in this section, we will discuss the geometry behind these systems. Doing so will give you some additional insight into dependent systems.

EXAMPLE 4

Solve the system.

$$\begin{array}{ll} x - 5y + 4z = 8 & (1) \\ 3x + y - 2z = 7 & (2) \\ -9x - 3y + 6z = 5 & (3) \end{array}$$

SOLUTION Multiplying equation (2) by 3 and adding the result to equation (3) produces

$$\begin{array}{ll} 9x + 3y - 6z = 21 & 3 \text{ times (2)} \\ \underline{-9x - 3y + 6z = 5} & (3) \\ 0 = 26 & \end{array}$$

In this case, all three variables have been eliminated, and we are left with a false statement. The two equations are inconsistent; there are no ordered triples that satisfy both equations. The solution set for the system is the empty set, $\varnothing$. If equations (2) and (3) have no ordered triples in common, then certainly (1), (2), and (3) do not either.

EXAMPLE 5

Solve the system.

$$x + 3y = 5 \qquad (1)$$
$$6y + z = 12 \qquad (2)$$
$$x - 2z = -10 \qquad (3)$$

SOLUTION It may be helpful to rewrite the system as

$$x + 3y \quad\;\; = \quad 5 \qquad (1)$$
$$6y + z = 12 \qquad (2)$$
$$x \quad\quad - 2z = -10 \qquad (3)$$

Equation (2) does not contain the variable x. If we multiply equation (3) by -1 and add the result to equation (1), we will be left with another equation that does not contain the variable x:

$$x + 3y \quad\quad = 5 \qquad (1)$$
$$\underline{-x \quad\quad + 2z = 10} \qquad -1 \text{ times (3)}$$
$$3y + 2z = 15 \qquad (4)$$

Equations (2) and (4) form a linear system in two variables. Multiplying equation (2) by -2 and adding the result to equation (4) eliminates the variable z:

$$6y + z = 12 \;\xrightarrow{\text{Multiply by } -2}\; -12y - 2z = -24$$
$$3y + 2z = 15 \;\xrightarrow[\text{No Change}]{}\; \underline{\quad 3y + 2z = \quad 15}$$
$$-9y + \quad\;\; = \;\; -9$$
$$y = 1$$

Using $y = 1$ in equation (4) and solving for z, we have

$$z = 6$$

Substituting $y = 1$ into equation (1) gives

$$x = 2$$

The ordered triple that satisfies all three equations is $(2, 1, 6)$.

The Geometry Behind Equations in Three Variables

We can graph an ordered triple on a coordinate system with three axes. The graph will be a point in space. The coordinate system is drawn in perspective; you have to imagine that the x-axis comes out of the paper and is perpendicular to both the y-axis and the z-axis. To graph the point $(3, 4, 5)$, we move 3 units in the x-direction, 4 units in the y-direction, and then 5 units in the z-direction, as shown in Figure 1.

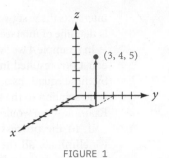

FIGURE 1

Although in actual practice it is sometimes difficult to graph equations in three variables, if we were to graph a linear equation in three variables, we would find that the graph was a plane in space. A system of three equations in three variables is represented by three planes in space.

There are a number of possible ways in which these three planes can intersect, some of which are shown below. And there are still other possibilities that are not among those shown.

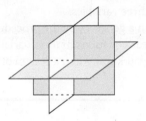

CASE 1   *The three planes have exactly one point in common. In this case we get one solution to our system, as in Examples 1, 2, and 5.*

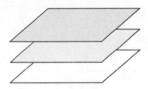

CASE 2   *The three planes have no points in common because they are all parallel to one another. The system they represent is an inconsistent system.*

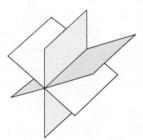

CASE 3   *The three planes intersect in a line. Any point on the line is a solution to the system of equations represented by the planes, so there is an infinite number of solutions to the system. This is an example of a dependent system.*

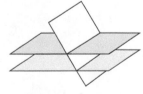

CASE 4   *Two of the planes are parallel; the third plane intersects each of the parallel planes. In this case, the three planes have no points in common. There is no solution to the system; it is an inconsistent system.*

In Example 3, we found that equations (1) and (2) were dependent equations. They represent the same plane. That is, they have all their points in common. But the system of equations that they came from has either no solution or an infinite number of solutions. It all depends on the third plane. If the third plane coincides with the first two, then the solution to the system is a plane. If the third plane is parallel to the first two, then there is no solution to the system. Finally, if the third plane intersects the first two but does not coincide with them, then the solution to the system is that line of intersection.

In Example 4 we found that trying to eliminate a variable from the second and third equations resulted in a false statement. This means that the two planes represented by these equations are parallel. It makes no difference where the third plane is; there is no solution to the system in Example 4. (If we were to graph the three planes from Example 4, we would obtain a diagram similar to Case 2 or Case 4.)

If, in the process of solving a system of linear equations in three variables, we eliminate all the variables from a pair of equations and are left with a false statement, we will say the system is inconsistent. If we eliminate all the variables and are left with a true statement, then we will say the system is a dependent one.

GETTING READY FOR CLASS

After reading through the preceding section, respond in your own words and in complete sentences.

A. What is an ordered triple of numbers?

B. Explain what it means for (1, 2, 3) to be a solution to a system of linear equations in three variables.

C. Explain in a general way the procedure you would use to solve a system of three linear equations in three variables.

D. How do you know when a system of linear equations in three variables has no solution?

Problem Set 4.2

Solve the following systems.

1. $x + y + z = 4$
$x - y + 2z = 1$
$x - y - 3z = -4$

2. $x - y - 2z = -1$
$x + y + z = 6$
$x + y - z = 4$

3. $x + y + z = 6$
$x - y + 2z = 7$
$2x - y - 4z = -9$

4. $x + y + z = 0$
$x + y - z = 6$
$x - y + 2z = -7$

5. $x + 2y + z = 3$
$2x - y + 2z = 6$
$3x + y - z = 5$

6. $2x + y - 3z = -14$
$x - 3y + 4z = 22$
$3x + 2y + z = 0$

7. $2x + 3y - 2z = 4$
$x + 3y - 3z = 4$
$3x - 6y + z = -3$

8. $4x + y - 2z = 0$
$2x - 3y + 3z = 9$
$-6x - 2y + z = 0$

9. $-x + 4y - 3z = 2$
$2x - 8y + 6z = 1$
$3x - y + z = 0$

10. $4x + 6y - 8z = 1$
$-6x - 9y + 12z = 0$
$x - 2y - 2z = 3$

11. $\frac{1}{2}x - y + z = 0$
$2x + \frac{1}{3}y + z = 2$
$x + y + z = -4$

12. $\frac{1}{3}x + \frac{1}{2}y + z = -1$
$x - y + \frac{1}{5}z = -1$
$x + y + z = -5$

13. $2x - y - 3z = 1$
$x + 2y + 4z = 3$
$4x - 2y - 6z = 2$

14. $3x + 2y + z = 3$
$x - 3y + z = 4$
$-6x - 4y - 2z = 1$

15. $2x - y + 3z = 4$
$x + 2y - z = -3$
$4x + 3y + 2z = -5$

16. $6x - 2y + z = 5$
$3x + y + 3z = 7$
$x + 4y - z = 4$

17. $x + y = 9$
$y + z = 7$
$x - z = 2$

18. $x - y = -3$
$x + z = 2$
$y - z = 7$

19. $2x + y = 2$
$y + z = 3$
$4x - z = 0$

20. $2x + y = 6$
$3y - 2z = -8$
$x + z = 5$

21. $2x - 3y = 0$
$6y - 4z = 1$
$x + 2z = 1$

22. $3x + 2y = 3$
$y + 2z = 2$
$6x - 4z = 1$

23. $x + y - z = 2$
$2x + y + 3z = 4$
$x - 2y + 2z = 6$

24. $x + 2y - 2z = 4$
$3x + 4y - z = -2$
$2x + 3y - 3z = -5$

25. $2x + 3y = -\frac{1}{2}$
$4x + 8z = 2$
$3y + 2z = -\frac{3}{4}$

26. $3x - 5y = 2$
$4x + 6z = \frac{1}{3}$
$5y - 7z = \frac{1}{6}$

341

27. $\dfrac{1}{3}x + \dfrac{1}{2}y - \dfrac{1}{6}z = 4$

$\dfrac{1}{4}x - \dfrac{3}{4}y + \dfrac{1}{2}z = \dfrac{3}{2}$

$\dfrac{1}{2}x - \dfrac{2}{3}y - \dfrac{1}{4}z = -\dfrac{16}{3}$

28. $-\dfrac{1}{4}x + \dfrac{3}{8}y + \dfrac{1}{2}z = -1$

$\dfrac{2}{3}x - \dfrac{1}{6}y - \dfrac{1}{2}z = 2$

$\dfrac{3}{4}x - \dfrac{1}{2}y - \dfrac{1}{8}z = 1$

29. $x - \dfrac{1}{2}y - \dfrac{1}{3}z = -\dfrac{4}{3}$

$\dfrac{1}{3}x \quad\quad - \dfrac{1}{2}z = 5$

$-\dfrac{1}{4}x + \dfrac{2}{3}y - \quad z = -\dfrac{3}{4}$

30. $x + \dfrac{1}{3}y - \dfrac{1}{2}z = -\dfrac{3}{2}$

$\dfrac{1}{2}x - \quad y + \dfrac{1}{3}z = 8$

$\dfrac{1}{3}x - \dfrac{1}{4}y - \quad z = -\dfrac{5}{6}$

31. Electric Current In the following diagram of an electrical circuit, x, y, and z represent the amount of current (in amperes) flowing across the 5-ohm, 20-ohm, and 10-ohm resistors, respectively. (In circuit diagrams, resistors are represented by ⏦ and potential differences by ⊣⊢.)

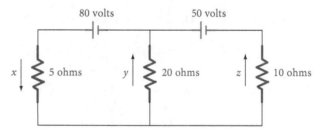

The system of equations used to find the three currents x, y, and z is

$$x - y - z = 0$$
$$5x + 20y = 80$$
$$20y - 10z = 50$$

Solve the system for all variables.

32. Cost of a Rental Car If a car rental company charges $10 a day and 8¢ a mile to rent one of its cars, then the cost z, in dollars, to rent a car for x days and drive y miles can be found from the equation

$$z = 10x + 0.08y$$

a. How much does it cost to rent a car for 2 days and drive it 200 miles under these conditions?

b. A second company charges $12 a day and 6¢ a mile for the same car. Write an equation that gives the cost z in dollars, to rent a car from this company for x days and drive it y miles.

c. A car is rented from each of the companies mentioned in **a.** and **b.** for 2 days. To find the mileage at which the cost of renting the cars from each of the two companies will be equal, solve the following system for y:

$$z = 10x + 0.08y$$
$$z = 12x + 0.06y$$
$$x = 2$$

33. **Movie Licensing** Assume that licensing fees for movies are the same over an entire month, but depend on the genre (rated G or PG or R). If, last month, the Crest Theater spent $4,200 to license 10 G-rated, 8 PG-rated, and 4 R-rated movies; the Apollo Theater spent $4,400 to license 5 G-rated, 6 PG-rated, and 8 R-rated movies; and the Shubert Theater spent $4,700 to license 1 G-rated, 4 PG-rated, and 12 R-rated movies, then the system of equations used to find the license costs for each genre is

$$10G + 8P + 4R = 4,200$$

$$5G + 6P + 8R = 4,400$$

$$G + 4P + 12R = 4,700$$

where G represents G-rated movies, P represents PG-rated movies, and R represents R-rated movies. Solve the system for all variables to find how much each type of license costs for that month.

34. **Movie Licensing** Assume that licensing fees for movies are the same over an entire month, but depend on the genre (rated G or PG or R). If, last month, the Fox Theater spent $4,250 to license 4 G-rated, 6 PG-rated, and 7 R-rated movies; the Oriental Theater spent $2,550 to license 1 G-rated, 5 PG-rated, and 4 R-rated movies; and the Uptown Theater spent $4,550 to license 8 G-rated, 8 PG-rated, and 5 R-rated movies, then the system of equations used to find the license costs for each genre is

$$4G + 6P + 7R = 4,250$$

$$G + 5P + 4R = 2,550$$

$$8G + 8P + 5R = 4,550$$

where G represents G-rated movies, P represents PG-rated movies, and R represents R-rated movies. Solve the system for all variables to find how much each type of license costs for that month.

GETTING READY FOR THE NEXT SECTION

Translate into symbols.

35. Two more than 3 times a number

36. One less than twice a number

Simplify.

37. $25 - \dfrac{385}{9}$

38. $0.30(12)$

39. $0.08(4,000)$

40. $500(1.5)$

41. $10(0.2x + 0.5y)$

42. $100(0.09x + 0.08y)$

Solve.

43. $x + (3x + 2) = 26$

44. $5x = 2,500$

Solve each system.

45. $-2y - 4z = -18$
    $-7y + 4z = 27$

46. $-x + 2y = 200$
    $4x - 2y = 1,300$

EXTENDING THE CONCEPTS

47. Solve the system:

$$\frac{1}{x} - \frac{2}{y} + \frac{1}{z} = 2$$

$$\frac{2}{x} - \frac{3}{y} + \frac{4}{z} = -1$$

$$\frac{3}{x} + \frac{4}{y} - \frac{3}{z} = 3$$

(*Hint:* Let $r = \dfrac{1}{x}$, $s = \dfrac{1}{y}$, and $t = \dfrac{1}{z}$.)

48. Solve the system:

$$\frac{2}{x} - \frac{3}{y} + \frac{4}{z} = -1$$

$$\frac{3}{x} + \frac{2}{y} - \frac{1}{z} = 2$$

$$\frac{5}{x} - \frac{4}{y} + \frac{2}{z} = 3$$

49. The height of an object thrown in the air is given by the function $h(t) = at^2 + bt + c$, where t is measured in seconds and $h(t)$ is measured in feet. After 1 second, the object reaches a height of 136 feet; after 2 seconds, the height is 184 feet; and after 3 seconds, the height is 200 feet. Find the values of a, b, and c.

50. The height of an object thrown in the air is given by the function $h(t) = at^2 + bt + c$, where t is measured in seconds and $h(t)$ is measured in feet. After 3 seconds, the object reaches a height of 60 feet; after 6 seconds, the height is 96 feet; and after 9 seconds, the height is 124 feet. Find the values of a, b, and c.

Solve each system for the solution (x, y, z, w).

51. $x + y + z + w = 10$

$x + 2y - z + w = 6$

$x - y - z + 2w = 4$

$x - 2y + z - 3w = -12$

52. $x + y + z + w = 16$

$x - y + 2z - w = 1$

$x + 3y - z - w = -2$

$x - 3y - 2z + 2w = -4$

4.3

OBJECTIVE

A Solve applications problems using a system of equations.

Suppose a chemist needs to mix a 20% alcohol solution with a 50% alcohol solution to obtain 12 gallons of a 30% alcohol solution. In order to calculate how much of each solution the chemist must begin with, we will need to set up and solve a system of linear equations in two variables.

Many times word problems involve more than one unknown quantity. If a problem is stated in terms of two unknowns, like the one above, and we represent each unknown quantity with a different variable, then we must write the relationships between the variables with two equations. The two equations written in terms of the two variables form a system of linear equations that we solve using the methods developed in this chapter. If we find a problem that relates three unknown quantities, then we need three equations to form a linear system we can solve.

 ## Blueprint for Problem Solving for Systems of Equations

Here is our Blueprint for Problem Solving, modified to fit the application problems that you will find in this section.

> ### BLUEPRINT FOR PROBLEM SOLVING
> #### Using a System of Equations ·
>
> **Step 1:** *Read* the problem, and then mentally *list* the items that are known and the items that are unknown.
>
> **Step 2:** *Assign variables* to each of the unknown items. That is, let $x =$ one of the unknown items and $y =$ the other unknown item (and $z =$ the third unknown item, if there is a third one). Then *translate* the other *information* in the problem to expressions involving the two (or three) variables.
>
> **Step 3:** *Reread* the problem, and then *write a system of equations*, using the items and variables listed in steps 1 and 2, that describes the situation.
>
> **Step 4:** *Solve the system* found in step 3.
>
> **Step 5:** *Write your answers* using complete sentences.
>
> **Step 6:** *Reread* the problem, and *check* your solution with the original words in the problem.

4.3 VIDEOS

EXAMPLE 1

One number is 2 more than 3 times another. Their sum is 26. Find the two numbers.

SOLUTION   Applying the steps from our Blueprint, we have:

Step 1: *Read and list.*
We know that we have two numbers whose sum is 26. One of them is 2 more than 3 times the other. The unknown quantities are the two numbers.

Step 2: *Assign variables and translate information.*
Let $x =$ one of the numbers and $y =$ the other number.

Step 3: *Write a system of equations.*
The first sentence in the problem translates into $y = 3x + 2$. The second sentence gives us a second equation: $x + y = 26$. Together, these two equations give us the following system of equations:

$$x + y = 26$$
$$y = 3x + 2$$

Step 4: *Solve the system.*
Substituting the expression for y from the second equation into the first and solving for x yields

$$x + (3x + 2) = 26$$
$$4x + 2 = 26$$
$$4x = 24$$
$$x = 6$$

Using $x = 6$ in $y = 3x + 2$ gives the second number:

$$y = 3(6) + 2$$
$$y = 20$$

Step 5: *Write answers.*
The two numbers are 6 and 20.

Step 6: *Reread and check.*
The sum of 6 and 20 is 26, and 20 is 2 more than 3 times 6.

EXAMPLE 2

Suppose 850 tickets were sold for a game for a total of $1,100. If adult tickets cost $1.50 and children's tickets cost $1.00, how many of each kind of ticket were sold?

SOLUTION

Step 1: *Read and list.*
The total number of tickets sold is 850. The total income from tickets is $1,100. Adult tickets are $1.50 each. Children's tickets are $1.00 each. We don't know how many of each type of ticket have been sold.

Step 2: *Assign variables and translate information.*
We let $x =$ the number of adult tickets and $y =$ the number of children's tickets.

Step 3: *Write a system of equations.*

The total number of tickets sold is 850, giving us our first equation.

$$x + y = 850$$

Because each adult ticket costs $1.50, and each children's ticket costs $1.00, and the total amount of money paid for tickets was $1,100, a second equation is

$$1.50x + 1.00y = 1,100$$

The same information can also be obtained by summarizing the problem with a table. One such table follows. Notice that the two equations we obtained previously are given by the two rows of the table.

	Adult Tickets	Children's Tickets	Total
Number	x	y	850
Value	$1.50x$	$1.00y$	1,100

Whether we use a table to summarize the information in the problem or just talk our way through the problem, the system of equations that describes the situation is

$$x + \quad y = \quad 850$$
$$1.50x + 1.00y = 1,100$$

Step 4: *Solve the system.*

If we multiply the second equation by 10 to clear it of decimals, we have the system

$$x + \quad y = \quad 850$$
$$15x + 10y = 11,000$$

Multiplying the first equation by -10 and adding the result to the second equation eliminates the variable y from the system:

$$-10x - 10y = -8,500$$
$$\underline{15x + 10y = \quad 11,000}$$
$$5x = \quad 2,500$$
$$x = \quad 500$$

The number of adult tickets sold was 500. To find the number of children's tickets, we substitute $x = 500$ into $x + y = 850$ to get

$$500 + y = 850$$
$$y = 350$$

Step 5: *Write answers.*

The number of children's tickets is 350, and the number of adult tickets is 500.

Step 6: *Reread and check.*

The total number of tickets is $350 + 500 = 850$. The amount of money from selling the two types of tickets is

350 children's tickets at $1.00 each is $350(1.00) = \$350$
500 adult tickets at $1.50 each is $\quad 500(1.50) = \$750$

The total income from ticket sales is $1,100

EXAMPLE 3

Suppose a person invests a total of $10,000 in two accounts. One account earns 8% annually, and the other earns 9% annually. If the total interest earned from both accounts in a year is $860, how much was invested in each account?

SOLUTION

Step 1: *Read and list.*

The total investment is $10,000 split between two accounts. One account earns 8% annually, and the other earns 9% annually. The interest from both accounts is $860 in 1 year. We don't know how much is in each account.

Step 2: *Assign variables and translate information.*

We let x equal the amount invested at 9% and y be the amount invested at 8%.

Step 3: *Write a system of equations.*

Because the total investment is $10,000, one relationship between x and y can be written as

$$x + y = 10{,}000$$

The total interest earned from both accounts is $860. The amount of interest earned on x dollars at 9% is $0.09x$, while the amount of interest earned on y dollars at 8% is $0.08y$. This relationship is represented by the equation

$$0.09x + 0.08y = 860$$

The two equations we have just written can also be found by first summarizing the information from the problem in a table. Again, the two rows of the table yield the two equations we found previously. Here is the table.

	Dollars at 9%	Dollars at 8%	Total
Number	x	y	10,000
Interest	$0.09x$	$0.08y$	860

The system of equations that describes this situation is given by

$$x + y = 10{,}000$$
$$0.09x + 0.08y = 860$$

Step 4: *Solve the system.*

Multiplying the second equation by 100 will clear it of decimals. The system that results after doing so is

$$x + y = 10{,}000$$
$$9x + 8y = 86{,}000$$

We can eliminate y from this system by multiplying the first equation by -8 and adding the result to the second equation.

$$-8x - 8y = -80{,}000$$
$$\underline{9x + 8y = 86{,}000}$$
$$x = 6{,}000$$

The amount of money invested at 9% is $6,000. Because the total investment was $10,000, the amount invested at 8% must be $4,000.

Step 5: *Write answers.*
The amount invested at 8% is $4,000, and the amount invested at 9% is $6,000.

Step 6: *Reread and check.*
The total investment is $4,000 + $6,000 = $10,000. The amount of interest earned from the two accounts is

In 1 year, $4,000 invested at 8% earns 0.08(4,000) = $320

In 1 year, $6,000 invested at 9% earns 0.09(6,000) = $540

The total interest from the two accounts is $860

EXAMPLE 4

How much 20% alcohol solution and 50% alcohol solution must be mixed to get 12 gallons of 30% alcohol solution?

SOLUTION   To solve this problem, we must first understand that a 20% alcohol solution is 20% alcohol and 80% water.

Step 1: *Read and list.*
We will mix two solutions to obtain 12 gallons of solution that is 30% alcohol. One of the solutions is 20% alcohol and the other 50% alcohol. We don't know how much of each solution we need.

Step 2: *Assign variables and translate information.*
Let x = the number of gallons of 20% alcohol solution needed, and y = the number of gallons of 50% alcohol solution needed.

Step 3: *Write a system of equations.*
Because we must end up with a total of 12 gallons of solution, one equation for the system is

$$x + y = 12$$

The amount of alcohol in the x gallons of 20% solution is $0.20x$, while the amount of alcohol in the y gallons of 50% solution is $0.50y$. Because the total amount of alcohol in the 20% and 50% solutions must add up to the amount of alcohol in the 12 gallons of 30% solution, the second equation in our system can be written as

$$0.20x + 0.50y = 0.30(12)$$

Again, let's make a table that summarizes the information we have to this point in the problem.

	20% Solution	50% Solution	Final Solution
Total number of gallons	x	y	12
Gallons of alcohol	$0.20x$	$0.50y$	$0.30(12)$

Our system of equations is

$$x + \quad y = 12$$
$$0.20x + 0.50y = 0.30(12) = 3.6$$

Step 4: *Solve the system.*

Multiplying the second equation by 10 gives us an equivalent system:

$$x + \ y = 12$$
$$2x + 5y = 36$$

Multiplying the top equation by -2 to eliminate the x-variable, we have

$$-2x - 2y = -24$$
$$\underline{2x + 5y = \quad 36}$$
$$3y = \quad 12$$
$$y = \quad 4$$

Substituting $y = 4$ into $x + y = 12$, we solve for x:

$$x + 4 = 12$$
$$x = 8$$

Step 5: *Write answers.*

It takes 8 gallons of 20% alcohol solution and 4 gallons of 50% alcohol solution to produce 12 gallons of 30% alcohol solution.

Step 6: *Reread and check.*

If we mix 8 gallons of 20% solution and 4 gallons of 50% solution, we end up with a total of 12 gallons of solution. To check the percentages we look for the total amount of alcohol in the two initial solutions and in the final solution.

In the initial solutions

The amount of alcohol in 8 gallons of 20% solution is $0.20(8) = 1.6$ gallons
The amount of alcohol in 4 gallons of 50% solution is $0.50(4) = 2.0$ gallons

The total amount of alcohol in the initial solutions is 3.6 gallons

In the final solution

The amount of alcohol in 12 gallons of 30% solution is $0.30(12) = 3.6$ gallons.

EXAMPLE 5

It takes 2 hours for a boat to travel 28 miles downstream (with the current). The same boat can travel 18 miles upstream (against the current) in 3 hours. What is the speed of the boat in still water, and what is the speed of the current of the river?

SOLUTION

Step 1: *Read and list.*

A boat travels 18 miles upstream and 28 miles downstream. The trip upstream takes 3 hours. The trip downstream takes 2 hours. We don't know the speed of the boat or the speed of the current.

Step 2: *Assign variables and translate information.*

Let $x =$ the speed of the boat in still water and let $y =$ the speed of the current. The average speed (rate) of the boat upstream is $x - y$, because it is traveling against the current. The rate of the boat downstream is $x + y$, because the boat is traveling with the current.

Step 3: *Write a system of equations.*

Putting the information into a table, we have

	d (distance, miles)	r (rate, mph)	t (time, h)
Upstream	18	$x - y$	3
Downstream	28	$x + y$	2

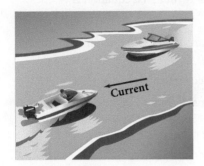

The formula for the relationship between distance d, rate r, and time t is $d = rt$ (the rate equation). Because $d = r \cdot t$, the system we need to solve the problem is

$$18 = (x - y) \cdot 3$$
$$28 = (x + y) \cdot 2$$

which is equivalent to

$$6 = x - y$$
$$14 = x + y$$

Step 4: *Solve the system.*

Adding the two equations, we have

$$20 = 2x$$
$$x = 10$$

Substituting $x = 10$ into $14 = x + y$, we see that

$$y = 4$$

Step 5: *Write answers.*

The speed of the boat in still water is 10 miles per hour; the speed of the current is 4 miles per hour.

Step 6: *Reread and check.*

The boat travels at $10 + 4 = 14$ miles per hour downstream, so in 2 hours it will travel $14 \cdot 2 = 28$ miles. The boat travels at $10 - 4 = 6$ miles per hour upstream, so in 3 hours it will travel $6 \cdot 3 = 18$ miles.

EXAMPLE 6

A coin collection consists of 14 coins with a total value of $1.35. If the coins are nickels, dimes, and quarters, and the number of nickels is 3 less than twice the number of dimes, how many of each coin is there in the collection?

SOLUTION This problem will require three variables and three equations.

Step 1: *Read and list.*

We have 14 coins with a total value of $1.35. The coins are nickels, dimes, and quarters. The number of nickels is 3 less than twice the number of dimes. We do not know how many of each coin we have.

Step 2: *Assign variables and translate information.*

Because we have three types of coins, we will have to use three variables. Let's let x = the number of nickels, y = the number of dimes, and z = the number of quarters.

Step 3: *Write a system of equations.*

Because the total number of coins is 14, our first equation is

$$x + y + z = 14$$

Because the number of nickels is 3 less than twice the number of dimes, a second equation is

$$x = 2y - 3 \qquad \text{which is equivalent to} \qquad x - 2y = -3$$

Our last equation is obtained by considering the value of each coin and the total value of the collection. Let's write the equation in terms of cents, so we won't have to clear it of decimals later.

$$5x + 10y + 25z = 135$$

Here is our system, with the equations numbered for reference:

$$
\begin{aligned}
x + \quad y + \quad z &= 14 \qquad (1) \\
x - \quad 2y \qquad\quad &= -3 \qquad (2) \\
5x + 10y + 25z &= 135 \qquad (3)
\end{aligned}
$$

Step 4: *Solve the system.*

Let's begin by eliminating x from the first and second equations, and the first and third equations. Adding -1 times the second equation to the first equation gives us an equation in only y and z. We call this equation (4).

$$3y + z = 17 \qquad (4)$$

Adding -5 times equation (1) to equation (3) gives us

$$5y + 20z = 65 \qquad (5)$$

We can eliminate z from equations (4) and (5) by adding -20 times (4) to (5). Here is the result:

$$-55y = -275$$
$$y = 5$$

Substituting $y = 5$ into equation (4) gives us $z = 2$. Substituting $y = 5$ and $z = 2$ into equation (1) gives us $x = 7$.

Step 5: *Write answers.*

The collection consists of 7 nickels, 5 dimes, and 2 quarters.

Step 6: *Reread and check.*

The total number of coins is $7 + 5 + 2 = 14$. The number of nickels, 7, is 3 less than twice the number of dimes, 5. To find the total value of the collection, we have

The value of the 7 nickels is	$7(0.05) = \$0.35$
The value of the 5 dimes is	$5(0.10) = \$0.50$
The value of the 2 quarters is	$2(0.25) = \$0.50$

The total value of the collection is $\$1.35$

If you go on to take a chemistry class, you may see the next example (or one much like it).

EXAMPLE 7

In a chemistry lab, students record the temperature of water at room temperature and find that it is 77° on the Fahrenheit temperature scale and 25° on the Celsius temperature scale. The water is then heated until it boils. The temperature of the boiling water is 212°F and 100°C. Assume that the relationship between the two temperature scales is a linear one, then use the preceding data to find the formula that gives the Celsius temperature C in terms of the Fahrenheit temperature F.

77°F

Corresponding Temperatures	
In Degrees Fahrenheit	In Degrees Celsius
77	25
212	100

SOLUTION   The data is summarized in the following table.

If we assume the relationship is linear, then the formula that relates the two temperature scales can be written in slope-intercept form as

$$C = mF + b$$

Substituting $C = 25$ and $F = 77$ into this formula gives us

$$25 = 77m + b$$

Substituting $C = 100$ and $F = 212$ into the formula yields

$$100 = 212m + b$$

Together, the two equations form a system of equations, which we can solve using the addition method.

$$25 = 77m + b \xrightarrow{\text{Multiply by } -1} -25 = -77m - b$$
$$100 = 212m + b \xrightarrow[\text{No change}]{} \underline{100 = 212m + b}$$
$$75 = 135m$$
$$m = \frac{75}{135} = \frac{5}{9}$$

To find the value of b, we substitute $m = \frac{5}{9}$ into $25 = 77m + b$ and solve for b.

$$25 = 77\left(\frac{5}{9}\right) + b$$

$$25 = \frac{385}{9} + b$$

$$b = 25 - \frac{385}{9} = \frac{225}{9} - \frac{385}{9} = -\frac{160}{9}$$

The equation that gives C in terms of F is

$$C = \frac{5}{9}F - \frac{160}{9}$$

GETTING READY FOR CLASS

After reading through the preceding section, respond in your own words and in complete sentences.

A. If you were to apply the Blueprint for Problem Solving from Section 2.3 to the examples in this section, what would be the first step?

B. If you were to apply the Blueprint for Problem Solving from Section 2.3 to the examples in this section, what would be the last step?

C. When working application problems involving boats moving in rivers, how does the current of the river affect the speed of the boat?

D. Write an application problem for which the solution depends on solving the system of equations:

$$x + y = 1,000$$
$$0.05x + 0.06y = 55$$

Problem Set 4.3

NUMBER PROBLEMS

1. One number is 3 more than twice another. The sum of the numbers is 18. Find the two numbers.

2. The sum of two numbers is 32. One of the numbers is 4 less than 5 times the other. Find the two numbers.

3. The difference of two numbers is 6. Twice the smaller is 4 more than the larger. Find the two numbers.

4. The larger of two numbers is 5 more than twice the smaller. If the smaller is subtracted from the larger, the result is 12. Find the two numbers.

5. The sum of three numbers is 8. Twice the smallest is 2 less than the largest, while the sum of the largest and smallest is 5. Use a linear system in three variables to find the three numbers.

6. The sum of three numbers is 14. The largest is 4 times the smallest, while the sum of the smallest and twice the largest is 18. Use a linear system in three variables to find the three numbers.

TICKET AND INTEREST PROBLEMS

7. A total of 925 tickets were sold for a game for a total of $1,150. If adult tickets sold for $2.00 and children's tickets sold for $1.00, how many of each kind of ticket were sold?

8. If tickets for a show cost $2.00 for adults and $1.50 for children, how many of each kind of ticket were sold if a total of 300 tickets were sold for $525?

9. Mr. Jones has $20,000 to invest. He invests part at 6% and the rest at 7%. If he earns $1,280 in interest after 1 year, how much did he invest at each rate?

10. A man invests $17,000 in two accounts. One account earns 5% interest per year and the other 6.5%. If his total interest after 1 year is $970, how much did he invest at each rate?

11. Susan invests twice as much money at 7.5% as she does at 6%. If her total interest after 1 year is $840, how much does she have invested at each rate?

12. A woman earns $1,350 in interest from two accounts in 1 year. If she has three times as much invested at 7% as she does at 6%, how much does she have in each account?

13. A man invests $2,200 in three accounts that pay 6%, 8%, and 9% in annual interest, respectively. He has three times as much invested at 9% as he does at 6%. If his total interest for the year is $178, how much is invested at each rate?

14. A student has money in three accounts that pay 5%, 7%, and 8% in annual interest. She has three times as much invested at 8% as she does at 5%. If the total amount she has invested is $1,600 and her interest for the year comes to $115, how much money does she have in each account?

MIXTURE PROBLEMS

15. How many gallons of 20% alcohol solution and 50% alcohol solution must be mixed to get 9 gallons of 30% alcohol solution?

16. How many ounces of 30% hydrochloric acid solution and 80% hydrochloric acid solution must be mixed to get 10 ounces of 50% hydrochloric acid solution?

17. A mixture of 16% disinfectant solution is to be made from 20% and 14% disinfectant solutions. How much of each solution should be used if 15 gallons of the 16% solution are needed?

18. How much 25% antifreeze and 50% antifreeze should be combined to give 40 gallons of 30% antifreeze?

19. Paul mixes nuts worth $1.55 per pound with oats worth $1.35 per pound to get 25 pounds of trail mix worth $1.45 per pound. How many pounds of nuts and how many pounds of oats did he use?

20. A chemist has three different acid solutions. The first acid solution contains 20% acid, the second contains 40%, and the third contains 60%. He wants to use all three solutions to obtain a mixture of 60 liters containing 50% acid, using twice as much of the 60% solution as the 40% solution. How many liters of each solution should be used?

RATE PROBLEMS

21. It takes a boat 2 hours to travel 24 miles downstream and 3 hours to travel 18 miles upstream. What is the speed of the boat in still water? What is the speed of the current of the river?

22. A boat on a river travels 20 miles downstream in only 2 hours. It takes the same boat 6 hours to travel 12 miles upstream. What are the speed of the boat and the speed of the current?

23. An airplane flying with the wind can cover a certain distance in 2 hours. The return trip against the wind takes $2\frac{1}{2}$ hours. How fast is the plane and what is the speed of the air, if the distance is 600 miles?

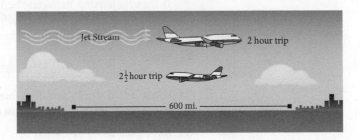

24. An airplane covers a distance of 1,500 miles in 3 hours when it flies with the wind and $3\frac{1}{3}$ hours when it flies against the wind. What is the speed of the plane in still air?

COIN PROBLEMS

25. Bob has 20 coins totaling $1.40. If he has only dimes and nickels, how many of each coin does he have?

26. If Amy has 15 coins totaling $2.70, and the coins are quarters and dimes, how many of each coin does she have?

27. A collection of nickels, dimes, and quarters consists of 9 coins with a total value of $1.20. If the number of dimes is equal to the number of nickels, find the number of each type of coin.

28. A coin collection consists of 12 coins with a total value of $1.20. If the collection consists only of nickels, dimes, and quarters, and the number of dimes is two more than twice the number of nickels, how many of each type of coin is in the collection?

29. A collection of nickels, dimes, and quarters amount to $10.00. If there are 140 coins in all and there are twice as many dimes as there are quarters, find the number of nickels.

30. A cash register contains a total of 95 coins consisting of pennies, nickels, dimes, and quarters. There are only 5 pennies and the total value of the coins is $12.05. Also, there are 5 more quarters than dimes. How many of each coin is in the cash register?

ADDITIONAL PROBLEMS

31. Price and Demand A manufacturing company finds that they can sell 300 items if the price per item is $2.00, and 400 items if the price is $1.50 per item. If the relationship between the number of items sold x and the price per item p is a linear one, find a formula that gives x in terms of p. Then use the formula to find the number of items they will sell if the price per item is $3.00.

32. **Price and Demand** A company manufactures and sells bracelets. They have found from past experience that they can sell 300 bracelets each week if the price per bracelet is $2.00, but only 150 bracelets are sold if the price is $2.50 per bracelet. If the relationship between the number of bracelets sold x and the price per bracelet p is a linear one, find a formula that gives x in terms of p. Then use the formula to find the number of bracelets they will sell at $3.00 each.

33. **Prices** If two drinks and seven tacos cost $51, and if six drinks and ten tacos cost $87, find the cost of each item (ignoring tax, service charge, etc.).

34. **Prices** If two drinks and four hamburgers cost $41 and if four drinks and five hamburgers cost $58, find the cost of each item (ignoring tax, service charge, etc.).

35. **Driving the Same** If CJ left at 2:00 pm driving north at 45 mph, and Edwin left at 2:30 pm driving north at the same 45 mph on the same route as CJ, determine when they will meet.

36. **Walking the Same** If Stephanie walks at 2 kilometers per hour, and Molly walks at 500 meters per 15 minutes on the same route going the same direction with both leaving at 7:00 am, determine when they will meet.

37. **Height of a Ball** A ball is tossed into the air so that the height after 1, 3, and 5 seconds is as given in the following table.

t (sec)	h (ft)
1	128
3	128
5	0

If the relationship between the height of the ball h and the time t is quadratic, then the relationship can be written as

$$h = at^2 + bt + c$$

Use the information in the table to write a system of three equations in three variables a, b, and c. Solve the system to find the exact relationship between h and t.

38. **Height of a Ball** A ball is tossed into the air and its height above the ground after 1, 3, and 4 seconds is recorded as shown in the following table.

t (sec)	h (ft)
1	96
3	64
4	0

The relationship between the height of the ball h and the time t is quadratic and can be written as

$$h = at^2 + bt + c$$

Use the information in the table to write a system of three equations in three variables a, b, and c. Solve the system to find the exact relationship between the variables h and t.

39. **Work Time** Breylor always works for B hours per day, Claire always works for C hours per day, and Nathan always works for N hours per day. A house took 55 total hours to paint, with Breylor working for 3 days, Claire working for 4 days, and Nathan working for 8 days. It took 51 total hours to tile a patio, with Breylor working for 5 days, Claire working for 1 day, and Nathan working for 6 days. It took 50 total hours to trim trees, with Breylor working for 2 days, Claire working for 6 days, and Nathan working for 7 days. Write a system of three equations in three variables B, C, and N, and solve to find the number of hours per day each person works.

40. **Work Time** Heather always works for H hours per day, Maddie always works for M hours per day, and Jackson always works for J hours per day. A house took 49 total hours to paint, with Heather working for 2 days, Maddie working for 4 days, and Jackson working for 5 days. It took 68 total hours to lay carpeting, with Heather working for 5 days, Maddie working for 6 days, and Jackson working for 4 days. It took 48 total hours to landscape, with Heather working for 4 days, Maddie working for 3 days, and Jackson working for 3 days. Write a system of three equations in three variables H, M, and J, and solve to find the number of hours per day each person works.

GETTING READY FOR THE NEXT SECTION

41. Does the graph of $x + y < 4$ include the boundary line?

42. Does the graph of $-x + y \leq 3$ include the boundary line?

43. Where do the graphs of the lines $x + y = 4$ and $x - 2y = 4$ intersect?

44. Where do the graphs of the line $x = -1$ and $x - 2y = 4$ intersect?

Solve.

45. $20x + 9,300 > 18,000$

46. $20x + 4,800 > 18,000$

EXTENDING THE CONCEPTS

47. **High School Dropout Rate** The high school dropout rates for males and females over a 24-year period are shown in the following table.

Year	Female Dropout Rate (%)	Male Dropout Rate (%)
Year 1	18	16
Year 3	16	14
Year 8	15	13
Year 13	13	16
Year 18	12	15
Year 24	13	14

Plotting the years along the horizontal axis and the dropout rates along the vertical axis, draw a line graph for these data. Draw the female line dashed and the male line solid for easier reading and comparison.

48. High School Dropout Rate Refer to the data in the preceding exercise to answer the following questions.

   a. Using the slope, determine the time interval when the decline in the female dropout rate was the steepest.

   b. Using the slope, determine the time interval when the increase in the male dropout rate was the steepest.

   c. What appears unusual about the time period from Year 8 to Year 13?

   d. Are the dropout rates for males and females generally increasing or generally decreasing?

   e. Do the dropout rates for males or females show the most consistent pattern?

49. High School Dropout Rate Refer to the data about high school dropout rates for males and females.

   a. Write a linear equation for the dropout rate of males, M, based on the year, x, for Year 8 and Year 13.

   b. Write a linear equation for the dropout rate of females, F, based on the year, x, for Year 8 and Year 13.

   c. Using your results from parts **a.** and **b.**, determine when the dropout rates for males and females were the same.

© iStockphoto/monkeybusinessimages

OBJECTIVES

A Graph the solution to a system of linear inequalities in two variables.

B Solve application problems using systems of inequalities.

4.4

Imagine you work at a local coffeehouse and your boss wants you to create a new house blend using the three most popular coffees: the dark roast, the medium roast, and the flavored blend. The dark roast coffee costs $20.00 per pound, the medium roast costs $14.50 per pound, and the flavored blend costs $16.00 per pound. Your boss wants the cost of the new blend to be at most $17.00 per pound and is requesting to see possible combinations to serve customers. As we have discussed, words like "at most" imply the use of inequalities. Working through this section will help us visualize solutions to systems of linear inequalities in two and three variables, such as the coffeehouse example.

Graphing Solutions to Linear Inequalities

In Section 3.4, we graphed linear inequalities in two variables. To review, we graph the boundary line, using a solid line if the boundary is part of the solution set and a broken line if the boundary is not part of the solution set. Then we test any point that is not on the boundary line in the original inequality. A true statement tells us that the point lies in the solution set, a false statement tells us the solution set is the other region.

Figure 1 shows the graph of the inequality $x + y < 4$. Note that the boundary is not included in the solution set, and is therefore drawn with a broken line. Figure 2 shows the graph of $-x + y \leq 3$. Note that the boundary is drawn with a solid line, because it is part of the solution set.

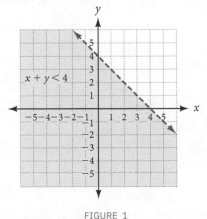

FIGURE 1

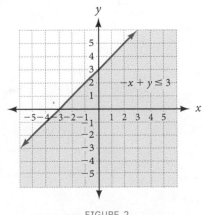

FIGURE 2

361

If we form a system of inequalities with the two inequalities, the solution set will be all the points common to both solution sets shown in Figures 1 and 2: It is the intersection of the two solution sets. Therefore, the solution set for the system of inequalities

$$x + y < 4$$
$$-x + y \leq 3$$

is all the ordered pairs that satisfy both inequalities. It is the set of points that are below the line $x + y = 4$, and also below (and including) the line $-x + y = 3$. The graph of the solution set to this system is shown in Figure 3. We have written the system in Figure 3 with the word *and* just to remind you that the solution set to a system of equations or inequalities is all the points that satisfy both equations or inequalities.

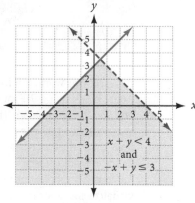

FIGURE 3

EXAMPLE 1

Graph the solution to the system of linear inequalities.

$$y < \frac{1}{2}x + 3$$

$$y \geq \frac{1}{2}x - 2$$

SOLUTION   Figures 4 and 5 show the solution set for each of the inequalities separately.

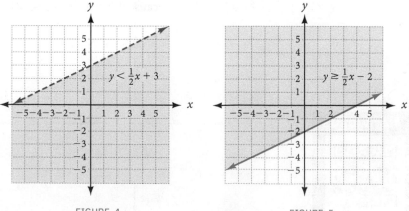

FIGURE 4   FIGURE 5

Figure 6 is the solution set to the system of inequalities. It is the region consisting of points whose coordinates satisfy both inequalities.

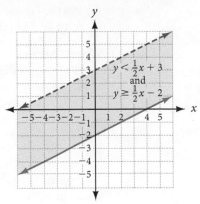

FIGURE 6

EXAMPLE 2

Graph the solution to the system of linear inequalities.

$$x + y < 4$$

$$x \geq 0$$

$$y \geq 0$$

SOLUTION   We graphed the first inequality, $x + y < 4$, in Figure 1 at the beginning of this section. The solution set to the inequality $x \geq 0$, shown in Figure 7, is all the points to the right of the y-axis; that is, all the points with x-coordinates that are greater than or equal to 0. Figure 8 shows the graph of $y \geq 0$. It consists of all points with y-coordinates greater than or equal to 0; that is, all points from the x-axis up.

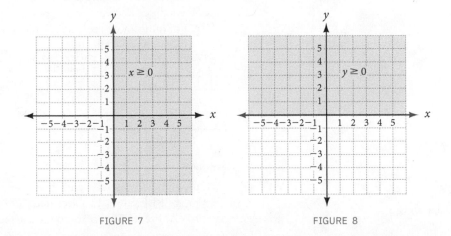

FIGURE 7   FIGURE 8

The regions shown in Figures 7 and 8 overlap in the first quadrant. Therefore, putting all three regions together we have the points in the first quadrant that are below the line $x + y = 4$. This region is shown in Figure 9, and it is the solution to our system of inequalities.

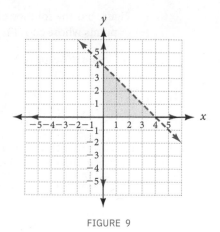

FIGURE 9

Extending the discussion in Example 2 we can name the points in each of the four quadrants using systems of inequalities.

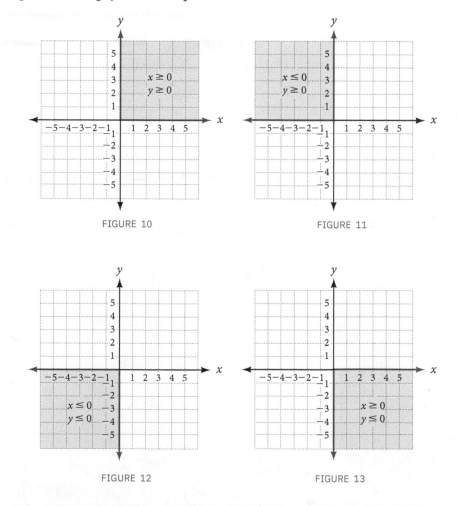

FIGURE 10

FIGURE 11

FIGURE 12

FIGURE 13

EXAMPLE 3

Graph the solution to the system of linear inequalities.

$$x \le 4$$
$$y \ge -3$$

SOLUTION   The solution to this system will consist of all points to the left of and including the vertical line $x = 4$ that intersect with all points above and including the horizontal line $y = -3$. The solution set is shown in Figure 14.

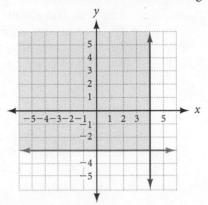

FIGURE 14

EXAMPLE 4

Graph the solution set for the following system.

$$x - 2y \le 4$$
$$x + y \le 4$$
$$x \ge -1$$

SOLUTION   We have three linear inequalities, representing three sections of the coordinate plane. The graph of the solution set for this system will be the intersection of these three sections. The graph of $x - 2y \le 4$ is the section above and including the boundary $x - 2y = 4$. The graph of $x + y \le 4$ is the section below and including the boundary line $x + y = 4$. The graph of $x \ge -1$ is all the points to the right of, and including, the vertical line $x = -1$. The intersection of these three graphs is shown in Figure 15.

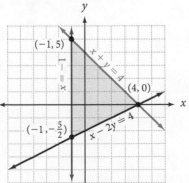

FIGURE 15

B Application Problems

EXAMPLE 5

A college basketball arena plans on charging $20 for certain seats and $15 for others. They want to bring in more than $18,000 from all ticket sales and have reserved at least 500 tickets at the $15 rate. Find a system of inequalities describing all possibilities and sketch the graph. If 620 tickets are sold for $15, at least how many tickets are sold for $20?

SOLUTION   Let x = the number of $20 tickets and y = the number of $15 tickets. We need to write a list of inequalities that describe this situation. That list will form our system of inequalities. First of all, we note that we cannot use negative numbers for either x or y. So, we have our first inequalities:

$$x \geq 0$$

$$y \geq 0$$

Next, we note that they are selling at least 500 tickets for $15, so we can replace our second inequality with $y \geq 500$. Now our system is

$$x \geq 0$$

$$y \geq 500$$

Now the amount of money brought in by selling $20 tickets is $20x$, and the amount of money brought in by selling $15 tickets is $15y$. It the total income from ticket sales is to be more than $18,000, then $20x + 15y$ must be greater than 18,000. This gives us our last inequality and completes our system.

$$20x + 15y > 18,000$$

$$x \geq 0$$

$$y \geq 500$$

We have used all the information in the problem to arrive at this system of inequalities. The solution set contains all the values of x and y that satisfy all the conditions given in the problem. Here is the graph of the solution set.

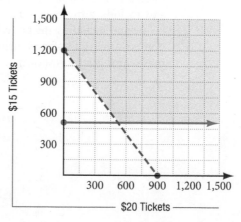

FIGURE 16

If 620 tickets are sold for $15, then we substitute 620 for y in our first inequality to obtain

$$20x + 15(620) > 18,000 \quad \text{Substitute 620 for } y.$$
$$20x + 9,300 > 18,000 \quad \text{Multiply.}$$
$$20x > 8,700 \quad \text{Add } -9,300 \text{ to each side.}$$
$$x > 435 \quad \text{Divide each side by 20.}$$

If they sell 620 tickets for $15 each, then they need to sell more than 435 tickets at $20 each to bring in more than $18,000.

GETTING READY FOR CLASS

After reading through the preceding section, respond in your own words and in complete sentences.

A. What is the solution set for a system of inequalities?

B. If an ordered pair satisfied only one inequality in a system of inequalities, is the ordered pair part of the solution set for that system? Explain.

C. Explain how you would graph the solution set for a system of three linear inequalities.

D. Give an example of an application problem that involves a system of linear inequalities.

Graph the solution set for each system of linear inequalities.

1. $x + y < 5$

$2x - y > 4$

2. $x + y < 5$

$2x - y < 4$

3. $y < \dfrac{1}{3}x + 4$

$y \geq \dfrac{1}{3}x - 3$

4. $y < 2x + 4$
$y \geq 2x - 3$

5. $x \geq -3$
$y < -2$

6. $x \leq 4$
$y \geq -2$

7. $1 \leq x \leq 3$
$2 \leq y \leq 4$

8. $-4 \leq x \leq -2$
$1 \leq y \leq 3$

9. $x + y \leq 4$
$x \geq 0$
$y \geq 0$

10. $x - y \leq 2$
$x \geq 0$
$y \leq 0$

11. $x + y \leq 3$
$x - 3y \leq 3$
$x \geq -2$

12. $x - y \leq 4$
$x + 2y \leq 4$
$x \geq -1$

13. $x + y \leq 2$
$-x + y \leq 2$
$y \geq -2$

14. $x - y \leq 3$
$-x - y \leq 3$
$y \leq -1$

15. $x + y < 5$
$y > x$
$y \geq 0$

16. $x + y < 5$
$y > x$
$x \geq 0$

17. $2x + 3y \leq 6$
$x \geq 0$
$y \geq 0$

18. $x + 2y \leq 10$
$3x + 2y \leq 12$
$x \geq 0$
$y \geq 0$

For each figure below, find a system of inequalities that describes the shaded region.

19.

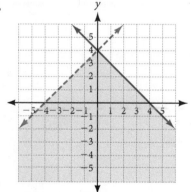

FIGURE 17

20.

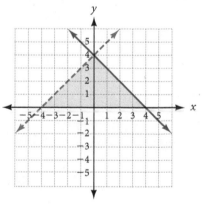

FIGURE 18

21.

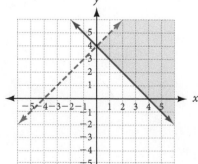

FIGURE 19

22.

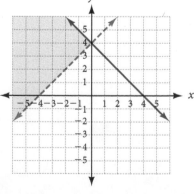

FIGURE 20

APPLYING THE CONCEPTS

23. **Office Supplies** An office worker wants to purchase some $0.55 postage stamps and also some $0.65 postage stamps totaling no more than $40. It is also desired to have at least twice as many $0.55 stamps and more than 15 $0.55 stamps.

    a. Find a system of inequalities describing all the possibilities and sketch the graph.

    b. If he purchases 20 $0.55 stamps, what is the maximum number of $0.65 stamps he can purchase?

24. **Inventory** A store sells two brands of DVD players. Customer demand indicates that it is necessary to stock at least twice as many DVD players of brand A as of brand B. At least 30 of brand A and 15 of brand B must be on hand. In the store, there is room for not more than 100 DVD players.

    a. Find a system of inequalities describing all possibilities, then sketch the graph.

    b. If there are 35 DVD players of brand A, what is the most number of brand B DVD players on hand?

25. **Work Time** Aaron will earn a one-time bonus check of at least $300 in addition to his usual $10 per hour, and Betsy's one-time bonus will be at least $200 in addition to her usual $12 per hour. They worked the same number of hours x.

    a. Find the number of hours where the boundary lines intersect.

    b. Explain whether or not it is possible for their checks to be $275.

26. **Work Time** Terry will earn a one-time bonus check of at least $200 in addition to her usual $9 per hour, and Jenny's one-time bonus will be at least $150 in addition to her usual $11 per hour. They worked the same number of hours x.

    a. Find the number of hours where the boundary lines intersect.

    b. Explain whether or not it is possible for their checks to be $175.

GETTING READY FOR THE NEXT SECTION

Simplify.

27. $1(4) - 3(2)$

28. $3(7) - (-2)(5)$

29. $1(0)(1) + 3(1)(4) + (-2)(2)(-1)$

30. $-4(0)(-2) - (-1)(1)(1) - 1(2)(3)$

31. $1[0 - (-1)] - 3(2 - 4) + (-2)(-2 - 0)$

32. $-3(-1 - 1) + 4[-2 - (-2)] - 5[2 - (-2)]$

Solve.

33. $x^2 - 2x = 8$

34. $2x^2 + 8 = -8x$

4.5

OBJECTIVES

A Find the value of a 2 × 2 determinant.

B Find the value of a 3 × 3 determinant.

C Write an equation as a determinant equation.

© iStockphoto/rappensuncle

In 1714, Daniel Gabriel Fahrenheit invented the first mercury thermometer. Ten years later, he introduced the scale to measure temperature that we use today. He based this scale on the temperature of the human body, which at the time was measured to be 100°F, although has since been adjusted to 98.6°F. In 1742, Anders Celsius invented the centigrade scale based on the freezing point of water at 0°C and the boiling point at 100°C. Since there were now two prominent scales to measure temperature, the linear formula $F = \frac{9}{5}C + 32$ was derived to convert Celsius temperatures to Fahrenheit temperatures. In this section, you will learn how a linear formula can also be written as a *determinant* equation. For example, $F = \frac{9}{5}C + 32$ is equivalent to the following determinant:

$$\begin{vmatrix} C & F & 1 \\ 5 & 41 & 1 \\ -10 & 14 & 1 \end{vmatrix} = 0$$

In this section, we will expand and evaluate *determinants*. The purpose of this section is simply to be able to find the value of a given determinant. As we will see in the next section, determinants are very useful in solving systems of linear equations. Before we apply determinants to systems of linear equations, however, we must practice calculating the value of some determinants.

 ## 2 × 2 Determinants

DEFINITION *2×2 determinant*

The value of the 2×2 (2 by 2) *determinant*

$$\begin{vmatrix} a & c \\ b & d \end{vmatrix}$$

is given by

$$\begin{vmatrix} a & c \\ b & d \end{vmatrix} = ad - bc$$

From the preceding definition, we see that a determinant is simply a square array of numbers with two vertical lines enclosing it. The value of a 2×2 determinant is found by cross-multiplying on the diagonals and then subtracting, a diagram of which looks like

$$\begin{vmatrix} a & c \\ b & d \end{vmatrix} = ad - bc$$

EXAMPLE 1

Find the value of the following 2×2 determinants:

a. $\begin{vmatrix} 1 & 2 \\ 3 & 4 \end{vmatrix} = 1(4) - 3(2) = 4 - 6 = -2$

b. $\begin{vmatrix} 3 & 5 \\ -2 & 7 \end{vmatrix} = 3(7) - (-2)5 = 21 + 10 = 31$

EXAMPLE 2

Solve for x if

$$\begin{vmatrix} x^2 & 2 \\ x & 1 \end{vmatrix} = 8$$

SOLUTION   We expand the determinant on the left side to get

$$x^2(1) - x(2) = 8$$
$$x^2 - 2x = 8$$
$$x^2 - 2x - 8 = 0$$
$$(x - 4)(x + 2) = 0$$
$$x - 4 = 0 \quad \text{or} \quad x + 2 = 0$$
$$x = 4 \qquad\qquad x = -2$$

B 3 × 3 Determinants

We now turn our attention to 3×3 determinants. A 3×3 determinant is also a square array of numbers, the value of which is given by the following definition.

DEFINITION   *3×3 determinant*

The value of the 3×3 determinant

$$\begin{vmatrix} a_1 & b_1 & c_1 \\ a_2 & b_2 & c_2 \\ a_3 & b_3 & c_3 \end{vmatrix}$$

is given by

$$\begin{vmatrix} a_1 & b_1 & c_1 \\ a_2 & b_2 & c_2 \\ a_3 & b_3 & c_3 \end{vmatrix} = a_1 b_2 c_3 + a_3 b_1 c_2 + a_2 b_3 c_1 - a_3 b_2 c_1 - a_1 b_3 c_2 - a_2 b_1 c_3$$

At first glance, the expansion of a 3×3 determinant looks a little complicated. There are actually two different methods used to find the six products in the preceding definition that simplify matters somewhat.

Method 1

We begin by writing the determinant with the first two columns repeated on the right:

$$\begin{vmatrix} a_1 & b_1 & c_1 \\ a_2 & b_2 & c_2 \\ a_3 & b_3 & c_3 \end{vmatrix} \begin{matrix} a_1 & b_1 \\ a_2 & b_2 \\ a_3 & b_3 \end{matrix}$$

The positive products in the definition come from multiplying down the three full diagonals:

$$\begin{vmatrix} a_1 & b_1 & c_1 \\ a_2 & b_2 & c_2 \\ a_3 & b_3 & c_3 \end{vmatrix} \begin{matrix} a_1 & b_1 \\ a_2 & b_2 \\ a_3 & b_3 \end{matrix}$$

The negative products come from multiplying up the three full diagonals:

$$\begin{vmatrix} a_1 & b_1 & c_1 \\ a_2 & b_2 & c_2 \\ a_3 & b_3 & c_3 \end{vmatrix} \begin{matrix} a_1 & b_1 \\ a_2 & b_2 \\ a_3 & b_3 \end{matrix}$$

EXAMPLE 3

Find the value of

$$\begin{vmatrix} 1 & 3 & -2 \\ 2 & 0 & 1 \\ 4 & -1 & 1 \end{vmatrix}$$

SOLUTION Repeating the first two columns and then finding the products down the diagonals and the products up the diagonals as given in Method 1, we have

$$\begin{vmatrix} 1 & 3 & -2 \\ 2 & 0 & 1 \\ 4 & -1 & 1 \end{vmatrix} \begin{matrix} 1 & 3 \\ 2 & 0 \\ 4 & -1 \end{matrix}$$

$$= 1(0)(1) + 3(1)(4) + (-2)(2)(-1) - 4(0)(-2) - (-1)(1)(1) - 1(2)(3)$$

$$= 0 + 12 + 4 - 0 - (-1) - 6$$

$$= 11$$

Method 2

The second method of evaluating a 3×3 determinant is called *expansion by minors*.

DEFINITION *minor*

The *minor* for an element in a 3×3 determinant is the determinant consisting of the elements remaining when the row and column to which the element belongs are deleted. For example, in the determinant

$$\begin{vmatrix} a_1 & b_1 & c_1 \\ a_2 & b_2 & c_2 \\ a_3 & b_3 & c_3 \end{vmatrix}$$

$$\text{Minor for element } a_1 = \begin{vmatrix} b_2 & c_2 \\ b_3 & c_3 \end{vmatrix}$$

$$\text{Minor for element } b_2 = \begin{vmatrix} a_1 & c_1 \\ a_3 & c_3 \end{vmatrix}$$

$$\text{Minor for element } c_3 = \begin{vmatrix} a_1 & b_1 \\ a_2 & b_2 \end{vmatrix}$$

Before we can evaluate a 3×3 determinant by Method 2, we must first define what is known as the sign array for a 3×3 determinant.

DEFINITION *sign array*

The *sign array* for a 3×3 determinant is a 3×3 array of signs in the following pattern:

$$\begin{vmatrix} + & - & + \\ - & + & - \\ + & - & + \end{vmatrix}$$

The sign array begins with a $+$ sign in the upper left-hand corner. The signs then alternate between $+$ and $-$ across every row and down every column.

NOTE This method of evaluating a determinant is actually more valuable than our first method, because it works with any size determinant from 3×3 to 4×4 to any higher order determinant. Method 1 works only on 3×3 determinants. It cannot be used on a 4×4 determinant.

HOW TO *Evaluate a 3×3 Determinant by Expansion of Minors*

We can evaluate a 3×3 determinant by expanding across any row or down any column as follows:

Step 1: Choose a row or column to expand about.

Step 2: Write the product of each element in the row or column chosen in Step 1 with its minor.

Step 3: Connect the three products in Step 2 with the signs in the corresponding row or column in the sign array.

To illustrate the procedure, we will use the same determinant we used in Example 3.

EXAMPLE 4

Expand across the first row:

$$\begin{vmatrix} 1 & 3 & -2 \\ 2 & 0 & 1 \\ 4 & -1 & 1 \end{vmatrix}$$

SOLUTION The products of the three elements in row 1 with their minors are

$$1\begin{vmatrix} 0 & 1 \\ -1 & 1 \end{vmatrix} \qquad 3\begin{vmatrix} 2 & 1 \\ 4 & 1 \end{vmatrix} \qquad (-2)\begin{vmatrix} 2 & 0 \\ 4 & -1 \end{vmatrix}$$

Connecting these three products with the signs from the first row of the sign array, we have

$$+1\begin{vmatrix} 0 & 1 \\ -1 & 1 \end{vmatrix} \quad -3\begin{vmatrix} 2 & 1 \\ 4 & 1 \end{vmatrix} \quad +(-2)\begin{vmatrix} 2 & 0 \\ 4 & -1 \end{vmatrix}$$

We complete the problem by evaluating each of the three 2×2 determinants and then simplifying the resulting expression:

$$+1[0 - (-1)] - 3(2 - 4) + (-2)(-2 - 0) = 1(1) - 3(-2) + (-2)(-2)$$
$$= 1 + 6 + 4$$
$$= 11$$

The results of Examples 3 and 4 match. It makes no difference which method we use — the value of a 3×3 determinant is unique.

EXAMPLE 5

Expand down column 2:

$$\begin{vmatrix} 2 & 3 & -2 \\ 1 & 4 & 1 \\ 1 & 5 & -1 \end{vmatrix}$$

SOLUTION We connect the products of elements in column 2 and their minors with the signs from the second column in the sign array:

$$\begin{vmatrix} 2 & 3 & -2 \\ 1 & 4 & 1 \\ 1 & 5 & -1 \end{vmatrix} = -3\begin{vmatrix} 1 & 1 \\ 1 & -1 \end{vmatrix} + 4\begin{vmatrix} 2 & -2 \\ 1 & -1 \end{vmatrix} - 5\begin{vmatrix} 2 & -2 \\ 1 & 1 \end{vmatrix}$$

$$= -3(-1 - 1) + 4[-2 - (-2)] - 5[2 - (-2)]$$
$$= -3(-2) + 4(0) - 5(4)$$
$$= 6 + 0 - 20$$
$$= -14$$

ⓒ Modeling: Equations from Determinants

EXAMPLE 6

Show that the following determinant equation is another way to write the equation $F = \frac{9}{5}C + 32$.

$$\begin{vmatrix} C & F & 1 \\ 5 & 41 & 1 \\ -10 & 14 & 1 \end{vmatrix} = 0$$

SOLUTION We expand the left side of the equation across the top row. Then we solve for F.

$$C(41 - 14) - F(5 + 10) + 1(70 + 410) = 0$$

$$27C - 15F + 480 = 0$$

$$15F = 27C + 480$$

$$F = \frac{27}{15}C + \frac{480}{15}$$

$$F = \frac{9}{5}C + 32$$

Using Technology

Your graphing calculator has built-in matrices, which we can use to find the determinants. Let's start by finding the determinant from Example 4 in your text.

EXAMPLE 7

Use your graphing calculator to evaluate this determinant.

$$\begin{vmatrix} 1 & 3 & -2 \\ 2 & 0 & 1 \\ 4 & -1 & 1 \end{vmatrix}$$

SOLUTION We first access the matrix screen on our calculator, then we tell the calculator we have a matrix with three rows and three columns. Once we have that set up, we can enter the numbers in the determinant above and ask the calculator to find the determinant. We start by accessing the [MATRIX] menu, to obtain a screen like this:

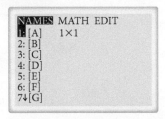

FIGURE 1

Next, scroll to EDIT, then select Matrix A to obtain the screen in Figure 2. The calculator will ask you for the dimensions of the matrix. It is a 3×3 matrix, so enter those numbers to see a screen like the one in Figure 3. Enter the numbers from the determinant in question to see a screen like the one in Figure 4.

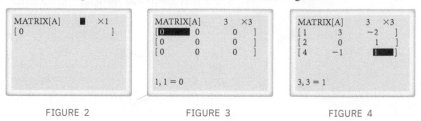

FIGURE 2   FIGURE 3   FIGURE 4

Go back to your home screen and then go to the [MATRIX] menu again and select MATH (Figure 5). Enter 1 to select det(. Next go back to the [MATRIX] menu and select Matrix A. Press [ENTER] twice to see a screen like the one in Figure 7.

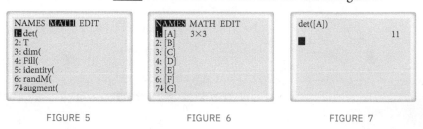

FIGURE 5   FIGURE 6   FIGURE 7

As you can see, the result is 11, which is consistent with the result we obtained in Example 4.

GETTING READY FOR CLASS

After reading through the preceding section, respond in your own words and in complete sentences.

A. What is a determinant?

B. How do you evaluate a 2×2 determinant?

C. How do you construct a sign array for a 3×3 determinant?

D. What is the minor for the first element in a 3×3 determinant?

Problem Set 4.5

Find the value of the following 2×2 determinants.

1. $\begin{vmatrix} 1 & 0 \\ 2 & 3 \end{vmatrix}$

2. $\begin{vmatrix} 5 & 4 \\ 3 & 2 \end{vmatrix}$

3. $\begin{vmatrix} 2 & 1 \\ 3 & 4 \end{vmatrix}$

4. $\begin{vmatrix} 4 & 1 \\ 5 & 2 \end{vmatrix}$

5. $\begin{vmatrix} 0 & 1 \\ 1 & 0 \end{vmatrix}$

6. $\begin{vmatrix} 1 & 0 \\ 0 & 1 \end{vmatrix}$

7. $\begin{vmatrix} -3 & 2 \\ 6 & -4 \end{vmatrix}$

8. $\begin{vmatrix} 8 & -3 \\ -2 & 5 \end{vmatrix}$

9. $\begin{vmatrix} -3 & -1 \\ 4 & -2 \end{vmatrix}$

10. $\begin{vmatrix} 5 & 3 \\ 7 & -6 \end{vmatrix}$

Solve each of the following for x.

11. $\begin{vmatrix} 2x & 1 \\ x & 3 \end{vmatrix} = 10$

12. $\begin{vmatrix} 3x & -2 \\ 2x & 3 \end{vmatrix} = 26$

13. $\begin{vmatrix} 1 & 2x \\ 2 & -3x \end{vmatrix} = 21$

14. $\begin{vmatrix} -5 & 4x \\ 1 & -x \end{vmatrix} = 27$

15. $\begin{vmatrix} 2x & -4 \\ 2 & x \end{vmatrix} = -8x$

16. $\begin{vmatrix} 3x & 2 \\ 2 & x \end{vmatrix} = -11x$

17. $\begin{vmatrix} x^2 & 3 \\ x & 1 \end{vmatrix} = 10$

18. $\begin{vmatrix} x^2 & -2 \\ x & 1 \end{vmatrix} = 35$

19. $\begin{vmatrix} x^2 & -4 \\ x & 1 \end{vmatrix} = 32$

20. $\begin{vmatrix} x^2 & 6 \\ x & 1 \end{vmatrix} = 72$

21. $\begin{vmatrix} x & 5 \\ 1 & x \end{vmatrix} = 4$

22. $\begin{vmatrix} 3x & 4 \\ 2 & x \end{vmatrix} = 10x$

Find the value of each of the following 3×3 determinants by using Method 1.

23. $\begin{vmatrix} 1 & 2 & 0 \\ 0 & 2 & 1 \\ 1 & 1 & 1 \end{vmatrix}$

24. $\begin{vmatrix} -1 & 0 & 2 \\ 3 & 0 & 1 \\ 0 & 1 & 3 \end{vmatrix}$

25. $\begin{vmatrix} 1 & 2 & 3 \\ 3 & 2 & 1 \\ 1 & 1 & 1 \end{vmatrix}$

26. $\begin{vmatrix} -1 & 2 & 0 \\ 3 & -2 & 1 \\ 0 & 5 & 4 \end{vmatrix}$

Find the value of each determinant by using Method 2 and expanding across the first row.

27. $\begin{vmatrix} 0 & 1 & 2 \\ 1 & 0 & 1 \\ -1 & 2 & 0 \end{vmatrix}$

28. $\begin{vmatrix} 3 & -2 & 1 \\ 0 & -1 & 0 \\ 2 & 0 & 1 \end{vmatrix}$

29. $\begin{vmatrix} 3 & 0 & 2 \\ 0 & -1 & -1 \\ 4 & 0 & 0 \end{vmatrix}$

30. $\begin{vmatrix} 1 & 1 & 1 \\ 1 & -1 & 1 \\ 1 & 1 & -1 \end{vmatrix}$

Find the value of each of the following determinants.

31. $\begin{vmatrix} 2 & -1 & 0 \\ 1 & 0 & -2 \\ 0 & 1 & 2 \end{vmatrix}$

32. $\begin{vmatrix} 5 & 0 & -4 \\ 0 & 1 & 3 \\ -1 & 2 & -1 \end{vmatrix}$

33. $\begin{vmatrix} 1 & 3 & 7 \\ -2 & 6 & 4 \\ 3 & 7 & -1 \end{vmatrix}$

34. $\begin{vmatrix} 2 & 1 & 5 \\ 6 & -3 & 4 \\ 8 & 9 & -2 \end{vmatrix}$

35. $\begin{vmatrix} -2 & 0 & 1 \\ 0 & 3 & 2 \\ 1 & 0 & -5 \end{vmatrix}$

36. $\begin{vmatrix} -1 & 1 & 1 \\ -2 & 2 & 2 \\ 5 & 7 & -4 \end{vmatrix}$

37. $\begin{vmatrix} 1 & 2 & 3 \\ 4 & 5 & 6 \\ 7 & 8 & 9 \end{vmatrix}$

38. $\begin{vmatrix} 4 & -2 & 1 \\ 0 & -5 & 4 \\ 1 & 2 & 1 \end{vmatrix}$

39. $\begin{vmatrix} -2 & 4 & -1 \\ 0 & 3 & 1 \\ -5 & -2 & 3 \end{vmatrix}$

40. $\begin{vmatrix} -3 & 2 & 4 \\ 1 & 2 & 3 \\ -1 & 1 & 5 \end{vmatrix}$

Find all values of x for which the given statement is true.

41. $\begin{vmatrix} x+3 & 2 & -3 \\ 0 & x-2 & 5 \\ 0 & 0 & x-4 \end{vmatrix} = 0$ **42.** $\begin{vmatrix} 1-x & -7 & 5 \\ 0 & 3+2x & -8 \\ 0 & 0 & 4x-1 \end{vmatrix} = 0$

43. $\begin{vmatrix} 2 & 2 & x \\ 4 & 4 & 4 \\ 3 & x & 3 \end{vmatrix} = 0$ **44.** $\begin{vmatrix} 1 & x & x^2 \\ 2 & 2 & 2 \\ 3 & 4 & 0 \end{vmatrix} = 0$

MODELING PRACTICE

45. Slope-Intercept Form Show that the following determinant equation is another way to write the slope-intercept form of the equation of a line.

$$\begin{vmatrix} y & x \\ m & 1 \end{vmatrix} = b$$

46. Consider the following determinant equation:

$$\begin{vmatrix} x & y & 1 \\ x_1 & y_1 & 1 \\ 1 & m & 0 \end{vmatrix} = 0$$

Expand the determinant and show that the resulting equation is equivalent to the point-slope formula for a line.

© Hal Bergman/iStockPhoto

47. Amusement Park Income From 1986 to 1990, the annual income of amusement parks was linearly increasing, after which time it remained fairly constant. The annual income y, in billions of dollars, may be found for one of these years by evaluating the following determinant equation, in which x represents the number of years after January 1, 1986.

$$\begin{vmatrix} x & -1.7 \\ 2 & 0.3 \end{vmatrix} = y$$

a. Write the determinant equation in slope-intercept form.

b. Use the equation from part **a.** to find the approximate income for amusement parks in the year 1988.

48. Military Enrollment From 1981, the enrollment of women in the United States armed forces was linearly increasing until 1990, after which it declined. The approximate number of women, w, enrolled in the armed forces from 1981 to 1990 may be found by evaluating the following determinant equation, in which x represents the number of years after January 1, 1981.

$$\begin{vmatrix} 6{,}509 & -2 \\ 85{,}709 & x \end{vmatrix} = w$$

Use this equation to determine the number of women enrolled in the armed forces in 1985.

USING TECHNOLOGY EXERCISES

49. Repeat Problem 37 on your graphing calculator.

50. Repeat Problem 36 on your graphing calculator.

Find the value of the following determinants.

51. $\begin{vmatrix} 1 & 3 & 2 & -4 \\ 0 & 4 & 1 & 0 \\ -2 & 1 & 3 & 0 \\ 2 & 3 & 4 & -1 \end{vmatrix}$

52. $\begin{vmatrix} 2 & 4 & -2 & -3 \\ 1 & 2 & 0 & 2 \\ -1 & 2 & 3 & -2 \\ 3 & 2 & 1 & -3 \end{vmatrix}$

GETTING READY FOR THE NEXT SECTION

Simplify.

53. $2(3) - 4(4)$

54. $2(5) - 4(-3)$

55. $1(-1)(-3)$

56. $-(-3)(2)(1)$

57. $\dfrac{2(3) - 4(4)}{2(5) - 4(-3)}$

58. $\dfrac{1(-2) - 2(-16) + 1(9)}{6(1) - (-5) + 1(2)}$

59. $6(1) - 1(-5) + 1(2)$

60. $-2(-14) + 3(-4) - 1(-10)$

Find the value of each determinant.

61. $\begin{vmatrix} 3 & -5 \\ 2 & 4 \end{vmatrix}$

62. $\begin{vmatrix} 3 & 2 \\ 2 & 1 \end{vmatrix}$

63. $\begin{vmatrix} 6 & 1 & 1 \\ 3 & -1 & 1 \\ -4 & 2 & -3 \end{vmatrix}$

64. $\begin{vmatrix} 1 & 1 & 0 \\ 2 & 0 & 1 \\ 0 & 1 & 2 \end{vmatrix}$

EXTENDING THE CONCEPTS

A 4×4 determinant can be evaluated only by using Method 2, expansion by minors; Method 1 will not work. Below is a 4×4 determinant and its associated sign array.

$$\begin{vmatrix} 2 & 0 & 1 & -3 \\ -1 & 2 & 0 & 1 \\ -3 & 0 & 1 & 0 \\ 1 & 1 & 0 & 0 \end{vmatrix} \qquad \begin{vmatrix} + & - & + & - \\ - & + & - & + \\ + & - & + & - \\ - & + & - & + \end{vmatrix}$$

4 × 4 determinant   *4 × 4 sign array*

65. Use expansion by minors to evaluate the preceding 4×4 determinant by expanding it across row 1.

66. Evaluate the preceding determinant by expanding it down column 4.

67. Use expansion by minors down column 3 to evaluate the preceding determinant.

68. Evaluate the preceding determinant by expanding it across row 4.

Find the value of the following determinants.

69.
$$\begin{vmatrix} 1 & 3 & 2 & -4 \\ 0 & 4 & 1 & 0 \\ -2 & 1 & 3 & 0 \\ 2 & 3 & 4 & -1 \end{vmatrix}$$

70.
$$\begin{vmatrix} 2 & 4 & -2 & -3 \\ 1 & 2 & 0 & -2 \\ -1 & 2 & 3 & -2 \\ 3 & 2 & -1 & -3 \end{vmatrix}$$

Find all the values of x for which the given statement is true.

71.
$$\begin{vmatrix} x+3 & 2 & -3 \\ 0 & x-2 & 5 \\ 0 & 0 & x-4 \end{vmatrix} = 0$$

72.
$$\begin{vmatrix} 1-x & -7 & 5 \\ 0 & 3+2x & -8 \\ 0 & 0 & 4x-1 \end{vmatrix} = 0$$

73.
$$\begin{vmatrix} 2 & 2 & x \\ 4 & 4 & 4 \\ 3 & x & 3 \end{vmatrix} = 0$$

74.
$$\begin{vmatrix} 1 & x & x^2 \\ 2 & 2 & 2 \\ 3 & 4 & 0 \end{vmatrix} = 0$$

CRAMER'S RULE

© iStockphoto/Tashi-Delek

OBJECTIVE

A Use Cramer's Rule to solve a system of linear equations.

4.6

We begin this section with a look at how determinants can be used to solve a system of linear equations in two variables. The method we use is called *Cramer's rule*. We state it here as a theorem without proof.

Cramer's Rule

> **THEOREM** *Cramer's Rule*
>
> The solution set to the system
>
> $$a_1x + b_1y = c_1$$
> $$a_2x + b_2y = c_2$$
>
> is given by
>
> $$x = \frac{D_x}{D}, \qquad y = \frac{D_y}{D}, \qquad (D \neq 0)$$
>
> where
>
> $$D = \begin{vmatrix} a_1 & b_1 \\ a_2 & b_2 \end{vmatrix} \quad D_x = \begin{vmatrix} c_1 & b_1 \\ c_2 & b_2 \end{vmatrix} \quad D_y = \begin{vmatrix} a_1 & c_1 \\ a_2 & c_2 \end{vmatrix}$$

The determinant D is made up of the coefficients of x and y in the original system. The determinants D_x and D_y are found by replacing the coefficients of x or y by the constant terms in the original system. Notice also that Cramer's rule does not apply if $D = 0$.

EXAMPLE 1

Use Cramer's rule to solve

$$2x - 3y = 4$$
$$4x + 5y = 3$$

4.6 VIDEOS

381

SOLUTION We begin by calculating the determinants D, D_x, and D_y.

$$D = \begin{vmatrix} 2 & -3 \\ 4 & 5 \end{vmatrix} = 2(5) - 4(-3) = 22$$

$$D_x = \begin{vmatrix} 4 & -3 \\ 3 & 5 \end{vmatrix} = 4(5) - 3(-3) = 29$$

$$D_y = \begin{vmatrix} 2 & 4 \\ 4 & 3 \end{vmatrix} = 2(3) - 4(4) = -10$$

$$x = \frac{D_x}{D} = \frac{29}{22} \quad \text{and} \quad y = \frac{D_y}{D} = \frac{-10}{22} = -\frac{5}{11}$$

The solution set for the system is $\left\{ \left(\frac{29}{22}, -\frac{5}{11} \right) \right\}$.

Cramer's rule can also be applied to systems of linear equations in three variables.

THEOREM *Also Cramer's Rule*
• •

The solution set to the system

$$a_1 x + b_1 y + c_1 z = d_1$$
$$a_2 x + b_2 y + c_2 z = d_2$$
$$a_3 x + b_3 y + c_3 z = d_3$$

is given by

$$x = \frac{D_x}{D}, \qquad y = \frac{D_y}{D}, \qquad z = \frac{D_z}{D}, \qquad \text{and } (D \neq 0)$$

where

$$D = \begin{vmatrix} a_1 & b_1 & c_1 \\ a_2 & b_2 & c_2 \\ a_3 & b_3 & c_3 \end{vmatrix} \qquad D_x = \begin{vmatrix} d_1 & b_1 & c_1 \\ d_2 & b_2 & c_2 \\ d_3 & b_3 & c_3 \end{vmatrix}$$

$$D_y = \begin{vmatrix} a_1 & d_1 & c_1 \\ a_2 & d_2 & c_2 \\ a_3 & d_3 & c_3 \end{vmatrix} \qquad D_z = \begin{vmatrix} a_1 & b_1 & d_1 \\ a_2 & b_2 & d_2 \\ a_3 & b_3 & d_3 \end{vmatrix}$$

Again, the determinant D consists of the coefficients of x, y, and z in the original system. The determinants D_x, D_y, and D_z are found by replacing the coefficients of x, y, and z, respectively, with the constant terms from the original system. If $D = 0$, there is no unique solution to the system.

EXAMPLE 2

Use Cramer's rule to solve

$$x + y + z = 6$$
$$2x - y + z = 3$$
$$x + 2y - 3z = -4$$

SOLUTION This is the same system used in Example 1 in Section 7.2, so we can compare Cramer's rule with our previous methods of solving a system in three variables. We begin by setting up and evaluating D, D_x, D_y, and D_z. (Recall that there are a number of ways to evaluate a 3×3 determinant. Because we have four of these determinants, we can use both Methods 1 and 2 from the previous section.) We evaluate D using Method 1 from Section 7.3.

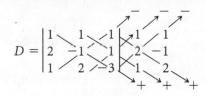

$$= 3 + 1 + 4 - (-1) - (2) - (-6) = 13$$

We evaluate D_x using Method 2 from Section 4.3 and expanding across row 1:

$$D_x = \begin{vmatrix} 6 & 1 & 1 \\ 3 & -1 & 1 \\ -4 & 2 & -3 \end{vmatrix} = 6\begin{vmatrix} -1 & 1 \\ 2 & -3 \end{vmatrix} - 1\begin{vmatrix} 3 & 1 \\ -4 & -3 \end{vmatrix} + 1\begin{vmatrix} 3 & -1 \\ -4 & 2 \end{vmatrix}$$

$$= 6(1) - (-5) + 1(2)$$

$$= 13$$

Find D_y by expanding across row 2:

$$D_y = \begin{vmatrix} 1 & 6 & 1 \\ 2 & 3 & 1 \\ 1 & -4 & -3 \end{vmatrix} = -2\begin{vmatrix} 6 & 1 \\ -4 & -3 \end{vmatrix} + 3\begin{vmatrix} 1 & 1 \\ 1 & -3 \end{vmatrix} - 1\begin{vmatrix} 1 & 6 \\ 1 & -4 \end{vmatrix}$$

$$= -2(-14) + 3(-4) - 1(-10)$$

$$= 26$$

Find D_z by expanding down column 1:

$$D_z = \begin{vmatrix} 1 & 1 & 6 \\ 2 & -1 & 3 \\ 1 & 2 & -4 \end{vmatrix} = 1\begin{vmatrix} -1 & 3 \\ 2 & -4 \end{vmatrix} - 2\begin{vmatrix} 1 & 6 \\ 2 & -4 \end{vmatrix} + 1\begin{vmatrix} 1 & 6 \\ -1 & 3 \end{vmatrix}$$

$$= 1(-2) - 2(-16) + 1(9)$$

$$= 39$$

NOTE We are solving each of these determinants by expanding about different rows or columns just to show the different ways these determinants can be evaluated.

$$x = \frac{D_x}{D} = \frac{13}{13} = 1 \qquad y = \frac{D_y}{D} = \frac{26}{13} = 2 \qquad z = \frac{D_z}{D} = \frac{39}{13} = 3$$

The solution set is $\{(1, 2, 3)\}$.

EXAMPLE 3

Use Cramer's rule to solve

$$x + y = -1$$
$$2x - z = 3$$
$$y + 2z = -1$$

SOLUTION It is helpful to rewrite the system using zeros for the coefficients of those variables not shown:

$$x + y + 0z = -1$$
$$2x + 0y - z = 3$$
$$0x + y + 2z = -1$$

The four determinants used in Cramer's rule are

$$D = \begin{vmatrix} 1 & 1 & 0 \\ 2 & 0 & -1 \\ 0 & 1 & 2 \end{vmatrix} = -3$$

$$D_x = \begin{vmatrix} -1 & 1 & 0 \\ 3 & 0 & -1 \\ -1 & 1 & 2 \end{vmatrix} = -6$$

$$D_y = \begin{vmatrix} 1 & -1 & 0 \\ 2 & 3 & -1 \\ 0 & -1 & 2 \end{vmatrix} = 9$$

$$D_z = \begin{vmatrix} 1 & 1 & -1 \\ 2 & 0 & 3 \\ 0 & 1 & -1 \end{vmatrix} = -3$$

$$x = \frac{D_x}{D} = \frac{-6}{-3} = 2 \qquad y = \frac{D_y}{D} = \frac{9}{-3} = -3 \qquad z = \frac{D_z}{D} = \frac{-3}{-3} = 1$$

The solution set is $\{(2, -3, 1)\}$.

Finally, we should mention the possible situations that can occur when the determinant D is 0, when we are using Cramer's rule.

If $D = 0$ and at least one of the other determinants, D_x or D_y (or D_z), is not 0, then the system is inconsistent. In this case, there is no solution to the system.

On the other hand, if $D = 0$ and both D_x and D_y (and D_z in a system of three equations in three variables) are 0, then the system will have infinitely many solutions that form either a line or a plane.

A Note on the History of Cramer's Rule Cramer's rule is named after the Swiss mathematician *Gabriel Cramer* (1704–1752). Cramer's rule appeared in the appendix of an algebraic work of his classifying curves, but the basic idea behind his now-famous rule was formulated earlier by Gottfried Leibniz (inventor of calculus) and Chinese mathematicians. It was actually Cramer's superior notation that helped to popularize the technique. He wrote on philosophy, law, and government, as well as mathematics; served in public office; and was an expert on cathedrals, often instructing workers about their repair and coordinating excavations to recover cathedral archives.

GETTING READY FOR CLASS

After reading through the preceding section, respond in your own words and in complete sentences.

A. When applying Cramer's rule, when will you use 2×2 determinants?

B. Why would it be impossible to use Cramer's rule if the determinant $D = 0$?

C. When applying Cramer's rule to solve a system of two linear equations in two variables, how many numbers should you obtain? How do these numbers relate to the system?

D. What will happen when you apply Cramer's rule to a system of equations made up of two parallel lines?

Problem Set 4.6

Solve each of the following systems using Cramer's rule.

1. $2x - 3y = 3$
$\quad 4x - 2y = 10$

2. $3x + y = -2$
$\quad -3x + 2y = -4$

3. $5x - 2y = 4$
$\quad -10x + 4y = 1$

4. $-4x + 3y = -11$
$\quad 5x + 4y = 6$

5. $4x - 7y = 3$
$\quad 5x + 2y = -3$

6. $3x - 4y = 7$
$\quad 6x - 2y = 5$

7. $9x - 8y = 4$
$\quad 2x + 3y = 6$

8. $4x - 7y = 10$
$\quad -3x + 2y = -9$

9. $3x + 2y = 6$
$\quad 4x - 5y = 8$

10. $-4x + 3y = 12$
$\quad 6x - 7y = 14$

11. $12x - 13y = 16$
$\quad 11x + 15y = 18$

12. $-13x + 15y = 17$
$\quad 12x - 14y = 19$

13. $x + y + z = -2$
$\quad x - y - z = 2$
$\quad 2x + 2y - z = 2$

14. $-x + y + 3z = 6$
$\quad x + y + 2z = 7$
$\quad 2x + 3y + z = 4$

15. $x + y - z = 2$
$\quad -x + y + z = 3$
$\quad x + y + z = 4$

16. $-x - y + z = 1$
$\quad x - y + z = 3$
$\quad x + y - z = 4$

17. $3x - y + 2z = 4$
$\quad 6x - 2y + 4z = 8$
$\quad x - 5y + 2z = 1$

18. $2x - 3y + z = 1$
$\quad 3x - y - z = 4$
$\quad 4x - 6y + 2z = 3$

19. $2x - y + 3z = 4$
$\quad x - 5y - 2z = 1$
$\quad -4x - 2y + z = 3$

20. $4x - y + 5z = 1$
$\quad 2x + 3y + 4z = 5$
$\quad x + y + 3z = 2$

21. $x + 2y - z = 4$
$\quad 2x + 3y + 2z = 5$
$\quad x - 3y + z = 6$

22. $3x + 2y + z = 6$
$\quad 2x + 3y - 2z = 4$
$\quad x - 2y + 3z = 8$

23. $3x - 4y + 2z = 5$
$2x - 3y + 4z = 7$
$4x + 2y - 3z = 6$

24. $5x - 3y - 4z = 3$
$4x - 5y + 3z = 5$
$3x + 4y - 5z = -4$

25. $x - 3z = 6$
$y + 2z = 8$
$x + 4y = 10$

26. $x - 5y = -6$
$y - 4z = -5$
$2x + 3z = -6$

27. $-x - 7y = 1$
$x + 3z = 11$
$2y + z = 10$

28. $x + y = 2$
$-x + 3z = 0$
$2y + z = 3$

29. $x - y = 2$
$3x + z = 11$
$y - 2z = -3$

30. $4x + 5y = -1$
$2y + 3z = -5$
$x + 2z = -1$

MODELING PRACTICE

31. Break-Even Point If a company has fixed costs of $100 per week and each item it produces costs $10 to manufacture, then the total cost y per week to produce x items is

$$y = 10x + 100$$

If the company sells each item it manufactures for $12, then the total amount of money y the company brings in for selling x items is

$$y = 12x$$

Use Cramer's rule to solve the system

$$y = 10x + 100$$

$$y = 12x$$

for x to find the number of items the company must sell per week to break even.

32. Break-Even Point Suppose a company has fixed costs of $200 per week and each item it produces costs $20 to manufacture.

a. Write an equation that gives the total cost per week y to manufacture x items.

b. If each item sells for $25, write an equation that gives the total amount of money y the company brings in for selling x items.

c. Use Cramer's rule to find the number of items the company must sell each week to break even.

GETTING READY FOR THE NEXT SECTION

Simplify.

33. $-2(-3) + 5$

34. $-2(9) - 4$

35. $-2(1) + 2$

36. $-2(2) + 7$

Solve each system.

37. $x - 3y = 9$
$\qquad y = -2$

38. $x + y - z = 2$
$\qquad y + z = 3$
$\qquad\qquad z = 2$

EXTENDING THE CONCEPTS

Solve for x and y using Cramer's rule. Your answers will contain the constants a and b.

39. $ax + by = -1$
$\quad bx + ay = 1$

40. $ax + y = b$
$\quad bx + y = a$

41. $a^2x + by = 1$
$\quad b^2x + ay = 1$

42. $ax + by = a$
$\quad bx + ay = a$

43. Name the system of equations for which Cramer's rule yields the following determinants.

$$D = \begin{vmatrix} 1 & 2 \\ 3 & 4 \end{vmatrix}, \qquad D_x = \begin{vmatrix} 1 & 2 \\ 0 & 4 \end{vmatrix}$$

44. Name the system of equations for which Cramer's rule yields the following determinants.

$$D_x = \begin{vmatrix} -2 & 3 \\ 6 & -4 \end{vmatrix}, \qquad D_y = \begin{vmatrix} 1 & -2 \\ 5 & 6 \end{vmatrix}$$

45. Name the system of equations for which Cramer's rule yields the following determinants.

$$D = \begin{vmatrix} 1 & -3 & 4 \\ 3 & 4 & 1 \\ 2 & -2 & -6 \end{vmatrix}, \quad D_z = \begin{vmatrix} 1 & -3 & 5 \\ 3 & 4 & 6 \\ 2 & -2 & -5 \end{vmatrix}$$

46. Name the system of equations for which Cramer's rule yields the following determinants.

$$D_x = \begin{vmatrix} -6 & 2 & 4 \\ 8 & 1 & -5 \\ -3 & 7 & 6 \end{vmatrix} \quad D_z = \begin{vmatrix} -3 & 2 & -6 \\ 2 & 1 & 8 \\ 4 & 7 & -3 \end{vmatrix}.$$

Use Cramer's Rule to solve the following systems.

47.
$$\begin{aligned} x + 2y \ - \ z + 3w &= 4 \\ 2x + \ y + 2z - 2w &= 5 \\ x - 3y + \ z \ - \ w &= 2 \\ -2x + \ y \ - \ z + 3w &= 3 \end{aligned}$$

48.
$$\begin{aligned} 3x - 2y \ - \ z \ - \ w &= -4 \\ 2x + \ y + 2z - 2w &= 3 \\ x - 3y + 3z + \ w &= -2 \\ -2x + 3y \ - \ z + 3w &= -1 \end{aligned}$$

49.
$$\begin{aligned} ax + \ y + \ z &= 1 \\ x + ay + \ z &= 1 \\ x + \ y + az &= 1 \end{aligned}$$

50.
$$\begin{aligned} ax + \ y + \ z &= a \\ x + ay + \ z &= a \\ x + \ y + az &= a \end{aligned}$$

© iStockphoto/laughingmango

OBJECTIVE

A Solve a system of linear equations using an augmented matrix.

Suppose you decide to sell homegrown zucchini and artichokes at two local farmer's market. For the first market, you price the zucchini at $0.75 each and the artichokes at $1.50 each, and you make a total of $21. For the second market, you increase your prices by a quarter each and end up selling the same quantity of each vegetable as you did at the first market. You make a total of $26 at the second market. The following system of equations shows how many of each vegetable you sold at the markets.

$$0.75x + 1.5y = 21$$
$$x + 1.75y = 26$$

In mathematics, a *matrix* is a rectangular array of elements considered as a whole. We can use matrices to represent systems of linear equations. To do so, we write the coefficients of the variables and the constant terms in the same position in the matrix as they occur in the system of equations. To show where the coefficients end and the constant terms begin, we use vertical lines instead of equal signs. For example, the system

$$2x + 5y = -4$$
$$x - 3y = 9$$

can be represented by the matrix

$$\begin{bmatrix} 2 & 5 & | & -4 \\ 1 & -3 & | & 9 \end{bmatrix}$$

which is called an *augmented matrix* because it includes both the coefficients of the variables and the constant terms.

To solve a system of linear equations by using the augmented matrix for that system, we need the following row operations as the tools of that solution process. The row operations tell us what we can do to an augmented matrix that may change the numbers in the matrix, but will always produce a matrix that represents a system of equations with the same solution as that of our original system.

4.7 VIDEOS

389

🅐 Row Operations

1. We can interchange any two rows of a matrix.
2. We can multiply any row by a nonzero constant.
3. We can add to any row a constant multiple of another row.

The three row operations are simply a list of the properties we use to solve systems of linear equations, translated to fit an augmented matrix. For instance, the second operation in our list is actually just another way to state the multiplication property of equality.

We solve a system of linear equations by first transforming the augmented matrix into a matrix that has 1s down the diagonal of the coefficient matrix, and 0s below it. For instance, we will solve the system

$$2x + 5y = -4$$
$$x - 3y = 9$$

by transforming the matrix

$$\begin{bmatrix} 2 & 5 & | & -4 \\ 1 & -3 & | & 9 \end{bmatrix}$$

using the row operations listed earlier to get a matrix of the form

$$\begin{bmatrix} 1 & - & | & - \\ 0 & 1 & | & - \end{bmatrix}$$

To accomplish this, we begin with the first column and try to produce a 1 in the first position and a 0 below it. Interchanging rows 1 and 2 gives us a 1 in the top position of the first column:

$$\begin{bmatrix} 1 & -3 & | & 9 \\ 2 & 5 & | & -4 \end{bmatrix} \longleftarrow \qquad \text{Interchange rows 1 and 2.}$$

Multiplying row 1 by -2 and adding the result to row 2 gives us a 0 where we want it.

$$\begin{bmatrix} 1 & -3 & | & 9 \\ 0 & 11 & | & -22 \end{bmatrix} \longleftarrow \qquad \text{Multiply row 1 by } -2 \text{ and add the result to row 2.}$$

Multiplying row 2 by $\dfrac{1}{11}$ gives us a 1 where we want it.

$$\begin{bmatrix} 1 & -3 & | & 9 \\ 0 & 1 & | & -2 \end{bmatrix} \longleftarrow \qquad \text{Multiply row 2 by } \dfrac{1}{11}.$$

Taking this last matrix and writing the system of equations it represents, we have

$$x - 3y = 9$$
$$y = -2$$

Substituting -2 for y in the top equation gives us

$$x = 3$$

The solution to our system is $(3, -2)$.

EXAMPLE 1

Solve the following system using an augmented matrix:

$$x + y - z = 2$$
$$2x + 3y - z = 7$$
$$3x - 2y + z = 9$$

SOLUTION We begin by writing the system in terms of an augmented matrix:

$$\left[\begin{array}{ccc|c} 1 & 1 & -1 & 2 \\ 2 & 3 & -1 & 7 \\ 3 & -2 & 1 & 9 \end{array}\right]$$

Next, we want to produce 0s in the second two positions of column 1:

$$\left[\begin{array}{ccc|c} 1 & 1 & -1 & 2 \\ 0 & 1 & 1 & 3 \\ 3 & -2 & 1 & 9 \end{array}\right] \longleftarrow \quad$$ Multiply row 1 by -2 and add the result to row 2.

$$\left[\begin{array}{ccc|c} 1 & 1 & -1 & 2 \\ 0 & 1 & 1 & 3 \\ 0 & -5 & 4 & 3 \end{array}\right] \longleftarrow \quad$$ Multiply row 1 by -3 and add the result to row 3.

Note that we could have done these two steps in one single step. As you become more familiar with this method of solving systems of equations, you will do just that.

$$\left[\begin{array}{ccc|c} 1 & 1 & -1 & 2 \\ 0 & 1 & 1 & 3 \\ 0 & 0 & 9 & 18 \end{array}\right] \longleftarrow \quad$$ Multiply row 2 by 5 and add the result to row 3.

$$\left[\begin{array}{ccc|c} 1 & 1 & -1 & 2 \\ 0 & 1 & 1 & 3 \\ 0 & 0 & 1 & 2 \end{array}\right] \longleftarrow \quad$$ Multiply row 3 by $\frac{1}{9}$.

Converting back to a system of equations, we have

$$x + y - z = 2$$
$$y + z = 3$$
$$z = 2$$

This system is equivalent to our first one, but much easier to solve. Substituting $z = 2$ into the second equation, we have

$$y = 1$$

Substituting $z = 2$ and $y = 1$ into the first equation, we have

$$x = 3$$

The solution to our original system is $(3, 1, 2)$. It satisfies each of our original equations. You can check this, if you want.

GETTING READY FOR CLASS

After reading through the preceding section, respond in your own words and in complete sentences.

A. What is a matrix?

B. What is an augmented matrix?

C. Name one elementary row operation for matrices.

D. What augmented matrix could be used to represent this system of equations:

$$x + y - z = 2$$
$$y + z = 3$$
$$z = 2$$

SPOTLIGHT ON SUCCESS

Breylor | student instructor

There are three ingredients in the good life:
learning, earning, and yearning.

—Christopher Morley

It can be hard to improve yourself in life, no matter what you are doing. To succeed and prosper, it is helpful to think about Christopher Morley's quote above. I love to learn new things and value what I earn from learning, but sometimes life can get busy and I think, "I know enough. I can slow down." The real key to improvement is perseverance and yearning for more. Training in martial arts is a passion of mine, and it continues to enthuse me to this day. However, obstacles often pop up that can distract me from my training. In the moment, I find it easy to think, "I'll just skip today." Then I ask myself, "Where would that thinking take me?" I strive for improvement, I yearn for it, and skipping a day of learning will not help me reach my goals and earn the success I seek. This thinking relates to all aspects of life, math included. Improvement only happens if someone has a desire to get better, or a yearning for the knowledge to come.

Problem Set 4.7

Solve the following systems of equations by using matrices.

1. $x + y = 5$
$3x - y = 3$

2. $x + y = -2$
$2x - y = -10$

3. $3x - 5y = 7$
$-x + y = -1$

4. $2x - y = 4$
$x + 3y = 9$

5. $2x - 8y = 6$
$3x - 8y = 13$

6. $3x - 6y = 3$
$-2x + 3y = -4$

7. $x + y + z = 4$
$x - y + 2z = 1$
$x - y - z = -2$

8. $x - y - 2z = -1$
$x + y + z = 6$
$x + y - z = 4$

9. $x + 2y + z = 3$
$2x - y + 2z = 6$
$3x + y - z = 5$

10. $x - 3y + 4z = -4$
$2x + y - 3z = 14$
$3x + 2y + z = 10$

11. $x + 2y = 3$
$y + z = 3$
$4x - z = 2$

12. $x + y = 2$
$3y - 2z = -8$
$x + z = 5$

13. $x + 3y = 7$
$3x - 4z = -8$
$5y - 2z = -5$

14. $x + 4y = 13$
$2x - 5z = -3$
$4y - 3z = 9$

Solve each system using matrices. Remember, multiplying a row by a nonzero constant will not change the solution to the system.

15. $\frac{1}{3}x + \frac{1}{5}y = 2$

$\frac{1}{3}x - \frac{1}{2}y = -\frac{1}{3}$

16. $\frac{1}{2}x + \frac{1}{3}y = 13$

$\frac{1}{5}x + \frac{1}{8}y = 5$

The following systems are inconsistent systems. In each case, the lines are parallel. Try solving each system using matrices and see what happens.

17. $2x - 3y = 4$
$4x - 6y = 4$

18. $10x - 15y = 5$
$-4x + 6y = -4$

The following systems contain dependent equations. In each case, the lines coincide. Try solving each system using matrices and see what happens.

19. $-6x + 4y = 8$
$-3x + 2y = 4$

20. $x + 2y = 5$
$-x - 2y = -5$

MAINTAINING YOUR SKILLS

For each of the following straight lines, identify the x-intercept, y-intercept, and slope, and sketch the graph.

21. $2x + y = 6$ **22.** $y = \dfrac{3}{2}x + 4$ **23.** $x = -2$

Find the equation for each line.

24. Give the equation of the line through $(-1, 3)$ that has slope $m = 2$.

25. Give the equation of the line through $(-3, 2)$ and $(4, -1)$.

26. Line l contains the point $(5, -3)$ and has a graph parallel to the graph of $2x - 5y = 10$. Find the equation for l.

27. Give the equation of the vertical line through $(4, -7)$.

State the domain and range for the following relations, and indicate which relations are also functions.

28. $\{(-2, 0), (-3, 0), (-2, 1)\}$ **29.** $y = x^2 - 9$

Let $f(x) = x - 2$, $g(x) = 3x + 4$ and $h(x) = 3x^2 - 2x - 8$, and find the following.

30. $f(3) + g(2)$ **31.** $h(0) + g(0)$

32. $f[g(2)]$ **33.** $g[f(2)]$

Solve the following variation problems.

34. **Direct Variation** Quantity y varies directly with the square of x. If y is 50 when x is 5, find y when x is 3.

35. **Joint Variation** Quantity z varies jointly with x and the cube of y. If z is 15 when x is 5 and y is 2, find z when x is 2 and y is 3.

The numbers in brackets refer to the section(s) in which the topic can be found.

EXAMPLES

1. The solution to the system

$$x + 2y = 4$$
$$x - y = 1$$

is the ordered pair (2, 1). It is the only ordered pair that satisfies both equations.

Systems of Linear Equations [4.1, 4.2]

A system of linear equations consists of two or more linear equations considered simultaneously. The solution set to a linear system in two variables is the set of ordered pairs that satisfy both equations. The solution set to a linear system in three variables consists of all the ordered triples that satisfy each equation in the system.

2. We can eliminate the y-variable from the system in Example 1 by multiplying both sides of the second equation by 2 and adding the result to the first equation:

$$
\begin{array}{l}
x + 2y = 4 \xrightarrow{\text{No Change}} \quad x + 2y = 4 \\
x - y = 1 \xrightarrow[\text{Multiply by 2}]{} \quad \underline{2x - 2y = 2} \\
\qquad\qquad\qquad\qquad\quad 3x \quad\;\; = 6 \\
\qquad\qquad\qquad\qquad\quad\; x \quad\;\; = 2
\end{array}
$$

Substituting $x = 2$ into either of the original two equations gives $y = 1$. The solution is (2, 1).

To Solve a System by the Addition Method [4.1]

Step 1: Look the system over to decide which variable will be easier to eliminate.

Step 2: Use the multiplication property of equality on each equation separately, if necessary, to ensure that the coefficients of the variable to be eliminated are opposites.

Step 3: Add the left and right sides of the system produced in step 2, and solve the resulting equation.

Step 4: Substitute the solution from step 3 back into any equation with both x- and y-variables, and solve.

Step 5: Check your solution in both equations if necessary.

3. We can apply the substitution method to the system in Example 1 by first solving the second equation for x to get

$$x = y + 1$$

Substituting this expression for x into the first equation we have

$$
\begin{array}{l}
(y + 1) + 2y = 4 \\
\qquad 3y + 1 = 4 \\
\qquad\quad 3y = 3 \\
\qquad\quad\; y = 1
\end{array}
$$

Using $y = 1$ in either of the original equations gives $x = 2$.

To Solve a System by the Substitution Method [4.1]

Step 1: Solve either of the equations for one of the variables (this step is not necessary if one of the equations has the correct form already).

Step 2: Substitute the results of step 1 into the other equation, and solve.

Step 3: Substitute the results of step 2 into an equation with both x- and y-variables, and solve. (The equation produced in step 1 is usually a good one to use.)

Step 4: Check your solution if necessary.

4. If the two lines are parallel, then the system will be inconsistent and the solution is $\varnothing$. If the two lines coincide, then the equations are dependent.

Inconsistent and Dependent Equations [4.1, 4.2]

A system of two linear equations that have no solutions in common is said to be an *inconsistent* system, whereas two linear equations that have all their solutions in common are said to be *dependent* equations.

5. The solution set for the system

$$x + y < 4$$
$$-x + y \leq 3$$

is shown below.

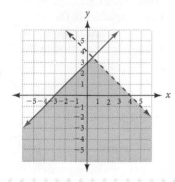

Systems of Linear Inequalities [4.4]

A system of linear inequalities is two or more linear inequalities considered at the same time. To find the solution set to the system, we graph each of the inequalities on the same coordinate system. The solution set is the region that is common to all the regions graphed.

Determinant [4.5]

6a. $\begin{vmatrix} 3 & 4 \\ -2 & 5 \end{vmatrix} = 15 - (-8) = 23$

The value of the **2 × 2** (2 by 2) **determinant**

$$\begin{vmatrix} a & c \\ b & d \end{vmatrix}$$

is given by

$$\begin{vmatrix} a & c \\ b & d \end{vmatrix} = ad - bc$$

6b. Expanding $\begin{vmatrix} 1 & 3 & -2 \\ 2 & 0 & 1 \\ 4 & -1 & 1 \end{vmatrix}$ across the first row gives us

$$1\begin{vmatrix} 0 & 1 \\ -1 & 1 \end{vmatrix} - 3\begin{vmatrix} 2 & 1 \\ 4 & 1 \end{vmatrix} - 2\begin{vmatrix} 2 & 0 \\ 4 & -1 \end{vmatrix}$$

$$= 1(1) - 3(-2) - 2(-2)$$

$$= 11$$

The value of the **3 × 3 determinant**

$$\begin{vmatrix} a_1 & b_1 & c_1 \\ a_2 & b_2 & c_2 \\ a_3 & b_3 & c_3 \end{vmatrix}$$

is given by

$$\begin{vmatrix} a_1 & b_1 & c_1 \\ a_2 & b_2 & c_2 \\ a_3 & b_3 & c_3 \end{vmatrix} = a_1b_2c_3 + a_3b_1c_2 + a_2b_3c_1 - a_3b_2c_1 - a_1b_3c_2 - a_2b_1c_3$$

Cramer's Rule [4.6]

7. For the system $x + y = 6$
$\qquad 3x - 2y = -2$

we have

$$D = \begin{vmatrix} 1 & 1 \\ 3 & -2 \end{vmatrix} = -5$$

$$D_x = \begin{vmatrix} 6 & 1 \\ -2 & -2 \end{vmatrix} = -10$$

$$x = \frac{-10}{-5} = 2$$

$$D_y = \begin{vmatrix} 1 & 6 \\ 3 & -2 \end{vmatrix} = -20$$

$$y = \frac{-20}{-5} = 4$$

The solution set to the system

$$a_1x + b_1y = c_1$$
$$a_2x + b_2y = c_2$$

is given by

$$x = \frac{D_x}{D}, \qquad y = \frac{D_y}{D}$$

where

$$D = \begin{vmatrix} a_1 & b_1 \\ a_2 & b_2 \end{vmatrix} \qquad D_x = \begin{vmatrix} c_1 & b_1 \\ c_2 & b_2 \end{vmatrix} \qquad D_y = \begin{vmatrix} a_1 & c_1 \\ a_2 & c_2 \end{vmatrix} \qquad (D \neq 0)$$

Augmented Matrices [4.7]

The system

$$2x + 5y = -4$$

$$x - 3y = 9$$

can be represented by the matrix

$$\begin{bmatrix} 2 & 5 & \bigm| & -4 \\ 1 & -3 & \bigm| & 9 \end{bmatrix}$$

which is called an ***augmented matrix*** because it includes both the coefficients of the variables and the constant terms.

Row Operations [4.7]

To solve a system of linear equations by using the augmented matrix for that system, we need the following row operations as the tools of that solution process. The row operations tell us what we can do to an augmented matrix that may change the numbers in the matrix, but will always produce a matrix that represents a system of equations with the same solution as that of our original system.

1. We can interchange any two rows of a matrix.

2. We can multiply any row by a nonzero constant.

3. We can add to any row a constant multiple of another row.

Chapter 4 Review

Solve each system by graphing both equations on the same set of axes and then reading the solution from the graph.

1. $2x - 2y = 6$
$\quad y = x - 3$

2. $x + y = 2$
$\quad x - y = 4$

Solve the following systems by the substitution method. [4.1]

3. $4x + 2y = -2$
$\quad -2x + y = 11$

4. $x - 5y = 10$
$\quad 5x + 3y = -6$

5. $2x - 2y = 2$
$\quad x - 3y = -7$

6. $-10x + 8y = -6$
$\quad y = \dfrac{5}{4}x$

7. $2x + 6y = -18$
$\quad x + 3y = -9$

8. $0.20x + 0.50y = 3.60$
$\quad y = 12 - x$

Solve the following systems by the elimination method. [4.1]

9. $6x - y = 7$
$\quad 3x + 2y = 16$

10. $\dfrac{1}{5}x + \dfrac{3}{5}y = 2$
$\quad \dfrac{1}{4}x + \dfrac{1}{2}y = 2$

11. $-3x + 2y = 5$
$\quad 9x - 6y = -14$

12. $\dfrac{1}{2}x - \dfrac{1}{3}y = \dfrac{5}{6}$
$\quad -\dfrac{2}{5}x + \dfrac{1}{2}y = -\dfrac{9}{10}$

Solve each of the following systems by eliminating the y variable. [4.1]

13. $3x - y = 7$
$\quad 10x - 5y = 25$

14. $4x - y = -6$
$\quad 3x + 2y = 7$

Solve each of the following systems by eliminating the x variable. [4.1]

15. $-x + 10y = 1$
$\quad -5x + 15y = -9$

16. $x - 3y = -4$
$\quad -3x + 9y = 12$

Solve each of the following systems by substitution or by elimination. [4.1]

17. $-5x + 3y = -15$
$\quad x = \dfrac{4}{5}y - 2$

18. $3x - 4y = 8$
$\quad y = \dfrac{3}{4}x - 2$

19. $\dfrac{1}{5}x - \dfrac{1}{10}y = -\dfrac{1}{5}$
$\quad \dfrac{2}{3}x - \dfrac{1}{2}y = -\dfrac{1}{6}$

20. $-\dfrac{1}{2}x + \dfrac{1}{6}y = -\dfrac{1}{3}$
$\quad -\dfrac{1}{8}x + \dfrac{1}{2}y = \dfrac{3}{4}$

21. Solve the system using the elimination method. [4.2]
$\quad -2x + 4y + 4z = 7$
$\quad -2x + 3y - z = 1$
$\quad -4x + 2y - 4z = -4$

Solve the following systems. [4.2]

22. $4x + y - 2z = 0$
$\quad 2x - 3y + 3z = 9$
$\quad -6x - 2y + z = 0$

23. $4x + 6y - 8z = 1$
$\quad -6x - 9y + 12z = 0$
$\quad x - 2y - 2z = 3$

24. $3x - 5y = 2$
$\quad 4x + 6z = \dfrac{1}{3}$
$\quad 5y - 7z = \dfrac{1}{6}$

25. $x + \dfrac{1}{3}y - \dfrac{1}{2}z = -\dfrac{3}{2}$
$\quad \dfrac{1}{2}x - y + \dfrac{1}{3}z = 8$
$\quad \dfrac{1}{3}x - \dfrac{1}{4}y - z = -\dfrac{5}{6}$

Solve the following systems of equations by using matrices. [4.3]

26. $3x + 5y = 3$
$\quad x = 4y + 1$

27. $5x + 2y = -25$
$\quad -3x + 2y = -1$

Solve each system using matrices. Remember, multiplying a row by a nonzero constant will not change the solution to a system. [4.3]

28. $\dfrac{1}{2}x + \dfrac{1}{3}y = 13$
$\quad \dfrac{1}{5}x + \dfrac{1}{8}y = 5$

29. $\dfrac{1}{3}x - \dfrac{5}{6}y = 16$
$\quad -\dfrac{1}{2}x + \dfrac{3}{4}y = -18$

30. Solve the system using augmented matrices. [4.3]

$5x - 2y - 2z = 3$

$3x - y - 4z = -7$

$x + 3y + 2z = 1$

31. Find the determinant. [4.4]

$$\begin{vmatrix} 4 & 7 \\ -2 & -3 \end{vmatrix}$$

Solve each of the following for x. [4.4]

32. $\begin{vmatrix} -5 & 4x \\ 1 & -x \end{vmatrix} = 27$ **33.** $\begin{vmatrix} 3x & -2 \\ x & 4 \end{vmatrix} = 21$

34. $\begin{vmatrix} x^2 & 6 \\ x & 1 \end{vmatrix} = 72$ **35.** $\begin{vmatrix} 3x & 4 \\ 2 & x \end{vmatrix} = 10x$

36. Find the determinant. [4.4]

$$\begin{vmatrix} 1 & -2 & -3 \\ 2 & 1 & 3 \\ 1 & -3 & 2 \end{vmatrix}$$

Find the value of each of the following determinants by expanding across any row or down any column. [4.4]

37. $\begin{vmatrix} 5 & 0 & -4 \\ 0 & 1 & 3 \\ -1 & 2 & -1 \end{vmatrix}$ **38.** $\begin{vmatrix} 2 & 1 & 5 \\ 6 & -3 & 4 \\ 8 & 9 & -2 \end{vmatrix}$

Solve the following systems using Cramer's rule. [4.4]

39. $-4x + 3y = -11$
     $5x + 4y = 6$

40. $-4x + 3y = 12$
     $6x - 7y = 14$

41. $-13x + 15y = 17$
     $12x - 14y = 19$

42. $2x - 3y + z = 1$
     $3x - y - z = 4$
     $4x - 6y + 2z = 3$

43. $4x - y + 5z = 1$
     $2x + 3y + 4z = 5$
     $x + y + 3z = 2$

44. $x + y = 2$
     $-x + 3z = 0$
     $2y + z = 3$

45. $4x + 5y = -1$
     $2y + 3z = -5$
     $x + 2z = -1$

46. Temperature Conversions Show that the following determinant equation is another way to write the equation $F = \frac{9}{5}C + 32$.

$$\begin{vmatrix} C & F & 1 \\ 5 & 41 & 1 \\ -10 & 14 & 1 \end{vmatrix} = 0$$

47. Military Enlistment From 1981, the enlistment of women in the United States armed forces was linearly increasing until 1990, after which it declined. The approximate number of women, w, enlisted in the armed forces from 1981 to 1990 may be found by evaluating the following determinant equation, in which x represents the number of years after January 1, 1981.

$$\begin{vmatrix} 6{,}509 & -2 \\ 85{,}709 & x \end{vmatrix} = w$$

Use this equation to determine the number of women enlisted in the armed forces in 1985.

Solve each word problem. [4.5]

48. Number Problem A number is 4 more than three times another. Their sum is 16. Find the two numbers.

49. Number Problem The difference of two positive numbers is 17. The larger number is one more than twice the smaller number. Find the numbers.

50. Interest Rates Ralph owes twice as much money at a 10% annual interest rate as he does at 12%. If his loan charges a total of $560 in 1 year, how much does he owe at each rate?

51. Ticket Cost There were 905 tickets sold for a basketball game for a total of $9,050. If adult tickets cost $12.00 and children's tickets cost $7.00, how many of each kind were sold?

52. Mixture How many ounces of 30% hydrochloric acid solution and 80% hydrochloric acid solution must be mixed to get 10 ounces of 50% hydrochloric acid solution?

53. Speed of a Boat A boat can travel 80 miles downstream in 8 hours. The same boat can travel 20 miles upstream in 5 hours. What is the speed of the boat in still water, and what is the speed of the current?

54. Coin Problem A collection of nickels, dimes, and quarters consists of 22 coins with a total value of $1.75. If the number of nickels is three times the number of dimes, how many of each coin are contained in the collection?

55. Price and Demand A company manufactures and sells bracelets. They have found from past experience that they can sell 300 bracelets each week if the price per bracelet is $2.00, but only 150 bracelets are sold if the price is $2.50 per bracelet. If the relationship between the number of bracelets sold x and the price per bracelet p is a linear one, find a formula that gives x in terms of p. Then use the formula to find the number of bracelets they will sell at $3.00 each.

56. The length of a rectangle is three inches less than twice the width. The perimeter is 36 inches. Find the length and width.

Find a system of inequalities that describes the shaded region. [4.6]

65.

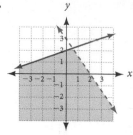

66.

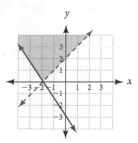

Find the equilibrium price and quantity for the following supply and demand equations.

67. $d(p) = 12{,}500 - 17p$
$s(p) = 2{,}500 + 3p$

68. $d(p) = 300{,}000 - 47p$
$s(p) = 3p$

Graph the solution set for each system of linear inequalities. [4.6]

57. $x - 2y \geq 4$
$2x + 3y < 6$

58. $y < -\dfrac{3}{4}x + 5$
$x \geq 0$
$y \geq 0$

59. $y < 2x + 4$
$y \geq 2x - 3$

60. $-4 \leq x \leq -2$
$1 \leq y \leq 3$

61. $x - y \leq 3$
$-x - y \leq 3$
$y \leq -1$

62. $x + y < 5$
$y > x$
$x \geq 0$

63. $x + 2y \leq 10$
$3x + 2y \leq 12$
$x \geq 0$
$y \geq 0$

64. $3x + y > 6$
$2x + 3y > 12$
$x > 0$
$y > 0$

Simplify each of the following.

1. $3^2 - 7^2$

2. $(3 - 7)^2$

3. $16 - 9 \div 3 - 6 \cdot 2$

4. $3(8 - 11)^3 + 2(2 - 6)^2$

5. $6\left(\dfrac{2}{3}\right) - 18\left(\dfrac{2}{3}\right)^2$

6. $18\left(\dfrac{5}{6}\right) - 72\left(\dfrac{5}{6}\right)^2$

Find the value of each expression when x is -3.

7. $x^2 + 10x - 25$

8. $(x + 5)(x - 5)$

Simplify each of the following by combining similar terms. [4.1]

9. $(8y^3 - 3y^2 + 7y + 2) - (-4y^3 + 6y^2 - 5y - 8)$

Multiply. [4.2]

10. $(2x + 3)(2x^2 - 4x - 1)$ **11.** $(2x - 5)(2x + 5)$

12. $\left(3x - \dfrac{1}{5}\right)\left(5x + \dfrac{1}{3}\right)$

Factor the following expressions. [4.3, 4.4, 4.5, 4.6]

13. $6x^4 + 14x^2 - 40$ **14.** $6ab + 3ay - 8bx - 4xy$

15. $y^3 - \dfrac{1}{64}$

Solve each equation [4.7]

16. $(x - 3)(x + 4) = 8$ **17.** $x^3 + 3x^2 - 16x - 48 = 0$

Simplify each of the following expressions.

18. $3(2x - 4) + 5(3x - 1)$ **19.** $7 - 2[4x - 5(x + 3)]$

20. $-6\left(\dfrac{2}{3}x + \dfrac{5}{2}y\right)$ **21.** $8\left(\dfrac{1}{4}x - \dfrac{3}{8}y\right)$

Solve.

22. $-3y + 6 = 7y - 4$ **23.** $-30 + 6(7x + 5) = 0$

24. $|7x + 2| + 4 = 3$ **25.** $\left|\dfrac{3}{2}x - 1\right| + 2 = 1$

Solve the following equations for y.

26. $y - 4 = -2(x + 1)$ **27.** $y + 2 = \dfrac{3}{4}(x + 8)$

Solve the following equations for x.

28. $ax + 1 = bx - 3$ **29.** $ax - 3 = cx - 4$

Solve each inequality and write your answers using interval notation.

30. $-4t \le 12$ **31.** $|3x - 1| \ge 5$

Let $A = \{0, 5, 10, 15\}$ and $B = \{1, 3, 5, 7\}$. Find the following.

32. $A \cup B$ **33.** $A \cap B$

Solve each of the following systems.

34. $-2x + 7y = -27$
$\quad\ -x + 3y = -11$

35. $6x + 3y = 13$
$\quad\ 10x + 5y = 7$

36. $3x + y = 1$
$\quad\ y = -3x + 1$

37. $5x + 2y = -z$
$\quad\ 3y - 4z = -22$
$\quad\ 7x + 2z = 8$

38. Graph on a rectangular coordinate system.
$$3x + y < -2$$

39. Find the determinant: $\begin{vmatrix} -2 & 0 & 1 \\ 1 & 2 & 1 \\ 3 & -1 & 0 \end{vmatrix}$

40. Graph the solution set to the system.
$$x + y < -3$$
$$y > -5$$

41. Graph the solution set to the system.
$$y < \frac{1}{4}x + 3$$
$$y \geq \frac{1}{4}x - 2$$

42. Find the slope of the line through $\left(\frac{3}{2}, -\frac{1}{5}\right)$ and $\left(\frac{3}{5}, \frac{7}{10}\right)$.

43. Find the slope and y-intercept of $2x - 7y = 14$.

44. Give the equation of a line with slope $-\frac{1}{2}$ and y-intercept $= 13$.

45. Find the equation of the line that is perpendicular to $4x - 3y = 17$ and contains the point $(0, -3)$.

46. Find the equation of the line through $(5, -6)$ and $(9, 2)$.

Let $f(x) = x - 1$ and $g(x) = 3 - x^2$. Find the following.

47. $f(-3)$ **48.** $g(-2) + 5$

49. $(f - g)(x)$ **50.** $(g \circ f)(x)$

Use the graph below to work Problems 51–54.

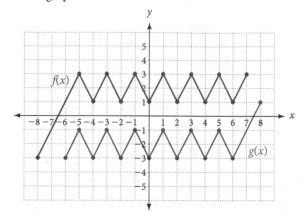

51. Find $f(2)$.   **52.** Find $(f + g)(-3)$.

53. Find $(g \circ f)(-8)$.   **54.** Find x if $g(x) = 1$.

Solve the following systems by the addition method. [4.1]

1. $2x - 5y = -8$

$3x + y = 5$

2. $4x - 7y = -2$

$-5x + 6y = -3$

3. $\dfrac{1}{3}x - \dfrac{1}{6}y = 3$

$-\dfrac{1}{5}x + \dfrac{1}{4}y = 0$

Solve the following systems by the substitution method. [4.1]

4. $2x - 5y = 14$
$y = 3x + 8$

5. $6x - 3y = 0$
$x + 2y = 5$

Solve each system. [4.2]

6. $2x - y + z = 9$
$x + y - 3z = -2$
$3x + y - z = 6$

7. $2x + 4y = 3$
$-4x - 8y = -6$

8. $2x - y + 3z = 2$
$x - 4y - z = 6$
$3x - 2y + z = 4$

Solve each word problem. [4.3]

9. Number Problem A number is 1 less than twice another. Their sum is 14. Find the two numbers.

10. Investing John invests twice as much money at 6% as he does at 5%. If his investments earn a total of $680 in one year, how much does he have invested at each rate?

11. Ticket Cost There were 750 tickets sold for a basketball game for a total of $1,090. If adult tickets cost $2.00 and children's tickets cost $1.00, how many of each kind were sold?

12. Mixture Problem How much 30% alcohol solution and 70% alcohol solution must be mixed to get 16 gallons of 60% solution?

13. Speed of a Boat A boat can travel 20 miles downstream in 2 hours. The same boat can travel 18 miles upstream in 3 hours. What is the speed of the boat in still water, and what is the speed of the current?

14. Coin Problem A collection of nickels, dimes, and quarters consists of 15 coins with a total value of $1.10. If the number of nickels is 1 less than 4 times the number of dimes, how many of each coin are contained in the collection?

Graph the solution set for each system of linear inequalities. [4.4]

15. $x + 4y \le 4$

$-3x + 2y > -12$

16. $y < -\dfrac{1}{2}x + 4$

$x \ge 0$

$y \ge 0$

17. Solve for x. [4.5]

$$\begin{vmatrix} 2 & 3x \\ -1 & 2x \end{vmatrix} = 4$$

18. Evaluate. [4.5]

$$\begin{vmatrix} 1 & 0 & -3 \\ 2 & 1 & 0 \\ 0 & 5 & 4 \end{vmatrix}$$

19. Use Cramer's rule to solve. [4.6]

$$2x + 4y = 3$$
$$-4x - 8y = -6$$

20. Set up the augmented matrix that corresponds to this system. [4.7]

$$2x - y + 3z = 1$$
$$x - 4y - z = 6$$
$$3x - 2y + z = 4$$

21. Write the augmented matrix that corresponds to this system. Then use that matrix to solve the system. [4.7]

$$x + y = 4$$
$$2x + 5y = 2$$

5

Rational Expressions and Rational Functions

© iStockphoto/BrianAJackson

If you have ever put yourself on a weight loss diet, you know that you lose more weight at the beginning of the diet than you do later. If we let $W(x)$ represent a person's weight after x weeks on the diet, then the rational function

$$W(x) = \frac{80(2x + 15)}{x + 6}$$

is a mathematical model of the person's weekly progress on a diet intended to take them from 200 pounds to about 160 pounds. Rational functions are good models for quantities that fall off rapidly to begin with, and then level off over time. The table shows some values for this function, along with the graph of this function.

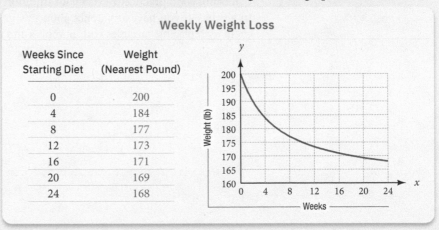

Weekly Weight Loss

Weeks Since Starting Diet	Weight (Nearest Pound)
0	200
4	184
8	177
12	173
16	171
20	169
24	168

As you progress through this chapter, you will acquire an intuitive feel for these types of functions, and as a result, you will see why they are good models for situations such as dieting.

405

STUDY SKILLS

The study skills for this chapter cover the way you approach new situations in mathematics. The first study skill is a point of view you hold about your natural instincts for what does and doesn't work in mathematics. The second study skill gives you a way of testing your instincts.

1. **Don't Let Your Intuition Fool You**   As you become more experienced and more successful in mathematics you will be able to trust your mathematical intuition. For now, though, it can get in the way of your success. For example, if you ask some students to "subtract 3 from −5" they will answer −2 or 2. Both answers are incorrect, even though they may seem intuitively true. Likewise, some students will expand $(a + b)^2$ and arrive at $a^2 + b^2$, which is incorrect. In both cases, intuition leads directly to the wrong answer.

2. **Test Properties of Which You Are Unsure**   From time to time, you will be in a situation where you would like to apply a property or rule, but you are not sure it is true. You can always test a property or statement by substituting numbers for variables. For instance, I always have students who rewrite $(x + 3)^2$ as $x^2 + 9$, thinking that the two expressions are equivalent. The fact that the two expressions are not equivalent becomes obvious when we substitute 10 for x in each one.

    When $x = 10$, the expression $(x + 3)^2$ is $(10 + 3)^2 = 13^2 = 169$

    When $x = 10$, the expression $x^2 + 9 = 10^2 + 9 = 100 + 9 = 109$

    When you test the equivalence of expressions by substituting numbers for the variable, make it easy on yourself by choosing numbers that are easy to work with, such as 10. Don't try to verify the equivalence of expressions by substituting 0, 1, or 2 for the variable, as using these numbers will occasionally give you false results.

It is not good practice to trust your intuition or instincts in every new situation in algebra. If you have any doubt about the generalizations you are making, test them by replacing variables with numbers and simplifying.

OBJECTIVES

A Reduce a rational expression to lowest terms.

B Find a rational function given the value of variables.

C Simplify a difference quotient.

© iStockphoto/faslooff

The Norwegian fjords are a popular location for wingsuit flying. Many of the thrill-seekers opt to leap, flip, or be thrown off the edge of a sheer cliff to begin their flight. A wingsuit flyer wears a special jumpsuit that adds surface area to his body to create lift during a flight. Carbon fiber is an extremely strong material in relation to its weight. Because of this quality, wingsuit flyers have utilized carbon fiber in a cutting-edge piece of equipment called a wingpack. The wing of the wingpack is made of rigid carbon fiber. Strapped to his back, the wingpack allows the flyer to travel longer horizontal distances than with the wingsuit alone. A rating scale called the glide ratio is used to categorize the equipment's aptitude for increasing the flyer's horizontal distance. The glide ratio compares the amount of lift L from a wingpack divided by the amount of drag D it causes while moving through the air. Therefore, this ratio can be written as $\frac{L}{D}$, and is an example of a rational expression. The wingpack has a glide ratio value of 6, whereas the wingsuit is only 2.5.

We will begin this section with the definition of a rational expression. We will then state the two basic properties associated with rational expressions and go on to apply one of the properties to reduce rational expressions to lowest terms.

Recall that a *rational number* is any number that can be expressed as the ratio of two integers:

$$\text{Rational numbers} = \left\{ \frac{a}{b} \mid a \text{ and } b \text{ are integers}, b \neq 0 \right\}$$

A *rational expression* is defined similarly as any expression that can be written as the ratio of two polynomials:

$$\text{Rational expressions} = \left\{ \frac{P}{Q} \mid P \text{ and } Q \text{ are polynominals}, Q \neq 0 \right\}$$

Some examples of rational expressions are

$$\frac{2x - 3}{x + 5} \qquad \frac{x^2 - 5x - 6}{x^2 - 1} \qquad \frac{a - b}{b - a}$$

Multiplying the numerator and denominator by the same nonzero expression may change the form of the rational expression, but it will always produce an expression equivalent to the original one. The same is true when dividing the numerator and denominator by the same nonzero quantity.

> **PROPERTY** *Properties of Rational Expressions*
>
> If P, Q, and K are polynomials with $Q \neq 0$ and $K \neq 0$, then
>
> $$\frac{P}{Q} = \frac{PK}{QK} \qquad \text{and} \qquad \frac{P}{Q} = \frac{\frac{P}{K}}{\frac{Q}{K}}$$

Ⓐ Reducing to Lowest Terms

The fraction $\frac{6}{8}$ can be written in lowest terms as $\frac{3}{4}$. The process is shown here:

$$\frac{6}{8} = \frac{3 \cdot 2}{4 \cdot 2} = \frac{3}{4}$$

Reducing $\frac{6}{8}$ to $\frac{3}{4}$ involves dividing the numerator and denominator by 2, the factor they have in common. Before dividing out the common factor 2, we must notice that the common factor *is* 2. (This may not be obvious because we are very familiar with the numbers 6 and 8 and therefore do not have to put much thought into finding what number divides both of them.)

We reduce rational expressions to lowest terms by first factoring the numerator and denominator and then dividing both numerator and denominator by any factors they have in common.

EXAMPLE 1

Reduce $\dfrac{x^2 - 9}{x - 3}$ to lowest terms.

SOLUTION   Factoring, we have

$$\frac{x^2 - 9}{x - 3} = \frac{(x + 3)(x - 3)}{x - 3}$$

The numerator and denominator have the factor $x - 3$ in common. Dividing the numerator and denominator by $x - 3$, we have

$$\frac{(x + 3)(x - 3)}{x - 3} = \frac{x + 3}{1} = x + 3$$

NOTE The lines drawn through the $(x - 3)$ in the numerator and denominator indicate that we have divided through by $(x - 3)$. As the problems become more involved, these lines will help keep track of which factors have been divided out and which have not.

For the problem in Example 1, there is an implied restriction on the variable x: It cannot be 3. If x were 3, the expression $\frac{(x^2 - 9)}{(x - 3)}$ would become $\frac{0}{0}$, an expression that we cannot associate with a real number. For all problems involving rational expressions, we restrict the variable to only those values that result in a nonzero denominator. When we state the relationship

$$\frac{x^2 - 9}{x - 3} = x + 3$$

we are assuming that it is true for all values of x except $x = 3$.

Here are some other examples of reducing rational expressions to lowest terms.

EXAMPLE 2

Reduce $\dfrac{y^2 - 5y - 6}{y^2 - 1}$ to lowest terms.

SOLUTION

$$\frac{y^2 - 5y - 6}{y^2 - 1} = \frac{(y - 6)(y + 1)}{(y - 1)(y + 1)}$$

$$= \frac{y - 6}{y - 1}$$

EXAMPLE 3

Reduce $\dfrac{2a^3 - 16}{4a^2 - 12a + 8}$ to lowest terms.

SOLUTION

$$\frac{2a^3 - 16}{4a^2 - 12a + 8} = \frac{2(a^3 - 8)}{4(a^2 - 3a + 2)}$$

$$= \frac{2(a - 2)(a^2 + 2a + 4)}{4(a - 2)(a - 1)}$$

$$= \frac{a^2 + 2a + 4}{2(a - 1)}$$

EXAMPLE 4

Reduce $\dfrac{x^2 - 3x + ax - 3a}{x^2 - ax - 3x + 3a}$ to lowest terms.

SOLUTION

$$\frac{x^2 - 3x + ax - 3a}{x^2 - ax - 3x + 3a} = \frac{x(x - 3) + a(x - 3)}{x(x - a) - 3(x - a)}$$

$$= \frac{(x - 3)(x + a)}{(x - a)(x - 3)}$$

$$= \frac{x + a}{x - a}$$

The answer to Example 4 cannot be reduced further. It is a fairly common mistake to attempt to divide out an x or an a in this last expression. Remember, we can divide out only the factors common to the numerator and denominator of a rational expression.

The next example involves what we call a trick. The trick is to reverse the order of the terms in a difference by factoring -1 from each term in either the numerator or the denominator. The next examples illustrate how this is done.

EXAMPLE 5

Reduce to lowest terms: $\dfrac{a - b}{b - a}$

SOLUTION   The relationship between $a - b$ and $b - a$ is that they are opposites. We can show this fact by factoring -1 from each term in the numerator:

$$\frac{a - b}{b - a} = \frac{-1(-a + b)}{b - a} \qquad \text{Factor } -1 \text{ from each term in the numerator.}$$

$$= \frac{-1(b - a)}{b - a} \qquad \text{Reverse the order of the terms in the numerator.}$$

$$= -1 \qquad \text{Divide out common factor } b - a.$$

EXAMPLE 6

Reduce to lowest terms: $\dfrac{x^2 - 25}{5 - x}$

SOLUTION   We begin by factoring the numerator:

$$\frac{x^2 - 25}{5 - x} = \frac{(x - 5)(x + 5)}{5 - x}$$

The factors $x - 5$ and $5 - x$ are similar but are not exactly the same. We can reverse the order of either by factoring -1 from it.

That is: $5 - x = -1(-5 + x) = -1(x - 5)$.

$$\frac{(x - 5)(x + 5)}{5 - x} = \frac{(x - 5)(x + 5)}{-1(x - 5)}$$

$$= \frac{x + 5}{-1}$$

$$= -(x + 5)$$

B Rational Functions

We can extend our knowledge of rational expressions to rational functions with the following definition:

> **DEFINITION** *rational function*
> •
>
> A *rational function* is any function that can be written in the form
>
> $$f(x) = \frac{P(x)}{Q(x)}$$
>
> where $P(x)$ and $Q(x)$ are polynomials and $Q(x) \neq 0$.

EXAMPLE 7

For the rational function $f(x) = \dfrac{x - 4}{x - 2}$, find $f(0), f(-4), f(4), f(-2)$, and $f(2)$.

SOLUTION To find these function values, we substitute the given value of x into the rational expression, and then simplify if possible.

$$f(0) = \frac{0 - 4}{0 - 2} = \frac{-4}{-2} = 2$$

$$f(-4) = \frac{-4 - 4}{-4 - 2} = \frac{-8}{-6} = \frac{4}{3}$$

$$f(4) = \frac{4 - 4}{4 - 2} = \frac{0}{2} = 0$$

$$f(-2) = \frac{-2 - 4}{-2 - 2} = \frac{-6}{-4} = \frac{3}{2}$$

$$f(2) = \frac{2 - 4}{2 - 2} = \frac{-2}{0} \qquad \text{Undefined}$$

Because the rational function in Example 7 is not defined when x is 2, the domain of that function does not include 2. We have more to say about the domain of a rational function next.

The Domain of a Rational Function

If the domain of a rational function is not specified, it is assumed to be all real numbers for which the function is defined. That is, the domain of the rational function

$$f(x) = \frac{P(x)}{Q(x)}$$

is all x for which $Q(x)$ is nonzero.

EXAMPLE 8

Find the domain for each function.

a. $f(x) = \dfrac{x - 4}{x - 2}$ **b.** $g(x) = \dfrac{x^2 + 5}{x + 1}$ **c.** $h(x) = \dfrac{x}{x^2 - 9}$

SOLUTION

a. The domain for $f(x) = \dfrac{x - 4}{x - 2}$ is $\{x \mid x \neq 2\}$.

b. The domain for $g(x) = \dfrac{x^2 + 5}{x + 1}$ is $\{x \mid x \neq -1\}$.

c. The domain for $h(x) = \dfrac{x}{x^2 - 9}$ is $\{x \mid x \neq -3, x \neq 3\}$.

Notice that, for these functions, $f(2), g(-1), h(-3)$, and $h(3)$ are all undefined, and that is why the domains are written as shown.

C Difference Quotients

The diagram in Figure 1 is an important diagram from calculus. Although it may look complicated, the point of it is simple: The slope of the line passing through the points P and Q is given by the formula

$$\text{Slope of line through } PQ = m = \frac{f(x) - f(a)}{x - a}$$

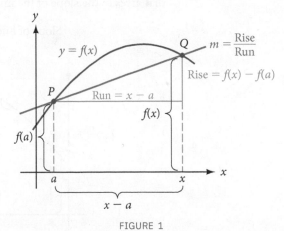

FIGURE 1

The expression $\frac{f(x) - f(a)}{x - a}$ is called a *difference quotient*. When $f(x)$ is a polynomial, it will be a rational expression.

EXAMPLE 9

If $f(x) = 3x - 5$, find $\dfrac{f(x) - f(a)}{x - a}$.

SOLUTION

$$\begin{aligned}
\frac{f(x) - f(a)}{x - a} &= \frac{(3x - 5) - (3a - 5)}{x - a} \\
&= \frac{3x - 3a}{x - a} \\
&= \frac{3(x - a)}{x - a} \\
&= 3
\end{aligned}$$

EXAMPLE 10

If $f(x) = x^2 - 4$, find $\dfrac{f(x) - f(a)}{x - a}$ and simplify.

SOLUTION   Because $f(x) = x^2 - 4$ and $f(a) = a^2 - 4$, we have

$$\begin{aligned}
\frac{f(x) - f(a)}{x - a} &= \frac{(x^2 - 4) - (a^2 - 4)}{x - a} \\
&= \frac{x^2 - 4 - a^2 + 4}{x - a} \\
&= \frac{x^2 - a^2}{x - a} \\
&= \frac{(x + a)(x - a)}{x - a} \qquad \text{Factor and divide out common factor.}\\
&= x + a
\end{aligned}$$

The diagram in Figure 2 is similar to the one in Figure 1. The main difference is in how we label the points. From Figure 2, we can see another difference quotient that gives us the slope of the line through the points P and Q.

$$\text{Slope of line through } PQ = m = \frac{f(x + h) - f(x)}{h}$$

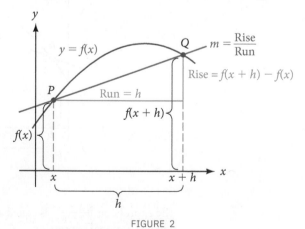

FIGURE 2

Examples 11 and 12 use the same functions used in Examples 9 and 10, but this time the new difference quotient is used.

EXAMPLE 11

If $f(x) = 3x - 5$, find $\dfrac{f(x + h) - f(x)}{h}$.

SOLUTION   The expression $f(x + h)$ is given by

$$f(x + h) = 3(x + h) - 5$$
$$= 3x + 3h - 5$$

Using this result gives us

$$\frac{f(x + h) - f(x)}{h} = \frac{(3x + 3h - 5) - (3x - 5)}{h}$$
$$= \frac{3h}{h}$$
$$= 3$$

EXAMPLE 12

If $f(x) = x^2 - 4$, find $\dfrac{f(x + h) - f(x)}{h}$.

SOLUTION   The expression $f(x + h)$ is given by

$$f(x + h) = (x + h)^2 - 4$$
$$= x^2 + 2xh + h^2 - 4$$

Using this result gives us

$$\frac{f(x + h) - f(x)}{h} = \frac{(x^2 + 2xh + h^2 - 4) - (x^2 - 4)}{h}$$
$$= \frac{2xh + h^2}{h}$$
$$= \frac{h(2x + h)}{h}$$
$$= 2x + h$$

GETTING READY FOR CLASS

After reading through the preceding section, respond in your own words and in complete sentences.

A. What is a rational expression?

B. Explain how to determine if a rational expression is in "lowest terms."

C. When is a rational expression undefined?

D. Explain the process we use to reduce a rational expression or a fraction to lowest terms.

Reduce to lowest terms.

1. $\dfrac{8}{12}$

2. $\dfrac{9}{12}$

3. $\dfrac{8x^3}{12x^2}$

4. $\dfrac{9x^4}{12x^2}$

5. $\dfrac{6x^2y}{3xy^2}$

6. $\dfrac{8x^3y}{4xy^3}$

7. $\dfrac{39a^3b^2}{26a^2b}$

8. $\dfrac{45a^2b^4}{18a^2b}$

9. If $g(x) = \dfrac{x+3}{x-1}$, find $g(0)$, $g(-3)$, $g(3)$, $g(-1)$, and $g(1)$, if possible.

10. If $g(x) = \dfrac{x-2}{x-1}$, find $g(0)$, $g(-2)$, $g(2)$, $g(-1)$, and $g(1)$, if possible.

11. If $h(t) = \dfrac{t-3}{t+1}$, find $h(0)$, $h(-3)$, $h(3)$, $h(-1)$, and $h(1)$, if possible.

12. If $h(t) = \dfrac{t-2}{t+1}$, find $h(0)$, $h(-2)$, $h(2)$, $h(-1)$, and $h(1)$, if possible.

State the domain for each rational function.

13. $f(x) = \dfrac{x-3}{x-1}$

14. $f(x) = \dfrac{x+4}{x-2}$

15. $g(x) = \dfrac{x^2-4}{x-2}$

16. $g(x) = \dfrac{x^2-9}{x-3}$

17. $h(t) = \dfrac{t-4}{t^2-16}$

18. $h(t) = \dfrac{t-5}{t^2-25}$

Reduce each rational expression to lowest terms.

19. $\dfrac{x^2-16}{6x+24}$

20. $\dfrac{12x-9y}{3x^2+3xy}$

21. $\dfrac{a^4-81}{a-3}$

22. $\dfrac{a^2-4a-12}{a^2+8a+12}$

23. $\dfrac{20y^2-45}{10y^2-5y-15}$

24. $\dfrac{20x^2-93x+34}{4x^2-9x-34}$

25. $\dfrac{12y-2xy-2x^2y}{6y-4xy-2x^2y}$

26. $\dfrac{250a+100ax+10ax^2}{50a-2ax^2}$

27. $\dfrac{(x-3)^2(x+2)}{(x+2)^2(x-3)}$

28. $\dfrac{(x-4)^3(x+3)}{(x+3)^2(x-4)}$

29. $\dfrac{x^3+1}{x^2-1}$

30. $\dfrac{x^3-1}{x^2-1}$

31. $\dfrac{4am-4an}{3n-3m}$

32. $\dfrac{ad-ad^2}{d-1}$

33. $\dfrac{ab-a+b-1}{ab+a+b+1}$

34. $\dfrac{6cd-4c-9d+6}{6d^2-13d+6}$

35. $\dfrac{21x^2-23x+6}{21x^2+x-10}$

36. $\dfrac{36x^2-11x-12}{20x^2-39x+18}$

37. $\dfrac{8x^2-6x-9}{8x^2-18x+9}$

38. $\dfrac{42x^2+23x-10}{14x^2+45x-14}$

39. $\dfrac{4x^2+29x+45}{8x^2-10x-63}$

40. $\dfrac{30x^2-61x+30}{60x^2+22x-60}$

41. $\dfrac{a^3+b^3}{a^2-b^2}$

42. $\dfrac{a^2-b^2}{a^3-b^3}$

43. $\dfrac{8x^4-8x}{4x^4+4x^3+4x^2}$

44. $\dfrac{6x^5-48x^3}{12x^3+24x^2+48x}$

45. $\dfrac{ax+2x+3a+6}{ay+2y-4a-8}$

46. $\dfrac{x^2-3ax-2x+6a}{x^2-3ax+2x-6a}$

47. $\dfrac{x^3+3x^2-4x-12}{x^2+x-6}$

48. $\dfrac{x^3+5x^2-4x-20}{x^2+7x+10}$

49. $\dfrac{x^3-8}{x^2-4}$

50. $\dfrac{y^2-9}{y^3+27}$

51. $\dfrac{8x^3-27}{4x^2-9}$

52. $\dfrac{25y^2-4}{125y^3+8}$

Refer to Examples 5 and 6 in this section, and reduce the following to lowest terms.

53. $\dfrac{x-4}{4-x}$

54. $\dfrac{6-x}{x-6}$

55. $\dfrac{y^2-36}{6-y}$

56. $\dfrac{1-y}{y^2-1}$

57. $\dfrac{1-9a^2}{9a^2-6a+1}$

58. $\dfrac{1-a^2}{a^2-2a+1}$

Simplify each expression.

59. $\dfrac{(3x-5)-(3a-5)}{x-a}$

60. $\dfrac{(2x+3)-(2a+3)}{x-a}$

61. $\dfrac{(x^2-4)-(a^2-4)}{x-a}$

62. $\dfrac{(x^2-1)-(a^2-1)}{x-a}$

For the functions below, evaluate

a. $\dfrac{f(x)-f(a)}{x-a}$

b. $\dfrac{f(x+h)-f(x)}{h}$

63. $f(x)=4x$

64. $f(x)=-3x$

65. $f(x)=5x+3$

66. $f(x)=6x-5$

67. $f(x)=x^2$

68. $f(x)=3x^2$

69. $f(x)=x^2+1$

70. $f(x)=x^2-3$

71. $f(x)=x^2-3x+4$

72. $f(x)=x^2+4x-7$

The graphs of two rational functions are given in Figures 3 and 4. Use the graphs to find the following.

73. **a.** $f(2)$ **b.** $f(-1)$ **c.** $f(0)$ **d.** $g(3)$

74. **a.** $g(6)$ **b.** $g(-1)$ **c.** $f(g(6))$ **d.** $g(f(-2))$

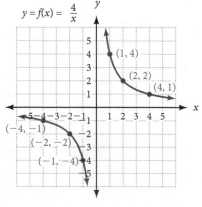

FIGURE 3

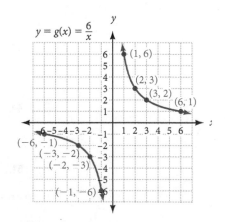

FIGURE 4

APPLYING THE CONCEPTS

75. Diet The following rational function is the one we mentioned in the introduction to this chapter. The quantity $W(x)$ is the weight (in pounds) of the person after x weeks of dieting. Use the function to fill in the table. Then compare your results with the graph in the chapter introduction.

$$W(x) = \frac{80(2x + 15)}{x + 6}$$

Weeks	Weight (lb)
x	$W(x)$
0	
1	
4	
12	
24	

76. Drag Racing The following rational function gives the speed $V(x)$, in miles per hour, of a dragster at each second x during a quarter-mile race.

Use the function to fill in the table.

$$V(x) = \frac{340x}{x + 3}$$

Time (sec)	Speed (mi/hr)
x	$V(x)$
0	
1	
2	
3	
4	
5	
6	

77. Test Scores The rational function $G(x)$ gives the grade, as a percent, when 100% is scored on each test x following a 70% earned on the first test. If 100% is scored on the next two tests, find the grade.

$$G(x) = \frac{70 + 100x}{x + 1}$$

78. Test Scores The rational function $G(x)$ gives the grade, as a percent, when 100% is scored on each test x following a 75% earned on the first test. If 100% is scored on the next test, find the grade.

$$G(x) = \frac{75 + 100x}{x + 1}$$

79. Chemistry A solution is created from 15 grams of sugar mixed into 200 milliliters of water. Another mixture of 8 grams per milliliter is added to the solution. Find the concentration of the solution if 50 milliliters of the mixture is added. Use the function

$$C(x) = \frac{15 + 8x}{200 + x}$$

where x is the amount of mixture added in milliliters.

80. Chemistry A solution is created from 15 grams of sugar mixed into 200 milliliters of water. Another mixture of 8 grams per milliliter is added to the solution. Find the concentration of the solution if 50 milliliters of the mixture is added. Use the function

$$C(x) = \frac{10 + 6x}{120 + x}$$

where x is the amount of mixture added in milliliters.

GETTING READY FOR THE NEXT SECTION ━━━

Multiply or divide, as indicated.

81. $\frac{6}{7} \cdot \frac{14}{18}$

82. $\frac{6}{8} \div \frac{3}{5}$

83. $5y^2 \cdot 4x^2$

84. $4y^3 \cdot 3x^2$

85. $9x^4 \cdot 8y^5$

86. $6x^4 \cdot 12y^5$

Factor.

87. $x^2 - 4$

88. $x^2 - 6x + 9$

89. $x^3 - x^2y$

90. $a^2 - 5a + 6$

91. $2y^2 - 2$

92. $xa + xb + ya + yb$

EXTENDING THE CONCEPTS ━━━

93. $\dfrac{(x + y)^3 - 8z^3}{(x + y - 2z)}$

94. $\dfrac{(x + y)^3 + 8z^3}{(x + y + 2z)}$

95. $\dfrac{a^3 + (b - c)^3}{a + b - c}$

96. $\dfrac{a^3 - (b - c)^3}{a - b + c}$

97. $\dfrac{x^2 - (y + 1)^2}{x + y + 1}$

98. $\dfrac{x^2 - (y - 1)^2}{x + y - 1}$

99. $\dfrac{(x^2 - 1)(x + 1)}{(x^2 - 2x + 1)^2}$

100. $\dfrac{x^4 - y^4}{(x^2 + 2xy + y^2)(x^2 + y^2)}$

MULTIPLICATION AND DIVISION OF RATIONAL EXPRESSIONS

 OBJECTIVE

A Multiply and divide rational expressions by factoring and then dividing out common factors.

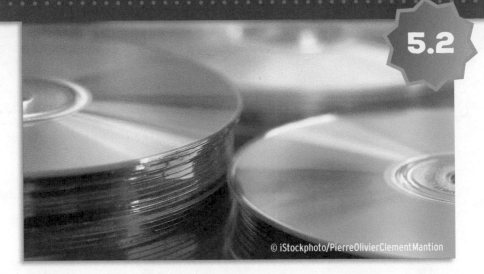

© iStockphoto/PierreOlivierClementMantion

If you have ever needed duplicates of a DVD, you know the amount you pay for the transfer depends on the number of duplications made: the more duplications you have made, the lower the charge per copy. The following demand function gives the price (in dollars) per DVD $p(x)$ a company charges for making x DVDs. As you can see, it is a rational function.

$$p(x) = \frac{2(x + 40)}{x + 5}$$

The graph in Figure 1 shows this function from $x = 0$ to $x = 100$. As you can see, the more copies that are made, the lower the price per copy.

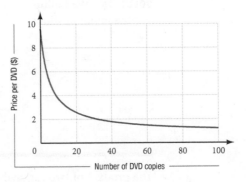

FIGURE 1

If we were interested in finding the revenue function for this situation, we would multiply the number of copies made x by the price per copy $p(x)$. This involves multiplication with a rational expression, which is one of the topics we cover in this section.

5.2 VIDEOS

A Multiplying and Dividing Rational Expressions

In Section 5.1, we found the process of reducing rational expressions to lowest terms to be the same process used in reducing fractions to lowest terms. The similarity also holds for the process of multiplication or division of rational expressions.

Multiplication with fractions is the simplest of the four basic operations. To multiply two fractions, we simply multiply numerators and multiply denominators. That is, if a, b, c, and d are real numbers, with $b \neq 0$ and $d \neq 0$, then

$$\frac{a}{b} \cdot \frac{c}{d} = \frac{ac}{bd}$$

EXAMPLE 1

Multiply $\frac{6}{7} \cdot \frac{14}{18}$.

SOLUTION

$$\frac{6}{7} \cdot \frac{14}{18} = \frac{6(14)}{7(18)} \qquad \text{Multiply numerators and denominators.}$$

$$= \frac{2 \cdot 3(2 \cdot 7)}{7(2 \cdot 3 \cdot 3)} \qquad \text{Factor.}$$

$$= \frac{2}{3} \qquad \text{Divide out common factors.}$$

In our next example, we multiply fractions whose numerators and denominators are monomials by multiplying numerators and multiplying denominators and then reducing to lowest terms. Here is how it looks.

EXAMPLE 2

Multiply $\frac{8x^3}{27y^8} \cdot \frac{9y^3}{12x^2}$.

SOLUTION   We multiply numerators and denominators without actually carrying out the multiplication:

$$\frac{8x^3}{27y^8} \cdot \frac{9y^3}{12x^2} = \frac{8 \cdot 9x^3 y^3}{27 \cdot 12x^2 y^8} \qquad \begin{array}{l} \text{Multiply numerators.} \\ \text{Multiply denominators.} \end{array}$$

$$= \frac{4 \cdot 2 \cdot 9x^3 y^3}{9 \cdot 3 \cdot 4 \cdot 3x^2 y^8} \qquad \text{Factor coefficients.}$$

$$= \frac{2x}{9y^5} \qquad \text{Divide out common factors.}$$

The product of two rational expressions is the product of their numerators over the product of their denominators.

Once again, we should mention that the little slashes we have drawn through the factors are simply used to denote the factors we have divided out of the numerator and denominator.

EXAMPLE 3

Multiply $\frac{x-3}{x^2-4} \cdot \frac{x+2}{x^2-6x+9}$.

SOLUTION We begin by multiplying numerators and denominators. We then factor all polynomials and divide out factors common to the numerator and denominator:

$$\frac{x-3}{x^2-4} \cdot \frac{x+2}{x^2-6x+9} = \frac{(x-3)(x+2)}{(x^2-4)(x^2-6x+9)} \qquad \text{Multiply.}$$

$$= \frac{(x-3)(x+2)}{(x+2)(x-2)(x-3)(x-3)} \qquad \text{Factor.}$$

$$= \frac{1}{(x-2)(x-3)}$$

The first two steps can be combined to save time. We can perform the multiplication and factoring steps together.

EXAMPLE 4

Multiply $\dfrac{2y^2-4y}{2y^2-2} \cdot \dfrac{y^2-2y-3}{y^2-5y+6}$.

SOLUTION

$$\frac{2y^2-4y}{2y^2-2} \cdot \frac{y^2-2y-3}{y^2-5y+6} = \frac{2y(y-2)(y-3)(y+1)}{2(y+1)(y-1)(y-3)(y-2)}$$

$$= \frac{y}{y-1}$$

Notice in both of the preceding examples that we did not actually multiply the polynomials as we did in Section 3.5. It would be senseless to do that because we would then have to factor each of the resulting products to reduce them to lowest terms.

The quotient of two rational expressions is the product of the first and the reciprocal of the second. That is, we find the quotient of two rational expressions the same way we find the quotient of two fractions. Here is an example that reviews division with fractions.

EXAMPLE 5

Divide $\dfrac{6}{8} \div \dfrac{3}{5}$.

SOLUTION

$$\frac{6}{8} \div \frac{3}{5} = \frac{6}{8} \cdot \frac{5}{3} \qquad \text{Write division in terms of multiplication.}$$

$$= \frac{6(5)}{8(3)} \qquad \text{Multiply numerators and denominators.}$$

$$= \frac{2 \cdot 3(5)}{2 \cdot 2 \cdot 2(3)} \qquad \text{Factor.}$$

$$= \frac{5}{4} \qquad \text{Divide out common factors.}$$

To divide one rational expression by another, we use the definition of division to multiply by the reciprocal of the expression that follows the division symbol.

EXAMPLE 6

Divide $\dfrac{8x^3}{5y^2} \div \dfrac{4x^2}{10y^6}$.

SOLUTION First, we rewrite the problem in terms of multiplication. Then we multiply.

$$\frac{8x^3}{5y^2} \div \frac{4x^2}{10y^6} = \frac{8x^3}{5y^2} \cdot \frac{10y^6}{4x^2}$$

$$= \frac{\overset{2}{8} \cdot \overset{2}{10} x^3 y^6}{4 \cdot 5 x^2 y^2}$$

$$= 4xy^4$$

EXAMPLE 7

Divide $\dfrac{x^2 - y^2}{x^2 - 2xy + y^2} \div \dfrac{x^3 + y^3}{x^3 - x^2y}$.

SOLUTION We begin by writing the problem as the product of the first and the reciprocal of the second and then proceed as in the previous two examples:

$$\frac{x^2 - y^2}{x^2 - 2xy + y^2} \div \frac{x^3 + y^3}{x^3 - x^2y}$$
Multiply by the reciprocal of the divisor.

$$= \frac{x^2 - y^2}{x^2 - 2xy + y^2} \cdot \frac{x^3 - x^2y}{x^3 + y^3}$$

$$= \frac{(x-y)(x+y)(x^2)(x-y)}{(x-y)(x-y)(x+y)(x^2-xy+y^2)}$$
Factor and multiply.

$$= \frac{x^2}{x^2 - xy + y^2}$$
Divide out common factors.

Here are some more examples of multiplication and division with rational expressions.

EXAMPLE 8

Perform the indicated operations.

$$\frac{a^2 - 8a + 15}{a + 4} \cdot \frac{a + 2}{a^2 - 5a + 6} \div \frac{a^2 - 3a - 10}{a^2 + 2a - 8}$$

SOLUTION First, we rewrite the division as multiplication by the reciprocal. Then we proceed as usual.

$$\frac{a^2 - 8a + 15}{a + 4} \cdot \frac{a + 2}{a^2 - 5a + 6} \div \frac{a^2 - 3a - 10}{a^2 + 2a - 8}$$
Change division to multiplication by the reciprocal.

$$= \frac{(a^2 - 8a + 15)(a + 2)(a^2 + 2a - 8)}{(a + 4)(a^2 - 5a + 6)(a^2 - 3a - 10)}$$
Factor.

$$= \frac{(a-5)(a-3)(a+2)(a+4)(a-2)}{(a+4)(a-3)(a-2)(a-5)(a+2)}$$
Divide out common factors.

$$= 1$$

Our next example involves factoring by grouping. As you may have noticed, working the problems in this chapter gives you a very detailed review of factoring.

EXAMPLE 9

Multiply $\dfrac{xa + xb + ya + yb}{xa - xb - ya + yb} \cdot \dfrac{xa + xb - ya - yb}{xa - xb + ya - yb}$.

SOLUTION We will factor each polynomial by grouping, which takes two steps.

$$\frac{xa + xb + ya + yb}{xa - xb - ya + yb} \cdot \frac{xa + xb - ya - yb}{xa - xb + ya - yb}$$

$$= \frac{x(a + b) + y(a + b)}{x(a - b) - y(a - b)} \cdot \frac{x(a + b) - y(a + b)}{x(a - b) + y(a - b)} \left.\right\} \quad \text{Factor by grouping.}$$

$$= \frac{(a + b)(x + y)(a + b)(x - y)}{(a - b)(x - y)(a - b)(x + y)}$$

$$= \frac{(a + b)^2}{(a - b)^2}$$

EXAMPLE 10

Multiply $(4x^2 - 36) \cdot \dfrac{12}{4x + 12}$.

SOLUTION We can think of $4x^2 - 36$ as having a denominator of 1. Thinking of it in this way allows us to proceed as we did in the previous examples.

$$(4x^2 - 36) \cdot \frac{12}{4x + 12}$$

$$= \frac{4x^2 - 36}{1} \cdot \frac{12}{4x + 12} \qquad \text{Write } 4x^2 - 36 \text{ with denominator 1.}$$

$$= \frac{4(x - 3)(x + 3)12}{4(x + 3)} \qquad \text{Factor.}$$

$$= 12(x - 3) \qquad \text{Divide out common factors.}$$

EXAMPLE 11

Multiply $3(x - 2)(x - 1) \cdot \dfrac{5}{x^2 - 3x + 2}$.

SOLUTION This problem is very similar to the problem in Example 10. Writing the first rational expression with a denominator of 1, we have

$$\frac{3(x - 2)(x - 1)}{1} \cdot \frac{5}{x^2 - 3x + 2} = \frac{3(x - 2)(x - 1)5}{(x - 2)(x - 1)}$$

$$= 3 \cdot 5$$

$$= 15$$

GETTING READY FOR CLASS

After reading through the preceding section, respond in your own words and in complete sentences.

A. Summarize the steps used to multiply fractions.

B. What is the first step in multiplying two rational expressions?

C. Why is factoring important when multiplying and dividing rational expressions?

D. How is division with rational expressions different than multiplication of rational expressions?

Problem Set 5.2

Perform the indicated operations.

1. $\dfrac{2}{9} \cdot \dfrac{3}{4}$

2. $\dfrac{5}{6} \cdot \dfrac{7}{8}$

3. $\dfrac{3}{4} \div \dfrac{1}{3}$

4. $\dfrac{3}{8} \div \dfrac{5}{4}$

5. $\dfrac{3}{7} \cdot \dfrac{14}{24} \div \dfrac{1}{2}$

6. $\dfrac{6}{5} \cdot \dfrac{10}{36} \div \dfrac{3}{4}$

7. $\dfrac{10x^2}{5y^2} \cdot \dfrac{15y^3}{2x^4}$

8. $\dfrac{8x^3}{7y^4} \cdot \dfrac{14y^6}{16x^2}$

9. $\dfrac{11a^2b}{5ab^2} \div \dfrac{22a^3b^2}{10ab^4}$

10. $\dfrac{8ab^3}{9a^2b} \div \dfrac{16a^2b^2}{18ab^3}$

11. $\dfrac{6x^2}{5y^3} \cdot \dfrac{11z^2}{2x^2} \div \dfrac{33z^5}{10y^8}$

12. $\dfrac{4x^3}{7y^2} \cdot \dfrac{6z^5}{5x^6} \div \dfrac{24z^2}{35x^6}$

Perform the indicated operations. Be sure to write all answers in lowest terms.

13. $\dfrac{x^2-9}{x^2-4} \cdot \dfrac{x-2}{x-3}$

14. $\dfrac{x^2-16}{x^2-25} \cdot \dfrac{x-5}{x-4}$

15. $\dfrac{y^2-1}{y+2} \cdot \dfrac{y^2+5y+6}{y^2+2y-3}$

16. $\dfrac{y-1}{y^2-y-6} \cdot \dfrac{y^2+5y+6}{y^2-1}$

17. $\dfrac{3x-12}{x^2-4} \cdot \dfrac{x^2+6x+8}{x-4}$

18. $\dfrac{x^2+5x+1}{4x-4} \cdot \dfrac{x-1}{x^2+5x+1}$

19. $\dfrac{xy}{xy+1} \div \dfrac{x}{y}$

20. $\dfrac{y}{x} \div \dfrac{xy}{xy-1}$

21. $\dfrac{1}{x^2-9} \div \dfrac{1}{x^2+9}$

22. $\dfrac{1}{x^2-9} \div \dfrac{1}{(x-3)^2}$

23. $\dfrac{y-3}{y^2-6y+9} \cdot \dfrac{y-3}{4}$

24. $\dfrac{y-3}{y^2-6y+9} \div \dfrac{y-3}{4}$

25. $\dfrac{5x+2y}{25x^2-5xy-6y^2} \cdot \dfrac{20x^2-7xy-3y^2}{4x+y}$

26. $\dfrac{7x+3y}{42x^2-17xy-15y^2} \cdot \dfrac{12x^2-4xy-5y^2}{2x+y}$

27. $\dfrac{a^2-5a+6}{a^2-2a-3} \div \dfrac{a-5}{a^2+3a+2}$

28. $\dfrac{a^2+7a+12}{a-5} \div \dfrac{a^2+9a+18}{a^2-7a+10}$

29. $\dfrac{x^2 + 2x - 3}{x^2 - 9} \cdot \dfrac{2x - 6}{x^2 - 1}$

30. $\dfrac{x^2 + 4x - 5}{x^2 - 25} \cdot \dfrac{3x - 15}{x^2 + 3x - 4}$

31. $\dfrac{x^2 + 6x + 5}{x^2 - 2x - 3} \div \dfrac{x^2 + 8x + 15}{x^2 - 9}$

32. $\dfrac{x^2 - 49}{x^2 - 6x - 7} \div \dfrac{x^2 + 10x + 21}{x^2 + 4x + 3}$

33. $\dfrac{x^3 + 8}{x^2 - 4} \cdot \dfrac{x^2 - 5x + 6}{x^2 - 2x + 4}$

34. $\dfrac{x^2 - 6x - 27}{x^2 + 3x + 9} \cdot \dfrac{x^3 - 27}{x^2 - 9}$

35. $\dfrac{4t^2 - 1}{6t^2 + t - 2} \div \dfrac{8t^3 + 1}{27t^3 + 8}$

36. $\dfrac{9t^2 - 1}{6t^2 + 7t - 3} \div \dfrac{27t^3 + 1}{8t^3 + 27}$

37. $\dfrac{2x^2 - 5x - 12}{4x^2 + 8x + 3} \div \dfrac{x^2 - 16}{2x^2 + 7x + 3}$

38. $\dfrac{x^2 - 2x + 1}{3x^2 + 7x - 20} \div \dfrac{x^2 + 3x - 4}{3x^2 - 2x - 5}$

39. $\dfrac{2a^2 - 21ab - 36b^2}{a^2 - 11ab - 12b^2} \div \dfrac{10a + 15b}{a^2 - b^2}$

40. $\dfrac{3a^2 + 7ab - 20b^2}{a^2 + 5ab + 4b^2} \div \dfrac{3a^2 - 17ab + 20b^2}{3a - 12b}$

41. $\dfrac{6c^2 - c - 15}{9c^2 - 25} \cdot \dfrac{15c^2 + 22c - 5}{6c^2 + 5c - 6}$

42. $\dfrac{m^2 + 4m - 21}{m^2 - 12m + 27} \cdot \dfrac{m^2 - 7m + 12}{m^2 + 3m - 28}$

43. $\dfrac{6a^2b + 2ab^2 - 20b^3}{4a^2b - 16b^3} \cdot \dfrac{10a^2 - 22ab + 4b^2}{27a^3 - 125b^3}$

44. $\dfrac{12a^2b - 3ab^2 - 42b^3}{9a^2 - 36b^2} \cdot \dfrac{6a^2 - 15ab + 6b^2}{8a^3b - b^4}$

45. $\dfrac{360x^3 - 490x}{36x^2 + 84x + 49} \cdot \dfrac{30x^2 + 83x + 56}{150x^3 + 65x^2 - 280x}$

46. $\dfrac{490x^2 - 640}{49x^2 - 112x + 64} \cdot \dfrac{28x^2 - 95x + 72}{56x^3 - 62x^2 - 144x}$

47. $\dfrac{x^5 - x^2}{5x^2 - 5x} \cdot \dfrac{10x^4 - 10x^2}{2x^4 + 2x^3 + 2x^2}$

48. $\dfrac{2x^4 - 16x}{3x^6 - 48x^2} \cdot \dfrac{6x^5 + 24x^3}{4x^4 + 8x^3 + 16x^2}$

49. $\dfrac{a^2 - 16b^2}{a^2 - 8ab + 16b^2} \cdot \dfrac{a^2 - 9ab + 20b^2}{a^2 - 7ab + 12b^2} \div \dfrac{a^2 - 25b^2}{a^2 - 6ab + 9b^2}$

50. $\dfrac{a^2 - 6ab + 9b^2}{a^2 - 4b^2} \cdot \dfrac{a^2 - 5ab + 6b^2}{(a - 3b)^2} \div \dfrac{a^2 - 9b^2}{a^2 - ab - 6b^2}$

51. $\dfrac{2y^2 - 7y - 15}{42y^2 - 29y - 5} \cdot \dfrac{12y^2 - 16y + 5}{7y^2 - 36y + 5} \div \dfrac{4y^2 - 9}{49y^2 - 1}$

52. $\dfrac{8y^2 + 18y - 5}{21y^2 - 16y + 3} \cdot \dfrac{35y^2 - 22y + 3}{6y^2 + 17y + 5} \div \dfrac{16y^2 - 1}{9y^2 - 1}$

53. $\dfrac{xy - 2x + 3y - 6}{xy + 2x - 4y - 8} \cdot \dfrac{xy + x - 4y - 4}{xy - x + 3y - 3}$

54. $\dfrac{ax + bx + 2a + 2b}{ax - 3a + bx - 3b} \cdot \dfrac{ax - bx - 3a + 3b}{ax - bx - 2a + 2b}$

55. $\dfrac{xy^2 - y^2 + 4xy - 4y}{xy - 3y + 4x - 12} \div \dfrac{xy^3 + 2xy^2 + y^3 + 2y^2}{xy^2 - 3y^2 + 2xy - 6y}$

56. $\dfrac{4xb - 8b + 12x - 24}{xb^2 + 3b^2 + 3xb + 9b} \div \dfrac{4xb - 8b - 8x + 16}{xb^2 + 3b^2 - 2xb - 6b}$

57. $\dfrac{2x^3 + 10x^2 - 8x - 40}{x^3 + 4x^2 - 9x - 36} \cdot \dfrac{x^2 + x - 12}{2x^2 + 14x + 20}$

58. $\dfrac{x^3 + 2x^2 - 9x - 18}{x^4 + 3x^3 - 4x^2 - 12x} \cdot \dfrac{x^3 + 5x^2 + 6x}{x^2 - x - 6}$

59. $\dfrac{w^3 - w^2x}{wy - w} \div \left(\dfrac{w - x}{y - 1} \right)^2$ **60.** $\dfrac{a^3 - a^2b}{ac - a} \div \left(\dfrac{a - b}{c - 1} \right)^2$

61. $\dfrac{mx + my + 2x + 2y}{6x^2 - 5xy - 4y^2} \div \dfrac{2mx - 4x + my - 2y}{3mx - 6x - 4my + 8y}$

62. $\dfrac{ax - 2a + 2xy - 4y}{ax + 2a - 2xy - 4y} \div \dfrac{ax + 2a + 2xy + 4y}{ax - 2a - 2xy + 4y}$

63. $(3x - 6) \cdot \dfrac{x}{x - 2}$

64. $(4x + 8) \cdot \dfrac{x}{x + 2}$

65. $(x^2 - 25) \cdot \dfrac{2}{x - 5}$

66. $(x^2 - 49) \cdot \dfrac{5}{x + 7}$

67. $(x^2 - 3x + 2) \cdot \dfrac{3}{3x - 3}$

68. $(x^2 - 3x + 2) \cdot \dfrac{-1}{x - 2}$

69. $(y - 3)(y - 4)(y + 3) \cdot \dfrac{-1}{y^2 - 9}$

70. $(y + 1)(y + 4)(y - 1) \cdot \dfrac{3}{y^2 - 1}$

71. $a(a + 5)(a - 5) \cdot \dfrac{a + 1}{a^2 + 5a}$

72. $a(a + 3)(a - 3) \cdot \dfrac{a - 1}{a^2 - 3a}$

PAYING ATTENTION TO INSTRUCTIONS

73. Work each problem according to the instructions given.

a. Simplify: $\dfrac{16 - 1}{64 - 1}$

b. Reduce: $\dfrac{25x^2 - 9}{125x^3 - 27}$

c. Multiply: $\dfrac{25x^2 - 9}{125x^3 - 27} \cdot \dfrac{5x - 3}{5x + 3}$

d. Divide: $\dfrac{25x^2 - 9}{125x^3 - 27} \div \dfrac{5x - 3}{25x^2 + 15x + 9}$

74. Work each problem according to the instructions given.

a. Simplify: $\dfrac{64 - 49}{64 + 112 + 49}$

b. Reduce: $\dfrac{9x^2 - 49}{9x^2 + 42x + 49}$

c. Multiply: $\dfrac{9x^2 - 49}{9x^2 + 42x + 49} \cdot \dfrac{3x + 7}{3x - 7}$

d. Divide: $\dfrac{9x^2 - 49}{9x^2 + 42x + 49} \div \dfrac{3x + 7}{3x - 7}$

GETTING READY FOR THE NEXT SECTION

Combine.

75. $\dfrac{4}{9} + \dfrac{2}{9}$ **76.** $\dfrac{3}{8} + \dfrac{1}{8}$ **77.** $\dfrac{3}{14} + \dfrac{7}{30}$ **78.** $\dfrac{3}{10} + \dfrac{11}{42}$

Multiply.

79. $-1(7 - x)$ **80.** $-1(3 - x)$

Factor.

81. $x^2 - 1$ **82.** $x^2 - 2x - 3$ **83.** $2x + 10$

84. $x^2 + 4x + 3$ **85.** $a^3 - b^3$ **86.** $8y^3 - 27$

EXTENDING THE CONCEPTS

87. $\dfrac{x^6 + y^6}{x^4 + 4x^2y^2 + 3y^4} \div \dfrac{x^4 + 3x^2y^2 + 2y^4}{x^4 + 5x^2y^2 + 6y^4}$

88. $\dfrac{x^2 + 9xy + 8y^2}{x^2 + 7xy - 8y^2} \div \dfrac{x^2 - y^2}{x^2 + 5xy - 6y^2}$

89. $\dfrac{a^2(2a + b) + 6a(2a + b) + 5(2a + b)}{3a^2(2a + b) - 2a(2a + b) + (2a + b)} \div \dfrac{a + 1}{a - 1}$

90. $\dfrac{2x^2(x - 3z) - 5x(x - 3z) + 2(x - 3z)}{4x^2(x - 3z) - 11x(x - 3z) + 6(x - 3z)} \div \dfrac{4x - 3}{4x + 1}$

91. $\dfrac{a^3 - a^2b}{ac - a} \div \left(\dfrac{a - b}{c - 1} \right)^2$

92. $\dfrac{p^3 + q^3}{q - p} \div \dfrac{(p + q)^2}{p^2 - q^2}$

© sxc.hu, johnnyberg, 2008

5.3

OBJECTIVES

A Add and subtract rational expressions that have the same denominator.

B Add and subtract rational expressions that have different denominators.

If a human faced off against an elephant in a hot dog bun-eating contest, who would win? One summer, three brave humans signed up to find out firsthand. Three elephants, Susie, Minnie, and Bunny, from Ringling Bros. and Barnum & Bailey Circus stood behind a table on which a tall pile of hot dog buns sat. A few feet away, three competitive eating contestants stood in front of their own table outfitted with stacks of buns and large cups of water in which they could dunk the buns to help their cause. When the announcer yelled "Go!" the two teams began devouring the buns. They had 6 minutes to eat as many as possible. The final tally: 41 dozen buns for the elephants, 15 dozen for the humans.

We can use this scenario to help us introduce the topic of this section: addition and subtraction of rational expressions. Suppose Susie the elephant ate x dozen hot dog buns, Minnie ate one more dozen than Susie, and Bunny ate twice as many dozen as Susie. What rational expression represents the total number of buns eaten by the three elephants? First, we can create a rational expression for each elephant by writing the amount eaten as a numerator over a denominator of 41 (dozen). Then we can add them together using our knowledge of combining fractions with the same denominator.

$$\underset{\underset{\downarrow}{\text{Susie}}}{} \quad \underset{\underset{\downarrow}{\text{Minnie}}}{} \qquad\qquad \underset{\underset{\downarrow}{\text{Bunny}}}{}$$

$$\text{Add } \frac{x}{41} + \frac{x+1}{41} + \frac{2x}{41} \text{ which equals } \frac{4x+1}{41}.$$

This section is concerned with addition and subtraction of rational expressions. In the first part of this section, we will look at addition of expressions that have the same denominator. In the second part of this section, we will look at addition of expressions that have different denominators.

5.3 VIDEOS

A Addition and Subtraction with the Same Denominator

To add two expressions that have the same denominator, we simply add numerators and put the sum over the common denominator. Because the process we use to add and subtract rational expressions is the same process used to add and subtract fractions, we will begin with an example involving fractions.

EXAMPLE 1

Add $\dfrac{4}{9} + \dfrac{2}{9}$.

SOLUTION We add fractions with the same denominator by using the distributive property. Here is a detailed look at the steps involved.

$$\frac{4}{9} + \frac{2}{9} = 4\left(\frac{1}{9}\right) + 2\left(\frac{1}{9}\right)$$

$$= (4 + 2)\left(\frac{1}{9}\right) \qquad \text{Distributive property}$$

$$= 6\left(\frac{1}{9}\right)$$

$$= \frac{6}{9}$$

$$= \frac{2}{3} \qquad \text{Divide numerator and denominator by common factor 3.}$$

Note that the important thing about the fractions in this example is that they each have a denominator of 9. If they did not have the same denominator, we could not have written them as two terms with a factor of $\frac{1}{9}$ in common. Without the $\frac{1}{9}$ common to each term, we couldn't apply the distributive property. Without the distributive property, we would not have been able to add the two fractions in this form.

In the following examples, we will not show all the steps we showed in Example 1. The steps are shown in Example 1 so you will see why both fractions must have the same denominator before we can add them. In practice, we simply add numerators and place the result over the common denominator.

We add and subtract rational expressions with the same denominator by combining numerators and writing the result over the common denominator. Then we reduce the result to lowest terms, if possible. Example 2 shows this process in detail. If you see the similarities between operations on rational numbers and operations on rational expressions, this chapter will look like an extension of rational numbers rather than a completely new set of topics.

EXAMPLE 2

Add $\dfrac{x}{x^2 - 1} + \dfrac{1}{x^2 - 1}$.

SOLUTION Because the denominators are the same, we simply add numerators:

$$\frac{x}{x^2 - 1} + \frac{1}{x^2 - 1} = \frac{x + 1}{x^2 - 1} \qquad \text{Add numerators.}$$

$$= \frac{x + 1}{(x - 1)(x + 1)} \qquad \text{Factor denominator.}$$

$$= \frac{1}{x - 1} \qquad \text{Divide out common factor } x + 1.$$

Our next example involves subtraction of rational expressions. Pay careful attention to what happens to the signs of the terms in the numerator of the second expression when we subtract it from the first expression.

EXAMPLE 3

Subtract $\dfrac{2x - 5}{x - 2} - \dfrac{x - 3}{x - 2}$.

SOLUTION Because each expression has the same denominator, we simply subtract the numerator in the second expression from the numerator in the first expression and write the difference over the common denominator $x - 2$. We must be careful, however, that we subtract both terms in the second numerator. To ensure that we do, we will enclose that numerator in parentheses.

$$\frac{2x - 5}{x - 2} - \frac{x - 3}{x - 2} = \frac{2x - 5 - (x - 3)}{x - 2} \qquad \text{Subtract numerators.}$$

$$= \frac{2x - 5 - x + 3}{x - 2} \qquad \text{Remove parentheses.}$$

$$= \frac{\cancel{x - 2}}{\cancel{x - 2}} \qquad \text{Combine similar terms in the numerator.}$$

$$= 1 \qquad \text{Reduce (or divide).}$$

Note the $+3$ in the numerator of the second step. It is a common mistake to write this as -3, by forgetting to subtract both terms in the numerator of the second expression. Whenever the expression we are subtracting has two or more terms in its numerator, we have to watch for this mistake.

Next we consider addition and subtraction of fractions and rational expressions that have different denominators.

B Addition and Subtraction with Different Denominators

Before we look at an example of addition of fractions with different denominators, we need to review the definition for the least common denominator (LCD).

> **DEFINITION** *least common denominator*
>
> The *least common denominator* for a set of denominators is the smallest expression that is divisible by each of the denominators.

The first step in combining two fractions is to find the LCD. Once we have the common denominator, we rewrite each fraction as an equivalent fraction with the common denominator. After that, we simply add or subtract as we did in our first three examples.

Example 4 is a review of the step-by-step procedure used to add two fractions with different denominators.

EXAMPLE 4

Add $\dfrac{3}{14} + \dfrac{7}{30}$.

SOLUTION

Step 1: Find the LCD.

To do this, we first factor both denominators into prime factors.

Factor 14: $14 = 2 \cdot 7$

Factor 30: $30 = 2 \cdot 3 \cdot 5$

Because the LCD must be divisible by 14, it must have factors of 2 and 7. It must also be divisible by 30 and, therefore, have factors of 2, 3, and 5. We do not need to repeat the 2 that appears in both the factors of 14 and those of 30. Therefore,

$$LCD = 2 \cdot 3 \cdot 5 \cdot 7 = 210$$

Step 2: Change to equivalent fractions.

Because we want each fraction to have a denominator of 210 and at the same time keep its original value, we multiply each by 1 in the appropriate form.

Change $\frac{3}{14}$ to a fraction with denominator 210:

$$\frac{3}{14} \cdot \frac{15}{15} = \frac{45}{210}$$

Change $\frac{7}{30}$ to a fraction with denominator 210:

$$\frac{7}{30} \cdot \frac{7}{7} = \frac{49}{210}$$

Step 3: Add numerators of equivalent fractions found in step 2:

$$\frac{45}{210} + \frac{49}{210} = \frac{94}{210}$$

Step 4: Reduce to lowest terms, if necessary:

$$\frac{94}{210} = \frac{47}{105}$$

The main idea in adding fractions is to write each fraction again with the LCD for a denominator. In doing so, we must be sure not to change the value of either of the original fractions.

EXAMPLE 5

Add $\dfrac{-2}{x^2 - 2x - 3} + \dfrac{3}{x^2 - 9}$.

SOLUTION

Step 1: *Factor each denominator and build the LCD from the factors:*

$$x^2 - 2x - 3 = (x - 3)(x + 1)$$
$$x^2 - 9 \quad = (x - 3)(x + 3)$$
$$LCD = (x - 3)(x + 3)(x + 1)$$

Step 2: *Change each rational expression to an equivalent expression that has the LCD for a denominator:*

$$\frac{-2}{x^2 - 2x - 3} = \frac{-2}{(x - 3)(x + 1)} \cdot \frac{(x + 3)}{(x + 3)} = \frac{-2x - 6}{(x - 3)(x + 3)(x + 1)}$$

$$\frac{3}{x^2 - 9} = \frac{3}{(x - 3)(x + 3)} \cdot \frac{(x + 1)}{(x + 1)} = \frac{3x + 3}{(x - 3)(x + 3)(x + 1)}$$

Step 3: *Add numerators of the rational expressions found in step 2:*

$$\frac{-2x - 6}{(x - 3)(x + 3)(x + 1)} + \frac{3x + 3}{(x - 3)(x + 3)(x + 1)} = \frac{x - 3}{(x - 3)(x + 3)(x + 1)}$$

Step 4: *Reduce to lowest terms by dividing out the common factor x − 3:*

$$\frac{\cancel{x - 3}}{(\cancel{x - 3})(x + 3)(x + 1)} = \frac{1}{(x + 3)(x + 1)}$$

EXAMPLE 6

Subtract $\dfrac{x + 4}{2x + 10} - \dfrac{5}{x^2 - 25}$.

SOLUTION   We begin by factoring each denominator:

$$\frac{x + 4}{2x + 10} - \frac{5}{x^2 - 25} = \frac{x + 4}{2(x + 5)} - \frac{5}{(x + 5)(x - 5)}$$

The LCD is $2(x + 5)(x - 5)$. Completing the problem, we have

$$= \frac{x + 4}{2(x + 5)} \cdot \frac{(x - 5)}{(x - 5)} - \frac{5}{(x + 5)(x - 5)} \cdot \frac{2}{2}$$

$$= \frac{x^2 - x - 20}{2(x + 5)(x - 5)} - \frac{10}{2(x + 5)(x - 5)}$$

$$= \frac{x^2 - x - 30}{2(x + 5)(x - 5)}$$

To see if this expression will reduce, we factor the numerator into $(x - 6)(x + 5)$.

$$= \frac{(x - 6)\cancel{(x + 5)}}{2\cancel{(x + 5)}(x - 5)}$$

$$= \frac{x - 6}{2(x - 5)}$$

EXAMPLE 7

Subtract $\dfrac{2x - 2}{x^2 + 4x + 3} - \dfrac{x - 1}{x^2 + 5x + 6}$.

SOLUTION We factor each denominator and build the LCD from those factors:

$$\frac{2x - 2}{x^2 + 4x + 3} - \frac{x - 1}{x^2 + 5x + 6}$$

$$= \frac{2x - 2}{(x + 3)(x + 1)} - \frac{x - 1}{(x + 3)(x + 2)}$$

$$= \frac{2x - 2}{(x + 3)(x + 1)} \cdot \frac{(x + 2)}{(x + 2)} - \frac{x - 1}{(x + 3)(x + 2)} \cdot \frac{(x + 1)}{(x + 1)} \qquad \text{The LCD is } (x + 1)(x + 2)(x + 3).$$

$$= \frac{2x^2 + 2x - 4}{(x + 1)(x + 2)(x + 3)} - \frac{x^2 - 1}{(x + 1)(x + 2)(x + 3)} \qquad \text{Multiply out each numerator.}$$

$$= \frac{(2x^2 + 2x - 4) - (x^2 - 1)}{(x + 1)(x + 2)(x + 3)} \qquad \text{Subtract numerators.}$$

$$= \frac{x^2 + 2x - 3}{(x + 1)(x + 2)(x + 3)} \qquad \text{Factor numerator to see if we can reduce.}$$

$$= \frac{(x + 3)(x - 1)}{(x + 1)(x + 2)(x + 3)} \qquad \text{Reduce.}$$

$$= \frac{x - 1}{(x + 1)(x + 2)}$$

EXAMPLE 8

Add $\dfrac{x^2}{x - 7} + \dfrac{6x + 7}{7 - x}$.

SOLUTION In Section 5.1, we were able to reverse the terms in a factor such as $7 - x$ by factoring -1 from each term. In a problem like this, the same result can be obtained by multiplying the numerator and denominator by -1:

$$\frac{x^2}{x - 7} + \frac{6x + 7}{7 - x} \cdot \frac{-1}{-1} = \frac{x^2}{x - 7} + \frac{-6x - 7}{x - 7}$$

$$= \frac{x^2 - 6x - 7}{x - 7} \qquad \text{Add numerators.}$$

$$= \frac{(x - 7)(x + 1)}{(x - 7)} \qquad \text{Factor numerator.}$$

$$= x + 1 \qquad \text{Divide out } x - 7.$$

For our next example, we will look at a problem in which we combine a whole number and a rational expression.

EXAMPLE 9

Subtract $2 - \dfrac{9}{3x + 1}$.

SOLUTION To subtract these two expressions, we think of 2 as a rational expression with a denominator of 1.

$$2 - \frac{9}{3x + 1} = \frac{2}{1} - \frac{9}{3x + 1}$$

The LCD is $3x + 1$. Multiplying the numerator and denominator of the first expression by $3x + 1$ gives us a rational expression equivalent to 2, but with a denominator of $3x + 1$.

$$\frac{2}{1} \cdot \frac{(3x + 1)}{(3x + 1)} - \frac{9}{3x + 1} = \frac{6x + 2 - 9}{3x + 1}$$

$$= \frac{6x - 7}{3x + 1}$$

The numerator and denominator of this last expression do not have any factors in common other than 1, so the expression is in lowest terms.

EXAMPLE 10

Write an expression for the sum of a number and twice its reciprocal. Then, simplify that expression.

SOLUTION If x is the number, then its reciprocal is $\frac{1}{x}$. Twice its reciprocal is $\frac{2}{x}$. The sum of the number and twice its reciprocal is

$$x + \frac{2}{x}$$

To combine these two expressions, we think of the first term x as a rational expression with a denominator of 1. The LCD is x:

$$x + \frac{2}{x} = \frac{x}{1} + \frac{2}{x}$$

$$= \frac{x}{1} \cdot \frac{x}{x} + \frac{2}{x}$$

$$= \frac{x^2 + 2}{x}$$

GETTING READY FOR CLASS

After reading through the preceding section, respond in your own words and in complete sentences.

A. Briefly describe how you would add two rational expressions that have the same denominator.

B. Why is factoring important in finding a least common denominator?

C. What is the last step in adding or subtracting two rational expressions?

D. Explain how you would change the fraction $\dfrac{5}{x - 3}$ to an equivalent fraction with denominator $x^2 - 9$.

Combine the following fractions.

1. $\dfrac{3}{4} + \dfrac{1}{2}$ **2.** $\dfrac{5}{6} + \dfrac{1}{3}$ **3.** $\dfrac{2}{5} - \dfrac{1}{15}$ **4.** $\dfrac{5}{8} - \dfrac{1}{4}$

5. $\dfrac{5}{6} + \dfrac{7}{8}$ **6.** $\dfrac{3}{4} + \dfrac{2}{3}$ **7.** $\dfrac{9}{48} - \dfrac{3}{54}$ **8.** $\dfrac{6}{28} - \dfrac{5}{42}$

9. $\dfrac{3}{4} - \dfrac{1}{8} + \dfrac{2}{3}$ **10.** $\dfrac{1}{3} - \dfrac{5}{6} + \dfrac{5}{12}$

Combine the following rational expressions. Reduce all answers to lowest terms.

11. $\dfrac{x}{x+3} + \dfrac{3}{x+3}$ **12.** $\dfrac{5x}{5x+2} + \dfrac{2}{5x+2}$ **13.** $\dfrac{4}{y-4} - \dfrac{y}{y-4}$

14. $\dfrac{8}{y+8} + \dfrac{y}{y+8}$ **15.** $\dfrac{x}{x^2-y^2} - \dfrac{y}{x^2-y^2}$ **16.** $\dfrac{x}{x^2-y^2} + \dfrac{y}{x^2-y^2}$

17. $\dfrac{2x-3}{x-2} - \dfrac{x-1}{x-2}$ **18.** $\dfrac{2x-4}{x+2} - \dfrac{x-6}{x+2}$ **19.** $\dfrac{1}{a} + \dfrac{2}{a^2} - \dfrac{3}{a^3}$

20. $\dfrac{3}{a} + \dfrac{2}{a^2} - \dfrac{1}{a^3}$ **21.** $\dfrac{3}{x} + \dfrac{4}{x^2}$ **22.** $\dfrac{7}{x^3} + \dfrac{2}{x}$

23. $\dfrac{5a}{2b} + \dfrac{3b}{a}$ **24.** $\dfrac{6a}{5b} + \dfrac{7b}{a}$ **25.** $\dfrac{2}{a^2 b} + \dfrac{3}{ab^2}$

26. $\dfrac{5}{a^3 b^2} + \dfrac{9}{ab}$ **27.** $\dfrac{7x-2}{2x+1} - \dfrac{5x-3}{2x+1}$ **28.** $\dfrac{7x-1}{3x+2} - \dfrac{4x-3}{3x+2}$

PAYING ATTENTION TO INSTRUCTIONS

29. Work each problem according to the instructions given.

  a. Multiply: $\dfrac{3}{8} \cdot \dfrac{1}{6}$ **b.** Divide: $\dfrac{3}{8} \div \dfrac{1}{6}$

  c. Add: $\dfrac{3}{8} + \dfrac{1}{6}$ **d.** Multiply: $\dfrac{x+3}{x-3} \cdot \dfrac{5x+15}{x^2-9}$

  e. Divide: $\dfrac{x+3}{x-3} \div \dfrac{5x+15}{x^2-9}$ **f.** Subtract: $\dfrac{x+3}{x-3} - \dfrac{5x+15}{x^2-9}$

30. Work each problem according to the instructions given.

  a. Multiply: $\dfrac{16}{49} \cdot \dfrac{1}{28}$ **b.** Divide: $\dfrac{16}{49} \div \dfrac{1}{28}$

  c. Subtract: $\dfrac{16}{49} - \dfrac{1}{28}$ **d.** Multiply: $\dfrac{3x-2}{3x+2} \cdot \dfrac{15x+6}{9x^2-4}$

  e. Divide: $\dfrac{3x-2}{3x+2} \div \dfrac{15x+6}{9x^2-4}$ **f.** Subtract: $\dfrac{3x+2}{3x-2} - \dfrac{15x+6}{9x^2-4}$

Combine the following rational expressions. Reduce all answers to lowest terms.

31. $\dfrac{3x+1}{2x-6} - \dfrac{x+2}{x-3}$ **32.** $\dfrac{x+1}{x-2} - \dfrac{4x+7}{5x-10}$

33. $\dfrac{6x+5}{5x-25} - \dfrac{x+2}{x-5}$ **34.** $\dfrac{4x+2}{3x+12} - \dfrac{x-2}{x+4}$

35. $\dfrac{x+1}{2x-2} - \dfrac{2}{x^2-1}$

36. $\dfrac{x+7}{2x+12} + \dfrac{6}{x^2-36}$

37. $\dfrac{1}{a-b} - \dfrac{3ab}{a^3-b^3}$

38. $\dfrac{1}{a+b} + \dfrac{3ab}{a^3+b^3}$

39. $\dfrac{1}{2y-3} - \dfrac{18y}{8y^3-27}$

40. $\dfrac{1}{3y-2} - \dfrac{18y}{27y^3-8}$

41. $\dfrac{x}{x^2-5x+6} - \dfrac{3}{3-x}$

42. $\dfrac{x}{x^2+4x+4} - \dfrac{2}{2+x}$

43. $\dfrac{2}{4t-5} + \dfrac{9}{8t^2-38t+35}$

44. $\dfrac{3}{2t-5} + \dfrac{21}{8t^2-14t-15}$

45. $\dfrac{1}{a^2-5a+6} + \dfrac{3}{a^2-a-2}$

46. $\dfrac{-3}{a^2+a-2} + \dfrac{5}{a^2-a-6}$

47. $\dfrac{1}{8x^3-1} - \dfrac{1}{4x^2-1}$

48. $\dfrac{1}{27x^3-1} - \dfrac{1}{9x^2-1}$

49. $\dfrac{4}{4x^2-9} - \dfrac{6}{8x^2-6x-9}$

50. $\dfrac{9}{9x^2+6x-8} - \dfrac{6}{9x^2-4}$

51. $\dfrac{4a}{a^2+6a+5} - \dfrac{3a}{a^2+5a+4}$

52. $\dfrac{3a}{a^2+7a+10} - \dfrac{2a}{a^2+6a+8}$

53. $\dfrac{2x-1}{x^2+x-6} - \dfrac{x+2}{x^2+5x+6}$

54. $\dfrac{4x+1}{x^2+5x+4} - \dfrac{x+3}{x^2+4x+3}$

55. $\dfrac{2x-8}{3x^2+8x+4} + \dfrac{x+3}{3x^2+5x+2}$

56. $\dfrac{5x+3}{2x^2+5x+3} - \dfrac{3x+9}{2x^2+7x+6}$

57. $\dfrac{2}{x^2+5x+6} - \dfrac{4}{x^2+4x+3} + \dfrac{3}{x^2+3x+2}$

58. $\dfrac{-5}{x^2+3x-4} + \dfrac{5}{x^2+2x-3} + \dfrac{1}{x^2+7x+12}$

59. $\dfrac{2x+8}{x^2+5x+6} - \dfrac{x+5}{x^2+4x+3} - \dfrac{x-1}{x^2+3x+2}$

60. $\dfrac{2x+11}{x^2+9x+20} - \dfrac{x+1}{x^2+7x+12} - \dfrac{x+6}{x^2+8x+15}$

61. $2 + \dfrac{3}{2x+1}$

62. $3 - \dfrac{2}{2x+3}$

63. $5 + \dfrac{2}{4-t}$

64. $7 + \dfrac{3}{5-t}$

65. $x - \dfrac{4}{2x+3}$

66. $x - \dfrac{5}{3x+4} + 1$

67. $\dfrac{x}{x+2} + \dfrac{1}{2x+4} - \dfrac{3}{x^2+2x}$

68. $\dfrac{x}{x+3} + \dfrac{7}{3x+9} - \dfrac{2}{x^2+3x}$

69. $\dfrac{1}{x} + \dfrac{x}{2x+4} - \dfrac{2}{x^2+2x}$

70. $\dfrac{1}{x} + \dfrac{x}{3x+9} - \dfrac{3}{x^2+3x}$

71. Let $f(x) = \dfrac{2}{x+4}$ and $g(x) = \dfrac{x-1}{x^2 + 3x - 4}$; find $f(x) + g(x)$

72. Let $f(t) = \dfrac{5}{3t-2}$ and $g(t) = \dfrac{t-3}{3t^2 + 7t - 6}$; find $f(t) - g(t)$

73. Let $f(x) = \dfrac{2x}{x^2 - x - 2}$ and $g(x) = \dfrac{5}{x^2 + x - 6}$; find $f(x) + g(x)$

74. Let $f(x) = \dfrac{7}{x^2 - x - 12}$ and $g(x) = \dfrac{5}{x^2 + x - 6}$; find $f(x) - g(x)$

APPLYING THE CONCEPTS

75. Optometry The formula

$$P = \frac{1}{a} + \frac{1}{b}$$

is used by optometrists to help determine how strong to make the lenses for a pair of eyeglasses. If a is 10 and b is 0.2, find the corresponding value of P.

76. Quadratic Formula Later in the book we will work with the quadratic formula. The derivation of the formula requires that you can add the fractions below. Add the fractions.

$$\frac{-c}{a} + \frac{b^2}{(2a)^2}$$

77. Number Problem Write an expression for the sum of a number and 4 times its reciprocal. Then, simplify that expression.

78. Number Problem Write an expression for the sum of a number and 3 times its reciprocal. Then, simplify that expression.

79. Number Problem Write an expression for the sum of the reciprocals of two consecutive integers. Then, simplify that expression.

80. Number Problem Write an expression for the sum of the reciprocals of two consecutive even integers. Then, simplify that expression.

GETTING READY FOR THE NEXT SECTION

Divide.

81. $\dfrac{3}{4} \div \dfrac{5}{8}$

82. $\dfrac{2}{3} \div \dfrac{5}{6}$

Multiply.

83. $x\left(1 + \dfrac{2}{x}\right)$ **84.** $3\left(x + \dfrac{1}{3}\right)$ **85.** $3x\left(\dfrac{1}{x} - \dfrac{1}{3}\right)$ **86.** $3x\left(\dfrac{1}{x} + \dfrac{1}{3}\right)$

Factor.

87. $x^2 - 4$

88. $x^2 - x - 6$

EXTENDING THE CONCEPTS

89. $\left(1 - \dfrac{1}{x}\right)\left(1 - \dfrac{1}{x+1}\right)\left(1 - \dfrac{1}{x+2}\right)\left(1 - \dfrac{1}{x+3}\right)$

90. $\left(1 + \dfrac{1}{x}\right)\left(1 + \dfrac{1}{x+1}\right)\left(1 + \dfrac{1}{x+2}\right)\left(1 + \dfrac{1}{x+3}\right)$

91. $\left(\dfrac{a^2 - b^2}{u^2 - v^2}\right)\left(\dfrac{av - au}{b - a}\right) + \left(\dfrac{a^2 - av}{u + v}\right)\left(\dfrac{1}{a}\right)$

92. $\left(\dfrac{6r^2}{r^2 - 1}\right)\left(\dfrac{r+1}{3}\right) - \dfrac{2r^2}{r - 1}$

93. $\dfrac{18x - 19}{4x^2 + 27x - 7} - \dfrac{12x - 41}{3x^2 + 17x - 28}$

94. $\dfrac{42 - 22y}{3y^2 - 13y - 10} - \dfrac{21 - 13y}{2y^2 - 9y - 5}$

95. $\left(\dfrac{1}{y^2 - 1} \div \dfrac{1}{y^2 + 1}\right)\left(\dfrac{y^3 + 1}{y^4 - 1}\right) + \dfrac{1}{(y + 1)^2(y - 1)}$

96. $\left(\dfrac{a^3 - 64}{a^2 - 16} \div \dfrac{a^2 - 4a + 16}{a^2 - 4} \div \dfrac{a^2 + 4a + 16}{a^3 + 64}\right) + 4 - a^2$

OBJECTIVE

A   Simplify complex fractions.

5.4

Every week in Portland, Oregon, bicyclists called zoobombers, young and old, convene on a street corner near the local zoo. Bicycles are decorated in streamers, strings of lights, and other ornate details to coincide with the often-costumed riders. Even the bikes themselves are known to cause a stir, falling into a variety of categories, such as swing bikes, mini bikes, tall bikes, and cruisers. Amidst laughter and shouts of joy, the bicyclists spend the evening speeding together down the steep hills of the neighborhood.

Imagine you were a spectator on the sidewalk of a street the zoobombers were speeding down. You hear the sound of a single screaming bicyclist zooming by because of the Doppler effect, where the pitch of the sound changes as the bicyclist approaches and then passes you. The formula for the sound frequency h you hear is

$$h = \frac{f}{1 + \frac{v}{s}}$$

where f is the actual frequency of the sound being produced, s is the speed of the sound (about 740 miles per hour), and v is the velocity of the bicyclist. This formula uses a complex rational expression, which is the focus of this section.

DEFINITION   *complex fractions*

A *complex fraction* is a fraction or rational expression that contains other fractions in its numerator or denominator.

Each of the following is a *complex fraction*:

$$\frac{\frac{1}{2}}{\frac{2}{3}} \qquad\qquad \frac{x + \frac{1}{y}}{y + \frac{1}{x}} \qquad\qquad \frac{\frac{a+1}{a^2-9}}{\frac{2}{a+3}}$$

Ⓐ Simplifying Complex Fractions

We will begin this section by simplifying the first of these complex fractions. Before we do, though, let's agree on some vocabulary. So that we won't have to use phrases such as "the numerator of the denominator," let's call the numerator of a complex fraction the *top* and the denominator of a complex fraction the *bottom*.

EXAMPLE 1

Simplify $\dfrac{\frac{3}{4}}{\frac{5}{8}}$.

SOLUTION   There are generally two methods that can be used to simplify complex fractions.

Method 1   We can multiply the numerator and denominator of the complex fractions by the LCD for both of the fractions, which in this case is 8.

$$\frac{\frac{3}{4}}{\frac{5}{8}} = \frac{\frac{3}{4} \cdot 8}{\frac{5}{8} \cdot 8} = \frac{6}{5}$$

Method 2   Instead of dividing by $\frac{5}{8}$ we can multiply by $\frac{8}{5}$.

$$\frac{\frac{3}{4}}{\frac{5}{8}} = \frac{3}{4} \cdot \frac{8}{5} = \frac{24}{20} = \frac{6}{5}$$

Here are some examples of complex fractions involving rational expressions. Most can be solved using either of the two methods shown in Example 1.

EXAMPLE 2

Simplify $\dfrac{\frac{1}{x} + \frac{1}{y}}{\frac{1}{x} - \frac{1}{y}}$.

SOLUTION   This problem is most easily solved using Method 1. We begin by multiplying both the numerator and denominator by the quantity xy, which is the LCD for all the fractions:

$$\frac{\frac{1}{x} + \frac{1}{y}}{\frac{1}{x} - \frac{1}{y}} = \frac{\left(\frac{1}{x} + \frac{1}{y}\right) \cdot xy}{\left(\frac{1}{x} - \frac{1}{y}\right) \cdot xy}$$

$$= \frac{\frac{1}{x}(xy) + \frac{1}{y}(xy)}{\frac{1}{x}(xy) - \frac{1}{y}(xy)}$$

Apply the distributive property to distribute xy over both terms in the numerator and denominator.

$$= \frac{y + x}{y - x}$$

EXAMPLE 3

Simplify $\dfrac{\dfrac{x-2}{x^2-9}}{\dfrac{x^2-4}{x+3}}$.

SOLUTION   Applying Method 2, we have

$$\dfrac{\dfrac{x-2}{x^2-9}}{\dfrac{x^2-4}{x+3}} = \dfrac{x-2}{x^2-9} \cdot \dfrac{x+3}{x^2-4}$$

$$= \dfrac{(x-2)(x+3)}{(x+3)(x-3)(x+2)(x-2)}$$

$$= \dfrac{1}{(x-3)(x+2)}$$

EXAMPLE 4

Simplify $\dfrac{1-\dfrac{4}{x^2}}{1-\dfrac{1}{x}-\dfrac{6}{x^2}}$.

SOLUTION   The simplest way to simplify this complex fraction is to multiply the numerator and denominator by the LCD, x^2:

$$\dfrac{1-\dfrac{4}{x^2}}{1-\dfrac{1}{x}-\dfrac{6}{x^2}} = \dfrac{x^2\left(1-\dfrac{4}{x^2}\right)}{x^2\left(1-\dfrac{1}{x}-\dfrac{6}{x^2}\right)} \qquad \text{Multiply numerator and denominator by } x^2.$$

$$= \dfrac{x^2 \cdot 1 - x^2 \cdot \dfrac{4}{x^2}}{x^2 \cdot 1 - x^2 \cdot \dfrac{1}{x} - x^2 \cdot \dfrac{6}{x^2}} \qquad \text{Distributive property.}$$

$$= \dfrac{x^2-4}{x^2-x-6} \qquad \text{Simplify.}$$

$$= \dfrac{(x-2)(x+2)}{(x-3)(x+2)} \qquad \text{Factor.}$$

$$= \dfrac{x-2}{x-3} \qquad \text{Reduce.}$$

EXAMPLE 5

Simplify $2 - \dfrac{3}{x+\dfrac{1}{3}}$.

SOLUTION   First, we simplify the expression that follows the subtraction sign.

$$2 - \dfrac{3}{x+\dfrac{1}{3}} = 2 - \dfrac{3 \cdot 3}{3\left(x+\dfrac{1}{3}\right)} = 2 - \dfrac{9}{3x+1}$$

Now we subtract by rewriting the first term, 2, with the LCD, $3x + 1$.

$$2 - \frac{9}{3x + 1} = \frac{2}{1} \cdot \frac{3x + 1}{3x + 1} - \frac{9}{3x + 1}$$

$$= \frac{6x + 2 - 9}{3x + 1} = \frac{6x - 7}{3x + 1}$$

GETTING READY FOR CLASS

Respond in your own words and in complete sentences.

A. What is a complex fraction?

B. Explain how a least common denominator can be used to simplify a complex fraction.

C. Explain how some complex fractions can be converted to division problems. When is it more efficient to convert a complex fraction to a division problem of rational expressions?

D. Which method of simplifying complex fractions do you prefer? Why?

Problem Set 5.4

Simplify each of the following as much as possible.

1. $\dfrac{\dfrac{3}{4}}{\dfrac{2}{3}}$

2. $\dfrac{\dfrac{5}{9}}{\dfrac{7}{12}}$

3. $\dfrac{\dfrac{1}{3} - \dfrac{1}{4}}{\dfrac{1}{2} + \dfrac{1}{8}}$

4. $\dfrac{\dfrac{1}{6} - \dfrac{1}{3}}{\dfrac{1}{4} - \dfrac{1}{8}}$

5. $\dfrac{3 + \dfrac{2}{5}}{1 - \dfrac{3}{7}}$

6. $\dfrac{2 + \dfrac{5}{6}}{1 - \dfrac{7}{8}}$

7. $\dfrac{\dfrac{1}{x}}{1 + \dfrac{1}{x}}$

8. $\dfrac{1 - \dfrac{1}{x}}{\dfrac{1}{x}}$

9. $\dfrac{1 + \dfrac{1}{a}}{1 - \dfrac{1}{a}}$

10. $\dfrac{1 - \dfrac{2}{a}}{1 - \dfrac{3}{a}}$

11. $\dfrac{\dfrac{1}{x} - \dfrac{1}{y}}{\dfrac{1}{x} + \dfrac{1}{y}}$

12. $\dfrac{\dfrac{1}{x} + \dfrac{2}{y}}{\dfrac{2}{x} + \dfrac{1}{y}}$

13. $\dfrac{\dfrac{x - 5}{x^2 - 4}}{\dfrac{x^2 - 25}{x + 2}}$

14. $\dfrac{\dfrac{3x + 1}{x^2 - 49}}{\dfrac{9x^2 - 1}{x - 7}}$

15. $\dfrac{\dfrac{4a}{2a^3 + 2}}{\dfrac{8a}{4a + 4}}$

16. $\dfrac{\dfrac{2a}{3a^3 - 3}}{\dfrac{4a}{6a - 6}}$

17. $\dfrac{1 - \dfrac{9}{x^2}}{1 - \dfrac{1}{x} - \dfrac{6}{x^2}}$

18. $\dfrac{4 - \dfrac{1}{x^2}}{4 + \dfrac{4}{x} + \dfrac{1}{x^2}}$

19. $\dfrac{2 + \dfrac{5}{a} - \dfrac{3}{a^2}}{2 - \dfrac{5}{a} + \dfrac{2}{a^2}}$

20. $\dfrac{3 + \dfrac{5}{a} - \dfrac{2}{a^2}}{3 - \dfrac{10}{a} + \dfrac{3}{a^2}}$

21. $\dfrac{2 + \dfrac{3}{x} - \dfrac{18}{x^2} - \dfrac{27}{x^3}}{2 + \dfrac{9}{x} + \dfrac{9}{x^2}}$

22. $\dfrac{3 + \dfrac{5}{x} - \dfrac{12}{x^2} - \dfrac{20}{x^3}}{3 + \dfrac{11}{x} + \dfrac{10}{x^2}}$

23. $\dfrac{1 + \dfrac{1}{x + 3}}{1 - \dfrac{1}{x + 3}}$

24. $\dfrac{1 + \dfrac{1}{x - 2}}{1 - \dfrac{1}{x - 2}}$

25. $\dfrac{1 + \dfrac{1}{x + 3}}{1 + \dfrac{7}{x - 3}}$

26. $\dfrac{1 + \dfrac{1}{x - 2}}{1 - \dfrac{3}{x + 2}}$

27. $\dfrac{1 - \dfrac{1}{a + 1}}{1 + \dfrac{1}{a - 1}}$

28. $\dfrac{\dfrac{1}{a-1}+1}{\dfrac{1}{a+1}-1}$

29. $\dfrac{\dfrac{1}{x+3}+\dfrac{1}{x-3}}{\dfrac{1}{x+3}-\dfrac{1}{x-3}}$

30. $\dfrac{\dfrac{1}{x+a}+\dfrac{1}{x-a}}{\dfrac{1}{x+a}-\dfrac{1}{x-a}}$

31. $\dfrac{\dfrac{y+1}{y-1}+\dfrac{y-1}{y+1}}{\dfrac{y+1}{y-1}-\dfrac{y-1}{y+1}}$

32. $\dfrac{\dfrac{y-1}{y+1}-\dfrac{y+1}{y-1}}{\dfrac{y-1}{y+1}+\dfrac{y+1}{y-1}}$

33. $1-\dfrac{x}{1-\dfrac{1}{x}}$

34. $x-\dfrac{1}{x-\dfrac{1}{2}}$

35. $1+\dfrac{1}{1+\dfrac{1}{1+1}}$

36. $1-\dfrac{1}{1-\dfrac{1}{1-\dfrac{1}{2}}}$

37. $\dfrac{1-\dfrac{1}{x+\dfrac{1}{2}}}{1+\dfrac{1}{x+\dfrac{1}{2}}}$

38. $\dfrac{2+\dfrac{1}{x-\dfrac{1}{3}}}{2-\dfrac{1}{x-\dfrac{1}{3}}}$

39. $\dfrac{\dfrac{1}{x+h}-\dfrac{1}{x}}{h}$

40. $\dfrac{\dfrac{1}{(x+h)^2}-\dfrac{1}{x^2}}{h}$

41. $\dfrac{\dfrac{3}{ab}+\dfrac{4}{bc}-\dfrac{2}{ac}}{\dfrac{5}{abc}}$

42. $\dfrac{\dfrac{x}{yz}-\dfrac{y}{xz}+\dfrac{z}{xy}}{\dfrac{1}{x^2y^2}-\dfrac{1}{x^2z^2}+\dfrac{1}{y^2z^2}}$

43. $\dfrac{\dfrac{t^2-2t-8}{t^2+7t+6}}{\dfrac{t^2-t-6}{t^2+2t+1}}$

44. $\dfrac{\dfrac{y^2-5y-14}{y^2+3y-10}}{\dfrac{y^2-8y+7}{y^2+6y+5}}$

45. $\dfrac{5+\dfrac{4}{b-1}}{\dfrac{7}{b+5}-\dfrac{3}{b-1}}$

46. $\dfrac{\dfrac{6}{x+5}-7}{\dfrac{8}{x+5}-\dfrac{9}{x+3}}$

47. $\dfrac{\dfrac{3}{x^2-x-6}}{\dfrac{2}{x+2}-\dfrac{4}{x-3}}$

48. $\dfrac{\dfrac{9}{a-7}+\dfrac{8}{2a+3}}{\dfrac{10}{2a^2-11a-21}}$

49. $\dfrac{\dfrac{1}{m-4}+\dfrac{1}{m-5}}{\dfrac{1}{m^2-9m+20}}$

50. $\dfrac{\dfrac{1}{k^2-7k+12}}{\dfrac{1}{k-3}+\dfrac{1}{k-4}}$

APPLYING THE CONCEPTS

51. Difference Quotient For each rational function below, find the difference quotient

$$\frac{f(x) - f(a)}{x - a}$$

a. $f(x) = \dfrac{4}{x}$ **b.** $f(x) = \dfrac{1}{x + 1}$ **c.** $f(x) = \dfrac{1}{x^2}$

52. Difference Quotient For each rational function below, find the difference quotient

$$\frac{f(x + h) - f(x)}{h}$$

a. $f(x) = \dfrac{4}{x}$ **b.** $f(x) = \dfrac{1}{x + 1}$ **c.** $f(x) = \dfrac{1}{x^2}$

53. Doppler Effect The change in the pitch of a sound (such as a train whistle) as an object passes is called the Doppler effect, named after C. J. Doppler (1803 – 1853). A person will *hear* a sound with a frequency, h, according to the formula

$$h = \frac{f}{1 + \dfrac{v}{s}}$$

where f is the actual frequency of the sound being produced, s is the speed of sound (about 740 miles per hour), and v is the velocity of the moving object.

a. Examine this fraction, and then explain why h and f approach the same value as v becomes smaller and smaller.

b. Solve this formula for v.

54. Work Problem A water storage tank has two drains. It can be shown that the time it takes to empty the tank if both drains are open is given by the formula

$$\frac{1}{\dfrac{1}{a} + \dfrac{1}{b}}$$

where a = time it takes for the first drain to empty the tank, and b = time for the second drain to empty the tank.

a. Simplify this complex fraction.

b. Find the amount of time needed to empty the tank using both drains if, used alone, the first drain empties the tank in 4 hours and the second drain can empty the tank in 3 hours.

55. Average Speed Average speed for a round trip can be given by the formula

$$v = \frac{2}{\dfrac{1}{v_1} + \dfrac{1}{v_2}}$$

where v_1 is the average speed to the destination and v_2 is the average speed for the return trip.

a. Simplify this complex fraction.

b. Find the average speed, to the nearest tenth, for the round trip if $v_1 = 35$ mph and $v_2 = 45$ mph.

GETTING READY FOR THE NEXT SECTION

Multiply.

56. $x(y - 2)$

57. $x(y - 1)$

58. $6\left(\dfrac{x}{2} - 3\right)$

59. $6\left(\dfrac{x}{3} + 1\right)$

60. $xab \cdot \dfrac{1}{x}$

61. $xab\left(\dfrac{1}{b} + \dfrac{1}{a}\right)$

Factor.

62. $y^2 - 25$

63. $x^2 - 3x + 2$

64. $xa + xb$

65. $xy - y$

Solve.

66. $5x - 4 = 6$

67. $y^2 + y - 20 = 2y$

EXTENDING THE CONCEPTS

68. $\dfrac{\left(\dfrac{1}{3}\right) - \left(\dfrac{1}{3}\right)^2}{1 - \dfrac{1}{3}}$

69. $\dfrac{\left(\dfrac{1}{2}\right) - \left(\dfrac{1}{2}\right)^2}{1 - \dfrac{1}{2}}$

70. $\dfrac{\left(\dfrac{1}{9}\right) - \dfrac{1}{9}\left(\dfrac{1}{3}\right)^4}{1 - \dfrac{1}{3}}$

71. $\dfrac{\left(\dfrac{1}{6}\right) - \dfrac{1}{6}\left(\dfrac{1}{2}\right)^4}{1 - \dfrac{1}{2}}$

72. $\dfrac{1 + \dfrac{1}{1 - \dfrac{a}{b}}}{1 - \dfrac{3}{1 - \dfrac{a}{b}}}$

73. $\dfrac{1 - \dfrac{1}{\dfrac{a}{b} + 2}}{1 + \dfrac{3}{\dfrac{a}{2b} + 1}}$

74. $\dfrac{a^{-1} + b^{-1}}{(ab)^{-1}}$

75. $\dfrac{(r^{-1} - s^{-1})^{-1}}{(rs)^{-2}}$

76. $\dfrac{(q^{-2} - t^{-2})^{-1}}{(t^{-1} - q^{-1})^{-1}}$

77. $\dfrac{(q^{-2} + t^{-2})^{-1}}{(t^{-1} + q^{-1})^{-1}}$

5.5

OBJECTIVES

A Solve an equation that contains rational expressions.

B Graph rational functions given the value of the variables.

© NASA JPL

The Mars500 Project consisted of six crewmembers who simulated a 520-day trip to Mars and back. The crew was locked inside a large facility that was built to look like the interior of a space shuttle. Isolated from the outside world, the men studied the effects of the mission on their own bodies. The mission included a trip to the planet for x number of days, y number of days to perform mock experiments on the planet, and then z number of days to fly back to Earth. We can set up the following rational expressions to represent the days in each stage of the mission:

Trip to Mars ↘   ↙ Trip to Earth

$$\frac{x}{520} + \frac{y}{520} + \frac{z}{520} = \text{Total days of mission}$$

↑
Experiments

Then if given the value for two of the variables, we can determine the value of the third variable using a rational equation. Equations that involve rational expressions are the focus of this section.

A Solving Equations with Rational Expressions

EXAMPLE 1

Solve $\frac{x}{2} - 3 = \frac{2}{3}$.

SOLUTION The LCD for 2 and 3 is 6. Multiplying both sides by 6, we have

$$6\left(\frac{x}{2} - 3\right) = 6\left(\frac{2}{3}\right)$$

$$6\left(\frac{x}{2}\right) - 6(3) = 6\left(\frac{2}{3}\right)$$

$$3x - 18 = 4$$

$$3x = 22$$

$$x = \frac{22}{3}$$

5.5 VIDEOS

Multiplying both sides of an equation by the LCD clears the equation of fractions because the LCD has the property that all the denominators divide it evenly.

EXAMPLE 2

Solve $\dfrac{6}{a-4} = \dfrac{3}{8}$.

SOLUTION The LCD for $a-4$ and 8 is $8(a-4)$. Multiplying both sides by this quantity yields

$$8(a-4) \cdot \frac{6}{a-4} = 8(a-4) \cdot \frac{3}{8}$$

$$48 = (a-4) \cdot 3$$

$$48 = 3a - 12$$

$$60 = 3a$$

$$20 = a$$

The solution set is 20, which checks in the original equation.

When we multiply both sides of an equation by an expression containing the variable, we must be sure to check our solutions. The multiplication property of equality does not allow multiplication by 0. If the expression we multiply by contains the variable, then it has the possibility of being 0. In the last example, we multiplied both sides by $8(a-4)$. This gives a restriction $a \neq 4$ for any solution we come up with.

EXAMPLE 3

Solve $\dfrac{x}{x-2} + \dfrac{2}{3} = \dfrac{2}{x-2}$.

SOLUTION The LCD is $3(x-2)$. We are assuming $x \neq 2$ when we multiply both sides of the equation by $3(x-2)$:

$$3(x-2) \cdot \left(\frac{x}{x-2} + \frac{2}{3} \right) = 3(x-2) \cdot \frac{2}{x-2}$$

$$3x + (x-2) \cdot 2 = 3 \cdot 2$$

$$3x + 2x - 4 = 6$$

$$5x - 4 = 6$$

$$5x = 10$$

$$x = 2$$

NOTE In the process of solving the equation, we multiplied both sides by $3(x-2)$, solved for x, and got $x = 2$ for our solution. But when x is 2, the quantity $3(x-2) = 3(2-2) = 3(0) = 0$, which means we multiplied both sides of our equation by 0, which is not allowed under the multiplication property of equality.

The only possible solution is $x = 2$. Checking this value back in the original equation gives

$$\frac{2}{2-2} + \frac{2}{3} \overset{?}{=} \frac{2}{2-2}$$

$$\frac{2}{0} + \frac{2}{3} \overset{?}{=} \frac{2}{0}$$

The first and last terms are undefined. The proposed solution, $x = 2$, does not check in the original equation. The solution set is the empty set. There is no solution to the original equation.

When the proposed solution to an equation is not actually a solution, it is called an *extraneous* solution. In the last example, $x = 2$ is an extraneous solution.

EXAMPLE 4

Solve $\dfrac{5}{x^2 - 3x + 2} - \dfrac{1}{x - 2} = \dfrac{1}{3x - 3}$.

SOLUTION Writing the equation again with the denominators in factored form, we have

$$\frac{5}{(x - 2)(x - 1)} - \frac{1}{x - 2} = \frac{1}{3(x - 1)}$$

The LCD is $3(x - 2)(x - 1)$. Multiplying through by the LCD, we have

$$3(x - 2)(x - 1)\,\frac{5}{(x - 2)(x - 1)} - 3(x - 2)(x - 1) \cdot \frac{1}{(x - 2)}$$

$$= 3(x - 2)(x - 1) \cdot \frac{1}{3(x - 1)}$$

$$3 \cdot 5 - 3(x - 1) \cdot 1 = (x - 2) \cdot 1$$

$$15 - 3x + 3 = x - 2$$

$$-3x + 18 = x - 2$$

$$-4x + 18 = -2$$

$$-4x = -20$$

$$x = 5$$

> **NOTE** We can check the proposed solution in any of the equations obtained before multiplying through by the LCD. We cannot check the proposed solution in an equation obtained after multiplying through by the LCD because, if we have multiplied by 0, the resulting equations will not be equivalent to the original one.

EXAMPLE 5

Checking the proposed solution $x = 5$ in the original equation yields a true statement. Try it and see.

$$\text{Solve } 3 + \frac{1}{x} = \frac{10}{x^2}.$$

SOLUTION To clear the equation of denominators, we multiply both sides by x^2:

$$x^2\left(3 + \frac{1}{x}\right) = x^2\left(\frac{10}{x^2}\right)$$

$$3(x^2) + \left(\frac{1}{x}\right)(x^2) = \left(\frac{10}{x^2}\right)(x^2)$$

$$3x^2 + x = 10$$

Rewrite in standard form, and solve:

$$3x^2 + x - 10 = 0$$

$$(3x - 5)(x + 2) = 0$$

$$3x - 5 = 0 \quad \text{or} \quad x + 2 = 0$$

$$x = \frac{5}{3} \qquad\qquad x = -2$$

The solution set is $\left[-2, \frac{5}{3}\right]$. Both solutions check in the original equation. Remember: We have to check all solutions any time we multiply both sides of the equation by an expression that contains the variable, just to be sure we haven't multiplied by 0.

EXAMPLE 6

Solve $\dfrac{y-4}{y^2-5y} = \dfrac{2}{y^2-25}$.

SOLUTION Factoring each denominator, we find the LCD is $y(y-5)(y+5)$. Multiplying each side of the equation by the LCD clears the equation of denominators and leads us to our possible solutions:

$$y(y-5)(y+5) \cdot \frac{y-4}{y(y-5)} = \frac{2}{(y-5)(y+5)} \cdot y(y-5)(y+5)$$

$$(y+5)(y-4) = 2y$$

$$y^2 + y - 20 = 2y \qquad \text{Multiply out the left side.}$$

$$y^2 - y - 20 = 0 \qquad \text{Add } -2y \text{ to each side.}$$

$$(y-5)(y+4) = 0$$

$$y - 5 = 0 \quad \text{or} \quad y + 4 = 0$$

$$y = 5 \qquad\qquad y = -4$$

The two possible solutions are 5 and -4. If we substitute -4 for y in the original equation, we find that it leads to a true statement. It is therefore a solution. On the other hand, if we substitute 5 for y in the original equation, we find that both sides of the equation are undefined. The only solution to our original equation is $y = -4$. The other possible solution, $y = 5$, is extraneous.

EXAMPLE 7

Solve for y: $x = \dfrac{y-4}{y-2}$.

SOLUTION To solve for y, we first multiply each side by $y - 2$ to obtain

$$x(y-2) = y - 4$$

$$xy - 2x = y - 4 \qquad \text{Distributive property}$$

$$xy - y = 2x - 4 \qquad \text{Collect all terms containing } y \text{ on the left side.}$$

$$y(x-1) = 2x - 4 \qquad \text{Factor } y \text{ from each term on the left side.}$$

$$y = \frac{2x-4}{x-1} \qquad \text{Divide each side by } x - 1.$$

EXAMPLE 8

Solve the formula $\dfrac{1}{x} = \dfrac{1}{b} + \dfrac{1}{a}$ for x.

SOLUTION We begin by multiplying both sides by the least common denominator xab. As you can see from our previous examples, multiplying both sides of an equation by the LCD is equivalent to multiplying each term of both sides by the LCD:

$$xab \cdot \frac{1}{x} = \frac{1}{b} \cdot xab + \frac{1}{a} \cdot xab$$

$$ab = xa + xb$$

$$ab = (a+b)x \qquad \text{Factor } x \text{ from the right side.}$$

$$\frac{ab}{a+b} = x$$

We know we are finished because the variable we were solving for is alone on one side of the equation and does not appear on the other side.

Graphing Rational Functions

In our next example, we investigate the graph of a rational function.

EXAMPLE 9

Graph the rational function $f(x) = \dfrac{6}{x-2}$.

SOLUTION   To find the y-intercept, we let x equal 0.

$$\text{When } x = 0: \qquad y = \frac{6}{0-2} = \frac{6}{-2} = -3 \qquad \text{\textit{y}-intercept}$$

The graph will not cross the x-axis. If it did, we would have a solution to the equation

$$0 = \frac{6}{x-2}$$

which has no solution because there is no number to divide 6 by to obtain 0.

The graph of our equation is shown in Figure 1 along with a table giving values of x and y that satisfy the equation. Notice that y is undefined when x is 2. This means that the graph will not cross the vertical line $x = 2$. (If it did, there would be a value of y for $x = 2$.) The line $x = 2$ is called a *vertical asymptote* of the graph. The graph will get very close to the vertical asymptote, but will never touch or cross it.

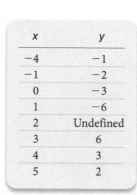

x	y
-4	-1
-1	-2
0	-3
1	-6
2	Undefined
3	6
4	3
5	2

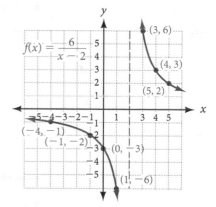

FIGURE 1 *The graph of $f(x) = \dfrac{6}{x-2}$*

If you were to graph $y = \frac{6}{x}$ on the coordinate system in Figure 1, you would see that the graph of $y = \frac{6}{x-2}$ is the graph of $y = \frac{6}{x}$ with all points shifted 2 units to the right.

USING TECHNOLOGY *More About Example 9*

We know the graph of $f(x) = \frac{6}{x-2}$ will not cross the vertical asymptote $x = 2$ because replacing x with 2 in the equation gives us an undefined expression, meaning there is no value of y to associate with $x = 2$. We can use a graphing calculator to explore the behavior of this function when x gets closer and closer to 2 by using the table function on the calculator. We want to put our own values for X into the table, so we set the independent variable to Ask. (On a TI-82/83, use the TBLSET key to set up the table.) To see how the function behaves as x gets close to 2, we let X take on values of 1.9, 1.99, and 1.999. Then we move to the other side of 2 and let X become 2.1, 2.01, and 2.001.

```
TABLE SETUP
  TblStart = 0
  ΔTbl = 1
Indpnt:  Auto  Ask
Depend:  Auto  Ask
```

```
Plot1  Plot2  Plot3
\Y₁ ■ 6/(X − 2)
\Y₂ =
\Y₃ =
\Y₄ =
\Y₅ =
\Y₆ =
\Y₇ =
```

The table will look like this:

X	Y₁
1.9	−60
1.99	−600
1.999	−6000
2.1	60
2.01	600
2.001	6000

Again, the calculator asks us for a table increment. Because we are inputting the x values ourselves, the increment value does not matter.

As you can see, the values in the table support the shape of the curve in Figure 1 around the vertical asymptote $x = 2$.

EXAMPLE 10

Graph: $g(x) = \dfrac{6}{x+2}$

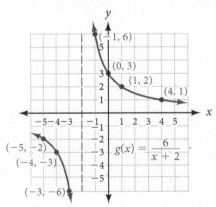

FIGURE 2 *The graph of $g(x) = \dfrac{6}{x+2}$*

SOLUTION The only difference between this equation and the equation in Example 9 is in the denominator. This graph will have the same shape as the graph in Example 9, but the vertical asymptote will be $x = -2$ instead of $x = 2$. Figure 2 shows the graph.

Notice that the graphs shown in Figures 1 and 2 are both graphs of functions because no vertical line will cross either graph in more than one place. Notice the similarities and differences in our two functions,

$$f(x) = \frac{6}{x - 2} \qquad \text{and} \qquad g(x) = \frac{6}{x + 2}$$

and their graphs. The vertical asymptotes shown in Figures 1 and 2 correspond to the fact that both $f(2)$ and $g(-2)$ are undefined. The domain for the function f is all real numbers except $x = 2$, while the domain for g is all real numbers except $x = -2$.

GETTING READY FOR CLASS

After reading through the preceding section, respond in your own words and in complete sentences.

A. Explain how a least common denominator can be used to simplify an equation.

B. What is an extraneous solution?

C. How does the location of the vertical asymptote in the graph of a rational function relate to the equation of the function?

D. What is the last step in solving an equation that contains rational expressions?

Problem Set 5.5

Solve each of the following equations.

1. $\frac{x}{5} + 4 = \frac{5}{3}$

2. $\frac{x}{5} = \frac{x}{2} - 9$

3. $\frac{a}{3} + 2 = \frac{4}{5}$

4. $\frac{a}{4} + \frac{1}{2} = \frac{2}{3}$

5. $\frac{y}{2} + \frac{y}{4} + \frac{y}{6} = 3$

6. $\frac{y}{3} - \frac{y}{6} + \frac{y}{2} = 1$

7. $\frac{5}{2x} = \frac{1}{x} + \frac{3}{4}$

8. $\frac{1}{2a} = \frac{2}{a} - \frac{3}{8}$

9. $\frac{1}{x} = \frac{1}{3} - \frac{2}{3x}$

10. $\frac{5}{2x} = \frac{2}{x} - \frac{1}{12}$

11. $\frac{2x}{x-3} + 2 = \frac{2}{x-3}$

12. $\frac{2}{x+5} = \frac{2}{5} - \frac{x}{x+5}$

13. $1 - \frac{1}{x} = \frac{12}{x^2}$

14. $2 + \frac{5}{x} = \frac{3}{x^2}$

15. $y - \frac{4}{3y} = -\frac{1}{3}$

16. $\frac{y}{2} - \frac{4}{y} = -\frac{7}{2}$

17. $\frac{x+2}{x+1} = \frac{1}{x+1} + 2$

18. $\frac{x+6}{x+3} = \frac{3}{x+3} + 2$

19. $\frac{3}{a-2} = \frac{2}{a-3}$

20. $\frac{5}{a+1} = \frac{4}{a+2}$

21. $6 - \frac{5}{x^2} = \frac{7}{x}$

22. $10 - \frac{3}{x^2} = -\frac{1}{x}$

23. $\frac{1}{x-1} - \frac{1}{x+1} = \frac{3x}{x^2-1}$

24. $\frac{5}{x-1} + \frac{2}{x-1} = \frac{4}{x+1}$

25. $\frac{2}{x-3} + \frac{x}{x^2-9} = \frac{4}{x+3}$

26. $\frac{2}{x+5} + \frac{3}{x+4} = \frac{2x}{x^2+9x+20}$

27. $\frac{3}{2} - \frac{1}{x-4} = \frac{-2}{2x-8}$

28. $\frac{2}{x} - \frac{1}{x+1} = \frac{-2}{5x+5}$

29. $\frac{t-4}{t^2-3t} = \frac{-2}{t^2-9}$

30. $\frac{t+3}{t^2-2t} = \frac{10}{t^2-4}$

31. $\frac{3}{y-4} - \frac{2}{y+1} = \frac{5}{y^2-3y-4}$

32. $\frac{1}{y+2} - \frac{2}{y-3} = \frac{-2y}{y^2-y-6}$

33. $\frac{2}{1+a} = \frac{3}{1-a} + \frac{5}{a}$

34. $\frac{1}{a+3} - \frac{a}{a^2-9} = \frac{2}{3-a}$

455

35. $\dfrac{3}{2x-6} - \dfrac{x+1}{4x-12} = 4$

36. $\dfrac{2x-3}{5x+10} + \dfrac{3x-2}{4x+8} = 1$

37. $\dfrac{y+2}{y^2-y} - \dfrac{6}{y^2-1} = 0$

38. $\dfrac{y+3}{y^2-y} - \dfrac{8}{y^2-1} = 0$

39. $\dfrac{4}{2x-6} - \dfrac{12}{4x+12} = \dfrac{12}{x^2-9}$

40. $\dfrac{1}{x+2} + \dfrac{1}{x-2} = \dfrac{4}{x^2-4}$

41. $\dfrac{2}{y^2-7y+12} - \dfrac{1}{y^2-9} = \dfrac{4}{y^2-y-12}$

42. $\dfrac{1}{y^2+5y+4} + \dfrac{3}{y^2-1} = \dfrac{-1}{y^2+3y-4}$

43. Let $f(x) = \dfrac{1}{x-3}$ and $g(x) = \dfrac{1}{x+3}$ and find x if

    a. $f(x) + g(x) = \dfrac{5}{8}$ **b.** $\dfrac{f(x)}{g(x)} = 5$ **c.** $f(x) = g(x)$

44. Let $f(x) = \dfrac{4}{x+2}$ and $g(x) = \dfrac{4}{x-2}$ and find x if

    a. $f(x) - g(x) = -\dfrac{4}{3}$ **b.** $\dfrac{g(x)}{f(x)} = -7$ **c.** $f(x) = -g(x)$

45. Solve each equation.

    a. $6x - 2 = 0$ **b.** $\dfrac{6}{x} - 2 = 0$ **c.** $\dfrac{x}{6} - 2 = -\dfrac{1}{2}$

    d. $\dfrac{6}{x} - 2 = -\dfrac{1}{2}$ **e.** $\dfrac{6}{x^2} + 6 = \dfrac{20}{x}$

46. Solve each equation.

    a. $5x - 2 = 0$ **b.** $5 - \dfrac{2}{x} = 0$ **c.** $\dfrac{x}{2} - 5 = -\dfrac{3}{4}$

    d. $\dfrac{2}{x} - 5 = -\dfrac{3}{4}$ **e.** $-\dfrac{3}{x} + \dfrac{2}{x^2} = 5$

PAYING ATTENTION TO INSTRUCTIONS

47. Work each problem according to the instructions given.

    a. Divide: $\dfrac{6}{x^2-2x-8} \div \dfrac{x+3}{x+2}$

    b. Add: $\dfrac{6}{x^2-2x-8} + \dfrac{x+3}{x+2}$

    c. Solve: $\dfrac{6}{x^2-2x-8} + \dfrac{x+3}{x+2} = 2$

48. Work each problem according to the instructions given.

   a. Divide: $\dfrac{-10}{x^2 - 25} \div \dfrac{x - 4}{x - 5}$

   b. Add: $\dfrac{-10}{x^2 - 25} + \dfrac{x - 4}{x - 5}$

   c. Solve: $\dfrac{-10}{x^2 - 25} + \dfrac{x - 4}{x - 5} = \dfrac{4}{5}$

49. Solve $\dfrac{1}{x} = \dfrac{1}{b} - \dfrac{1}{a}$ for x.   **50.** Solve $\dfrac{1}{x} = \dfrac{1}{a} - \dfrac{1}{b}$ for x.

Solve for y.

51. $x = \dfrac{y - 3}{y - 1}$ **52.** $x = \dfrac{y - 2}{y - 3}$ **53.** $x = \dfrac{2y + 1}{3y + 1}$ **54.** $x = \dfrac{3y + 2}{5y + 1}$

Graph each function. Show the vertical asymptote.

55. $f(x) = \dfrac{1}{x - 3}$ **56.** $f(x) = \dfrac{1}{x + 3}$ **57.** $f(x) = \dfrac{4}{x + 2}$ **58.** $f(x) = \dfrac{4}{x - 2}$

59. $g(x) = \dfrac{2}{x - 4}$ **60.** $g(x) = \dfrac{2}{x + 4}$ **61.** $g(x) = \dfrac{6}{x + 1}$ **62.** $g(x) = \dfrac{6}{x - 1}$

APPLYING THE CONCEPTS

63. Geometry From plane geometry and the principle of similar triangles, the relationship between y_1, y_2, and h shown in Figure 3 can be expressed as

$$\frac{1}{h} = \frac{1}{y_1} + \frac{1}{y_2}$$

Two poles are 12 feet high and 8 feet high. If a cable is attached to the top of each one and stretched to the bottom of the other, what is the height above the ground at which the two wires will meet?

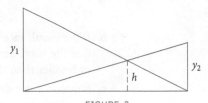

FIGURE 3

64. Kayak Race In a kayak race, the participants must paddle a kayak 450 meters down a river and then return 450 meters up the river to the starting point (Figure 4). Susan has correctly deduced that the total time t (in seconds) depends on the speed c (in meters per second) of the water according to the following expression:

$$t = \frac{450}{v + c} + \frac{450}{v - c}$$

where v is the speed of the kayak relative to the water (the speed of the kayak in still water).

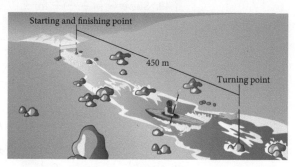

FIGURE 4

a. Fill in the following table.

Time	Speed of Kayak Relative to the Water	Current of the River
t(sec)	v(m/sec)	c(m/sec)
240		1
300		2
	4	3
	3	1
540	3	
	3	3

b. If the kayak race were conducted in the still waters of a lake, do you think that the total time of a given participant would be greater than, equal to, or smaller than the time in the river? Justify your answer.

c. Suppose Peter can paddle his kayak at 4.1 meters per second and that the speed of the current is 4.1 meters per second. What will happen when Peter makes the turn and tries to come back up the river? How does this situation show up in the equation for total time?

65. Resistance In an electrical circuit, when two resistors R_1 and R_2 are connected in parallel, the combined resistance R is given by

$$\frac{1}{R} = \frac{1}{R_1} + \frac{1}{R_2}$$

Find the total resistance when one resistor has a resistance of 12 ohms and the other has a resistance of 6 ohms.

66. Resistance In an electrical circuit, when two resistors R_1 and R_2 are connected in parallel, the combined resistance R is given by

$$\frac{1}{R} = \frac{1}{R_1} + \frac{1}{R_2}$$

Find the total resistance when one resistor has a resistance of 9 ohms and the other has a resistance of 3 ohms.

GETTING READY FOR THE NEXT SECTION

Multiply.

67. $39.3 \cdot 60$

68. $1,100 \cdot 60 \cdot 60$

Divide. Round to the nearest tenth, if necessary.

69. $65,000 \div 5,280$

70. $3,960,000 \div 5,280$

Multiply.

71. $2x\left(\dfrac{1}{x} + \dfrac{1}{2x}\right)$

72. $3x\left(\dfrac{1}{x} + \dfrac{1}{3x}\right)$

Solve.

73. $12(x + 3) + 12(x - 3) = 3(x^2 - 9)$ **74.** $40 + 2x = 60 - 3x$

75. $\dfrac{1}{10} - \dfrac{1}{12} = \dfrac{1}{x}$

76. $\dfrac{1}{x} + \dfrac{1}{2x} = 2$

EXTENDING THE CONCEPTS

77. $\dfrac{12}{x} + \dfrac{8}{x^2} - \dfrac{75}{x^3} - \dfrac{50}{x^4} = 0$

78. $\dfrac{45}{x} + \dfrac{18}{x^2} - \dfrac{80}{x^3} - \dfrac{32}{x^4} = 0$

79. $\dfrac{1}{x^3} - \dfrac{1}{3x^2} - \dfrac{1}{4x} + \dfrac{1}{12} = 0$

80. $\dfrac{1}{x^3} - \dfrac{1}{2x^2} - \dfrac{1}{9x} + \dfrac{1}{18} = 0$

81. Solve for x. $\dfrac{2}{x} + \dfrac{4}{x + a} = \dfrac{-6}{a - x}$

82. Solve for x. $\dfrac{1}{b - x} - \dfrac{1}{x} = \dfrac{-2}{b + x}$

83. Solve for v. $\dfrac{s - vt}{t^2} = -16$

84. Solve for r. $A = P\left(1 + \dfrac{r}{n}\right)$

85. Solve for f. $\dfrac{1}{p} = \dfrac{1}{f} + \dfrac{1}{g}$

86. Solve for p. $h = \dfrac{v^2}{2g} + \dfrac{p}{c}$

5.6

© iStockphoto/YuriyS

OBJECTIVES

A Solve applications where solutions depend on solving an equation containing rational expressions.

B Graph rational functions with binomials in the numerator and denominator.

In May 2010, an American adventurer fulfilled a lifelong dream of crossing the English Channel by balloon—helium balloons, that is. Jonathan Trappe is a registered pilot for lighter-than-air systems. He strapped himself to a chair outfitted with safety equipment and hung from the bundle of large colorful balloons. Then he took to the skies from an English field. He floated over the White Cliffs of Dover, continued across the English Channel, and landed in a French cabbage patch without incident.

Imagine yourself trying to re-create Trappe's helium balloon flight across the English Channel. The flight's distance was approximately 21 miles, and the complete trip took 5 hours. If the speed that Trappe completed his flight was 3 miles per hour faster than yours, how fast was your speed? This is an example of an application that involves a rational equation. We will work similar problems in this section.

Applications Involving Rational Expressions

EXAMPLE 1

One number is twice another. The sum of their reciprocals is 2. Find the numbers.

SOLUTION Let x = the smaller number. The larger number is $2x$. Their reciprocals are $\frac{1}{x}$ and $\frac{1}{2x}$. The equation that describes the situation is

$$\frac{1}{x} + \frac{1}{2x} = 2$$

Multiplying both sides by the LCD $2x$, we have

$$2x \cdot \frac{1}{x} + 2x \cdot \frac{1}{2x} = 2x(2)$$

$$2 + 1 = 4x$$

$$3 = 4x$$

$$x = \frac{3}{4}$$

5.6 VIDEOS

The smaller number is $\frac{3}{4}$. The larger is $2\left(\frac{3}{4}\right) = \frac{6}{4} = \frac{3}{2}$. Adding their reciprocals, we have

$$\frac{4}{3} + \frac{2}{3} = \frac{6}{3} = 2$$

The sum of the reciprocals of $\frac{3}{4}$ and $\frac{3}{2}$ is 2.

EXAMPLE 2

Two families from the same neighborhood plan a ski trip together. The first family is driving a newer vehicle and makes the 455-mile trip at a speed 5 miles per hour faster than the second family, who is traveling in an older vehicle. The second family takes a half-hour longer to make the trip. What are the speeds of the two families?

SOLUTION   The following table will be helpful in finding the equation necessary to solve this problem.

	d(distance)	r(rate)	t(time)
First Family			
Second Family			

If we let x be the speed of the second family, then the speed of the first family will be $x + 5$. Both families travel the same distance of 455 miles. Putting this information into the table we have

	d	r	t
First Family	455	$x + 5$	
Second Family	455	x	

To fill in the last two spaces in the table, we use the relationship $d = r \cdot t$. Since the last column of the table is the time, we solve the equation $d = r \cdot t$ for t and get

$$t = \frac{d}{r}$$

Taking the distance and dividing by the rate (speed) for each family, we complete the table.

	d	r	t
First Family	455	$x + 5$	$\frac{455}{x + 5}$
Second Family	455	x	$\frac{455}{x}$

Reading the problem again, we find that the time for the second family is longer than the time for the first family by one-half hour. In other words, the time for the second family can be found by adding one-half hour to the time for the first family, or

$$\frac{455}{x + 5} + \frac{1}{2} = \frac{455}{x}$$

Multiplying both sides by the LCD of $2x(x + 5)$ gives

$$2x \cdot (455) + x(x + 5) \cdot 1 = 455 \cdot 2(x + 5)$$

$$910x + x^2 + 5x = 910x + 4550$$

$$x^2 + 5x - 4550 = 0$$

$$(x + 70)(x - 65) = 0$$

$$x = -70 \quad \text{or} \quad x = 65$$

Since we cannot have a negative speed, the only solution is $x = 65$. Then

$$x + 5 = 65 + 5 = 70$$

The speed of the first family is 70 miles per hour, and the speed of the second family is 65 miles per hour.

EXAMPLE 3

The current of a river is 3 miles per hour. It takes a motorboat a total of 3 hours to travel 12 miles upstream and return 12 miles downstream. What is the speed of the boat in still water?

SOLUTION   This time we let $x =$ the speed of the boat in still water. Then, we fill in as much of the table as possible using the information given in the problem. For instance, because we let $x =$ the speed of the boat in still water, the rate upstream (against the current) must be $x - 3$. The rate downstream (with the current) is $x + 3$.

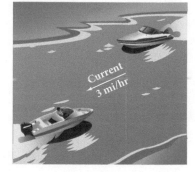

	d	r	t
Upstream	12	$x - 3$	
Downstream	12	$x + 3$	

The last two boxes can be filled in using the relationship

$$t = \frac{d}{r}$$

	d	r	t
Upstream	12	$x - 3$	$\dfrac{12}{x - 3}$
Downstream	12	$x + 3$	$\dfrac{12}{x + 3}$

The total time for the trip up and back is 3 hours:

$$\text{Time upstream} + \text{Time downstream} = \text{Total time}$$

$$\frac{12}{x-3} + \frac{12}{x+3} = 3$$

Multiplying both sides by $(x - 3)(x + 3)$, we have

$$12(x + 3) + 12(x - 3) = 3(x^2 - 9)$$

$$12x + 36 + 12x - 36 = 3x^2 - 27$$

$$3x^2 - 24x - 27 = 0$$

$$x^2 - 8x - 9 = 0 \qquad \text{Divide both sides by 3.}$$

$$(x - 9)(x + 1) = 0$$

$$x = 9 \qquad \text{or} \qquad x = -1$$

The speed of the motorboat in still water is 9 miles per hour. (We don't use $x = -1$ because the speed of the motorboat cannot be a negative number.)

EXAMPLE 4

An inlet pipe can fill a pool in 10 hours, while the drain can empty it in 12 hours. If the pool is empty and both the inlet pipe and drain are open, how long will it take to fill the pool?

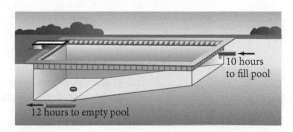
10 hours to fill pool
12 hours to empty pool

SOLUTION It is helpful to think in terms of how much work is done by each pipe in 1 hour.

Let $x =$ the time it takes to fill the pool with both pipes open.

If the inlet pipe can fill the pool in 10 hours, then in 1 hour it is $\frac{1}{10}$ full. If the outlet pipe empties the pool in 12 hours, then in 1 hour it is $\frac{1}{12}$ empty. If the pool can be filled in x hours with both the inlet pipe and the drain open, then in 1 hour it is $\frac{1}{x}$ full when both pipes are open.

Here is the equation:

In 1 hour

$$\begin{bmatrix} \text{Amount filled} \\ \text{by inlet pipe} \end{bmatrix} - \begin{bmatrix} \text{Amount emptied} \\ \text{by the drain} \end{bmatrix} = \begin{bmatrix} \text{Fraction of pool} \\ \text{filled with both pipes} \end{bmatrix}$$

$$\frac{1}{10} - \frac{1}{12} = \frac{1}{x}$$

Multiplying through by $60x$, we have

$$60x \cdot \frac{1}{10} - 60x \cdot \frac{1}{12} = 60x \cdot \frac{1}{x}$$

$$6x - 5x = 60$$

$$x = 60$$

It takes 60 hours to fill the pool if both the inlet pipe and the drain are open.

B More About Graphing Rational Functions

We continue our investigation of the graphs of rational functions by considering the graph of a rational function with binomials in the numerator and denominator.

EXAMPLE 5

Graph the rational function $y = \frac{x - 4}{x - 2}$.

SOLUTION   In addition to making a table to find some points on the graph, we can analyze the graph as follows:

1. The graph will have a y-intercept of 2, because when $x = 0$, $y = \frac{-4}{-2} = 2$.

2. To find the x-intercept, we let $y = 0$ to get

    $$0 = \frac{x - 4}{x - 2}$$

    The only way this expression can be 0 is if the numerator is 0, which happens when $x = 4$. (If you want to solve this equation, multiply both sides by $x - 2$. You will get the same solution, $x = 4$.)

3. The graph will have a vertical asymptote at $x = 2$, because $x = 2$ will make the denominator of the function 0, meaning y is undefined when x is 2.

4. The graph will have a *horizontal asymptote* at $y = 1$ because for very large values of x, $\frac{x-4}{x-2}$ is very close to 1. The larger x is, the closer $\frac{x-4}{x-2}$ is to 1. The same is true for very small values of x, such as $-1{,}000$ and $-10{,}000$.

Putting this information together with the ordered pairs in the table next to the figure, we have the graph shown in Figure 1.

x	y
-1	$\frac{5}{3}$
0	2
1	3
2	Undefined
3	-1
4	0
5	$\frac{1}{3}$

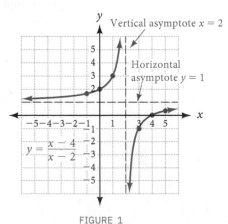

FIGURE 1

USING TECHNOLOGY *More About Example 5*

In the previous section, we used technology to explore the graph of a rational function around a vertical asymptote. This time, we are going to explore the graph near the horizontal asymptote. In Figure 1, the horizontal asymptote is at $y = 1$. To show that the graph approaches this line as x becomes very large, we use the table function on our graphing calculator, with X taking values of 100, 1,000, and 10,000. To show that the graph approaches the line $y = 1$ on the left side of the coordinate system, we let X become -100, $-1,000$ and $-10,000$.

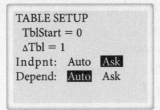

The table will look like this:

X	Y_1	
100	.97959	
1000	.998	
10000	.9998	
−100	1.0196	
−1000	1.002	
−100000	1.0002	

As you can see, as x becomes very large in the positive direction, the graph approaches the line $y = 1$ from below. As x becomes very small in the negative direction, the graph approaches the line $y = 1$ from above.

GETTING READY FOR CLASS

After reading through the preceding section, respond in your own words and in complete sentences.

A. Briefly list the steps in the Blueprint for Problem Solving that you have used previously to solve application problems.

B. Write an application problem for which the solution depends on solving the equation $\frac{1}{2} + \frac{1}{3} = \frac{1}{x}$.

C. One number is twice another, write an expression for the sum of their reciprocals.

D. Write a formula for the relationship between distance, rate, and time.

Problem Set 5.6

Solve the following word problems. Be sure to show the equation in each case.

NUMBER PROBLEMS

1. One number is 3 times another. The sum of their reciprocals is $\frac{20}{3}$. Find the numbers.

2. One number is 3 times another. The sum of their reciprocals is $\frac{4}{9}$. Find the numbers.

3. The sum of a number and its reciprocal is $\frac{10}{3}$. Find the number.

4. The sum of a number and twice its reciprocal is $\frac{27}{5}$. Find the number.

5. The sum of the reciprocals of two consecutive integers is $\frac{7}{12}$. Find the two integers.

6. Find two consecutive even integers, the sum of whose reciprocals is $\frac{3}{4}$.

7. If a certain number is added to the numerator and denominator of $\frac{7}{9}$, the result is $\frac{5}{6}$. Find the number.

8. Find the number you would add to both the numerator and denominator of $\frac{8}{11}$ so that the result would be $\frac{6}{7}$.

9. The speed of a boat in still water is 5 miles per hour. If the boat travels 3 miles downstream in the same amount of time it takes to travel 1.5 miles upstream, what is the speed of the current?

   a. Let x be the speed of the current. Complete the distance and rate columns in the table.

   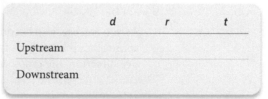

	d	r	t
Upstream			
Downstream			

   b. Now use the distance and rate information to complete the time column.

   c. What does the problem tell us about the two times? Use this fact to write an equation involving the two expressions for time.

   d. Solve the equation. Write your answer as a complete sentence.

10. A boat, which moves at 18 miles per hour in still water, travels 14 miles downstream in the same amount of time it takes to travel 10 miles upstream. Find the speed of the current.

    a. Let x be the speed of the current. Complete the distance and rate columns in the table.

    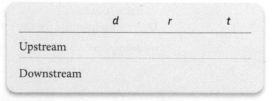

	d	r	t
Upstream			
Downstream			

    b. Now use the distance and rate information to complete the time column.

    c. What does the problem tell us about the two times? Use this fact to write an equation involving the two expressions for time.

    d. Solve the equation. Write your answer as a complete sentence.

RATE PROBLEMS

11. The current of a river is 2 miles per hour. A boat travels to a point 8 miles upstream and back again in 3 hours. What is the speed of the boat in still water?

12. A motorboat travels at 4 miles per hour in still water. It goes 12 miles upstream and 12 miles back again in a total of 8 hours. Find the speed of the current of the river.

13. Train A has a speed 15 miles per hour greater than that of train B. If train A travels 150 miles in the same time that train B travels 120 miles, what are the speeds of the two trains?

    a. Let x be the speed of train B. Complete the distance and rate columns in the table.

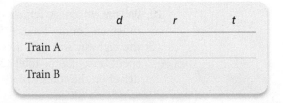

	d	r	t
Train A			
Train B			

    b. Now use the distance and rate information to complete the time column.

    c. What does the problem tell us about the two times? Use this fact to write an equation involving the two expressions for time.

    d. Solve the equation. Write your answer as a complete sentence.

14. A train travels 30 miles per hour faster than a car. If the train covers 120 miles in the same time the car covers 80 miles, what are the speeds of each of them?

    a. Let x be the speed of the car. Complete the distance and rate columns in the table.

	d	r	t
Car			
Train			

    b. Now use the distance and rate information to complete the time column.

    c. What does the problem tell us about the two times? Use this fact to write an equation involving the two expressions for time.

    d. Solve the equation. Write your answer as a complete sentence.

15. A small airplane flies 810 miles from Los Angeles to Portland, OR, with an average speed of 270 miles per hour. An hour and a half after the plane leaves, a Boeing 747 leaves Los Angeles for Portland. Both planes arrive in Portland at the same time. What was the average speed of the 747?

16. Lou leaves for a cross-country excursion on a bicycle traveling at 20 miles per hour. His friends are driving the trip and will meet him at several rest stops along the way. The first stop is scheduled 30 miles from the original starting point. If the people driving leave 15 minutes after Lou from the same place, how fast will they have to drive to reach the first rest stop at the same time as Lou?

17. A tour bus leaves Sacramento every Friday evening at 5:00 P.M. for a 270-mile trip to Las Vegas. This week, however, the bus leaves at 5:30 P.M. To arrive in Las Vegas on time, the driver drives 6 miles per hour faster than usual. What is the bus' usual speed?

18. A bakery delivery truck leaves the bakery at 5:00 A.M. each morning on its 140-mile route. One day the driver gets a late start and does not leave the bakery until 5:30 A.M. To finish her route on time the driver drives 5 miles per hour faster than usual. At what speed does she usually drive?

19. **Uniform Motion** A motorcycle drove 120 miles on the highway and then drove 30 miles on dirt roads. The rate of speed the motorcyclist was able to travel on the highway was two times the rate of speed for the dirt roads. The time spent traveling the total 150 miles was 3 hours. Find the rate of speed of the motorcyclist on the dirt roads.

20. **Uniform Motion** A vehicle drove 90 miles on the highway and then drove 40 miles on county roads. The rate of speed the driver was able to travel on the highway was 1.25 times the rate of speed for the county roads. The time spent traveling the total 130 miles was 2 hours. Find the rate of speed of the driver on the county roads.

WORK PROBLEMS

21. A water tank can be filled by an inlet pipe in 8 hours. It takes twice that long for the outlet pipe to empty the tank. How long will it take to fill the tank if both pipes are open?

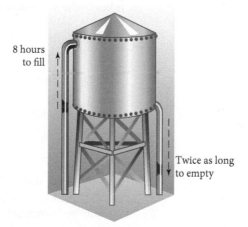

8 hours to fill

Twice as long to empty

22. A sink can be filled from the faucet in 5 minutes. It takes only 3 minutes to empty the sink when the drain is open. If the sink is full and both the faucet and the drain are open, how long will it take to empty the sink?

23. It takes 10 hours to fill a pool with the inlet pipe. It can be emptied in 15 hours with the outlet pipe. If the pool is half full to begin with, how long will it take to fill it from there if both pipes are open?

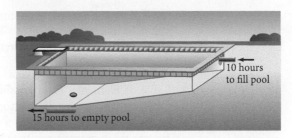

24. A sink is one-quarter full when both the faucet and the drain are opened. The faucet alone can fill the sink in 6 minutes, while it takes 8 minutes to empty it with the drain. How long will it take to fill the remaining three quarters of the sink?

25. A sink has two faucets: one for hot water and one for cold water. The sink can be filled by a cold water faucet in 3.5 minutes. If both faucets are open, the sink is filled in 2.1 minutes. How long does it take to fill the sink with just the hot water faucet open?

26. A water tank is being filled by two inlet pipes. Pipe A can fill the tank in $4\frac{1}{2}$ hours, but both pipes together can fill the tank in 2 hours. How long does it take to fill the tank using only pipe B?

27. Shawn can replace trim in a room of a house in 6 hours. Working alone, Jerry can replace the trim in 9 hours. If Shawn and Jerry work together, how long will it take them to replace the trim?

28. Sandy and Tim, working together, can mow a field in 75 minutes. Tim can mow twice as fast as Sandy. If Tim had to work alone, how long would it take him to mow?

MISCELLANEOUS PROBLEMS

29. Rhind Papyrus Nearly 4,000 years ago, Egyptians worked mathematical exercises involving reciprocals. The Rhind Papyrus contains a wealth of such problems, and one of them is as follows:

> "A quantity and its two thirds are added together, one third of this is added, then one third of the sum is taken, and the result is 10."

Write an equation and solve this exercise.

30. Photography For clear photographs, a camera must be properly focused. Professional photographers use a mathematical relationship relating the distance from the camera lens to the object being photographed, a; the distance from the lens to the film, b; and the focal length of the lens, f. These quantities, a, b, and f, are related by the equation

$$\frac{1}{a} + \frac{1}{b} = \frac{1}{f}$$

A camera has a focal length of 3 inches. If the lens is 5 inches from the film, how far should the lens be placed from the object being photographed for the camera to be perfectly focused?

31. Length Conversions Computations for certain formulas may have you convert standard units to metric units. There are 30.48 centimeters in 1 foot. How many feet are in 110 cm? Round to the nearest tenth.

32. Length Conversions Computations for certain formulas may have you convert standard units to metric units. There are 0.914 meters in 1 yard. How many yards are in 15 m? Round to the nearest tenth.

The Periodic Table If you take a chemistry class, you will work with the Periodic Table of Elements. Figure 3 shows three of the elements listed in the periodic table. As you can see, the bottom number in each figure is the molecular weight of the element. In chemistry, a mole is the amount of a substance that will give the weight in grams equal to the molecular weight. For example, 1 mole of lead is 207.2 grams.

FIGURE 3

33. Chemistry For the element carbon, 1 mole = 12.01 grams.

   a. To the nearest gram, how many grams of carbon are in 2.5 moles of carbon?

   b. How many moles of carbon are in 39 grams of carbon? Round to the nearest hundredth.

34. Chemistry For the element sulfur, 1 mole = 32.07 grams.

   a. How many grams of sulfur are in 3 moles of sulfur?

   b. How many moles of sulfur are found in 80.2 grams of sulfur?

Graph each rational function. In each case, show the vertical asymptote, the horizontal asymptote, and any intercepts that exist.

35. $f(x) = \dfrac{x-3}{x-1}$ **36.** $f(x) = \dfrac{x+4}{x-2}$ **37.** $f(x) = \dfrac{x+3}{x-1}$

38. $f(x) = \dfrac{x-2}{x-1}$ **39.** $g(x) = \dfrac{x-3}{x+1}$ **40.** $g(x) = \dfrac{x-2}{x+1}$

GETTING READY FOR THE NEXT SECTION

Divide.

41. $\dfrac{10x^2}{5x^2}$ **42.** $\dfrac{-15x^4}{5x^2}$ **43.** $\dfrac{4x^4y^3}{-2x^2y}$

44. $\dfrac{10a^4b^2}{4a^2b^2}$ **45.** $4{,}628 \div 25$ **46.** $7{,}546 \div 35$

Multiply.

47. $2x^2(2x-4)$ **48.** $3x^2(x-2)$

49. $(2x-4)(2x^2+4x+5)$ **50.** $(x-2)(3x^2+6x+15)$

Subtract.

51. $(2x^2 - 7x + 9) - (2x^2 - 4x)$ **52.** $(x^2 - 6xy - 7y^2) - (x^2 + xy)$

Factor.

53. $x^2 - a^2$ **54.** $x^2 - 1$

55. $x^2 - 6xy - 7y^2$ **56.** $2x^2 - 5xy + 3y^2$

SPOTLIGHT ON SUCCESS

Edwin | student instructor

You never fail until you stop trying.

—Albert Einstein

Coming to the United States at the age of 10 and not knowing how to speak English was a very difficult hurdle to overcome. However, with hard work and dedication I was able to rise above those obstacles. When I came to the U.S. our school did not have a strong English development program as it was known at that time, English as a Second Language (ESL). The approach back then was "sink or swim." When my self-esteem was low, my mom and my three older sisters were always there for me and they would always encourage me to do well. My mom was a single parent, and her number one priority was that we would receive a good education. My mother's perseverance is what has made me the person I am today.

The day everything began to change for me was when I was in eighth grade. I was sitting in my biology class not paying attention to the teacher because I was really focusing on a piece of paper on the wall. It said, "You never fail until you stop trying." I read it over and over, trying to digest what the quote meant. With my limited English I was doing my best to translate what it meant in my native language. It finally clicked! I memorized the quote and began to apply it to my academics and to real-life situations.

I was able to finish junior high successfully. Going to high school was a lot easier and I ended up with very good grades and eventually I was accepted to an excellent college. I was never the smartest student on campus, but I always did well because I never quit. I earned my college degree and now I teach at a dual-immersion elementary school. I have that same quote in my classroom and I constantly remind my students to never stop trying.

OBJECTIVES

A Divide a polynomial by a monomial.

B Divide a polynomial by a polynomial.

First Bank of San Luis Obispo charges \$2.00 per month and \$0.15 per check for a regular checking account. So, if you write x checks in one month, the total monthly cost of the checking account will be $C(x) = 2.00 + 0.15x$. From this formula, we see that the more checks we write in a month, the more we pay for the account. But it is also true that the more checks we write in a month, the lower the cost per check. To find the cost per check, we use the *average cost* function. To find the average cost function, we divide the total cost by the number of checks written.

$$\text{Average Cost} = \overline{C}(x) = \frac{C(x)}{x} = \frac{2.00 + 0.15x}{x}$$

This last expression gives us the average cost per check for each of the x checks written. To work with this last expression, we need to know something about division with polynomials, and that is what we will cover in this section.

We begin this section by considering division of a polynomial by a monomial. This is the simplest kind of polynomial division. The rest of the section is devoted to division of a polynomial by a polynomial. This kind of division is similar to long division with whole numbers.

Ⓐ Dividing a Polynomial by a Monomial

To divide a polynomial by a monomial, we use the definition of division and apply the distributive property. The following example illustrates the procedure.

EXAMPLE 1

Divide $\dfrac{10x^5 - 15x^4 + 20x^3}{5x^2}$.

SOLUTION

$$= (10x^5 - 15x^4 + 20x^3) \cdot \frac{1}{5x^2}$$
Dividing by $5x^2$ is the same as multiplying by $\frac{1}{5x^2}$.

$$= 10x^5 \cdot \frac{1}{5x^2} - 15x^4 \cdot \frac{1}{5x^2} + 20x^3 \cdot \frac{1}{5x^2}$$
Distributive property

$$= \frac{10x^5}{5x^2} - \frac{15x^4}{5x^2} + \frac{20x^3}{5x^2}$$
Multiplying by $\frac{1}{5x^2}$ is the same as multiplying by $5x^2$.

$$= 2x^3 - 3x^2 + 4x$$

5.7 VIDEOS

Notice that division of a polynomial by a monomial is accomplished by dividing each term of the polynomial by the monomial. The first two steps are usually not shown in a problem like this. They are part of Example 1 to justify distributing $5x^2$ under all three terms of the polynomial $10x^5 - 15x^4 + 20x^3$.

Here are some more examples of this kind of division.

EXAMPLE 2

Divide $\dfrac{8x^3y^5 - 16x^2y^2 + 4x^4y^3}{-2x^2y}$. Write the result with positive exponents.

SOLUTION

$$\frac{8x^3y^5 - 16x^2y^2 + 4x^4y^3}{-2x^2y} = \frac{8x^3y^5}{-2x^2y} + \frac{-16x^2y^2}{-2x^2y} + \frac{4x^4y^3}{-2x^2y}$$

$$= -4xy^4 + 8y - 2x^2y^2$$

EXAMPLE 3

Divide $\dfrac{10a^4b^2 + 8ab^3 - 12a^3b + 6ab}{4a^2b^2}$. Write the result with positive exponents.

SOLUTION

$$\frac{10a^4b^2 + 8ab^3 - 12a^3b + 6ab}{4a^2b^2} = \frac{10a^4b^2}{4a^2b^2} + \frac{8ab^3}{4a^2b^2} - \frac{12a^3b}{4a^2b^2} + \frac{6ab}{4a^2b^2}$$

$$= \frac{5a^2}{2} + \frac{2b}{a} - \frac{3a}{b} + \frac{3}{2ab}$$

Notice in Example 3 that the result is not a polynomial because of the last three terms. If we were to write each as a product, some of the variables would have negative exponents. For example, the second term would be

$$\frac{2b}{a} = 2a^{-1}b$$

The divisor in each of the preceding examples was a monomial. We now want to turn our attention to division of polynomials in which the divisor has two or more terms.

B Dividing a Polynomial by a Polynomial

EXAMPLE 4

Divide: $\dfrac{x^2 - 6xy - 7y^2}{x + y}$

SOLUTION In this case, we can factor the numerator and perform division by simply dividing out common factors, just like we did in previous sections:

$$\frac{x^2 - 6xy - 7y^2}{x + y} = \frac{(x + y)(x - 7y)}{x + y}$$

$$= x - 7y$$

Long Division

For the type of division shown in Example 4, the denominator must be a factor of the numerator. When the denominator is not a factor of the numerator, or in the case where we can't factor the numerator, the method used in Example 4 won't work. We need to develop a new method for these cases. Because this new method is very similar to *long division* with whole numbers, we will review the method of long division here.

EXAMPLE 5

Divide $25\overline{)4{,}628}$.

SOLUTION

$$
\begin{array}{r}
1 \\
25\overline{)4{,}628} \\
2\,5 \\
\hline
2\,1
\end{array}
$$

Estimate 25 into 46.

Multiply $1 \times 25 = 25$.

Subtract $46 - 25 = 21$.

$$
\begin{array}{r}
1 \\
25\overline{)4{,}628} \\
2\,5\downarrow \\
\hline
2\,12
\end{array}
$$

Bring down the 2.

These are the four basic steps in long division: estimate, multiply, subtract, and bring down the next term. To complete the problem, we simply perform the same four steps:

$$
\begin{array}{r}
18 \\
25\overline{)4{,}628} \\
2\,5 \\
\hline
2\,12 \\
2\,00\downarrow \\
\hline
128
\end{array}
$$

8 is the estimate.

Multiply to get 200.

Subtract to get 12, then bring down the 8.

One more time:

$$
\begin{array}{r}
185 \\
25\overline{)4{,}628} \\
2\,5 \\
\hline
2\,12 \\
2\,00\downarrow \\
\hline
128 \\
125 \\
\hline
3
\end{array}
$$

5 is the estimate.

Multiply to get 125.

Subtract to get 3.

Because 3 is less than 25 and we have no more terms to bring down, we have our answer:

$$\frac{4{,}628}{25} = 185 + \frac{3}{25}$$

To check our answer, we multiply 185 by 25 and then add 3 to the result:

$$25(185) + 3 = 4{,}625 + 3 = 4{,}628$$

Long division with polynomials is similar to long division with whole numbers. Both use the same four basic steps: estimate, multiply, subtract, and bring down the next term. We use long division with polynomials when the denominator has two or more terms and is not a factor of the numerator. Here is an example.

EXAMPLE 6

Divide $\dfrac{2x^2 - 7x + 9}{x - 2}$.

SOLUTION

$$
\begin{array}{r}
2x \\
x - 2 \overline{)\ 2x^2 - 7x + 9} \\
\ \ \ -\ \ \ \ + \\
\overline{\ \ \ \cancel{+}\ 2x^2\ \cancel{+}\ 4x} \\
-3x
\end{array}
$$

Estimate $2x^2 \div x = 2x$.

Multiply $2x(x - 2) = 2x^2 - 4x$.
Subtract $(2x^2 - 7x) - (2x^2 - 4x) = -3x$.

$$
\begin{array}{r}
2x \\
x - 2 \overline{)\ 2x^2 - 7x + 9} \\
\ \ \ -\ \ \ \ +\ \ \ \ \downarrow \\
\overline{\ \ \ \cancel{+}\ 2x^2\ \cancel{+}\ 4x\ \ \downarrow} \\
-3x + 9
\end{array}
$$

Bring down the 9.

Notice we change the signs on $2x^2 - 4x$ and add in the subtraction step. Subtracting a polynomial is equivalent to adding its opposite.

We repeat the four steps.

$$
\begin{array}{r}
2x - 3 \\
x - 2 \overline{)\ 2x^2 - 7x + 9} \\
\ \ \ -\ \ \ \ + \\
\overline{\ \ \ \cancel{+}\ 2x^2\ \cancel{+}\ 4x} \\
-3x + 9 \\
+\ \ \ \ - \\
\overline{\ \ \ \cancel{+}3x\ \cancel{+}\ 6} \\
3
\end{array}
$$

-3 is the estimate: $-3x \div x = -3$.

Multiply $-3(x - 2) = -3x + 6$.
Subtract $(-3x + 9) - (-3x + 6) = 3$.

Because we have no other term to bring down, we have our answer:

$$
\frac{2x^2 - 7x + 9}{x - 2} = 2x - 3 + \frac{3}{x - 2}
$$

To check, we multiply $(2x - 3)(x - 2)$ to get $2x^2 - 7x + 6$; then, adding the remainder 3 to this result, we have $2x^2 - 7x + 9$.

In setting up a division problem involving two polynomials, we must remember two things: (1) Both polynomials should be in decreasing powers of the variable, and (2) neither should skip any powers from the highest power down to the constant term. If there are any missing terms, they can be filled in using a coefficient of 0.

EXAMPLE 7

Divide $2x - 4 \overline{)4x^3 - 6x - 11}$.

SOLUTION   Because the trinomial is missing a term in x^2, we can fill it in with $0x^2$:

$$
4x^3 - 6x - 11 = 4x^3 + 0x^2 - 6x - 11
$$

Adding $0x^2$ does not change our original problem.

$$
\begin{array}{r}
2x^2 + 4x + 5 \\
2x - 4 \overline{) 4x^3 + 0x^2 - 6x - 11 } \\
\end{array}
$$

NOTE Adding the $0x^2$ term gives us a column in which to write $-8x^2$.

$$\frac{4x^3 - 6x - 11}{2x - 4} = 2x^2 + 4x + 5 + \frac{9}{2x - 4}$$

To check this result, we multiply $2x - 4$ and $2x^2 + 4x + 5$:

$$
\begin{array}{r}
2x^2 + 4x + 5 \\
\times 2x - 4 \\
\hline
4x^3 + 8x^2 + 10x \\
+ - 8x^2 - 16x - 20 \\
\hline
4x^3 - 6x - 20 \\
\end{array}
$$

Adding 9 (the remainder) to this result gives us the polynomial $4x^3 - 6x - 11$. Our answer checks.

For our next example, let's do Example 4 again, but this time using long division.

EXAMPLE 8

Divide $\dfrac{x^2 - 6xy - 7y^2}{x + y}$.

SOLUTION

$$
\begin{array}{r}
x - 7y \\
x + y \overline{) x^2 - 6xy - 7y^2 } \\
\end{array}
$$

In this case, the remainder is 0, and we have

$$\frac{x^2 - 6xy - 7y^2}{x + y} = x - 7y$$

which is easy to check because

$$(x + y)(x - 7y) = x^2 - 6xy - 7y^2$$

EXAMPLE 9

Factor $x^3 + 9x^2 + 26x + 24$ completely if $x + 2$ is one of its factors.

SOLUTION   Because $x + 2$ is one of the factors of the polynomial we are trying to factor, it must divide that polynomial evenly — that is, without a remainder. Therefore, we begin by dividing the polynomial by $x + 2$:

$$
\begin{array}{r}
x^2 + 7x + 12 \\
x + 2 \overline{)\; x^3 + 9x^2 + 26x + 24} \\
\underline{\cancel{+}\, x^3 \cancel{+}\, 2x^2} \qquad\qquad \\
+ 7x^2 + 26x \qquad \\
\underline{\cancel{+}\, 7x^2 \cancel{+}\, 14x} \qquad \\
+ 12x + 24 \\
\underline{\cancel{+}\, 12x \cancel{+}\, 24} \\
0
\end{array}
$$

Now we know that the polynomial we are trying to factor is equal to the product of $x + 2$ *and* $x^2 + 7x + 12$. To factor completely, we simply factor $x^2 + 7x + 12$:

$$x^3 + 9x^2 + 26x + 24 = (x + 2)(x^2 + 7x + 12)$$

$$= (x + 2)(x + 3)(x + 4)$$

GETTING READY FOR CLASS

After reading through the preceding section, respond in your own words and in complete sentences.

A. What are the four steps used in long division with polynomials?

B. What does it mean to have a remainder of 0?

C. When must long division be performed, and when can factoring be used to divide polynomials?

D. What property of real numbers is the key to dividing a polynomial by a monomial?

Problem Set 5.7

Find the following quotients.

1. $\dfrac{4x^3 - 8x^2 + 6x}{2x}$

2. $\dfrac{6x^3 + 12x^2 - 9x}{3x}$

3. $\dfrac{10x^4 + 15x^3 - 20x^2}{-5x^2}$

4. $\dfrac{12x^5 - 18x^4 - 6x^3}{6x^3}$

5. $\dfrac{8y^5 + 10y^3 - 6y}{4y^3}$

6. $\dfrac{6y^4 - 3y^3 + 18y^2}{9y^2}$

7. $\dfrac{5x^3 - 8x^2 - 6x}{-2x^2}$

8. $\dfrac{-9x^5 + 10x^3 - 12x}{-6x^4}$

9. $\dfrac{28a^3b^5 + 42a^4b^3}{7a^2b^2}$

10. $\dfrac{a^2b + ab^2}{ab}$

11. $\dfrac{10x^3y^2 - 20x^2y^3 - 30x^3y^3}{-10x^2y}$

12. $\dfrac{9x^4y^4 + 18x^3y^4 - 27x^2y^4}{-9xy^3}$

Divide by factoring numerators and then dividing out common factors.

13. $\dfrac{x^2 - x - 6}{x - 3}$

14. $\dfrac{x^2 - x - 6}{x + 2}$

15. $\dfrac{2a^2 - 3a - 9}{2a + 3}$

16. $\dfrac{2a^2 + 3a - 9}{2a - 3}$

17. $\dfrac{5x^2 - 14xy - 24y^2}{x - 4y}$

18. $\dfrac{5x^2 - 26xy - 24y^2}{5x + 4y}$

19. $\dfrac{x^3 - y^3}{x - y}$

20. $\dfrac{x^3 + 8}{x + 2}$

21. $\dfrac{y^4 - 16}{y - 2}$

22. $\dfrac{y^4 - 81}{y - 3}$

23. $\dfrac{x^3 + 2x^2 - 25x - 50}{x - 5}$

24. $\dfrac{x^3 + 2x^2 - 25x - 50}{x + 5}$

25. $\dfrac{4x^3 + 12x^2 - 9x - 27}{x + 3}$

26. $\dfrac{9x^3 + 18x^2 - 4x - 8}{x + 2}$

Divide using the long division method.

27. $\dfrac{x^2 - 5x - 7}{x + 2}$

28. $\dfrac{x^2 + 4x - 8}{x - 3}$

29. $\dfrac{6x^2 + 7x - 18}{3x - 4}$

30. $\dfrac{8x^2 - 26x - 9}{2x - 7}$

31. $\dfrac{2x^3 - 3x^2 - 4x + 5}{x + 1}$

32. $\dfrac{3x^3 - 5x^2 + 2x - 1}{x - 2}$

33. $\dfrac{2y^3 - 9y^2 - 17y + 39}{2y - 3}$

34. $\dfrac{3y^3 - 19y^2 + 17y + 4}{3y - 4}$

35. $\dfrac{2x^3 - 9x^2 + 11x - 6}{2x^2 - 3x + 2}$

36. $\dfrac{6x^3 + 7x^2 - x + 3}{3x^2 - x + 1}$

37. $\dfrac{6y^3 - 8y + 5}{2y - 4}$

38. $\dfrac{9y^3 - 6y^2 + 8}{3y - 3}$

39. $\dfrac{a^4 - 2a + 5}{a - 2}$

40. $\dfrac{a^4 + a^3 - 1}{a + 2}$

41. $\dfrac{y^4 - 16}{y - 2}$

42. $\dfrac{y^4 - 81}{y - 3}$

43. $\dfrac{x^4 + x^3 - 3x^2 - x + 2}{x^2 + 3x + 2}$

44. $\dfrac{2x^4 + x^3 + 4x - 3}{2x^2 - x + 3}$

45. Factor $x^3 + 6x^2 + 11x + 6$ completely if one of its factors is $x + 3$.

46. Factor $x^3 + 10x^2 + 29x + 20$ completely if one of its factors is $x + 4$.

47. Factor $x^3 + 5x^2 - 2x - 24$ completely if one of its factors is $x + 3$.

48. Factor $x^3 + 3x^2 - 10x - 24$ completely if one of its factors is $x + 2$.

49. Problems 21 and 41 are the same problem. Are the two answers you obtained equivalent?

50. Problems 22 and 42 are the same problem. Are the two answers you obtained equivalent?

51. Find $P(-2)$ if $P(x) = x^2 - 5x - 7$. Compare it with the remainder in Problem 27.

52. Find $P(3)$ if $P(x) = x^2 + 4x - 8$. Compare it with the remainder in Problem 28.

APPLYING THE CONCEPTS

53. The Factor Theorem The factor theorem of algebra states that if $x - a$ is a factor of a polynomial, $P(x)$, then $P(a) = 0$. Verify the following.
   a. That $x - 2$ is a factor of $P(x) = x^3 - 3x^2 + 5x - 6$, and that $P(2) = 0$
   b. That $x - 5$ is a factor of $P(x) = x^4 - 5x^3 - x^2 + 6x - 5$, and that $P(5) = 0$

54. The Remainder Theorem The remainder theorem of algebra states that if a polynomial, $P(x)$, is divided by $x - a$, then the remainder is $P(a)$. Verify the remainder theorem by showing that when $P(x) = x^2 - x + 3$ is divided by $x - 2$ the remainder is 5, and that $P(2) = 5$.

55. Checking Account First Bank of San Luis Obispo charges $2.00 per month and $0.15 per check for a regular checking account. As we mentioned in the introduction to this section, the total monthly cost of this account is $C(x) = 2.00 + 0.15x$. To find the average cost of each of the x checks, we divide the total cost by the number of checks written. That is,

$$\overline{C}(x) = \frac{C(x)}{x}$$

a. Use the total cost function to fill in the following table.

x	1	5	10	15	20
$C(x)$					

b. Find the formula for the average cost function, $\overline{C}(x)$.

c. Use the average cost function to fill in the following table.

x	1	5	10	15	20
$\overline{C}(x)$					

d. What happens to the average cost as more checks are written?

e. Give the domain and range of each of the functions.

56. Average Cost A company that manufactures memory sticks uses the function $C(x) = 200 + 2x$ to represent the daily cost of producing x memory sticks.

a. Find the average cost function, $\overline{C}(x)$.

b. Use the average cost function to fill in the following table:

x	1	5	10	15	20
$\overline{C}(x)$					

c. What happens to the average cost as more items are produced?

d. Graph the function $y = \overline{C}(x)$ for $x > 0$.

e. What is the domain of this function?

f. What is the range of this function?

57. Average Cost For long-distance service, a particular phone company charges a monthly fee of $4.95 plus $0.07 per minute of calling time used. The relationship between the number of minutes of calling time used, m, and the amount of the monthly phone bill $T(m)$ is given by the function $T(m) = 4.95 + 0.07m$.

a. Find the total cost when 100, 400, and 500 minutes of calling time is used in 1 month.

b. Find a formula for the average cost per minute function $\overline{T}(m)$.

c. Find the average cost per minute of calling time used when 100, 400, and 500 minutes are used in 1 month.

58. Average Cost A company manufactures electric pencil sharpeners. Each month they have fixed costs of \$40,000 and variable costs of \$8.50 per sharpener. Therefore, the total monthly cost to manufacture x sharpeners is given by the function $C(x) = 40,000 + 8.5x$.

a. Find the total cost to manufacture 1,000, 5,000, and 10,000 sharpeners a month.

b. Write an expression for the average cost per sharpener function $\overline{C}(x)$.

c. Find the average cost per sharpener to manufacture 1,000, 5,000, and 10,000 sharpeners per month.

MAINTAINING YOUR SKILLS

Reviewing these problems will help clarify the different methods we have used in this chapter.

Perform the indicated operations.

59. $\dfrac{2a + 10}{a^3} \cdot \dfrac{a^2}{3a + 15}$

60. $\dfrac{4a + 8}{a^2 - a - 6} \div \dfrac{a^2 + 7a + 12}{a^2 - 9}$

61. $(x^2 - 9)\left(\dfrac{x + 2}{x + 3}\right)$

62. $\dfrac{1}{x + 4} + \dfrac{8}{x^2 - 16}$

63. $\dfrac{2x - 7}{x - 2} - \dfrac{x - 5}{x - 2}$

64. $2 + \dfrac{25}{5x - 1}$

Simplify each expression.

65. $\dfrac{\dfrac{1}{x} - \dfrac{1}{3}}{\dfrac{1}{x} + \dfrac{1}{3}}$

66. $\dfrac{1 - \dfrac{9}{x^2}}{1 - \dfrac{1}{x} - \dfrac{6}{x^2}}$

Solve each equation.

67. $\dfrac{x}{x - 3} + \dfrac{3}{2} = \dfrac{3}{x - 3}$

68. $1 - \dfrac{3}{x} = \dfrac{-2}{x^2}$

EXTENDING THE CONCEPTS

69. $\dfrac{4x^5 - x^4 - 20x^3 + 8x^2 - 15}{x^2 - 5}$

70. $\dfrac{4x^5 + 2x^4 + x^3 - 20x^2 - 10x - 5}{x^3 - 5}$

71. $\dfrac{0.5x^3 - 0.3x^2 + 0.22x + 0.06}{x + 0.2}$

72. $\dfrac{0.6x^3 - 1.1x^2 - 0.1x + 0.6}{0.3x + 0.2}$

73. $\dfrac{3x^2 + x - 10}{2x + 4}$

74. $\dfrac{2x^2 - x - 10}{3x + 6}$

75. $\dfrac{2x^2 + \dfrac{1}{3}x + \dfrac{5}{3}}{3x - 1}$

76. $\dfrac{x^2 + \dfrac{3}{5}x + \dfrac{8}{5}}{5x - 2}$

Rational Numbers and Expressions [5.1]

1. $\frac{3}{4}$ is a rational number. $\frac{x-3}{x^2-9}$ is a rational expression.

A *rational number* is any number that can be expressed as the ratio of two integers:

$$\text{Rational numbers} = \left\{ \frac{a}{b} \, \middle| \, a \text{ and } b \text{ are integers, } b \neq 0 \right\}$$

A *rational expression* is any quantity that can be expressed as the ratio of two polynomials:

$$\text{Rational expressions} = \left\{ \frac{P}{Q} \, \middle| \, P \text{ and } Q \text{ are polynomials, } Q \neq 0 \right\}$$

Properties of Rational Expressions [5.1]

If P, Q, and K are polynomials with $Q \neq 0$ and $K \neq 0$, then

$$\frac{P}{Q} = \frac{PK}{QK} \qquad \text{and} \qquad \frac{P}{Q} = \frac{\frac{P}{K}}{\frac{Q}{K}}$$

which is to say that multiplying or dividing the numerator and denominator of a rational expression by the same nonzero quantity always produces an equivalent rational expression.

Reducing to Lowest Terms [5.1]

2. $\dfrac{x-3}{x^2-9} = \dfrac{x-3}{(x-3)(x+3)}$

$= \dfrac{1}{x+3}$

To reduce a rational expression to lowest terms, we first factor the numerator and denominator and then divide the numerator and denominator by any factors they have in common.

Multiplication [5.2]

3. $\dfrac{x+1}{x^2-4} \cdot \dfrac{x+2}{3x+3}$

$= \dfrac{(x+1)(x+2)}{(x-2)(x+2)(3)(x+1)}$

$= \dfrac{1}{3(x-2)}$

To multiply two rational numbers or rational expressions, multiply numerators and multiply denominators. In symbols,

$$\frac{P}{Q} \cdot \frac{R}{S} = \frac{PR}{QS} \qquad (Q \neq 0 \text{ and } S \neq 0)$$

In practice, we don't really multiply, but rather, we factor and then divide out common factors.

Division [5.2]

4. $\dfrac{x^2-y^2}{x^3+y^3} \div \dfrac{x-y}{x^2-xy+y^2}$

$= \dfrac{x^2-y^2}{x^3+y^3} \cdot \dfrac{x^2-xy+y^2}{x-y}$

$= \dfrac{(x+y)(x-y)(x^2-xy+y^2)}{(x+y)(x^2-xy+y^2)(x-y)}$

$= 1$

To divide one rational expression by another, we use the definition of division to rewrite our division problem as an equivalent multiplication problem. To divide by a rational expression, we multiply by its reciprocal. In symbols,

$$\frac{P}{Q} \div \frac{R}{S} = \frac{P}{Q} \cdot \frac{S}{R} = \frac{PS}{QR} \qquad (Q \neq 0, S \neq 0, R \neq 0)$$

Least Common Denominator [5.3]

5. The LCD for $\frac{2}{x-3}$ and $\frac{3}{5}$ is $5(x-3)$.

The *least common denominator*, LCD, for a set of denominators is the smallest quantity divisible by each of the denominators.

Addition and Subtraction [5.3]

6. $\dfrac{2}{x-3} + \dfrac{3}{5}$

$= \dfrac{2}{x-3} \cdot \dfrac{5}{5} + \dfrac{3}{5} \cdot \dfrac{x-3}{x-3}$

$= \dfrac{3x+1}{5(x-3)}$

If P, Q, and R represent polynomials, $R \neq 0$, then

$$\frac{P}{R} + \frac{Q}{R} = \frac{P+Q}{R} \quad \text{and} \quad \frac{P}{R} - \frac{Q}{R} = \frac{P-Q}{R}$$

When adding or subtracting rational expressions with different denominators, we must find the LCD for all denominators and change each rational expression to an equivalent expression that has the LCD.

Complex Fractions [5.4]

7. $\dfrac{\dfrac{1}{x} + \dfrac{1}{y}}{\dfrac{1}{x} - \dfrac{1}{y}} = \dfrac{xy\left(\dfrac{1}{x} + \dfrac{1}{y}\right)}{xy\left(\dfrac{1}{x} - \dfrac{1}{y}\right)}$

$= \dfrac{y+x}{y-x}$

A rational expression that contains, in its numerator or denominator, other rational expressions is called a *complex fraction*. One method of simplifying a complex fraction is to multiply the numerator and denominator by the LCD for all denominators.

Equations Involving Rational Expressions [5.5]

8. Solve $\dfrac{x}{2} + 3 = \dfrac{1}{3}$.

$6\left(\dfrac{x}{2}\right) + 6 \cdot 3 = 6 \cdot \dfrac{1}{3}$

$3x + 18 = 2$

$x = -\dfrac{16}{3}$

To solve an equation involving rational expressions, we first find the LCD for all denominators appearing on either side of the equation. We then multiply both sides by the LCD to clear the equation of all fractions and solve as usual.

Dividing a Polynomial by a Monomial [5.7]

9. $\dfrac{15x^3 - 20x^2 + 10x}{5x}$

$= 3x^2 - 4x + 2$

To divide a polynomial by a monomial, divide each term of the polynomial by the monomial.

Long Division with Polynomials [5.7]

10.
$$\begin{array}{r} x - 2 \\ x - 3 \overline{)\ x^2 - 5x + 8} \\ \underline{\mp x^2 \pm 3x} \\ -2x + 8 \\ \underline{\pm 2x \mp 6} \\ 2 \end{array}$$

If division with polynomials cannot be accomplished by dividing out factors common to the numerator and denominator, then we use a process similar to long division with whole numbers. The steps in the process are estimate, multiply, subtract, and bring down the next term.

COMMON MISTAKES

1. Attempting to divide the numerator and denominator of a rational expression by a quantity that is not a factor of both. Like this:

$$\frac{x^2 - \overset{3}{9}x + \overset{2}{20}}{x^2 - \underset{1}{3}x - \underset{1}{10}} \quad \text{Mistake}$$

This makes no sense at all. The numerator and denominator must be factored completely before any factors they have in common can be recognized:

$$\frac{x^2 - 9x + 20}{x^2 - 3x - 10} = \frac{(x - 5)(x - 4)}{(x - 5)(x + 2)}$$

$$= \frac{x - 4}{x + 2}$$

2. Forgetting to check solutions to equations involving rational expressions. When we multiply both sides of an equation by a quantity containing the variable, we must be sure to check for extraneous solutions.

Reduce each rational expression to lowest terms. [5.1]

1. $\dfrac{42x^2 + 23x - 10}{14x^2 + 45x - 14}$

2. $\dfrac{6x^2 + xy - 12y^2}{16y^2 - 9x^2}$

3. $\dfrac{y^2 - 9}{y^3 + 27}$

4. $\dfrac{x + 3}{x^3 + 27}$

5. $\dfrac{2x^2 - 3xy + y^2}{2x - y}$

6. $\dfrac{3x^2 - 4x - 4}{3x^2 - x - 2}$

Simplify each expression. [5.1]

7. $\dfrac{(2x + 3) - (2a + 3)}{x - a}$

8. $\dfrac{(x^2 - 2x) - (a^2 - 2a)}{x - a}$

State the domain for each rational function. [5.1]

9. $g(x) = \dfrac{x^2 - 9}{x - 3}$

10. $h(t) = \dfrac{t - 5}{t^2 - 25}$

11. Average Speed A person riding a Ferris wheel with a diameter of 75 feet travels once around the wheel in 70 seconds. What is the average speed of the rider in feet per minute? Use 3.14 as an approximation for π and round your answer to the nearest tenth. [5.1]

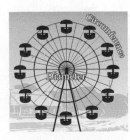

For the functions in Problems 12 and 13, evaluate the given difference quotients. [5.1]

a. $\dfrac{f(x + h) - f(x)}{h}$

b. $\dfrac{f(x) - f(a)}{x - a}$

12. $f(x) = 3x + 5$

13. $f(x) = x^2 + 2x - 8$

Multiply and divide as indicated. [5.2]

14. $\dfrac{x^2 - 16}{3x + 12} \cdot \dfrac{3(x + 3)^2}{x^2 - x - 12}$

15. $\dfrac{y}{x} \div \dfrac{xy}{xy - 1}$

16. $\dfrac{1}{x^2 - 9} \div \dfrac{1}{(x - 3)^2}$

17. $\dfrac{x^4 - 81}{2x^2 - 9x + 9} \div \dfrac{x^2 + 9}{10x - 15}$

18. $\dfrac{7x + 3y}{42x^2 - 17xy - 15y^2} \cdot \dfrac{12x^2 - 4xy - 5y^2}{2x + y}$

19. $\dfrac{490x^2 - 640}{49x^2 - 112x + 64} \cdot \dfrac{28x^2 - 95x + 72}{56x^3 - 62x^2 - 144x}$

20. $\dfrac{4x^2 + 6x - 4}{2x^2 - 11x + 5} \div \dfrac{x^3 - 8}{x^2 - 7x + 10}$

21. $\dfrac{k^2 - 1}{k^3 + 1} \div \dfrac{k^3 - 1}{k - 1} \cdot (k^2 + k + 1)$

22. $\dfrac{r^3 - s^3}{s - r} \cdot \dfrac{(r + s)^2}{r^2 + s^2}$

Add and subtract as indicated. [5.3]

23. $\dfrac{3}{14} + \dfrac{9}{20}$

24. $\dfrac{1}{2} - \dfrac{7}{24} + \dfrac{5}{6}$

25. $\dfrac{8}{y - 8} - \dfrac{y}{y - 8}$

26. $\dfrac{2x - 4}{x + 2} - \dfrac{x - 6}{x + 2}$

27. $\dfrac{a + 7}{a^2 - 81} + \dfrac{2}{a^2 - 81}$

28. $\dfrac{3}{x - 1} + \dfrac{6}{2x - 1}$

29. $\dfrac{4x}{x^2 + 2x - 3} - \dfrac{2x}{x^2 - 1}$

30. $\dfrac{2x - 1}{x^2 - 1} - \dfrac{x + 4}{x^2 + 4x + 3}$

Find and simplify $f(x) - g(x)$ for the following functions. [5.3]

31. $f(x) = \dfrac{x + 1}{x - 2}, g(x) = \dfrac{4x + 7}{5x - 10}$

Find the quotient.

32. $\dfrac{9t^2 - 1}{6t^2 + 7t - 3} \div \dfrac{27t^3 + 1}{8t^3 + 27}$

Combine the following rational expressions. Reduce all answers to lowest terms. [5.3]

33. $\dfrac{3a}{a^2 + 7a + 10} - \dfrac{2a}{a^2 + 6a + 8}$

34. $\dfrac{5x + 3}{2x^2 + 5x + 3} - \dfrac{3x + 9}{2x^2 + 7x + 6}$

35. $\dfrac{x}{x + 3} + \dfrac{7}{3x + 9} - \dfrac{2}{x^2 + 3x}$

36. $\dfrac{1}{x} + \dfrac{x}{3x + 9} - \dfrac{3}{x^2 + 3x}$

Simplify each complex rational expression. [5.4]

37. $\dfrac{1 - \dfrac{1}{x}}{\dfrac{1}{x}}$

38. $\dfrac{\dfrac{1}{x} + \dfrac{2}{y}}{\dfrac{2}{x} + \dfrac{1}{y}}$

39. $\dfrac{4 - x^{-2}}{4 + 4x^{-1} + x^{-2}}$

40. $\dfrac{(x + a)^{-1} + (x - a)^{-1}}{(x + a)^{-1} - (x - a)^{-1}}$

41. $\dfrac{4 - \dfrac{1}{a + 2}}{4 + \dfrac{1}{a + 2}}$

42. $\dfrac{1 - \dfrac{16}{x^2}}{1 - \dfrac{1}{x} - \dfrac{12}{x^2}}$

43. $x - \dfrac{1}{x - \dfrac{1}{2}}$

44. $\dfrac{2 + \dfrac{1}{x - \dfrac{1}{3}}}{2 - \dfrac{1}{x - \dfrac{1}{3}}}$

45. $\dfrac{\dfrac{1}{x + 7} - \dfrac{1}{x}}{7}$

46. $\dfrac{1}{(x + 7)^2} - \dfrac{1}{x^2}$

Solve each of the following equations. [5.5]

47. $\dfrac{y}{2} - \dfrac{4}{y} = -\dfrac{7}{2}$

48. $\dfrac{x + 6}{x + 3} = \dfrac{3}{x + 3} + 2$

49. $5 - \dfrac{2}{x} = \dfrac{9}{5}$

50. $\dfrac{3x}{x + 2} + \dfrac{3}{2} = \dfrac{9}{x + 1}$

51. $\dfrac{4}{x + 5} + \dfrac{x + 6}{8x} = \dfrac{7}{8}$

52. $1 = \dfrac{11}{x} - \dfrac{30}{x^2}$

53. $\dfrac{y + 3}{y^2 - y} - \dfrac{8}{y^2 - 1} = 0$

54. $\dfrac{1}{x + 2} + \dfrac{1}{x - 2} = \dfrac{4}{x^2 - 4}$

55. Let $f(x) = \dfrac{4}{x + 2}$ and $g(x) = \dfrac{4}{x - 2}$. Find x if

  a. $f(x) - g(x) = -\dfrac{4}{3}$

  b. $\dfrac{g(x)}{f(x)} = -7$

  c. $f(x) = -g(x)$

Graph each function. Show the vertical asymptote. [5.5]

56. $f(x) = \dfrac{1}{x + 3}$ **57.** $f(x) = \dfrac{4}{x - 2}$

For each function given, find the location of any vertical asymptotes. Also give the domain of the function. (Do not graph.) [5.5]

58. $y(x) = \dfrac{x + 3}{x^2 - 5x + 4}$ **59.** $y(x) = \dfrac{7}{3x^2 + 5x - 2}$

Solve the following applications. Be sure to show the equation in each case. [5.6]

60. Number Problem What number must be added to the denominator of $\frac{23}{63}$ to make the result $\frac{1}{3}$?

61. Speed of a Boat The current of a river is 7 miles per hour. It takes a motorboat a total of 11 hours to travel 36 miles upstream and return 36 miles downstream. What is the speed of the boat in still water?

62. Filling a Tub An inlet pipe can fill a tub in 5 minutes, and the drain can empty it in 6 minutes. If both the inlet pipe and the drain are left open, how long will it take to fill the tub?

Graph each rational function. In each case, show the vertical asymptote, the horizontal asymptote, and any intercepts that exist. [5.6]

63. $f(x) = \dfrac{x - 2}{x - 1}$ **64.** $g(x) = \dfrac{x}{x + 2}$

Divide. [5.7]

65. $\dfrac{y^4 - 81}{y - 3}$ **66.** $\dfrac{x^3 + 2x^2 - 25x - 50}{x + 5}$

67. $\dfrac{8x^4y^2 - 20x^5y + 28x^2y^3}{4x^2y}$ **68.** $\dfrac{9x^3 - 9x^2 + 2x - 12}{3x - 5}$

Chapter 5 Cumulative Review

Simplify.

1. $\left(-\dfrac{5}{3}\right)^3$

2. $\left(\dfrac{8}{7}\right)^{-2}$

3. $|-18 - 32| - 90$

4. $3^3 - 6(5^2 - 6^2)$

5. $81 \div 18 \cdot 2$

6. $36 \div 12 \cdot 4$

7. $\dfrac{2^4 - 5}{11(6^2 - 5^2)}$

8. $\dfrac{-5^2(4 - 4 \cdot 2)}{19 + 3^4}$

9. Subtract $\dfrac{4}{5}$ from the product of -3 and $\dfrac{7}{15}$.

10. Write in symbols: The difference of $2a$ and $9b$ is greater than their sum.

Let $P(x) = 14x - 0.01x^2$ and find the following.

11. $P(10)$

12. $P(-10)$

Let $Q(x) = \dfrac{5}{4}x^2 + \dfrac{1}{5}x - 33$ and find the following.

13. $Q(10)$

14. $Q(-10)$

Simplify each of the following expressions.

15. $-4(4x + 5) - 11x$

16. $(x + 5)^2 - (x - 5)^2$

Simplify.

17. $\dfrac{5}{y^2 + y - 6} - \dfrac{4}{y^2 - 4}$

18. $\dfrac{x}{y^2 - x^2} + \dfrac{y}{y^2 - x^2}$

Multiply.

19. $\left(5t^2 + \dfrac{1}{4}\right)\left(4t^2 - \dfrac{1}{5}\right)$

20. $\dfrac{x^2 - 81}{x^2 + 5x + 6} \cdot \dfrac{x + 2}{x + 9}$

Divide.

21. $\dfrac{9x^{7n} - 6x^{4n}}{3x^{2n}}$

22. $\dfrac{2a^5 + a^2 - 8}{a - 1}$

Reduce to lowest terms. Write the answer with positive exponents only.

23. $\dfrac{x^{-6}}{x^{-11}}$

24. $\left(\dfrac{x^{-5}y^{-8}}{x^2 y^{-6}}\right)^{-1}$

25. $\dfrac{x^3 + 3x^2 - 25x - 75}{x^2 - 2x - 15}$

26. $\dfrac{\dfrac{7a}{4a^3 - 4}}{\dfrac{14a}{8a - 8}}$

Find the equation of the line through the two points.

27. $(5, 4)$ and $(10, 1)$

28. $(3, -1)$ and $(11, -1)$

Factor each of the following expressions.

29. 396

30. $x^2 + 3x - 40$

31. $x^2 + 2xy - 36 + y^2$

32. $y^3 - \dfrac{64}{125}x^3$

Solve the following equations.

33. $\dfrac{9}{2}x - 6 = 57$

34. $9y + 27 = 2y - 8$

35. $\dfrac{2}{3}(9x - 1) - \dfrac{1}{3} = -7$

36. $\dfrac{3}{7}(14x - 4) + \dfrac{5}{7} = 8$

37. $\dfrac{4}{y - 1} = \dfrac{7}{y + 6}$

38. $3 - \dfrac{16}{x} = \dfrac{12}{x^2}$

Solve each system.

39. $2x + 5y = 17$
$\quad\;\; 9x + 3y = 2x + 16$

40. $3x - 2y = 5$
$\quad\;\; 2x + 9y = 6$

Solve each inequality, and graph the solution.

41. $-3(x - 2) \le -2(x - 6)$

42. $|x - 4| + 3 < 5$

Graph on a rectangular coordinate system.

43. $-6x + 5y = -15$ **44.** $3x + y < -4$

Use the graph to work Problems 45–48.

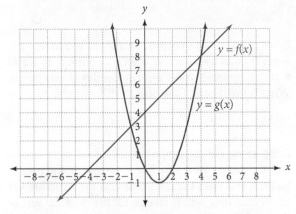

45. Find $(f - g)(3)$.   **46.** Find $(f + g)(-2)$.

47. Find $(f \circ g)(-1)$.   **48.** Find $(g \circ f)(-1)$.

49. **Unit Analysis**   Convert 105.6 feet/minute into miles/hour.

50. **Piecewise Functions**   Graph the following piecewise function.

$$f(x) = \begin{cases} -4 & \text{if } x \leq -1 \\ 2x & \text{if } x > -1 \end{cases}$$

Reduce to lowest terms. [5.1]

1. $\dfrac{x^2 - y^2}{x - y}$

2. $\dfrac{2x^2 - 5x + 3}{2x^2 - x - 3}$

Multiply and divide as indicated. [5.2]

3. $\dfrac{a^2 - 16}{5a - 15} \cdot \dfrac{10(a - 3)^2}{a^2 - 7a + 12}$

4. $\dfrac{a^4 - 81}{a^2 + 9} \div \dfrac{a^2 - 8a + 15}{4a - 20}$

5. $\dfrac{x^3 - 8}{2x^2 - 9x + 10} \div \dfrac{x^2 + 2x + 4}{2x^2 + x - 15}$

Add and subtract as indicated. [5.3]

6. $\dfrac{4}{21} + \dfrac{6}{35}$

7. $\dfrac{3}{4} - \dfrac{1}{2} + \dfrac{5}{8}$

8. $\dfrac{a}{a^2 - 9} + \dfrac{3}{a^2 - 9}$

9. $\dfrac{1}{x} + \dfrac{2}{x - 3}$

10. $\dfrac{4x}{x^2 + 6x + 5} - \dfrac{3x}{x^2 + 5x + 4}$

11. $\dfrac{2x + 8}{x^2 + 4x + 3} - \dfrac{x + 4}{x^2 + 5x + 6}$

Simplify each complex fraction. [5.4]

12. $\dfrac{3 - \dfrac{1}{a + 3}}{3 + \dfrac{1}{a + 3}}$

13. $\dfrac{1 - \dfrac{9}{x^2}}{1 + \dfrac{1}{x} - \dfrac{6}{x^2}}$

Solve each of the following equations. [5.5]

14. $\dfrac{1}{x} + 3 = \dfrac{4}{3}$

15. $\dfrac{x}{x - 3} + 3 = \dfrac{3}{x - 3}$

16. $\dfrac{y + 3}{2y} + \dfrac{5}{y - 1} = \dfrac{1}{2}$

17. $1 - \dfrac{1}{x} = \dfrac{6}{x^2}$

18. Graph $f(x) = \dfrac{x + 4}{x - 1}$.

Solve the following applications. Be sure to show the equation in each case. [5.6]

19. Number Problem What number must be subtracted from the denominator of $\frac{10}{23}$ to make the result $\frac{1}{3}$?

20. Speed of a Boat The current of a river is 2 miles per hour. It takes a motorboat a total of 3 hours to travel 8 miles upstream and return 8 miles downstream. What is the speed of the boat in still water?

21. Filling a Pool An inlet pipe can fill a pool in 10 hours, and the drain can empty it in 15 hours. If the pool is half full and both the inlet pipe and the drain are left open, how long will it take to fill the pool the rest of the way?

22. Unit Analysis The top of Mount Whitney, the highest point in California, is 14,494 feet above sea level. Give this height in miles to the nearest tenth of a mile.

23. Unit Analysis A bullet fired from a gun travels a distance of 4,750 feet in 3.2 seconds. Find the average speed of the bullet in miles per hour. Round to the nearest whole number.

Divide. [5.7]

24. $\dfrac{24x^3y + 12x^2y^2 - 16xy^3}{4xy}$

25. $\dfrac{2x^3 - 9x^2 + 10}{2x - 1}$

6

Rational Exponents and Roots

© iStockphoto/trait2lumiere

Ecology and conservation are topics that interest most college students. If our rivers and oceans are to be preserved for future generations, we need to work to eliminate pollution from our waters. If a river is flowing at 1 meter per second and a pollutant is entering the river at a constant rate, the shape of the pollution plume can often be modeled by the simple equation

$$y = \sqrt{x}$$

The following table and graph were produced from the equation.

Width of a Pollutant Plume

Distance from Source (meters)	Width of Plume (meters)
x	y
0	0
1	1
4	2
9	3
16	4

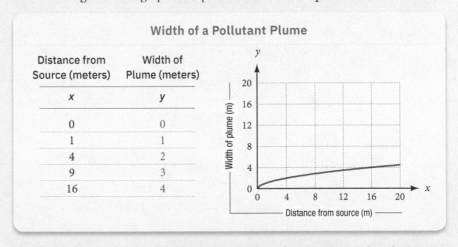

To visualize how the graph models the pollutant plume, imagine that the river is flowing from left to right, parallel to the *x*-axis, with the *x*-axis as one of its banks. The pollutant is entering the river from the bank at (0, 0).

By modeling pollution with mathematics, we can use our knowledge of mathematics to help control and eliminate pollution.

491

STUDY SKILLS

This is the last chapter in which we will mention study skills. You know by now what works best for you and what you have to do to achieve your goals for this course. From now on, it is simply a matter of sticking with the things that work for you and avoiding the things that do not. It seems simple, but as with anything that takes effort, it is up to you to see that you maintain the skills that get you where you want to be in the course.

If you intend to take more classes in mathematics and want to ensure your success in those classes, then you can work toward this goal: ***Become the type of student who can learn mathematics on his or her own.*** Most people who have degrees in mathematics were students who could learn mathematics on their own. This doesn't mean that you have to learn it all on your own; it simply means that if you have to, you can learn it on your own. Attaining this goal gives you independence and puts you in control of your success in any math class you take.

OBJECTIVES

A Simplify radical expressions involving numbers using the definition for square roots.

B Simplify radical expressions involving variables.

C Simplify expressions with rational exponents.

Figure 1 shows a square in which each of the four sides is 1 inch long. To find the square of the length of the diagonal c, we apply the Pythagorean theorem:

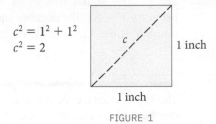

$$c^2 = 1^2 + 1^2$$
$$c^2 = 2$$

1 inch

c

1 inch

1 inch

FIGURE 1

Because we know that c is positive and that its square is 2, we call c the *positive square root* of 2, and we write $c = \sqrt{2}$. Associating numbers, such as $\sqrt{2}$, with the diagonal of a square or rectangle allows us to analyze some interesting items from geometry. One particularly interesting geometric object that we will study in this section is shown in Figure 2. It is constructed from a right triangle, and the length of the diagonal is found from the Pythagorean theorem. We will come back to this figure at the end of this section.

Some people believe Dutch painter Piet Mondrian used the golden ratio in several of his works. He is said to believe that using ratios and straight lines enabled his paintings to achieve harmony. Look at the illustration at the beginning of this section, which was inspired by Mondrian's *Composition with yellow, blue, and red, 1937–42*. Can you see how he used golden rectangles to create this work of art?

The Golden Rectangle

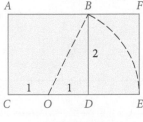

FIGURE 2

🅐 Radical Expressions

In Chapter 1, we developed notation (exponents) to give us the square, cube, or any other power of a number. For instance, if we wanted the square of 3, we wrote $3^2 = 9$. If we wanted the cube of 3, we wrote $3^3 = 27$. In this section, we will develop notation that will take us in the reverse direction, that is, from the square of a number, say 25, back to the original number, 5.

> **DEFINITION**   *positive square root*
>
> ●
>
> If x is a nonnegative real number, then the expression $\sqrt{x}$ is called the *positive square root* of x and is such that
>
> $$(\sqrt{x})^2 = x$$
>
> *In words:* $\sqrt{x}$ is the positive number we square to get x.

The negative square root of x, $-\sqrt{x}$, is defined in a similar manner.

EXAMPLE 1

The positive square root of 64 is 8 because 8 is the positive number with the property $8^2 = 64$. The negative square root of 64 is -8 because -8 is the negative number whose square is 64. We can summarize both these facts by saying

$$\sqrt{64} = 8 \qquad \text{and} \qquad -\sqrt{64} = -8$$

The higher roots, cube roots, fourth roots, and so on, are defined by definitions similar to that of square roots.

> **DEFINITION**
>
> ●
>
> If x is a real number and n is a positive integer, then
>
> Positive square root of x, $\sqrt{x}$, is such that $(\sqrt{x})^2 = x$ $x \geq 0$
>
> Cube root of x, $\sqrt[3]{x}$, is such that $(\sqrt[3]{x})^3 = x$
>
> Positive fourth root of x, $\sqrt[4]{x}$, is such that $(\sqrt[4]{x})^4 = x$ $x \geq 0$
>
> Fifth root of x, $\sqrt[5]{x}$, is such that $(\sqrt[5]{x})^5 = x$
>
> ⋮   ⋮
>
> The nth root of x, $\sqrt[n]{x}$, is such that $(\sqrt[n]{x})^n = x$ $x \geq 0$ if n is even

NOTE It is a common mistake to assume that an expression like $\sqrt{25}$ indicates both square roots, 5 and -5. The expression $\sqrt{25}$ indicates only the positive square root of 25, which is 5. If we want the negative square root, we must use a negative sign: $-\sqrt{25} = -5$.

NOTE We have restricted the even roots in this definition to nonnegative numbers. Even roots of negative numbers exist, but are not represented by real numbers. That is, $\sqrt{-4}$ is not a real number because there is no real number whose square is -4.

The following is a table of the most common roots used in this book. Any of the roots that are unfamiliar should be memorized.

Square Roots		Cube Roots	Fourth Roots
$\sqrt{0} = 0$ $\quad$ $\sqrt{49} = 7$		$\sqrt[3]{0} = 0$	$\sqrt[4]{0} = 0$
$\sqrt{1} = 1$ $\quad$ $\sqrt{64} = 8$		$\sqrt[3]{1} = 1$	$\sqrt[4]{1} = 1$
$\sqrt{4} = 2$ $\quad$ $\sqrt{81} = 9$		$\sqrt[3]{8} = 2$	$\sqrt[4]{16} = 2$
$\sqrt{9} = 3$ $\quad$ $\sqrt{100} = 10$		$\sqrt[3]{27} = 3$	$\sqrt[4]{81} = 3$
$\sqrt{16} = 4$ $\quad$ $\sqrt{121} = 11$		$\sqrt[3]{64} = 4$	
$\sqrt{25} = 5$ $\quad$ $\sqrt{144} = 12$		$\sqrt[3]{125} = 5$	
$\sqrt{36} = 6$ $\quad$ $\sqrt{169} = 13$			

Notation An expression like $\sqrt[3]{8}$ that involves a root is called a *radical expression*. In the expression $\sqrt[3]{8}$, the ^{3} is called the *index*, the $\sqrt{}$ is the *radical sign*, and 8 is called the *radicand*. The index of a radical must be a positive integer greater than 1. If no index is written, it is assumed to be 2.

Roots and Negative Numbers

When dealing with negative numbers and radicals, the only restriction concerns negative numbers under even roots. We can have negative signs in front of radicals and negative numbers under odd roots and still obtain real numbers. Here are some examples to help clarify this. In the last section of this chapter, we will see how to deal with even roots of negative numbers.

EXAMPLE

Simplify each expression, if possible.

2. $\sqrt[3]{-8} = -2$ because $(-2)^3 = -8$.

3. $\sqrt{-4}$ is not a real number because there is no real number whose square is -4.

4. $-\sqrt{25} = -5$, because -5 is the negative square root of 25.

5. $\sqrt[5]{-32} = -2$ because $(-2)^5 = -32$.

6. $\sqrt[4]{-81}$ is not a real number because there is no real number we can raise to the fourth power and obtain -81.

B Variables Under a Radical

From the preceding examples, it is clear that we must be careful that we do not try to take an even root of a negative number. For this reason, we will assume that all variables appearing under a radical sign represent nonnegative numbers.

EXAMPLE

Assume all variables represent nonnegative numbers, and simplify each expression as much as possible.

7. $\sqrt{25a^4b^6} = 5a^2b^3$ because $(5a^2b^3)^2 = 25a^4b^6$.

8. $\sqrt[3]{x^6y^{12}} = x^2y^4$ because $(x^2y^4)^3 = x^6y^{12}$.

9. $\sqrt[4]{81r^8s^{20}} = 3r^2s^5$ because $(3r^2s^5)^4 = 81r^8s^{20}$.

⚙ Rational Numbers as Exponents

We will now develop a second kind of notation involving exponents that will allow us to designate square roots, cube roots, and so on in another way.

Consider the equation $x = 8^{1/3}$. Although we have not encountered fractional exponents before, let's assume that all the properties of exponents hold in this case. Cubing both sides of the equation, we have

$$x^3 = (8^{1/3})^3$$
$$x^3 = 8^{(1/3)(3)}$$
$$x^3 = 8^1$$
$$x^3 = 8$$

The last line tells us that x is the number whose cube is 8. It must be true, then, that x is the cube root of 8, $x = \sqrt[3]{8}$. Because we started with $x = 8^{1/3}$, it follows that

$$8^{1/3} = \sqrt[3]{8}$$

It seems reasonable, then, to define fractional exponents as indicating roots. Here is the formal definition.

> **DEFINITION**
> •
> If x is a real number and n is a positive integer greater than 1, then
> $$x^{1/n} = \sqrt[n]{x} \qquad (x \geq 0 \text{ when } n \text{ is even})$$
> *In words:*   The quantity $x^{1/n}$ is the nth root of x.

With this definition, we have a way of representing roots with exponents. Here are some examples.

EXAMPLE

Write each expression as a root and then simplify, if possible.

10. $8^{1/3} = \sqrt[3]{8} = 2$

11. $36^{1/2} = \sqrt{36} = 6$

12. $-25^{1/2} = -\sqrt{25} = -5$

13. $(-25)^{1/2} = \sqrt{-25}$, which is not a real number

14. $\left(\dfrac{4}{9}\right)^{1/2} = \sqrt{\dfrac{4}{9}} = \dfrac{2}{3}$

The properties of exponents developed in Chapter 1 were applied to integer exponents only. We will now extend these properties to include rational exponents also. We do so without proof.

> **PROPERTY** *Properties of Exponents*
>
> If a and b are real numbers and r and s are rational numbers, and a and b are nonnegative whenever r and s indicate even roots, then
>
> **1.** $a^r \cdot a^s = a^{r+s}$ **4.** $a^{-r} = \dfrac{1}{a^r}$ $(a \neq 0)$
>
> **2.** $(a^r)^s = a^{rs}$ **5.** $\left(\dfrac{a}{b}\right)^r = \dfrac{a^r}{b^r}$ $(b \neq 0)$
>
> **3.** $(ab)^r = a^r b^r$ **6.** $\dfrac{a^r}{a^s} = a^{r-s}$ $(a \neq 0)$

Sometimes rational exponents can simplify our work with radicals. Here are Examples 8 and 9 again, but this time we will work them using rational exponents.

EXAMPLE

Write each radical with a rational exponent, then simplify.

15. $\sqrt[3]{x^6 y^{12}} = (x^6 y^{12})^{1/3}$

$\qquad\qquad = (x^6)^{1/3}(y^{12})^{1/3}$

$\qquad\qquad = x^2 y^4$

16. $\sqrt[4]{81 r^8 s^{20}} = (81 r^8 s^{20})^{1/4}$

$\qquad\qquad = 81^{1/4}(r^8)^{1/4}(s^{20})^{1/4}$

$\qquad\qquad = 3 r^2 s^5$

So far, the numerators of all the rational exponents we have encountered have been 1. The next theorem extends the work we can do with rational exponents to rational exponents with numerators other than 1.

We can extend our properties of exponents with the following theorem.

> ### Rational Exponent Theorem
>
> If a is a nonnegative real number, m is an integer, and n is a positive integer, then
> $$a^{m/n} = (a^{1/n})^m = (a^m)^{1/n}$$

Proof We can prove the Rational exponent theorem using the properties of exponents. Because $m/n = m(1/n)$, we have

$$a^{m/n} = a^{m(1/n)} \qquad\Big|\qquad a^{m/n} = a^{(1/n)(m)}$$

$$= (a^m)^{1/n} \qquad\Big|\qquad = (a^{1/n})^m$$

Here are some examples that illustrate how we use this theorem.

EXAMPLE

Simplify as much as possible.

17. $8^{2/3} = (8^{1/3})^2$ — Rational exponent theorem

$= 2^2$ — Definition of fractional exponents

$= 4$ — The square of 2 is 4.

18. $25^{(3/2)} = (25^{1/2})^3$ — Rational exponent theorem

$= 5^3$ — Definition of fractional exponents

$= 125$ — The cube of 5 is 125.

19. $9^{-3/2} = (9^{1/2})^{-3}$ — Rational exponent theorem

$= 3^{-3}$ — Definition of fractional exponents

$= \dfrac{1}{3^3}$ — Property 4 for exponents

$= \dfrac{1}{27}$ — The cube of 3 is 27.

20. $\left(\dfrac{27}{8}\right)^{-4/3} = \left[\left(\dfrac{27}{8}\right)^{1/3}\right]^{-4}$ — Rational exponent theorem

$= \left(\dfrac{3}{2}\right)^{-4}$ — Definition of fractional exponents

$= \left(\dfrac{2}{3}\right)^{4}$ — Property 4 for exponents

$= \dfrac{16}{81}$ — The fourth power of $\frac{2}{3}$ is $\frac{16}{81}$.

NOTE On a scientific calculator, Example 17 would look like this:

$8 \boxed{y^x} \boxed{(} 2 \boxed{\div} 3 \boxed{)} \boxed{=}$

The following examples show the application of the properties of exponents to rational exponents.

EXAMPLE

Assume all variables represent positive quantities, and simplify as much as possible.

21. $x^{1/3} \cdot x^{5/6} = x^{1/3 + 5/6}$ — Property 1

$= x^{2/6 + 5/6}$ — LCD is 6

$= x^{7/6}$ — Add fractions.

22. $(y^{2/3})^{3/4} = y^{(2/3)(3/4)}$ — Property 2

$= y^{1/2}$ — Multiply fractions: $\frac{2}{3} \cdot \frac{3}{4} = \frac{6}{12} = \frac{1}{2}$.

23. $\dfrac{z^{1/3}}{z^{1/4}} = z^{1/3 - 1/4}$ — Property 6

$= z^{4/12 - 3/12}$ — LCD is 12

$= z^{1/12}$ — Subtract fractions.

24. $\left(\dfrac{a^{-1/3}}{b^{1/2}}\right)^6 = \dfrac{(a^{-1/3})^6}{(b^{1/2})^6}$ Property 5

$\qquad\qquad = \dfrac{a^{-2}}{b^3}$ Property 2

$\qquad\qquad = \dfrac{1}{a^2 b^3}$ Property 4

25. $\dfrac{(x^{-3}y^{1/2})^4}{x^{10}y^{3/2}} = \dfrac{(x^{-3})^4(y^{1/2})^4}{x^{10}y^{3/2}}$ Property 3

$\qquad\qquad = \dfrac{x^{-12}y^2}{x^{10}y^{3/2}}$ Property 2

$\qquad\qquad = x^{-22}y^{1/2}$ Property 6

$\qquad\qquad = \dfrac{y^{1/2}}{x^{22}}$ Property 4

FACTS FROM GEOMETRY   *The Pythagorean Theorem (Again) and the Golden Rectangle*

Now that we have had some experience working with square roots, we can rewrite the Pythagorean theorem using a square root. If triangle *ABC* is a right triangle with $C = 90°$, then the length of the longest side is the *positive square root* of the sum of the squares of the other two sides (see Figure 3).

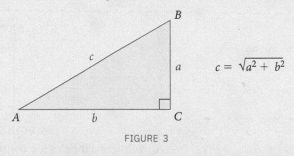

$$c = \sqrt{a^2 + b^2}$$

FIGURE 3

In the introduction to this chapter, we mentioned the golden rectangle. Its origins can be traced back over 2,000 years to the Greek civilization that produced Pythagoras, Socrates, Plato, Aristotle, and Euclid. The most important mathematical work to come from that Greek civilization was Euclid's *Elements,* an elegantly written summary of all that was known about geometry at that time in history. Euclid's *Elements,* according to Howard Eves, an authority on the history of mathematics, exercised a greater influence on scientific thinking than any other work. Here is how we construct a golden rectangle from a square of side 2, using the same method that Euclid used in his *Elements.*

Constructing a Golden Rectangle from a Square of Side 2

Step 1: Draw a square with a side of length 2. Connect the midpoint of side *CD* to corner *B*. (Note that we have labeled the midpoint of segment *CD* with the letter *O*.)

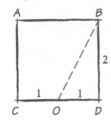

Step 2: Drop the diagonal from step 1 down so it aligns with side *CD*.

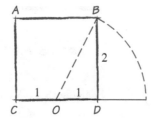

Step 3: Form rectangle *ACEF*. This is a golden rectangle.

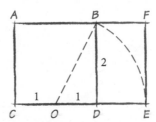

All golden rectangles are constructed from squares. Every golden rectangle, no matter how large or small it is, will have the same shape. To associate a number with the shape of the golden rectangle, we use the ratio of its length to its width. This ratio is called the *golden ratio*. To calculate the golden ratio, we must first find the length of the diagonal we used to construct the golden rectangle. Figure 4 shows the golden rectangle we constructed from a square of side 2. The length of the diagonal *OB* is found by applying the Pythagorean theorem to triangle *OBD*.

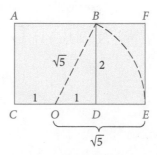

FIGURE 4

The length of segment OE is equal to the length of diagonal OB; both are $\sqrt{5}$. Because the distance from C to O is 1, the length CE of the golden rectangle is $1 + \sqrt{5}$. Now we can find the golden ratio:

$$\text{Golden ratio} = \frac{\text{length}}{\text{width}} = \frac{CE}{EF} = \frac{1 + \sqrt{5}}{2}$$

USING TECHNOLOGY   *Graphing Calculators:*
                        A Word of Caution

Some graphing calculators give surprising results when evaluating expressions such as $(-8)^{2/3}$. As you know from reading this section, the expression $(-8)^{2/3}$ simplifies to 4, either by taking the cube root first and then squaring the result, or by squaring the base first and then taking the cube root of the result. Here are three different ways to evaluate this expression on your calculator:

1. $(-8)^\wedge(2/3)$ To evaluate $(-8)^{2/3}$
2. $((-8)^\wedge 2)^\wedge(1/3)$ To evaluate $((-8)^2)^{1/3}$
3. $((-8)^\wedge(1/3))^\wedge 2$ To evaluate $((-8)^{1/3})^2$

Note any differences in the results.
   Next, graph each of the following functions, one at a time.

1. $Y_1 = X^{2/3}$ **2.** $Y_2 = (X^2)^{1/3}$ **3.** $Y_3 = (X^{1/3})^2$

The correct graph is shown in Figure 5. Note which of your graphs match the correct graph.
   Different calculators evaluate exponential expressions in different ways. You should use the method (or methods) that gave you the correct graph.

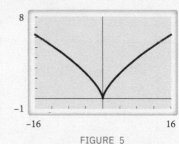

FIGURE 5

GETTING READY FOR CLASS

After reading through the preceding section, respond in your own words and in complete sentences.

A. Every real number has two square roots. Explain the notation we use to tell them apart. Use the square roots of 3 for examples.
B. Explain why a square root of -4 is not a real number.
C. We use the notation $\sqrt{2}$ to represent the positive square root of 2. Explain why there isn't a simpler way to express the positive square root of 2.
D. For the expression $a^{m/n}$, explain the significance of the numerator m and the significance of the denominator n in the exponent.

Problem Set 6.1

Find each of the following roots, if possible.

1. $\sqrt{144}$ **2.** $-\sqrt{144}$ **3.** $\sqrt{-144}$ **4.** $\sqrt{-49}$

5. $-\sqrt{49}$ **6.** $\sqrt{49}$ **7.** $\sqrt[3]{-27}$ **8.** $-\sqrt[3]{27}$

9. $\sqrt[4]{16}$ **10.** $-\sqrt[4]{16}$ **11.** $\sqrt[4]{-16}$ **12.** $-\sqrt[4]{-16}$

13. $\sqrt{0.04}$ **14.** $\sqrt{0.81}$ **15.** $\sqrt[3]{0.008}$ **16.** $\sqrt[3]{0.125}$

Simplify each expression. Assume all variables represent nonnegative numbers.

17. $\sqrt{36a^8}$ **18.** $\sqrt{49a^{10}}$ **19.** $\sqrt[3]{27a^{12}}$ **20.** $\sqrt[3]{8a^{15}}$

21. $\sqrt[3]{x^3y^6}$ **22.** $\sqrt[3]{x^6y^3}$ **23.** $\sqrt[5]{32x^{10}y^5}$ **24.** $\sqrt[5]{32x^5y^{10}}$

25. $\sqrt[4]{16a^{12}b^{20}}$ **26.** $\sqrt[4]{81a^{24}b^8}$

Use the definition of rational exponents to write each of the following with the appropriate root. Then simplify.

27. $36^{1/2}$ **28.** $49^{1/2}$ **29.** $-9^{1/2}$ **30.** $-16^{1/2}$

31. $8^{1/3}$ **32.** $-8^{1/3}$ **33.** $(-8)^{1/3}$ **34.** $-27^{1/3}$

35. $32^{1/5}$ **36.** $81^{1/4}$ **37.** $\left(\dfrac{81}{25}\right)^{1/2}$ **38.** $\left(\dfrac{9}{16}\right)^{1/2}$

39. $\left(\dfrac{64}{125}\right)^{1/3}$ **40.** $\left(\dfrac{8}{27}\right)^{1/3}$

Use the Rational Exponent Theorem to simplify each of the following as much as possible.

41. $27^{2/3}$ **42.** $8^{4/3}$ **43.** $25^{3/2}$ **44.** $9^{3/2}$

45. $16^{3/4}$ **46.** $81^{3/4}$

Simplify each expression. Remember, negative exponents give reciprocals.

47. $27^{-1/3}$ **48.** $9^{-1/2}$ **49.** $81^{-3/4}$ **50.** $4^{-3/2}$

51. $\left(\dfrac{25}{36}\right)^{-1/2}$ **52.** $\left(\dfrac{16}{49}\right)^{-1/2}$ **53.** $\left(\dfrac{81}{16}\right)^{-3/4}$ **54.** $\left(\dfrac{27}{8}\right)^{-2/3}$

55. $16^{1/2} + 27^{1/3}$ **56.** $25^{1/2} + 100^{1/2}$ **57.** $8^{-2/3} + 4^{-1/2}$ **58.** $49^{-1/2} + 25^{-1/2}$

Use the properties of exponents to simplify each of the following as much as possible. Assume all bases are positive.

59. $x^{3/5} \cdot x^{1/5}$ **60.** $x^{3/4} \cdot x^{5/4}$ **61.** $(a^{3/4})^{4/3}$ **62.** $(a^{2/3})^{3/4}$

63. $\dfrac{x^{1/5}}{x^{3/5}}$ **64.** $\dfrac{x^{2/7}}{x^{5/7}}$ **65.** $\dfrac{x^{5/6}}{x^{2/3}}$ **66.** $\dfrac{x^{7/8}}{x^{8/7}}$

67. $(x^{3/5}y^{5/6}z^{1/3})^{3/5}$ **68.** $(x^{3/4}y^{1/8}z^{5/6})^{4/5}$ **69.** $\dfrac{a^{3/4}b^2}{a^{7/8}b^{1/4}}$ **70.** $\dfrac{a^{1/3}b^4}{a^{3/5}b^{1/3}}$

71. $\dfrac{(y^{2/3})^{3/4}}{(y^{1/3})^{3/5}}$ **72.** $\dfrac{(y^{5/4})^{2/5}}{(y^{1/4})^{4/3}}$ **73.** $\left(\dfrac{a^{-1/4}}{b^{1/2}}\right)^8$ **74.** $\left(\dfrac{a^{-1/5}}{b^{1/3}}\right)^{15}$

Simplify. (Assume all variables are nonnegative.)

75. a. $\sqrt{25}$ **b.** $\sqrt{0.25}$ **c.** $\sqrt{2500}$ **d.** $\sqrt{0.0025}$

76. a. $\sqrt[3]{8}$ **b.** $\sqrt[3]{0.008}$ **c.** $\sqrt[3]{8,000}$ **d.** $\sqrt[3]{8 \times 10^{-6}}$

77. a. $\sqrt{16a^4b^8}$ **b.** $\sqrt[3]{16a^4b^8}$ **c.** $\sqrt[4]{16a^4b^8}$

78. a. $\sqrt{64x^5y^{10}}$ **b.** $\sqrt[3]{64x^5y^{10}}$ **c.** $\sqrt[4]{64x^5y^{10}}$

79. Show that the expression $(a^{1/2} + b^{1/2})^2$ is not equal to $a + b$ by replacing a with 9 and b with 4 in both expressions and then simplifying each.

80. Show that the statement $(a^2 + b^2)^{1/2} = a + b$ is not, in general, true by replacing a with 3 and b with 4 and then simplifying both sides.

81. You may have noticed, if you have been using a calculator to find roots, that you can find the fourth root of a number by pressing the square root button twice. Written in symbols, this fact looks like this:

$$\sqrt{\sqrt{a}} = \sqrt[4]{a} \qquad (a \geq 0)$$

Show that this statement is true by rewriting each side with exponents instead of radical notation and then simplifying the left side.

82. Show that the statement is true by rewriting each side with exponents instead of radical notation and then simplifying the left side.

$$\sqrt[3]{\sqrt{a}} = \sqrt[6]{a} \qquad (a \geq 0)$$

APPLYING THE CONCEPTS

83. Maximum Speed The maximum speed (v) that an automobile can travel around a curve of radius r without skidding is given by the equation

$$v = \left(\frac{5r}{2}\right)^{1/2}$$

where v is in miles per hour and r is measured in feet. What is the maximum speed a car can travel around a curve with a radius of 250 feet without skidding?

84. Relativity The equation

$$L = \left(1 - \frac{v^2}{c^2}\right)^{1/2}$$

gives the relativistic length of a 1-foot ruler traveling with velocity v. Find L if

$$\frac{v}{c} = \frac{3}{5}$$

85. Volume The length s of a side of a cube can be expressed in terms of the cube's volume as

$$s = V^{1/3}$$

What is the length of each side of a cube that has a volume of 1,728 cubic inches?

86. Volume The length s of a side of a cube can be expressed in terms of the cube's volume as

$$s = V^{1/3}$$

What is the length of each side of a cube that has a volume of 512 cubic inches?

87. Economics The number of units N of a finished product produced from L units of labor input and C units of capital input can be expressed as

$$N = 30L^{3/4}C^{1/4}$$

Find the units of finished product that used 625 units of labor and 256 units of capital.

88. Economics The number of units N of a finished product produced from L units of labor input and C units of capital input can be expressed as

$$N = 30L^{\frac{3}{4}}C^{\frac{1}{4}}$$

Find the units of finished product that used 256 units of labor and 81 units of capital.

89. Atmospheric Pressure Atmospheric pressure (in millibars) can be expressed using the formula

$$P = 1000(0.5)^{\frac{x}{5.6}}$$

where x is elevation in kilometers. What is the approximate atmospheric pressure at the top of Pikes Peak, 4.3 kilometers in elevation?

90. Atmospheric Pressure Atmospheric pressure (in millibars) can be expressed using the formula

$$P = 1000(0.5)^{\frac{x}{5.6}}$$

where x is elevation in kilometers. What is the approximate atmospheric pressure at the top of Picard Peak, 2.7 kilometers in elevation?

91. Golden Ratio The golden ratio is the ratio of the length to the width in any golden rectangle. The exact value of this number is $\frac{1 + \sqrt{5}}{2}$. Use a calculator to find a decimal approximation to this number and round it to the nearest thousandth.

92. Golden Ratio The reciprocal of the golden ratio is $\frac{2}{1 + \sqrt{5}}$. Find a decimal approximation to this number that is accurate to the nearest thousandth.

93. Sequences Find the next term in the following sequence. Then explain how this sequence is related to the Fibonacci sequence.

$$\frac{3}{2}, \frac{5}{3}, \frac{8}{5}, \cdots$$

94. Sequences Write the first 10 terms in the sequence shown in Problem 93. Then find a decimal approximation to each of the 10 terms, rounding each to the nearest thousandth.

95. Chemistry Figure 6 shows part of a model of a magnesium oxide (MgO) crystal. Each corner of the square is at the center of one oxygen ion (O^{2-}), and the center of the middle ion is at the center of the square. The radius for each oxygen ion is 150 picometers (pm), and the radius for each magnesium ion (Mg^{2+}) is 60 picometers.

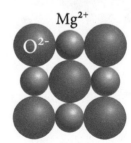

FIGURE 6

   a. Find the length of the side of the square. Write your answer in picometers.

   b. Find the length of the diagonal of the square. Write your answer in picometers.

   c. If 1 meter is 10^{12} picometers, give the length of the diagonal of the square in meters.

96. Geometry The length of each side of the cube shown in Figure 7 is 1 inch.

   a. Find the length of the diagonal CH.

   b. Find the length of the diagonal CF.

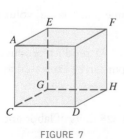

FIGURE 7

GETTING READY FOR THE NEXT SECTION

Simplify. Assume all variable are positive real numbers.

97. $\sqrt{25}$ **98.** $\sqrt{4}$ **99.** $\sqrt{6^2}$ **100.** $\sqrt{3^2}$

101. $\sqrt{16x^4y^2}$ **102.** $\sqrt{4x^6y^8}$ **103.** $\sqrt{(5y)^2}$ **104.** $\sqrt{(8x^3)^2}$

105. $\sqrt[3]{27}$ **106.** $\sqrt[3]{-8}$ **107.** $\sqrt[3]{2^3}$ **108.** $\sqrt[3]{(-5)^3}$

109. $\sqrt[3]{8a^3b^3}$ **110.** $\sqrt[3]{64a^6b^3}$

Fill in the blank.

111. $50 = \underline{\quad} \cdot 2$ **112.** $12 = \underline{\quad} \cdot 3$

113. $48x^4y^3 = \underline{\quad} \cdot y$ **114.** $40a^5b^4 = \underline{\quad} \cdot 5a^2b$

115. $12x^7y^6 = \underline{\quad} \cdot 3x$ **116.** $54a^6b^2c^4 = \underline{\quad} \cdot 2b^2c$

EXTENDING THE CONCEPTS

Graph the equation $y = x^{3/4}$, and then use the Trace and Zoom features to approximate each of the following to the nearest tenth.

117. $2^{3/4}$ **118.** $3^{3/4}$ **119.** $10^{3/4}$ **120.** $16^{3/4}$

Graph $y = x^{3/4}$ and $y = x^{4/3}$ on the same coordinate system, using a window with X from -4 to 4 and Y from -1 to 4. Use the graphs to answer the following questions.

121. Where do the two graphs intersect?

122. For what values to x is $x^{3/4} \geq x^{4/3}$?

Carbon-14 dating is used extensively in science to find the age of fossils. If at one time a fossilized substance contains an amount A of carbon-14, then t years later the amount of carbon-14 it contains is given by the formula

$$A \cdot 2^{-t/5,600}$$

Use this formula and a calculator to solve the following problems.

123. A fossilized substance contains 3 micrograms of carbon-14. How much carbon-14 will be left at the end of

a. 5,000 years?   **b.** 10,000 years?

c. 56,000 years?   **d.** 112,000 years?

124. A fossilized substance contains 5 micrograms of carbon-14. How much carbon-14 will be left at the end of

a. 500 years?   **b.** 5,000 years?

c. 56,000 years?   **d.** 112,000 years?

A Multiply expressions containing rational exponents.

B Divide expressions containing rational exponents.

C Factor expressions containing rational exponents.

D Combine two expressions by adding or subtracting rational expressions.

6.2

© iStockphoto/alexis84

Suppose you purchased 10 silver proof coin sets in 1997 for $21 each, for a total investment of $210. Three years later, in 2000, you find that each set is worth $30, which means that your 10 sets have a total value of $300.

You can calculate the annual rate of return on this investment using a formula that involves rational exponents. The annual rate of return will tell you at what interest rate you would have to invest your original $210 for it to be worth $300, 3 years later. As you will see at the end of this section, the annual rate of return on this investment is 12.6%, which is a good return on your money.

In this section, we will look at multiplication, division, factoring, and simplification of some expressions that resemble polynomials but contain rational exponents. The problems in this section will be of particular interest to you if you are planning to take either an engineering calculus class or a business calculus class. As was the case in the previous section, we will assume all variables represent nonnegative real numbers. That way we will not have to worry about the possibility of introducing undefined terms—even roots of negative numbers—into any of our examples. Let's begin this section with a look at multiplication of expressions containing rational exponents.

Multiplication of Expressions Containing Rational Exponents

EXAMPLE 1

Multiply $x^{2/3}(x^{4/3} - x^{1/3})$.

SOLUTION   Applying the distributive property and then simplifying the resulting terms, we have

$$x^{2/3}(x^{4/3} - x^{1/3}) = x^{2/3}x^{4/3} - x^{2/3}x^{1/3} \qquad \text{Distributive property}$$

$$= x^{6/3} - x^{3/3} \qquad \text{Add exponents.}$$

$$= x^2 - x \qquad \text{Simplify.}$$

EXAMPLE 2

Multiply $(x^{2/3} - 3)(x^{2/3} + 5)$.

SOLUTION   Applying the FOIL method, we multiply as if we were multiplying two binomials:

$$(x^{2/3} - 3)(x^{2/3} + 5) = x^{2/3}x^{2/3} + 5x^{2/3} - 3x^{2/3} - 15$$

$$= x^{4/3} + 2x^{2/3} - 15$$

EXAMPLE 3

Multiply $(3a^{1/3} - 2b^{1/3})(4a^{1/3} - b^{1/3})$.

SOLUTION   Again, we use the FOIL method to multiply:

$$(3a^{1/3} - 2b^{1/3})(4a^{1/3} - b^{1/3}) = 3a^{1/3}4a^{1/3} - 3a^{1/3}b^{1/3} - 2b^{1/3}4a^{1/3} + 2b^{1/3}b^{1/3}$$

$$= 12a^{2/3} - 11a^{1/3}b^{1/3} + 2b^{2/3}$$

EXAMPLE 4

Expand $(t^{1/2} - 5)^2$.

SOLUTION   We can use the definition of exponents and the FOIL method:

$$(t^{1/2} - 5)^2 = (t^{1/2} - 5)(t^{1/2} - 5)$$

$$= t^{1/2}t^{1/2} - 5t^{1/2} - 5t^{1/2} + 25$$

$$= t - 10t^{1/2} + 25$$

We can obtain the same result by using the formula for the square of a binomial, $(a - b)^2 = a^2 - 2ab + b^2$:

$$(t^{1/2} - 5)^2 = (t^{1/2})^2 - 2t^{1/2} \cdot 5 + 5^2$$

$$= t - 10t^{1/2} + 25$$

EXAMPLE 5

Multiply $(x^{3/2} - 2^{3/2})(x^{3/2} + 2^{3/2})$.

SOLUTION   This product has the form $(a - b)(a + b)$, which will result in the difference of two squares, $a^2 - b^2$:

$$(x^{3/2} - 2^{3/2})(x^{3/2} + 2^{3/2}) = (x^{3/2})^2 - (2^{3/2})^2$$

$$= x^3 - 2^3$$

$$= x^3 - 8$$

EXAMPLE 6

Multiply $(a^{1/3} - b^{1/3})(a^{2/3} + a^{1/3}b^{1/3} + b^{2/3})$.

SOLUTION   We can find this product by multiplying in columns:

$$
\begin{array}{r}
a^{2/3} + a^{1/3}b^{1/3} + b^{2/3} \\
a^{1/3} - b^{1/3} \\
\hline
a\ + a^{2/3}b^{1/3} + a^{1/3}b^{2/3} \\
-\ a^{2/3}b^{1/3}\ - a^{1/3}b^{2/3} -\qquad b \\
\hline
a - b
\end{array}
$$

The product is $a - b$.

B Division of Expressions Containing Rational Exponents

Our next example involves division with expressions that contain rational exponents. As you will see, this kind of division is very similar to division of a polynomial by a monomial.

EXAMPLE 7

Divide $\dfrac{15x^{2/3}y^{1/3} - 20x^{4/3}y^{2/3}}{5x^{1/3}y^{1/3}}$.

SOLUTION   We can approach this problem in the same way we approached division by a monomial. We simply divide each term in the numerator by the term in the denominator:

$$\frac{15x^{2/3}y^{1/3} - 20x^{4/3}y^{2/3}}{5x^{1/3}y^{1/3}} = \frac{15x^{2/3}y^{1/3}}{5x^{1/3}y^{1/3}} - \frac{20x^{4/3}y^{2/3}}{5x^{1/3}y^{1/3}}$$

$$= 3x^{1/3} - 4xy^{1/3}$$

C Factoring Expressions Containing Rational Exponents

The next three examples involve factoring. In the first example, we are told what to factor from each term of an expression.

EXAMPLE 8

Factor $3(x - 2)^{1/3}$ from $12(x - 2)^{4/3} - 9(x - 2)^{1/3}$.

SOLUTION   This solution is similar to factoring out the greatest common factor:

$$12(x - 2)^{4/3} - 9(x - 2)^{1/3} = 3(x - 2)^{1/3}[4(x - 2) - 3]$$

$$= 3(x - 2)^{1/3}(4x - 11)$$

Although an expression containing rational exponents is not a polynomial—remember, a polynomial must have exponents that are whole numbers—we are going to treat the following expressions as if they were polynomials.

EXAMPLE 9

Factor $x^{2/3} - 3x^{1/3} - 10$ as if it were a trinomial.

SOLUTION   We can think of $x^{2/3} - 3x^{1/3} - 10$ as if it is a trinomial in which the variable is $x^{1/3}$. To see this, replace $x^{1/3}$ with y to get

$$y^2 - 3y - 10$$

Because this trinomial in y factors as $(y - 5)(y + 2)$, we can factor our original expression similarly:

$$x^{2/3} - 3x^{1/3} - 10 = (x^{1/3} - 5)(x^{1/3} + 2)$$

Remember, with factoring, we can always multiply our factors to check that we have factored correctly.

EXAMPLE 10

Factor $6x^{2/5} + 11x^{1/5} - 10$ as if it were a trinomial.

SOLUTION   We can think of the expression in question as a trinomial in $x^{1/5}$.

$$6x^{2/5} + 11x^{1/5} - 10 = (3x^{1/5} - 2)(2x^{1/5} + 5)$$

Ⓓ Combining Two Expressions

In our next example, we combine two expressions by applying the methods we used to add and subtract fractions or rational expressions in Chapter 4.

EXAMPLE 11

Subtract $(x^2 + 4)^{1/2} - \dfrac{x^2}{(x^2 + 4)^{1/2}}$.

SOLUTION   To combine these two expressions, we need to find the least common denominator (LCD), change to equivalent fractions, and subtract numerators. The LCD is $(x^2 + 4)^{1/2}$.

$$(x^2 + 4)^{1/2} - \frac{x^2}{(x^2 + 4)^{1/2}} = \frac{(x^2 + 4)^{1/2}}{1} \cdot \frac{(x^2 + 4)^{1/2}}{(x^2 + 4)^{1/2}} - \frac{x^2}{(x^2 + 4)^{1/2}}$$

$$= \frac{x^2 + 4 - x^2}{(x^2 + 4)^{1/2}}$$

$$= \frac{4}{(x^2 + 4)^{1/2}}$$

EXAMPLE 12

If you purchase an investment for P dollars and t years later it is worth A dollars, then the annual rate of return on that investment is given by the formula

$$r = \left(\frac{A}{P} \right)^{1/t} - 1$$

Find the annual rate of return on a coin collection that was purchased for $210 and sold 3 years later for $300.

SOLUTION   Using $A = 300$, $P = 210$, and $t = 3$ in the formula, we have

$$r = \left(\frac{300}{210} \right)^{1/3} - 1$$

The easiest way to simplify this expression is with a calculator.

$$\boxed{(}\, 300 \,\boxed{\div}\, 210 \,\boxed{)}\,\boxed{\wedge}\,\boxed{(}\,\boxed{(}\, 1 \,\boxed{\div}\, 3 \,\boxed{)}\,\boxed{-}\, 1 \,\boxed{\text{ENTER}}$$

Allowing three decimal places, the result is 0.126. The annual rate of return on the coin collection is approximately 12.6%. To do as well with a savings account, we would have to invest the original $210 in an account that paid 12.6% compounded annually.

GETTING READY FOR CLASS

After reading through the preceding section, respond in your own words and in complete sentences.

A. When multiplying expressions with fractional exponents, when do we add the fractional exponents?

B. Is it possible to multiply two expressions with fractional exponents and end up with an expression containing integer exponents? Support your answer with examples.

C. Write an application modeled by the equation $r = \left(\frac{1,000}{600} \right)^{1/8} - 1$.

D. When can you use the FOIL method with expressions that contain rational exponents?

Problem Set 6.2

Multiply. (Assume all variables in this problem set represent nonnegative real numbers.)

1. $x^{2/3}(x^{1/3} + x^{4/3})$

2. $x^{2/5}(x^{3/5} - x^{8/5})$

3. $a^{1/2}(a^{3/2} - a^{1/2})$

4. $a^{1/4}(a^{3/4} + a^{7/4})$

5. $2x^{1/3}(3x^{8/3} - 4x^{5/3} + 5x^{2/3})$

6. $5x^{1/2}(4x^{5/2} + 3x^{3/2} + 2x^{1/2})$

7. $4x^{1/2}y^{3/5}(3x^{3/2}y^{-3/5} - 9x^{-1/2}y^{7/5})$

8. $3x^{4/5}y^{1/3}(4x^{6/5}y^{-1/3} - 12x^{-4/5}y^{5/3})$

9. $(x^{2/3} - 4)(x^{2/3} + 2)$

10. $(x^{2/3} - 5)(x^{2/3} + 2)$

11. $(a^{1/2} - 3)(a^{1/2} - 7)$

12. $(a^{1/2} - 6)(a^{1/2} - 2)$

13. $(4y^{1/3} - 3)(5y^{1/3} + 2)$

14. $(5y^{1/3} - 2)(4y^{1/3} + 3)$

15. $(5x^{2/3} + 3y^{1/2})(2x^{2/3} + 3y^{1/2})$

16. $(4x^{2/3} - 2y^{1/2})(5x^{2/3} - 3y^{1/2})$

17. $(t^{1/2} + 5)^2$

18. $(t^{1/2} - 3)^2$

19. $(x^{3/2} + 4)^2$

20. $(x^{3/2} - 6)^2$

21. $(a^{1/2} - b^{1/2})^2$

22. $(a^{1/2} + b^{1/2})^2$

23. $(2x^{1/2} - 3y^{1/2})^2$

24. $(5x^{1/2} + 4y^{1/2})^2$

25. $(a^{1/2} - 3^{1/2})(a^{1/2} + 3^{1/2})$

26. $(a^{1/2} - 5^{1/2})(a^{1/2} + 5^{1/2})$

27. $(x^{3/2} + y^{3/2})(x^{3/2} - y^{3/2})$

28. $(x^{5/2} + y^{5/2})(x^{5/2} - y^{5/2})$

29. $(t^{1/2} - 2^{3/2})(t^{1/2} + 2^{3/2})$

30. $(t^{1/2} - 5^{3/2})(t^{1/2} + 5^{3/2})$

31. $(2x^{3/2} + 3^{1/2})(2x^{3/2} - 3^{1/2})$

32. $(3x^{1/2} + 2^{3/2})(3x^{1/2} - 2^{3/2})$

33. $(x^{1/3} + y^{1/3})(x^{2/3} - x^{1/3}y^{1/3} + y^{2/3})$

34. $(x^{1/3} - y^{1/3})(x^{2/3} + x^{1/3}y^{1/3} + y^{2/3})$

35. $(a^{1/3} - 2)(a^{2/3} + 2a^{1/3} + 4)$

36. $(a^{1/3} + 3)(a^{2/3} - 3a^{1/3} + 9)$

37. $(2x^{1/3} + 1)(4x^{2/3} - 2x^{1/3} + 1)$

38. $(3x^{1/3} - 1)(9x^{2/3} + 3x^{1/3} + 1)$

39. $(t^{1/4} - 1)(t^{1/4} + 1)(t^{1/2} + 1)$

40. $(t^{1/4} - 2)(t^{1/4} + 2)(t^{1/2} + 4)$

Divide.

41. $\dfrac{18x^{3/4} + 27x^{1/4}}{9x^{1/4}}$

42. $\dfrac{25x^{1/4} + 30x^{3/4}}{5x^{1/4}}$

43. $\dfrac{12x^{2/3}y^{1/3} - 16x^{1/3}y^{2/3}}{4x^{1/3}y^{1/3}}$

44. $\dfrac{12x^{4/3}y^{1/3} - 18x^{1/3}y^{4/3}}{6x^{1/3}y^{1/3}}$

45. $\dfrac{21a^{7/5}b^{3/5} - 14a^{2/5}b^{8/5}}{7a^{2/5}b^{3/5}}$

46. $\dfrac{24a^{9/5}b^{3/5} - 16a^{4/5}b^{8/5}}{8a^{4/5}b^{3/5}}$

47. Factor $3(x - 2)^{1/2}$ from $12(x - 2)^{3/2} - 9(x - 2)^{1/2}$.

48. Factor $4(x + 1)^{1/3}$ from $4(x + 1)^{4/3} + 8(x + 1)^{1/3}$.

49. Factor $5(x - 3)^{7/5}$ from $5(x - 3)^{12/5} - 15(x - 3)^{7/5}$.

50. Factor $6(x + 3)^{8/7}$ from $6(x + 3)^{15/7} - 12(x + 3)^{8/7}$.

51. Factor $3(x + 1)^{1/2}$ from $9x(x + 1)^{3/2} + 6(x + 1)^{1/2}$.

52. Factor $4x(x + 1)^{1/2}$ from $4x^2(x + 1)^{1/2} + 8x(x + 1)^{3/2}$.

Factor each of the following as if it were a trinomial.

53. $x^{2/3} - 5x^{1/3} + 6$ **54.** $x^{2/3} - x^{1/3} - 6$

55. $a^{2/5} - 2a^{1/5} - 8$ **56.** $a^{2/5} + 2a^{1/5} - 8$

57. $2y^{2/3} - 5y^{1/3} - 3$ **58.** $3y^{2/3} + 5y^{1/3} - 2$

59. $9t^{2/5} - 25$ **60.** $16t^{2/5} - 49$

61. $4x^{2/7} + 20x^{1/7} + 25$ **62.** $25x^{2/7} - 20x^{1/7} + 4$

Simplify each of the following to a single fraction.

63. $\dfrac{3}{x^{1/2}} + x^{1/2}$ **64.** $\dfrac{2}{x^{1/2}} - x^{1/2}$

65. $x^{2/3} + \dfrac{5}{x^{1/3}}$ **66.** $x^{3/4} - \dfrac{7}{x^{1/4}}$

67. $\dfrac{3x^2}{(x^3 + 1)^{1/2}} + (x^3 + 1)^{1/2}$ **68.** $\dfrac{x^3}{(x^2 - 1)^{1/2}} + 2x(x^2 - 1)^{1/2}$

69. $\dfrac{x^2}{(x^2 + 4)^{1/2}} - (x^2 + 4)^{1/2}$ **70.** $\dfrac{x^5}{(x^2 - 2)^{1/2}} + 4x^3(x^2 - 2)^{1/2}$

Use a calculator to find approximations to each of the following. Round your answers for Problems 75 and 76 to three places past the decimal point.

71. $16^{0.25}$ **72.** $81^{0.25}$

73. $9^{1.5}$ **74.** $32^{0.4}$

75. $\left(\dfrac{1}{2}\right)^{1/5}$ **76.** $\left(\dfrac{1}{2}\right)^{1/10}$

APPLYING THE CONCEPTS

77. Investing A coin collection is purchased as an investment for $500 and sold 4 years later for $900. Find the annual rate of return on the investment.

78. Investing An investor buys stock in a company for $800. Five years later, the same stock is worth $1,600. Find the annual rate of return on the stock.

79. Investing Find the annual rate of return on a home that is purchased for $60,000 and is sold 5 years later for $80,000.

80. Investing Find the annual rate of return on a home that is purchased for $75,000 and is sold 10 years later for $150,000.

81. Falling Object If an object is dropped, the time, in seconds, that it takes to fall d feet is given by the equation $t = \left(\dfrac{d}{16}\right)^{1/2}$. How long does it take for a ball that is dropped off a 200-foot cliff to hit the bottom?

82. Stopping Distance Police can measure the skid marks left by a car and calculate the speed the car was traveling before the driver slammed on the brakes. If it takes d feet to stop, then the initial speed of the car, in miles per hour, is given approximately by $v = \left(\dfrac{d}{k}\right)^{1/2}$, where k is a constant that depends on the particular car. According to *Consumer Reports*, the 1998 Mercury Mountaineer takes 147 feet to stop at 60 mph on dry pavement. This allows us to calculate the value of k for this sport utility vehicle as $k = .0408$. Suppose it took a 1998 Mercury Mountaineer 240 feet to stop on dry pavement. How fast was it initially going?

83. **Kepler's Law** Kepler's third law of planetary motion says the period T of the orbit of a planet or satellite is related to the radius r of its orbit by the equation $T = kr^{3/2}$, where k is a constant that depends on the planetary system. Earth is about 93 million miles from the Sun and orbits the Sun in 1 year. Therefore, if we measure r in millions of miles and T in years, we can calculate the value of the constant k as $k \approx .001115$. Mars is about 141.4 million miles from the Sun. How long does it take for Mars to complete one orbit?

84. **Population Growth** If a population of P_0 grows to a size of P in t years, the annual percentage growth rate is given by the equation $r = \left(\frac{P}{P_0}\right)^{1/t} - 1$. In 1970, the population of Algeria was 13.9 million. In 1980, the population had grown to 18.8 million people. What is the annual growth rate of the population? Express your answer as a percentage rounded to two decimal places.

GETTING READY FOR THE NEXT SECTION

The following problems review material we covered in sections 6.1 and 6.2. Simplify.

85. $\dfrac{x^2 - 9}{x^4 - 81}$

86. $\dfrac{6 - a - a^2}{3 - 2a - a^2}$

87. $\dfrac{15x^2y - 20x^4y^2}{5xy}$

88. $\dfrac{12x^3y^2 - 24x^2y^3}{6xy}$

Divide using long division. [6.2]

89. $\dfrac{10x^2 + 7x - 12}{2x + 3}$

90. $\dfrac{6x^2 - x - 35}{2x - 5}$

91. $\dfrac{x^3 - 125}{x - 5}$

92. $\dfrac{x^3 + 64}{x + 4}$

EXTENDING THE CONCEPTS

93. Choose values for x and y to show the following: $(x^{1/2} + y^{1/2})^2 \neq x + y$.

94. Choose values for x and y to show the following: $(x^2 + y^2)^{1/2} \neq x + y$.

Write the following with positive exponents and simplify.

95. $\dfrac{x + y}{x^{-1} + y^{-1}}$

96. $\dfrac{x^2 - y^2}{x^{-1} - y^{-1}}$

Write the following with rational exponents and simplify.

97. $\sqrt[5]{x^3} \cdot \sqrt[4]{x}$

98. $\sqrt[6]{x^5} \cdot \sqrt[3]{x^2}$

99. $\sqrt[4]{\sqrt[3]{m}}$

100. $\sqrt[3]{\sqrt{k}}$

OBJECTIVES

A Write radical expressions in simplified form.

B Rationalize a denominator that contains only one term.

C Simplify expressions using the absolute value property for radicals.

D Construct a spiral of roots.

© iStockphoto/joingate

6.3

Earlier in this chapter, we showed how the Pythagorean theorem can be used to construct a golden rectangle. In a similar manner, the Pythagorean theorem can be used to construct the attractive spiral shown here.

This spiral is called the Spiral of Roots because each of the diagonals is the positive square root of one of the positive integers. At the end of this section, we will use the Pythagorean theorem and some of the material in this section to construct this spiral.

In this section, we will use radical notation instead of rational exponents. We will begin by stating two properties of radicals. Following this, we will give a definition for simplified form for radical expressions. The examples in this section show how we use the properties of radicals to write radical expressions in simplified form.

The Spiral of Roots

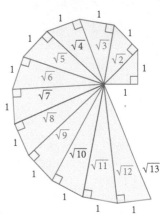

Simplified Form

Here are the first two properties of radicals. For these two properties, we will assume a and b are nonnegative real numbers whenever n is an even number.

PROPERTY *Property 1 for Radicals*

· ·

$$\sqrt[n]{ab} = \sqrt[n]{a}\sqrt[n]{b}$$

In words: The nth root of a product is the product of the nth roots.

Proof of Property 1

$$\sqrt[n]{ab} = (ab)^{1/n} \qquad \text{Definition of fractional exponents}$$

$$= a^{1/n}b^{1/n} \qquad \text{Exponents distribute over products}$$

$$= \sqrt[n]{a}\sqrt[n]{b} \qquad \text{Definition of fractional exponents}$$

NOTE There is not a property for radicals that says the nth root of a sum is the sum of the nth roots. That is,

$$\sqrt[n]{a+b} \neq \sqrt[n]{a} + \sqrt[n]{b}$$

PROPERTY *Property 2 for Radicals*
• •

$$\sqrt[n]{\frac{a}{b}} = \frac{\sqrt[n]{a}}{\sqrt[n]{b}} = \quad (b \neq 0)$$

In words: The nth root of a quotient is the quotient of the nth roots.

The proof of Property 2 is similar to the proof of Property 1.

These two properties of radicals allow us to change the form of and simplify radical expressions without changing their value.

RULE *Simplified Form for Radical Expressions*
• •

A radical expression is in *simplified form* if

1. None of the factors of the radicand (the quantity under the radical sign) can be written as powers greater than or equal to the index — that is, no perfect squares can be factors of the quantity under a square root sign, no perfect cubes can be factors of what is under a cube root sign, and so forth.
2. There are no fractions under the radical sign.
3. There are no radicals in the denominator.

Satisfying the first condition for simplified form actually amounts to taking as much out from under the radical sign as possible. The following examples illustrate the first condition for simplified form.

EXAMPLE 1

Write $\sqrt{50}$ in simplified form.

SOLUTION The largest perfect square that divides 50 is 25. We write 50 as $25 \cdot 2$ and apply Property 1 for radicals:

$$\sqrt{50} = \sqrt{25 \cdot 2} \qquad 50 = 25 \cdot 2$$
$$= \sqrt{25}\sqrt{2} \qquad \text{Property 1}$$
$$= 5\sqrt{2} \qquad \sqrt{25} = 5$$

We have taken as much as possible out from under the radical sign — in this case, factoring 25 from 50 and then writing $\sqrt{25}$ as 5.

As we progress through this chapter, you will see more and more expressions that involve the product of a number and a radical. Here are some examples:

$$3\sqrt{2} \qquad \frac{1}{2}\sqrt{5} \qquad 5\sqrt{7} \qquad 3x\sqrt{2x} \qquad 2ab\sqrt{5a}$$

All of these are products. The first expression, $3\sqrt{2}$, is the product of 3 and $\sqrt{2}$. That is,

$$3\sqrt{2} = 3 \cdot \sqrt{2}$$

The 3 and the $\sqrt{2}$ are not stuck together in some mysterious way. The expression $3\sqrt{2}$ is simply the product of two numbers, one of which is rational, and the other is irrational.

EXAMPLE 2

Write in simplified form: $\sqrt{48x^4y^3}$, where $x, y \geq 0$

SOLUTION The largest perfect square that is a factor of the radicand is $16x^4y^2$. Applying Property 1 again, we have

$$\sqrt{48x^4y^3} = \sqrt{16x^4y^2 \cdot 3y}$$
$$= \sqrt{16x^4y^2}\sqrt{3y}$$
$$= 4x^2y\sqrt{3y}$$

EXAMPLE 3

Write $\sqrt[3]{40a^5b^4}$ in simplified form.

SOLUTION We now want to factor the largest perfect cube from the radicand. We write $40a^5b^4$ as $8a^3b^3 \cdot 5a^2b$ and proceed as we did in Examples 1 and 2.

$$\sqrt[3]{40a^5b^4} = \sqrt[3]{8a^3b^3 \cdot 5a^2b}$$
$$= \sqrt[3]{8a^3b^3}\sqrt[3]{5a^2b}$$
$$= 2ab\sqrt[3]{5a^2b}$$

Our next examples involve fractions and simplified form for radicals.

EXAMPLE 4

Simplify each expression.

a. $\dfrac{\sqrt{12}}{6}$ **b.** $\dfrac{5\sqrt{18}}{15}$ **c.** $\dfrac{6 + \sqrt{8}}{2}$ **d.** $\dfrac{-1 + \sqrt{45}}{2}$

SOLUTION In each case, we simplify the radical first, then we factor and reduce to lowest terms.

a.
$$\frac{\sqrt{12}}{6} = \frac{2\sqrt{3}}{6} \qquad \text{Simplify the radical } \sqrt{12} = \sqrt{4 \cdot 3} = \sqrt{4}\sqrt{3} = 2\sqrt{3}$$
$$= \frac{2\sqrt{3}}{2 \cdot 3} \qquad \text{Factor denominator.}$$
$$= \frac{\sqrt{3}}{3} \qquad \text{Divide out common factors.}$$

b.
$$\frac{5\sqrt{18}}{15} = \frac{5 \cdot 3\sqrt{2}}{15} \qquad \sqrt{18} = \sqrt{9 \cdot 2} = \sqrt{9}\sqrt{2} = 3\sqrt{2}$$
$$= \frac{5 \cdot 3\sqrt{2}}{3 \cdot 5} \qquad \text{Factor denominator.}$$
$$= \sqrt{2} \qquad \text{Divide out common factors.}$$

c.
$$\frac{6 + \sqrt{8}}{2} = \frac{6 + 2\sqrt{2}}{2} \qquad \sqrt{8} = \sqrt{4 \cdot 2} = \sqrt{4}\sqrt{2} = 2\sqrt{2}$$
$$= \frac{2(3 + \sqrt{2})}{2} \qquad \text{Factor numerator.}$$
$$= 3 + \sqrt{2} \qquad \text{Divide out common factors.}$$

d.
$$\frac{-1 + \sqrt{45}}{2} = \frac{-1 + 3\sqrt{5}}{2} \qquad \sqrt{45} = \sqrt{9 \cdot 5} = \sqrt{9}\sqrt{5} = 3\sqrt{5}$$

This expression cannot be simplified further because $-1 + 3\sqrt{5}$ and 2 have no factors in common.

B Rationalizing the Denominator

EXAMPLE 5

Simplify $\sqrt{\dfrac{3}{4}}$.

SOLUTION   Applying Property 2 for radicals, we have

$$\sqrt{\frac{3}{4}} = \frac{\sqrt{3}}{\sqrt{4}} \qquad \text{Property 2}$$

$$= \frac{\sqrt{3}}{2} \qquad \sqrt{4} = 2$$

The last expression is in simplified form because it satisfies all three conditions for simplified form.

EXAMPLE 6

Write $\sqrt{\dfrac{5}{6}}$ in simplified form.

SOLUTION   Proceeding as in Example 5, we have

$$\sqrt{\frac{5}{6}} = \frac{\sqrt{5}}{\sqrt{6}}$$

The resulting expression satisfies the second condition for simplified form because neither radical contains a fraction. It does, however, violate Condition 3 because it has a radical in the denominator. Getting rid of the radical in the denominator is called *rationalizing the denominator* and is accomplished, in this case, by multiplying the numerator and denominator by $\sqrt{6}$:

$$\frac{\sqrt{5}}{\sqrt{6}} = \frac{\sqrt{5}}{\sqrt{6}} \cdot \frac{\sqrt{6}}{\sqrt{6}}$$

$$= \frac{\sqrt{30}}{\sqrt{6^2}}$$

$$= \frac{\sqrt{30}}{6}$$

EXAMPLE

Rationalize the denominator.

7. $\dfrac{4}{\sqrt{3}} = \dfrac{4}{\sqrt{3}} \cdot \dfrac{\sqrt{3}}{\sqrt{3}}$

$\quad = \dfrac{4\sqrt{3}}{\sqrt{3^2}}$

$\quad = \dfrac{4\sqrt{3}}{3}$

8. $\dfrac{2\sqrt{3x}}{\sqrt{5y}} = \dfrac{2\sqrt{3x}}{\sqrt{5y}} \cdot \dfrac{\sqrt{5y}}{\sqrt{5y}}$

$\quad = \dfrac{2\sqrt{15xy}}{\sqrt{(5y)^2}}$

$\quad = \dfrac{2\sqrt{15xy}}{5y}$

When the denominator involves a cube root, we must multiply by a radical that will produce a perfect cube under the cube root sign in the denominator, as Example 9 illustrates.

EXAMPLE 9

Rationalize the denominator in $\dfrac{7}{\sqrt[3]{4}}$.

SOLUTION Because $4 = 2^2$, we can multiply both numerator and denominator by $\sqrt[3]{2}$ and obtain $\sqrt[3]{2^3}$ in the denominator.

$$\frac{7}{\sqrt[3]{4}} = \frac{7}{\sqrt[3]{2^2}}$$

$$= \frac{7}{\sqrt[3]{2^2}} \cdot \frac{\sqrt[3]{2}}{\sqrt[3]{2}}$$

$$= \frac{7\sqrt[3]{2}}{\sqrt[3]{2^3}}$$

$$= \frac{7\sqrt[3]{2}}{2}$$

EXAMPLE 10

Simplify $\sqrt{\dfrac{12x^5y^3}{5z}}$.

SOLUTION We use Property 2 to write the numerator and denominator as two separate radicals:

$$\sqrt{\frac{12x^5y^3}{5z}} = \frac{\sqrt{12x^5y^3}}{\sqrt{5z}}$$

Simplifying the numerator, we have

$$\frac{\sqrt{12x^5y^3}}{\sqrt{5z}} = \frac{\sqrt{4x^4y^2}\sqrt{3xy}}{\sqrt{5z}}$$

$$= \frac{2x^2y\sqrt{3xy}}{\sqrt{5z}}$$

To rationalize the denominator, we multiply the numerator and denominator by $\sqrt{5z}$:

$$\frac{2x^2y\sqrt{3xy}}{\sqrt{5z}} \cdot \frac{\sqrt{5z}}{\sqrt{5z}} = \frac{2x^2y\sqrt{15xyz}}{\sqrt{(5z)^2}}$$

$$= \frac{2x^2y\sqrt{15xyz}}{5z}$$

C Square Root of a Perfect Square

So far in this chapter, we have assumed that all our variables are nonnegative when they appear under a square root symbol. There are times, however, when this is not the case.

Consider the following two statements:

$$\sqrt{3^2} = \sqrt{9} = 3 \qquad \text{and} \qquad \sqrt{(-3)^2} = \sqrt{9} = 3$$

Whether we operate on 3 or -3, the result is the same: Both expressions simplify to 3. The other operation we have worked with in the past that produces the same result is absolute value. That is,

$$|3| = 3 \quad \text{and} \quad |-3| = 3$$

This leads us to the next property of radicals.

PROPERTY   *Property 3 for Radicals*
• •

If a is a real number, then $\sqrt{a^2} = |a|$.

The result of this discussion and Property 3 is simply this:

If we know a is positive, then $\sqrt{a^2} = a$.

If we know a is negative, then $\sqrt{a^2} = |a|$.

If we don't know if a is positive or negative, then $\sqrt{a^2} = |a|$.

EXAMPLE

Simplify each expression. Do *not* assume the variables represent positive numbers.

11. $\sqrt{9x^2} = 3|x|$

12. $\sqrt{x^3} = |x|\sqrt{x}$

13. $\sqrt{x^2 - 6x + 9} = \sqrt{(x-3)^2} = |x - 3|$

14. $\sqrt{x^3 - 5x^2} = \sqrt{x^2(x-5)} = |x|\sqrt{x-5}$

As you can see, we must use absolute value symbols when we take a square root of a perfect square, unless we know that the base of the perfect square is a positive number. The same idea holds for higher even roots, but not for odd roots. With odd roots, no absolute value symbols are necessary.

EXAMPLE

Simplify each expression.

15. $\sqrt[3]{(-2)^3} = \sqrt[3]{-8} = -2$

16. $\sqrt[3]{(-5)^3} = \sqrt[3]{-125} = -5$

We can extend this discussion to all roots as follows:

PROPERTY   *Extending Property 3 for Radicals*
• •

If a is a real number, then

$$\sqrt[n]{a^n} = |a| \qquad \text{if} \qquad n \text{ is even}$$

$$\sqrt[n]{a^n} = a \qquad \text{if} \qquad n \text{ is odd}$$

ⓓ The Spiral of Roots

To visualize the square roots of the positive integers, we can construct the spiral of roots that we mentioned in the introduction to this section. To begin, we draw two line segments, each of length 1, at right angles to each other. Then we use the Pythagorean theorem to find the length of the diagonal. Figure 1 illustrates this procedure.

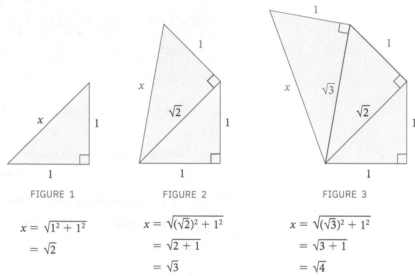

FIGURE 1   FIGURE 2   FIGURE 3

$$x = \sqrt{1^2 + 1^2}$$
$$= \sqrt{2}$$

$$x = \sqrt{(\sqrt{2})^2 + 1^2}$$
$$= \sqrt{2 + 1}$$
$$= \sqrt{3}$$

$$x = \sqrt{(\sqrt{3})^2 + 1^2}$$
$$= \sqrt{3 + 1}$$
$$= \sqrt{4}$$

Next, we construct a second triangle by connecting a line segment of length 1 to the end of the first diagonal so that the angle formed is a right angle. We find the length of the second diagonal using the Pythagorean theorem. Figure 2 illustrates this procedure. Continuing to draw new triangles by connecting line segments of length 1 to the end of each new diagonal, so that the angle formed is a right angle, the spiral of roots begins to appear (Figure 3).

The Spiral of Roots and Function Notation

Looking over the diagrams and calculations in the preceding discussion, we see that each diagonal in the spiral of roots is found by using the length of the previous diagonal.

First diagonal: $\sqrt{1^2 + 1^2} = \sqrt{2}$

Second diagonal: $\sqrt{(\sqrt{2})^2 + 1^2} = \sqrt{3}$

Third diagonal: $\sqrt{(\sqrt{3})^2 + 1^2} = \sqrt{4}$

Fourth diagonal: $\sqrt{(\sqrt{4})^2 + 1^2} = \sqrt{5}$

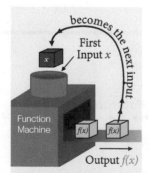

A process like this one, in which the answer to one calculation is used to find the answer to the next calculation, is called a *recursive* process. In this particular case, we can use function notation to model the process. If we let *x* represent the length of any diagonal, then the length of the next diagonal is given by

$$f(x) = \sqrt{x^2 + 1}$$

To begin the process of finding the diagonals, we let $x = 1$:

$$f(1) = \sqrt{1^2 + 1} = \sqrt{2}$$

To find the next diagonal, we substitue $\sqrt{2}$ for x to obtain

$$f[f(1)] = f(\sqrt{2}) = \sqrt{(\sqrt{2})^2 + 1} = \sqrt{3}$$

$$f(f[f(1)]) = f(\sqrt{3}) = \sqrt{(\sqrt{3})^2 + 1} = \sqrt{4}$$

We can describe this process of finding the diagonals of the spiral of roots very concisely this way:

$$f(1), f[f(1)], f(f[f(1)]), \ldots \qquad \text{where } f(x) = \sqrt{x^2 + 1}$$

This sequence of function values is a special case of a general category of similar sequences that are closely connected to *fractals* and *chaos,* two topics in mathematics that are currently receiving a good deal of attention.

USING TECHNOLOGY

As our preceding discussion indicates, the length of each diagonal in the spiral of roots is used to calculate the length of the next diagonal. The ANS key on a graphing calculator can be used effectively in a situation like this. To begin, we store the number 1 in the variable ANS. Next, we key in the fomula used to produce each diagonal using ANS for the variable. After that, it is simply a matter of pressing ENTER as many times as we like to produce the lengths of as many diagonals as we like. Here is a summary of what we do:

Enter This	Display Shows
1 ENTER	1.000
$\sqrt{\ }$ (ANS^2 + 1) ENTER	1.414
ENTER	1.732
ENTER	2.000
ENTER	2.236

If you continue to press the ENTER key, you will produce decimal approximations for as many of the diagonals in the spiral of roots as you like.

GETTING READY FOR CLASS

After reading through the preceding section, respond in your own words and in complete sentences.

A. Explain why this statement is false: "The square root of a sum is the sum of the square roots."

B. What is simplified form for an expression that contains a square root?

C. Why is it not necessarily true that $\sqrt{a^2} = a$?

D. What does it mean to rationalize the denominator in an expression?

Problem Set 6.3

Use Property 1 for radicals to write each of the following expressions in simplified form. (Assume all variables are nonnegative through Problem 70.)

1. $\sqrt{8}$ **2.** $\sqrt{32}$ **3.** $\sqrt{98}$ **4.** $\sqrt{75}$

5. $\sqrt{288}$ **6.** $\sqrt{128}$ **7.** $\sqrt{80}$ **8.** $\sqrt{200}$

9. $\sqrt{48}$ **10.** $\sqrt{27}$ **11.** $\sqrt{675}$ **12.** $\sqrt{972}$

13. $\sqrt[3]{54}$ **14.** $\sqrt[3]{24}$ **15.** $\sqrt[3]{128}$ **16.** $\sqrt[3]{162}$

17. $\sqrt[3]{432}$ **18.** $\sqrt[3]{1,536}$ **19.** $\sqrt[5]{64}$ **20.** $\sqrt[4]{48}$

21. $\sqrt{18x^3}$ **22.** $\sqrt{27x^5}$ **23.** $\sqrt[4]{32y^7}$ **24.** $\sqrt[5]{32y^7}$

25. $\sqrt[3]{40x^4y^7}$ **26.** $\sqrt[3]{128x^6y^2}$ **27.** $\sqrt{48a^2b^3c^4}$ **28.** $\sqrt{72a^4b^3c^2}$

29. $\sqrt[3]{48a^2b^3c^4}$ **30.** $\sqrt[3]{72a^4b^3c^2}$ **31.** $\sqrt[5]{64x^8y^{12}}$ **32.** $\sqrt[4]{32x^9y^{10}}$

33. $\sqrt[5]{243x^7y^{10}z^5}$ **34.** $\sqrt[5]{64x^8y^4z^{11}}$

Substitute the given numbers into the expression $\sqrt{b^2 - 4ac}$, and then simplify.

35. $a = 2, b = -6, c = 3$ **36.** $a = 6, b = 7, c = -5$

37. $a = 1, b = 2, c = 6$ **38.** $a = 2, b = 5, c = 3$

39. $a = \dfrac{1}{2}, b = -\dfrac{1}{2}, c = -\dfrac{5}{4}$ **40.** $a = \dfrac{7}{4}, b = -\dfrac{3}{4}, c = -2$

41. Simplify each expression.

a. $\dfrac{\sqrt{20}}{4}$ b. $\dfrac{3\sqrt{20}}{15}$ c. $\dfrac{4 + \sqrt{12}}{2}$ d. $\dfrac{2 + \sqrt{9}}{5}$

42. Simplify each expression.

a. $\dfrac{\sqrt{12}}{4}$ b. $\dfrac{2\sqrt{32}}{8}$ c. $\dfrac{9 + \sqrt{27}}{3}$ d. $\dfrac{-6 - \sqrt{64}}{2}$

43. Simplify each expression.

a. $\dfrac{10 + \sqrt{75}}{5}$ b. $\dfrac{-6 + \sqrt{45}}{3}$ c. $\dfrac{-2 - \sqrt{27}}{6}$

44. Simplify each expression.

a. $\dfrac{12 - \sqrt{12}}{6}$ b. $\dfrac{-4 - \sqrt{8}}{2}$ c. $\dfrac{6 - \sqrt{48}}{8}$

Rationalize the denominator in each of the following expressions.

45. $\dfrac{2}{\sqrt{3}}$ **46.** $\dfrac{3}{\sqrt{2}}$ **47.** $\dfrac{5}{\sqrt{6}}$ **48.** $\dfrac{7}{\sqrt{5}}$

49. $\sqrt{\dfrac{1}{2}}$ **50.** $\sqrt{\dfrac{1}{3}}$ **51.** $\sqrt{\dfrac{1}{5}}$ **52.** $\sqrt{\dfrac{1}{6}}$

53. $\dfrac{4}{\sqrt[3]{2}}$ **54.** $\dfrac{5}{\sqrt[3]{3}}$ **55.** $\dfrac{2}{\sqrt[3]{9}}$ **56.** $\dfrac{3}{\sqrt[3]{4}}$

57. $\sqrt[4]{\dfrac{3}{2x^2}}$ **58.** $\sqrt[4]{\dfrac{5}{3x^2}}$ **59.** $\sqrt[4]{\dfrac{8}{y}}$ **60.** $\sqrt[4]{\dfrac{27}{y}}$

61. $\sqrt[3]{\dfrac{4x}{3y}}$ **62.** $\sqrt[3]{\dfrac{7x}{6y}}$ **63.** $\sqrt[3]{\dfrac{2x}{9y}}$ **64.** $\sqrt[3]{\dfrac{5x}{4y}}$

Write each of the following in simplified form.

65. $\sqrt{\dfrac{27x^3}{5y}}$ **66.** $\sqrt{\dfrac{12x^5}{7y}}$ **67.** $\sqrt{\dfrac{75x^3y^2}{2z}}$ **68.** $\sqrt{\dfrac{50x^2y^3}{3z}}$

Rationalize the denominator.

69. a. $\dfrac{1}{\sqrt{2}}$ **b.** $\dfrac{1}{\sqrt[3]{2}}$ **c.** $\dfrac{1}{\sqrt[4]{2}}$

70. a. $\dfrac{1}{\sqrt{3}}$ **b.** $\dfrac{1}{\sqrt[3]{9}}$ **c,** $\dfrac{1}{\sqrt[4]{27}}$

Simplify each expression. Do *not* assume the variables represent positive numbers.

71. $\sqrt{25x^2}$ **72.** $\sqrt{49x^2}$ **73.** $\sqrt{27x^3y^2}$ **74.** $\sqrt{40x^3y^2}$

75. $\sqrt{x^2 - 10x + 25}$ **76.** $\sqrt{x^2 - 16x + 64}$

77. $\sqrt{4x^2 + 12x + 9}$ **78.** $\sqrt{16x^2 + 40x + 25}$

79. $\sqrt{4a^4 + 16a^3 + 16a^2}$ **80.** $\sqrt{9a^4 + 18a^3 + 9a^2}$

81. $\sqrt{4x^3 - 8x^2}$ **82.** $\sqrt{18x^3 - 9x^2}$

83. Show that the statement $\sqrt{a + b} = \sqrt{a} + \sqrt{b}$ is not true by replacing a with 9 and b with 16 and simplifying both sides.

84. Find a pair of values for a and b that will make the statement $\sqrt{a + b} = \sqrt{a} + \sqrt{b}$ true.

APPLYING THE CONCEPTS

85. Diagonal Distance The distance d between opposite corners of a rectangular room with length l and width w is given by

$$d = \sqrt{l^2 + w^2}$$

How far is it between opposite corners of a living room that measures 10 by 15 feet?

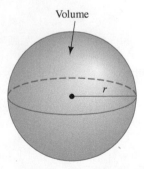

86. Radius of a Sphere The radius r of a sphere with volume V can be found by using the formula

$$r = \sqrt[3]{\dfrac{3V}{4\pi}}$$

Find the radius of a sphere with volume 9 cubic feet. Write your answer in simplified form. (Use $\dfrac{22}{7}$ for π.)

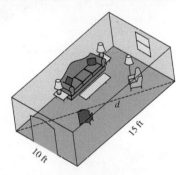

Volume

87. Distance to the Horizon If you are at a point k miles above the surface of Earth, the distance you can see, in miles, is approximated by the equation $d = \sqrt{8000k + k^2}$.

a. How far can you see from a point that is 1 mile above the surface of Earth?

b. How far can you see from a point that is 2 miles above the surface of Earth?

c. How far can you see from a point that is 3 miles above the surface of Earth?

88. **Investing** If you invest P dollars and you want the investment to grow to A dollars in t years, the interest rate that must be earned if interest is compounded annually is given by the formula

$$r = \sqrt[t]{\frac{A}{P}} - 1$$

If you invest \$4,000 and want to have \$7,000 in 8 years, what interest rate must be earned?

89. **Tsunami Speed** The speed s of a tsunami can be found by the formula $s = \sqrt{gd}$, where g is acceleration due to gravity and d is the water depth where the disturbance occurred. If $g = 32$ feet per square second (ft/s²) and $d = 9,000$ feet, what is the approximate speed of the tsunami in feet per second?

90. **Tsunami Speed** The speed s of a tsunami can be found by the formula $s = \sqrt{gd}$, where g is acceleration due to gravity and d is the water depth where the disturbance occurred. If $g = 32$ feet per square second (ft/s²) and $d = 4,000$ feet, what is the approximate speed of the tsunami in feet per second?

91. **Body Surface Area** A simple body surface area formula for a human body is given by

$$B = \sqrt{\frac{mh}{3,600}}$$

where B is the body surface area in square meters, m is mass in kilograms, and h is height in centimeters. What is the body surface area of a person with a mass of 86 kilograms and height of 183 centimeters?

92. **Body Surface Area** A simple body surface area formula for a human body is given by

$$B = \sqrt{\frac{mh}{3,600}}$$

where B is the body surface area in square meters, m is mass in kilograms, and h is height in centimeters. What is the body surface area of a person with a mass of 75 kilograms and height of 170 centimeters?

93. **Spiral of Roots** Construct your own spiral of roots by using a ruler. Draw the first triangle by using two 1-inch lines. The first diagonal will have a length of $\sqrt{2}$ inches. Each new triangle will be formed by drawing a 1-inch line segment at the end of the previous diagonal so the angle formed is 90°.

94. **Spiral of Roots** Construct a spiral of roots by using line segments of length 2 inches. The length of the first diagonal will be $2\sqrt{2}$ inches. The length of the second diagonal will be $2\sqrt{3}$ inches.

95. **Spiral of Roots** If $f(x) = \sqrt{x^2 + 1}$, find the first six terms in the following sequence. Use your results to predict the value of the 10th term and the 100th term.

$$f(1), f[f(1)], f(f[f(1)]), \ldots$$

96. **Spiral of Roots** If $f(x) = \sqrt{x^2 + 4}$, find the first six terms in the following sequence. Use your results to predict the value of the 10th term and the 100th term. (The numbers in this sequence are the lengths of the diagonals of the spiral you drew in Problem 93.)

$$f(2), f[f(2)], f(f[f(2)]), \ldots$$

GETTING READY FOR THE NEXT SECTION

Simplify the following.

97. $5x - 4x + 6x$

98. $12x + 8x - 7x$

99. $35xy^2 - 8xy^2$

100. $20a^2b + 33a^2b$

101. $\dfrac{1}{2}x + \dfrac{1}{3}x$

102. $\dfrac{2}{3}x + \dfrac{5}{8}x$

Write in simplified form for radicals.

103. $\sqrt{18}$

104. $\sqrt{8}$

105. $\sqrt{75xy^3}$

106. $\sqrt{12xy}$

107. $\sqrt[3]{8a^4b^2}$

108. $\sqrt[3]{27ab^2}$

EXTENDING THE CONCEPTS

Factor each radicand into the product of prime factors. Then simplify each radical.

109. $\sqrt[3]{8,640}$

110. $\sqrt{8,640}$

111. $\sqrt[3]{10,584}$

112. $\sqrt{10,584}$

Assume a is a positive number, and rationalize each denominator.

113. $\dfrac{1}{\sqrt[10]{a^3}}$

114. $\dfrac{1}{\sqrt[12]{a^7}}$

115. $\dfrac{1}{\sqrt[20]{a^{11}}}$

116. $\dfrac{1}{\sqrt[15]{a^{13}}}$

117. Show that the two expressions $\sqrt{x^2 + 1}$ and $x + 1$ are not, in general, equal to each other by graphing $y = \sqrt{x^2 + 1}$ and $y = x + 1$ in the same viewing window.

118. Show that the two expressions $\sqrt{x^2 + 9}$ and $x + 3$ are not, in general, equal to each other by graphing $y = \sqrt{x^2 + 9}$ and $y = x + 3$ in the same viewing window.

119. Approximately how far apart are the graphs in Problem 117 when $x = 2$?

120. Approximately how far apart are the graphs in Problem 118 when $x = 2$?

121. For what value of x are the expressions $\sqrt{x^2 + 1}$ and $x + 1$ equal?

122. For what value of x are the expressions $\sqrt{x^2 + 9}$ and $x + 3$ equal?

OBJECTIVES

A Add and subtract radicals.

© Luca Giarelli

6.4

As the Sun sets on a high mountain valley called Val Camonica in northern Italy, shadows bring out ancient rock carvings on the surrounding cliffs. Archaeologists believe the Camuni tribe may have lit the carvings with candlelight, thus making them "move" as the candle flame flickered. The location of the nearby rock walls would also make a storyteller's voice echo. This recurring echo may have provided a cinematic experience for the tribe.

The echo's time delay is the time it takes for the sound of the storyteller's voice to travel to the nearby rock wall, bounce off it, and be audible to the storyteller and audience. To determine this time delay, we need to know the distance of the storyteller to the nearby wall. Suppose the storyteller was standing $12\sqrt{3}$ meters from the opposing rock wall. We can use the following formula for the echo's time delay t:

$$t = \frac{d}{s}$$

where d is the doubled distance (distance of the storyteller to the wall and back), and s is the speed of sound. Since the formula calls for us to double the distance, we need to know how to add $12\sqrt{3}$ and $12\sqrt{3}$ together. In this section, we will learn how to apply our knowledge of similar terms to the addition and subtraction of radical expressions such as these.

A Combining Similar Terms

In Chapter 1, we found we could add similar terms when combining polynomials. The same idea applies to addition and subtraction of radical expressions.

> **DEFINITION** *similar radicals*
>
> Two radicals are said to be *similar radicals* if they have the same index and the same radicand.

The expressions $5\sqrt[3]{7}$ and $-8\sqrt[3]{7}$ are similar since the index is 3 in both cases and the radicands are 7. The expressions $3\sqrt[4]{5}$ and $7\sqrt[3]{5}$ are not similar because they have different indices, and the expressions $2\sqrt[5]{8}$ and $3\sqrt[5]{9}$ are not similar because the radicands are not the same.

We add and subtract radical expressions in the same way we add and subtract polynomials — by combining similar terms under the distributive property.

EXAMPLE 1

Combine $5\sqrt{3} - 4\sqrt{3} + 6\sqrt{3}$.

SOLUTION   All three radicals are similar. We apply the distributive property to get

$$5\sqrt{3} - 4\sqrt{3} + 6\sqrt{3} = (5 - 4 + 6)\sqrt{3}$$
$$= 7\sqrt{3}$$

EXAMPLE 2

Combine $3\sqrt{8} + 5\sqrt{18}$.

SOLUTION   The two radicals do not seem to be similar. We must write each in simplified form before applying the distributive property.

$$3\sqrt{8} + 5\sqrt{18} = 3\sqrt{4 \cdot 2} + 5\sqrt{9 \cdot 2}$$
$$= 3\sqrt{4}\,\sqrt{2} + 5\sqrt{9}\,\sqrt{2}$$
$$= 3 \cdot 2\,\sqrt{2} + 5 \cdot 3\,\sqrt{2}$$
$$= 6\,\sqrt{2} + 15\,\sqrt{2}$$
$$= (6 + 15)\,\sqrt{2}$$
$$= 21\,\sqrt{2}$$

The result of Example 2 can be generalized to the following rule for sums and differences of radical expressions.

> **RULE**
> •
> To add or subtract radical expressions, put each in simplified form and apply the distributive property, if possible. We can add only similar radicals. We must write each expression in simplified form for radicals before we can tell if the radicals are similar.

EXAMPLE 3

Combine $7\sqrt{75xy^3} - 4y\sqrt{12xy}$, where $x, y \geq 0$.

SOLUTION   We write each expression in simplified form and combine similar radicals:

$$7\sqrt{75xy^3} - 4y\sqrt{12xy} = 7\sqrt{25y^2}\,\sqrt{3xy} - 4y\sqrt{4}\,\sqrt{3xy}$$
$$= 35y\sqrt{3xy} - 8y\sqrt{3xy}$$
$$= (35y - 8y)\sqrt{3xy}$$
$$= 27y\sqrt{3xy}$$

EXAMPLE 4

Combine $10\sqrt[3]{8a^4b^2} + 11a\sqrt[3]{27ab^2}$.

SOLUTION Writing each radical in simplified form and combining similar terms, we have

$$10\sqrt[3]{8a^4b^2} + 11a\sqrt[3]{27ab^2} = 10\sqrt[3]{8a^3}\sqrt[3]{ab^2} + 11a\sqrt[3]{27}\sqrt[3]{ab^2}$$
$$= 20a\sqrt[3]{ab^2} + 33a\sqrt[3]{ab^2}$$
$$= 53a\sqrt[3]{ab^2}$$

EXAMPLE 5

Combine $\dfrac{\sqrt{3}}{2} + \dfrac{1}{\sqrt{3}}$.

SOLUTION We begin by writing the second term in simplified form.

$$\frac{\sqrt{3}}{2} + \frac{1}{\sqrt{3}} = \frac{\sqrt{3}}{2} + \frac{1}{\sqrt{3}} \cdot \frac{\sqrt{3}}{\sqrt{3}}$$
$$= \frac{\sqrt{3}}{2} + \frac{\sqrt{3}}{3}$$
$$= \frac{1}{2}\sqrt{3} + \frac{1}{3}\sqrt{3}$$
$$= \left(\frac{1}{2} + \frac{1}{3}\right)\sqrt{3}$$
$$= \frac{5}{6}\sqrt{3} = \frac{5\sqrt{3}}{6}$$

EXAMPLE 6

Construct a golden rectangle from a square of side 4. Then show that the ratio of the length to the width is the golden ratio $\dfrac{1 + \sqrt{5}}{2}$.

SOLUTION Figure 1 shows the golden rectangle constructed from a square of side 4. The length of the diagonal OB is found from the Pythagorean theorem.

$$OB = \sqrt{2^2 + 4^2} = \sqrt{4 + 16} = \sqrt{20} = 2\sqrt{5}$$

The ratio of the length to the width for the rectangle is the golden ratio.

$$\text{Golden ratio} = \frac{CE}{EF} = \frac{2 + 2\sqrt{5}}{4} = \frac{2(1 + \sqrt{5})}{2 \cdot 2} = \frac{1 + \sqrt{5}}{2}$$

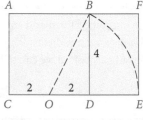

FIGURE 1

As you can see, showing that the ratio of length to width in this rectangle is the golden ratio depends on our ability to write $\sqrt{20}$ as $2\sqrt{5}$ and our ability to reduce to lowest terms by factoring and then dividing out the common factor 2 from the numerator and denominator.

GETTING READY FOR CLASS

After reading through the preceding section, respond in your own words and in complete sentences.

A. What are similar radicals?

B. When can we add two radical expressions?

C. What is the first step when adding or subtracting expressions with radicals?

D. What is the golden ratio, and where does it come from?

Problem Set 6.4

Combine the following expressions. (Assume any variables under an even root are nonnegative.)

1. $3\sqrt{5} + 4\sqrt{5}$

2. $6\sqrt{3} - 5\sqrt{3}$

3. $3x\sqrt{7} - 4x\sqrt{7}$

4. $6y\sqrt{a} + 7y\sqrt{a}$

5. $5\sqrt[3]{10} - 4\sqrt[3]{10}$

6. $6\sqrt[4]{2} + 9\sqrt[4]{2}$

7. $8\sqrt[5]{6} - 2\sqrt[5]{6} + 3\sqrt[5]{6}$

8. $7\sqrt[6]{7} - \sqrt[6]{7} + 4\sqrt[6]{7}$

9. $3x\sqrt{2} - 4x\sqrt{2} + x\sqrt{2}$

10. $5x\sqrt{6} - 3x\sqrt{6} - 2x\sqrt{6}$

11. $\sqrt{20} - \sqrt{80} + \sqrt{45}$

12. $\sqrt{8} - \sqrt{32} - \sqrt{18}$

13. $4\sqrt{8} - 2\sqrt{50} - 5\sqrt{72}$

14. $\sqrt{48} - 3\sqrt{27} + 2\sqrt{75}$

15. $5x\sqrt{8} + 3\sqrt{32x^2} - 5\sqrt{50x^2}$

16. $2\sqrt{50x^2} - 8x\sqrt{18} - 3\sqrt{72x^2}$

17. $5\sqrt[3]{16} - 4\sqrt[3]{54}$

18. $\sqrt[3]{81} + 3\sqrt[3]{24}$

19. $\sqrt[3]{x^4y^2} + 7x\sqrt[3]{xy^2}$

20. $2\sqrt[3]{x^8y^6} - 3y^2\sqrt[3]{8x^8}$

21. $5a^2\sqrt{27ab^3} - 6b\sqrt{12a^5b}$

22. $9a\sqrt{20a^3b^2} + 7b\sqrt{45a^5}$

23. $b\sqrt[3]{24a^5b} + 3a\sqrt[3]{81a^2b^4}$

24. $7\sqrt[3]{a^4b^3c^2} - 6ab\sqrt[3]{ac^2}$

25. $5x\sqrt[4]{3y^5} + y\sqrt[4]{243x^4y} + \sqrt[4]{48x^4y^5}$

26. $x\sqrt[4]{5xy^8} + y\sqrt[4]{405x^5y^4} + y^2\sqrt[4]{80x^5}$

27. $\dfrac{\sqrt{2}}{2} + \dfrac{1}{\sqrt{2}}$

28. $\dfrac{\sqrt{3}}{3} + \dfrac{1}{\sqrt{3}}$

29. $\dfrac{\sqrt{5}}{3} + \dfrac{1}{\sqrt{5}}$

30. $\dfrac{\sqrt{6}}{2} + \dfrac{1}{\sqrt{6}}$

31. $\sqrt{x} - \dfrac{1}{\sqrt{x}}$

32. $\sqrt{x} + \dfrac{1}{\sqrt{x}}$

33. $\dfrac{\sqrt{18}}{6} + \sqrt{\dfrac{1}{2}} + \dfrac{\sqrt{2}}{2}$

34. $\dfrac{\sqrt{12}}{6} + \sqrt{\dfrac{1}{3}} + \dfrac{\sqrt{3}}{3}$

35. $\sqrt{6} - \sqrt{\dfrac{2}{3}} + \sqrt{\dfrac{1}{6}}$

36. $\sqrt{15} - \sqrt{\dfrac{3}{5}} + \sqrt{\dfrac{5}{3}}$

37. $\sqrt[3]{25} + \dfrac{3}{\sqrt[3]{5}}$

38. $\sqrt[4]{8} + \dfrac{1}{\sqrt[4]{2}}$

39. Use a calculator to find a decimal approximation for $\sqrt{12}$ and for $2\sqrt{3}$.

40. Use a calculator to find decimal approximations for $\sqrt{50}$ and $5\sqrt{2}$.

41. Use a calculator to find a decimal approximation for $\sqrt{8} + \sqrt{18}$. Is it equal to the decimal approximation for $\sqrt{26}$ or $\sqrt{50}$?

42. Use a calculator to find a decimal approximation for $\sqrt{3} + \sqrt{12}$. Is it equal to the decimal approximation for $\sqrt{15}$ or $\sqrt{27}$?

Each of the following statements is false. Correct the right side of each one to make the statement true.

43. $3\sqrt{2x} + 5\sqrt{2x} = 8\sqrt{4x}$

44. $5\sqrt{3} - 7\sqrt{3} = -2\sqrt{9}$

45. $\sqrt{9 + 16} = 3 + 4$

46. $\sqrt{36 + 64} = 6 + 8$

APPLYING THE CONCEPTS

47. Golden Rectangle Construct a golden rectangle from a square of side 8. Then show that the ratio of the length to the width is the golden ratio $\frac{1+\sqrt{5}}{2}$.

48. Golden Rectangle Construct a golden rectangle from a square of side 10. Then show that the ratio of the length to the width is the golden ratio $\frac{1+\sqrt{5}}{2}$.

49. Golden Rectangle Use a ruler to construct a golden rectangle from a square of side 1 inch. Then show that the ratio of the length to the width is the golden ratio.

50. Golden Rectangle Use a ruler to construct a golden rectangle from a square of side $\frac{2}{3}$ inch. Then show that the ratio of the length to the width is the golden ratio.

51. Golden Rectangle To show that all golden rectangles have the same ratio of length to width, construct a golden rectangle from a square of side $2x$. Then show that the ratio of the length to the width is the golden ratio.

52. Golden Rectangle To show that all golden rectangles have the same ratio of length to width, construct a golden rectangle from a square of side x. Then show that the ratio of the length to the width is the golden ratio.

53. Isosceles Right Triangles A triangle is isosceles if it has two equal sides, and a triangle is a right triangle if it has a right angle in it. Sketch an isosceles right triangle, and find the ratio of the hypotenuse to a leg.

54. Equilateral Triangles A triangle is equilateral if it has three equal sides. The triangle in the figure is equilateral with each side of length $2x$. Find the ratio of the height to a side.

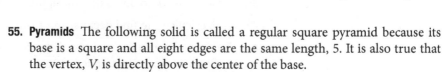

55. Pyramids The following solid is called a regular square pyramid because its base is a square and all eight edges are the same length, 5. It is also true that the vertex, V, is directly above the center of the base.

a. Find the ratio of a diagonal of the base to the length of a side.

b. Find the ratio of the area of the base to the diagonal of the base.

c. Find the ratio of the area of the base to the perimeter of the base.

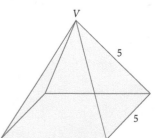

56. Pyramids Refer to this diagram of a square pyramid. Find the ratio of the height h of the pyramid to the altitude a.

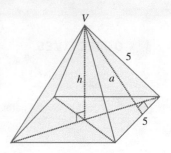

GETTING READY FOR THE NEXT SECTION

Simplify the following.

57. $3 \cdot 2$

58. $5 \cdot 7$

59. $(x + y)(4x - y)$

60. $(2x + y)(x - y)$

61. $(x + 3)^2$

62. $(3x - 2y)^2$

63. $(x - 2)(x + 2)$

64. $(2x + 5)(2x - 5)$

Simplify the following expressions.

65. $2\sqrt{18}$

66. $5\sqrt{36}$

67. $(\sqrt{6})^2$

68. $(\sqrt{2})^2$

69. $(3\sqrt{x})^2$

70. $(2\sqrt{y})^2$

Rationalize the denominator.

71. $\dfrac{\sqrt{3}}{\sqrt{2}}$

72. $\dfrac{\sqrt{5}}{\sqrt{6}}$

EXTENDING THE CONCEPTS

Combine expressions, if possible. Assume all variables represent positive numbers.

73. $\dfrac{\sqrt{36a^2b^4c}}{2} - \dfrac{b\sqrt{16a^2b^2c}}{3}$

74. $\dfrac{y\sqrt{18xyz}}{5} + \dfrac{\sqrt{32xy^3z}}{4}$

75. $\dfrac{-b + \sqrt{b^2 - 4ac}}{2a} + \dfrac{-b - \sqrt{b^2 - 4ac}}{2a}$

76. $\dfrac{c - \sqrt{c^2 - 4ad}}{2a} + \dfrac{c + \sqrt{c^2 - 4ad}}{2a}$

77. $\sqrt{\dfrac{7}{4}} + \dfrac{\sqrt{7}}{2}$

78. $\sqrt{\dfrac{5}{9}} - \dfrac{-2\sqrt{5}}{3}$

79. $-5a\sqrt{75b^3} + \sqrt{18a^2b} + 7b\sqrt{108a^2b} - a\sqrt{50b}$

80. $3\sqrt{24x^3y} - y\sqrt{27x} + 11x\sqrt{54x^2y} - y\sqrt{48x}$

OBJECTIVES

A Multiply expressions containing radicals.

B Rationalize a denominator containing two terms.

© sxc.hu, cempey, 2006

6.5

The insects of a London park now have a new hotel in which to stay. The city held a unique competition that requested participants to build a structure that could house multiple varieties of bugs at once. A 1.5-meter-tall box with a façade mimicking the multi-polygon design of a dragonfly's wing, also called voronoi, won first prize. The designers stacked 25 layers of 20mm-thick birch plywood front to back, with each layer having the voronoi design. Each polygonal hole caters to a specific type of insect, and is loosely stuffed with recycled or organic materials. Beetles, bees, butterflies, spiders, and more are all encouraged to make their homes in this new Bug Hotel.

A pentagon is one of the shapes present in the Bug Hotel's façade. We can find the area A of a regular pentagon with a side length t by using the formula

$$A = \frac{t^2\sqrt{25 + 10\sqrt{5}}}{4}$$

This formula uses a radical expression to find the pentagon's area.

In this section, we will look at multiplication and division of expressions like this one that contain radicals. As you will see, multiplication of expressions that contain radicals is very similar to multiplication of polynomials. The division problems in this section are just an extension of the work we did previously when we rationalized denominators.

Ⓐ Multiplication of Radical Expressions

We have worked with the golden rectangle more than once in this chapter. The following is one such golden rectangle.

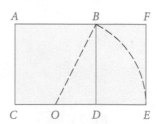

By now you know that, in any golden rectangle constructed from a square (of any size), the ratio of the length to the width will be

$$\frac{1 + \sqrt{5}}{2}$$

which we call the golden ratio. What is interesting is that the smaller rectangle on the right, *BFED*, is also a golden rectangle. We will use the mathematics developed in this section to confirm this fact.

EXAMPLE 1

Multiply $(3\sqrt{5})(2\sqrt{7})$.

SOLUTION We can rearrange the order and grouping of the numbers in this product by applying the commutative and associative properties. Following this, we apply Property 1 for radicals and multiply:

$$(3\sqrt{5})(2\sqrt{7}) = (3 \cdot 2)(\sqrt{5}\sqrt{7}) \qquad \text{Commutative and associative properties}$$

$$= (3 \cdot 2)(\sqrt{5 \cdot 7}) \qquad \text{Property 1 for radicals}$$

$$= 6\sqrt{35} \qquad \text{Multiplication}$$

In practice, it is not necessary to show the first two steps.

EXAMPLE 2

Multiply $\sqrt{3}(2\sqrt{6} - 5\sqrt{12})$.

SOLUTION Applying the distributive property, we have

$$\sqrt{3}(2\sqrt{6} - 5\sqrt{12}) = \sqrt{3} \cdot 2\sqrt{6} - \sqrt{3} \cdot 5\sqrt{12}$$

$$= 2\sqrt{18} - 5\sqrt{36}$$

Writing each radical in simplified form gives

$$2\sqrt{18} - 5\sqrt{36} = 2\sqrt{9}\sqrt{2} - 5\sqrt{36}$$

$$= 6\sqrt{2} - 30$$

EXAMPLE 3

Multiply $(\sqrt{3} + \sqrt{5})(4\sqrt{3} - \sqrt{5})$.

SOLUTION The same principle that applies when multiplying two binomials applies to this product. We must multiply each term in the first expression by each term in the second one. Any convenient method can be used. Let's use the FOIL method.

$$(\sqrt{3} + \sqrt{5})(4\sqrt{3} - \sqrt{5}) = \overset{\text{F}}{\sqrt{3} \cdot 4\sqrt{3}} - \overset{\text{O}}{\sqrt{3} \cdot \sqrt{5}} + \overset{\text{I}}{\sqrt{5} \cdot 4\sqrt{3}} - \overset{\text{L}}{\sqrt{5} \cdot \sqrt{5}}$$

$$= 4 \cdot 3 - \sqrt{15} + 4\sqrt{15} - 5$$

$$= 12 + 3\sqrt{15} - 5$$

$$= 7 + 3\sqrt{15}$$

EXAMPLE 4

Expand and simplify $(\sqrt{x} + 3)^2$.

SOLUTION 1   We can write this problem as a multiplication problem and proceed as we did in Example 3:

$$(\sqrt{x} + 3)^2 = (\sqrt{x} + 3)(\sqrt{x} + 3)$$

$$\overset{\text{F}\quad\text{O}\quad\text{I}\quad\text{L}}{= \sqrt{x} \cdot \sqrt{x} + 3\sqrt{x} + 3\sqrt{x} + 3 \cdot 3}$$

$$= x + 3\sqrt{x} + 3\sqrt{x} + 9$$

$$= x + 6\sqrt{x} + 9$$

SOLUTION 2   We can obtain the same result by applying the formula for the square of a sum: $(a + b)^2 = a^2 + 2ab + b^2$.

$$(\sqrt{x} + 3)^2 = (\sqrt{x})^2 + 2(\sqrt{x})(3) + 3^2$$

$$= x + 6\sqrt{x} + 9$$

EXAMPLE 5

Expand $(3\sqrt{x} - 2\sqrt{y})^2$ and simplify the result.

SOLUTION   Let's apply the formula for the square of a difference, $(a - b)^2 = a^2 - 2ab + b^2$.

$$(3\sqrt{x} - 2\sqrt{y})^2 = (3\sqrt{x})^2 - 2(3\sqrt{x})(2\sqrt{y}) + (2\sqrt{y})^2$$

$$= 9x - 12\sqrt{xy} + 4y$$

EXAMPLE 6

Expand and simplify $(\sqrt{x + 2} - 1)^2$.

SOLUTION   Applying the formula $(a - b)^2 = a^2 - 2ab + b^2$, we have

$$(\sqrt{x + 2} - 1)^2 = (\sqrt{x + 2})^2 - 2\sqrt{x + 2}(1) + 1^2$$

$$= x + 2 - 2\sqrt{x + 2} + 1$$

$$= x + 3 - 2\sqrt{x + 2}$$

EXAMPLE 7

Multiply $(\sqrt{6} + \sqrt{2})(\sqrt{6} - \sqrt{2})$.

SOLUTION   We notice the product is of the form $(a + b)(a - b)$, which always gives the difference of two squares, $a^2 - b^2$:

$$(\sqrt{6} + \sqrt{2})(\sqrt{6} - \sqrt{2}) = (\sqrt{6})^2 - (\sqrt{2})^2$$

$$= 6 - 2$$

$$= 4$$

In Example 7, the two expressions $(\sqrt{6} + \sqrt{2})$ and $(\sqrt{6} - \sqrt{2})$ are called *conjugates*. In general, the conjugate of $\sqrt{a} + \sqrt{b}$ is $\sqrt{a} - \sqrt{b}$. If a and b are integers, multiplying conjugates of this form always produces a rational number. That is, if a and b are positive integers, then

$$(\sqrt{a} + \sqrt{b})(\sqrt{a} - \sqrt{b}) = \sqrt{a}\sqrt{a} - \sqrt{a}\sqrt{b} + \sqrt{a}\sqrt{b} - \sqrt{b}\sqrt{b}$$

$$= a - \sqrt{ab} + \sqrt{ab} - b$$

$$= a - b$$

which is rational if a and b are rational.

B Division of Radical Expressions

Division with radical expressions is the same as rationalizing the denominator. In Section 6.2, we were able to divide $\sqrt{3}$ by $\sqrt{2}$ by rationalizing the denominator:

$$\frac{\sqrt{3}}{\sqrt{2}} = \frac{\sqrt{3}}{\sqrt{2}} \cdot \frac{\sqrt{2}}{\sqrt{2}} = \frac{\sqrt{6}}{2}$$

We can accomplish the same result with expressions such as

$$\frac{6}{\sqrt{5} - \sqrt{3}}$$

by multiplying the numerator and denominator by the conjugate of the denominator.

EXAMPLE 8

Divide $\dfrac{6}{\sqrt{5} - \sqrt{3}}$. (Rationalize the denominator.)

SOLUTION Because the product of two conjugates is a rational number, we multiply the numerator and denominator by the conjugate of the denominator.

$$\frac{6}{\sqrt{5} - \sqrt{3}} = \frac{6}{\sqrt{5} - \sqrt{3}} \cdot \frac{(\sqrt{5} + \sqrt{3})}{(\sqrt{5} + \sqrt{3})}$$

$$= \frac{6\sqrt{5} + 6\sqrt{3}}{(\sqrt{5})^2 - (\sqrt{3})^2}$$

$$= \frac{6\sqrt{5} + 6\sqrt{3}}{5 - 3}$$

$$= \frac{6\sqrt{5} + 6\sqrt{3}}{2}$$

The numerator and denominator of this last expression have a factor of 2 in common. We can reduce to lowest terms by factoring 2 from the numerator and then dividing both the numerator and denominator by 2:

$$= \frac{2(3\sqrt{5} + 3\sqrt{3})}{2}$$

$$= 3\sqrt{5} + 3\sqrt{3}$$

EXAMPLE 9

Rationalize the denominator $\dfrac{\sqrt{5} - 2}{\sqrt{5} + 2}$.

SOLUTION To rationalize the denominator, we multiply the numerator and denominator by the conjugate of the denominator:

$$\frac{\sqrt{5} - 2}{\sqrt{5} + 2} = \frac{\sqrt{5} - 2}{\sqrt{5} + 2} \cdot \frac{(\sqrt{5} - 2)}{(\sqrt{5} - 2)}$$

$$= \frac{5 - 2\sqrt{5} - 2\sqrt{5} + 4}{(\sqrt{5})^2 - 2^2}$$

$$= \frac{9 - 4\sqrt{5}}{5 - 4}$$

$$= \frac{9 - 4\sqrt{5}}{1}$$

$$= 9 - 4\sqrt{5}$$

EXAMPLE 10

A golden rectangle constructed from a square of side 2 is shown in Figure 1. Show that the smaller rectangle *BDEF* is also a golden rectangle by finding the ratio of its length to its width.

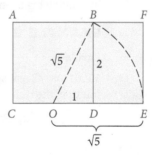

FIGURE 1

SOLUTION First, find expressions for the length and width of the smaller rectangle.

$$\text{Length} = EF = 2$$
$$\text{Width} = DE = \sqrt{5} - 1$$

Next, we find the ratio of length to width.

$$\text{Ratio of length to width} = \frac{EF}{DE} = \frac{2}{\sqrt{5} - 1}$$

To show that the small rectangle is a golden rectangle, we must show that the ratio of length to width is the golden ratio. We do so by rationalizing the denominator.

$$\frac{2}{\sqrt{5} - 1} = \frac{2}{\sqrt{5} - 1} \cdot \frac{\sqrt{5} + 1}{\sqrt{5} + 1}$$

$$= \frac{2(\sqrt{5} + 1)}{5 - 1}$$

$$= \frac{2(\sqrt{5} + 1)}{4}$$

$$= \frac{\sqrt{5} + 1}{2} \qquad \text{Divide out common factor 2.}$$

Because addition is commutative, this last expression is the golden ratio. Therefore, the small rectangle in Figure 1 is a golden rectangle.

GETTING READY FOR CLASS

After reading through the preceding section, respond in your own words and in complete sentences.

A. Explain why $(\sqrt{5} + \sqrt{2})^2 \neq 5 + 2$.

B. Explain in words how you would rationalize the denominator in the expression $\frac{\sqrt{3}}{\sqrt{5} - \sqrt{2}}$.

C. What are conjugates?

D. What result is guaranteed when multiplying radical expressions that are conjugates?

Problem Set 6.5

Multiply. (Assume all expressions appearing under a square root symbol represent nonnegative numbers throughout this problem set.)

1. $\sqrt{6}\sqrt{3}$

2. $\sqrt{6}\sqrt{2}$

3. $(2\sqrt{3})(5\sqrt{7})$

4. $(3\sqrt{5})(2\sqrt{7})$

5. $(4\sqrt{6})(2\sqrt{15})(3\sqrt{10})$

6. $(4\sqrt{35})(2\sqrt{21})(5\sqrt{15})$

7. $(3\sqrt[3]{3})(6\sqrt[3]{9})$

8. $(2\sqrt[3]{2})(6\sqrt[3]{4})$

9. $\sqrt{3}(\sqrt{2} - 3\sqrt{3})$

10. $\sqrt{2}(5\sqrt{3} + 4\sqrt{2})$

11. $6\sqrt[3]{4}(2\sqrt[3]{2} + 1)$

12. $7\sqrt[3]{5}(3\sqrt[3]{25} - 2)$

13. $(\sqrt{3} + \sqrt{2})(3\sqrt{3} - \sqrt{2})$

14. $(\sqrt{5} - \sqrt{2})(3\sqrt{5} + 2\sqrt{2})$

15. $(\sqrt{x} + 5)(\sqrt{x} - 3)$

16. $(\sqrt{x} + 4)(\sqrt{x} + 2)$

17. $(3\sqrt{6} + 4\sqrt{2})(\sqrt{6} + 2\sqrt{2})$

18. $(\sqrt{7} - 3\sqrt{3})(2\sqrt{7} - 4\sqrt{3})$

19. $(\sqrt{3} + 4)^2$

20. $(\sqrt{5} - 2)^2$

21. $(\sqrt{x} - 3)^2$

22. $(\sqrt{x} + 4)^2$

23. $(2\sqrt{a} - 3\sqrt{b})^2$

24. $(5\sqrt{a} - 2\sqrt{b})^2$

25. $(\sqrt{x - 4} + 2)^2$

26. $(\sqrt{x - 3} + 2)^2$

27. $(\sqrt{x - 5} - 3)^2$

28. $(\sqrt{x - 3} - 4)^2$

29. $(\sqrt{3} - \sqrt{2})(\sqrt{3} + \sqrt{2})$

30. $(\sqrt{5} - \sqrt{2})(\sqrt{5} + \sqrt{2})$

31. $(\sqrt{a} + 7)(\sqrt{a} - 7)$

32. $(\sqrt{a} + 5)(\sqrt{a} - 5)$

33. $(5 - \sqrt{x})(5 + \sqrt{x})$

34. $(3 - \sqrt{x})(3 + \sqrt{x})$

35. $(\sqrt{x - 4} + 2)(\sqrt{x - 4} - 2)$

36. $(\sqrt{x + 3} + 5)(\sqrt{x + 3} - 5)$

37. $(\sqrt{3} + 1)^3$

38. $(\sqrt{5} - 2)^3$

Rationalize the denominator in each of the following.

39. $\dfrac{\sqrt{2}}{\sqrt{6} - \sqrt{2}}$

40. $\dfrac{\sqrt{5}}{\sqrt{5} + \sqrt{3}}$

41. $\dfrac{\sqrt{5}}{\sqrt{5} + 1}$

42. $\dfrac{\sqrt{7}}{\sqrt{7} - 1}$

43. $\dfrac{\sqrt{x}}{\sqrt{x} - 3}$

44. $\dfrac{\sqrt{x}}{\sqrt{x} + 2}$

45. $\dfrac{\sqrt{5}}{2\sqrt{5} - 3}$

46. $\dfrac{\sqrt{7}}{3\sqrt{7} - 2}$

47. $\dfrac{3}{\sqrt{x} - \sqrt{y}}$

48. $\dfrac{2}{\sqrt{x} + \sqrt{y}}$

49. $\dfrac{\sqrt{6} + \sqrt{2}}{\sqrt{6} - \sqrt{2}}$

50. $\dfrac{\sqrt{5} - \sqrt{3}}{\sqrt{5} + \sqrt{3}}$

51. $\dfrac{\sqrt{7} - 2}{\sqrt{7} + 2}$

52. $\dfrac{\sqrt{11} + 3}{\sqrt{11} - 3}$

PAYING ATTENTION TO INSTRUCTIONS

53. Work each problem according to the instructions given.

  a. Add: $(\sqrt{x} + 2) + (\sqrt{x} - 2)$ **b.** Multiply: $(\sqrt{x} + 2)(\sqrt{x} - 2)$

  c. Square: $(\sqrt{x} + 2)^2$ **d.** Divide: $\dfrac{\sqrt{x} + 2}{\sqrt{x} - 2}$

54. Work each problem according to the instructions given.

  a. Add: $(\sqrt{x} - 3) + (\sqrt{x} + 3)$ **b.** Multiply: $(\sqrt{x} - 3)(\sqrt{x} + 3)$

  c. Square: $(\sqrt{x} + 3)^2$ **d.** Divide: $\dfrac{\sqrt{x} + 3}{\sqrt{x} - 3}$

55. Work each problem according to the instructions given.

  a. Add: $(5 + \sqrt{2}) + (5 - \sqrt{2})$ **b.** Multiply: $(5 + \sqrt{2})(5 - \sqrt{2})$

  c. Square: $(5 + \sqrt{2})^2$ **d.** Divide: $\dfrac{5 + \sqrt{2}}{5 - \sqrt{2}}$

56. Work each problem according to the instructions given.

  a. Add: $(2 + \sqrt{3}) + (2 - \sqrt{3})$ **b.** Multiply: $(2 + \sqrt{3})(2 - \sqrt{3})$

  c. Square: $(2 + \sqrt{3})^2$ **d.** Divide: $\dfrac{2 + \sqrt{3}}{2 - \sqrt{3}}$

57. Work each problem according to the instructions given.

  a. Add: $\sqrt{2} + (\sqrt{6} + \sqrt{2})$ **b.** Multiply: $\sqrt{2}(\sqrt{6} + \sqrt{2})$

  c. Divide: $\dfrac{\sqrt{6} + \sqrt{2}}{\sqrt{2}}$ **d.** Divide: $\dfrac{\sqrt{2}}{\sqrt{6} + \sqrt{2}}$

58. Work each problem according to the instructions given.

  a. Add: $\sqrt{5} + (\sqrt{5} + \sqrt{10})$ **b.** Multiply: $\sqrt{5}(\sqrt{5} + \sqrt{10})$

  c. Divide: $\dfrac{\sqrt{5} + \sqrt{10}}{\sqrt{5}}$ **d.** Divide: $\dfrac{\sqrt{5}}{\sqrt{5} + \sqrt{10}}$

59. Work each problem according to the instructions given.

  a. Add: $\left(\dfrac{1 + \sqrt{5}}{2}\right) + \left(\dfrac{1 - \sqrt{5}}{2}\right)$ **b.** Multiply: $\left(\dfrac{1 + \sqrt{5}}{2}\right)\left(\dfrac{1 - \sqrt{5}}{2}\right)$

60. Work each problem according to the instructions given.

  a. Add: $\left(\dfrac{1 + \sqrt{3}}{2}\right) + \left(\dfrac{1 - \sqrt{3}}{2}\right)$ **b.** Multiply: $\left(\dfrac{1 + \sqrt{3}}{2}\right)\left(\dfrac{1 - \sqrt{3}}{2}\right)$

61. Show that the product below is 5:

$$(\sqrt[3]{2} + \sqrt[3]{3})(\sqrt[3]{4} - \sqrt[3]{6} + \sqrt[3]{9})$$

62. Show that the product below is $x + 8$:

$$(\sqrt[3]{x} + 2)(\sqrt[3]{x^2} - 2\sqrt[3]{x} + 4)$$

Each of the following statements below is false. Correct the right side of each one to make it true.

63. $5(2\sqrt{3}) = 10\sqrt{15}$ **64.** $3(2\sqrt{x}) = 6\sqrt{3x}$ **65.** $(\sqrt{x} + 3)^2 = x + 9$

66. $(\sqrt{x} - 7)^2 = x - 49$ **67.** $(5\sqrt{3})^2 = 15$ **68.** $(3\sqrt{5})^2 = 15$

APPLYING THE CONCEPTS

69. Gravity If an object is dropped from the top of a 100-foot building, the amount of time t (in seconds) that it takes for the object to be h feet from the ground is given by the formula

$$t = \frac{\sqrt{100 - h}}{4}$$

How long does it take before the object is 50 feet from the ground? How long does it take to reach the ground? (When it is on the ground, h is 0.)

70. Gravity Use the formula given in Problem 69 to determine h if t is 1.25 seconds.

71. Golden Rectangle Rectangle $ACEF$ in Figure 2 is a golden rectangle. If side AC is 6 inches, show that the smaller rectangle $BDEF$ is also a golden rectangle.

72. Golden Rectangle Rectangle $ACEF$ in Figure 2 is a golden rectangle. If side AC is 1 inch, show that the smaller rectangle $BDEF$ is also a golden rectangle.

73. Golden Rectangle If side AC in Figure 2 is $2x$, show that rectangle $BDEF$ is a golden rectangle.

74. Golden Rectangle If side AC in Figure 2 is x, show that rectangle $BDEF$ is a golden rectangle.

GETTING READY FOR THE NEXT SECTION

Simplify.

75. $(t + 5)^2$ **76.** $(x - 4)^2$ **77.** $\sqrt{x} \cdot \sqrt{x}$ **78.** $\sqrt{3x} \cdot \sqrt{3x}$

Solve.

79. $3x + 4 = 5^2$ **80.** $4x - 7 = 3^2$

81. $t^2 + 7t + 12 = 0$ **82.** $x^2 - 3x - 10 = 0$

83. $t^2 + 10t + 25 = t + 7$ **84.** $x^2 - 4x + 4 = x - 2$

85. $(x + 4)^2 = x + 6$ **86.** $(x - 6)^2 = x - 4$

87. Is $x = 7$ a solution to $\sqrt{3x + 4} = 5$?

88. Is $x = 4$ a solution to $\sqrt{4x - 7} = -3$?

89. Is $t = -6$ a solution to $t + 5 = \sqrt{t + 7}$?

90. Is $t = -3$ a solution to $t + 5 = \sqrt{t + 7}$?

EXTENDING THE CONCEPTS

Simplify.

91. $\sqrt{\dfrac{2y}{x}} \cdot \sqrt{\dfrac{x^2}{8}}$

92. $\sqrt{\dfrac{3a}{b^3}} \cdot \sqrt{\dfrac{ab}{27}}$

93. $\sqrt[3]{\dfrac{16a^7}{b^4}} \cdot \sqrt[3]{\dfrac{b}{2a}}$

94. $\sqrt[3]{\dfrac{3y}{x^7}} \cdot \sqrt[3]{\dfrac{x}{81y^4}}$

95. $\sqrt[5]{\dfrac{x^2y}{z^7}} \cdot \sqrt[5]{\dfrac{y^9z^3}{x^7}}$

96. $\sqrt[5]{\dfrac{a^3b}{c^8}} \cdot \sqrt[5]{\dfrac{b^4c^4}{a^8}}$

Rationalize the denominator.

97. $\dfrac{3x}{\sqrt{x} - \sqrt{xy}}$

98. $\dfrac{2a}{\sqrt{a} - \sqrt{ab}}$

99. $\dfrac{\sqrt{a}}{\sqrt{ab} - \sqrt{a}}$

100. $\dfrac{2\sqrt{x}}{3\sqrt{x} - 2\sqrt{y}}$

101. $\dfrac{\sqrt{x - 4}}{\sqrt{x - 4} + 2}$

102. $\dfrac{\sqrt{x + 3}}{\sqrt{x + 3} + 5}$

103. $\dfrac{\sqrt{x + 3} + \sqrt{x - 3}}{\sqrt{x + 3} - \sqrt{x - 3}}$

104. $\dfrac{\sqrt{x + 5} + \sqrt{x - 5}}{\sqrt{x + 5} - \sqrt{x - 5}}$

105. $\dfrac{1}{\sqrt[3]{x} + 2}$

106. $\dfrac{1}{\sqrt[3]{x} - 2}$

107. $\dfrac{1}{\sqrt[3]{3} + \sqrt[3]{2}}$

108. $\dfrac{1}{\sqrt[3]{3} - \sqrt[3]{2}}$

OBJECTIVES

A Solve equations containing radicals.

B Graph simple square root and cube functions in two variables.

© bewinca, 2007

6.6

The Three Gorges Dam stretches across the Yangtze River in China. Every year, the Chinese government spends more than a million dollars to remove tons of domestic garbage from the river. The garbage has the potential to block and jam the shipping locks in the dam. It may also damage boats and pollute the water. Approximately 150 million people live upstream from the dam. Unfortunately, it is common practice to allow household waste and agricultural pollutants to run off into the river. In this section, we will work more with radicals and solving equations that contain them. For instance, if a pollutant enters the river at a constant rate, its plume disperses down the river and may be represented by the equation $y = \sqrt{x}$. We will learn how to solve for the variables in this equation. The first step is to eliminate the radical. To do so, we need to introduce an additional property.

 Solving Equations with Radicals

This section is concerned with solving equations that involve one or more radicals. The first step in solving an equation that contains a radical is to eliminate the radical from the equation. To do so, we need an additional property.

> **PROPERTY** *Squaring Property of Equality*
>
> If both sides of an equation are squared, the solutions to the original equation are solutions to the resulting equation.

We will never lose solutions to our equations by squaring both sides. We may, however, introduce *extraneous solutions*. Extraneous solutions satisfy the equation obtained by squaring both sides of the original equation, but do not satisfy the original equation.

We know that if two real numbers a and b are equal, then so are their squares:

$$\text{If } a = b \quad \text{then} \quad a^2 = b^2$$

On the other hand, extraneous solutions are introduced when we square opposites. That is, even though opposites are not equal, their squares are. For example,

$$5 = -5 \qquad \text{A false statement}$$

$$(5)^2 = (-5)^2 \qquad \text{Square both sides.}$$

$$25 = 25 \qquad \text{A true statement}$$

We are free to square both sides of an equation any time it is convenient. We must be aware, however, that doing so may introduce extraneous solutions. We must, therefore, check all our solutions in the original equation if at any time we square both sides of the original equation.

EXAMPLE 1

Solve for x: $\sqrt{3x + 4} = 5$.

SOLUTION   We square both sides and proceed as usual:

$$\sqrt{3x + 4} = 5$$

$$(\sqrt{3x + 4})^2 = 5^2$$

$$3x + 4 = 25$$

$$3x = 21$$

$$x = 7$$

Checking $x = 7$ in the original equation, we have

$$\sqrt{3(7) + 4} \stackrel{?}{=} 5$$

$$\sqrt{21 + 4} \stackrel{?}{=} 5$$

$$\sqrt{25} \stackrel{?}{=} 5$$

$$5 = 5$$

The solution $x = 7$ satisfies the original equation.

EXAMPLE 2

Solve $\sqrt{4x - 7} = -3$.

SOLUTION   Squaring both sides, we have

$$\sqrt{4x - 7} = -3$$

$$(\sqrt{4x - 7})^2 = (-3)^2$$

$$4x - 7 = 9$$

$$4x = 16$$

$$x = 4$$

Checking $x = 4$ in the original equation gives

$$\sqrt{4(4) - 7} \stackrel{?}{=} -3$$

$$\sqrt{16 - 7} \stackrel{?}{=} -3$$

$$\sqrt{9} \stackrel{?}{=} -3$$

$$3 = -3 \qquad \text{A false statement}$$

NOTE   The fact that there is no solution to the equation in Example 2 was obvious to begin with. Notice that the left side of the equation is the positive square root of $4x - 7$, which must be a positive number or 0. The right side of the equation is -3. Because we cannot have a number that is either positive or zero equal to a negative number, there is no solution to the equation.

The solution $x = 4$ produces a false statement when checked in the original equation. Because $x = 4$ was the only possible solution, there is no solution to the original equation. The possible solution $x = 4$ is an extraneous solution. It satisfies the equation obtained by squaring both sides of the original equation, but does not satisfy the original equation.

EXAMPLE 3

Solve $\sqrt{5x - 1} + 3 = 7$.

SOLUTION We must isolate the radical on the left side of the equation. If we attempt to square both sides without doing so, the resulting equation will also contain a radical. Adding -3 to both sides, we have

$$\sqrt{5x - 1} + 3 = 7$$
$$\sqrt{5x - 1} = 4$$

We can now square both sides and proceed as usual:

$$(\sqrt{5x - 1})^2 = 4^2$$
$$5x - 1 = 16$$
$$5x = 17$$
$$x = \frac{17}{5}$$

Checking $x = \frac{17}{5}$, we have

$$\sqrt{5\left(\frac{17}{5}\right) - 1} + 3 \stackrel{?}{=} 7$$
$$\sqrt{17 - 1} + 3 \stackrel{?}{=} 7$$
$$\sqrt{16} + 3 \stackrel{?}{=} 7$$
$$4 + 3 \stackrel{?}{=} 7$$
$$7 = 7$$

EXAMPLE 4

Solve $t + 5 = \sqrt{t + 7}$.

SOLUTION This time, squaring both sides of the equation results in a quadratic equation:

$$(t + 5)^2 = (\sqrt{t + 7})^2 \qquad \text{Square both sides.}$$
$$t^2 + 10t + 25 = t + 7$$
$$t^2 + 9t + 18 = 0 \qquad \text{Standard form}$$
$$(t + 3)(t + 6) = 0 \qquad \text{Factor the left side.}$$
$$t + 3 = 0 \quad \text{or} \quad t + 6 = 0 \qquad \text{Set factors equal to 0.}$$
$$t = -3 \qquad\qquad t = -6$$

We must check each solution in the original equation:

$$\text{Check } t = -3 \qquad\qquad \text{Check } t = -6$$

$$-3 + 5 \overset{?}{=} \sqrt{-3 + 7} \qquad\qquad -6 + 5 \overset{?}{=} \sqrt{-6 + 7}$$

$$2 \overset{?}{=} \sqrt{4} \qquad\qquad\qquad -1 \overset{?}{=} \sqrt{1}$$

$$2 = 2 \quad \text{A true statement} \qquad -1 = 1 \quad \text{A false statement}$$

Because $t = -6$ does not check, our only solution is $t = -3$.

EXAMPLE 5

Solve $\sqrt{x - 3} = \sqrt{x} - 3$.

SOLUTION   We begin by squaring both sides. Note carefully what happens when we square the right side of the equation, and compare the square of the right side with the square of the left side. You must convince yourself that these results are correct. (The note in the margin will help if you are having trouble convincing yourself that what is written below is true.)

$$(\sqrt{x - 3})^2 = (\sqrt{x} - 3)^2$$

$$x - 3 = x - 6\sqrt{x} + 9$$

Now we still have a radical in our equation, so we will have to square both sides again. Before we do, though, let's isolate the remaining radical.

$$x - 3 = x - 6\sqrt{x} + 9$$

$$-3 = -6\sqrt{x} + 9 \qquad \text{Add } -x \text{ to each side.}$$

$$-12 = -6\sqrt{x} \qquad \text{Add } -9 \text{ to each side.}$$

$$2 = \sqrt{x} \qquad \text{Divide each side by } -6.$$

$$4 = x \qquad \text{Square each side.}$$

Our only possible solution is $x = 4$, which we check in our original equation as follows:

$$\sqrt{4 - 3} \overset{?}{=} \sqrt{4} - 3$$

$$\sqrt{1} \overset{?}{=} 2 - 3$$

$$1 = -1 \qquad \text{A false statement}$$

Substituting 4 for x in the original equation yields a false statement. Because 4 was our only possible solution, there is no solution to our equation.

Here is another example of an equation for which we must apply our squaring property twice before all radicals are eliminated.

NOTE   It is very important that you realize that the square of $(\sqrt{x} - 3)$ is not $x + 9$. Remember, when we square a difference with two terms, we use the formula

$$(a - b)^2 = a^2 - 2ab + b^2$$

Applying this formula to $(\sqrt{x} - 3)^2$ we have

$$(\sqrt{x} - 3)^2 =$$
$$(\sqrt{x})^2 - 2(\sqrt{x})(3) + 3^2$$
$$= x - 6\sqrt{x} + 9$$

EXAMPLE 6

Solve $\sqrt{x+1} = 1 - \sqrt{2x}$.

SOLUTION This equation has two separate terms involving radical signs.
Squaring both sides gives

$$x + 1 = 1 - 2\sqrt{2x} + 2x$$

$$-x = -2\sqrt{2x} \qquad \text{Add } -2x \text{ and } -1 \text{ to both sides.}$$

$$x^2 = 4(2x) \qquad \text{Square both sides.}$$

$$x^2 - 8x = 0 \qquad \text{Standard form}$$

Our equation is a quadratic equation in standard form. To solve for x, we factor the
left side and set each factor equal to 0:

$$x(x - 8) = 0 \qquad \text{Factor left side.}$$

$$x = 0 \quad \text{or} \quad x - 8 = 0$$

$$x = 8 \qquad \text{Set factors equal to 0.}$$

Because we squared both sides of our equation, we have the possibility that one
or both of the solutions are extraneous. We must check each one in the original
equation:

Check $x = 8$	Check $x = 0$
$\sqrt{8 + 1} \stackrel{?}{=} 1 - \sqrt{2 \cdot 8}$	$\sqrt{0 + 1} \stackrel{?}{=} 1 - \sqrt{2 \cdot 0}$
$\sqrt{9} \stackrel{?}{=} 1 - \sqrt{16}$	$\sqrt{1} \stackrel{?}{=} 1 - \sqrt{0}$
$3 \stackrel{?}{=} 1 - 4$	$1 \stackrel{?}{=} 1 - 0$
$3 = -3$ A false statement	$1 = 1$ A true statement

Because $x = 8$ does not check, it is an extraneous solution. Our only solution is
$x = 0$.

EXAMPLE 7

Solve $\sqrt{x+1} = \sqrt{x+2} - 1$.

SOLUTION Squaring both sides, we have

$$(\sqrt{x+1})^2 = (\sqrt{x+2} - 1)^2$$

$$x + 1 = x + 2 - 2\sqrt{x+2} + 1$$

Once again, we are left with a radical in our equation. Before we square each side
again, we must isolate the radical on the right side of the equation.

$$x + 1 = x + 3 - 2\sqrt{x+2} \qquad \text{Simplify the right side.}$$

$$1 = 3 - 2\sqrt{x+2} \qquad \text{Add } -x \text{ to each side.}$$

$$-2 = -2\sqrt{x+2} \qquad \text{Add } -3 \text{ to each side.}$$

$$1 = \sqrt{x+2} \qquad \text{Divide each side by } -2.$$

$$1 = x + 2 \qquad \text{Square both sides.}$$

$$-1 = x \qquad \text{Add } -2 \text{ to each side.}$$

Checking our only possible solution, $x = -1$, in our original equation, we have

$$\sqrt{-1 + 1} \overset{?}{=} \sqrt{-1 + 2} - 1$$

$$\sqrt{0} \overset{?}{=} \sqrt{1} - 1$$

$$0 \overset{?}{=} 1 - 1$$

$$0 = 0 \qquad \text{A true statement}$$

Our solution checks.

It is also possible to raise both sides of an equation to powers greater than 2. We only need to check for extraneous solutions when we raise both sides of an equation to an even power. Raising both sides of an equation to an odd power will not produce extraneous solutions.

EXAMPLE 8

Solve $\sqrt[3]{4x + 5} = 3$.

SOLUTION   Cubing both sides, we have

$$(\sqrt[3]{4x + 5})^3 = 3^3$$

$$4x + 5 = 27$$

$$4x = 22$$

$$x = \frac{22}{4}$$

$$x = \frac{11}{2}$$

We do not need to check $x = \frac{11}{2}$ because we raised both sides to an odd power.

B Graphing Simple Radical Functions

We end this section by looking at graphs of some equations that contain radicals.

EXAMPLE 9

Graph $y = \sqrt{x}$ and $y = \sqrt[3]{x}$.

SOLUTION   The graphs are shown in Figures 1 and 2. Notice that the graph of $y = \sqrt{x}$ appears in the first quadrant only, because in the equation $y = \sqrt{x}$, x and y cannot be negative.

The graph of $y = \sqrt[3]{x}$ appears in quadrants 1 and 3 because the cube root of a positive number is also a positive number, and the cube root of a negative number is a negative number. That is, when x is positive, y will be positive, and when x is negative, y will be negative.

The graphs of both equations will contain the origin, because $y = 0$ when $x = 0$ in both equations.

x	y
−4	Undefined
−1	Undefined
0	0
1	1
4	2
9	3
16	4

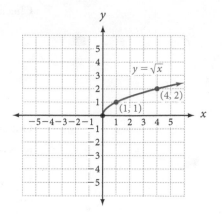

FIGURE 1

x	y
−27	−3
−8	−2
−1	−1
0	0
1	1
8	2
27	3

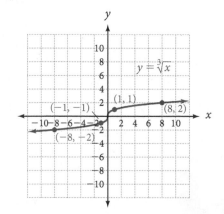

FIGURE 2

GETTING READY FOR CLASS

Respond in your own words and in complete sentences.

A. What is the squaring property of equality?

B. Under what conditions do we obtain extraneous solutions to equations that contain radical expressions?

C. If we have raised both sides of an equation to a power, when is it not necessary to check for extraneous solutions?

D. When will we need to apply the squaring property of equality twice in the process of solving an equation containing radicals?

Problem Set 6.6

Solve each of the following equations.

1. $\sqrt{2x + 1} = 3$

2. $\sqrt{3x + 1} = 4$

3. $\sqrt{4x + 1} = -5$

4. $\sqrt{6x + 1} = -5$

5. $\sqrt{2y - 1} = 3$

6. $\sqrt{3y - 1} = 2$

7. $\sqrt{5x - 7} = -1$

8. $\sqrt{8x + 3} = -6$

9. $\sqrt{2x - 3} - 2 = 4$

10. $\sqrt{3x + 1} - 4 = 1$

11. $\sqrt{4a + 1} + 3 = 2$

12. $\sqrt{5a - 3} + 6 = 2$

13. $\sqrt[4]{3x + 1} = 2$

14. $\sqrt[4]{4x + 1} = 3$

15. $\sqrt[3]{2x - 5} = 1$

16. $\sqrt[3]{5x + 7} = 2$

17. $\sqrt[3]{3a + 5} = -3$

18. $\sqrt[3]{2a + 7} = -2$

19. $\sqrt{y - 3} = y - 3$

20. $\sqrt{y + 3} = y - 3$

21. $\sqrt{a + 2} = a + 2$

22. $\sqrt{a + 10} = a - 2$

23. $\sqrt{a + 3} = a + 1$

24. $\sqrt{x + 15} = x - 5$

25. $\sqrt{x + 7} = x + 5$

26. $\sqrt{3x + 52} = x + 8$

27. $\sqrt{2x + 2} = x + 1$

28. $\sqrt{3x + 4} = -\sqrt{2x + 3}$

29. $\sqrt{4a + 7} = -\sqrt{a + 2}$

30. $\sqrt{7a - 1} = \sqrt{2a + 4}$

31. $\sqrt[4]{5x - 8} = \sqrt[4]{4x - 1}$

32. $\sqrt[4]{6x + 7} = \sqrt[4]{x + 2}$

33. $x + 1 = \sqrt{5x + 1}$

34. $x - 1 = \sqrt{6x + 1}$

35. $t + 5 = \sqrt{2t + 9}$

36. $t + 7 = \sqrt{2t + 13}$

37. $\sqrt{y - 8} = \sqrt{8 - y}$

38. $\sqrt{2y + 5} = \sqrt{5y + 2}$

39. $\sqrt[3]{3x + 5} = \sqrt[3]{5 - 2x}$

40. $\sqrt[3]{4x + 9} = \sqrt[3]{3 - 2x}$

The following equations will require that you square both sides twice before all the radicals are eliminated. Solve each equation using the methods shown in Examples 5, 6, and 7.

41. $\sqrt{x-8} = \sqrt{x} - 2$

42. $\sqrt{x+3} = \sqrt{x} - 3$

43. $\sqrt{x+1} = \sqrt{x} + 1$

44. $\sqrt{x-1} = \sqrt{x} - 1$

45. $\sqrt{x+8} = \sqrt{x-4} + 2$

46. $\sqrt{x+5} = \sqrt{x-3} + 2$

47. $\sqrt{x-5} - 3 = \sqrt{x-8}$

48. $\sqrt{x-3} - 4 = \sqrt{x-3}$

49. Solve each equation.

  a. $\sqrt{y} - 4 = 6$

  b. $\sqrt{y-4} = 6$

  c. $\sqrt{y-4} = -6$

  d. $\sqrt{y-4} = y - 6$

50. Solve each equation.

  a. $\sqrt{2y} + 15 = 7$

  b. $\sqrt{2y+15} = 7$

  c. $\sqrt{2y+15} = y$

  d. $\sqrt{2y+15} = y + 6$

51. Solve each equation.

  a. $x - 3 = 0$

  b. $\sqrt{x} - 3 = 0$

  c. $\sqrt{x-3} = 0$

  d. $\sqrt{x} + 3 = 0$

  e. $\sqrt{x} + 3 = 5$

  f. $\sqrt{x} + 3 = -5$

  g. $x - 3 = \sqrt{5-x}$

52. Solve each equation.

  a. $x - 2 = 0$

  b. $\sqrt{x} - 2 = 0$

  c. $\sqrt{x} + 2 = 0$

  d. $\sqrt{x+2} = 0$

  e. $\sqrt{x} + 2 = 7$

  f. $x - 2 = \sqrt{2x-1}$

APPLYING THE CONCEPTS

53. Solving a Formula Solve the following formula for h:

$$t = \frac{\sqrt{100-h}}{4}$$

54. Solving a Formula Solve the following formula for h:

$$t = \sqrt{\frac{2h-40t}{g}}$$

1 sec

55. Pendulum Clock The length of time (T) in seconds it takes the pendulum of a clock to swing through one complete cycle is given by the formula

$$T = 2\pi\sqrt{\frac{L}{32}}$$

where L is the length, in feet, of the pendulum, and π is approximately $\frac{22}{7}$. How long must the pendulum be if one complete cycle takes 2 seconds?

56. Pollution A long straight river, 100 meters wide, is flowing at 1 meter per second. A pollutant is entering the river at a constant rate from one of its banks. As the pollutant disperses in the water, it forms a plume that is modeled by the equation $y = \sqrt{x}$. Use this information to answer the following questions.

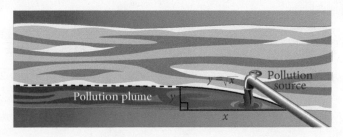

a. How wide is the plume 25 meters down river from the source of the pollution?

b. How wide is the plume 100 meters down river from the source of the pollution?

c. How far down river from the source of the pollution does the plume reach halfway across the river?

d. How far down the river from the source of the pollution does the plume reach the other side of the river?

57. Voltage Voltage V (measured in volts) is related to power P, in watts, and resistance R, in ohms, through the formula $V = \sqrt{PR}$. What is the power in a circuit with a voltage of 60 volts and resistance of 90 ohms?

58. Voltage Voltage V (measured in volts) is related to power P, in watts, and resistance R, in ohms, through the formula $V = \sqrt{PR}$. What is the power in a circuit with a voltage of 80 volts and resistance of 100 ohms?

59. Sail Area-Displacement Ratio The area A (in square feet) of the sails of a sailboat is related to the displacement d of the hull (in cubic feet) by the formula $A = 18\sqrt[3]{d^2}$. If a sailboat has 648 square feet of sail, what is the displacement of the hull?

60. Sail Area-Displacement Ratio The area A (in square feet) of the sails of a sailboat is related to the displacement d of the hull (in cubic feet) by the formula $A = 18\sqrt[3]{d^2}$. If a sailboat has 450 square feet of sail, what is the displacement of the hull?

61. Pythagorean Theorem A ladder that is 18 feet long is placed against a building so its base is 6 feet from the building. Find the height at which the ladder will touch the side of the building. Use the formula $L = \sqrt{b^2 + h^2}$, where L is the length of the ladder, b is the distance of the ladder base from the building, and h is the height of the building where the ladder touches.

62. Pythagorean Theorem A ladder that is 15 feet long is placed against a building so its base is 5 feet from the building. Find the height at which the ladder will touch the side of the building. Use the formula $L = \sqrt{b^2 + h^2}$, where L is the length of the ladder, b is the distance of the ladder base from the building, and h is the height of the building where the ladder touches.

Graph each equation.

63. $y = 2\sqrt{x}$ **64.** $y = -2\sqrt{x}$ **65.** $y = \sqrt{x} - 2$ **66.** $y = \sqrt{x} + 2$

67. $y = \sqrt{x - 2}$ **68.** $y = \sqrt{x + 2}$ **69.** $y = 3\sqrt[3]{x}$ **70.** $y = -3\sqrt[3]{x}$

71. $y = \sqrt[3]{x} + 3$ **72.** $y = \sqrt[3]{x} - 3$ **73.** $y = \sqrt[3]{x + 3}$ **74.** $y = \sqrt[3]{x - 3}$

GETTING READY FOR THE NEXT SECTION

Simplify.

75. $\sqrt{25}$ **76.** $\sqrt{49}$ **77.** $\sqrt{12}$ **78.** $\sqrt{50}$

79. $(-1)^{15}$ **80.** $(-1)^{20}$ **81.** $(-1)^{50}$ **82.** $(-1)^{5}$

Solve.

83. $3x = 12$ **84.** $4 = 8y$ **85.** $4x - 3 = 5$ **86.** $7 = 2y - 1$

Perform the indicated operation.

87. $(3 + 4x) + (7 - 6x)$ **88.** $(2 - 5x) + (-1 + 7x)$

89. $(7 + 3x) - (5 + 6x)$ **90.** $(5 - 2x) - (9 - 4x)$

91. $(3 - 4x)(2 + 5x)$ **92.** $(8 + x)(7 - 3x)$

93. $2x(4 - 6x)$ **94.** $3x(7 + 2x)$

95. $(2 + 3x)^2$ **96.** $(3 + 5x)^2$

97. $(2 - 3x)(2 + 3x)$ **98.** $(4 - 5x)(4 + 5x)$

EXTENDING THE CONCEPTS

99. $\dfrac{x}{3\sqrt{2x-3}} - \dfrac{1}{\sqrt{2x-3}} = \dfrac{1}{3}$

100. $\dfrac{x}{5\sqrt{2x+10}} + \dfrac{1}{\sqrt{2x+10}} = \dfrac{1}{5}$

101. $x + 1 = \sqrt[3]{4x+4}$

102. $x - 1 = \sqrt[3]{4x-4}$

Solve for y in terms of x.

103. $y + 2 = \sqrt{x^2 + (y-2)^2}$

104. $y + \dfrac{1}{2} = \sqrt{x^2 + \left(y - \dfrac{1}{2}\right)^2}$

105. Use your Y variables list, or write a program, to graph the family of curves $Y = \sqrt{X} + B$ for $B = -3, -2, -1, 0, 1, 2,$ and 3.

106. Use your Y variables list, or write a program, to graph the family of curves $Y = \sqrt{X + B}$ for $B = -3, -2, -1, 0, 1, 2,$ and 3.

© iStockphoto/afinocchiaro

 OBJECTIVES

A   Simplify square roots of negative numbers.

B   Simplify powers of i.

C   Solve for unknown variables by equating real parts and equating imaginary parts of two complex numbers.

D   Add and subtract complex numbers.

E   Multiply complex numbers.

F   Divide complex numbers.

In 2012, two deadly earthquakes struck the Emilia Romagna region of northern Italy. Widespread soil liquefaction damaged much of the region's buildings and infrastructure. Earthquake liquefaction occurs when shaking from a seismic event turns the once-solid ground to a consistency similar to quicksand. Buildings and bridges can literally sink into the ground! Some civil engineers study soil liquefaction using a special group of numbers called complex numbers, which is the focus of this section.

The equation $x^2 = -9$ has no real number solutions because the square of a real number is always positive. We have been unable to work with square roots of negative numbers like $\sqrt{-25}$ and $\sqrt{-16}$ for the same reason. Complex numbers allow us to expand our work with radicals to include square roots of negative numbers and to solve equations like $x^2 = -9$ and $x^2 = -64$.

Square Roots of Negative Numbers

Our work with complex numbers is based on the following definition.

> **DEFINITION**   *the number i*
> . . . .
> The ***number i*** is such that $i = \sqrt{-1}$ (which is the same as saying $i^2 = -1$).

The number i, as we have defined it here, is not a real number. Because of the way we have defined i, we can use it to simplify square roots of negative numbers.

> ***Square Roots of Negative Numbers***
> . . . .
> If a is a positive number, then $\sqrt{-a}$ can always be written as $i\sqrt{a}$. That is,
> $$\sqrt{-a} = i\sqrt{a} \qquad \text{if } a \text{ is a positive number.}$$

To justify our rule, we simply square the quantity $i\sqrt{a}$ to obtain $-a$. Here is what it looks like when we do so:

$$(i\sqrt{a})^2 = i^2 \cdot (\sqrt{a})^2$$
$$= -1 \cdot a$$
$$= -a$$

Here are some examples that illustrate the use of our new rule.

EXAMPLE

Write each square root in terms of the number i.

1. $\sqrt{-25} = i\sqrt{25} = i \cdot 5 = 5i$ **2.** $\sqrt{-49} = i\sqrt{49} = i \cdot 7 = 7i$

3. $\sqrt{-12} = i\sqrt{12} = i \cdot 2\sqrt{3} = 2i\sqrt{3}$ **4.** $\sqrt{-17} = i\sqrt{17}$

B Powers of i

If we assume all the properties of exponents hold when the base is i, we can write any power of i as i, -1, $-i$, or 1. Using the fact that $i^2 = -1$, we have

$$i^1 = i$$
$$i^2 = -1$$
$$i^3 = i^2 \cdot i = -1(i) = -i$$
$$i^4 = i^2 \cdot i^2 = -1(-1) = 1$$

Because $i^4 = 1$, i^5 will simplify to i, and we will begin repeating the sequence i, -1, $-i$, 1 as we simplify higher powers of i: Any power of i simplifies to i, -1, $-i$, or 1. The easiest way to simplify higher powers of i is to write them in terms of i^2. For instance, to simplify i^{21}, we would write it as

$$(i^2)^{10} \cdot i \quad \text{because} \quad 2 \cdot 10 + 1 = 21$$

Then, because $i^2 = -1$, we have

$$(-1)^{10} \cdot i = 1 \cdot i = i$$

EXAMPLE

Simplify as much as possible.

5. $i^{30} = (i^2)^{15} = (-1)^{15} = -1$

6. $i^{11} = (i^2)^5 \cdot i = (-1)^5 \cdot i = (-1)i = -i$

7. $i^{40} = (i^2)^{20} = (-1)^{20} = 1$

NOTE In Examples 3 and 4, we wrote i before the radical simply to avoid confusion. If we were to write the answer to 3 as $2\sqrt{3i}$, some people would think the i was under the radical sign, but it is not.

> **DEFINITION** *complex number*
>
> A *complex number* is any number that can be put in the form
>
> $$a + bi$$
>
> where a and b are real numbers and $i = \sqrt{-1}$. The form $a + bi$ is called *standard form* for complex numbers. The number a is called the *real part* of the complex number. The number b is called the *imaginary part* of the complex number.

Every real number is a complex number. For example, 8 can be written as $8 + 0i$. Likewise, $-\frac{1}{2}, \pi, \sqrt{3}$, and 29 are complex numbers because they can all be written in the form $a + bi$:

$$-\frac{1}{2} = -\frac{1}{2} + 0i \qquad \pi = \pi + 0i \qquad \sqrt{3} = \sqrt{3} + 0i \qquad -9 = -9 + 0i$$

The rest of the complex numbers that are not real numbers are divided into two additional categories: *compound numbers* and *pure imaginary numbers*. The diagram below shows all three subsets of the complex numbers, along with examples of the type of numbers that fall into those subsets.

Subsets of the Complex Numbers

All numbers of the form $a + bi$ fall into one of the following categories. Each category is a subset of the complex numbers.

Real Numbers	Compound Numbers	Pure Imaginary Numbers
When $a \neq 0$ and $b = 0$ Examples include: $-10, 0, 1, \sqrt{3}, \frac{5}{8}, \pi$	When neither a nor b is 0 Examples include: $5 + 4i, \frac{1}{3} + 4i, \sqrt{5} - i, -6 + i\sqrt{5}$	When $a = 0$ and $b \neq 0$ Examples include: $-4i, i\sqrt{3}, -5i\sqrt{7}, \frac{3}{4}i$

©2009 James Robert Metz

NOTE See Section 1.1 for a review of the subsets of real numbers.

Note: The definition for compound numbers is from Jim Metz of Kapiolani Community College in Hawaii. Some textbooks use the label *imaginary numbers* to represent both the compound numbers and the pure imaginary numbers. In those books, the pure imaginary numbers are a subset of the imaginary numbers. We like the definition from Mr. Metz because it keeps the three subsets from overlapping.

© Equality for Complex Numbers

Two complex numbers are equal if and only if their real parts are equal and their imaginary parts are equal. That is, for real numbers a, b, c, and d,

$$a + bi = c + di \qquad \text{if and only if} \qquad a = c \qquad \text{and} \qquad b = d$$

EXAMPLE 8

Find x and y if $3x + 4i = 12 - 8yi$.

SOLUTION Because the two complex numbers are equal, their real parts are equal and their imaginary parts are equal:

$$3x = 12 \quad \text{and} \quad 4 = -8y$$
$$x = 4 \qquad\qquad y = -\frac{1}{2}$$

EXAMPLE 9

Find x and y if $(4x - 3) + 7i = 5 + (2y - 1)i$.

SOLUTION The real parts are $4x - 3$ and 5. The imaginary parts are 7 and $2y - 1$:

$$4x - 3 = 5 \quad \text{and} \quad 7 = 2y - 1$$
$$4x = 8 \qquad\qquad 8 = 2y$$
$$x = 2 \qquad\qquad y = 4$$

D Addition and Subtraction of Complex Numbers

To add two complex numbers, add their real parts and their imaginary parts. That is, if a, b, c, and d are real numbers, then

$$(a + bi) + (c + di) = (a + c) + (b + d)i$$

If we assume that the commutative, associative, and distributive properties hold for the number i, then the definition of addition is simply an extension of these properties.

We define subtraction in a similar manner. If a, b, c, and d are real numbers, then

$$(a + bi) - (c + di) = (a - c) + (b - d)i$$

EXAMPLE

Add or subtract as indicated.

10. $(3 + 4i) + (7 - 6i) = (3 + 7) + (4 - 6)i = 10 - 2i$

11. $(7 + 3i) - (5 + 6i) = (7 - 5) + (3 - 6)i = 2 - 3i$

12. $(5 - 2i) - (9 - 4i) = (5 - 9) + (-2 + 4)i = -4 + 2i$

E Multiplication of Complex Numbers

Because complex numbers have the same form as binomials, we find the product of two complex numbers the same way we find the product of two binomials.

EXAMPLE 13

Multiply $(3 - 4i)(2 + 5i)$.

SOLUTION Multiplying each term in the second complex number by each term in the first, we have

$$\overset{\text{F} \qquad \text{O} \qquad \text{I} \qquad \text{L}}{(3 - 4i)(2 + 5i) = 3 \cdot 2 + 3 \cdot 5i - 2 \cdot 4i - 4i(5i)}$$

$$= 6 + 15i - 8i - 20i^2$$

Combining similar terms and using the fact that $i^2 = -1$, we can simplify as follows:

$$6 + 15i - 8i - 20i^2 = 6 + 7i - 20(-1)$$

$$= 6 + 7i + 20$$

$$= 26 + 7i$$

The product of the complex numbers $3 - 4i$ and $2 + 5i$ is the complex number $26 + 7i$.

EXAMPLE 14

Multiply $2i(4 - 6i)$.

SOLUTION Applying the distributive property gives us

$$2i(4 - 6i) = 2i \cdot 4 - 2i \cdot 6i$$

$$= 8i - 12i^2$$

$$= 12 + 8i$$

EXAMPLE 15

Expand $(3 + 5i)^2$.

SOLUTION We treat this like the square of a binomial. Remember, $(a + b)^2 = a^2 + 2ab + b^2$:

$$(3 + 5i)^2 = 3^2 + 2(3)(5i) + (5i)^2$$

$$= 9 + 30i + 25i^2$$

$$= 9 + 30i - 25$$

$$= -16 + 30i$$

EXAMPLE 16

Multiply $(2 - 3i)(2 + 3i)$.

SOLUTION This product has the form $(a - b)(a + b)$, which we know results in the difference of two squares, $a^2 - b^2$:

$$(2 - 3i)(2 + 3i) = 2^2 - (3i)^2$$

$$= 4 - 9i^2$$

$$= 4 + 9$$

$$= 13$$

The product of the two complex numbers $2 - 3i$ and $2 + 3i$ is the real number 13. The two complex numbers $2 - 3i$ and $2 + 3i$ are called complex conjugates. The fact that their product is a real number is very useful.

> **DEFINITION** *complex conjugates*
>
> The complex numbers $a + bi$ and $a - bi$ are called *complex conjugates*. One important property they have is that their product is the real number $a^2 + b^2$. Here's why:
>
> $$(a + bi)(a - bi) = a^2 - (bi)^2$$
> $$= a^2 - b^2 i^2$$
> $$= a^2 - b^2(-1)$$
> $$= a^2 + b^2$$

F Division with Complex Numbers

The fact that the product of two complex conjugates is a real number is the key to division with complex numbers.

EXAMPLE 17

Divide $\dfrac{2 + i}{3 - 2i}$.

SOLUTION We want a complex number in standard form that is equivalent to the quotient $\frac{2+i}{3-2i}$. We need to eliminate i from the denominator. Multiplying the numerator and denominator by $3 + 2i$ will give us what we want:

$$\frac{2 + i}{3 - 2i} = \frac{2 + i}{3 - 2i} \cdot \frac{(3 + 2i)}{(3 + 2i)}$$
$$= \frac{6 + 4i + 3i + 2i^2}{9 - 4i^2}$$
$$= \frac{6 + 7i - 2}{9 + 4}$$
$$= \frac{4 + 7i}{13}$$
$$= \frac{4}{13} + \frac{7}{13}i$$

Dividing the complex number $2 + i$ by $3 - 2i$ gives the complex number $\frac{4}{13} + \frac{7}{13}i$.

EXAMPLE 18

Divide $\dfrac{7 - 4i}{i}$.

SOLUTION The conjugate of the denominator is $-i$. Multiplying the numerator and denominator by this number, we have

$$\frac{7 - 4i}{i} = \frac{7 - 4i}{i} \cdot \frac{-i}{-i}$$
$$= \frac{-7i + 4i^2}{-i^2}$$
$$= \frac{-7i + 4(-1)}{-(-1)}$$
$$= -4 - 7i$$

After reading through the preceding section, respond in your own words and in complete sentences.

A. What is the number *i*?

B. What is a complex number?

C. What kind of number will always result when we multiply complex conjugates?

D. Explain how to divide complex numbers.

SPOTLIGHT ON SUCCESS

Aaron | student instructor

Sometimes you have to take a step back in order to get a running start forward.

—Anonymous

As a high school senior I was encouraged to go to college immediately after graduating. I earned good grades in high school and I knew that I would have a pretty good group of schools to pick from. Even though I felt like "more school" was not quite what I wanted, the counselors had so much faith and had done this process so many times that it was almost too easy to get the applications out. I sent out applications to schools I knew I could get into and a "dream school."

One night in my email inbox there was a letter of acceptance from my dream school. There was just one problem with getting into this school. It was going to be difficult and I still had senioritis. Going into my first quarter of college was as exciting and difficult as I knew it would be. But after my first quarter I could see that this was not the time for me to be here. I was interested in the subject matter but I could not find my motivating purpose like I had in high school. Instead of dropping out completely, I decided a community college would be a good way for me to stay on track. Without necessarily knowing my direction, I could take the general education classes and get those out of the way while figuring out exactly what and where I felt was a good place for me to be.

Now I know what I want to go to school for and the next time I walk onto a four-year campus it will be on my terms with my reasons for being there driving me to succeed. I encourage everyone to continue school after high school, even if you have no clue as to what you want to study. There are always stepping stones, like community colleges, that can help you get a clearer picture of what you want to strive for.

Problem Set 6.7

Write the following in terms of i, and simplify as much as possible.

1. $\sqrt{-36}$ **2.** $\sqrt{-49}$ **3.** $-\sqrt{-25}$ **4.** $-\sqrt{-81}$

5. $\sqrt{-72}$ **6.** $\sqrt{-48}$ **7.** $-\sqrt{-12}$ **8.** $-\sqrt{-75}$

Write each of the following as i, -1, $-i$, or 1.

9. i^{28} **10.** i^{31} **11.** i^{26} **12.** i^{37}

13. i^{75} **14.** i^{42}

Find x and y so that each of the following equations is true.

15. $2x + 3yi = 6 - 3i$ **16.** $4x - 2yi = 4 + 8i$

17. $2 - 5i = -x + 10yi$ **18.** $4 + 7i = 6x - 14yi$

19. $2x + 10i = -16 - 2yi$ **20.** $4x - 5i = -2 + 3yi$

21. $(2x - 4) - 3i = 10 - 6yi$ **22.** $(4x - 3) - 2i = 8 + yi$

23. $(7x - 1) + 4i = 2 + (5y + 2)i$ **24.** $(5x + 2) - 7i = 4 + (2y + 1)i$

Combine the following complex numbers.

25. $(2 + 3i) + (3 + 6i)$ **26.** $(4 + i) + (3 + 2i)$

27. $(3 - 5i) + (2 + 4i)$ **28.** $(7 + 2i) + (3 - 4i)$

29. $(5 + 2i) - (3 + 6i)$ **30.** $(6 + 7i) - (4 + i)$

31. $(3 - 5i) - (2 + i)$ **32.** $(7 - 3i) - (4 + 10i)$

33. $[(3 + 2i) - (6 + i)] + (5 + i)$ **34.** $[(4 - 5i) - (2 + i)] + (2 + 5i)$

35. $[(7 - i) - (2 + 4i)] - (6 + 2i)$ **36.** $[(3 - i) - (4 + 7i)] - (3 - 4i)$

37. $(3 + 2i) - [(3 - 4i) - (6 + 2i)]$ **38.** $(7 - 4i) - [(-2 + i) - (3 + 7i)]$

39. $(4 - 9i) + [(2 - 7i) - (4 + 8i)]$ **40.** $(10 - 2i) - [(2 + i) - (3 - i)]$

Find the following products.

41. $3i(4 + 5i)$ **42.** $2i(3 + 4i)$ **43.** $6i(4 - 3i)$

44. $11i(2 - i)$ **45.** $(3 + 2i)(4 + i)$ **46.** $(2 - 4i)(3 + i)$

47. $(4 + 9i)(3 - i)$ **48.** $(5 - 2i)(1 + i)$ **49.** $(1 + i)^3$

50. $(1 - i)^3$ **51.** $(2 - i)^3$ **52.** $(2 + i)^3$

53. $(2 + 5i)^2$ **54.** $(3 + 2i)^2$ **55.** $(1 - i)^2$

56. $(1 + i)^2$ **57.** $(3 - 4i)^2$ **58.** $(6 - 5i)^2$

59. $(2 + i)(2 - i)$ **60.** $(3 + i)(3 - i)$ **61.** $(6 - 2i)(6 + 2i)$

62. $(5 + 4i)(5 - 4i)$ **63.** $(2 + 3i)(2 - 3i)$ **64.** $(2 - 7i)(2 + 7i)$

65. $(10 + 8i)(10 - 8i)$ **66.** $(11 - 7i)(11 + 7i)$

Find the following quotients. Write all answers in standard form for complex numbers.

67. $\dfrac{2 - 3i}{i}$

68. $\dfrac{3 + 4i}{i}$

69. $\dfrac{5 + 2i}{-i}$

70. $\dfrac{4 - 3i}{-i}$

71. $\dfrac{4}{2 - 3i}$

72. $\dfrac{3}{4 - 5i}$

73. $\dfrac{6}{-3 + 2i}$

74. $\dfrac{-1}{-2 - 5i}$

75. $\dfrac{2 + 3i}{2 - 3i}$

76. $\dfrac{4 - 7i}{4 + 7i}$

77. $\dfrac{5 + 4i}{3 + 6i}$

78. $\dfrac{2 + i}{5 - 6i}$

APPLYING THE CONCEPTS

79. Electrical Circuits Complex numbers may be applied to electrical circuits. Electrical engineers use the fact that resistance R to electrical flow of the electrical current I and the voltage V are related by the formula $V = RI$. (Voltage is measured in volts, resistance in ohms, and current in amperes.) Find the resistance to electrical flow in a circuit that has a voltage $V = (80 + 20i)$ volts and current $I = (-6 + 2i)$ amps.

80. Electrical Circuits Refer to the information about electrical circuits in Problem 79, and find the current in a circuit that has a resistance of $(4 + 10i)$ ohms and a voltage of $(5 - 7i)$ volts.

MAINTAINING YOUR SKILLS

The following problems review material we covered in Sections 5.5 and 5.6.

Solve each equation. [5.5]

81. $\dfrac{t}{3} - \dfrac{1}{2} = -1$

82. $\dfrac{x}{x - 2} + \dfrac{2}{3} = \dfrac{2}{x - 2}$

83. $2 + \dfrac{5}{y} = \dfrac{3}{y^2}$

84. $1 - \dfrac{1}{y} = \dfrac{12}{y^2}$

Solve each application problem. [5.6]

85. The sum of a number and its reciprocal is $\dfrac{41}{20}$. Find the number.

86. It takes an inlet pipe 8 hours to fill a tank. The drain can empty the tank in 6 hours. If the tank is full and both the inlet pipe and drain are open, how long will it take to drain the tank?

EXTENDING THE CONCEPTS

87. Show that $-i$ and $\dfrac{1}{i}$ (the opposite and reciprocal of i) are the same number.

88. Show that $i^{2n + 1}$ is the same as i for all positive even integers.

89. Show that $x = 1 + i$ is a solution to the equation $x^2 - 2x + 2 = 0$.

90. Show that $x = 1 - i$ is a solution to the equation $x^2 - 2x + 2 = 0$.

91. Show that $x = 2 + i$ is a solution to the equation $x^3 - 11x + 20 = 0$.

92. Show that $x = 2 - i$ is a solution to the equation $x^3 - 11x + 20 = 0$.

Chapter 6 Summary

The numbers in brackets refer to the section(s) in which the topic can be found.

EXAMPLES

Square Roots [6.1]

1. The number 49 has two square roots, 7 and -7. They are written like this:

$$\sqrt{49} = 7 \qquad -\sqrt{49} = -7$$

Every positive real number x has two square roots. The *positive square root* of x is written $\sqrt{x}$, and the *negative square root* of x is written $-\sqrt{x}$. Both the positive and the negative square roots of x are numbers we square to get x; that is,

$$\left. \begin{array}{c} (\sqrt{x})^2 = x \\ (-\sqrt{x})^2 = x \end{array} \right\} \text{ for } x \geq 0$$

and

Higher Roots [6.1]

2. $\sqrt[3]{8} = 2$
$\sqrt[3]{-27} = -3$

In the expression $\sqrt[n]{a}$, n is the *index*, a is the *radicand*, and $\sqrt{}$ is the *radical sign*. The expression $\sqrt[n]{a}$ is such that

$$(\sqrt[n]{a})^n = a \qquad a \geq 0 \text{ when } n \text{ is even}$$

Rational Exponents [6.1, 6.2]

3. $25^{1/2} = \sqrt{25} = 5$
$8^{2/3} = (\sqrt[3]{8})^2 = 2^2 = 4$
$9^{3/2} = (\sqrt{9})^3 = 3^3 = 27$

Rational exponents are used to indicate roots. The relationship between rational exponents and roots is as follows:

$$a^{1/n} = \sqrt[n]{a} \qquad \text{and} \qquad a^{m/n} = (a^{1/n})^m = (a^m)^{1/n}$$

$$a \geq 0 \text{ when } n \text{ is even}$$

Properties of Radicals [6.3]

4. $\sqrt{4 \cdot 5} = \sqrt{4}\sqrt{5} = 2\sqrt{5}$

$\sqrt{\dfrac{7}{9}} = \dfrac{\sqrt{7}}{\sqrt{9}} = \dfrac{\sqrt{7}}{3}$

If a and b are nonnegative real numbers whenever n is even, then

1. $\sqrt[n]{ab} = \sqrt[n]{a}\,\sqrt[n]{b}$

2. $\sqrt[n]{\dfrac{a}{b}} = \dfrac{\sqrt[n]{a}}{\sqrt[n]{b}} \qquad (b \neq 0)$

Simplified Form for Radicals [6.2, 6.3]

5. $\sqrt{\dfrac{4}{5}} = \dfrac{\sqrt{4}}{\sqrt{5}}$

$= \dfrac{2}{\sqrt{5}} \cdot \dfrac{\sqrt{5}}{\sqrt{5}}$

$= \dfrac{2\sqrt{5}}{5}$

A radical expression is said to be in *simplified form*:

1. If there is no factor of the radicand that can be written as a power greater than or equal to the index;

2. If there are no fractions under the radical sign; and

3. If there are no radicals in the denominator.

6. $5\sqrt{3} - 7\sqrt{3} = (5 - 7)\sqrt{3}$
$= -2\sqrt{3}$

$\sqrt{20} + \sqrt{45} = 2\sqrt{5} + 3\sqrt{5}$
$= (2 + 3)\sqrt{5}$
$= 5\sqrt{5}$

Addition and Subtraction of Radical Expressions [6.4]

We add and subtract radical expressions by using the distributive property to combine similar radicals. Similar radicals are radicals with the same index and the same radicand.

Multiplication of Radical Expressions [6.5]

7. $(\sqrt{x} + 2)(\sqrt{x} + 3)$
$= \sqrt{x}\,\sqrt{x} + 3\sqrt{x} + 2\sqrt{x} + 2 \cdot 3$
$= x + 5\sqrt{x} + 6$

We multiply radical expressions in the same way that we multiply polynomials. We can use the distributive property and the FOIL method.

Rationalizing the Denominator [6.3, 6.5]

8. $\dfrac{3}{\sqrt{2}} = \dfrac{3}{\sqrt{2}} \cdot \dfrac{\sqrt{2}}{\sqrt{2}} = \dfrac{3\sqrt{2}}{2}$

$\dfrac{3}{\sqrt{5} - \sqrt{3}} = \dfrac{3}{\sqrt{5} - \sqrt{3}} \cdot \dfrac{\sqrt{5} + \sqrt{3}}{\sqrt{5} + \sqrt{3}}$
$= \dfrac{3\sqrt{5} + 3\sqrt{3}}{5 - 3}$
$= \dfrac{3\sqrt{5} + 3\sqrt{3}}{2}$

When a fraction contains a square root in the denominator, we rationalize the denominator by multiplying the numerator and denominator by

1. The square root itself if there is only one term in the denominator, or

2. The conjugate of the denominator if there are two terms in the denominator.

Rationalizing the denominator is also called division of radical expressions.

Squaring Property of Equality [6.6]

9. $\sqrt{2x + 1} = 3$
$(\sqrt{2x + 1})^2 = 3^2$
$2x + 1 = 9$
$x = 4$

We may square both sides of an equation any time it is convenient to do so, as long as we check all resulting solutions in the original equation.

Complex Numbers [6.7]

10. $3 + 4i$ is a complex number.

Addition
$(3 + 4i) + (2 - 5i) = 5 - i$

Multiplication
$(3 + 4i)(2 - 5i)$
$= 6 - 15i + 8i - 20i^2$
$= 6 - 7i + 20$
$= 26 - 7i$

Division
$\dfrac{2}{3 + 4i} = \dfrac{2}{3 + 4i} \cdot \dfrac{3 - 4i}{3 - 4i}$
$= \dfrac{6 - 8i}{9 + 16}$
$= \dfrac{6}{25} - \dfrac{8}{25}i$

A *complex number* is any number that can be put in the form

$$a + bi$$

where a and b are real numbers and $i = \sqrt{-1}$. The *real part* of the complex number is a, and b is the *imaginary part*.

If a, b, c, and d are real numbers, then we have the following definitions associated with complex numbers:

1. Equality
$$a + bi = c + di \quad \text{if and only if} \quad a = c \text{ and } b = d$$

2. Addition and subtraction
$$(a + bi) + (c + di) = (a + c) + (b + d)i$$
$$(a + bi) - (c + di) = (a - c) + (b - d)i$$

3. Multiplication
$$(a + bi)(c + di) = (ac - bd) + (ad + bc)i$$

4. Division is similar to rationalizing the denominator.

Find each of the following roots, if possible. [6.1]

1. $\sqrt{49}$

2. $-\sqrt[3]{27}$

3. $-\sqrt[4]{-16}$

4. $\sqrt{0.81}$

Simplify each of the following. (Assume all variable bases are positive integers and all variable exponents are positive real numbers throughout this test.) [6.1]

5. $27^{-2/3}$

6. $\left(\dfrac{144}{49}\right)^{-1/2}$

7. $x^{2/3} \cdot x^{1/5}$

8. $\dfrac{(a^{1/4}b)^{-2}}{(a^{3/2}b^6)^{-1}}$

9. $\sqrt{49x^8y^{12}}$

10. $\sqrt[3]{27x^6y^{12}}$

11. $\dfrac{(49a^8b^{10})^{1/2}}{(27a^{15}b^3)^{1/3}}$

12. $\dfrac{(x^{1/n}y^{2/n})^n}{(x^{1/n}y^n)^{n^2}}$

Convert the following radicals to expressions with rational exponents and simplify where possible. Assume all variables represent positive numbers. [6.1]

13. $\sqrt[10]{y^2}$

14. $\sqrt[12]{b^3}$

Use the product property for radicals to write each of the following expressions in simplified form. (Assume all variables are nonnegative.) [6.3]

15. $\sqrt[3]{72a^4b^3c^2}$

16. $\sqrt[5]{64x^8y^4z^{11}}$

17. $\sqrt[3]{-128x^{15}y^{20}}$

18. $\sqrt[4]{48x^{13}y^{23}}$

Write in simplified form. [6.2]

19. $\sqrt{50x^3y^7}$

20. $\sqrt[3]{135x^5y^7}$

21. $\sqrt{\dfrac{5}{7}}$

22. $\sqrt{\dfrac{12a^3b^5}{5c^2}}$

23. $\sqrt{\dfrac{12x^5}{7y}}$

24. $\sqrt{\dfrac{50x^2y^3}{3z}}$

Rationalize the denominator. [6.3]

25. $\sqrt{\dfrac{1}{6}}$

26. $\dfrac{5}{\sqrt[3]{3}}$

27. $\dfrac{3}{\sqrt{5}-2}$

28. $\dfrac{\sqrt{x}-\sqrt{5}}{\sqrt{x}+\sqrt{5}}$

29. $\sqrt[3]{\dfrac{5x}{4y}}$

30. $\sqrt[4]{\dfrac{5}{27y}}$

Combine. [6.4]

31. $\dfrac{3}{x^{1/3}} + x^{2/3}$

32. $\dfrac{x^4}{(x^4-6)^{1/2}} - (x^4-6)^{1/2}$

33. $7\sqrt[6]{7} - \sqrt[6]{7} + 4\sqrt[6]{7}$

34. $5x\sqrt{6} - 3x\sqrt{6} - 2x\sqrt{6}$

35. $2\sqrt{12} - 2\sqrt{48}$

36. $3\sqrt[3]{40a^3b^5} - a\sqrt[3]{5b^5}$

37. $\sqrt{x} + \dfrac{1}{\sqrt{x}}$

38. $\sqrt{15} - \sqrt{\dfrac{3}{5}} + \sqrt{\dfrac{5}{3}}$

39. $\sqrt[3]{49} + \dfrac{4}{\sqrt[3]{7}}$

40. $x\sqrt[4]{5xy^8} + y\sqrt[4]{405x^5y^4} + y^2\sqrt[4]{80x^5}$

Multiply. [6.5]

41. $3a^{5/2}(2a^{3/2} - a^{7/2})$

42. $(3a^{4/5} - 2)^2$

43. $(\sqrt{x} - 3)(\sqrt{x} + 5)$

44. $(4\sqrt{3} - \sqrt{5})^2$

Rationalize the denominator in each of the following. [6.4]

45. $\dfrac{\sqrt{7}}{3\sqrt{7}-2}$

46. $\dfrac{\sqrt{5}-\sqrt{3}}{\sqrt{5}+\sqrt{3}}$

Solve for x. [6.5]

47. $\sqrt{x + 7} = x - 5$

48. $\sqrt[3]{5x + 4} = 4$

49. $\sqrt{x + 8} = \sqrt{x - 16} + 4$

50. $\sqrt{3x + 4} = -\sqrt{2x + 3}$

51. $\sqrt[3]{4x + 9} = \sqrt[3]{3 - 2x}$

Graph. [6.6]

52. $y = \sqrt{x - 3}$

53. $y = \sqrt[3]{x} - 1$

Match each equation with its graph. [6.6]

54. $y = \sqrt[3]{x} \quad 2$

55. $y = \sqrt{x} \quad 2$

56. $y = \sqrt{x}$

57. $y = \sqrt[3]{x - 2}$

58. $y = \sqrt{x} + 2$

59. $y = -\sqrt{x} - 2$

A.

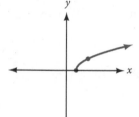

B.

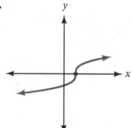

C.

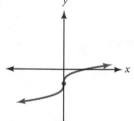

D.

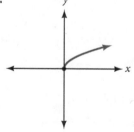

E.

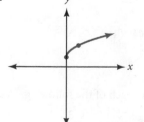

F.

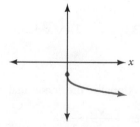

Write the following in terms of i, and simplify as much as possible.

60. $-\sqrt{-81}$

61. $-\sqrt{-75}$

Find x and y so that each of the following equations is true. [6.7]

62. $4x - 5i = -2 + 3yi$

63. $(4x - 3) - 2i = 8 + yi$

64. $9 - (2x - 3)i = (4y - 3) + 5i$

Perform the indicated operations. [6.7]

65. $(5 + 3i) - [(4 - 2i) - (3 + 7i)]$

66. $[(4 - 5i) - (2 + i)] + (2 + 5i)$

67. $(3 - 2i)(4 + 3i)$

68. $(5 - 3i)^2$

69. $\dfrac{4i - 1}{4i + 1}$

70. Show that i^{39} can be written as $-i$. [6.6]

Simplify.

1. $2 - [1 + 4(8 - 9)]$

2. $3 + 21 \div 7 + 4 \cdot 2$

3. $5 + (7 - 3)^2 - (4 + 1)^2$

4. $\sqrt[4]{128}$

5. $27^{2/3} + 25^{1/2}$

6. $\dfrac{1 - \dfrac{4}{5}}{1 + \dfrac{1}{5}}$

7. $\left(-\dfrac{8}{125}\right)^{-1/3}$

8. $32 + [4 - 9 \div 3(6 - 9)]$

Reduce.

9. $\dfrac{104}{117}$

10. $\dfrac{18x^2 - 21xy - 15y^2}{6x + 3y}$

11. $\dfrac{x^2 + x - 2}{x^2 + 5x + 6}$

12. $\dfrac{4a^8b^{-2}}{16a^3b^{-8}}$

Multiply or divide.

13. $(3\sqrt{x} - 5)(\sqrt{x} + 1)$

14. $(8 + 2i)(6 - i)$

15. $\dfrac{3 - i}{3 - 2i}$

16. $\dfrac{\sqrt{5}}{2 - \sqrt{5}}$

17. $(4x + 2y)(7x - y)$

18. $(16x^2 - 9) \cdot \dfrac{x - 7}{4x - 3}$

19. $\dfrac{24x^7y^{-3}}{5x^2y^4} \div \dfrac{6x^2y^2}{25x^3y}$

20. $\dfrac{\sqrt{x} - \sqrt{y}}{\sqrt{x} + \sqrt{y}}$

Solve.

21. $\dfrac{1}{2} + \dfrac{2}{a + 3} = \dfrac{1}{a + 3}$

22. $\sqrt{5 - x} = x - 5$

23. $(3x + 4)^2 = 7$

24. $0.02x^2 + 0.07x = 0.04$

25. $\sqrt{x - 7} = 7 - \sqrt{x}$

26. $3x^3 - 4x = -4x^2$

27. $\dfrac{1}{10}x^2 + \dfrac{2}{5}x = \dfrac{1}{2}$

28. $\dfrac{1}{8}(16x - 2) + \dfrac{1}{4} = 6$

29. $|8x - 9| - 4 = 3$

30. $\sqrt[3]{15 - 4x} = -1$

Solve each inequality, and graph the solution.

31. $-1 \le \dfrac{x}{5} - 5 \le 1$

32. $|7x + 2| > 3$

Solve each system.

33. $-9x + 6y = 30$
$4x - 5y = -25$

34. $2x + 5y = -23$
$7x - 3y = 22$

Graph on a rectangular coordinate system.

35. $2x + 3y = 3$

36. $-x + 2y < -2$

Find the slope of the line passing through the points.

37. $(-5, 7)$ and $(-8, -5)$

38. $(6, 13)$ and $(5, 2)$

Rationalize the denominator.

39. $\dfrac{6}{\sqrt[4]{27}}$

40. $\dfrac{5}{\sqrt{7} - 3}$

Factor completely.

41. $81a^4 - 256b^4$

42. $24a^4 - 44a^2 - 28$

43. Inverse Variation y varies inversely with the square root of x. If $y = -1$ when $x = 9$, find y when $x = 16$.

44. Joint Variation z varies jointly with x and the cube of y. If $z = -32$ when $x = 4$, and $y = 2$, find z when $x = 2$ and $y = 3$.

Let $f(x) = 2x + 2$ and $g(x) = x^2 + x$. Find the following.

45. $f(b)$

46. $(g - f)(3)$

47. $(g \circ f)(1)$

48. $(f \circ g)(-1)$

The chart shows the yearly sales for the top frozen pizza retailers. Use the information to answer Questions 49 and 50.

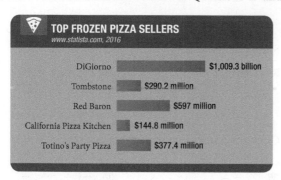

49. What is the difference in sales between California Pizza Kitchen and Red Baron?

50. What is the difference in sales between DiGiorno and Totinos?

51. Unit Analysis Convert 211.2 feet/minute into miles/hour.

52. Translations Sketch the basic graph and the graph of the given function by hand on the same axes. Label the coordinates of two or three points on the graph of the given function.

$$G(t) = 2|t - 5| - 1$$

Chapter 6 Test

Simplify each of the following. (Assume all variable bases are positive integers and all variable exponents are positive real numbers throughout this test.) [6.1]

1. $27^{-2/3}$

2. $\left(\dfrac{25}{49}\right)^{-1/2}$

3. $a^{3/4} \cdot a^{-1/3}$

4. $\dfrac{(x^{2/3}y^{-3})^{1/2}}{(x^{3/4}y^{1/2})^{-1}}$

5. $\sqrt{49x^8y^{10}}$

6. $\sqrt[5]{32x^{10}y^{20}}$

7. $\dfrac{(36a^8b^4)^{1/2}}{(27a^9b^6)^{1/3}}$

8. $\dfrac{(x^ny^{1/n})^n}{(x^{1/n}y^n)^{n^2}}$

Multiply. [6.1]

9. $2a^{1/2}(3a^{3/2} - 5a^{1/2})$ **10.** $(4a^{3/2} - 5)^2$

Factor. [6.1]

11. $3x^{2/3} + 5x^{1/3} - 2$ **12.** $9x^{2/3} - 49$

Combine. [6.3]

13. $\dfrac{4}{x^{1/2}} + x^{1/2}$ **14.** $\dfrac{x^2}{(x^2 - 3)^{1/2}} - (x^2 - 3)^{1/2}$

Write in simplified form. [6.2]

15. $\sqrt{125x^3y^5}$ **16.** $\sqrt[3]{40x^7y^8}$

17. $\sqrt{\dfrac{2}{3}}$ **18.** $\sqrt{\dfrac{12a^4b^3}{5c}}$

Combine. [6.3]

19. $3\sqrt{12} - 4\sqrt{27}$ **20.** $\sqrt[3]{24a^3b^3} - 5a\sqrt[3]{3b^3}$

Multiply. [6.4]

21. $(\sqrt{x} + 7)(\sqrt{x} - 4)$ **22.** $(3\sqrt{2} - \sqrt{3})^2$

Rationalize the denominator. [6.4]

23. $\dfrac{5}{\sqrt{3} - 1}$ **24.** $\dfrac{\sqrt{x} - \sqrt{2}}{\sqrt{x} + \sqrt{2}}$

Solve for x. [6.5]

25. $\sqrt{3x + 1} = x - 3$ **26.** $\sqrt[3]{2x + 7} = -1$

27. $\sqrt{x + 3} = \sqrt{x + 4} - 1$

Graph. [6.5]

28. $y = \sqrt{x - 2}$ **29.** $y = \sqrt[3]{x} + 3$

30. Solve for x and y so that the following equation is true [6.6]:

$$(2x + 5) - 4i = 6 - (y - 3)i$$

Perform the indicated operations. [6.6]

31. $(3 + 2i) - [(7 - i) - (4 + 3i)]$

32. $(2 - 3i)(4 + 3i)$

33. $(5 - 4i)^2$

34. $\dfrac{2 - 3i}{2 + 3i}$

35. Show that i^{38} can be written as -1. [6.6}

7

Quadratic Functions

Chapter Outline

© Fir0002/Flagstaffotos/http://www.gnu.org/licenses/old-licenses/fdl-1.2.html

If you have been to the circus or the county fair recently, you may have witnessed one of the more spectacular acts, the human cannonball. The human cannonball shown in the photograph will reach a height of 70 feet and travel a distance of 160 feet before landing in a safety net. In this chapter, we use this information to derive the equation

$$f(x) = -\frac{7}{640}(x - 80)^2 + 70 \quad \text{for } 0 \le x \le 160$$

which describes the path flown by this particular cannonball. The table and graph below were constructed from this equation.

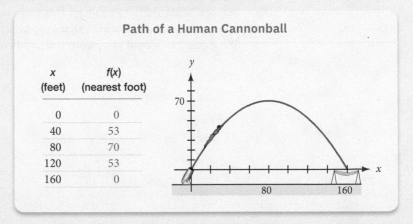

Path of a Human Cannonball

x (feet)	f(x) (nearest foot)
0	0
40	53
80	70
120	53
160	0

All objects that are projected into the air, whether they are basketballs, bullets, arrows, or coins, follow parabolic paths like the one shown in the graph. Studying the material in this chapter will give you a more mathematical understanding of the world around you.

If you have made it this far, then you have the study skills necessary to be successful in this course. For the remaining chapters, this introductory section will be titled "Success Skills." Success skills are more general in nature and will help you with all your classes and ensure your success in college as well.

Let's start with a question:

> *Question:* What quality is most important for success in any college course?
> *Answer:* Independence. You want to become an independent learner.

We all know people like this. They are generally happy. They don't worry about getting the right instructor, or whether or not things work out every time. They have a confidence that comes from knowing that they are responsible for their success or failure in the goals they set for themselves.

Here are some of the qualities of independent learners:

- ► Intend to succeed.
- ► Don't let setbacks deter them.
- ► Know their resources.
    - ▷ Instructor's office hours
    - ▷ Math lab
    - ▷ Student Solutions Manual
    - ▷ Group study
    - ▷ Internet
- ► Don't mistake activity for achievement.
- ► Have a positive attitude.

There are other traits as well. The first step in becoming an independent learner is doing a little self-evaluation and then making a list of traits that you would like to acquire. What skills do you have that align with those of an independent learner? What attributes do you have that keep you from being an independent learner? What qualities would you like to develop that you don't have now?

COMPLETING THE SQUARE

© sxc.hu, abejo, 2009

OBJECTIVES

A Solve quadratic equations by taking the square root of both sides.

B Solve quadratic equations by completing the square.

C Solve for missing parts of right triangles using quadratic equations.

Table 1 is taken from the trail map given to skiers at the Northstar at Tahoe Ski Resort in Lake Tahoe, California. The table gives the length of each chair lift at Northstar, along with the change in elevation from the beginning of the lift to the end of the lift.

Right triangles are good mathematical models for chair lifts. In this section, we will use our knowledge of right triangles, along with the new material developed in the section, to solve problems involving chair lifts and a variety of other examples.

TABLE 1 From the Trail Map for the Northstar at Tahoe Ski Resort

Lift	Lift Information	
	Vertical Rise (feet)	Length (feet)
Big Springs Gondola	480	4,100
Bear Paw Double	120	790
Echo Triple	710	4,890
Aspen Express Quad	900	5,100
Forest Double	1,170	5,750
Lookout Double	960	4,330
Comstock Express Quad	1,250	5,900
Rendezvous Triple	650	2,900
Schaffer Camp Triple	1,860	6,150
Chipmunk Tow Lift	28	280
Bear Cub Tow Lift	120	750

7.1 VIDEOS

The Square Root Method

In this section, we will develop the first of our new methods of solving quadratic equations. The new method is called *completing the square*. Completing the square on a quadratic equation allows us to obtain solutions, regardless of whether the equation can be factored. Before we solve equations by completing the square, we need to learn how to solve equations by taking square roots of both sides.

Consider the equation

$$x^2 = 16$$

We could solve it by writing it in standard form, factoring the left side, and proceeding as we did in Chapter 2. We can shorten our work considerably, however, if we simply notice that x must be either the positive square root of 16 or the negative square root of 16. That is,

$$\text{If} \qquad x^2 = 16$$
$$\text{Then} \qquad x = \sqrt{16} \quad \text{or} \quad x = -\sqrt{16}$$
$$x = 4 \qquad \text{or} \qquad x = -4$$

We can generalize this result as follows.

> **PROPERTY** *Square Root Property for Equations*
> ●
> If $a^2 = b$, where b is a real number, then $a = \sqrt{b}$ or $a = -\sqrt{b}$.

Notation The expression $a = \sqrt{b}$ or $a = -\sqrt{b}$ can be written in shorthand form as $a = \pm\sqrt{b}$. The symbol $\pm$ is read "plus or minus."

We can apply the Square Root Property for Equations to some fairly complicated quadratic equations.

EXAMPLE 1

Solve $(2x - 3)^2 = 25$.

SOLUTION

$$(2x - 3)^2 = 25$$

$2x - 3 = \pm\sqrt{25}$	Square Root Property for Equations
$2x - 3 = \pm 5$	$\sqrt{25} = 5$
$2x = 3 \pm 5$	Add 3 to both sides.
$x = \dfrac{3 \pm 5}{2}$	Divide both sides by 2.

The last equation can be written as two separate statements:

$$x = \frac{3 + 5}{2} \quad \text{or} \quad x = \frac{3 - 5}{2}$$
$$= \frac{8}{2} \qquad\qquad = \frac{-2}{2}$$
$$= 4 \qquad\qquad = -1$$

The solution set is $4, -1$.

Notice that we could have solved the equation in Example 1 by expanding the left side, writing the resulting equation in standard form, and then factoring. The problem would look like this:

$$(2x - 3)^2 = 25 \qquad \text{Original equation}$$

$$4x^2 - 12x + 9 = 25 \qquad \text{Expand the left side.}$$

$$4x^2 - 12x - 16 = 0 \qquad \text{Add } -25 \text{ to each side.}$$

$$4(x^2 - 3x - 4) = 0 \qquad \text{Begin factoring.}$$

$$4(x - 4)(x + 1) = 0 \qquad \text{Factor completely.}$$

$$x - 4 = 0 \quad \text{or} \quad x + 1 = 0 \qquad \text{Set variable factors equal to 0.}$$

$$x = 4 \qquad\qquad x = -1$$

As you can see, solving the equation by factoring leads to the same two solutions.

EXAMPLE 2

Solve for x: $(3x - 1)^2 = -12$

SOLUTION

$$(3x - 1)^2 = -12$$

$$3x - 1 = \pm\sqrt{-12} \qquad \text{Square Root Property for Equations}$$

$$3x - 1 = \pm 2i\sqrt{3} \qquad \sqrt{-12} = 2i\sqrt{3}$$

$$3x = 1 \pm 2i\sqrt{3} \qquad \text{Add 1 to both sides.}$$

$$x = \frac{1 \pm 2i\sqrt{3}}{3} \qquad \text{Divide both sides by 3.}$$

The solution set is $\left\{ \dfrac{1 + 2i\sqrt{3}}{3}, \dfrac{1 - 2i\sqrt{3}}{3} \right\}$.

Both solutions are complex. Here is a check of the first solution:

When $\qquad\qquad\qquad\qquad\qquad\qquad x = \dfrac{1 + 2i\sqrt{3}}{3}$

the equation $\qquad\qquad\qquad\qquad (3x - 1)^2 = -12$

becomes $\qquad\qquad\qquad \left(3 \cdot \dfrac{1 + 2i\sqrt{3}}{3} - 1 \right)^2 \overset{?}{=} -12$

or $\qquad\qquad\qquad\qquad (1 + 2i\sqrt{3} - 1)^2 \overset{?}{=} -12$

$$(2i\sqrt{3})^2 \overset{?}{=} -12$$

$$4 \cdot i^2 \cdot 3 \overset{?}{=} -12$$

$$12(-1) \overset{?}{=} -12$$

$$-12 = -12$$

NOTE We cannot solve the equation in Example 2 by factoring. If we expand the left side and write the resulting equation in standard form, we are left with a quadratic equation that does not factor:

$$(3x - 1)^2 = -12$$

Equation from Example 2

$$9x^2 - 6x + 1 = -12$$

Expand the left side.

$$9x^2 - 6x + 13 = 0$$

Standard form, but not factorable

EXAMPLE 3

Solve $x^2 + 6x + 9 = 12$.

SOLUTION We can solve this equation as we have the equations in Examples 1 and 2 if we first write the left side as $(x + 3)^2$.

$$x^2 + 6x + 9 = 12 \qquad \text{Original equation}$$

$$(x + 3)^2 = 12 \qquad \text{Write } x^2 + 6x + 9 \text{ as } (x + 3)^2$$

$$x + 3 = \pm 2\sqrt{3} \qquad \text{Square Root Property for Equations}$$

$$x = -3 \pm 2\sqrt{3} \qquad \text{Add } -3 \text{ to each side.}$$

We have two irrational solutions: $-3 + 2\sqrt{3}$ and $-3 - 2\sqrt{3}$. What is important about this problem, however, is the fact that the equation was easy to solve because the left side was a perfect square trinomial.

B Method of Completing the Square

The method of completing the square is simply a way of transforming any quadratic equation into an equation of the form found in the preceding three examples.

The key to understanding the method of completing the square lies in recognizing the relationship between the last two terms of any perfect square trinomial whose leading coefficient is 1.

Consider the following list of perfect square trinomials and their corresponding binomial squares:

$$x^2 - 6x + 9 = (x - 3)^2$$

$$x^2 + 8x + 16 = (x + 4)^2$$

$$x^2 - 10x + 25 = (x - 5)^2$$

$$x^2 + 12x + 36 = (x + 6)^2$$

In each case, the leading coefficient is 1. A more important observation comes from noticing the relationship between the linear and constant terms (middle and last terms) in each trinomial. Observe that the constant term in each case is the square of half the coefficient of x in the middle term. For example, in the last expression, the constant term 36 is the square of half of 12, where 12 is the coefficient of x in the middle term. (Notice also that the second terms in all the binomials on the right side are half the coefficients of the middle terms of the trinomials on the left side.) We can use these observations to build our own perfect square trinomials and, in doing so, solve some quadratic equations.

NOTE The equation in Example 4 can be solved quickly by factoring:

$$x^2 - 6x + 5 = 0$$
$$(x - 5)(x - 1) = 0$$
$$x - 5 = 0 \quad \text{or} \quad x - 1 = 0$$
$$x = 5 \qquad\qquad x = 1$$

The reason we didn't solve it by factoring is we want to practice completing the square on some simple equations.

EXAMPLE 4

Solve $x^2 - 6x + 5 = 0$ by completing the square.

SOLUTION We begin by adding -5 to both sides of the equation. We want just $x^2 - 6x$ on the left side so that we can add on our own final term to get a perfect square trinomial:

$$x^2 - 6x + 5 = 0$$

$$x^2 - 6x \qquad = -5 \qquad \text{Add } -5 \text{ to both sides.}$$

Now we can add 9 to both sides and the left side will be a perfect square:

$$x^2 - 6x + 9 = -5 + 9$$

$$(x - 3)^2 = 4$$

The final line is in the form of the equations we solved previously:

$$x - 3 = \pm 2$$

$$x = 3 \pm 2 \qquad \text{Add 3 to both sides.}$$

$$x = 3 + 2 \quad \text{or} \quad x = 3 - 2$$

$$x = 5 \qquad\qquad x = 1$$

The two solutions are 5 and 1.

EXAMPLE 5

Solve by completing the square: $x^2 + 5x - 2 = 0$

SOLUTION   We must begin by adding 2 to both sides. (The left side of the equation, as it is, is not a perfect square because it does not have the correct constant term. We will simply "move" that term to the other side and use our own constant term.)

$$x^2 + 5x = 2 \qquad \text{Add 2 to each side.}$$

We complete the square by adding the square of half the coefficient of the linear term to both sides:

$$x^2 + 5x + \frac{25}{4} = 2 + \frac{25}{4} \qquad \text{Half of 5 is } \frac{5}{2}, \text{ the square of which is } \frac{25}{4}$$

$$\left(x + \frac{5}{2}\right)^2 = \frac{33}{4} \qquad 2 + \frac{25}{4} = \frac{8}{4} + \frac{25}{4} = \frac{33}{4}$$

$$x + \frac{5}{2} = \pm\sqrt{\frac{33}{4}} \qquad \text{Square Root Property for Equations}$$

$$x + \frac{5}{2} = \pm\frac{\sqrt{33}}{2} \qquad \text{Simplify the radical.}$$

$$x = -\frac{5}{2} \pm \frac{\sqrt{33}}{2} \qquad \text{Add } -\frac{5}{2} \text{ to both sides.}$$

$$x = \frac{-5 \pm \sqrt{33}}{2}$$

The solution set is $\left\{ \dfrac{-5 + \sqrt{33}}{2}, \dfrac{-5 - \sqrt{33}}{2} \right\}$.

We can use a calculator to get decimal approximations to these solutions. If $\sqrt{33} \approx 5.74$, then

$$\frac{-5 + 5.74}{2} = 0.37$$

$$\frac{-5 - 5.74}{2} = -5.37$$

EXAMPLE 6

Solve for x: $3x^2 - 8x + 7 = 0$

SOLUTION

$$3x^2 - 8x + 7 = 0$$

$$3x^2 - 8x = -7 \qquad \text{Add } -7 \text{ to both sides.}$$

We cannot complete the square on the left side because the leading coefficient is not 1.

We take an extra step and divide both sides by 3:

$$\frac{3x^2}{3} - \frac{8x}{3} = -\frac{7}{3}$$

$$x^2 - \frac{8}{3}x = -\frac{7}{3}$$

Half of $\frac{8}{3}$ is $\frac{4}{3}$, the square of which is $\frac{16}{9}$:

$$x^2 - \frac{8}{3}x + \frac{16}{9} = -\frac{7}{3} + \frac{16}{9} \qquad \text{Add } \frac{16}{9} \text{ to both sides.}$$

$$\left(x - \frac{4}{3}\right)^2 = -\frac{5}{9} \qquad \text{Simplify right side.}$$

$$x - \frac{4}{3} = \pm\sqrt{-\frac{5}{9}} \qquad \text{Square Root Property for Equations}$$

$$x - \frac{4}{3} = \pm\frac{i\sqrt{5}}{3} \qquad \sqrt{-\frac{5}{9}} = \frac{\sqrt{-5}}{3} = \frac{i\sqrt{5}}{3}$$

$$x = \frac{4}{3} \pm \frac{i\sqrt{5}}{3} \qquad \text{Add } \frac{4}{3} \text{ to both sides.}$$

$$x = \frac{4 \pm i\sqrt{5}}{3}$$

The solution set is $\left\{\dfrac{4 + i\sqrt{5}}{3}, \dfrac{4 - i\sqrt{5}}{3}\right\}$.

HOW TO *Solve a Quadratic Equation by Completing the Square*

To summarize the method used in the preceding two examples, we list the following steps:

Step 1: Write the equation in the form $ax^2 + bx = c$.

Step 2: If the leading coefficient is not 1, divide both sides by the coefficient so that the resulting equation has a leading coefficient of 1. That is, if $a \neq 1$, then divide both sides by a.

Step 3: Add the square of half the coefficient of the linear term to both sides of the equation.

Step 4: Write the left side of the equation as the square of a binomial, and simplify the right side if possible.

Step 5: Apply the Square Root Property for Equations, and solve as usual.

Quadratic Equations and Right Triangles

FACTS FROM GEOMETRY   *More Special Triangles*

The triangles shown in Figures 1 and 2 occur frequently in mathematics.

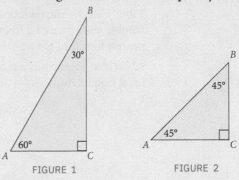

FIGURE 1   FIGURE 2

Note that both of the triangles are right triangles. We refer to the triangle in Figure 1 as a 30°–60°–90° triangle, and the triangle in Figure 2 as a 45°–45°–90° triangle.

EXAMPLE 7

If the shortest side in a 30°–60°–90° triangle is 1 inch, find the lengths of the other two sides.

SOLUTION   In Figure 3, triangle ABC is a 30°–60°–90° triangle in which the shortest side, AC, is 1 inch long. Triangle DBC is also a 30°–60°–90° triangle in which the shortest side, DC, is 1 inch long.

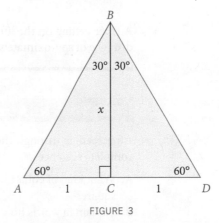

FIGURE 3

Notice that the large triangle ABD is an equilateral triangle because each of its interior angles is 60°. Each side of triangle ABD is 2 inches long. Side AB in triangle ABC is therefore 2 inches. To find the length of side BC, we use the Pythagorean theorem.

$$BC^2 + AC^2 = AB^2$$
$$x^2 + 1^2 = 2^2$$
$$x^2 + 1 = 4$$
$$x^2 = 3$$
$$x = \sqrt{3} \text{ inches}$$

Note that we write only the positive square root because x is the length of a side in a triangle and is therefore a positive number.

EXAMPLE 8

Table 1 in the introduction to this section gives the vertical rise of the Forest Double chair lift as 1,170 feet and the length of the chair lift as 5,750 feet. To the nearest foot, find the horizontal distance covered by a person riding this lift.

SOLUTION   Figure 4 is a model of the Forest Double chair lift. A rider gets on the lift at point A and exits at point B. The length of the lift is AB.

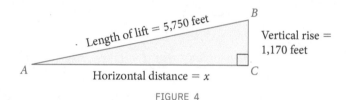

FIGURE 4

To find the horizontal distance covered by a person riding the chair lift, we use the Pythagorean theorem.

$$5{,}750^2 = x^2 + 1{,}170^2 \qquad \text{Pythagorean theorem}$$
$$33{,}062{,}500 = x^2 + 1{,}368{,}900 \qquad \text{Simplify squares.}$$
$$x^2 = 33{,}062{,}500 - 1{,}368{,}900 \qquad \text{Solve for } x^2.$$
$$x^2 = 31{,}693{,}600 \qquad \text{Simplify the right side.}$$
$$x = \sqrt{31{,}693{,}600} \qquad \text{Square Root Property for Equations}$$
$$= 5{,}630 \text{ feet} \qquad \text{To the nearest foot}$$

A rider getting on the lift at point A and riding to point B will cover a horizontal distance of approximately 5,630 feet.

GETTING READY FOR CLASS

After reading through the preceding section, respond in your own words and in complete sentences.

A. What kind of equation do we solve using the method of completing the square?
B. Explain in words how you would complete the square on $x^2 - 16x = 4$.
C. What is the relationship between the shortest side and the longest side in a $30° - 60° - 90°$ triangle?
D. What two expressions together are equivalent to $x = \pm 4$?

Problem Set 7.1

Solve the following equations.

1. $x^2 = 25$ **2.** $x^2 = 16$ **3.** $a^2 = -9$ **4.** $a^2 = -49$

5. $y^2 = \dfrac{3}{4}$ **6.** $y^2 = \dfrac{5}{9}$ **7.** $x^2 + 12 = 0$ **8.** $x^2 + 8 = 0$

9. $4a^2 - 45 = 0$ **10.** $9a^2 - 20 = 0$ **11.** $(2y - 1)^2 = 25$ **12.** $(3y + 7)^2 = 1$

13. $(2a + 3)^2 = -9$ **14.** $(3a - 5)^2 = -49$

15. $(5x + 2)^2 = -8$ **16.** $(6x - 7)^2 = -75$

17. $x^2 + 8x + 16 = -27$ **18.** $x^2 - 12x + 36 = -8$

19. $4a^2 - 12a + 9 = -4$ **20.** $9a^2 - 12a + 4 = -9$

Copy each of the following, and fill in the blanks so the left side of each is a perfect square trinomial. That is, complete the square.

21. $x^2 + 12x + \underline{\hspace{1em}} = (x + \underline{\hspace{1em}})^2$ **22.** $x^2 + 6x + \underline{\hspace{1em}} = (x + \underline{\hspace{1em}})^2$

23. $x^2 - 4x + \underline{\hspace{1em}} = (x - \underline{\hspace{1em}})^2$ **24.** $x^2 - 2x + \underline{\hspace{1em}} = (x - \underline{\hspace{1em}})^2$

25. $a^2 - 10a + \underline{\hspace{1em}} = (a - \underline{\hspace{1em}})^2$ **26.** $a^2 - 8a + \underline{\hspace{1em}} = (a - \underline{\hspace{1em}})^2$

27. $x^2 + 5x + \underline{\hspace{1em}} = (x + \underline{\hspace{1em}})^2$ **28.** $x^2 + 3x + \underline{\hspace{1em}} = (x + \underline{\hspace{1em}})^2$

29. $y^2 - 7y + \underline{\hspace{1em}} = (y - \underline{\hspace{1em}})^2$ **30.** $y^2 - y + \underline{\hspace{1em}} = (y - \underline{\hspace{1em}})^2$

31. $x^2 + \dfrac{1}{2}x + \underline{\hspace{1em}} = (x + \underline{\hspace{1em}})^2$ **32.** $x^2 - \dfrac{3}{4}x + \underline{\hspace{1em}} = (x - \underline{\hspace{1em}})^2$

33. $x^2 + \dfrac{2}{3}x + \underline{\hspace{1em}} = (x + \underline{\hspace{1em}})^2$ **34.** $x^2 - \dfrac{4}{5}x + \underline{\hspace{1em}} = (x - \underline{\hspace{1em}})^2$

Solve each of the following quadratic equations by completing the square.

35. $x^2 + 4x = 12$ **36.** $x^2 - 2x = 8$ **37.** $x^2 + 12x = -27$

38. $x^2 - 6x = 16$ **39.** $a^2 - 2a + 5 = 0$ **40.** $a^2 + 10a + 22 = 0$

41. $y^2 - 8y + 1 = 0$ **42.** $y^2 + 6y - 1 = 0$ **43.** $x^2 - 5x - 3 = 0$

44. $x^2 - 5x - 2 = 0$ **45.** $2x^2 - 4x - 8 = 0$ **46.** $3x^2 - 9x - 12 = 0$

47. $3t^2 - 8t + 1 = 0$ **48.** $5t^2 + 12t - 1 = 0$ **49.** $4x^2 - 3x + 5 = 0$

50. $7x^2 - 5x + 2 = 0$ **51.** $3x^2 + 4x - 1 = 0$ **52.** $2x^2 + 6x - 1 = 0$

53. $2x^2 - 10x = 11$ **54.** $25x^2 - 20x = 1$ **55.** $4x^2 - 10x + 11 = 0$

56. $4x^2 - 6x + 1 = 0$

57. For the equation $x^2 = -9$
  a. Can it be solved by factoring? **b.** Solve it.

58. For the equation $x^2 - 10x + 18 = 0$
  a. Can it be solved by factoring? **b.** Solve it.

59. Solve the equation $x^2 - 6x = 0$
  a. by factoring **b.** by completing the square

60. Solve the equation $x^2 + ax = 0$
  a. by factoring **b.** by completing the square

61. Solve the equation $x^2 + 2x = 35$
  a. by factoring **b.** by completing the square

62. Solve the equation $8x^2 - 10x - 25 = 0$

    a. by factoring   **b.** by completing the square

63. Is $x = -3 + \sqrt{2}$ a solution to $x^2 - 6x = 7$?

64. Is $x = 2 - \sqrt{5}$ a solution to $x^2 - 4x = 1$?

65. Solve each equation.

    a. $5x - 7 = 0$ **b.** $5x - 7 = 8$ **c.** $(5x - 7)^2 = 8$

    d. $\sqrt{5x - 7} = 8$ **e.** $\dfrac{5}{2} - \dfrac{7}{2x} = \dfrac{4}{x}$

66. Solve each equation.

    a. $5x + 11 = 0$ **b.** $5x + 11 = 9$ **c.** $(5x + 11)^2 = 9$

    d. $\sqrt{5x + 11} = 9$ **e.** $\dfrac{5}{3} - \dfrac{11}{3x} = \dfrac{3}{x}$

APPLYING THE CONCEPTS

67. Geometry If the shortest side in a 30°–60°–90° triangle is $\frac{1}{2}$ inch long, find the lengths of the other two sides.

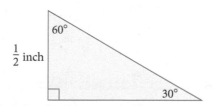

68. Geometry If the length of the longest side of a 30°–60°–90° triangle is x, find the lengths of the other two sides in terms of x.

69. Geometry If the length of the shorter sides of a 45°–45°–90° triangle is 1 inch, find the length of the hypotenuse.

70. Geometry If the length of the shorter sides of a 45°–45°–90° triangle is x, find the length of the hypotenuse in terms of x.

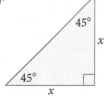

71. Chair Lift Use Table 1 from the introduction to this section to find the horizontal distance covered by a person riding the Bear Paw Double chair lift. Round your answer to the nearest foot.

72. Fermat's Last Theorem As mentioned in a previous chapter, the postage stamp shows Fermat's last theorem, which states that if n is an integer greater than 2, then there are no positive integers x, y, and z that will make the formula $x^n + y^n = z^n$ true. Use the formula $x^n + y^n = z^n$ to

    a. find z if $n = 2$, $x = 6$, and $y = 8$ **b.** find y if $n = 2$, $x = 5$, and $z = 13$

73. **Interest Rate** Suppose a deposit of $3,000 in a savings account that paid an annual interest rate r (compounded yearly) is worth $3,456 after 2 years. Using the formula $A = P(1 + r)^t$, we have

$$3,456 = 3,000(1 + r)^2$$

Solve for r to find the annual interest rate.

74. **Compound Interest** The formula for the amount (A) of money in an account earning r% interest compounded annually for t years after initially depositing a principal (P) is

$$A = P(1 + r)^t$$

Find r if there is $840.50 in an account 2 years after depositing $800.

75. **Compound Interest** The formula for the amount (A) of money in an account earning r% interest compounded annually for t years after initially depositing a principal (P) is

$$A = P(1 + r)^t$$

Find r if there is $10,612.08 in an account 3 years after depositing $10,000.

76. **Planetary Motion** Johannes Kepler discovered the formula $T^2 = R^3/0.000075$ in 1618, where R is the semi-major axis distance in astronomical units (1 AU = the mean distance between Earth and the Sun) and T is the period in Earth days. At the time, Kepler thought that Jupiter had a semi-major axis distance of 5.2 AU. Use this information to find the period of Jupiter (the number of Earth days in Jupiter's orbit).

77. **Planetary Motion** Johannes Kepler discovered the formula $R^3 = 0.000075T^2$ in 1618, where R is the semi-major axis distance in astronomical units (1 AU = the mean distance between Earth and the Sun) and T is the period in Earth days. At the time, Kepler thought that Venus had a period of 224.7 Earth days. Use this information to find the semi-major axis length for Venus.

78. **Distance in a Computer Chip** If the base of a rectangular chip has width 3 cm and a variable length x cm, and the height is twice the length, then find a formula for the distance along the base diagonal.

79. **Distance in a Computer Chip** If the base of a rectangular chip has width 3 cm and a variable length x cm, and the height is twice the length, then find a formula for the distance along the grand diagonal.

80. **Special Triangles** In Figure 5, triangle ABC has angles 45° and 30°, and height x. Find the lengths of sides AB, BC, and AC, in terms of x.

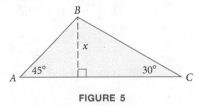

FIGURE 5

81. **Length of an Escalator** An escalator in a department store is made to carry people a vertical distance of 20 feet between floors. How long is the escalator if it makes an angle of 45° with the ground? (See Figure 6.)

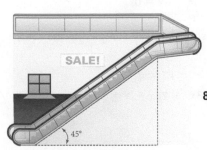

FIGURE 6

82. **Dimensions of a Tent** A two-person tent is to be made so that the height at the center is 4 feet. If the sides of the tent are to meet the ground at an angle of 60° and the tent is to be 6 feet in length, how many square feet of material will be needed to make the tent? (Figure 7; assume that the tent has a floor and is closed at both ends.) Give your answer to the nearest tenth of a square foot.

FIGURE 7

GETTING READY FOR THE NEXT SECTION

Simplify.

83. $49 - 4(6)(-5)$

84. $49 - 4(6)(2)$

85. $(-27)^2 - 4(0.1)(1,700)$

86. $25 - 4(4)(-10)$

87. $-7 + \dfrac{169}{12}$

88. $-7 - \dfrac{169}{12}$

Factor.

89. $27t^3 - 8$

90. $125t^3 + 1$

EXTENDING THE CONCEPTS

Solve for x.

91. $(x + a)^2 + (x - a)^2 = 10a^2$

92. $(ax + 1)^2 + (ax - 1)^2 = 10$

Assume a is a positive number and solve for x by completing the square on x.

93. $x^2 + 2ax = -a^2$

94. $x^2 + 2ax = -4a^2$

95. $x^2 + 2ax = 0$

96. $x^2 + ax = 0$

Assume p and q are positive numbers and solve for x by completing the square on x.

97. $x^2 + px + q = 0$

98. $x^2 - px + q = 0$

99. $3x^2 + px + q = 0$

100. $3x^2 + 2px + q = 0$

7.2

OBJECTIVES

A Solve quadratic equations using the quadratic formula.

B Solve application problems using the quadratic formula.

© iStockphoto/Lemon_tm

If you go on to take a business course or an economics course, you will find yourself spending lots of time with the three expressions that form the mathematical foundation of business: profit, revenue, and cost. Many times these expressions are given as polynomials, the topic of this section. The relationship between the three equations is known as the profit equation:

$$\text{Profit} = \text{Revenue} - \text{Cost}$$

$$P(x) = R(x) - C(x)$$

The table and graph below were produced on a graphing calculator. They give numerical and graphical descriptions of revenue, profit, and cost for a company that manufactures and sells smartphone cases according to the equations

$$R(x) = 11.5x - 0.05x^2 \quad \text{and} \quad C(x) = 200 + 2x$$

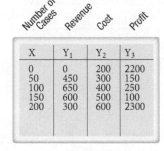

| Number of Cases | Revenue | Cost | Profit |
X	Y_1	Y_2	Y_3
0	0	200	2200
50	450	300	150
100	650	400	250
150	600	500	100
200	300	600	2300

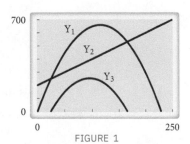

FIGURE 1

By studying the material in this section, you will get a more thorough look at the equations and relationships that are emphasized in business and economics.

In this section, we will use the method of completing the square from the preceding section to derive the quadratic formula. The *quadratic formula* is a very useful tool in mathematics. It allows us to solve all types of quadratic equations.

7.2 VIDEOS

The Quadratic Formula

> **THEOREM** *The Quadratic Theorem*
>
> For any quadratic equation in the form $ax^2 + bx + c = 0$, $a \neq 0$, the two solutions are
>
> $$x = \frac{-b + \sqrt{b^2 - 4ac}}{2a} \quad \text{and} \quad x = \frac{-b - \sqrt{b^2 - 4ac}}{2a}$$

Proof We will prove the quadratic theorem by completing the square on $ax^2 + bx + c = 0$:

$$ax^2 + bx + c = 0$$

$$ax^2 + bx = -c \qquad \text{Add } -c \text{ to both sides.}$$

$$x^2 + \frac{b}{a}x = -\frac{c}{a} \qquad \text{Divide both sides by } a.$$

To complete the square on the left side, we add the square of $\frac{1}{2}$ of $\frac{b}{a}$ to both sides $\left(\frac{1}{2} \text{ of } \frac{b}{a} \text{ is } \frac{b}{2a}\right)$.

$$x^2 + \frac{b}{a}x + \left(\frac{b}{2a}\right)^2 = -\frac{c}{a} + \left(\frac{b}{2a}\right)^2$$

We now simplify the right side as a separate step. We combine the two terms by writing each with the least common denominator $4a^2$:

$$-\frac{c}{a} + \left(\frac{b}{2a}\right)^2 = -\frac{c}{a} + \frac{b^2}{4a^2} = \frac{4a}{4a}\left(\frac{-c}{a}\right) + \frac{b^2}{4a^2} = \frac{-4ac + b^2}{4a^2}$$

It is convenient to write this last expression as

$$\frac{b^2 - 4ac}{4a^2}$$

Continuing with the proof, we have

$$x^2 + \frac{b}{a}x + \left(\frac{b}{2a}\right)^2 = \frac{b^2 - 4ac}{4a^2}$$

$$\left(x + \frac{b}{2a}\right)^2 = \frac{b^2 - 4ac}{4a^2} \qquad \text{Write left side as a binomial square.}$$

$$x + \frac{b}{2a} = \pm\frac{\sqrt{b^2 - 4ac}}{2a} \qquad \text{Square Root Property for Equations}$$

$$x = -\frac{b}{2a} \pm \frac{\sqrt{b^2 - 4ac}}{2a} \qquad \text{Add } -\frac{b}{2a} \text{ to both sides.}$$

$$= \frac{-b \pm \sqrt{b^2 - 4ac}}{2a}$$

Our proof is now complete. What we have is this: If our equation is in the form $ax^2 + bx + c = 0$ (standard form), where $a \neq 0$, the two solutions are always given by the formula

$$x = \frac{-b \pm \sqrt{b^2 - 4ac}}{2a}$$

This formula is known as the *quadratic formula*. If we substitute the coefficients a, b, and c of any quadratic equation in standard form into the formula, we need only perform some basic arithmetic to arrive at the solution set.

EXAMPLE 1

Solve $x^2 - 5x - 6 = 0$ by using the quadratic formula.

SOLUTION To use the quadratic formula, we must make sure the equation is in standard form; identify a, b, and c; substitute them into the formula; and work out the arithmetic.

For the equation $x^2 - 5x - 6 = 0$, $a = 1$, $b = -5$, and $c = -6$:

$$x = \frac{-b \pm \sqrt{b^2 - 4ac}}{2a}$$

$$= \frac{-(-5) \pm \sqrt{(-5)^2 - 4(1)(-6)}}{2(1)}$$

$$= \frac{5 \pm \sqrt{49}}{2}$$

$$= \frac{5 \pm 7}{2}$$

$$x = \frac{5 + 7}{2} \quad \text{or} \quad x = \frac{5 - 7}{2}$$

$$x = \frac{12}{2} \quad\quad\quad\quad x = -\frac{2}{2}$$

$$x = 6 \quad\quad\quad\quad x = -1$$

The two solutions are 6 and -1.

> **NOTE** Whenever the solutions to our quadratic equations turn out to be rational numbers, as in Example 1, it means the original equation could have been solved by factoring. (We didn't solve the equation in Example 1 by factoring because we were trying to get some practice with the quadratic formula.)

EXAMPLE 2

Solve for x: $2x^2 = -4x + 3$.

SOLUTION Before we can identify a, b, and c, we must write the equation in standard form. To do so, we add $4x$ and -3 to each side of the equation.

$$2x^2 = -4x + 3$$

$$2x^2 + 4x - 3 = 0 \qquad \text{Add } 4x \text{ and } -3 \text{ to each side.}$$

Now that the equation is in standard form, we see that $a = 2$, $b = 4$, and $c = -3$. Using the quadratic formula, we have

$$x = \frac{-b \pm \sqrt{b^2 - 4ac}}{2a}$$

$$= \frac{-4 \pm \sqrt{4^2 - 4(2)(-3)}}{2(2)}$$

$$= \frac{-4 \pm \sqrt{40}}{4}$$

$$= \frac{-4 \pm 2\sqrt{10}}{4}$$

We can reduce the final expression in the preceding equation to lowest terms by factoring 2 from the numerator and denominator and then dividing it out.

$$x = \frac{2(-2 \pm \sqrt{10})}{2 \cdot 2}$$

$$= \frac{-2 \pm \sqrt{10}}{2}$$

Our two solutions are $\dfrac{-2 + \sqrt{10}}{2}$ and $\dfrac{-2 - \sqrt{10}}{2}$.

EXAMPLE 3

Solve $x^2 - 6x = -7$.

SOLUTION We begin by writing the equation in standard form:

$$x^2 - 6x = -7$$

$$x^2 - 6x + 7 = 0 \qquad \text{Add 7 to each side.}$$

Using $a = 1$, $b = -6$, and $c = 7$ in the quadratic formula

$$x = \frac{-b \pm \sqrt{b^2 - 4ac}}{2a}$$

we have

$$x = \frac{-(-6) \pm \sqrt{(-6)^2 - 4(1)(7)}}{2(1)}$$

$$= \frac{6 \pm \sqrt{36 - 28}}{2}$$

$$= \frac{6 \pm \sqrt{8}}{2}$$

$$= \frac{6 \pm 2\sqrt{2}}{2}$$

The two terms in the numerator have a 2 in common. We reduce to lowest terms by factoring the 2 from the numerator and then dividing the numerator and denominator by 2.

$$= \frac{2(3 \pm \sqrt{2})}{2}$$

$$= 3 \pm \sqrt{2}$$

The two solutions are $3 + \sqrt{2}$ and $3 - \sqrt{2}$. This time, let's check our solutions in the original equation, $x^2 - 6x = -7$.

Checking $x = 3 + \sqrt{2}$, we have:

$$(3 + \sqrt{2})^2 - 6(3 + \sqrt{2}) \stackrel{?}{=} -7$$

$$9 + 6\sqrt{2} + 2 - 18 - 6\sqrt{2} = -7 \qquad \text{Multiply.}$$

$$11 - 18 + 6\sqrt{2} - 6\sqrt{2} = -7 \qquad \text{Add 9 and 2.}$$

$$-7 + 0 = -7 \qquad \text{Subtraction}$$

$$-7 = -7 \qquad \text{A true statement}$$

Checking $x = 3 - \sqrt{2}$, we have:

$$(3 - \sqrt{2})^2 - 6(3 - \sqrt{2}) \stackrel{?}{=} -7$$

$$9 - 6\sqrt{2} + 2 - 18 + 6\sqrt{2} = -7 \qquad \text{Multiply.}$$

$$11 - 18 - 6\sqrt{2} + 6\sqrt{2} = -7 \qquad \text{Add 9 and 2.}$$

$$-7 + 0 = -7 \qquad \text{Subtraction}$$

$$-7 = -7 \qquad \text{A true statement}$$

As you can see, both solutions yield true statements when used in place of the variable in the original equation.

EXAMPLE 4

Solve for x: $\frac{1}{10}x^2 - \frac{1}{5}x = -\frac{1}{2}$.

SOLUTION It will be easier to apply the quadratic formula if we clear the equation of fractions. Multiplying both sides of the equation by the LCD, 10, gives us:

$$x^2 - 2x = -5$$

Next, we add 5 to both sides to put the equation into standard form:

$$x^2 - 2x + 5 = 0 \qquad \text{Add 5 to both sides.}$$

Applying the quadratic formula with $a = 1$, $b = -2$, and $c = 5$, we have:

$$x = \frac{-(-2) \pm \sqrt{(-2)^2 - 4(1)(5)}}{2(1)} = \frac{2 \pm \sqrt{-16}}{2} = \frac{2 \pm 4i}{2}$$

Dividing the numerator and denominator by 2, we have the two solutions:

$$x = 1 \pm 2i$$

The two solutions are $1 + 2i$ and $1 - 2i$.

EXAMPLE 5

Solve $(2x - 3)(2x - 1) = -4$.

SOLUTION We multiply the binomials on the left side and then add 4 to each side to write the equation in standard form. From there, we identify a, b, and c and apply the quadratic formula.

$$(2x - 3)(2x - 1) = -4$$
$$4x^2 - 8x + 3 = -4 \qquad \text{Multiply binomials on left side.}$$
$$4x^2 - 8x + 7 = 0 \qquad \text{Add 4 to each side.}$$

Placing $a = 4$, $b = -8$, and $c = 7$ in the quadratic formula, we have

$$x = \frac{-(-8) \pm \sqrt{(-8)^2 - 4(4)(7)}}{2(4)}$$

$$= \frac{8 \pm \sqrt{64 - 112}}{8}$$

$$= \frac{8 \pm \sqrt{-48}}{8}$$

$$= \frac{8 \pm 4i\sqrt{3}}{8} \qquad \sqrt{-48} = i\sqrt{48} = i\sqrt{16}\sqrt{3} = 4i\sqrt{3}$$

To reduce this final expression to lowest terms, we factor a 4 from the numerator and then divide the numerator and denominator by 4.

$$= \frac{4(2 \pm i\sqrt{3})}{4 \cdot 2}$$

$$= \frac{2 \pm i\sqrt{3}}{2}$$

NOTE It would be a mistake to try to reduce this final expression further. Sometimes first-year algebra students will try to divide the 2 in the denominator into the 2 in the numerator, which is a mistake. Remember, when we reduce to lowest terms, we do so by dividing the numerator and denominator by any factors they have in common. In this case, 2 is not a factor of the numerator. This expression is in lowest terms.

Although the equation in our next example is not a quadratic equation, we solve it by using both factoring and the quadratic formula.

EXAMPLE 6

Solve $27t^3 - 8 = 0$.

SOLUTION It would be a mistake to add 8 to each side of this equation and then take the cube root of each side because we would lose two of our solutions. Instead, we factor the left side, and then set the factors equal to 0:

$$27t^3 - 8 = 0 \qquad \text{Equation in standard form}$$

$$(3t - 2)(9t^2 + 6t + 4) = 0 \qquad \text{Factor as the difference of two cubes.}$$

$$3t - 2 = 0 \quad \text{or} \quad 9t^2 + 6t + 4 = 0 \qquad \text{Set each factor equal to 0.}$$

The first equation leads to a solution of $t = \frac{2}{3}$. The second equation does not factor, so we use the quadratic formula with $a = 9$, $b = 6$, and $c = 4$:

$$t = \frac{-6 \pm \sqrt{36 - 4(9)(4)}}{2(9)}$$

$$= \frac{-6 \pm \sqrt{36 - 144}}{18}$$

$$= \frac{-6 \pm \sqrt{-108}}{18}$$

$$= \frac{-6 \pm 6i\sqrt{3}}{18} \qquad \sqrt{-108} = i\sqrt{36 \cdot 3} = 6i\sqrt{3}$$

$$= \frac{6(-1 \pm i\sqrt{3})}{6 \cdot 3} \qquad \text{Factor 6 from the numerator and denominator.}$$

$$= \frac{-1 \pm i\sqrt{3}}{3} \qquad \text{Divide out common factor 6.}$$

The three solutions to our original equation are

$$\frac{2}{3}, \qquad \frac{-1 + i\sqrt{3}}{3}, \qquad \text{and} \qquad \frac{-1 - i\sqrt{3}}{3}$$

B Applications

EXAMPLE 7

If an object is thrown downward with an initial velocity of 20 feet per second, the distance $s(t)$, in feet, it travels in t seconds is given by the function $s(t) = 20t + 16t^2$. How long does it take the object to fall 40 feet?

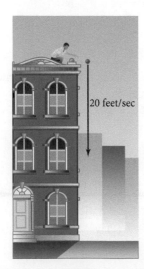

20 feet/sec

SOLUTION We let $s(t) = 40$, and solve for t:

When $\qquad\qquad s(t) = 40$

the function $\qquad\qquad s(t) = 20t + 16t^2$

becomes $\qquad\qquad 40 = 20t + 16t^2$

or $\qquad 16t^2 + 20t - 40 = 0$

$\qquad\qquad 4t^2 + 5t - 10 = 0 \qquad \text{Divide by 4.}$

Using the quadratic formula, we have

$$t = \frac{-5 \pm \sqrt{25 - 4(4)(-10)}}{2(4)}$$

$$= \frac{-5 \pm \sqrt{185}}{8}$$

$$= \frac{-5 + \sqrt{185}}{8} \quad \text{or} \quad t = \frac{-5 - \sqrt{185}}{8}$$

The second solution is impossible because it is a negative number and time t must be positive. It takes

$$t = \frac{-5 + \sqrt{185}}{8} \quad \text{or approximately} \quad \frac{-5 + 13.60}{8} \approx 1.08 \text{ seconds}$$

for the object to fall 40 feet.

The relationship between profit, revenue, and cost is given by the formula

$$P(x) = R(x) - C(x)$$

where $P(x)$ is the profit, $R(x)$ is the total revenue, and $C(x)$ is the total cost of producing and selling x items.

EXAMPLE 8

A company produces and sells copies of an accounting program for home computers. The total weekly cost (in dollars) to produce x copies of the program is $C(x) = 8x + 500$, and the weekly revenue from selling all x copies of the program is $R(x) = 35x - 0.1x^2$. How many copies of the program must be sold each week for the weekly profit to be $1,200?

SOLUTION Substituting the given expressions for $R(x)$ and $C(x)$ in the equation $P(x) = R(x) - C(x)$, we have a polynomial in x that represents the weekly profit $P(x)$:

$$P(x) = R(x) - C(x)$$

$$= 35x - 0.1x^2 - (8x + 500)$$

$$= 35x - 0.1x^2 - 8x - 500$$

$$= -500 + 27x - 0.1x^2$$

Setting this expression equal to 1,200, we have a quadratic equation to solve that gives us the number of copies x that need to be sold each week to bring in a profit of $1,200:

$$1,200 = -500 + 27x - 0.1x^2$$

We can write this equation in standard form by adding the opposite of each term on the right side of the equation to both sides of the equation. Doing so produces the following equation:

$$0.1x^2 - 27x + 1,700 = 0$$

Applying the quadratic formula to this equation with $a = 0.1$, $b = -27$, and $c = 1,700$, we have

$$x = \frac{27 \pm \sqrt{(-27)^2 - 4(0.1)(1,700)}}{2(0.1)}$$

$$= \frac{27 \pm \sqrt{729 - 680}}{0.2}$$

$$= \frac{27 \pm \sqrt{49}}{0.2}$$

$$= \frac{27 \pm 7}{0.2}$$

Writing this last expression as two separate expressions, we have our two solutions:

$$x = \frac{27 + 7}{0.2} \quad \text{or} \quad x = \frac{27 - 7}{0.2}$$

$$= \frac{34}{0.2} \qquad\qquad = \frac{20}{0.2}$$

$$= 170 \qquad\qquad\quad = 100$$

The weekly profit will be $1,200 if the company produces and sells 100 copies or 170 copies of the program.

What is interesting about this last example is that it has rational solutions, meaning it could have been solved by factoring. But looking back at the equation, factoring does not seem like a reasonable method of solution because the coefficients are either very large or very small. So, there are times when using the quadratic formula is a faster method of solution, even though the equation you are solving is factorable.

GETTING READY FOR CLASS

After reading through the preceding section, respond in your own words and in complete sentences.

A. What is the quadratic formula?

B. Under what circumstances should the quadratic formula be applied?

C. When would the quadratic formula result in complex solutions?

D. When will the quadratic formula result in only one solution?

USING TECHNOLOGY *Graphing Calculators*

More About Example 7

We can solve the problem discussed in Example 7 by graphing the function $Y_1 = 20X + 16X^2$ in a window with X from 0 to 2 (because X is taking the place of t and we know t is a positive quantity) and Y from 0 to 50 (because we are looking for X when Y_1 is 40). Graphing Y_1 gives a graph similar to the graph in Figure 2. Using the Zoom and Trace features at $Y_1 = 40$ gives us X = 1.08 to the nearest hundredth, matching the results we obtained by solving the original equation algebraically.

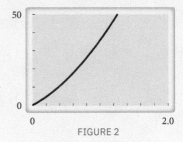

FIGURE 2

More About Example 8

To visualize the functions in Example 8, we set up our calculator this way:

$$Y_1 = 35X - .1X^2 \qquad \text{Revenue function}$$

$$Y_2 = 8X + 500 \qquad \text{Cost function}$$

$$Y_3 = Y_1 - Y_2 \qquad \text{Profit function}$$

Window: X from 0 to 350, Y from 0 to 3,500

Graphing these functions produces graphs similar to the ones shown in Figure 3. The lowest graph is the graph of the profit function. Using the Zoom and Trace features on the lowest graph at $Y_3 = 1,200$ produces two corresponding values of X, 170 and 100, which match the results in Example 8.

We will continue this discussion of the relationship between graphs of functions and solutions to equations in the Using Technology material in the next section.

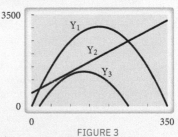

FIGURE 3

Problem Set 7.2

Solve each equation. Use factoring or the quadratic formula, whichever is appropriate. (Try factoring first. If you have any difficulty factoring, then go right to the quadratic formula.)

1. $x^2 + 5x + 6 = 0$ **2.** $x^2 + 5x - 6 = 0$ **3.** $a^2 - 4a + 1 = 0$

4. $a^2 + 4a + 1 = 0$ **5.** $\frac{1}{6}x^2 - \frac{1}{2}x + \frac{1}{3} = 0$ **6.** $\frac{1}{6}x^2 + \frac{1}{2}x + \frac{1}{3} = 0$

7. $\frac{x^2}{2} + 1 = \frac{2x}{3}$ **8.** $\frac{x^2}{2} + \frac{2}{3} = -\frac{2x}{3}$ **9.** $y^2 - 5y = 0$

10. $2y^2 + 10y = 0$ **11.** $30x^2 + 40x = 0$ **12.** $50x^2 - 20x = 0$

13. $\frac{2t^2}{3} - t = -\frac{1}{6}$ **14.** $\frac{t^2}{3} - \frac{t}{2} = -\frac{3}{2}$

15. $0.01x^2 + 0.06x - 0.08 = 0$ **16.** $0.02x^2 - 0.03x + 0.05 = 0$

17. $2x + 3 = -2x^2$ **18.** $2x - 3 = 3x^2$

19. $100x^2 - 200x + 100 = 0$ **20.** $100x^2 - 600x + 900 = 0$

21. $\frac{1}{2}r^2 = \frac{1}{6}r - \frac{2}{3}$ **22.** $\frac{1}{4}r^2 = \frac{2}{5}r + \frac{1}{10}$

23. $(x - 3)(x - 5) = 1$ **24.** $(x - 3)(x + 1) = -6$

25. $(x + 3)^2 + (x - 8)(x - 1) = 16$ **26.** $(x - 4)^2 + (x + 2)(x + 1) = 9$

27. $\frac{x^2}{3} - \frac{5x}{6} = \frac{1}{2}$ **28.** $\frac{x^2}{6} + \frac{5}{6} = -\frac{x}{3}$

Multiply both sides of each equation by its LCD. Then solve the resulting equation.

29. $\frac{1}{x + 1} - \frac{1}{x} = \frac{1}{2}$ **30.** $\frac{1}{x + 1} + \frac{1}{x} = \frac{1}{3}$ **31.** $\frac{1}{y - 1} + \frac{1}{y + 1} = 1$

32. $\frac{2}{y + 2} + \frac{3}{y - 2} = 1$ **33.** $\frac{1}{x + 2} + \frac{1}{x + 3} = 1$ **34.** $\frac{1}{x + 3} + \frac{1}{x + 4} = 1$

35. $\frac{6}{r^2 - 1} - \frac{1}{2} = \frac{1}{r + 1}$ **36.** $2 + \frac{5}{r - 1} = \frac{12}{(r - 1)^2}$

Solve each equation. In each case, you will have three solutions.

37. $x^3 - 8 = 0$ **38.** $x^3 - 27 = 0$ **39.** $8a^3 + 27 = 0$

40. $27a^3 + 8 = 0$ **41.** $125t^3 - 1 = 0$ **42.** $64t^3 + 1 = 0$

Each of the following equations has three solutions. Look for the greatest common factor, then use the quadratic formula to find all solutions.

43. $2x^3 + 2x^2 + 3x = 0$ **44.** $6x^3 - 4x^2 + 6x = 0$ **45.** $3y^4 = 6y^3 - 6y^2$

46. $4y^4 = 16y^3 - 20y^2$ **47.** $6t^5 + 4t^4 = -2t^3$ **48.** $8t^5 + 2t^4 = -10t^3$

49. Which two of the expressions below are equivalent?

     a. $\frac{6 + 2\sqrt{3}}{4}$ **b.** $\frac{3 + \sqrt{3}}{2}$ **c.** $6 + \frac{\sqrt{3}}{2}$

50. Which two of the expressions below are equivalent?

   a. $\dfrac{8 - 4\sqrt{2}}{4}$ **b.** $2 - 4\sqrt{3}$ **c.** $2 - \sqrt{2}$

51. Solve $3x^2 - 5x = 0$

   a. by factoring   **b.** by the quadratic formula

52. Solve $3x^2 + 23x - 70 = 0$

   a. by factoring   **b.** by the quadratic formula

53. Can the equation $x^2 - 4x + 7 = 0$ be solved by factoring? Solve it.

54. Can the equation $x^2 = 5$ be solved by factoring? Solve it.

55. Is $x = -1 + i$ a solution to $x^2 + 2x = -2$?

56. Is $x = 2 + 2i$ a solution to $(x - 2)^2 = -4$?

APPLYING THE CONCEPTS

57. Falling Object An object is thrown downward with an initial velocity of 5 feet per second. The relationship between the distance s it travels and time t is given by $s = 5t + 16t^2$. How long does it take the object to fall 74 feet?

58. Coin Toss A coin is tossed upward with an initial velocity of 32 feet per second from a height of 16 feet above the ground. The equation giving the object's height h at any time t is $h = 16 + 32t - 16t^2$. Does the object ever reach a height of 32 feet?

59. Profit The total cost (in dollars) for a company to manufacture and sell x items per week is $C = 60x + 300$, whereas the revenue brought in by selling all x items is $R = 100x - 0.5x^2$. How many items must be sold to obtain a weekly profit of \$300?

60. Profit Suppose a company manufactures and sells x picture frames each month with a total cost of $C = 1{,}200 + 3.5x$ dollars. If the revenue obtained by selling x frames is $R = 9x - 0.002x^2$, find the number of frames the company must sell each month if its monthly profit is to be \$2,300.

GETTING READY FOR THE NEXT SECTION

Find the value of $b^2 - 4ac$ when

61. $a = 1, b = -3, c = -40$ **62.** $a = 2, b = 3, c = 4$

63. $a = 4, b = 12, c = 9$ **64.** $a = -3, b = 8, c = -1$

Solve.

65. $k^2 - 144 = 0$ **66.** $36 - 20k = 0$

Multiply.

67. $(x - 3)(x + 2)$ **68.** $(t - 5)(t + 5)$

69. $(x - 3)(x - 3)(x + 2)$ **70.** $(t - 5)(t + 5)(t - 3)$

EXTENDING THE CONCEPTS

So far, all the equations we have solved have had coefficients that were rational numbers. Here are some equations that have irrational coefficients and some that have complex coefficients. Solve each equation. (Remember, $i^2 = -1$.)

71. $x^2 + \sqrt{3}x - 6 = 0$

72. $x^2 - \sqrt{5}x - 5 = 0$

73. $\sqrt{2}x^2 + 2x - \sqrt{2} = 0$

74. $\sqrt{7}x^2 + 2\sqrt{2}x - \sqrt{7} = 0$

75. $x^2 + ix + 2 = 0$

76. $x^2 + 3ix - 2 = 0$

77. $ix^2 + 3x + 4i = 0$

78. $4ix^2 + 5x + 9i = 0$

OBJECTIVES

A Find the number and types of solutions to a quadratic equation by using the discriminant.

B Find an unknown constant in a quadratic equation so that there is exactly one solution.

C Find an equation given its solutions and multiplicities.

© iStockphoto/fredrocko

7.3

A shortstop can jump straight up with an initial velocity of 2 feet per second. If he wants to catch a line drive hit over his head, he needs to be able to jump 5 feet off the ground. The equation that models the jump is $h = -16t^2 + 2t$. Can he make the out? Problems like this can be answered without having to name the solutions. We can use an expression called the *discriminant* to determine if there are any real solutions to this equation. In this section, we will define the discriminant and use it to find the number and types of solutions an equation may have. We will also use the zero-factor property to build equations from their solutions, which may help us decide the initial velocity the shortstop will need to catch the line drive.

A The Discriminant

The quadratic formula

$$x = \frac{-b \pm \sqrt{b^2 - 4ac}}{2a}$$

gives the solutions to any quadratic equation in standard form. There are times, when working with quadratic equations, that it is important only to know what kind of solutions the equation has.

> **DEFINITION** *discriminant*
> ·······································
> The expression under the radical in the quadratic formula is called the *discriminant*:
>
> $$\text{Discriminant} = D = b^2 - 4ac$$

The discriminant indicates the number and type of solutions to a quadratic equation when the original equation has integer coefficients. For example, if we were to use the quadratic formula to solve the equation $2x^2 + 2x + 3 = 0$, we would find the discriminant to be

$$b^2 - 4ac = 2^2 - 4(2)(3) = -20$$

Because the discriminant appears under a square root symbol, we have the square root of a negative number in the quadratic formula. Our solutions would therefore be complex numbers. Similarly, if the discriminant were 0, the quadratic formula would yield

$$x = \frac{-b \pm \sqrt{0}}{2a} = \frac{-b \pm 0}{2a} = \frac{-b}{2a}$$

and the equation would have one rational solution, the number $\frac{-b}{2a}$.

The following table gives the relationship between the discriminant and the type of solutions to the equation.

For the equation $ax^2 + bx + c = 0$ where a, b, and c are integers and $a \neq 0$:

If the Discriminant $b^2 - 4ac$ is	Then the Equation Will Have
Negative	Two complex solutions containing i
Zero	One rational solution
A positive number that is also a perfect square	Two rational solutions
A positive number that is not a perfect square	Two irrational solutions

In the second and third cases, when the discriminant is 0 or a positive perfect square, the solutions are rational numbers. The quadratic equations in these two cases are the ones that can be factored.

EXAMPLES

For each equation, give the number and kind of solutions.

1. $x^2 - 3x - 40 = 0$

SOLUTION Using $a = 1$, $b = -3$, and $c = -40$ in $b^2 - 4ac$, we have

$$(-3)^2 - 4(1)(-40) = 9 + 160 = 169.$$

The discriminant is a perfect square. The equation therefore has two rational solutions.

2. $2x^2 - 3x + 4 = 0$

SOLUTION Using $a = 2$, $b = -3$, and $c = 4$, we have

$$b^2 - 4ac = (-3)^2 - 4(2)(4) = 9 - 32 = -23$$

The discriminant is negative, implying the equation has two complex solutions that contain i.

3. $4x^2 - 12x + 9 = 0$

SOLUTION Using $a = 4$, $b = -12$, and $c = 9$, the discriminant is

$$b^2 - 4ac = (-12)^2 - 4(4)(9) = 144 - 144 = 0$$

Because the discriminant is 0, the equation will have one rational solution.

4. $x^2 + 6x = 8$

SOLUTION We must first put the equation in standard form by adding -8 to each side. If we do so, the resulting equation is

$$x^2 + 6x - 8 = 0$$

Now we identify a, b, and c as 1, 6, and -8, respectively:

$$b^2 - 4ac = 6^2 - 4(1)(-8) = 36 + 32 = 68$$

The discriminant is a positive number, but not a perfect square. The equation will therefore have two irrational solutions.

B Finding an Unknown Constant

EXAMPLE 5

Find an appropriate k so that the equation $4x^2 - kx = -9$ has exactly one rational solution.

SOLUTION   We begin by writing the equation in standard form:

$$4x^2 - kx + 9 = 0$$

Using $a = 4$, $b = -k$, and $c = 9$, we have

$$b^2 - 4ac = (-k)^2 - 4(4)(9)$$
$$= k^2 - 144$$

An equation has exactly one rational solution when the discriminant is 0. We set the discriminant equal to 0 and solve:

$$k^2 - 144 = 0$$
$$k^2 = 144$$
$$k = \pm\, 12$$

Choosing k to be 12 or -12 will result in an equation with one rational solution.

C Building Equations from Their Solutions

Suppose we know that the solutions to an equation are $x = 3$ and $x = -2$. We can find equations with these solutions by using the zero-factor property. First, let's write our solutions as equations with 0 on the right side:

If	$x = 3$	First solution
then	$x - 3 = 0$	Add -3 to each side.
and if	$x = -2$	Second solution
then	$x + 2 = 0$	Add 2 to each side.

Now, because both $x - 3$ and $x + 2$ are 0, their product must be 0 also. We can therefore write

$(x - 3)(x + 2) = 0$	Zero-factor property
$x^2 - x - 6 = 0$	Multiply out the left side.

Many other equations have 3 and -2 as solutions. For example, any constant multiple of $x^2 - x - 6 = 0$, such as $5x^2 - 5x - 30 = 0$, also has 3 and -2 as solutions. Similarly, any equation built from positive integer powers of the factors $x - 3$ and $x + 2$ will also have 3 and -2 as solutions. One such equation is

$$(x - 3)^2(x + 2) = 0$$
$$(x^2 - 6x + 9)(x + 2) = 0$$
$$x^3 - 4x^2 - 3x + 18 = 0$$

In mathematics, we distinguish between the solutions to this last equation and those to the equation $x^2 - x - 6 = 0$ by saying $x = 3$ is a solution of *multiplicity* 2 in the equation $x^3 - 4x^2 - 3x + 18 = 0$, and a solution of *multiplicity* 1 in the equation $x^2 - x - 6 = 0$.

EXAMPLE 6

Find an equation that has solutions $t = 5$, $t = -5$, and $t = 3$.

SOLUTION First, we use the given solutions to write equations that have 0 on their right sides:

$$\text{If} \qquad t = 5 \qquad t = -5 \qquad t = 3$$

$$\text{then} \qquad t - 5 = 0 \qquad t + 5 = 0 \qquad t - 3 = 0$$

Since $t - 5$, $t + 5$, and $t - 3$ are all 0, their product is also 0 by the zero-factor property. An equation with solutions of 5, -5, and 3 is

$(t - 5)(t + 5)(t - 3) = 0$	Zero-factor property
$(t^2 - 25)(t - 3) = 0$	Multiply first two binomials.
$t^3 - 3t^2 - 25t + 75 = 0$	Complete the multiplication.

The last line, $t^3 - 3t^2 - 25t + 75 = 0$, gives us an equation with solutions of 5, -5, and 3. Remember, many other equations have these same solutions.

USING TECHNOLOGY   *Order of Operations*

Solving Equations

Now that we have explored the relationship between equations and their solutions, we can look at how a graphing calculator can be used in the solution process. To begin, let's solve the equation $x^2 = x + 2$ using techniques from algebra: writing it in standard form, factoring, and then setting each factor equal to 0.

$x^2 - x - 2 = 0$	Standard form
$(x - 2)(x + 1) = 0$	Factor.
$x - 2 = 0$ or $x + 1 = 0$	Set each factor equal to 0.
$x = 2 \qquad\qquad x = -1$	Solve.

Our original equation, $x^2 = x + 2$, has two solutions: $x = 2$ and $x = -1$. To solve the equation using a graphing calculator, we need to associate it with an equation (or equations) in two variables. One way to do this is to associate the left side with the equation $y = x^2$ and the right side of the equation with $y = x + 2$. To do so, we set up the functions list in our calculator this way:

$$Y_1 = X^2$$

$$Y_2 = X + 2$$

Window: X from -5 to 5, Y from -5 to 5

Graphing these functions in this window will produce a graph similar to the one shown in Figure 1.

If we use the Trace feature to find the coordinates of the points of intesection, we find that the two curves intersect at $(-1, 1)$ and $(2, 4)$. We note that the x-coordinates of these two points match the solutions to the equation $x^2 = x + 2$, which we found using algebraic techniques. This makes sense because if two graphs intersect at a point (x, y), then the coordinates of that point satisfy both equations. If a point (x, y) satisfies both $y = x^2$ and $y = x + 2$, then for that particular point, $x^2 = x + 2$. From this, we conclude that the x-coordinates of the points of intersection are solutions to our original equation. Here is a summary of what we have discovered:

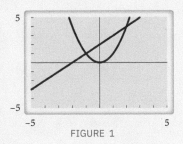

FIGURE 1

Conclusion 1   If the graphs of two functions $y = f(x)$ and $y = g(x)$ intersect in the coordinate plane, then the x-coordinates of the points of intersection are solutions to the equation $f(x) = g(x)$.

A second method of solving our original equation, $x^2 = x + 2$, graphically requires the use of one function instead of two. To begin, we write the equation in standard form as $x^2 - x - 2 = 0$. Next, we graph the function $y = x^2 - x - 2$. The x-intercepts of the graph are the points with y-coordinates of 0. They therefore satisfy the equation $0 = x^2 - x - 2$, which is equivalent to our original equation. The graph in Figure 2 shows $Y_1 = X^2 - X - 2$ in a window with X from -5 to 5 and Y from -5 to 5.

Using the Trace feature, we find that the x-intercepts of the graph are $x = -1$ and $x = 2$, which match the solutions to our original equation $x^2 = x + 2$. We can summarize the relationship between solutions to an equation and the intercepts of its associated graph this way:

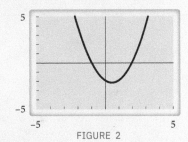

FIGURE 2

Conclusion 2   If $y = f(x)$ is a function, then any x-intercept on the graph of $y = f(x)$ is a solution to the equation $f(x) = 0$.

EXAMPLE 7

Find an equation with solutions $x = -\dfrac{2}{3}$ and $x = \dfrac{4}{5}$.

SOLUTION   The solution $x = -\frac{2}{3}$ can be rewritten as $3x + 2 = 0$ as follows:

$$x = -\frac{2}{3} \qquad \text{The first solution}$$

$$3x = -2 \qquad \text{Multiply each side by 3.}$$

$$3x + 2 = 0 \qquad \text{Add 2 to each side.}$$

Similarly, the solution $x = \frac{4}{5}$ can be rewritten as $5x - 4 = 0$:

$$x = \frac{4}{5} \qquad \text{The second solution}$$

$$5x = 4 \qquad \text{Multiply each side by 5.}$$

$$5x - 4 = 0 \qquad \text{Add } -4 \text{ to each side.}$$

Because both $3x + 2$ and $5x - 4$ are 0, their product is 0 also, giving us the equation we are looking for:

$$(3x + 2)(5x - 4) = 0 \qquad \text{Zero-factor property}$$
$$15x^2 - 2x - 8 = 0 \qquad \text{Multiplication}$$

GETTING READY FOR CLASS

After reading through the preceding section, respond in your own words and in complete sentences.

A. What is the discriminant?

B. What kind of solutions do we get to a quadratic equation when the discriminant is negative?

C. What does it mean for a solution to have multiplicity 3?

D. When will a quadratic equation have two rational solutions?

Problem Set 7.3

Use the discriminant to find the number and kind of solutions for each of the following equations.

1. $x^2 - 6x + 5 = 0$

2. $x^2 - x - 12 = 0$

3. $4x^2 - 4x = -1$

4. $9x^2 + 12x = -4$

5. $x^2 + x - 1 = 0$

6. $x^2 - 2x + 3 = 0$

7. $2y^2 = 3y + 1$

8. $3y^2 = 4y - 2$

9. $x^2 - 9 = 0$

10. $4x^2 - 81 = 0$

11. $5a^2 - 4a = 5$

12. $3a = 4a^2 - 5$

Determine k so that each of the following has exactly one rational solution.

13. $x^2 - kx + 25 = 0$

14. $x^2 + kx + 25 = 0$

15. $x^2 = kx - 36$

16. $x^2 = kx - 49$

17. $4x^2 - 12x + k = 0$

18. $9x^2 + 30x + k = 0$

19. $kx^2 - 40x = 25$

20. $kx^2 - 2x = -1$

21. $3x^2 - kx + 2 = 0$

22. $5x^2 + kx + 1 = 0$

For each of the following problems, find an equation that has the given solutions.

23. $x = 5, x = 2$

24. $x = -5, x = -2$

25. $t = -3, t = 6$

26. $t = -4, t = 2$

27. $y = 2, y = -2, y = 4$

28. $y = 1, y = -1, y = 3$

29. $x = \dfrac{1}{2}, x = 3$

30. $x = \dfrac{1}{3}, x = 5$

31. $t = -\dfrac{3}{4}, t = 3$

32. $t = -\dfrac{4}{5}, t = 2$

33. $x = 3, x = -3, x = \dfrac{5}{6}$

34. $x = 5, x = -5, x = \dfrac{2}{3}$

35. $a = -\dfrac{1}{2}, a = \dfrac{3}{5}$

36. $a = -\dfrac{1}{3}, a = \dfrac{4}{7}$

37. $x = -\dfrac{2}{3}, x = \dfrac{2}{3}, x = 1$

38. $x = -\dfrac{4}{5}, x = \dfrac{4}{5}, x = -1$

39. $x = 2, x = -2, x = 3, x = -3$

40. $x = 1, x = -1, x = 5, x = -5$

41. $x = \sqrt{7}, x = -\sqrt{7}$

42. $x = -\sqrt{3}, x = \sqrt{3}$

43. $x = 5i, x = -5i$

44. $x = -2i, x = 2i$

45. $x = 1 + i, x = 1 - i$

46. $x = 2 + 3i, x = 2 - 3i$

47. $x = -2 - 3i, x = -2 + 3i$

48. $x = -1 + i, x = -1 - i$

49. Find an equation that has a solution of $x = 3$ of multiplicity 1 and a solution $x = -5$ of multiplicity 2.

50. Find an equation that has a solution of $x = 5$ of multiplicity 1 and a solution $x = -3$ of multiplicity 2.

51. Find an equation that has solutions $x = 3$ and $x = -3$, both of multiplicity 2.

52. Find an equation that has solutions $x = 4$ and $x = -4$, both of multiplicity 2.

53. Find all solutions to $x^3 + 6x^2 + 11x + 6 = 0$, if $x = -3$ is one of its solutions.

54. Find all solutions to $x^3 + 10x^2 + 29x + 20 = 0$, if $x = -4$ is one of its solutions.

55. One solution to $y^3 + 5y^2 - 2y - 24 = 0$ is $y = -3$. Find all solutions.

56. One solution to $y^3 + 3y^2 - 10y - 24 = 0$ is $y = -2$. Find all solutions.

57. If $x = 3$ is one solution to $x^3 - 5x^2 + 8x = 6$, find the other solutions.

58. If $x = 2$ is one solution to $x^3 - 6x^2 + 13x = 10$, find the other solutions.

59. Find all solutions to $t^3 = 13t^2 - 65t + 125$, if $t = 5$ is one of the solutions.

60. Find all solutions to $t^3 = 8t^2 - 25t + 26$, if $t = 2$ is one of the solutions.

GETTING READY FOR THE NEXT SECTION

Simplify.

61. $(x + 3)^2 - 2(x + 3) - 8$

62. $(x - 2)^2 - 3(x - 2) - 10$

63. $(2a - 3)^2 - 9(2a - 3) + 20$

64. $(3a - 2)^2 + 2(3a - 2) - 3$

65. $2(4a + 2)^2 - 3(4a + 2) - 20$

66. $6(2a + 4)^2 - (2a + 4) - 2$

Solve.

67. $x^2 = \dfrac{1}{4}$

68. $x^2 = -2$

69. $\sqrt{x} = -3$

70. $\sqrt{x} = 2$

71. $x + 3 = 4$

72. $x + 3 = -2$

73. $y^2 - 2y - 8 = 0$

74. $y^2 + y - 6 = 0$

75. $4y^2 + 7y - 2 = 0$

76. $6x^2 - 13x - 5 = 0$

EXTENDING THE CONCEPTS

77. Find all solutions to $x^4 + x^3 - x^2 + x - 2 = 0$, if $x = -2$ is one solution.

78. Find all solutions to $x^4 - x^3 + 2x^2 - 4x - 8 = 0$, if $x = 2$ is one solution.

79. Find all solutions to $x^3 + 3ax^2 + 3a^2x + a^3 = 0$, if $x = -a$ is one solution.

80. If $x = -2a$ is one solution to $x^3 + 6ax^2 + 12a^2x + 8a^3 = 0$, find the other solutions.

Find all solutions to the following equations. Solve using algebra and by graphing. If rounding is necessary, round to the nearest hundredth.

81. $x^2 = 4x + 5$

82. $4x^2 = 8x + 5$

83. $x^2 - 1 = 2x$

84. $4x^2 - 1 = 4x$

Find all solutions to each equation. If rounding is necessary, round to the nearest hundredth.

85. $2x^3 - x^2 - 2x + 1 = 0$

86. $3x^3 - 2x^2 - 3x + 2 = 0$

87. $2x^3 + 2 = x^2 + 4x$

88. $3x^3 - 9x = 2x^2 - 6$

Each of the following equations has only one real solution. Find it by graphing. Then use long division and the quadratic formula to find the remaining solutions.

89. $3x^3 - 8x^2 + 10x - 4 = 0$

90. $10x^3 + 6x^2 + x - 2 = 0$

OBJECTIVES

A Solve equations that are reducible to a quadratic equation.

B Solve application problems using equations quadratic in form.

© iStockphoto/Sportlibrary

7.4

An Olympic platform diver jumps off a platform of height h with an upward velocity of about 8 feet per second. The time it takes her to reach the water is modeled by the equation $16t^2 - 8t - h = 0$. We can use this equation to calculate how long before she passes the platform. We can also use it to determine how long until she reaches the water. Let's put our knowledge of quadratic equations to work to solve a variety of equations, which can be converted to quadratic equations. The next few examples will illustrate these techniques.

Equations Quadratic in Form

EXAMPLE 1

Solve $(x + 3)^2 - 2(x + 3) - 8 = 0$.

SOLUTION We can see that this equation is quadratic in form by replacing $x + 3$ with another variable, say, y. Replacing $x + 3$ with y, we have

$$y^2 - 2y - 8 = 0$$

We can solve this equation by factoring the left side and then setting each factor equal to 0.

$$y^2 - 2y - 8 = 0$$

$$(y - 4)(y + 2) = 0 \qquad \text{Factor.}$$

$$y - 4 = 0 \quad \text{or} \quad y + 2 = 0 \quad \text{Set factors to 0.}$$

$$y = 4 \qquad\qquad y = -2$$

Because our original equation was written in terms of the variable x, we want our solutions in terms of x also. Replacing y with $x + 3$ and then solving for x, we have

$$x + 3 = 4 \quad \text{or} \quad x + 3 = -2$$

$$x = 1 \qquad\qquad x = -5$$

The solutions to our original equation are 1 and -5.

7.4 VIDEOS

The method we have just shown lends itself well to other types of equations that are quadratic in form, as we will see. In this example, however, there is another method that works just as well. Let's solve our original equation again, but this time, let's begin by expanding $(x + 3)^2$ and $2(x + 3)$.

$$(x + 3)^2 - 2(x + 3) - 8 = 0$$

$$x^2 + 6x + 9 - 2x - 6 - 8 = 0 \qquad \text{Multiply.}$$

$$x^2 + 4x - 5 = 0 \qquad \text{Combine similar terms.}$$

$$(x - 1)(x + 5) = 0 \qquad \text{Factor.}$$

$$x - 1 = 0 \quad \text{or} \quad x + 5 = 0 \qquad \text{Set factors to 0.}$$

$$x = 1 \qquad\qquad x = -5$$

As you can see, either method produces the same result.

EXAMPLE 2

Solve $4x^4 + 7x^2 = 2$.

SOLUTION   This equation is quadratic in x^2. We can make it easier to look at by using the substitution $y = x^2$. (The choice of the letter y is arbitrary. We could just as easily use the substitution $m = x^2$.) Making the substitution $y = x^2$ and then solving the resulting equation, we have

$$4y^2 + 7y = 2$$

$$4y^2 + 7y - 2 = 0 \qquad \text{Standard form}$$

$$(4y - 1)(y + 2) = 0 \qquad \text{Factor.}$$

$$4y - 1 = 0 \quad \text{or} \quad y + 2 = 0 \qquad \text{Set factors to 0.}$$

$$y = \frac{1}{4} \qquad\qquad y = -2$$

Now we replace y with x^2 to solve for x:

$$x^2 = \frac{1}{4} \quad \text{or} \quad x^2 = -2$$

$$x = \pm\sqrt{\frac{1}{4}} \qquad x = \pm\sqrt{-2} \qquad \text{Square Root Property for Equations}$$

$$x = \pm\frac{1}{2} \qquad\qquad = \pm i\sqrt{2}$$

The solution set is $\left\{ \frac{1}{2}, -\frac{1}{2}, i\sqrt{2}, -i\sqrt{2} \right\}$.

EXAMPLE 3

Solve for x: $x + \sqrt{x} - 6 = 0$

SOLUTION   To see that this equation is quadratic in form, we have to notice that $(\sqrt{x})^2 = x$. That is, the equation can be rewritten as

$$(\sqrt{x})^2 + \sqrt{x} - 6 = 0$$

Replacing $\sqrt{x}$ with y and solving as usual, we have

$$y^2 + y - 6 = 0$$
$$(y + 3)(y - 2) = 0$$
$$y + 3 = 0 \quad \text{or} \quad y - 2 = 0$$
$$y = -3 \qquad\qquad y = 2$$

Again, to find x, we replace y with x and solve:

$$\sqrt{x} = -3 \quad \text{or} \quad \sqrt{x} = 2$$
$$x = 9 \qquad\qquad x = 4 \quad \text{Square both sides of each equation.}$$

Because we squared both sides of each equation, we have the possibility of obtaining extraneous solutions. We have to check both solutions in our original equation.

When	$x = 9$	When	$x = 4$
the equation	$x + \sqrt{x} - 6 = 0$	the equation	$x + \sqrt{x} - 6 = 0$
becomes	$9 + \sqrt{9} - 6 \overset{?}{=} 0$	becomes	$4 + \sqrt{4} - 6 \overset{?}{=} 0$
	$9 + 3 - 6 \overset{?}{=} 0$		$4 + 2 - 6 \overset{?}{=} 0$
	$6 \neq 0$		$0 = 0$

This means 9 is extraneous.

This means 4 is a solution.

The only solution to the equation $x + \sqrt{x} - 6 = 0$ is $x = 4$.

We should note here that the two possible solutions, 9 and 4, to the equation in Example 3 can be obtained by another method. Instead of substituting for x, we can isolate it on one side of the equation and then square both sides to clear the equation of radicals.

$$x + \sqrt{x} - 6 = 0$$
$$\sqrt{x} = -x + 6 \qquad\qquad \text{Isolate } \sqrt{x}.$$
$$x = x^2 - 12x + 36 \qquad\qquad \text{Square both sides.}$$
$$0 = x^2 - 13x + 36 \qquad\qquad \text{Add } -x \text{ to both sides.}$$
$$0 = (x - 4)(x - 9) \qquad\qquad \text{Factor.}$$
$$x - 4 = 0 \quad \text{or} \quad x - 9 = 0$$
$$x = 4 \qquad\qquad x = 9$$

We obtain the same two possible solutions. Because we squared both sides of the equation to find them, we would have to check each one in the original equation. As was the case in Example 3, only $x = 4$ is a solution; $x = 9$ is extraneous.

Ⓑ Applications

EXAMPLE 4

If an object is tossed into the air with an upward velocity of 12 feet per second from the top of a building h feet high, the time it takes for the object to hit the ground is given by the formula

$$16t^2 - 12t - h = 0$$

Solve this formula for t.

SOLUTION The formula is in standard form and is quadratic in t. The coefficients a, b, and c that we need to apply to the quadratic formula are $a = 16$, $b = -12$, and $c = -h$. Substituting these quantities into the quadratic formula, we have

$$t = \frac{12 \pm \sqrt{144 - 4(16)(-h)}}{2(16)}$$

$$= \frac{12 \pm \sqrt{144 + 64h}}{32}$$

We can factor the perfect square 16 from the two terms under the radical and simplify our radical somewhat:

$$t = \frac{12 \pm \sqrt{16(9 + 4h)}}{32}$$

$$= \frac{12 \pm 4\sqrt{9 + 4h}}{32}$$

Now we can reduce to lowest terms by factoring a 4 from the numerator and denominator:

$$t = \frac{4(3 \pm \sqrt{9 + 4h})}{4 \cdot 8}$$

$$= \frac{3 \pm \sqrt{9 + 4h}}{8}$$

If we were given a value of h, we would find that one of the solutions to this last formula would be a negative number. Because time is always measured in positive units, we wouldn't use that solution.

USING TECHNOLOGY   *Graphing Calculators*

More About Example 1

As we mentioned before, algebraic expressions entered into a graphing calculator do not have to be simplified to be evaluated. This fact also applies to equations. We can graph the equation $y = (x + 3)^2 - 2(x + 3) - 8$ to assist us in solving the equation in Example 1. The graph is shown in Figure 1. Using the Zoom and Trace features at the x-intercepts gives us $x = 1$ and $x = -5$ as the solutions to the equation $0 = (x + 3)^2 - 2(x + 3) - 8$.

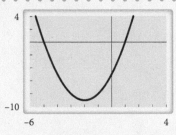

FIGURE 1

More About Example 2

Figure 2 shows the graph of $y = 4x^4 + 7x^2 - 2$. As we expect, the x-intercepts give the real number solutions to the equation $0 = 4x^4 + 7x^2 - 2$. The complex solutions do not appear on the graph.

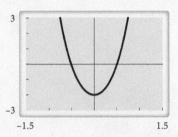

FIGURE 2

More About Example 3

In solving the equation in Example 3, we found that one of the possible solutions was an extraneous solution. If we solve the equation $x + \sqrt{x} - 6 = 0$ by graphing the function $y = x + \sqrt{x} - 6$, we find that the extraneous solution, 9, is not an x-intercept. Figure 3 shows that the only solution to the equation occurs at the x-intercept 4.

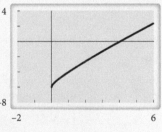

FIGURE 3

GETTING READY FOR CLASS

After reading through the preceding section, respond in your own words and in complete sentences.

A. What does it mean for an equation to be quadratic in form?

B. What are all the circumstances in solving equations (that we have studied) in which it is necessary to check for extraneous solutions?

C. How would you start to solve the equation $x + \sqrt{x} - 6 = 0$?

D. Is 9 a solution to $x + \sqrt{x} - 6 = 0$?

Problem Set 7.4

Solve each equation.

1. $(x - 3)^2 + 3(x - 3) + 2 = 0$ **2.** $(x + 4)^2 - (x + 4) - 6 = 0$

3. $2(x + 4)^2 + 5(x + 4) - 12 = 0$ **4.** $3(x - 5)^2 + 14(x - 5) - 5 = 0$

5. $x^4 - 6x^2 - 27 = 0$ **6.** $x^4 + 2x^2 - 8 = 0$

7. $x^4 + 9x^2 = -20$ **8.** $x^4 - 11x^2 = -30$

9. $(2a - 3)^2 - 9(2a - 3) = -20$ **10.** $(3a - 2)^2 + 2(3a - 2) = 3$

11. $2(4a + 2)^2 = 3(4a + 2) + 20$ **12.** $6(2a + 4)^2 = (2a + 4) + 2$

13. $6t^4 = -t^2 + 5$ **14.** $3t^4 = -2t^2 + 8$

15. $9x^4 - 49 = 0$ **16.** $25x^4 - 9 = 0$

Solve each of the following equations. Remember, if you square both sides of an equation in the process of solving it, you have to check all solutions in the original equation.

17. $x - 7\sqrt{x} + 10 = 0$ **18.** $x - 6\sqrt{x} + 8 = 0$

19. $t - 2\sqrt{t} - 15 = 0$ **20.** $t - 3\sqrt{t} - 10 = 0$

21. $6x + 11\sqrt{x} = 35$ **22.** $2x + \sqrt{x} = 15$

23. $(a - 2) - 11\sqrt{a - 2} + 30 = 0$ **24.** $(a - 3) - 9\sqrt{a - 3} + 20 = 0$

25. $(2x + 1) - 8\sqrt{2x + 1} + 15 = 0$ **26.** $(2x - 3) - 7\sqrt{2x - 3} + 12 = 0$

27. Solve the formula $16t^2 - vt - h = 0$ for t.

28. Solve the formula $16t^2 + vt + h = 0$ for t.

29. Solve the formula $kx^2 + 8x + 4 = 0$ for x.

30. Solve the formula $k^2x^2 + kx + 4 = 0$ for x.

31. Solve $x^2 + 2xy + y^2 = 0$ for x by using the quadratic formula with $a = 1$, $b = 2y$, and $c = y^2$.

32. Solve $x^2 - 2xy + y^2 = 0$ for x by using the quadratic formula with $a = 1$, $b = -2y$, $c = y^2$.

APPLYING THE CONCEPTS

For Problems 33 and 34, t is in seconds.

33. Falling Object An object is tossed into the air with an upward velocity of 8 feet per second from the top of a building h feet high. The time it takes for the object to hit the ground is given by the formula $16t^2 - 8t - h = 0$. Solve this formula for t.

34. Falling Object An object is tossed into the air with an upward velocity of 6 feet per second from the top of a building h feet high. The time it takes for the object to hit the ground is given by the formula $16t^2 - 6t - h = 0$. Solve this formula for t.

35. Saint Louis Arch The shape of the famous "Gateway to the West" arch in Saint Louis can be modeled by a parabola. The equation for one such parabola is:

$$y = -\frac{1}{150}x^2 + \frac{21}{5}x$$

a. Sketch the graph of the arch's equation on a coordinate axis.

b. Approximately how far do you have to walk to get from one side of the arch to the other?

36. Area In the following diagram, $ABCD$ is a rectangle with diagonal AC. Find its area.

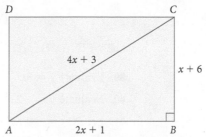

37. Area and Perimeter A total of 160 yards of fencing is to be used to enclose part of a lot that borders a river. This situation is shown in the following diagram.

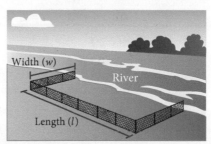

a. Write an equation that gives the relationship between the length and width and the 160 yards of fencing.

b. The formula for the area that is enclosed by the fencing and the river is $A = lw$. Solve the equation in part **a** for l, and then use the result to write the area in terms of w only.

c. Make a table that gives at least five possible values of w and associated area A.

d. From the pattern in your table shown in part **c**, what is the largest area that can be enclosed by the 160 yards of fencing? (Try some other table values if necessary.)

38. Area and Perimeter Rework all four parts of the preceding problem if it is desired to have an opening 2 yards wide in one of the shorter sides, as shown in the diagram.

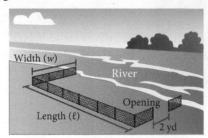

Width (w)

River

Opening

Length (ℓ)

2 yd

GETTING READY FOR THE NEXT SECTION

39. Evaluate $y = 3x^2 - 6x + 1$ for $x = 1$.

40. Evaluate $y = -2x^2 + 6x - 5$ for $x = \frac{3}{2}$.

41. Let $P(x) = -0.1x^2 + 27x - 500$ and find $P(135)$.

42. Let $P(x) = -0.1x^2 + 12x - 400$ and find $P(600)$.

Solve.

43. $0 = a(80)^2 + 70$

44. $0 = a(80)^2 + 90$

45. $x^2 - 6x + 5 = 0$

46. $x^2 - 3x - 4 = 0$

47. $-x^2 - 2x + 3 = 0$

48. $-x^2 + 4x + 12 = 0$

49. $2x^2 - 6x + 5 = 0$

50. $x^2 - 4x + 5 = 0$

Fill in the blanks to complete the square.

51. $x^2 - 6x + \square = (x - \square)^2$

52. $x^2 - 10x + \square = (x - \square)^2$

53. $y^2 + 2y + \square = (y + \square)^2$

54. $y^2 - 12y + \square = (x - \square)^2$

EXTENDING THE CONCEPTS

Solve each equation.

55. $x^6 - 9x^3 + 8 = 0$

56. $x^6 + 26x^3 - 27 = 0$

57. $x^8 - 17x^4 + 16 = 0$

58. $x^8 - 82x^3 + 81 = 0$

59. $\left(\frac{x-1}{x-3}\right)^2 - 2\left(\frac{x-1}{x-3}\right) = 8$

60. $\left(\frac{x+2}{x+5}\right)^2 - 5\left(\frac{x+2}{x+5}\right) = -6$

61. $(x^2 - x)^2 - 4(x^2 - x) - 12 = 0$

62. $(z^2 - z)^2 - (z^2 - z) - 6 = 0$

63. $4(t^2 + 1)^2 - 7(t^2 + 1) = 2$

64. $4(m^2 - 1)^2 - 12(m^2 - 1) = -8$

65. $(2r^2 - 1)^2 + 11(2r^2 - 1) = -30$

66. $(k^2 + 2)^2 + 13(k^2 + 2) + 12 = 0$

67. $6\left(\frac{x}{x+1}\right)^2 = 5\left(\frac{x}{x+1}\right) - 1$

68. $6\left(\frac{2x}{x+3}\right)^2 = 7\left(\frac{2x}{x+3}\right) - 2$

Find the x- and y-intercepts.

69. $y = x^3 - 4x$

70. $y = x^4 - 10x^2 + 9$

71. $y = 3x^3 + x^2 - 27x - 9$

OBJECTIVES

A   Graph a quadratic function.

B   Solve application problems using information from a graph.

C   Find an equation from its graph.

© iStockphoto/littlehenrabi

7.5

Participants of parkour, also called freerunning, make their own paths of travel by overcoming whatever physical obstacle stands before them. Their intense, and sometimes dangerous training requires them to climb walls, flip off ramps, leap over great distances, and land with body rolls to ease the impact on their bones and joints. The art of parkour has a French origin and has been practiced for nearly a century; however, its American popularity has only begun to grow in the past decade.

Imagine a freerunner as a falling object. He has just leaped over a railing at the top of a 10-foot-high brick wall. If he is falling downward at an initial velocity of 5 feet per second, the relationship between the distance s traveled and the time t is $s = 5t + 16t^2$. Problems like this one can be found by graphing a parabola. In this section, we will continue our work with parabolas and look at several ways to graph them.

 Graphing Quadratic Functions

The solution set to the equation

$$y = x^2 - 3$$

consists of ordered pairs. One method of graphing the solution set is to find a number of ordered pairs that satisfy the equation and to graph them. We can obtain some ordered pairs that are solutions to $y = x^2 - 3$ by use of a table as follows:

x	$y = x^2 - 3$	y	Solutions
-3	$y = (-3)^2 - 3 = 9 - 3 = 6$	6	$(-3, 6)$
-2	$y = (-2)^2 - 3 = 4 - 3 = 1$	1	$(-2, 1)$
-1	$y = (-1)^2 - 3 = 1 - 3 = -2$	-2	$(-1, -2)$
0	$y = 0^2 - 3 = 0 - 3 = -3$	-3	$(0, -3)$
1	$y = 1^2 - 3 = 1 - 3 = -2$	-2	$(1, -2)$
2	$y = 2^2 - 3 = 4 - 3 = 1$	1	$(2, 1)$
3	$y = 3^2 - 3 = 9 - 3 = 6$	6	$(3, 6)$

7.5 VIDEOS

Graphing these solutions and then connecting them with a smooth curve, we have the graph of $y = x^2 - 3$. (See Figure 1.)

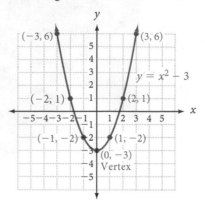

FIGURE 1

This graph is an example of a *parabola*. All equations of the form $y = ax^2 + bx + c$, $a \neq 0$, have parabolas for graphs.

Although it is always possible to graph parabolas by making a table of values of x and y that satisfy the equation, there are other methods that are faster and, in some cases, more accurate.

The important points associated with the graph of a parabola are the highest (or lowest) point on the graph and the x-intercepts. The y-intercepts can also be useful.

Intercepts for Parabolas

The graph of the equation $y = ax^2 + bx + c$ crosses the y-axis at $y = c$, because substituting $x = 0$ into $y = ax^2 + bx + c$ yields $y = c$.

Because the graph crosses the x-axis when $y = 0$, the x-intercepts are those values of x that are solutions to the quadratic equation $0 = ax^2 + bx + c$.

The Vertex of a Parabola

The highest or lowest point on a parabola is called the *vertex*. The vertex for the graph of $y = ax^2 + bx + c$ will always occur when

$$x = \frac{-b}{2a}$$

To see this, we must transform the right side of $y = ax^2 + bx + c$ into an expression that contains x in just one of its terms. This is accomplished by completing the square on the first two terms. Here is what it looks like:

$$y = ax^2 + bx + c$$
$$= a\left(x^2 + \frac{b}{a}x\right) + c$$
$$= a\left[x^2 + \frac{b}{a}x + \left(\frac{b}{2a}\right)^2\right] + c - a\left(\frac{b}{2a}\right)^2$$
$$= a\left(x + \frac{b}{2a}\right)^2 + \frac{4ac - b^2}{4a}$$

It may not look like it, but this last line indicates that the vertex of the graph of $y = ax^2 + bx + c$ has an x-coordinate of $\frac{-b}{2a}$. Because a, b, and c are constants, the only quantity that is varying in the last expression is the x in $\left(x + \frac{b}{2a}\right)^2$. Because the quantity $\left(x + \frac{b}{2a}\right)^2$ is the square of $x + \frac{b}{2a}$, the smallest it will ever be is 0, and that will happen when $x = \frac{-b}{2a}$.

We can use the vertex point along with the x- and y-intercepts to sketch the graph of any equation of the form $y = ax^2 + bx + c$. Here is a summary of the preceding information.

> ### Graphing Parabolas I
>
> The graph of $y = ax^2 + bx + c$, $a \neq 0$, will be a parabola with
> 1. A y-intercept at $y = c$
> 2. x-intercepts (if they exist) at
> $$x = \frac{-b \pm \sqrt{b^2 - 4ac}}{2a}$$
> 3. A vertex when $x = \dfrac{-b}{2a}$

EXAMPLE 1

Sketch the graph of $y = x^2 - 6x + 5$.

SOLUTION To find the x-intercepts, we let $y = 0$ and solve for x:

$$0 = x^2 - 6x + 5$$
$$0 = (x - 5)(x - 1)$$
$$x = 5 \quad \text{or} \quad x = 1$$

To find the coordinates of the vertex, we first find

$$x = \frac{-b}{2a} = \frac{-(-6)}{2(1)} = 3$$

The x-coordinate of the vertex is 3. To find the y-coordinate, we substitute 3 for x in our original equation:

$$y = 3^2 - 6(3) + 5 = 9 - 18 + 5 = -4$$

The graph crosses the x-axis at 1 and 5 and has its vertex at $(3, -4)$. Plotting these points and connecting them with a smooth curve, we have the graph shown in Figure 2.

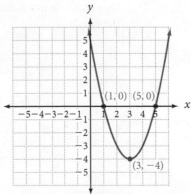

FIGURE 2

The graph is a parabola that opens up, so we say the graph is *concave up*. The vertex is the lowest point on the graph. (Note that the graph crosses the y-axis at 5, which is the value of y we obtain when we let $x = 0$.)

Finding the Vertex by Completing the Square

Another way to locate the vertex of the parabola in Example 1 is by completing the square on the first two terms on the right side of the equation $y = x^2 - 6x + 5$. In this case, we would do so by adding 9 to and subtracting 9 from the right side of the equation. This amounts to adding 0 to the equation, so we know we haven't changed its solutions. This is what it looks like:

$$y = (x^2 - 6x \qquad) + 5$$

$$= (x^2 - 6x + 9) + 5 - 9$$

$$= (x - 3)^2 - 4$$

You may have to look at this last equation a while to see this, but when $x = 3$, then $y = (x - 3)^2 - 4 = 0^2 - 4 = -4$ is the smallest y will ever be. That is why the vertex is at $(3, -4)$. As a matter of fact, this is the same kind of reasoning we used when we derived the formula $x = -\frac{b}{2a}$ for the x-coordinate of the vertex.

EXAMPLE 2

Graph $y = -x^2 - 2x + 3$.

SOLUTION To find the x-intercepts, we let $y = 0$:

$$0 = -x^2 - 2x + 3$$

$$0 = x^2 + 2x - 3 \qquad \text{Multiply each side by } -1.$$

$$0 = (x + 3)(x - 1)$$

$$x = -3 \qquad \text{or} \qquad x = 1$$

The x-coordinate of the vertex is given by

$$x = \frac{-b}{2a} = \frac{-(-2)}{2(-1)} = \frac{2}{-2} = -1$$

To find the y-coordinate of the vertex, we substitute -1 for x in our original equation to get

$$y = -(-1)^2 - 2(-1) + 3 = -1 + 2 + 3 = 4$$

Our parabola has x-intercepts at -3 and 1, and a vertex at $(-1, 4)$. Figure 3 shows the graph.

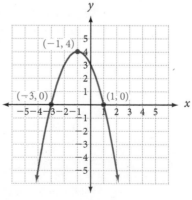

FIGURE 3

We say the graph is *concave down* because it opens downward. Again, we could have obtained the coordinates of the vertex by completing the square on the first two terms on the right side of our equation. To do so, we must first factor -1 from the first two terms. (Remember, the leading coefficient must be 1 to complete the square.) When we complete the square, we add 1 inside the parentheses, which actually decreases the right side of the equation by -1 because everything in the parentheses is multiplied by -1. To make up for it, we add 1 outside the parentheses.

$$y = -1(x^2 + 2x \quad) + 3$$
$$= -1(x^2 + 2x + 1) + 3 + 1$$
$$= -1(x + 1)^2 + 4$$

The last line tells us that the *largest* value of y will be 4, and that will occur when $x = -1$.

EXAMPLE 3

Graph $y = 3x^2 - 6x + 1$.

SOLUTION   To find the x-intercepts, we let $y = 0$ and solve for x:

$$0 = 3x^2 - 6x + 1$$

Because the right side of this equation does not factor, we can look at the discriminant to see what kind of solutions are possible. The discriminant for this equation is

$$b^2 - 4ac = 36 - 4(3)(1) = 24$$

Because the discriminant is a positive number but not a perfect square, the equation will have irrational solutions. This means that the x-intercepts are irrational numbers and will have to be approximated with decimals using the quadratic formula. Rather than use the quadratic formula, we will find some other points on the graph, but let's find the vertex first.

Here are both methods of finding the vertex:

Using the formula that gives us the x-coordinate of the vertex, we have

$$x = \frac{-b}{2a} = \frac{-(-6)}{2(3)} = 1$$

Substituting 1 for x in the equation gives us the y-coordinate of the vertex:

$$y = 3 \cdot 1^2 - 6 \cdot 1 + 1 = -2$$

To complete the square on the right side of the equation, we factor 3 from the first two terms, add 1 inside the parentheses, and add -3 outside the parentheses (this amounts to adding 0 to the right side):

$$y = 3(x^2 - 2x \quad) + 1$$
$$= 3(x^2 - 2x + 1) + 1 - 3$$
$$= 3(x - 1)^2 - 2$$

In either case, the vertex is $(1, -2)$.

If we can find two points, one on each side of the vertex, we can sketch the graph. Let's let $x = 0$ and $x = 2$, because each of these numbers is the same distance from $x = 1$, and $x = 0$ will give us the y-intercept.

When $x = 0$ | When $x = 2$

$$y = 3(0)^2 - 6(0) + 1 \qquad y = 3(2)^2 - 6(2) + 1$$
$$= 0 - 0 + 1 \qquad\qquad = 12 - 12 + 1$$
$$= 1 \qquad\qquad\qquad = 1$$

The two points just found are (0, 1) and (2, 1). Plotting these two points along with the vertex (1, −2), we have the graph shown in Figure 4.

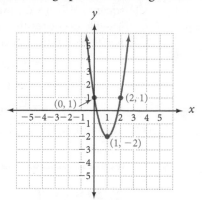

FIGURE 4

EXAMPLE 4

Graph $y = -2x^2 + 6x - 5$.

SOLUTION   Letting $y = 0$, we have

$$0 = -2x^2 + 6x - 5$$

Again, the right side of this equation does not factor. The discriminant is $b^2 - 4ac = 36 - 4(-2)(-5) = -4$, which indicates that the solutions are complex numbers. This means that our original equation does not have x-intercepts. The graph does not cross the x-axis.

Let's find the vertex.

Using our formula for the x-coordinate of the vertex, we have

$$x = \frac{-b}{2a} = \frac{-6}{2(-2)} = \frac{6}{4} = \frac{3}{2}$$

To find the y-coordinate, we let $x = \frac{3}{2}$:

$$y = -2\left(\frac{3}{2}\right)^2 + 6\left(\frac{3}{2}\right) - 5$$

$$= \frac{-18}{4} + \frac{18}{2} - 5$$

$$= \frac{-18 + 36 - 20}{4}$$

$$= -\frac{1}{2}$$

Finding the vertex by completing the square is a more complicated matter. To make the coefficient of x^2 a 1, we must factor -2 from the first two terms. To complete the square inside the parentheses, we add $\frac{9}{4}$. Since each term inside the parentheses is multiplied by -2, we add $\frac{9}{2}$ outside the parentheses so that the net result is the same as adding 0 to the right side:

$$y = -2(x^2 - 3x \qquad) - 5$$

$$= -2\left(x^2 - 3x + \frac{9}{4}\right) - 5 + \frac{9}{2}$$

$$= -2\left(x - \frac{3}{2}\right)^2 - \frac{1}{2}$$

The vertex is $\left(\frac{3}{2}, -\frac{1}{2}\right)$. Because this is the only point we have so far, we must find two others. Let's let $x = 3$ and $x = 0$, because each point is the same distance from $x = \frac{3}{2}$ and on either side:

When $x = 3$ | | When $x = 0$
|||

$$y = -2(3)^2 + 6(3) - 5 \qquad\qquad y = -2(0)^2 + 6(0) - 5$$
$$= -18 + 18 - 5 \qquad\qquad\qquad = 0 + 0 - 5$$
$$= -5 \qquad\qquad\qquad\qquad = -5$$

The two additional points on the graph are $(3, -5)$ and $(0, -5)$. Figure 5 shows the graph.

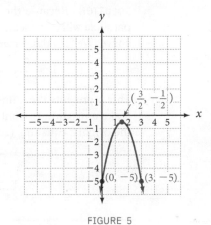

FIGURE 5

The graph is concave down. The vertex is the highest point on the graph.

By looking at the equations and graphs in Examples 1 through 4, we can conclude that the graph of $y = ax^2 + bx + c$ will be concave up when a is positive, and concave down when a is negative. Taking this even further, if $a > 0$, then the vertex is the lowest point on the graph, and if $a < 0$, the vertex is the highest point on the graph. Finally, if we complete the square on x in the equation $y = ax^2 + bx + c$, $a \neq 0$, we can rewrite the equation of our parabola as $y = a(x - h)^2 + k$. When the equation is in this form, the vertex is at the point (h, k). Here is a summary:

Graphing Parabolas II

The graph of

$$y = a(x - h)^2 + k, a \neq 0$$

will be a parabola with a vertex at (h, k). The vertex will be the highest point on the graph when $a < 0$, and the lowest point on the graph when $a > 0$.

B Applications

EXAMPLE 5

A company selling copies of an accounting program for home computers finds that it will make a weekly profit of P dollars from selling x copies of the program, according to the equation

$$P(x) = -0.1x^2 + 27x - 500$$

How many copies of the program should it sell to make the largest possible profit, and what is the largest possible profit?

SOLUTION   Because the coefficient of x^2 is negative, we know the graph of this parabola will be concave down, meaning that the vertex is the highest point of the curve. We find the vertex by first finding its x-coordinate:

$$x = \frac{-b}{2a} = \frac{-27}{2(-0.1)} = \frac{27}{0.2} = 135$$

This represents the number of programs the company needs to sell each week to make a maximum profit. To find the maximum profit, we substitute 135 for x in the original equation. (A calculator is helpful for these kinds of calculations.)

$$P(135) = -0.1(135)^2 + 27(135) - 500$$

$$= -0.1(18{,}225) + 3{,}645 - 500$$

$$= -1{,}822.5 + 3{,}645 - 500$$

$$= 1{,}322.5$$

The maximum weekly profit is $1,322.50 and is obtained by selling 135 copies of the program a week.

EXAMPLE 6

An art supply store finds that they can sell x sketch pads each week at p dollars each, according to the equation $x = 900 - 300p$. Graph the revenue equation $R = xp$. Then use the graph to find the price p that will bring in the maximum revenue. Finally, find the maximum revenue.

SOLUTION   As it stands, the revenue equation contains three variables. Because we are asked to find the value of p that gives us the maximum value of R, we rewrite the equation using just the variables R and p. Because $x = 900 - 300p$, we have

$$R = xp = (900 - 300p)p$$

The graph of this equation is shown in Figure 6. The graph appears in the first quadrant only, because R and p are both positive quantities.

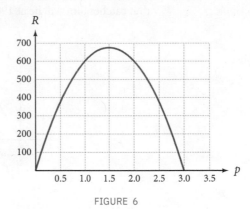

FIGURE 6

From the graph, we see that the maximum value of R occurs when $p = \$1.50$. We can calculate the maximum value of R from the equation:

When $p = 1.5$

the equation $R = (900 - 300p)p$

becomes $R = (900 - 300 \cdot 1.5)1.5$

$= (900 - 450)1.5$

$= 450 \cdot 1.5$

$= 675$

The maximum revenue is $675. It is obtained by setting the price of each sketch pad at $p = \$1.50$.

USING TECHNOLOGY   *Graphing Calculators*

If you have been using a graphing calculator for some of the material in this course, you are well aware that your calculator can draw all the graphs in this section very easily. It is important, however, that you be able to recognize and sketch the graph of any parabola by hand. It is a skill that all successful intermediate algebra students should possess, even if they are proficient in the use of a graphing calculator. My suggestion is that you work the problems in this section and problem set without your calculator. Then use your calculator to check your results.

C **Finding the Equation from the Graph**

EXAMPLE 7 ━━━━━━━━━━━━━━━━━━

In 2011, David Smith, Jr., also known as "The Bullet," was shot from a cannon in Milan, Italy. As a human cannonball, he reached a height of 75 feet before landing in a net 190 feet from the cannon. Sketch the graph of his path, and then find the equation of the graph.

SOLUTION We assume that the path taken by the human cannonball is a parabola. If the origin of the coordinate system is at the opening of the cannon, then the net that catches him will be at 190 on the x-axis. Figure 7 shows the graph.

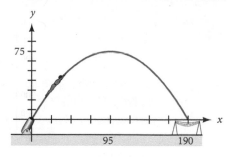

FIGURE 7

Because the curve is a parabola, we know the equation will have the form

$$y = a(x - h)^2 + k$$

Because the vertex of the parabola is at (95, 75), we can fill in two of the three constants in our equation, giving us

$$y = a(x - 95)^2 + 75$$

To find a, we note that the landing point will be (190, 0). Substituting the coordinates of this point into the equation, we solve for a:

$$0 = a(190 - 95)^2 + 75$$

$$0 = a(95)^2 + 75$$

$$0 = 9{,}025a + 75$$

$$a = -\frac{75}{9{,}025} = -\frac{3}{361}$$

The equation that describes the path of the human cannonball is

$$y = -\frac{3}{361}(x - 95)^2 + 75 \text{ for } 0 \le x \le 190$$

USING TECHNOLOGY *Graphing Calculators*

Graph the equation found in Example 7 on a graphing calculator using the window shown here. (We will use this graph later in the book to find the angle between the cannon and the horizontal.)

Window: X from 0 to 200, increment 20
Y from 0 to 80, increment 10

On the TI-83, an increment of 20 for X means Xscl = 20.

GETTING READY FOR CLASS

After reading through the preceding section, respond in your own words and in complete sentences.

A. What is a parabola?

B. What part of the equation of a parabola determines whether the graph is concave up or concave down?

C. Suppose $f(x) = ax^2 + bx + c$ is the equation of a parabola. Explain how $f(4) = 1$ relates to the graph of the parabola.

D. A line can be graphed with two points. How many points are necessary to get a reasonable sketch of a parabola? Explain.

Problem Set 7.5

For each of the following equations, give the x-intercepts and the coordinates of the vertex, and sketch the graph.

1. $y = x^2 + 2x - 3$ **2.** $y = x^2 - 2x - 3$ **3.** $y = -x^2 - 4x + 5$

4. $y = x^2 + 4x - 5$ **5.** $y = x^2 - 1$ **6.** $y = x^2 - 4$

7. $y = -x^2 + 9$ **8.** $y = -x^2 + 1$ **9.** $y = 2x^2 - 4x - 6$

10. $y = 2x^2 + 4x - 6$ **11.** $y = x^2 - 2x - 4$ **12.** $y = x^2 - 2x - 2$

Graph each parabola. Label the vertex and any intercepts that exist.

13. $y = 2(x - 1)^2 + 3$ **14.** $y = 2(x + 1)^2 - 3$

15. $f(x) = -(x + 2)^2 + 4$ **16.** $f(x) = -(x - 3)^2 + 1$

17. $g(x) = \dfrac{1}{2}(x - 2)^2 - 4$ **18.** $g(x) = \dfrac{1}{3}(x - 3)^2 - 3$

19. $f(x) = -2(x - 4)^2 - 1$ **20.** $f(x) = -4(x - 1)^2 + 4$

Find the vertex and any two convenient points to sketch the graphs of the following equations.

21. $y = x^2 - 4x - 4$ **22.** $y = x^2 - 2x + 3$ **23.** $y = -x^2 + 2x - 5$

24. $y = -x^2 + 4x - 2$ **25.** $f(x) = x^2 + 1$ **26.** $f(x) = x^2 + 4$

27. $y = -x^2 - 3$ **28.** $y = -x^2 - 2$ **29.** $g(x) = 3x^2 + 4x + 1$

30. $g(x) = 2x^2 + 4x + 3$

For each of the following equations, find the coordinates of the vertex, and indicate whether the vertex is the highest point on the graph or the lowest point on the graph. (Do not graph.)

31. $y = x^2 - 6x + 5$ **32.** $y = -x^2 + 6x - 5$ **33.** $y = -x^2 + 2x + 8$

34. $y = x^2 - 2x - 8$ **35.** $y = 12 + 4x - x^2$ **36.** $y = -12 - 4x + x^2$

37. $y = -x^2 - 8x$ **38.** $y = x^2 + 8x$

APPLYING THE CONCEPTS

39. Maximum Profit A company finds that it can make a profit of P dollars each month by selling x patterns, according to the formula $P(x) = -0.002x^2 + 3.5x - 800$. How many patterns must it sell each month to have a maximum profit? What is the maximum profit?

40. Maximum Profit A company selling picture frames finds that it can make a profit of P dollars each month by selling x frames, according to the formula $P(x) = -0.002x^2 + 5.5x - 1{,}200$. How many frames must it sell each month to have a maximum profit? What is the maximum profit?

41. Maximum Height Chaudra is tossing a softball into the air with an underhand motion. The distance of the ball above her hand at any time is given by the function

$$h(t) = 32t - 16t^2 \quad \text{for} \quad 0 \le t \le 2$$

where $h(t)$ is the height of the ball (in feet) and t is the time (in seconds). Find the times at which the ball is in her hand, and the maximum height of the ball.

42. **Maximum Area** Justin wants to fence three sides of a rectangular exercise yard for his dog. The fourth side of the exercise yard will be a side of the house. He has 80 feet of fencing available. Find the dimensions of the exercise yard that will enclose the maximum area.

43. **Maximum Revenue** A company that manufactures smartphone cases knows that the number of cases x it can sell each week is related to the price p of each case by the equation $x = 1,200 - 100p$. Graph the revenue equation $R = xp$. Then use the graph to find the price p that will bring in the maximum revenue. Finally, find the maximum revenue.

44. **Maximum Revenue** A company that manufactures earphones for smartphones finds that it can sell x earphone pairs each day at p dollars per pair, according to the equation $x = 800 - 100p$. Graph the revenue equation $R = xp$. Then use the graph to find the price p that will bring in the maximum revenue. Finally, find the maximum revenue.

45. **Maximum Revenue** The relationship between the number of calculators x a company sells each day and the price p of each calculator is given by the equation $x = 1,700 - 100p$. Graph the revenue equation $R = xp$, and use the graph to find the price p that will bring in the maximum revenue. Then find the maximum revenue.

46. **Maximum Revenue** The relationship between the number x of pencil sharpeners a company sells each week and the price p of each sharpener is given by the equation $x = 1,800 - 100p$. Graph the revenue equation $R = xp$, and use the graph to find the price p that will bring in the maximum revenue. Then find the maximum revenue.

47. **Human Cannonball** A human cannonball is shot from a cannon at the county fair. She reaches a height of 60 feet before landing in a net 180 feet from the cannon. Sketch the graph of her path, and then find the equation of the graph.

48. Interpreting Graphs The graph below shows the different paths taken by the human cannonball when his velocity out of the cannon is 50 miles/hour, and his cannon is inclined at varying angles.

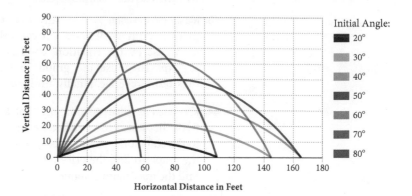

Initial Velocity: 50 miles per hour
Angle: 20°, 30°, 40°, 50°, 60°, 70°, 80°

a. If his landing net is placed 104 feet from the cannon, at what angle should the cannon be inclined so that he lands in the net?

b. Approximately where do you think he would land if the cannon was inclined at 45°?

c. If the cannon was inclined at 45°, approximately what height do you think he would attain?

d. Do you think there is another angle for which he would travel the same distance he travels at 80°? Give an estimate of that angle.

e. The fact that every landing point can come from two different paths makes us think that the equations that give us the landing points must be what type of equations?

GETTING READY FOR THE NEXT SECTION

Solve.

49. $x^2 - 2x - 8 = 0$

50. $x^2 - x - 12 = 0$

51. $6x^2 - x = 2$

52. $3x^2 - 5x = 2$

53. $x^2 - 6x + 9 = 0$

54. $x^2 + 8x + 16 = 0$

EXTENDING THE CONCEPTS

Finding the Equation from the Graph For each of the following problems, the graph is a parabola. In each case, find an equation in the form $y = a(x - h)^2 + k$ that describes the graph.

55.

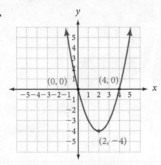

56.

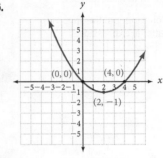

57. Comparing Expressions, Equations, and Functions Four problems follow. The solution to part **c** is shown in Figure 8. Solve the other three problems, and then explain how the solutions to the four problems are related.

a. Factor the expression $x^2 - 4$.

b. Solve the equation $x^2 - 4 = 0$.

c. Graph the function $y = x^2 - 4$.

d. If $f(x) = x^2 - 4$, find the value of x for which $f(x) = 0$.

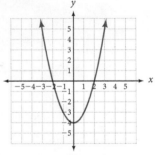

FIGURE 8

58. Comparing Expressions, Equations, and Functions Four problems are shown here. The solution to part **c** is shown in Figure 9. Solve the other three problems, and then explain how the solutions to the four problems are related.

a. Factor the expression $x^2 + 2x - 3$.

b. Solve the equation $x^2 + 2x - 3 = 0$.

c. Graph the function $y = x^2 + 2x - 3$.

d. If $f(x) = x^2 + 2x - 3$, find the value of x for which $f(x) = 0$.

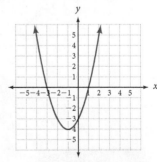

FIGURE 9

 OBJECTIVES

A Solve quadratic and polynomial inequalities and graph the solution set.

B Solve inequalities involving rational expressions.

© iStockphoto/dell640

Let's say the manufacturer of ski goggles knows that the weekly revenue produced by selling x goggles is given by the equation $R = 13{,}000p - 100p^2$, where p is the price of each set of goggles (in dollars). To find the price that needs to be charged for each set of goggles if the weekly revenue is to be at least \$400,000, we need to understand quadratic inequalities, which is the focus of this section.

 ## Quadratic Inequalities

Quadratic inequalities in one variable are inequalities of the form

$$ax^2 + bx + c < 0 \qquad ax^2 + bx + c > 0$$
$$ax^2 + bx + c \leq 0 \qquad ax^2 + bx + c \geq 0$$

where a, b, and c are constants, with $a \neq 0$. The technique we will use to solve inequalities of this type involves graphing. Suppose, for example, we want to find the solution set for the inequality $x^2 - x - 6 > 0$. We begin by factoring the left side to obtain

$$(x - 3)(x + 2) > 0$$

We have two real numbers $x - 3$ and $x + 2$ whose product $(x - 3)(x + 2)$ is greater than zero. That is, their product is positive. The only way the product can be positive is either if both factors, $(x - 3)$ and $(x + 2)$, are positive or if they are both negative. To help visualize where $x - 3$ is positive and where it is negative, we draw a real number line and label it accordingly:

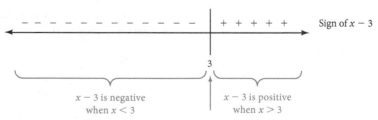

Here is a similar diagram showing where the factor $x + 2$ is positive and where it is negative:

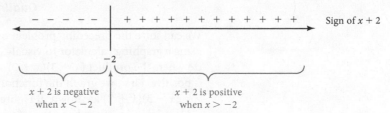

Drawing the two number lines together and eliminating the unnecessary numbers, we have

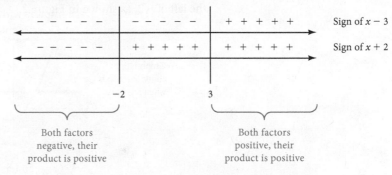

We can see from the preceding diagram that the graph of the solution to $x^2 - x - 6 > 0$ is

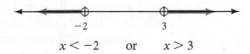

$$x < -2 \quad \text{or} \quad x > 3$$

USING TECHNOLOGY *Graphical Solutions to Quadratic Inequalities*

We can solve the preceding problem by using a graphing calculator to visualize where the product $(x - 3)(x + 2)$ is positive. First, we graph the function $y = (x - 3)(x + 2)$ as shown in Figure 1.

Next, we observe where the graph is above the x-axis. As you can see, the graph is above the x-axis to the right of 3 and to the left of -2, as shown in Figure 2.

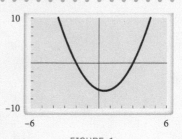

FIGURE 1

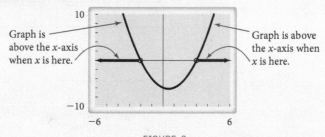

Graph is above the x-axis when x is here.

Graph is above the x-axis when x is here.

FIGURE 2

When the graph is above the x-axis, we have points whose y-coordinates are positive. Because these y-coordinates are the same as the expression $(x - 3)(x + 2)$, the values of x for which the graph of $y = (x - 3)(x + 2)$ is above the x-axis are the values of x for which the inequality $(x - 3)(x + 2) > 0$ is true. Our solution set is therefore

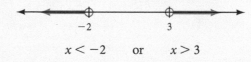

$$x < -2 \quad \text{or} \quad x > 3$$

EXAMPLE 1

Solve for x: $x^2 - 2x - 8 \le 0$.

ALGEBRAIC SOLUTION   We begin by factoring:

$$x^2 - 2x - 8 \le 0$$
$$(x - 4)(x + 2) \le 0$$

The product $(x - 4)(x + 2)$ is negative or zero. The factors must have opposite signs. We draw a diagram showing where each factor is positive and where each factor is negative:

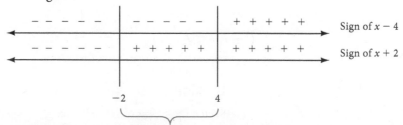

From the diagram, we have the graph of the solution set:

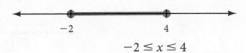

$$-2 \leq x \leq 4$$

GRAPHICAL SOLUTION   To solve this inequality with a graphing calculator, we graph the function $y = (x - 4)(x + 2)$ and observe where the graph is below the x-axis. These points have negative y-coordinates, which means that the product $(x - 4)(x + 2)$ is negative for these points. Figure 3 shows the graph of $y = (x - 4)(x + 2)$, along with the region on the x-axis where the graph contains points with negative y-coordinates.

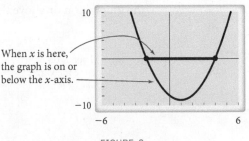

When x is here, the graph is on or below the x-axis.

FIGURE 3

As you can see, the graph is below the x-axis when x is between -2 and 4. Because our original inequality includes the possibility that $(x - 4)(x + 2)$ is 0, we include the endpoints, -2 and 4, with our solution set.

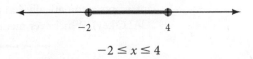

$$-2 \leq x \leq 4$$

EXAMPLE 2

Solve for x: $6x^2 - x \geq 2$.

ALGEBRAIC SOLUTION

$$6x^2 - x \geq 2$$

$$6x^2 - x - 2 \geq 0 \quad \leftarrow \text{ Standard form}$$

$$(3x - 2)(2x + 1) \geq 0$$

The product is positive or zero, so the factors must agree in sign. Here is the diagram showing where that occurs:

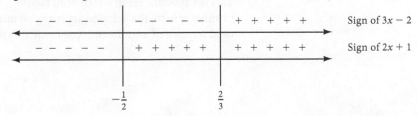

Because the factors agree in sign below $-\frac{1}{2}$ and above $\frac{2}{3}$, the graph of the solution set is

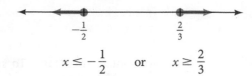

$$x \le -\frac{1}{2} \quad \text{or} \quad x \ge \frac{2}{3}$$

GRAPHICAL SOLUTION To solve this inequality with a graphing calculator, we graph the function $y = (3x - 2)(2x + 1)$ and observe where the graph is above the x-axis. These are the points that have positive y-coordinates, which means that the product $(3x - 2)(2x + 1)$ is positive for these points. Figure 4 shows the graph of $y = (3x - 2)(2x + 1)$, along with the regions on the x-axis where the graph is on or above the x-axis.

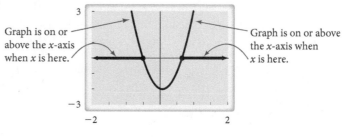

FIGURE 4

To find the points where the graph crosses the x-axis, we need to use either the Trace and Zoom features to zoom in on each point, or the calculator function that finds the intercepts automatically (on the TI-82/83) this is the root/zero function under the CALC key). Whichever method we use, we will obtain the following result:

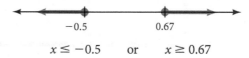

$$x \le -0.5 \quad \text{or} \quad x \ge 0.67$$

EXAMPLE 3

Solve $x^2 - 6x + 9 \ge 0$.

ALGEBRAIC SOLUTION

$$x^2 - 6x + 9 \ge 0$$
$$(x - 3)^2 \ge 0$$

This is a special case in which both factors are the same. Because $(x - 3)^2$ is always positive or zero, the solution set is all real numbers. That is, any real number that is used in place of x in the original inequality will produce a true statement.

GRAPHICAL SOLUTION   The graph of $y = (x - 3)^2$ is shown in Figure 5.

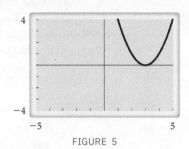

FIGURE 5

Notice that it touches the x-axis at 3 and is above the x-axis everywhere else. This means that every point on the graph has a y-coordinate greater than or equal to 0, no matter the value of x. The conclusion that we draw from the graph is that the inequality $(x - 3)^2 \geq 0$ is true for all values of x.

B Inequalities Involving Rational Expressions

Our next two examples involve inequalities that contain rational expressions.

EXAMPLE 4

Solve: $\dfrac{x - 4}{x + 1} \leq 0$.

SOLUTION   The inequality indicates that the quotient of $(x - 4)$ and $(x + 1)$ is negative or 0 (less than or equal to 0). We can use the same reasoning we used to solve the first three examples, because quotients are positive or negative under the same conditions that products are positive or negative. Here is the diagram that shows where each factor is positive and where each factor is negative:

Between -1 and 4 the factors have opposite signs, making the quotient negative. Thus, the region between -1 and 4 is where the solutions lie, because the original inequality indicates the quotient $\frac{x-4}{x+1}$ is negative. The solution set and its graph are shown here:

$$-1 < x \leq 4$$

Notice that the left endpoint is open—that is, it is not included in the solution set—because $x = -1$ would make the denominator in the original inequality 0. It is important to check all endpoints of solution sets to inequalities that involve rational expressions.

EXAMPLE 5

Solve: $\dfrac{3}{x-2} - \dfrac{2}{x-3} > 0$.

SOLUTION   We begin by adding the two rational expressions on the left side. The common denominator is $(x-2)(x-3)$:

$$\frac{3}{x-2} \cdot \frac{(x-3)}{(x-3)} - \frac{2}{x-3} \cdot \frac{(x-2)}{(x-2)} > 0$$

$$\frac{3x - 9 - 2x + 4}{(x-2)(x-3)} > 0$$

$$\frac{x-5}{(x-2)(x-3)} > 0$$

This time the quotient involves three factors. Here is the diagram that shows the signs of the three factors:

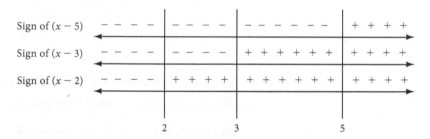

The original inequality indicates that the quotient is positive. For this to happen, either all three factors must be positive, or exactly two factors must be negative. Looking back at the diagram, we see the regions that satisfy these conditions are between 2 and 3 or above 5. Here is our solution set:

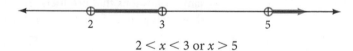

$$2 < x < 3 \text{ or } x > 5$$

GETTING READY FOR CLASS

After reading through the preceding section, respond in your own words and in complete sentences.

A. What is the first step in solving a quadratic inequality?

B. How do you show that the endpoint of a line segment is not part of the graph of a quadratic inequality?

C. How would you use the graph of $y = ax^2 + bx + c$ to help you find the graph of $ax^2 + bx + c < 0$?

D. Can a quadratic inequality have exactly one solution? Give an example.

Problem Set 7.6

Solve each of the following inequalities, and graph the solution set.

1. $x^2 + x - 6 > 0$ **2.** $x^2 + x - 6 < 0$ **3.** $x^2 - x - 12 \leq 0$

4. $x^2 - x - 12 \geq 0$ **5.** $x^2 + 5x \geq -6$ **6.** $x^2 - 5x > 6$

7. $6x^2 < 5x - 1$ **8.** $4x^2 \geq -5x + 6$ **9.** $x^2 - 9 < 0$

10. $x^2 - 16 \geq 0$ **11.** $4x^2 - 9 \geq 0$ **12.** $9x^2 - 4 < 0$

13. $2x^2 - x - 3 < 0$ **14.** $3x^2 + x - 10 \geq 0$ **15.** $x^2 - 4x + 4 \geq 0$

16. $x^2 - 4x + 4 < 0$ **17.** $x^2 - 10x + 25 < 0$ **18.** $x^2 - 10x + 25 > 0$

19. $(x - 2)(x - 3)(x - 4) > 0$ **20.** $(x - 2)(x - 3)(x - 4) < 0$

21. $(x + 1)(x + 2)(x + 3) \leq 0$ **22.** $(x + 1)(x + 2)(x + 3) \geq 0$

23. $\dfrac{x - 1}{x + 4} \leq 0$ **24.** $\dfrac{x + 4}{x - 1} \leq 0$

25. $\dfrac{3x}{x + 6} - \dfrac{8}{x + 6} < 0$ **26.** $\dfrac{5x}{x + 1} - \dfrac{3}{x + 1} < 0$

27. $\dfrac{4}{x - 6} + 1 > 0$ **28.** $\dfrac{2}{x - 3} + 1 \geq 0$

29. $\dfrac{x - 2}{(x + 3)(x - 4)} < 0$ **30.** $\dfrac{x - 1}{(x + 2)(x - 5)} < 0$

31. $\dfrac{2}{x - 4} - \dfrac{1}{x - 3} > 0$ **32.** $\dfrac{4}{x + 3} - \dfrac{3}{x + 2} > 0$

33. $\dfrac{x + 7}{2x + 12} + \dfrac{6}{x^2 - 36} \leq 0$ **34.** $\dfrac{x + 1}{2x - 2} - \dfrac{2}{x^2 - 1} \leq 0$

35. The graph of $y = x^2 - 4$ is shown in Figure 6. Use the graph to write the solution set for each of the following:

     a. $x^2 - 4 < 0$ **b.** $x^2 - 4 > 0$ **c.** $x^2 - 4 = 0$

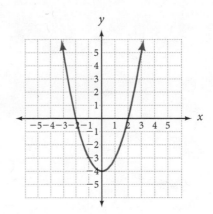

FIGURE 6

635

36. The graph of $y = 4 - x^2$ is shown in Figure 7. Use the graph to write the solution set for each of the following:

a. $4 - x^2 < 0$ **b.** $4 - x^2 > 0$ **c.** $4 - x^2 = 0$

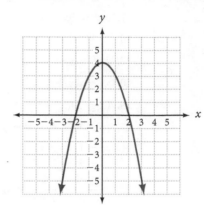

FIGURE 7

37. The graph of $y = x^2 - 3x - 10$ is shown in Figure 8. Use the graph to write the solution set for each of the following:

a. $x^2 - 3x - 10 < 0$ **b.** $x^2 - 3x - 10 > 0$ **c.** $x^2 - 3x - 10 = 0$

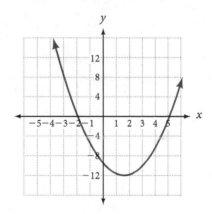

FIGURE 8

38. The graph of $y = x^2 + x - 12$ is shown in Figure 9. Use the graph to write the solution set for each of the following:

a. $x^2 + x - 12 < 0$ **b.** $x^2 + x - 12 > 0$ **c.** $x^2 + x - 12 = 0$

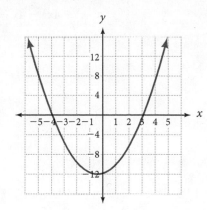

FIGURE 9

39. The graph of $y = x^3 - 3x^2 - x + 3$ is shown in Figure 10. Use the graph to write the solution set for each of the following:

a. $x^3 - 3x^2 - x + 3 < 0$ **b.** $x^3 - 3x^2 - x + 3 > 0$

c. $x^3 - 3x^2 - x + 3 = 0$

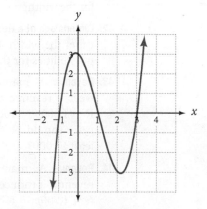

FIGURE 10

40. The graph of $y = x^3 + 4x^2 - 4x - 16$ is shown in Figure 11. Use the graph to write the solution set for each of the following:

a. $x^3 + 4x^2 - 4x - 16 < 0$ **b.** $x^3 + 4x^2 - 4x - 16 > 0$

c. $x^3 + 4x^2 - 4x - 16 = 0$

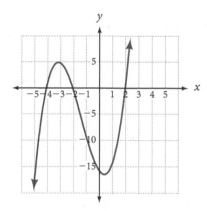

FIGURE 11

APPLYING THE CONCEPTS

41. Dimensions of a Rectangle The length of a rectangle is 3 inches more than twice the width. If the area is to be at least 44 square inches, what are the possibilities for the width?

42. Dimensions of a Rectangle The length of a rectangle is 5 inches less than three times the width. If the area is to be less than 12 square inches, what are the possibilities for the width?

43. Revenue A manufacturer of portable wireless speakers knows that the weekly revenue produced by selling x speakers is given by the equation $R = 1{,}300p - 100p^2$, where p is the price of each speaker (in dollars). What price should be charged for each speaker if the weekly revenue is to be at least $4,000?

44. Revenue A manufacturer of small calculators knows that the weekly revenue produced by selling x calculators is given by the equation $R = 1{,}700p - 100p^2$, where p is the price of each calculator (in dollars). What price should be charged for each calculator if the revenue is to be at least $7,000 each week?

45. Union Dues A labor union has 10,000 members. For every $10 increase in union dues, membership is decreased by 200 people. If the current dues are $100, what should be the new dues (to the nearest multiple of $10) so income from dues is greatest, and what is that income? *Hint:* Because Income = (membership)(dues), we can let x = the number of $10 increases in dues, and then this will give us income of $y = (10{,}000 - 200x)(100 + 10x)$.

46. Bookstore Receipts The owner of a used bookstore charges $2 for quality paperbacks and usually sells 40 per day. For every 10-cent increase in the price of these paperbacks, he thinks that he will sell two fewer per day. What is the price he should charge (to the nearest 10 cents) for these books to maximize his income, and what would be that income? *Hint:* Let x = the number of 10-cent increases in price.

47. **Quick Lube** The owner of a quick oil-change business charges $20 per oil change and has 40 customers per day. If each increase of $2 results in 2 fewer daily customers, what price should the owner charge (to the nearest $2) for an oil change if the income from this business is to be as great as possible?

48. **Computer Sales** A computer manufacturer charges $2,200 for its basic model and sells 1,500 computers per month at this price. For every $200 increase in price, it is believed that 75 fewer computers will be sold. What price should the company place on its basic model of computer (to the nearest $100) to have the greatest income?

MAINTAINING YOUR SKILLS

Use a calculator to evaluate, give answers to 4 decimal places.

49. $\dfrac{50{,}000}{32{,}000}$

50. $\dfrac{2.4362}{1.9758} - 1$

51. $\dfrac{1}{2}\left(\dfrac{4.5926}{1.3876} - 2\right)$

52. $1 + \dfrac{0.06}{12}$

Solve each equation.

53. $2\sqrt{3t - 1} = 2$

54. $\sqrt{4t + 5} + 7 = 3$

55. $\sqrt{x + 3} = x - 3$

56. $\sqrt{x + 3} = \sqrt{x} - 3$

Graph each equation.

57. $y = \sqrt[3]{x - 1}$

58. $y = \sqrt[3]{x} - 1$

EXTENDING THE CONCEPTS

Graph the solution set for each inequality.

59. $x^2 - 2x - 1 < 0$

60. $x^2 - 6x + 7 < 0$

61. $x^2 - 8x + 13 > 0$

62. $x^2 - 10x + 18 > 0$

63. $x^3 + 2x^2 - 9x - 18 < 0$

64. $x^3 + 2x^2 - 9x - 18 \geq 0$

65. $12x^3 - 8x^2 - 3x + 2 > 0$

66. $3a^3 + 5a^2 - 27a - 45 \leq 0$

67. $4x^2 + 19x - 30 > 0$

68. $4x^2 - 29x + 30 \leq 0$

69. $36x^2 + 81x < 0$

70. $36x^2 > 40 - 81x$

SPOTLIGHT ON SUCCESS

CJ | student instructor

We are what we repeatedly do.
Excellence, then, is not an act, but a habit.

—Aristotle

Something that has worked for me in college, in addition to completing the assigned homework, is working on some extra problems from each section. Working on these extra problems is a great habit to get into because it helps further your understanding of the material, and you see the many different types of problems that can arise. If you have completed every problem that your book offers, and you still don't feel confident that you have a full grasp of the material, look for more problems. Many problems can be found online or in other books. Your professors may even have some problems that they would suggest doing for extra practice. The biggest benefit to working all the problems in the course's assigned textbook is that often teachers will choose problems either straight from the book or ones similar to problems that were not assigned for tests. Doing this will ensure that you do your best in all your classes.

EXAMPLES

The Square Root Property for Equations [7.1]

1. If $(x - 3)^2 = 25$
then $x - 3 = \pm 5$
$x = 3 \pm 5$
$x = 8$ or $x = -2$

If $a^2 = b$, where b is a real number, then

$$a = \sqrt{b} \qquad \text{or} \qquad a = -\sqrt{b}$$

which can be written as $a = \pm\sqrt{b}$.

To Solve a Quadratic Equation by Completing the Square [7.1]

2. Solve $x^2 - 6x - 6 = 0$
$x^2 - 6x = 6$
$x^2 - 6x + 9 = 6 + 9$
$(x - 3)^2 = 15$
$x - 3 = \pm\sqrt{15}$
$x = 3 \pm \sqrt{15}$

Step 1: Write the equation in the form $ax^2 + bx = c$.

Step 2: If $a \neq 1$, divide through by the constant a so the coefficient of x^2 is 1.

Step 3: Complete the square on the left side by adding the square of $\frac{1}{2}$ and the coefficient of x to both sides.

Step 4: Write the left side of the equation as the square of a binomial. Simplify the right side if possible.

Step 5: Apply the square root property for equations, and solve as usual.

The Quadratic Theorem [7.2]

3. If $2x^2 + 3x - 4 = 0$, then

$$x = \frac{-3 \pm \sqrt{9 - 4(2)(-4)}}{2(2)}$$

$$= \frac{-3 \pm \sqrt{41}}{4}$$

For any quadratic equation in the form $ax^2 + bx + c = 0$, $a \neq 0$, the two solutions are

$$x = \frac{-b \pm \sqrt{b^2 - 4ac}}{2a}$$

This last equation is known as the *quadratic formula*.

The Discriminant [7.3]

4. The discriminant for
$x^2 + 6x + 9 = 0$
is $D = 36 - 4(1)(9) = 0$, which means the equation has one rational solution.

The expression $b^2 - 4ac$ that appears under the radical sign in the quadratic formula is known as the *discriminant*.

We can classify the solutions to $ax^2 + bx + c = 0$:

The solutions are	When the discriminant is
Two complex numbers containing i	Negative
One rational number	Zero
Two rational numbers	A positive perfect square
Two irrational numbers	A positive number, but not a perfect square

Equations Quadratic in Form [7.4]

5. The equation $x^4 - x^2 - 12 = 0$ is quadratic in x^2. Letting $y = x^2$, we have

$$y^2 - y - 12 = 0$$
$$(y - 4)(y + 3) = 0$$
$$y = 4 \quad \text{or} \quad y = -3$$

Resubstituting x^2 for y, we have

$$x^2 = 4 \quad \text{or} \quad x^2 = -3$$
$$x = \pm 2 \qquad x = \pm i\sqrt{3}$$

There are a variety of equations whose form is quadratic. We solve most of them by making a substitution so that the equation becomes quadratic, and then solving the equation by factoring the quadratic formula. For example,

The equation	*is quadratic in*
$(2x - 3)^2 + 5(2x - 3) - 6 = 0$	$2x - 3$
$4x^4 - 7x^2 - 2 = 0$	x^2
$2x - 7\sqrt{x} + 3 = 0$	$\sqrt{x}$

Graphing Parabolas [7.5]

6. The graph of $y = x^2 - 4$ will be a parabola. It will cross the x-axis at 2 and -2, and the vertex will be $(0, -4)$.

The graph of any equation of the form

$$y = ax^2 + bx + c \qquad a \neq 0$$

is a *parabola*. The graph is *concave up* if $a > 0$ and *concave down* if $a < 0$. The highest or lowest point on the graph is called the *vertex* and always has an x-coordinate of

$$x = \frac{-b}{2a}.$$

Quadratic Inequalities [1.6]

7. Solve $x^2 - 2x - 8 > 0$. We factor and draw the sign diagram:

$$(x - 4)(x + 2) > 0$$

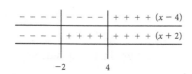

The solution is $x < -2$ or $x > 4$.

We solve quadratic inequalities by manipulating the inequality to get 0 on the right side and then factoring the left side. We then make a diagram that indicates where the factors are positive and where they are negative. From this sign diagram and the original inequality, we graph the appropriate solution set.

Chapter 7 Review

Solve each equation. [7.1, 7.2]

1. $a^2 = -49$

2. $x^2 + 8 = 0$

3. $(3x + 4)^2 = 25$

4. $(2x - 3)^2 = -18$

5. $y^2 - 6y + 9 = 36$

6. $(x - 4)(x + 7) = -10$

7. $125x^3 - 64 = 0$

8. $\dfrac{1}{a} - \dfrac{1}{3} = \dfrac{1}{a - 2}$

9. $\dfrac{t^2}{3} - \dfrac{t}{2} = -\dfrac{3}{2}$

10. $0.02x^2 - 0.03x + 0.05 = 0$

11. $\dfrac{x^2}{6} + \dfrac{5}{6} = -\dfrac{x}{3}$

12. $\sqrt{x} = x - 2$

Fill in the blanks so that the left side of each is a perfect square trinomial. That is, complete the square. [7.1]

13. $a^2 - 8a + \underline{\quad} = (a - \underline{\quad})^2$

14. $x^2 + 3x + \underline{\quad} = (x + \underline{\quad})^2$

15. Solve the formula $36(r - 4)^2 = A$ for r. [7.1]

Solve the following by completing the square. [7.1]

16. $x^2 - 8x = -13$

17. $3x^2 - 9x - 12 = 0$

18. $7x^2 - 5x + 2 = 0$

Find the x-intercepts of the given function.

19. $f(x) = x^2 + 5x - 6$

20. $f(x) = x^2 - 5x + 1$

21. $f(x) = x^2 + 3x + 1$

22. $f(x) = x^2 - 6x + 12$

23. If the length of the longest side of $30° - 60° - 90°$ triangle is 12 cm, find the lengths of the other two sides. [7.1]

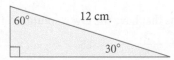

24. Projectile Motion An object projected upward with an initial velocity of 48 feet per second will rise and fall according to the equation $s(t) = 48t - 16t^2$, where s is its distance above the ground at time t. At what times will the object be 20 feet above the ground? [7.2]

25. Revenue The total weekly cost for a company to make x high-quality watches is given by the formula $C(x) = 10x + 300$. If the weekly revenue from selling all x watches is $R(x) = 161x - 0.5x^2$, how many watches must it sell a week to break even? [7.2]

26. Find k so that $3x^2 + 18x = -k$ has one rational solution. [7.3]

27. Find k so that $9x^2 + 30x + k = 0$ has one rational solution. [7.3]

28. Use the discriminant to identify the number and kind of solutions to $3x^2 = -16x + 35$. [7.3]

29. Use the discriminant to identify the number and kind of solutions to $x^2 - 2x + 3 = 0$. [7.3]

Solve each equation. In each case you will have three solutions. [7.2]

30. $x^3 - 27 = 0$

31. $64t^3 + 1 = 0$

Each of the following equations has three solutions. Look for the greatest common factor; then use the quadratic formula to find all solutions. [7.2]

32. $4y^4 = 16y^3 - 20y^2$

33. $8t^5 + 2t^4 = -10t^3$

Find equations that have the given solutions. [7.3]

34. $x = 3, x = -\dfrac{3}{5}$

35. $x = -4, x = 4, x = 3$

36. $x = \dfrac{-3 \pm 2i}{6}$

37. $x = 2 \pm \sqrt{5}$

38. Find an equation that has a solution of $x = 5$ of multiplicity 1 and a solution $x = -3$ of multiplicity 2.

39. Find all solutions to $t^3 = 8t^2 - 25t + 26$, if $t = 2$ is one of the solutions.

Solve each equation. [7.4]

40. $6(2a + 4)^2 = (2a + 4) + 2$

41. $3t^4 = -2t^2 + 8$

42. $4x^4 - 11x^2 - 45 = 0$

43. $(3x + 2)^2 - 4(3x + 2) - x - 6 = 0$

44. $2t - 5\sqrt{t} - 3 = 0$

45. $2x + \sqrt{x} = 15$

46. $(2x - 3) - 7\sqrt{2x - 3} + 12 = 0$

47. Projectile Motion An object is tossed into the air with an upward velocity of 12 feet per second from the top of a building h feet high. The time it takes for the object to hit the ground is given by the formula $16t^2 - 12t - h = 0$. Solve this formula for t. [7.4]

For each of the following functions, give the x-intercepts and the coordinates of the vertex, and sketch the graph. [7.5, 7.6]

48. $f(x) = 2x^2 + 4x - 6$

49. $f(x) = 2x^2 - 20x + 48$

50. $f(x) = -(x - 3)^2 + 1$

51. $f(x) = -4(x - 1)^2 + 4$

Find the vertex and any two convenient points to sketch the graphs of the following functions. [7.5]

52. $f(x) = -x^2 - 2$

53. $g(x) = -3x^2 + 6x$

Sketch the graph of each of the following. Give the coordinates of the vertex in each case. [7.5]

54. $y = x^2 - 4x - 6$

55. $y = -x^2 - 2x + 4$

56. Find an equation in the form $y = a(x - h)^2 + k$ that describes the graph of the parabola. [7.6]

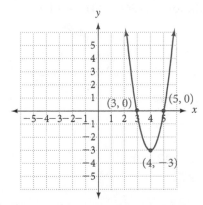

57. Profit Find the maximum weekly profit for a company with weekly costs of $C = 4x + 200$ and weekly revenue of $R = 26x - 0.4x^2$. [7.6]

Solve each inequality and graph the solution. [7.7]

58. $x^2 + 3x - 10 \leq 0$

59. $3x^2 - 8x > 3$

60. $x^2 - 16 \geq 0$

61. $9x^2 - 4 < 0$

62. $(x - 2)(x - 3)(x - 4) < 0$

63. $(x + 1)(x + 2)(x + 3) \geq 0$

64. $\dfrac{2}{x - 3} + 1 \geq 0$

65. $\dfrac{x - 1}{(x + 2)(x - 5)} < 0$

66. $\dfrac{x + 1}{2x - 2} - \dfrac{2}{x^2 - 1} \leq 0$ **67.** $\dfrac{x + 1}{2x + 4} \geq \dfrac{2}{x^2 - 4}$

69. Dimensions of a Rectangle The length of a rectangle is 5 inches less than three times the width. If the area is to be less than 12 square inches, what are the possibilities for the width?

70. Revenue A manufacturer of small calculators knows that the weekly revenue produced by selling x calculators is given by the equation $R = 1{,}700p - 100p^2$, where p is the price of each calculator (in dollars). What price should be charged for each calculator if the revenue is to be at least $7,000 each week?

68. The graph of $y = x^3 + 2x^2 - 5x - 6$ is shown below.

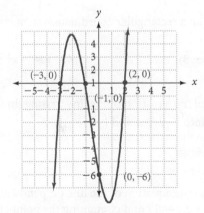

Use the graph to write the solution set for each of the following. [7.7]

a. $x^3 + 2x^2 - 5x - 6 < 0$

b. $x^3 + 2x^2 - 5x - 6 > 0$

c. $x^3 + 2x^2 - 5x - 6 = 0$

Simplify.

1. $6 - 36 \div 6 + 2 \cdot 2$ **2.** $5(13 - 17)^2 - 2(16 - 17)^2$

3. $\left(-\dfrac{5}{2}\right)^3$ **4.** $\left(-\dfrac{3}{4}\right)^2$

5. $3 + 2x - 5(3x - 2)$ **6.** $7 - 2[4x - 3(x + 1)]$

7. $\left(\dfrac{x^{-3}y^{-3}}{x^{-2}y^4}\right)^{-1}$ **8.** $\left(\dfrac{a^{-2}b^{-1}}{a^5b^{-2}}\right)^{-3}$

9. $\sqrt[3]{54}$ **10.** $\sqrt{32x^5}$

11. $64^{-2/3} - 16^{-1/2}$ **12.** $\left(\dfrac{125}{8}\right)^{-2/3}$

13. $\dfrac{\frac{4}{7} - 1}{\frac{4}{7} + 1}$ **14.** $\dfrac{\frac{1}{8} + 3}{1 - \frac{3}{2}}$

Reduce.

15. $\dfrac{3x^2 - 17xy - 28y^2}{3x + 4y}$ **16.** $\dfrac{x^2 - x - 20}{x + 4}$

Divide.

17. $\dfrac{-4 + i}{i}$ **18.** $\dfrac{4 + 2i}{5 - i}$

Solve.

19. $\dfrac{5}{4}x + 3 = 8$ **20.** $\dfrac{1}{3}a + \dfrac{3}{4} = 5$

21. $|a| + 4 = 7$ **22.** $|a + 1| = -2$

23. $\dfrac{x}{4} + \dfrac{15}{4x} = -2$ **24.** $-\dfrac{1}{3} + \dfrac{a}{a + 2} = \dfrac{2}{a + 2}$

25. $(2x - 3)^2 = 8$ **26.** $4x^3 + x = 9x^2$

27. $\dfrac{1}{10}x^2 - \dfrac{1}{2}x + \dfrac{2}{5} = 0$ **28.** $0.09a^2 + 0.02a = -0.04$

29. $\sqrt{y + 4} = y + 4$ **30.** $\sqrt{x - 5} = 5 - \sqrt{x}$

31. $x + 5\sqrt{x} - 24 = 0$ **32.** $x^{2/3} - x^{1/3} = 30$

Solve each inequality and graph the solution.

33. $-5 \le \dfrac{1}{3}x + 7 \le 1$ **34.** $|3x + 2| \ge 1$

35. $x^2 - 3x < 18$ **36.** $\dfrac{x - 4}{x + 2} \ge 0$

Solve each system.

37. $\begin{aligned} 7x - 3y &= 43 \\ -14x + 6y &= -5 \end{aligned}$ **38.** $\begin{aligned} -2x + 4y &= 8 \\ 3x - 6y &= 8 \end{aligned}$

Graph on a rectangular coordinate system.

39. $2x + 3y = 12$ **40.** $y = -x^2 + 4x - 3$

41. Find the equation of the line passing through the points $\left(\dfrac{1}{2}, \dfrac{13}{4}\right)$ and $\left(-\dfrac{3}{2}, \dfrac{1}{4}\right)$.

42. Find the equation of the line perpendicular to $2x + 8y = 17$ and containing the point $(1, 2)$.

Factor completely.

43. $x^2 - 6x + 9 - y^2$ **44.** $(x - 3)^2 + 5(x - 4) - 1$

Rationalize the denominator.

45. $\dfrac{6}{\sqrt[3]{4}}$

46. $\dfrac{\sqrt{3}}{\sqrt{3}-\sqrt{5}}$

If A = {1, 4, 9, 16} and B = {1, 2, 4, 8}, find the following.

47. $\{x \mid x \in A \text{ and } x \in B\}$

48. $\{x \mid x \notin A \text{ and } x \in B\}$

49. Geometry Find all three angles in a triangle if the smallest angle is three-tenths the largest angle and the remaining angle is 20° more than the smallest angle.

50. Investing A total of $6,000 is invested in two accounts. One account earns 3% per year, and the other earns 2% per year. If the total interest for the first year is $145, how much was invested in each account?

51. Direct Variation y varies directly with the square of x. If $y = -4$ when $x = -\frac{2}{3}$, find y when $x = \frac{5}{9}$.

52. Inverse Variation w varies inversely with the square root of c. If w is 8 when c is 4, find w when c is 16.

The chart shows what other types of phones iPhone users had before purchasing the iPhone. If 426,350 people were surveyed, use the information to answer Questions 53 and 54. Round to the nearest whole number.

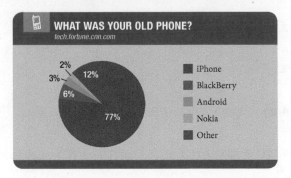

53. How many people formerly owned another iPhone?

54. How many people formerly owned either a Blackberry or an Android?

55. Unit Analysis Convert 422.4 feet/minute into miles/hour.

56. Piecewise Functions Graph the piecewise function.

$$f(x) = \begin{cases} \dfrac{1}{4}x - 4 & \text{if } x \le 4 \\ -3x + 9 & \text{if } x > 4 \end{cases}$$

Chapter 7 Test

Solve each equation. [7.1, 7.2]

1. $(2x + 4)^2 = 25$

2. $(2x - 6)^2 = -8$

3. $y^2 - 10y + 25 = -4$

4. $(y + 1)(y - 3) = -6$

5. $8t^3 - 125 = 0$

6. $\dfrac{1}{a + 2} - \dfrac{1}{3} = \dfrac{1}{a}$

7. Solve the formula $64(1 + r)^2 = A$ for r. [7.1]

8. Solve $x^2 - 4x = -2$ by completing the square. [7.1]

9. Projectile Motion An object projected upward with an initial velocity of 32 feet per second will rise and fall according to the equation $s(t) = 32t - 16t^2$, where s is its distance above the ground at time t. At what times will the object be 12 feet above the ground? [7.2]

10. Revenue The total weekly cost for a company to make x ceramic coffee cups is given by the formula $C(x) = 2x + 100$. If the weekly revenue from selling all x cups is $R(x) = 25x - 0.2x^2$, how many cups must it sell a week to make a profit of $200 a week? [7.2]

11. Find k so that $kx^2 = 12x - 4$ has one rational solution. [7.3]

12. Use the discriminant to identify the number and kind of solutions to $2x^2 - 5x = 7$. [7.3]

Find equations that have the given solutions. [7.3]

13. $x = 5, x = -\dfrac{2}{3}$

14. $x = 2, x = -2, x = 7$

Solve each equation. [7.4]

15. $4x^4 - 7x^2 - 2 = 0$

16. $(2t + 1)^2 - 5(2t + 1) + 6 = 0$

17. $2t - 7\sqrt{t} + 3 = 0$

18. Projectile Motion An object is tossed into the air with an upward velocity of 14 feet per second from the top of a building h feet high. The time it takes for the object to hit the ground is given by the formula $16t^2 - 14t - h = 0$. Solve this formula for t. [7.4]

Sketch the graph of each of the following equations. Give the coordinates of the vertex in each case. [7.5]

19. $y = x^2 - 2x - 3$

20. $y = -x^2 + 2x + 8$

Graph each of the following inequalities. [7.6]

21. $x^2 - x - 6 \leq 0$

22. $2x^2 + 5x > 3$

23. Profit Find the maximum weekly profit for a company with weekly costs of $C = 5x + 100$ and weekly revenue of $R = 25x - 0.1x^2$. [7.5]

8

Exponential and Logarithmic Functions

Chapter Outline

© istockphoto.com/aldomurillo

If you have had any problems with or had testing done on your thyroid gland, then you may have come in contact with radioactive iodine-131. Like all radioactive elements, iodine-131 decays naturally. The half-life of iodine-131 is 8 days, which means that every 8 days a sample of iodine-131 will decrease to half of its original amount. The following table and graph show what happens to a 1,600-microgram sample of iodine-131 over time.

Iodine-131 as a Function of Time

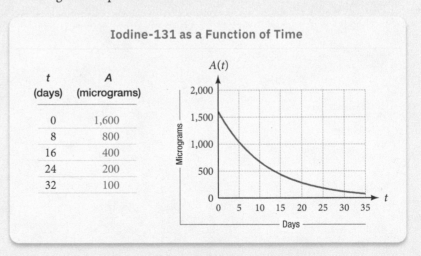

t (days)	A (micrograms)
0	1,600
8	800
16	400
24	200
32	100

The function represented by the information in the table and graph is

$$A(t) = 1{,}600 \cdot 2^{-t/8}$$

It is one of the types of functions we will study in this chapter.

Never mistake activity for achievement.
— John Wooden, legendary UCLA basketball coach

You may think that the John Wooden quote above has to do with being productive and efficient, or using your time wisely, but it is really about being honest with yourself. I have had students come to me after failing a test saying, "I can't understand why I got such a low grade after I put so much time in studying." One student even had help from a tutor and felt she understood everything that we covered. After asking her a few questions, it became clear that she spent all her time studying with a tutor and the tutor was doing most of the work. The tutor could work all the homework problems, but the student could not. She has mistaken activity for achievement.

Can you think of situations in your life when you are mistaking activity for achievement?

How would you describe someone who is mistaking activity for achievement in the way they study for their math class?

Which of the following best describes the idea behind the John Wooden quote?
► Always be efficient.
► Don't kid yourself.
► Take responsibility for your own success.
► Study with purpose.

EXPONENTIAL FUNCTIONS

OBJECTIVES

A Find function values for exponential functions.

B Graph exponential functions.

C Solve compound interest problems involving exponential functions.

To obtain an intuitive idea of how exponential functions behave, we can consider the heights attained by a bouncing ball. When a tennis ball is dropped from any height, the first bounce will reach a height that is $\frac{2}{3}$ of the original height. The second bounce will reach $\frac{2}{3}$ of the height of the first bounce, and so on, as shown in Figure 1.

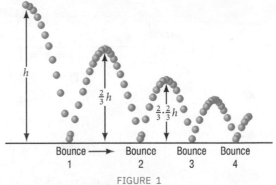

FIGURE 1

If the ball is initially dropped from a height of 1 meter, then during the first bounce it will reach a height of $\frac{2}{3}$ meter. The height of the second bounce will reach $\frac{2}{3}$ of the height reached on the first bounce. The maximum height of any bounce is $\frac{2}{3}$ of the height of the previous bounce.

Initial height:	$h = 1$
Bounce 1:	$h = \frac{2}{3}(1) = \frac{2}{3}$
Bounce 2:	$h = \frac{2}{3}\left(\frac{2}{3}\right) = \left(\frac{2}{3}\right)^2$
Bounce 3:	$h = \frac{2}{3}\left(\frac{2}{3}\right)^2 = \left(\frac{2}{3}\right)^3$
Bounce 4:	$h = \frac{2}{3}\left(\frac{2}{3}\right)^3 = \left(\frac{2}{3}\right)^4$
$\vdots$	$\vdots$
Bounce n:	$h = \frac{2}{3}\left(\frac{2}{3}\right)^{n-1} = \left(\frac{2}{3}\right)^n$

8.1 VIDEOS

651

This last equation is exponential in form. We classify all exponential functions together with the following definition.

> **DEFINITION** *exponential function*
>
> An *exponential function* is any function that can be written in the form
> $$f(x) = b^x$$
> where b is a positive real number other than 1.

Each of the following is an exponential function:

$$f(x) = 2^x \qquad y = 3^x \qquad f(x) = \left(\frac{1}{4}\right)^x$$

Ⓐ Evaluating Exponential Functions

The first step in becoming familiar with exponential functions is to find some values for specific exponential functions.

EXAMPLE 1

If the exponential functions f and g are defined by

$$f(x) = 2^x \quad \text{and} \quad g(x) = 3^x$$

then

$$f(0) = 2^0 = 1 \qquad\qquad g(0) = 3^0 = 1$$
$$f(1) = 2^1 = 2 \qquad\qquad g(1) = 3^1 = 3$$
$$f(2) = 2^2 = 4 \qquad\qquad g(2) = 3^2 = 9$$
$$f(3) = 2^3 = 8 \qquad\qquad g(3) = 3^3 = 27$$
$$f(-2) = 2^{-2} = \frac{1}{2^2} = \frac{1}{4} \qquad g(-2) = 3^{-2} = \frac{1}{3^2} = \frac{1}{9}$$
$$f(-3) = 2^{-3} = \frac{1}{2^3} = \frac{1}{8} \qquad g(-3) = 3^{-3} = \frac{1}{3^3} = \frac{1}{27}$$

In the introduction to this chapter, we indicated that the half-life of iodine-131 is 8 days, which means that every 8 days a sample of iodine-131 will decrease to half of its original amount. If we start with A_0 micrograms of iodine-131, then after t days the sample will contain

$$A(t) = A_0 \cdot 2^{-t/8}$$

micrograms of iodine-131.

EXAMPLE 2

A patient is administered a 1,200-microgram dose of iodine-131. How much iodine-131 will be in the patient's system after 10 days, and after 16 days?

SOLUTION The initial amount of iodine-131 is $A_0 = 1,200$, so the function that gives the amount left in the patient's system after t days is

$$A(t) = 1,200 \cdot 2^{-t/8}$$

After 10 days, the amount left in the patient's system is

$$A(10) = 1,200 \cdot 2^{-10/8} = 1,200 \cdot 2^{-1.25} \approx 504.5 \text{ microgram}$$

After 16 days, the amount left in the patient's system is

$$A(16) = 1,200 \cdot 2^{-16/8} = 1,200 \cdot 2^{-2} = 300 \text{ micrograms}$$

 NOTE Recall that the symbol $\approx$ is read "is approximately equal to."

We will now turn our attention to the graphs of exponential functions. Because the notation y is easier to use when graphing, and $y = f(x)$, for convenience we will write the exponential functions as

$$y = b^x$$

Ⓑ Graphing Exponential Functions

EXAMPLE 3

Sketch the graph of the exponential function $y = 2^x$.

SOLUTION Using the results of Example 1, we produce the following table. Graphing the ordered pairs given in the table and connecting them with a smooth curve, we have the graph of $y = 2^x$ shown in Figure 2.

x	y
-3	$\frac{1}{8}$
-2	$\frac{1}{4}$
0	1
1	2
2	4
3	8

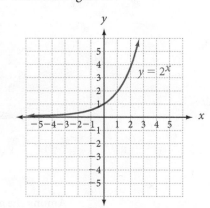

FIGURE 2

Notice that the graph does not cross the x-axis. It *approaches* the x-axis — in fact, we can get it as close to the x-axis as we want without it actually intersecting the x-axis. For the graph of $y = 2^x$ to intersect the x-axis, we would have to find a value of x that would make $2^x = 0$. Because no such value of x exists, the graph of $y = 2^x$ cannot intersect the x-axis.

EXAMPLE 4

Sketch the graph of $y = \left(\dfrac{1}{3}\right)^x$.

SOLUTION The table beside Figure 3 gives some ordered pairs that satisfy the equation. Using the ordered pairs from the table, we have the graph shown in Figure 3.

x	y
−3	27
−2	9
−1	3
0	1
1	$\frac{1}{3}$
2	$\frac{1}{9}$
3	$\frac{1}{27}$

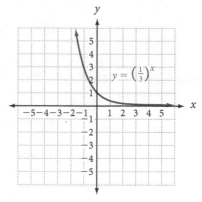

FIGURE 3

The graphs of all exponential functions have two things in common: (1) Each crosses the y-axis at $(0, 1)$ because $b^0 = 1$; and (2) none can cross the x-axis because $b^x = 0$ is impossible due to the restrictions on b.

Figures 4 and 5 show some families of exponential curves to help you become more familiar with them on an intuitive level.

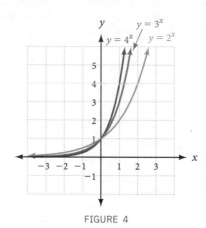

FIGURE 4

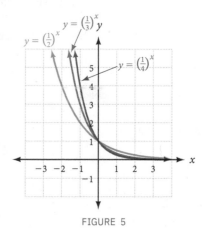

FIGURE 5

Among the many applications of exponential functions are the applications having to do with interest-bearing accounts. The next section will present the details.

C Compound Interest

If P dollars are deposited in an account with annual interest rate r, compounded n times per year, then the amount of money in the account after t years is given by the formula

$$A(t) = P\left(1 + \frac{r}{n}\right)^{nt}$$

EXAMPLE 5

Suppose you deposit $500 in an account with an annual interest rate of 4% compounded quarterly. Find an equation that gives the amount of money in the account after t years. Then find

a. The amount of money in the account after 5 years.

b. The number of years it will take for the account to contain $1,000.

SOLUTION   First, we note that $P = 500$ and $r = 0.04$. Interest that is compounded quarterly is compounded 4 times a year, giving us $n = 4$. Substituting these numbers into the preceding formula, we have our function

$$1 + 0.0444t = 500(1.02)^{4t}$$

a. To find the amount after 5 years, we let $t = 5$:

$$A(5) = 500(1.01)^{4 \cdot 5} = 500(1.01)^{20} \approx \$610.10$$

Our answer is found on a calculator, and then rounded to the nearest cent.

b. To see how long it will take for this account to total $1,000, we graph the equation $Y_1 = 500(1.01)^{4X}$ on a graphing calculator, and then look to see where it intersects the line $Y_2 = 1,000$. The two graphs are shown in Figure 6.

Using Zoom and Trace, or the Intersect function on the graphing calculator, we find that the two curves intersect at $X \approx 17.42$ and $Y = 1,000$. This means that our account will contain $1,000 after the money has been on deposit for 17.42 years.

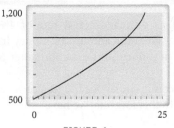

FIGURE 6

The Natural Exponential Function

A commonly occurring exponential function is based on a special number we denote with the letter e. The number e is a number like π. It is irrational and occurs in many formulas that describe the world around us. Like π, it can be approximated with a decimal number. Whereas π is approximately 3.1416, e is approximately 2.7183. (If you have a calculator with a key labeled e^x, you can use it to find e^1 to find a more accurate approximation to e.) We cannot give a more precise definition of the *number e* without using some of the topics taught in calculus. For the work we are going to do with the number e, we only need to know that it is an irrational number that is approximately 2.7183.

Here are a table and graph (Figure 7) for the natural exponential function

$$y = f(x) = e^x$$

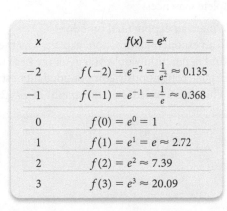

x	$f(x) = e^x$
-2	$f(-2) = e^{-2} = \frac{1}{e^2} \approx 0.135$
-1	$f(-1) = e^{-1} = \frac{1}{e} \approx 0.368$
0	$f(0) = e^0 = 1$
1	$f(1) = e^1 = e \approx 2.72$
2	$f(2) = e^2 \approx 7.39$
3	$f(3) = e^3 \approx 20.09$

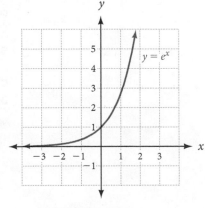

FIGURE 7

Continuing our discussion of natural exponential functions relating to interest-bearing accounts: In Example 5, we worked with the formula

$$A(t) = P\left(1 + \frac{r}{n}\right)^{nt}$$

that gives the amount of money in an account if P dollars are deposited for t years at annual interest rate r, compounded n times per year. In Example 5, the number of compounding periods was 4. What would happen if we let the number of compounding periods become larger and larger, so that we compounded the interest every day, then every hour, then every second, and so on? If we take this as far as it can go, we end up compounding the interest every moment. When this happens, we have an account with interest that is compounded continuously, and the amount of money in such an account depends on the number e. Here are the details.

Continuously Compounded Interest If P dollars are deposited in an account with annual interest rate r, compounded continuously, then the amount of money in the account after t years is given by the formula

$$A(t) = Pe^{rt}$$

EXAMPLE 6

Suppose you deposit $500 in an account with an annual interest rate of 4% compounded continuously. Find an equation that gives the amount of money in the account after t years. Then find the amount of money in the account after 5 years.

SOLUTION   Because the interest is compounded continuously, we use the formula $A(t) = Pe^{rt}$. Substituting $P = 500$ and $r = 0.04$ into this formula, we have

$$A(t) = 500e^{0.04t}$$

After 5 years, this account will contain

$$A(5) = 500e^{0.04 \cdot 5} = 500e^{0.2} \approx \$610.70$$

to the nearest cent. Compare this result with the answer to Example 5a.

GETTING READY FOR CLASS

After reading through the preceding section, respond in your own words and in complete sentences.

A. What is an exponential function?

B. In an exponential function, explain why the base b cannot equal 1. (What kind of function would you get if the base was equal to 1?)

C. Explain continuously compounded interest.

D. What characteristics do the graphs of $y = 2^x$ and $y = \left(\frac{1}{2}\right)^x$ have in common?

Problem Set 8.1

Let $f(x) = 3^x$ and $g(x) = \left(\frac{1}{2}\right)^x$, and evaluate each of the following.

1. $g(0)$ **2.** $f(0)$ **3.** $g(-1)$ **4.** $g(-4)$

5. $f(-3)$ **6.** $f(-1)$ **7.** $f(2) + g(-2)$ **8.** $f(2) - g(-2)$

Let $f(x) = 4^x$ and $g(x) = \left(\frac{1}{3}\right)^x$. Evaluate each of the following.

9. $f(-1) + g(1)$ **10.** $f(2) + g(-2)$ **11.** $\dfrac{f(-2)}{g(1)}$ **12.** $f(3) - g(2)$

Graph each of the following functions.

13. $y = 4^x$ **14.** $y = 2^{-x}$ **15.** $y = 3^{-x}$ **16.** $y = \left(\frac{1}{3}\right)^{-x}$

17. $y = 2^{x+1}$ **18.** $y = 2^{x-3}$ **19.** $y = e^x$ **20.** $y = e^{-x}$

21. $y = \left(\frac{1}{3}\right)^x$ **22.** $y = \left(\frac{1}{2}\right)^{-x}$ **23.** $y = 3^{x+2}$ **24.** $y = 2 \cdot 3^{-x}$

Graph each of the following functions on the same coordinate system for positive values of x only.

25. $y = 2x, y = x^2, y = 2^x$ **26.** $y = 3x, y = x^3, y = 3^x$

27. On a graphing calculator, graph the family of curves $y = b^x$, $b = 2, 4, 6, 8$.

28. On a graphing calculator, graph the family of curves $y = b^x$, $b = \frac{1}{2}, \frac{1}{4}, \frac{1}{6}, \frac{1}{8}$.

APPLYING THE CONCEPTS

29. Bouncing Ball Suppose the ball mentioned in the introduction to this section is dropped from a height of 6 feet above the ground. Find an exponential equation that gives the height h the ball will attain during the nth bounce. How high will it bounce on the fifth bounce?

30. Bouncing Ball A golf ball is manufactured so that if it is dropped from A feet above the ground onto a hard surface, the maximum height of each bounce will be one-half of the height of the previous bounce. Find an exponential equation that gives the height h the ball will attain during the nth bounce. If the ball is dropped from 10 feet above the ground onto a hard surface, how high will it bounce on the eighth bounce?

31. Exponential Decay Twinkies on the shelf of a convenience store lose their fresh tastiness over time. We say that the taste quality is 1 when the Twinkies are first put on the shelf at the store, and that the quality of tastiness declines according to the function $Q(t) = 0.85^t$ (t in days). Graph this function on a graphing calculator, and determine when the taste quality will be one-half of its original value.

32. Exponential Growth Automobiles built before 1993 use Freon in their air conditioners. The federal government now prohibits the manufacture of Freon. Because the supply of Freon is decreasing, the price per pound is increasing exponentially. Current estimates put the formula for the price per pound of Freon at $p(t) = 1.89(1.25)^t$, where t is the number of years since 1990. Find the price of Freon in 1995 and 1990. How much will Freon cost in the year 2020?

33. Compound Interest Suppose you deposit $1,200 in an account with an annual interest rate of 6% compounded quarterly.

   a. Find an equation that gives the amount of money in the account after t years.

   b. Find the amount of money in the account after 8 years.

   c. How many years will it take for the account to contain $2,400?

   d. If the interest were compounded continuously, how much money would the account contain after 8 years?

34. Compound Interest Suppose you deposit $500 in an account with an annual interest rate of 8% compounded monthly.

   a. Find an equation that gives the amount of money in the account after t years.

   b. Find the amount of money in the account after 5 years.

   c. How many years will it take for the account to contain $1,000?

   d. If the interest were compounded continuously, how much money would the account contain after 5 years?

Declining-Balance Depreciation The declining-balance method of depreciation is an accounting method businesses use to deduct most of the cost of new equipment during the first few years after purchase. Unlike other methods, the declining-balance formula does not consider salvage value.

35. Value of a Crane The function

$$V(t) = 450,000 \, (1 - 0.30)^t,$$

where V is value and t is time in years, can be used to find the value of a crane for the first 6 years of use.

   a. What is the value of the crane after 3 years and 6 months?

   b. State the domain of this function.

   c. Sketch the graph of this function.

   d. State the range of this function.

   e. After how many years will the crane be worth only $85,000?

36. Value of a Printing Press The function $V(t) = 375,000(1 - 0.25)^t$, where V is value and t is time in years, can be used to find the value of a printing press during the first 7 years of use.

   a. What is the value of the printing press after 4 years and 9 months?

   b. State the domain of this function.

   c. Sketch the graph of this function.

   d. State the range of this function.

   e. After how many years will the printing press be worth only $65,000?

37. Bacteria Growth Suppose it takes 12 hours for a certain strain of bacteria to reproduce by dividing in half. If 50 bacteria are present to begin with, then the total number present after x days will be $f(x) = 50 \cdot 4^x$. Find the total number present after 1 day, 2 days, and 3 days.

38. Bacteria Growth Suppose it takes 1 day for a certain strain of bacteria to reproduce by dividing in half. If 100 bacteria are present to begin with, then the total number present after x days will be $f(x) = 100 \cdot 2^x$. Find the total number present after 1 day, 2 days, 3 days, and 4 days. How many days must elapse before over 100,000 bacteria are present?

39. **Value of a Painting** A painting is purchased as an investment for $150. If the painting's value doubles every 3 years, then its value is given by the function

$$V(t) = 150 \cdot 2^{t/3} \text{ for } t \geq 0$$

where t is the number of years since it was purchased, and $V(t)$ is its value (in dollars) at that time. Graph this function.

40. **Value of a Painting** A painting is purchased as an investment for $125. If the painting's value doubles every 5 years, then its value is given by the function

$$V(t) = 125 \cdot 2^{t/5} \text{ for } t \geq 0$$

where t is the number of years since it was purchased, and $V(t)$ is its value (in dollars) at that time. Graph this function.

41. **Cost Increase** The cost of a can of Coca-Cola® in 1960 was $0.10. The exponential function that models the cost of a Coca-Cola by year is given below, where t is the number of years since 1960.

$$C(t) = 0.10e^{0.0576t}$$

   a. What was the expected cost of a can of Coca-Cola in 1985?

   b. What was the expected cost of a can of Coca-Cola in 2000?

   c. What is the expected cost of a can of Coca-Cola in 2020?

   d. What is the expected cost of a can of Coca-Cola in 2050?

42. **Airline Travel** The number of airline passengers in 1990 was 466 million. The number of passengers traveling by airplane each year has increased exponentially according to the model, $P(t) = 466 \cdot 1.035^t$, where t is the number of years since 1990 (U.S. Census Bureau).

   a. How many passengers traveled in 1997?

   b. How many passengers will travel in 2015?

43. **Bankruptcy Model** In 1997, there were a total of 1,316,999 bankruptcies filed under the Bankruptcy Reform Act (Administrative Office of the U.S. Courts, Statistical Tables for the Federal Judiciary). The model for the number of bankruptcies filed is $B(t) = 0.798 \cdot 1.164^t$, where t is the number of years since 1994 and B is the number of bankruptcies filed in terms of millions. How close was the model in predicting the actual number of bankruptcies filed in 1997?

44. **Value of a Car** As a car ages, its value decreases. The value of a particular car with an original purchase price of $25,600 is modeled by the following function, where c is the value at time t (*Kelly Blue Book*).

$$c(t) = 25,600(1 - 0.22)^t$$

   a. What is the value of the car when it is 3 years old?

   b. What is the total depreciation amount after 4 years?

45. **Bacteria Decay** You are conducting a biology experiment and begin with 5 million cells, but some of those cells are dying each minute. The rate of death of the cells is modeled by the function $A(t) = A_0 \cdot e^{-0.598t}$, where A_0 is the original number of cells, t is time in minutes, and A is the number of cells remaining after t minutes.

   a. How may cells remain after 5 minutes?

   b. How many cells remain after 10 minutes?

   c. How many cells remain after 20 minutes?

46. Health Care In 1990, $699 billion was spent on health care expenditures. The amount of money, E, in billions spent on health care expenditures can be estimated using the function $E(t) = 78.16(1.11)^t$, where t is time in years since 1970 (U.S. Census Bureau).

a. How close was the estimate determined by the function in estimating the actual amount of money spent on health care expenditures in 1990?

b. What are the expected health care expenditures in 2018, 2019, and 2020?

GETTING READY FOR THE NEXT SECTION

Solve each equation for y.

47. $x = 2y - 3$

48. $x = \dfrac{y + 7}{5}$

49. $x = y^2 - 3$

50. $x = (y + 4)^3$

51. $x = \dfrac{y - 4}{y - 2}$

52. $x = \dfrac{y + 5}{y - 3}$

53. $x = \sqrt{y - 3}$

54. $x = \sqrt{y} + 5$

EXTENDING THE CONCEPTS

55. Reading Graphs The graphs of two exponential functions are given in Figures 8 and 9. Use the graphs to find the following.

a. $f(0)$ **b.** $f(-1)$ **c.** $f(1)$ **d.** $g(0)$

e. $g(1)$ **f.** $g(-1)$ **g.** $f(g(0))$ **h.** $g(f(0))$

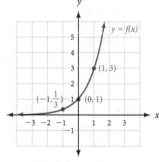

FIGURE 8

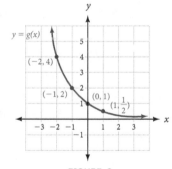

FIGURE 9

56. **Drag Racing** A dragster is equipped with a computer that measures its racing speed. Table 1 gives the speed of the dragster every second during one particular race. Figure 10 is a line graph constructed from the data in Table 1. The graph of the function $s(t) = 250(1 - 1.5^{-t})$ contains the first point and the last point shown in Figure 10. That is, both (0, 0) and (6, 228.1) satisfy the function. Graph the function to see how close it comes to the other points in Figure 10.

Table 1	
Elapsed Time (sec)	Speed (mi/hr)
0	0.0
1	72.7
2	129.9
3	162.8
4	192.2
5	212.4
6	228.1

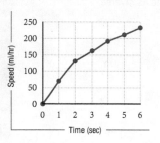

FIGURE 10

THE INVERSE OF A FUNCTION

© istockphoto.com/Rostislav_Sedlacek

OBJECTIVES

A Find the equation of the inverse of a function.

B Sketch a graph of a function and its inverse.

The following diagram (Figure 1) shows the route Justin takes to school. He leaves his home and drives 3 miles east, and then turns left and drives 2 miles north. When he leaves school to drive home, he drives the same two segments, but in the reverse order and the opposite direction; that is, he drives 2 miles south, turns right, and drives 3 miles west. When he arrives home from school, he is right where he started. His route home "undoes" his route to school, leaving him where he began.

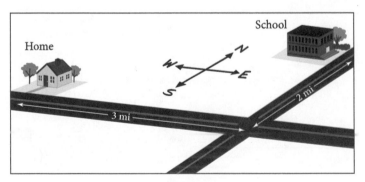

FIGURE 1

As you will see, the relationship between a function and its inverse function is similar to the relationship between Justin's route from home to school and his route from school to home.

Suppose the function f is given by

$$f = (1, 4), (2, 5), (3, 6), (4, 7)$$

The inverse of f is obtained by reversing the order of the coordinates in each ordered pair in f. The inverse of f is the relation given by

$$g = (4, 1), (5, 2), (6, 3), (7, 4)$$

It is obvious that the domain of f is now the range of g, and the range of f is now the domain of g. Every function (or relation) has an inverse that is obtained from the original function by interchanging the components of each ordered pair.

8.2 VIDEOS

Finding the Inverse Relation

Suppose a function f is defined with an equation instead of a list of ordered pairs. We can obtain the equation of the inverse of f by interchanging the role of x and y in the equation for f.

EXAMPLE 1

If the function f is defined by $f(x) = 2x - 3$, find the equation that represents the inverse of f.

SOLUTION   Because the inverse of f is obtained by interchanging the components of all the ordered pairs belonging to f, and each ordered pair in f satisfies the equation $y = 2x - 3$, we simply exchange x and y in the equation $y = 2x - 3$ to get the formula for the inverse of f:

$$x = 2y - 3$$

We now solve this equation for y in terms of x:

$$x + 3 = 2y$$
$$\frac{x + 3}{2} = y$$
$$y = \frac{x + 3}{2}$$

The last line gives the equation that defines the inverse of f. Let's compare the graphs of f and its inverse as given here. (See Figure 2.)

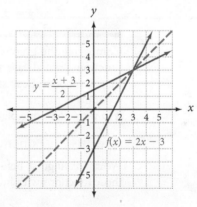

FIGURE 2

The graphs of f and its inverse have symmetry about the line $y = x$. This is a reasonable result since the one function was obtained from the other by interchanging x and y in the equation. The ordered pairs (a, b) and (b, a) always have symmetry about the line $y = x$.

B Graph a Function with Its Inverse

EXAMPLE 2

Graph the function $y = x^2 - 2$ and its inverse. Give the equation for the inverse.

SOLUTION We can obtain the graph of the inverse of $y = x^2 - 2$ by graphing $y = x^2 - 2$ by the usual methods, and then reflecting the graph about the line $y = x$.

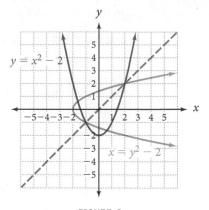

FIGURE 3

The equation that corresponds to the inverse of $y = x^2 - 2$ is obtained by interchanging x and y to get $x = y^2 - 2$.

We can solve the equation $x = y^2 - 2$ for y in terms of x as follows:

$$x = y^2 - 2$$
$$x + 2 = y^2$$
$$y = \pm\sqrt{x + 2}$$

Comparing the graphs from Examples 1 and 2, we observe that the inverse of a function is not always a function. In Example 1, both f and its inverse have graphs that are nonvertical straight lines and therefore both represent functions. In Example 2, the inverse of function f is not a function, since a vertical line crosses it in more than one place.

One-to-One Functions

We can distinguish between those functions with inverses that are also functions and those functions with inverses that are not functions with the following definition.

DEFINITION *one-to-one functions*

A function is a *one-to-one function* if every element in the range comes from exactly one element in the domain.

This definition indicates that a one-to-one function will yield a set of ordered pairs in which no two different ordered pairs have the same second coordinates. For example, the function

$$f = \{(2, 3), (-1, 3), (5, 8)\}$$

is not one-to-one because the element 3 in the range comes from both 2 and -1 in the domain. On the other hand, the function

$$g = \{(5, 7), (3, -1), (4, 2)\}$$

is a one-to-one function because every element in the range comes from only one element in the domain.

Horizontal Line Test

If we have the graph of a function, we can determine if the function is one-to-one with the following test. If a horizontal line crosses the graph of a function in more than one place, then the function is not a one-to-one function because the points at which the horizontal line crosses the graph will be points with the same y-coordinates, but different x-coordinates. Therefore, the function will have an element in the range (the y-coordinate) that comes from more than one element in the domain (the x-coordinates).

Of the functions we have covered previously, all the linear functions and exponential functions are one-to-one functions because no horizontal lines can be found that will cross their graphs in more than one place.

Functions Whose Inverses Are Also Functions

Because one-to-one functions do not repeat second coordinates, when we reverse the order of the ordered pairs in a one-to-one function, we obtain a relation in which no two ordered pairs have the same first coordinate—by definition, this relation must be a function. In other words, every one-to-one function has an inverse that is itself a function. Because of this, we can use function notation to represent that inverse.

Inverse Function Notation

If $y = f(x)$ is a one-to-one function, then the inverse of f is also a function and can be denoted by $y = f^{-1}(x)$.

NOTE The notation f^{-1} does not represent the reciprocal of f. That is, the -1 in this notation is not an exponent. The notation f^{-1} is defined as representing the inverse function for a one-to-one function.

To illustrate, in Example 1 we found that the inverse of $f(x) = 2x - 3$ was the function $y = \frac{x+3}{2}$. We can write this inverse function with inverse function notation as

$$f^{-1}(x) = \frac{x + 3}{2}$$

On the other hand, the inverse of the function in Example 2 is not itself a function, so we do not use the notation $f^{-1}(x)$ to represent it.

EXAMPLE 3

Find the inverse of $g(x) = \frac{x - 4}{x - 2}$.

SOLUTION To find the inverse for g, we begin by replacing $g(x)$ with y to obtain

$$y = \frac{x - 4}{x - 2} \qquad \text{The original function}$$

To find an equation for the inverse, we exchange x and y.

$$x = \frac{y - 4}{y - 2} \qquad \text{The inverse of the original function}$$

To solve for y, we first multiply each side by $y - 2$ to obtain

$$x(y - 2) = y - 4$$

$$xy - 2x = y - 4 \qquad \text{Distributive property}$$

$$xy - y = 2x - 4 \qquad \text{Collect all terms containing } y \text{ on the left side.}$$

$$y(x - 1) = 2x - 4 \qquad \text{Factor } y \text{ from each term on the left side.}$$

$$y = \frac{2x - 4}{x - 1} \qquad \text{Divide each side by } x - 1.$$

Because our original function is one-to-one, as verified by the graph in Figure 4, its inverse is also a function. Therefore, we can use inverse function notation to write

$$g^{-1}(x) = \frac{2x - 4}{x - 1}$$

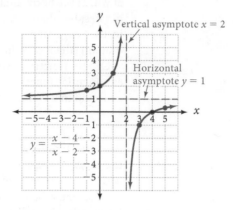

FIGURE 4

EXAMPLE 4

Graph the function $y = 2^x$ and its inverse $x = 2^y$.

SOLUTION We graphed $y = 2^x$ in the preceding section. We simply reflect its graph about the line $y = x$ to obtain the graph of its inverse, $x = 2^y$.

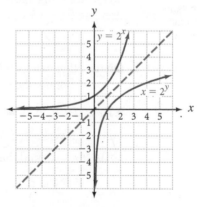

FIGURE 5

As you can see from the graph, $x = 2^y$ is a function. We do not have the mathematical tools to solve this equation for y, however. Therefore, we are unable to use the inverse function notation to represent this function. In the next section, we will give a definition that solves this problem. For now, we simply leave the equation as $x = 2^y$.

Functions, Relations, and Inverses: A Summary

Here are some of the things we know about functions, relations, and their inverses:

1. Every function is a relation, but not every relation is a function.

2. Every function has an inverse, but only one-to-one functions have inverses that are also functions.

3. The domain of a function is the range of its inverse, and the range of a function is the domain of its inverse.

4. If $y = f(x)$ is a one-to-one function, then we can use the notation $y = f^{-1}(x)$ to represent its inverse function.

5. The graph of a function and its inverse have symmetry about the line $y = x$.

6. If (a, b) belongs to the function f, then the point (b, a) belongs to its inverse.

GETTING READY FOR CLASS

After reading through the preceding section, respond in your own words and in complete sentences.

A. What is the inverse of a function?

B. What is the relationship between the graph of a function and the graph of its inverse?

C. Explain why only one-to-one functions have inverses that are also functions.

D. Describe the vertical line test, and explain the difference between the vertical line test and the horizontal line test.

Problem Set 8.2

For each of the following one-to-one functions, find the equation of the inverse. Write the inverse using the notation $f^{-1}(x)$.

1. $f(x) = 3x - 1$ **2.** $f(x) = 2x - 5$ **3.** $f(x) = x^3$

4. $f(x) = x^3 - 2$ **5.** $f(x) = \dfrac{x - 3}{x - 1}$ **6.** $f(x) = \dfrac{x - 2}{x - 3}$

7. $f(x) = \dfrac{x - 3}{4}$ **8.** $f(x) = \dfrac{x + 7}{2}$ **9.** $f(x) = \dfrac{1}{2}x - 3$

10. $f(x) = \dfrac{1}{3}x + 1$ **11.** $f(x) = \dfrac{2}{3}x - 3$ **12.** $f(x) = -\dfrac{1}{2}x + 4$

13. $f(x) = x^3 - 4$ **14.** $f(x) = -3x^3 + 2$ **15.** $f(x) = \dfrac{4x - 3}{2x + 1}$

16. $f(x) = \dfrac{3x - 5}{4x + 3}$ **17.** $f(x) = \dfrac{2x + 1}{3x + 1}$ **18.** $f(x) = \dfrac{3x + 2}{5x + 1}$

For each of the following relations, sketch the graph of the relation and its inverse, and write an equation for the inverse.

19. $y = 2x - 1$ **20.** $y = 3x + 1$ **21.** $y = x^2 - 3$ **22.** $y = x^2 + 1$

23. $y = x^2 - 2x - 3$ **24.** $y = x^2 + 2x - 3$

25. $y = 3^x$ **26.** $y = \left(\dfrac{1}{2}\right)^x$ **27.** $y = 4$ **28.** $y = -2$

29. $y = \dfrac{1}{2}x^3$ **30.** $y = x^3 - 2$ **31.** $y = \dfrac{1}{2}x + 2$ **32.** $y = \dfrac{1}{3}x - 1$

33. $y = \sqrt{x + 2}$ **34.** $y = \sqrt{x} + 2$

35. Determine if the following functions are one-to-one.

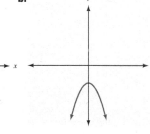

36. Could the following tables of values represent ordered pairs from one-to-one functions? Explain your answer.

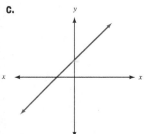

a.

x	y
−2	5
−1	4
0	3
1	4
2	5

b.

x	y
1.5	0.1
2.0	0.2
2.5	0.3
3.0	0.4
3.5	0.5

37. If $f(x) = 3x - 2$, then $f^{-1}(x) = \dfrac{x + 2}{3}$. Use these two functions to find

a. $f(2)$ **b.** $f^{-1}(2)$ **c.** $f[f^{-1}(2)]$ **d.** $f^{-1}[f(2)]$

38. If $f(x) = \frac{1}{2}x + 5$, then $f^{-1}(x) = 2x - 10$. Use these two functions to find

    a. $f(-4)$ **b.** $f^{-1}(-4)$ **c.** $f[f^{-1}(-4)]$ **d.** $f^{-1}[f(-4)]$

39. Let $f(x) = \frac{1}{x}$, and find $f^{-1}(x)$.

40. Let $f(x) = \frac{a}{x}$, and find $f^{-1}(x)$. (a is a real number constant.)

APPLYING THE CONCEPTS

Inverse Functions in Words Inverses may also be found by *inverse reasoning*. For example, to find the inverse of $f(x) = 3x + 2$, first list, in order, the operations done to variable x:

    Step 1: Multiply by 3.
    Step 2: Add 2.

Then, to find the inverse, simply apply the inverse operations, in reverse order, to the variable x. That is:

    Step 3: Subtract 2.
    Step 4: Divide by 3.

The inverse function then becomes $f^{-1}(x) = \frac{x-2}{3}$.

41. Use this method of "inverse reasoning" to find the inverse of the *function* $f(x) = \frac{x}{7} - 2$.

42. **Inverse Functions in Words** Use *inverse reasoning* to find the following inverses:

    a. $f(x) = 2x + 7$ **b.** $f(x) = \sqrt{x} - 9$ **c.** $f(x) = x^3 - 4$ **d.** $f(x) = \sqrt{x^3 - 4}$

43. **Reading Tables** Evaluate each of the following functions using the functions defined by Tables 1 and 2.

    a. $f[g(-3)]$ **b.** $g[f(-6)]$ **c.** $g[f(2)]$
    d. $f[g(3)]$ **e.** $f[g(-2)]$ **f.** $g[f(3)]$

    g. What can you conclude about the relationship between functions f and g?

TABLE 1			TABLE 2	
x	$f(x)$		x	$g(x)$
-6	3		-3	2
2	-3		-2	3
3	-2		3	-6
6	4		4	6

44. **Reading Tables** Use the functions defined in Tables 1 and 2 in Problem 43 to answer the following questions.

    a. What are the domain and range of f?
    b. What are the domain and range of g?
    c. How are the domain and range of f related to the domain and range of g?
    d. Is f a one-to-one function?
    e. Is g a one-to-one function?

45. Social Security A function that models the billions of dollars of Social Security payments (as shown in the chart) per year is $s(t) = 16t + 249.4$, where t is time in years since 1990 (U.S. Census Bureau).

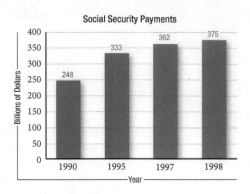

Social Security Payments

a. Use the model to estimate the amount of Social Security payments to be paid in 2020.

b. Write the inverse of the function.

c. Using the inverse function, estimate the year in which payments will reach $507 billion.

46. Families The function for the percentage of one-parent families (as shown in the following chart) is $f(x) = 0.417x + 24$, when x is the time in years since 1990 (U.S. Census Bureau).

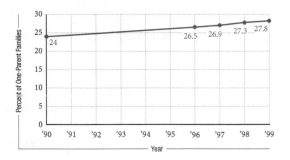

a. Use the function to predict the percentage of families with one parent in the year 2010.

b. Determine the inverse of the function, and estimate the year in which approximately 29% of the families are one-parent families.

47. Speed The fastest type of plane, a rocket plane, can travel at a speed of 4,520 miles per hour. The function $f(m) = \frac{22m}{15}$ converts miles per hour, m, to feet per second (*World Book Encyclopedia*).

a. Use the function to convert the speed of the rocket plane to feet per second.

b. Write the inverse of the function.

c. Using the inverse function, convert 2 feet per second to miles per hour.

48. **Speed** A Lockheed SR-71A airplane set a world record (as reported by the Air Force Armament Museum in 1996) with an absolute speed record of 2,193.167 miles per hour. The function $s(h) = 0.4468424h$ converts miles per hour, h, to meters per second, s.
    a. What is the absolute speed of the Lockheed SR-71A in meters per second?
    b. What is the inverse of this function?
    c. Using the inverse function, determine the speed of an airplane in miles per hour that flies 150 meters per second.

GETTING READY FOR THE NEXT SECTION

Simplify.

49. 3^{-2}

50. 2^3

Solve.

51. $2 = 3x$ 52. $3 = 5x$ 53. $4 = x^3$ 54. $12 = x^2$

Fill in the boxes to make each statement true.

55. $8 = 2^{\square}$

56. $27 = 3^{\square}$

57. $10{,}000 = 10^{\square}$

58. $1{,}000 = 10^{\square}$

59. $81 = 3^{\square}$

60. $81 = 9^{\square}$

61. $6 = 6^{\square}$

62. $1 = 5^{\square}$

EXTENDING THE CONCEPTS

For each of the following functions, find $f^{-1}(x)$. Then show that $f[f^{-1}(x)] = x$.

63. $f(x) = 3x + 5$

64. $f(x) = 6 - 8x$

65. $f(x) = x^3 + 1$

66. $f(x) = x^3 - 8$

67. $f(x) = \dfrac{x - 4}{x - 2}$

68. $f(x) = \dfrac{x - 3}{x - 1}$

69. Reading Graphs The graphs of a function and its inverse are shown in Figure 6. Use the graphs to find the following.

a. $f(0)$ **b.** $f(1)$ **c.** $f(2)$ **d.** $f^{-1}(1)$

e. $f^{-1}(2)$ **f.** $f^{-1}(5)$ **g.** $f^{-1}[f(2)]$ **h.** $f[f^{-1}(5)]$

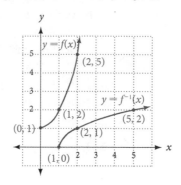

FIGURE 6

70. Domain from a Graph The function f is defined by the equation $f(x) = \sqrt{x + 4}$ for $x \geq -4$, meaning the domain is all x in the interval $[-4, \infty)$.

a. Graph the function f.

b. On the same set of axes, graph the line $y = x$ and the inverse function f^{-1}.

c. State the domain for the inverse function for f^{-1}.

d. Find the equation for $f^{-1}(x)$.

LOGARITHMS ARE EXPONENTS

OBJECTIVES

A Convert between logarithmic form and exponential form.

B Solve simple logarithmic equations using the definition of logarithm.

C Sketch the graph of logarithmic functions.

D Simplify expressions using logarithms.

E Solve applications involving logarithms.

© istockphoto.com/Claudiad

In January 2010, ABC News reported that an earthquake had occurred in Haiti, causing massive destruction. They reported the strength of the quake by indicating that it measured 7.0 on the Richter scale. For comparison, Table 1 gives the Richter magnitude of a number of other earthquakes.

Although the size of the numbers in the table do not seem to be very different, the intensity of the earthquakes they measure can be very different. For example, the 2017 Mexico earthquake was more than 10 times stronger than the 2010 earthquake in Haiti. The reason behind this is that the Richter scale is a *logarithmic scale*.

TABLE 1 Earthquakes

Year	Earthquake	Richter Magnitude
2009	Sumatra, Indonesia	7.5
2010	Haiti	7.0
2011	Turkey	7.2
2015	Nepal	7.8
2016	Italy	6.2
2017	Mexico	8.1

In this section, we start our work with logarithms, which will give you an understanding of the Richter scale. Let's begin.

Logarithmic Functions

As you know from your work in the previous sections, equations of the form

$$y = b^x \quad b > 0, b \neq 1$$

are called exponential functions. Because the equation of the inverse of a function can be obtained by exchanging x and y in the equation of the original function, the inverse of an exponential function must have the form

$$x = b^y \quad b > 0, b \neq 1$$

8.3 VIDEOS

Now, this last equation is actually the equation of a logarithmic function, as the following definition indicates:

DEFINITION

The expression $y = \log_b x$ is read "y is the logarithm to the base b of x" and is equivalent to the expression

$$x = b^y \qquad b > 0, b \neq 1$$

In words, we say "y is the number we raise b to in order to get x."

Notation   When an expression is in the form $x = b^y$, it is said to be in exponential form. On the other hand, if an expression is in the form $y = \log_b x$, it is said to be in logarithmic form.

Here are some equivalent statements written in both forms.

Exponential Form		Logarithmic Form
$8 = 2^3$	$\Leftrightarrow$	$\log_2 8 = 3$
$25 = 5^2$	$\Leftrightarrow$	$\log_5 25 = 2$
$0.1 = 10^{-1}$	$\Leftrightarrow$	$\log_{10} 0.1 = -1$
$\frac{1}{8} = 2^{-3}$	$\Leftrightarrow$	$\log_2 \frac{1}{8} = -3$
$r = z^s$	$\Leftrightarrow$	$\log_z r = s$

B Logarithmic Equations

EXAMPLE 1

Solve for x: $\log_3 x = -2$.

SOLUTION   In exponential form, the equation looks like this:

$$x = 3^{-2}$$

or

$$x = \frac{1}{9}$$

The solution is $\frac{1}{9}$.

EXAMPLE 2

Solve $\log_x 4 = 3$.

SOLUTION   Again, we use the definition of logarithms to write the expression in exponential form:

$$4 = x^3$$

Taking the cube root of both sides, we have

$$\sqrt[3]{4} = \sqrt[3]{x^3}$$

$$x = \sqrt[3]{4}$$

The solution set is $\{\sqrt[3]{4}\}$.

EXAMPLE 3

Solve $\log_8 4 = x$.

SOLUTION   We write the expression again in exponential form:

$$4 = 8^x$$

Because both 4 and 8 can be written as powers of 2, we write them in terms of powers of 2:

$$2^2 = (2^3)^x$$
$$2^2 = 2^{3x}$$

The only way the left and right sides of this last line can be equal is if the exponents are equal — that is, if

$$2 = 3x$$

or

$$x = \frac{2}{3}$$

The solution is $\frac{2}{3}$. We check as follows:

$$\log_8 4 = \frac{2}{3} \Leftrightarrow 4 = 8^{2/3}$$
$$4 = (\sqrt[3]{8})^2$$
$$4 = 2^2$$
$$4 = 4$$

The solution checks when used in the original equation.

🅒 Graphing Logarithmic Functions

Graphing logarithmic functions can be done using the graphs of exponential functions and the fact that the graphs of inverse functions have symmetry about the line $y = x$. Here's an example to illustrate.

EXAMPLE 4

Graph the equation $y = \log_2 x$.

SOLUTION   The equation $y = \log_2 x$ is, by definition, equivalent to the exponential equation

$$x = 2^y$$

which is the equation of the inverse of the function

$$y = 2^x$$

The graph of $y = 2^x$ was given in Figure 2 of Section 8.1. We simply reflect the graph of $y = 2^x$ about the line $y = x$ to get the graph of $x = 2^y$, which is also the graph of $y = \log_2 x$. (See Figure 1 on the next page.)

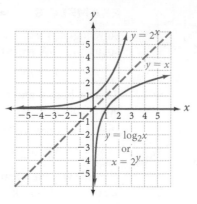

FIGURE 1

It is apparent from the graph that $y = \log_2 x$ is a function, because no vertical line will cross its graph in more than one place. The same is true for all logarithmic equations of the form $y = \log_b x$, where b is a positive number other than 1. Note also that the graph of $y = \log_b x$ will always appear to the right of the y-axis, meaning that x will always be positive in the expression $y = \log_b x$.

Ⓓ Two Special Identities

If b is a positive real number other than 1, then each of the following is a consequence of the definition of a logarithm:

$$(1)\ b^{\log_b x} = x \qquad \text{and} \qquad (2)\ \log_b b^x = x$$

The justifications for these identities are similar. Let's consider only the first one. Consider the expression

$$y = \log_b x$$

By definition, it is equivalent to

$$x = b^y$$

Substituting $\log_b x$ for y in the last line gives us

$$x = b^{\log_b x}$$

The next examples in this section show how these two special properties can be used to simplify expressions involving logarithms.

EXAMPLE 5

Simplify the following logarithmic expressions.

a. $\log_2 8$ **b.** $\log_{10} 10{,}000$ **c.** $\log_b b$

d. $\log_b 1$ **e.** $\log_4 (\log_5 5)$

SOLUTION

a. Substitute 2^3 for 8:

$$\log_2 8 = \log_2 2^3$$
$$= 3$$

b. 10,000 can be written as 10^4:

$$\log_{10} 10{,}000 = \log_{10} 10^4$$
$$= 4$$

c. Because $b^1 = b$, we have

$$\log_b b = \log_b b^1$$
$$= 1$$

d. Because $1 = b^0$, we have

$$\log_b 1 = \log_b b^0$$
$$= 0$$

e. Because $\log_5 5 = 1$,

$$\log_4(\log_5 5) = \log_4 1$$
$$= 0$$

🄴 Application

As we mentioned in the introduction to this section, one application of logarithms is in measuring the magnitude of an earthquake. If an earthquake has a shock wave T times greater than the smallest shock wave that can be measured on a seismograph, then the magnitude M of the earthquake, as measured on the Richter scale, is given by the formula

$$M = \log_{10} T$$

(When we talk about the size of a shock wave, we are talking about its amplitude. The amplitude of a wave is half the difference between its highest point and its lowest point.)

To illustrate the discussion, an earthquake that produces a shock wave that is 10,000 times greater than the smallest shock wave measurable on a seismograph will have a magnitude M on the Richter scale of

$$M = \log_{10} 10{,}000 = 4$$

EXAMPLE 6

If an earthquake has a magnitude of $M = 5$ on the Richter scale, what can you say about the size of its shock wave?

SOLUTION To answer this question, we put $M = 5$ into the formula $M = \log_{10} T$ to obtain

$$5 = \log_{10} T$$

Writing this expression in exponential form, we have

$$T = 10^5 = 100{,}000$$

We can say that an earthquake that measures 5 on the Richter scale has a shock wave 100,000 times greater than the smallest shock wave measurable on a seismograph.

From Example 6 and the discussion that preceded it, we find that an earthquake of magnitude 5 has a shock wave that is 10 times greater than an earthquake of magnitude 4, because 100,000 is 10 times 10,000.

GETTING READY FOR CLASS

After reading through the preceding section, respond in your own words and in complete sentences.

A. What is a logarithm?

B. What is the relationship between $y = 2^x$ and $y = \log_2 x$? How are their graphs related?

C. Will the graph of $y = \log_b x$ ever appear in the second or third quadrants? Explain why or why not.

D. Explain why $\log_2 0 = x$ has no solution for x.

Problem Set 8.3

Write each of the following expressions in logarithmic form.

1. $2^4 = 16$
2. $3^2 = 9$
3. $125 = 5^3$
4. $16 = 4^2$

5. $0.01 = 10^{-2}$
6. $0.001 = 10^{-3}$
7. $2^{-5} = \dfrac{1}{32}$
8. $4^{-2} = \dfrac{1}{16}$

9. $\left(\dfrac{1}{2}\right)^{-3} = 8$
10. $\left(\dfrac{1}{3}\right)^{-2} = 9$
11. $27 = 3^3$
12. $81 = 3^4$

Write each of the following expressions in exponential form.

13. $\log_{10} 100 = 2$
14. $\log_2 8 = 3$
15. $\log_2 64 = 6$

16. $\log_2 32 = 5$
17. $\log_8 1 = 0$
18. $\log_9 9 = 1$

19. $\log_{10} 0.001 = -3$
20. $\log_{10} 0.0001 = -4$
21. $\log_6 36 = 2$

22. $\log_7 49 = 2$
23. $\log_5 \dfrac{1}{25} = -2$
24. $\log_3 \dfrac{1}{81} = -4$

Solve each of the following equations for x.

25. $\log_3 x = 2$
26. $\log_4 x = 3$
27. $\log_5 x = -3$
28. $\log_2 x = -4$

29. $\log_2 16 = x$
30. $\log_3 27 = x$
31. $\log_8 2 = x$
32. $\log_{25} 5 = x$

33. $\log_x 4 = 2$
34. $\log_x 16 = 4$
35. $\log_x 5 = 3$
36. $\log_x 8 = 2$

37. $\log_5 25 = x$
38. $\log_5 x = -2$
39. $\log_x 36 = 2$
40. $\log_x \dfrac{1}{25} = 2$

41. $\log_8 4 = x$
42. $\log_{16} 8 = x$
43. $\log_9 \dfrac{1}{3} = x$
44. $\log_{27} 9 = x$

45. $\log_8 x = -2$
46. $\log_{36} \dfrac{1}{6} = x$

Sketch the graph of each of the following logarithmic equations.

47. $y = \log_3 x$
48. $y = \log_{1/2} x$
49. $y = \log_{1/3} x$
50. $y = \log_4 x$

51. $y = \log_5 x$
52. $y = \log_{1/5} x$
53. $y = \log_{10} x$
54. $y = \log_{1/4} x$

Each of the following graphs has an equation of the form $y = b^x$ or $y = \log_b x$. Find the equation for each graph.

55.

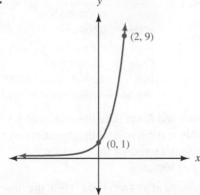

56.

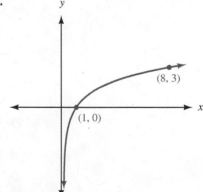

679

57. **58.**

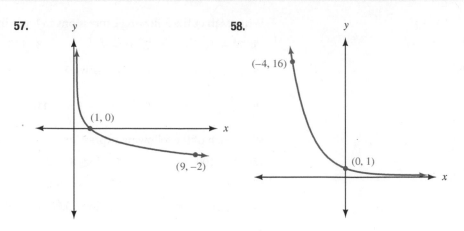

Simplify each of the following.

59. $\log_2 16$ **60.** $\log_3 9$ **61.** $\log_{25} 125$ **62.** $\log_9 27$

63. $\log_{10} 1{,}000$ **64.** $\log_{10} 10{,}000$ **65.** $\log_3 3$ **66.** $\log_4 4$

67. $\log_5 1$ **68.** $\log_{10} 1$ **69.** $\log_{17} 1$ **70.** $\log_4 8$

71. $\log_{16} 4$ **72.** $\log_{10} 0.0001$ **73.** $\log_{100} 1000$ **74.** $\log_{32} 16$

75. $\log_3 (\log_2 8)$ **76.** $\log_5 (\log_{32} 2)$ **77.** $\log_{1/2} (\log_3 81)$ **78.** $\log_9 (\log_8 2)$

79. $\log_3 (\log_6 6)$ **80.** $\log_5 (\log_3 3)$ **81.** $\log_4 [\log_2(\log_2 16)]$

82. $\log_4 [\log_3(\log_2 8)]$

APPLYING THE CONCEPTS

83. Metric System The metric system uses logical and systematic prefixes for multiplication. For instance, to multiply a unit by 100, the prefix "hecto" is applied, so a hectometer is equal to 100 meters. For each of the prefixes in the following table, find the logarithm, base 10, of the multiplying factor.

Prefix	Multiplying Factor	$\log_{10}$ (Multiplying Factor)
Nano	0.000 000 001	
Micro	0.000 001	
Deci	0.1	
Giga	1,000,000,000	
Peta	1,000,000,000,000,000	

84. Domain and Range Use the graphs of $y = 2^x$ and $y = \log_2 x$ shown in Figure 1 of this section to find the domain and range for each function. Explain how the domain and range found for $y = 2^x$ relate to the domain and range found for $y = \log_2 x$.

85. Magnitude of an Earthquake Find the magnitude M of an earthquake with a shock wave that measures T $= 100$ on a seismograph.

86. Magnitude of an Earthquake Find the magnitude M of an earthquake with a shock wave that measures T $= 100{,}000$ on a seismograph.

87. Shock Wave If an earthquake has a magnitude of 8 on the Richter scale, how many times greater is its shock wave than the smallest shock wave measurable on a seismograph?

88. Shock Wave If the 2010 Haiti earthquake had a magnitude of 7 on the Richter scale, how many times greater was its shock wave than the smallest shock wave measurable on a seismograph?

Earthquake The table below categorizes earthquake by the magnitude and identifies the average annual occurrence.

Earthquakes		
Descriptor	Magnitude	Average Annual Occurrence
Great	≥8.0	1
Major	7–7.9	18
Strong	6–6.9	120
Moderate	5–5.9	800
Light	4–4.9	6,200
Minor	3–3.9	49,000
Very Minor	2–2.9	1,000 per day
Very Minor	1–1.9	8,000 per day

SOURCE: *USGS National Earthquake Information.*

89. What is the average number of earthquakes that occur per year when the number of times the associated shock wave is greater than the smallest measurable shock wave, T, is 1,000,000?

90. What is the average number of earthquakes that occur per year when $T = 1{,}000{,}000$ or greater?

GETTING READY FOR THE NEXT SECTION

Simplify.

91. $8^{2/3}$

92. $27^{2/3}$

Solve.

93. $(x + 2)(x) = 2^3$

94. $(x + 3)(x) = 2^2$

95. $\dfrac{x - 2}{x + 1} = 9$

96. $\dfrac{x + 1}{x - 4} = 25$

Write in exponential form.

97. $\log_2 [(x + 2)(x)] = 3$

98. $\log_4 [x(x - 6)] = 2$

99. $\log_3 \left(\dfrac{x - 2}{x + 1} \right) = 4$

100. $\log_3 \left(\dfrac{x - 1}{x - 4} \right) = 2$

EXTENDING THE CONCEPTS

101. The graph of the exponential function $y = f(x) = b^x$ is shown below. Use the graph to complete parts a through d.

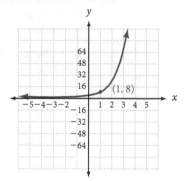

a. Fill in the table.

x	$f(x)$
-1	
0	
1	
2	

b. Fill in the table.

x	$f^{-1}(x)$
	-1
	0
	1
	2

c. Find the equation for $f(x)$.
d. Find the equation for $f^{-1}(x)$.

102. The graph of the exponential function $y = f(x) = b^x$ is shown below. Use the graph to complete parts a through d.

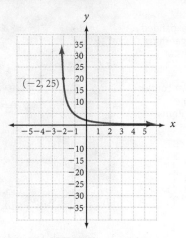

(−2, 25)

a. Fill in the table.

x	$f(x)$
−1	
0	
1	
2	

b. Fill in the table.

x	$f(x)$
	−1
	0
	1
	2

c. Find the equation for $f(x)$.

d. Find the equation for $f^{-1}(x)$.

PROPERTIES OF LOGARITHMS

OBJECTIVES

A Convert between expanded form and single logarithms using the properties of logarithms.

B Solve equations that contain logarithms using the properties of logarithms.

© istockphoto.com/LightFieldStudios

8.4

If we search for a definition of the word *decibel*, we find the following: A unit used to express relative difference in power or intensity, usually between two acoustic or electric signals, equal to ten times the common logarithm of the ratio of the two levels.

Decibels	Comparable to
10	A light whisper
20	Quiet conversation
30	Normal conversation
40	Light traffic
50	Typewriter, loud conversation
60	Noisy office
70	Normal traffic, quiet train
80	Rock music, subway
90	Heavy traffic, thunder
100	Jet plane at takeoff

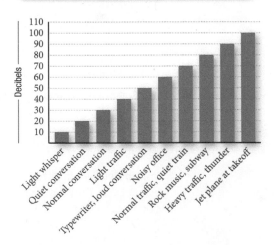

The precise definition for a *decibel* is

$$D = 10 \log_{10}\left(\frac{I}{I_0}\right)$$

where I is the intensity of the sound being measured, and I_0 is the intensity of the least audible sound. (Sound intensity is related to the amplitude of the sound wave that models the sound and is given in units of watts per meter^2.) In this section, we will see that the preceding formula can also be written as

$$D = 10(\log_{10} I - \log_{10} I_0)$$

The rules we use to rewrite expressions containing logarithms are called the *properties of logarithms*. There are three of them.

Ⓐ Properties of Logarithms

For the following three properties, x, y, and b are all positive real numbers, $b \neq 1$, and r is any real number.

RULE *Property 1*

$$\log_b(xy) = \log_b x + \log_b y$$

In words: The logarithm of a **product** is the **sum** of the logarithms.

RULE *Property 2*

$$\log_b\left(\frac{x}{y}\right) = \log_b x - \log_b y$$

In words: The logarithm of a **quotient** is the **difference** of the logarithms.

RULE *Property 3*

$$\log_b x^r = r \log_b x$$

In words: The logarithm of a number raised to a *power* is the *product* of the power and the logarithm of the number.

Proof of Property 1 To prove Property 1, we simply apply the first identity for logarithms given in the preceding section:

$$b^{\log_b xy} = xy = (b^{\log_b x})(b^{\log_b y}) = b^{\log_b x + \log_b y}$$

Because the first and last expressions are equal and the bases are the same, the exponents $\log_b xy$ and $\log_b x + \log_b y$ must be equal. Therefore,

$$\log_b xy = \log_b x + \log_b y$$

The proofs of Properties 2 and 3 proceed in much the same manner, so we will omit them here. The examples that follow show how the three properties can be used.

EXAMPLE 1

Expand, using the properties of logarithms: $\log_5 \dfrac{3xy}{z}$

SOLUTION   Applying Property 2, we can write the quotient of $3xy$ and z in terms of a difference:

$$\log_5 \frac{3xy}{z} = \log_5 3xy - \log_5 z$$

Applying Property 1 to the product $3xy$, we write it in terms of addition:

$$\log_5 \frac{3xy}{z} = \log_5 3 + \log_5 x + \log_5 y - \log_5 z$$

EXAMPLE 2

Expand, using the properties of logarithms:

$$\log_2 \frac{x^4}{\sqrt{y} \cdot z^3}$$

SOLUTION   We write $\sqrt{y}$ as $y^{1/2}$ and apply the properties:

$$\log_2 \frac{x^4}{\sqrt{y} \cdot z^3} = \log_2 \frac{x^4}{y^{1/2} z^3} \qquad \sqrt{y} = y^{1/2}$$

$$= \log_2 x^4 - \log_2(y^{1/2} \cdot z^3) \qquad \text{Property 2}$$

$$= \log_2 x^4 - (\log_2 y^{1/2} + \log_2 z^3) \qquad \text{Property 1}$$

$$= \log_2 x^4 - \log_2 y^{1/2} - \log_2 z^3 \qquad \text{Remove parentheses and distribute} -1.$$

$$= 4 \log_2 x - \frac{1}{2} \log_2 y - 3 \log_2 z \qquad \text{Property 3}$$

We can also use the three properties to write an expression in expanded form as just one logarithm.

EXAMPLE 3

Write as a single logarithm:

$$2 \log_{10} a + 3 \log_{10} b - \frac{1}{3} \log_{10} c$$

SOLUTION   We begin by applying Property 3:

$$2 \log_{10} a + 3 \log_{10} b - \frac{1}{3} \log_{10} c = \log_{10} a^2 + \log_{10} b^3 - \log_{10} c^{1/3} \qquad \text{Property 3}$$

$$= \log_{10} (a^2 \cdot b^3) - \log_{10} c^{1/3} \qquad \text{Property 1}$$

$$= \log_{10} \frac{a^2 b^3}{c^{1/3}} \qquad \text{Property 2}$$

$$= \log_{10} \frac{a^2 b^3}{\sqrt[3]{c}} \qquad c^{1/3} = \sqrt[3]{c}$$

B Equations with Logarithms

The properties of logarithms along with the definition of logarithms are useful in solving equations that involve logarithms.

EXAMPLE 4

Solve for x: $\log_2(x + 2) + \log_2 x = 3$

SOLUTION   Applying Property 1 to the left side of the equation allows us to write it as a single logarithm:

$$\log_2(x + 2) + \log_2 x = 3$$

$$\log_2[(x + 2)(x)] = 3$$

The last line can be written in exponential form using the definition of logarithms:

$$(x + 2)(x) = 2^3$$

Solve as usual:

$$x^2 + 2x = 8$$

$$x^2 + 2x - 8 = 0$$

$$(x + 4)(x - 2) = 0$$

$$x + 4 = 0 \quad \text{or} \quad x - 2 = 0$$

$$x = -4 \qquad\qquad x = 2$$

In the previous section, we noted the fact that x in the expression $y = \log_b x$ cannot be a negative number. Because substitution of $x = -4$ into the original equation gives

$$\log_2(-2) + \log_2(-4) = 3$$

which contains logarithms of negative numbers, we cannot use -4 as a solution. The solution set is 2.

GETTING READY FOR CLASS

After reading through the preceding section, respond in your own words and in complete sentences.

A. Explain why the following statement is false: "The logarithm of a product is the product of the logarithms."

B. Explain why the following statement is false: "The logarithm of a quotient is the quotient of the logarithms."

C. Explain the difference between $\log_b m + \log_b n$ and $\log_b(m + n)$. Are they equivalent?

D. Explain the difference between $\log_b(mn)$ and $(\log_b m)(\log_b n)$. Are they equivalent?

Problem Set 8.4

Use the three properties of logarithms given in this section to expand each expression as much as possible.

1. $\log_3 4x$ **2.** $\log_2 5x$ **3.** $\log_6 \dfrac{5}{x}$ **4.** $\log_3 \dfrac{x}{5}$

5. $\log_2 y^5$ **6.** $\log_7 y^3$ **7.** $\log_9 \sqrt[3]{z}$ **8.** $\log_8 \sqrt{z}$

9. $\log_6 x^2 y^4$ **10.** $\log_{10} x^2 y^4$ **11.** $\log_5 \sqrt{x} \cdot y^4$ **12.** $\log_8 \sqrt[3]{xy^6}$

13. $\log_b \dfrac{xy}{z}$ **14.** $\log_b \dfrac{3x}{y}$ **15.** $\log_{10} \dfrac{4}{xy}$ **16.** $\log_{10} \dfrac{5}{4y}$

17. $\log_{10} \dfrac{x^2 y}{\sqrt{z}}$ **18.** $\log_{10} \dfrac{\sqrt{x} \cdot y}{z^3}$ **19.** $\log_{10} \dfrac{x^3 \sqrt{y}}{z^4}$ **20.** $\log_{10} \dfrac{x^4 \sqrt[3]{y}}{\sqrt{z}}$

21. $\log_b \sqrt[3]{\dfrac{x^2 y}{z^4}}$ **22.** $\log_b \sqrt[4]{\dfrac{x^4 y^3}{z^5}}$ **23.** $\log_3 \sqrt[3]{\dfrac{x^2 y}{z^6}}$ **24.** $\log_8 \sqrt[4]{\dfrac{x^5 y^6}{z^3}}$

25. $\log_a \dfrac{4x^5}{9a^2}$ **26.** $\log_b \dfrac{16b^2}{25y^3}$

Write each expression as a single logarithm.

27. $\log_b x + \log_b z$

28. $\log_b x - \log_b z$

29. $2 \log_3 x - 3 \log_3 y$

30. $4 \log_2 x + 5 \log_2 y$

31. $\dfrac{1}{2} \log_{10} x + \dfrac{1}{3} \log_{10} y$

32. $\dfrac{1}{3} \log_{10} x - \dfrac{1}{4} \log_{10} y$

33. $3 \log_2 x + \dfrac{1}{2} \log_2 y - \log_2 z$

34. $2 \log_3 x + 3 \log_3 y - \log_3 z$

35. $\dfrac{1}{2} \log_2 x - 3 \log_2 y - 4 \log_2 z$

36. $3 \log_{10} x - \log_{10} y - \log_{10} z$

37. $\dfrac{3}{2} \log_{10} x - \dfrac{3}{4} \log_{10} y - \dfrac{4}{5} \log_{10} z$

38. $3 \log_{10} x - \dfrac{4}{3} \log_{10} y - 5 \log_{10} z$

39. $\dfrac{1}{2} \log_5 x + \dfrac{2}{3} \log_5 y - 4 \log_5 z$

40. $\dfrac{1}{4} \log_7 x + 5 \log_7 y - \dfrac{1}{3} \log_7 z$

41. $\log_3(x^2 - 16) - 2 \log_3(x + 4)$

42. $\log_4(x^2 - x - 6) - \log_4(x^2 - 9)$

Solve each of the following equations.

43. $\log_2 x + \log_2 3 = 1$

44. $\log_3 x + \log_3 3 = 1$

45. $\log_3 x - \log_3 2 = 2$

46. $\log_3 x + \log_3 2 = 2$

47. $\log_3 x + \log_3(x - 2) = 1$

48. $\log_6 x + \log_6(x - 1) = 1$

49. $\log_3(x + 3) - \log_3(x - 1) = 1$

50. $\log_4(x - 2) - \log_4(x + 1) = 1$

51. $\log_2 x + \log_2(x - 2) = 3$

52. $\log_4 x + \log_4(x + 6) = 2$

53. $\log_8 x + \log_8(x - 3) = \dfrac{2}{3}$

54. $\log_{27} x + \log_{27}(x + 8) = \dfrac{2}{3}$

55. $\log_3(x + 2) - \log_3 x = 1$

56. $\log_2(x + 3) - \log_2(x - 3) = 2$

57. $\log_2(x + 1) + \log_2(x + 2) = 1$

58. $\log_3 x + \log_3(x + 6) = 3$

59. $\log_9 \sqrt{x} + \log_9 \sqrt{2x + 3} = \dfrac{1}{2}$

60. $\log_8 \sqrt{x} + \log_8 \sqrt{5x + 2} = \dfrac{2}{3}$

61. $4 \log_3 x - \log_3 x^2 = 6$

62. $9 \log_4 x - \log_4 x^3 = 12$

63. $\log_5 \sqrt{x} + \log_5 \sqrt{6x + 5} = 1$

64. $\log_2 \sqrt{x} + \log_2 \sqrt{6x + 5} = 1$

APPLYING THE CONCEPTS

65. Decibel Formula Use the properties of logarithms to rewrite the decibel formula $D = 10 \log_{10}\left(\frac{I}{I_0}\right)$ as

$$D = 10(\log_{10} I - \log_{10} I_0).$$

66. Decibel Formula In the decibel formula $D = 10 \log_{10}\left(\frac{I}{I_0}\right)$, the threshold of hearing, I_0, is

$$I_0 = 10^{-12} \text{ watts/meter}^2$$

Substitute 10^{-12} for I_0 in the decibel formula, then show that it simplifies to

$$D = 10(\log_{10} I + 12)$$

67. Finding Logarithms If $\log_{10} 8 = 0.903$ and $\log_{10} 5 = 0.699$, find the following without using a calculator.

    a. $\log_{10} 40$ **b.** $\log_{10} 320$ **c.** $\log_{10} 1{,}600$

68. Matching Match each expression in the first column with an equivalent expression in the second column:

    a. $\log_2(ab)$ **i.** b

    b. $\log_2\left(\frac{a}{b}\right)$ **ii.** 2

    c. $\log_5 a^b$ **iii.** $\log_2 a + \log_2 b$

    d. $\log_a b^a$ **iv.** $\log_2 a - \log_2 b$

    e. $\log_a a^b$ **v.** $a \log_a b$

    f. $\log_3 9$ **vi.** $b \log_5 a$

69. Henderson–Hasselbalch Formula Doctors use the Henderson–Hasselbalch formula to calculate the pH of a person's blood. pH is a measure of the acidity and/or the alkalinity of a solution. This formula is represented as

$$\text{pH} = 6.1 + \log_{10}\left(\frac{x}{y}\right)$$

where x is the base concentration and y is the acidic concentration. Rewrite the Henderson–Hasselbalch formula so that the logarithm of a quotient is not involved.

70. Henderson–Hasselbalch Formula Refer to the information in the preceding problem about the Henderson–Hasselbalch formula. If most people have a blood pH of 7.4, use the Henderson–Hasselbalch formula to find the ratio of $\frac{x}{y}$ for an average person.

71. Food Processing The formula $M = 0.21(\log_{10} a - \log_{10} b)$ is used in the food processing industry to find the number of minutes M of heat processing a certain food should undergo at 250°F to reduce the probability of survival of *Clostridium botulinum* spores. The letter a represents the number of spores per can before heating, and b represents the number of spores per can after heating. Find M if $a = 1$ and $b = 10^{-12}$. Then find M using the same values for a and b in the formula $M = 0.21 \log_{10} \frac{a}{b}$.

72. Acoustic Powers The formula $N = \log_{10} \frac{P_1}{P_2}$ is used in radio electronics to find the ratio of the acoustic powers of two electric circuits in terms of their electric powers. Find N if P_1 is 100 and P_2 is 1. Then use the same two values of P_1 and P_2 to find N in the formula $N = \log_{10} P_1 - \log_{10} P_2$.

GETTING READY FOR THE NEXT SECTION

Simplify.

73. 5^0 **74.** 4^1 **75.** $\log_3 3$ **76.** $\log_5 5$

77. $\log_b b^4$ **78.** $\log_a a^k$

Use a calculator to find each of the following. Write your answer in scientific notation with the first number in each answer rounded to the nearest tenth.

79. $10^{-5.6}$ **80.** $10^{-4.1}$

Divide and round to the nearest whole number.

81. $\dfrac{2.00 \times 10^8}{3.96 \times 10^6}$ **82.** $\dfrac{3.25 \times 10^{12}}{1.72 \times 10^{10}}$

EXTENDING THE CONCEPTS

83. Use properties of logarithms to evaluate $\log_3 153 - \log_3 17$.

84. Use properties of logarithms to evaluate $\log_2 120 - \log_2 15$.

Given that $\log_b x = u$ and $\log_b y = v$, express each logarithm in terms of u and v.

85. $\log_b \left(x^2 \sqrt{y} \right)$ **86.** $\log_b \sqrt{\dfrac{x^5}{y^3}}$

87. $\log_b \sqrt{x^5 y^7}$ **88.** $\log_b \left(\sqrt{x} \cdot \sqrt[3]{y} \right)$

89. Let $f(x) = \log_x 5$. Find $f(25) - f(5)$.

90. Let $g(x) = \log_x 4$. For what value of x is $g(x) = -2$?

©istockphoto.com/Nuttanun

OBJECTIVES

A Find common logarithms using a calculator.

B Find a number given its common logarithm using a calculator.

C Solve applications involving common logarithms.

D Simplify expressions containing natural logarithms.

8.5

Acid rain was first discovered in the 1960s by Gene Likens and his research team, who studied the damage caused by acid rain to Hubbard Brook in New Hampshire. Acid rain is rain with a pH of 5.6 and below. As you will see as you work your way through this section, pH is defined in terms of common logarithms — one of the topics we present in this section. So, when you are finished with this section, you will have a more detailed knowledge of pH and acid rain.

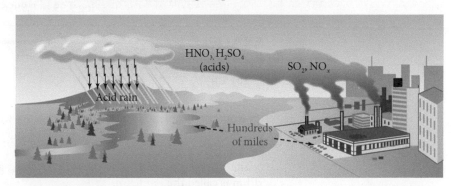

HNO_3, H_2SO_4 (acids)

SO_2, NO_x

Acid rain

Hundreds of miles

Two kinds of logarithms occur more frequently than other logarithms. Logarithms with a base of 10 are very common because our number system is a base-10 number system. For this reason, we call base-10 logarithms *common logarithms*.

> **DEFINITION** *common logarithms*
>
> A *common logarithm* is a logarithm with a base of 10. Because common logarithms are used so frequently, it is customary, in order to save time, to omit notating the base. That is,
>
> $$\log_{10} x = \log x$$
>
> When the base is not shown, it is assumed to be 10.

Common Logarithms

Common logarithms of powers of 10 are simple to evaluate. We need only recognize that $\log 10 = \log_{10} 10 = 1$ and apply the third property of logarithms: $\log_b x^r = r \log_b x$.

$$
\begin{aligned}
\log 1{,}000 &= \log 10^3 &=&\ 3 \log 10 =&\ 3(1) =&\ 3 \\
\log 100 &= \log 10^2 &=&\ 2 \log 10 =&\ 2(1) =&\ 2 \\
\log 10 &= \log 10^1 &=&\ 1 \log 10 =&\ 1(1) =&\ 1 \\
\log 1 &= \log 10^0 &=&\ 0 \log 10 =&\ 0(1) =&\ 0 \\
\log 0.1 &= \log 10^{-1} &=&\ -1 \log 10 =&\ -1(1) =&\ -1 \\
\log 0.01 &= \log 10^{-2} &=&\ -2 \log 10 =&\ -2(1) =&\ -2 \\
\log 0.001 &= \log 10^{-3} &=&\ -3 \log 10 =&\ -3(1) =&\ -3
\end{aligned}
$$

To find common logarithms of numbers that are not powers of 10, we use a calculator with a $\boxed{\log}$ key.

Check the following logarithms to be sure you know how to use your calculator. (These answers have been rounded to the nearest ten-thousandth.)

$$\log 7.02 \approx 0.8463$$
$$\log 1.39 \approx 0.1430$$
$$\log 6.00 \approx 0.7782$$
$$\log 9.99 \approx 0.9996$$

EXAMPLE 1

Use a calculator to find log 2,760.

SOLUTION

$$\log 2{,}760 \approx 3.4409$$

To work this problem on a scientific calculator, we simply enter the number 2,760 and press the key labeled $\boxed{\log}$. On a graphing calculator, we press the $\boxed{\log}$ key first, then 2,760.

The 3 in the answer is called the *characteristic*, and the decimal part of the logarithm is called the *mantissa*.

EXAMPLE 2

Find log 0.0391.

SOLUTION $\log 0.0391 \approx -1.4078$

EXAMPLE 3

Find log 0.00523.

SOLUTION $\log 0.00523 \approx -2.2815$

B Antilogarithms

EXAMPLE 4

Find x if $\log x = 3.8774$.

SOLUTION   We are looking for the number whose logarithm is 3.8774. On a scientific calculator, we enter 3.8774 and press the key labeled $\boxed{10^x}$. On a graphing calculator, we press $\boxed{10^x}$ first, then 3.8774. The result is 7,540 to four significant digits. Here's why:

$$\text{If} \qquad \log x = 3.8774$$
$$\text{then} \qquad x = 10^{3.8774}$$
$$\approx 7{,}540$$

The number 7,540 is called the *antilogarithm* or just *antilog* of 3.8774. That is, 7,540 is the number whose logarithm is 3.8774.

EXAMPLE 5

Find x if $\log x = -2.4179$.

SOLUTION   Using the $\boxed{10^x}$ key, the result is 0.00382.

$$\text{If} \qquad \log x = -2.4179$$
$$\text{then} \qquad x = 10^{-2.4179}$$
$$\approx 0.00382$$

The antilog of -2.4179 is 0.00382. That is, the logarithm of 0.00382 is -2.4179.

C Applications of Logarithms

In Section 8.3, we found that the magnitude M of an earthquake that produces a shock wave T times larger than the smallest shock wave that can be measured on a seismograph is given by the formula

$$M = \log_{10} T$$

We can rewrite this formula using our shorthand notation for common logarithms as

$$M = \log T$$

EXAMPLE 6

The San Francisco earthquake of 1906 is estimated to have measured 8.3 on the Richter scale. The San Fernando earthquake of 1971 measured 6.6 on the Richter scale. Find T for each earthquake, and then give some indication of how much stronger the 1906 earthquake was than the 1971 earthquake.

SOLUTION   For the 1906 earthquake:

If $\log T = 8.3$, then $T = 2.00 \times 10^8$.

For the 1971 earthquake:

If $\log T = 6.6$, then $T = 3.98 \times 10^6$.

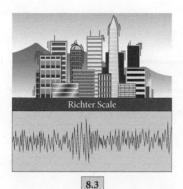

Richter Scale

8.3

Dividing the two values of T and rounding our answer to the nearest whole number, we have

$$\frac{2.00 \times 10^8}{3.98 \times 10^6} \approx 50$$

The shock wave for the 1906 earthquake was approximately 50 times larger than the shock wave for the 1971 earthquake.

In chemistry, the pH of a solution is the measure of the acidity of the solution. The definition for pH involves common logarithms. Here it is:

$$pH = -\log[H^+]$$

where $[H^+]$ is the concentration of the hydrogen ion in moles per liter. The range for pH is from 0 to 14. Pure water, a neutral solution, has a pH of 7. An acidic solution, such as vinegar, will have a pH less than 7, and an alkaline solution, such as ammonia, has a pH above 7.

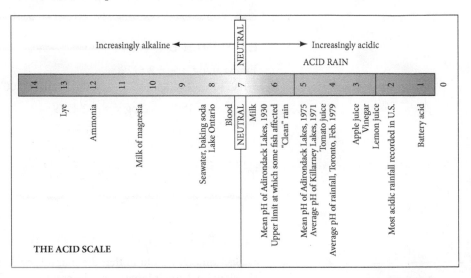

EXAMPLE 7

Normal rainwater has a pH of 5.6. What is the concentration of the hydrogen ion in normal rainwater?

SOLUTION   Substituting 5.6 for pH in the formula $pH = -\log[H^+]$, we have

$5.6 = -\log[H^+]$	Substitution
$\log[H^+] = -5.6$	Isolate the logarithm.
$[H^+] = 10^{-5.6}$	Write in exponential form.
$\approx 2.5 \times 10^{-6}$ moles per liter	Answer in scientific notation.

EXAMPLE 8

The concentration of the hydrogen ion in a sample of acid rain known to kill fish is 3.2×10^{-5} mole per liter. Find the pH of this acid rain to the nearest tenth.

SOLUTION   Substituting 3.2×10^{-5} for $[H^+]$ in the formula $pH = -\log[H^+]$, we have

$$pH = -\log[3.2 \times 10^{-5}] \qquad \text{Substitution}$$
$$\approx -(-4.5) \qquad \text{Evaluate the logarithm.}$$
$$\approx 4.5 \qquad \text{Simplify.}$$

D Natural Logarithms

> **DEFINITION** *natural logarithms*
> ·
>
> A *natural logarithm* is a logarithm with a base of e. The natural logarithm of x is denoted by $\ln x$. That is,
>
> $$\ln x = \log_e x$$

We can assume that all our properties of exponents and logarithms hold for expressions with a base of e, because e is a real number. Here are some examples intended to make you more familiar with the number e and natural logarithms.

EXAMPLE 9

Simplify each of the following expressions.

a. $e^0 = 1$

b. $e^1 = e$

c. $\ln e = 1$ In exponential form, $e^1 = e$

d. $\ln 1 = 0$ In exponential form, $e^0 = 1$

e. $\ln e^3 = 3$

f. $\ln e^{-4} = -4$

g. $\ln e^t = t$

EXAMPLE 10

Use the properties of logarithms to expand the expression $\ln Ae^{5t}$.

SOLUTION   Because the properties of logarithms hold for natural logarithms, we have

$$\ln Ae^{5t} = \ln A + \ln e^{5t}$$
$$= \ln A + 5t \ln e$$
$$= \ln A + 5t \qquad \text{Because } \ln e = 1$$

EXAMPLE 11

If $\ln 2 = 0.6931$ and $\ln 3 = 1.0986$, find

a. $\ln 6$ **b.** $\ln 0.5$ **c.** $\ln 8$

SOLUTION

a. Because $6 = 2 \cdot 3$, we have

$$\begin{aligned}
\ln 6 &= \ln 2 \cdot 3 \\
&= \ln 2 + \ln 3 \\
&= 0.6931 + 1.0986 \\
&= 1.7917
\end{aligned}$$

b. Writing 0.5 as $\dfrac{1}{2}$ and applying Property 2 for logarithms gives us

$$\begin{aligned}
\ln 0.5 &= \ln \frac{1}{2} \\
&= \ln 1 - \ln 2 \\
&= 0 - 0.6931 \\
&= -0.6931
\end{aligned}$$

c. Writing 8 as 2^3 and applying Property 3 for logarithms, we have

$$\begin{aligned}
\ln 8 &= \ln 2^3 \\
&= 3 \ln 2 \\
&= 3(0.6931) \\
&= 2.0793
\end{aligned}$$

GETTING READY FOR CLASS

After reading through the preceding section, respond in your own words and in complete sentences.

A. What is a common logarithm?

B. What is a natural logarithm?

C. Is e a rational number? Explain.

D. Find $\ln e$, and explain how you arrived at your answer.

Problem Set 8.5

Find the following logarithms.

1. log 378   **2.** log 426   **3.** log 37.8   **4.** log 42,600

5. log 3,780   **6.** log 0.4260   **7.** log 0.0378   **8.** log 0.0426

9. log 37,800   **10.** log 4,900   **11.** log 600   **12.** log 900

13. log 2,010   **14.** log 10,200   **15.** log 0.00971   **16.** log 0.0312

17. log 0.0314   **18.** log 0.00052   **19.** log 0.399   **20.** log 0.111

Find x in the following equations.

21. $\log x = 2.8802$ **22.** $\log x = 4.8802$ **23.** $\log x = -2.1198$

24. $\log x = -3.1198$ **25.** $\log x = 3.1553$ **26.** $\log x = 5.5911$

27. $\log x = -5.3497$ **28.** $\log x = -1.5670$ **29.** $\log x = -7.0372$

30. $\log x = -4.2000$ **31.** $\log x = 10$ **32.** $\log x = -1$

33. $\log x = -10$ **34.** $\log x = 1$ **35.** $\log x = 20$

36. $\log x = -20$ **37.** $\log x = -2$ **38.** $\log x = 4$

39. $\log x = \log_2 8$ **40.** $\log x = \log_3 9$ **41.** $\ln x = -1$

42. $\ln x = 4$ **43.** $\log x = 2 \log 5$ **44.** $\log x = -\log 4$

45. $\ln x = -3 \ln 2$ **46.** $\ln x = 5 \ln 3$

Simplify each of the following expressions.

47. $\ln e$ **48.** $\ln 1$ **49.** $\ln e^5$ **50.** $\ln e^{-3}$

51. $\ln e^x$ **52.** $\ln e^y$ **53.** $\log 10,000$ **54.** $\log 0.001$

55. $\ln \dfrac{1}{e^3}$ **56.** $\ln \sqrt{e}$ **57.** $\log \sqrt{1000}$ **58.** $\log \sqrt[3]{10,000}$

Use the properties of logarithms to expand each of the following expressions.

59. $\ln 10e^{3t}$ **60.** $\ln 10e^{4t}$ **61.** $\ln Ae^{-2t}$

62. $\ln Ae^{-3t}$ **63.** $\log [100(1.01)^{3t}]$ **64.** $\log \left[\dfrac{1}{10} (1.5)^{t+2} \right]$

65. $\ln (Pe^{rt})$ **66.** $\ln \left(\dfrac{1}{2} e^{-kt} \right)$ **67.** $-\log (4.2 \times 10^{-3})$

68. $-\log (5.7 \times 10^{-10})$

If $\ln 2 = 0.6931$, $\ln 3 = 1.0986$, and $\ln 5 = 1.6094$, find each of the following.

69. $\ln 15$ **70.** $\ln 10$ **71.** $\ln \dfrac{1}{3}$ **72.** $\ln \dfrac{1}{5}$

73. $\ln 9$ **74.** $\ln 25$ **75.** $\ln 16$ **76.** $\ln 81$

APPLYING THE CONCEPTS

77. Atomic Bomb Tests The formula for determining the magnitude, M, of an earthquake on the Richter scale is $M = \log_{10} T$, where T is the number of times the shock wave is greater than the smallest measurable shock wave. The Bikini Atoll in the Pacific Ocean was used as a location for atomic bomb tests by the United States government in the 1950s. One such test resulted in an earthquake measurement of 5.0 on the Richter scale. Compare the 1906 San Francisco earthquake of estimated magnitude 8.3 on the Richter scale to this atomic bomb test. Use the shock wave T for purposes of comparison.

78. Atomic Bomb Tests Today's nuclear weapons are 1,000 times more powerful than the atomic bombs tested in the Bikini Atoll mentioned in Problem 77. Use the shock wave T to determine the Richter scale measurement of a nuclear test today.

79. Getting Close to e Use a calculator to complete the following table.

x	$(1 + x)^{1/x}$
1	
0.5	
0.1	
0.01	
0.001	
0.0001	
0.00001	

What number does the expression $(1 + x)^{1/x}$ seem to approach as x gets closer and closer to zero?

80. Getting Close to e Use a calculator to complete the following table.

x	$\left(1 + \frac{1}{x}\right)^{x}$
1	
10	
50	
100	
500	
1,000	
10,000	
1,000,000	

What number does the expression $\left(1 + \frac{1}{x}\right)^{x}$ seem to approach as x gets larger and larger?

81. University Enrollment The percentage of students enrolled in a university who are between the ages of 25 and 34 can be modeled by the formula $s = 5 \ln x$, where s is the percentage of students and x is the number of years since 1989. Predict the year in which approximately 15% of students enrolled in a university are between the ages of 25 and 34.

82. **Memory** A class of students take a test on the mathematics concept of solving quadratic equations. That class agrees to take a similar form of the test each month for the next 6 months to test their memory of the topic since instruction. The function of the average score earned each month on the test is $m(x) = 75 - 5 \ln(x + 1)$, where x represents time in months. Complete the table to indicate the average score earned by the class at each month.

Time, x	Score, m
0	
1	
2	
3	
4	
5	
6	

Use the following figure to solve Problems 83–86.

pH Scale

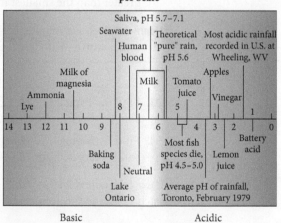

83. **pH** Find the pH of orange juice if the concentration of the hydrogen ion in the juice is $[H^+] = 6.50 \times 10^{-4}$.

84. **pH** Find the pH of milk if the concentration of the hydrogen ions in milk is $[H^+] = 1.88 \times 10^{-6}$.

85. **pH** Find the concentration of hydrogen ions in a glass of wine if the pH is 4.75.

86. **pH** Find the concentration of hydrogen ions in a bottle of vinegar if the pH is 5.75.

The Richter Scale Find the relative size T of the shock wave of earthquakes with the following magnitudes, as measured on the Richter scale.

87. 5.5   88. 6.6   89. 8.3   90. 8.7

91. Earthquake The chart below is a partial listing of earthquakes that were recorded in Canada during one year. Complete the chart by computing the magnitude on the Richter scale, M, or the number of times the associated shock wave is larger than the smallest measurable shock wave, T.

Location	Date	Magnitude M	Shock wave T
Moresby Island	Jan. 23	4.0	
Vancouver Island	Apr. 30		1.99×10^5
Quebec City	June 29	3.2	
Mould Bay	Nov. 13	5.2	
St. Lawrence	Dec. 14		5.01×10^3

SOURCE: *National Resources Canada, National Earthquake Hazards Program.*

92. Earthquake On April 25, 2015, an earthquake with a magnitude of 7.8 on the Richter scale hit Nepal. By what factor was this earthquake's shock wave greater than the smallest measurable shock wave?

Depreciation The annual rate of depreciation r on a car that is purchased for P dollars and is worth W dollars t years later can be found from the formula.

$$\log(1 - r) = \frac{1}{t} \log \frac{W}{P}$$

93. Find the annual rate of depreciation on a car that is purchased for $9,000 and sold 5 years later for $4,500.

94. Find the annual rate of depreciation on a car that is purchased for $9,000 and sold 4 years later for $3,000.

Two cars depreciate in value according to the following depreciation tables. In each case, find the annual rate of depreciation.

95.

Age in Years	Value in Dollars
New	7,550
5	5,750

96.

Age in Years	Value in Dollars
New	7,550
3	5,750

GETTING READY FOR THE NEXT SECTION

Solve.

97. $5(2x + 1) = 12$ **98.** $4(3x - 2) = 21$

Use a calculator to evaluate, give answers to 4 decimal places.

99. $\dfrac{100{,}000}{32{,}000}$ **100.** $\dfrac{1.4982}{6.5681} + 3$

101. $\dfrac{1}{2}\left(\dfrac{-0.6931}{1.4289} + 3\right)$ **102.** $1 + \dfrac{0.04}{52}$

Use the power rule to rewrite the following logarithms.

103. $\log 1.05^t$ **104.** $\log 1.033^t$

Use identities to simplify.

105. $\ln e^{0.05t}$ **106.** $\ln e^{-0.000121t}$

OBJECTIVES

A Solve exponential equations.

B Calculate logarithms using the change-of-base property.

C Solve application problems involving logarithms or exponential equations.

© istockphoto.com/gemenacom

8.6

For items involved in exponential growth, the time it takes for a quantity to double is called the *doubling time*. For example, if you invest $5,000 in an account that pays 5% annual interest, compounded quarterly, you may want to know how long it will take for your money to double in value. You can find this doubling time if you can solve the equation

$$10{,}000 = 5{,}000 \,(1.0125)^{4t}$$

As you will see as you progress through this section, logarithms are the key to solving equations of this type.

 Exponential Equations

Logarithms are very important in solving equations in which the variable appears as an exponent. The equation

$$5^x = 12$$

is an example of one such equation. Equations of this form are called *exponential equations*. Because the quantities 5^x and 12 are equal, so are their common logarithms. We begin our solution by taking the logarithm of both sides:

$$\log 5^x = \log 12$$

We now apply Property 3 for logarithms, $\log x^r = r \log x$, to turn x from an exponent into a coefficient:

$$x \log 5 = \log 12$$

Dividing both sides by $\log 5$ gives us

$$x = \frac{\log 12}{\log 5}$$

If we want a decimal approximation to the solution, we can find $\log 12$ and $\log 5$ on a calculator and divide:

$$x \approx \frac{1.0792}{0.6990}$$

$$\approx 1.5439$$

The complete problem looks like this:

$$5^x = 12$$

$$\log 5^x = \log 12$$

$$x \log 5 = \log 12$$

$$x = \frac{\log 12}{\log 5}$$

$$\approx \frac{1.0792}{0.6990}$$

$$\approx 1.5439$$

Here is another example of solving an exponential equation using logarithms.

EXAMPLE 1

Solve for x: $25^{2x+1} = 15$

SOLUTION   Taking the logarithm of both sides and then writing the exponent $(2x + 1)$ as a coefficient, we proceed as follows:

$$25^{2x+1} = 15$$

$$\log 25^{2x+1} = \log 15 \qquad \text{Take the log of both sides.}$$

$$(2x + 1)\log 25 = \log 15 \qquad \text{Property 3}$$

$$2x + 1 = \frac{\log 15}{\log 25} \qquad \text{Divide by log 25.}$$

$$2x = \frac{\log 15}{\log 25} - 1 \qquad \text{Add } -1 \text{ to both sides.}$$

$$x = \frac{1}{2}\left(\frac{\log 15}{\log 25} - 1\right) \qquad \text{Multiply both sides by } \tfrac{1}{2}.$$

Using a calculator, we can write a decimal approximation to the answer:

$$x \approx \frac{1}{2}\left(\frac{1.1761}{1.3979} - 1\right)$$

$$\approx \frac{1}{2}(0.8413 - 1)$$

$$\approx \frac{1}{2}(-0.1587)$$

$$\approx -0.079$$

If you invest P dollars in an account with an annual interest rate r that is compounded n times a year, then t years later the amount of money in that account will be

$$A = P\left(1 + \frac{r}{n}\right)^{nt}$$

EXAMPLE 2

How long does it take for $5,000 to double if it is deposited in an account that yields 5% interest compounded once a year?

SOLUTION Substituting $P = 5{,}000$, $r = 0.05$, $n = 1$, and $A = 10{,}000$ into our formula, we have

$$10{,}000 = 5{,}000(1 + 0.05)^t$$

$$10{,}000 = 5{,}000(1.05)^t$$

$$2 = (1.05)^t \qquad \text{Divide by 5,000.}$$

This is an exponential equation. We solve by taking the logarithm of both sides:

$$\log 2 = \log(1.05)^t$$

$$= t \log 1.05$$

Dividing both sides by $\log 1.05$, we have

$$t = \frac{\log 2}{\log 1.05}$$

$$\approx 14.2$$

It takes a little over 14 years for $5,000 to double if it earns 5% interest per year, compounded once a year.

B Change of Base

There is a fourth property of logarithms we have not yet considered. This last property allows us to change from one base to another and is therefore called the *change-of-base property*.

> **PROPERTY** *Property 4 (Change of Base)*
>
> If a and b are both positive numbers other than 1, and if $x > 0$, then
>
> $$\log_a x = \frac{\log_b x}{\log_b a}$$
>
> $\uparrow$ Base a $\qquad$ $\uparrow$ Base b

The logarithm on the left side has a base of a, and both logarithms on the right side have a base of b. This allows us to change from base a to any other base b that is a positive number other than 1. Here is a proof of Property 4 for logarithms.

Proof We begin by writing the identity

$$a^{\log_a x} = x$$

Taking the logarithm base b of both sides and writing the exponent $\log_a x$ as a coefficient, we have

$$\log_b a^{\log_a x} = \log_b x$$

$$\log_a x \log_b a = \log_b x$$

Dividing both sides by $\log_b a$, we have the desired result:

$$\frac{\log_a x \log_b a}{\log_b a} = \frac{\log_b x}{\log_b a}$$

$$\log_a x = \frac{\log_b x}{\log_b a}$$

We can use this property to find logarithms we could not otherwise compute on our calculators — that is, logarithms with bases other than 10 or e. The next example illustrates the use of this property.

EXAMPLE 3

Find $\log_8 24$.

SOLUTION   Because we do not have base-8 logarithms on our calculators, we can change this expression to an equivalent expression that contains only base-10 logarithms:

$$\log_8 24 = \frac{\log 24}{\log 8} \qquad \text{Property 4}$$

Don't be confused. We did not just drop the base; we changed to base 10. We could have written the last line like this:

$$\log_8 24 = \frac{\log_{10} 24}{\log_{10} 8}$$

From our calculators, we write

$$\log_8 24 \approx \frac{1.3802}{0.9031}$$

$$\approx 1.5283$$

Application

EXAMPLE 4

Suppose that the population in a small city is 32,000 in the beginning of 2010 and that the city council assumes that the population size t years later can be estimated by the equation

$$P = 32{,}000e^{0.05t}$$

Approximately when will the city have a population of 50,000?

SOLUTION   We substitute 50,000 for P in the equation and solve for t:

$$50{,}000 = 32{,}000e^{0.05t}$$

$$1.5625 = e^{0.05t} \qquad \frac{50{,}000}{32{,}000} = 1.5625$$

To solve this equation for t, we can take the natural logarithm of each side:

$$\ln 1.5625 = \ln e^{0.05t}$$

$$= 0.05t \ln e \qquad \text{Property 3 for logarithms}$$

$$= 0.05t \qquad \text{Because } \ln e = 1$$

$$t = \frac{\ln 1.5625}{0.05} \qquad \text{Divide each side by 0.05.}$$

$$\approx 8.93 \text{ years}$$

We can estimate that the population will reach 50,000 toward the end of 2018.

USING TECHNOLOGY   *Graphing Calculators*

We can evaluate many logarithmic expressions on a graphing calculator by using the fact that logarithmic functions and exponential functions are inverses.

EXAMPLE 5

Evaluate the logarithmic expression $\log_3 7$ from the graph of an exponential function.

SOLUTION   First, we let $\log_3 7 = x$. Next, we write this expression in exponential form as $3^x = 7$. We can solve this equation graphically by finding the intersection of the graphs $Y_1 = 3^x$ and $Y_2 = 7$, as shown in Figure 1.

Using the calculator, we find the two graphs intersect at $(1.77, 7)$. Therefore, $\log_3 7 = 1.77$ to the nearest hundredth. We can check our work by evaluating the expression $3^{1.77}$ on our calculator with the key strokes

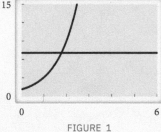

FIGURE 1

$$3 \;\boxed{\wedge}\; 1.77 \;\boxed{\text{ENTER}}$$

The result is 6.99 to the nearest hundredth, which seems reasonable since 1.77 is accurate to the nearest hundredth. To get a result closer to 7, we would need to find the intersection of the two graphs more accurately.

GETTING READY FOR CLASS

After reading through the preceding section, respond in your own words and in complete sentences.

A. What is an exponential equation?

B. How do logarithms help us solve exponential equations?

C. What is the change-of-base property?

D. Write an application modeled by the equation
$A = 10,000 \left(1 + \frac{0.08}{2}\right)^{2 \cdot 5}$.

SPOTLIGHT ON SUCCESS

Shelby | student instructor

The price of success is hard work, dedication to the job at hand, and the determination that whether we win or lose, we have applied the best of ourselves to the task at hand.

—Vince Lombardi

My earliest memory of math is from elementary school, having to continuously take the division and times tables test because I couldn't finish it in the allotted time. I have never been naturally gifted at math, but I enjoy a challenge. I wouldn't allow my setbacks to stop me from succeeding in my classes. As a high school freshman I struggled in my geometry class, and as a senior, I thrived in my AP Calculus class. The difference in those short four years was the effort I put in, as well as having great teachers who provided the necessary knowledge and support along the way.

However, to continue to thrive in math, I had to realize that my success wasn't solely in the hands of the teacher. I saw students fail classes taught by some of the best teachers on campus. It all stemmed from the personal goals each student held. There were multiple times when I told myself, "You can do this. You can learn the concepts." I'm not sure I would have moved past pre-algebra without those personal words of encouragement.

Having confidence in yourself is necessary to be successful in all aspects of life. It's easy to give up on something that doesn't come naturally, but the reward of achieving something you thought to be impossible is worth the hardship.

Problem Set 8.6

Solve each exponential equation. Use a calculator to write the answer in decimal form.

1. $3^x = 5$

2. $4^x = 3$

3. $5^x = 3$

4. $3^x = 4$

5. $5^{-x} = 12$

6. $7^{-x} = 8$

7. $12^{-x} = 5$

8. $8^{-x} = 7$

9. $8^{x+1} = 4$

10. $9^{x+1} = 3$

11. $4^{x-1} = 4$

12. $3^{x-1} = 9$

13. $3^{2x+1} = 2$

14. $2^{2x+1} = 3$

15. $3^{1-2x} = 2$

16. $2^{1-2x} = 3$

17. $15^{3x-4} = 10$

18. $10^{3x-4} = 15$

19. $6^{5-2x} = 4$

20. $9^{7-3x} = 5$

21. $3^{-4x} = 81$

22. $2^{5x} = \dfrac{1}{16}$

23. $5^{3x-2} = 15$

24. $7^{4x+3} = 200$

25. $100e^{3t} = 250$

26. $150e^{0.065t} = 400$

27. $1200\left(1 + \dfrac{0.072}{4}\right)^{4t} = 25000$

28. $2700\left(1 + \dfrac{0.086}{12}\right)^{12t} = 10000$

29. $50e^{-0.0742t} = 32$

30. $19e^{-0.000243t} = 12$

Use the change-of-base property and a calculator to find a decimal approximation to each of the following logarithms.

31. $\log_8 16$

32. $\log_9 27$

33. $\log_{16} 8$

34. $\log_{27} 9$

35. $\log_7 15$

36. $\log_3 12$

37. $\log_{15} 7$

38. $\log_{12} 3$

39. $\log_8 240$

40. $\log_6 180$

41. $\log_4 321$

42. $\log_5 462$

Find a decimal approximation to each of the following natural logarithms.

43. $\ln 345$

44. $\ln 3{,}450$

45. $\ln 0.345$

46. $\ln 0.0345$

47. $\ln 10$

48. $\ln 100$

49. $\ln 45{,}000$

50. $\ln 450{,}000$

APPLYING THE CONCEPTS

51. Compound Interest How long will it take for $500 to double if it is invested at 6% annual interest compounded 2 times a year?

52. Compound Interest How long will it take for $500 to double if it is invested at 6% annual interest compounded 12 times a year?

53. Compound Interest How long will it take for $1,000 to triple if it is invested at 12% annual interest compounded 6 times a year?

54. Compound Interest How long will it take for $1,000 to become $4,000 if it is invested at 12% annual interest compounded 6 times a year?

55. Doubling Time How long does it take for an amount of money P to double itself if it is invested at 8% interest compounded 4 times a year?

56. Tripling Time How long does it take for an amount of money P to triple itself if it is invested at 8% interest compounded 4 times a year?

57. Tripling Time If a $25 investment is worth $75 today, how long ago must that $25 have been invested at 6% interest compounded twice a year?

58. Doubling Time If a $25 investment is worth $50 today, how long ago must that $25 have been invested at 6% interest compounded twice a year?

Recall from Section 8.1 that if P dollars are invested in an account with annual interest rate r, compounded continuously, then the amount of money in the account after t years is given by the formula

$$A(t) = Pe^{rt}$$

59. **Continuously Compounded Interest** Repeat Problem 51 if the interest is compounded continuously.

60. **Continuously Compounded Interest** Repeat Problem 54 if the interest is compounded continuously.

61. **Continuously Compounded Interest** How long will it take $500 to triple if it is invested at 6% annual interest, compounded continuously?

62. **Continuously Compounded Interest** How long will it take $500 to triple if it is invested at 12% annual interest, compounded continuously?

63. **Continuously Compounded Interest** How long will it take for $1,000 to be worth $2,500 at 8% interest, compounded continuously?

64. **Continuously Compounded Interest** How long will it take for $1,000 to be worth $5,000 at 8% interest, compounded continuously?

65. **Exponential Growth** Suppose that the population in a small city is 32,000 at the beginning of 2015 and that the city council assumes that the population size t years later can be estimated by the equation

$$P(t) = 32,000e^{0.05t}$$

Approximately when will the city have a population of 64,000?

66. **Exponential Growth** Suppose the population of a city is given by the equation

$$P(t) = 100,000e^{0.05t}$$

where t is the number of years from the present time. How large is the population now? (*Now* corresponds to a certain value of t. Once you realize what that value of t is, the problem becomes very simple.)

67. **Airline Travel** The number of airline passengers in 1990 was 466 million. The number of passengers traveling by airplane each year has increased exponentially according to the model, $P(t) = 466 \cdot 1.035^t$, where t is the number of years since 1990 (U.S. Census Bureau). In what year is it predicted that 900 million passengers will travel by airline?

68. **Bankruptcy Model** In 1997, there were a total of 1,316,999 bankruptcies filed under the Bankruptcy Reform Act. The model for the number of bankruptcies filed is $B(t) = 0.798 \cdot 1.164^t$, where t is the number of years since 1994 and B is the number of bankruptcies filed in terms of millions (Administrative Office of the U.S. Courts, Statistical Tables for the Federal Judiciary). In what year is it predicted that 12 million bankruptcies will be filed?

69. **Health Care** In 1990, $699 billion was spent on health care expenditures. The amount of money, E, in billions spent on health care expenditures can be estimated using the function $E(t) = 78.16(1.11)^t$, where t is time in years since 1970 (U.S. Census Bureau). In what year was it estimated that $800 billion will be spent on health care expenditures?

70. Value of a Car As a car ages, its value decreases. The value of a particular car with an original purchase price of $25,600 is modeled by the function $c(t) = 25,600(1 - 0.22)^t$, where c is the value at time t (*Kelly Blue Book*). How old is the car when its value is $10,000?

71. Compound Interest In 2017, the average cost of attending a public university was $25,290 per year (U.S. Department of Education, National Center for Educational Statistics). If John's parents deposited that amount in an account in 2001 at an interest rate of 7% compounded semi-annually, how long will it take for the money to double?

72. Carbon Dating Scientists use carbon-14 dating to find the age of fossils and other artifacts. The amount of carbon-14 in an organism will yield information concerning its age. A formula used in carbon-14 dating is $A(t) = A_0 \cdot 2^{-t/5600}$, where A_0 is the amount of carbon originally in the organism, t is time in years, and A is the amount of carbon remaining after t years. Determine the number of years since an organism died if it originally contained 1,000 gram of carbon-14 and it currently contains 600 gram of carbon-14.

73. Cost Increase The cost of a can of Coca-Cola® in 1960 was $0.10. The function that models the cost of a Coca-Cola by year is $C(t) = 0.10e^{0.0576t}$, where t is the number of years since 1960. In what year is it expected that a can of Coca-Cola will cost $1.00?

74. Online Banking Use The number of households using online banking services increased from 754,000 in 2010 to 12,980,000 in 2015. The formula $H(t) = 0.76e^{0.55t}$ models the number of households, H, in millions when time is t years since 2010. In what year is it estimated that 50,000,000 households will use online banking services?

75. Half-life The Nobelium-259 isotope has a half-life of 58 minutes. If the initial amount is 500 grams, then find:

a. the amount remaining after 290 minutes.

b. how much time is needed before only 1 gram remains.

76. Half-life The Fermium-253 isotope has a half-life of 3 days. If the initial amount is 200 grams, then find:

a. the amount remaining after 15 days.

b. how much time is needed before only 0.5 gram remains.

MAINTAINING YOUR SKILLS

The following problems review material covered in Section 7.5.

Find the vertex for each of the following parabolas, and then indicate if it is the highest or lowest point on the graph.

77. $y = 2x^2 + 8x - 15$ **78.** $y = 3x^2 - 9x - 10$

79. $y = 12x - 4x^2$ **80.** $y = 18x - 6x^2$

81. Maximum Height An object is projected into the air with an initial upward velocity of 64 feet per second. Its height h at any time t is given by the formula $h = 64t - 16t^2$. Find the time at which the object reaches its maximum height. Then, find the maximum height.

82. **Maximum Height** An object is projected into the air with an initial upward velocity of 64 feet per second from the top of a building 40 feet high. If the height h of the object t seconds after it is projected into the air is $h = 40 + 64t - 16t^2$, find the time at which the object reaches its maximum height. Then, find the maximum height it attains.

EXTENDING THE CONCEPTS

83. Solve the formula $A = Pe^{rt}$ for t.

84. Solve the formula $A = Pe^{-rt}$ for t.

85. Solve the formula $A = P2^{-kt}$ for t.

86. Solve the formula $A = P2^{kt}$ for t.

87. Solve the formula $A = P(l - r)^t$ for t.

88. Solve the formula $A = P(l + r)^t$ for t.

Simplify each expression. You should not need a calculator.

89. $(\log_2 3)(\log_3 8)$

90. $(\log_3 5)(\log_5 81)$

91. $(\log_{25} 9)(\log_{81} 5)$

92. $(\log_6 16)(\log_{32} 36)$

93. $(\log_{125} 81)(\log_{243} 25)$

94. $(\log_{27} 16)(\log_{64} 81)$

Exponential Functions [8.1]

1. For the exponential function $f(x) = 2^x$,
$$f(0) = 2^0 = 1$$
$$f(1) = 2^1 = 2$$
$$f(2) = 2^2 = 4$$
$$f(3) = 2^3 = 8$$

Any function of the form
$$f(x) = b^x$$
where $b > 0$ and $b \neq 1$, is an *exponential function*.

One-to-One Functions [8.2]

2. The function $f(x) = x^2$ is not one-to-one because 9, which is in the range, comes from both 3 and -3 in the domain.

A function is a *one-to-one function* if every element in the range comes from exactly one element in the domain.

Inverse Functions [8.2]

3. The inverse of $f(x) = 2x - 3$ is
$$f^{-1}(x) = \frac{x + 3}{2}$$

The *inverse* of a function is obtained by reversing the order of the coordinates of the ordered pairs belonging to the function. Only one-to-one functions have inverses that are also functions.

Definition of Logarithms [8.3]

4. The definition allows us to write expressions like
$$y = \log_3 27$$
equivalently in exponential form as
$$3^y = 27$$
which makes it apparent that y is 3.

If b is a positive number not equal to 1, then the expression
$$y = \log_b x$$
is equivalent to $x = b^y$; that is, in the expression $y = \log_b x$, y is the number to which we raise b in order to get x. Expressions written in the form $y = \log_b x$ are said to be in *logarithmic form*. Expressions like $x = b^y$ are in *exponential form*.

Two Special Identities [8.3]

5. Examples of the two special identities are
$$5^{\log_5 12} = 12$$
and
$$\log_8 8^3 = 3$$

For $b > 0$, $b \neq 1$, the following two expressions hold for all positive real numbers x:

(1) $b^{\log_b x} = x$

(2) $\log_b b^x = x$

Properties of Logarithms [8.4]

6. We can rewrite the expression
$$\log_{10} \frac{45^6}{273}$$
using the properties of logarithms, as
$$6 \log_{10} 45 - \log_{10} 273$$

If x, y, and b are positive real numbers, $b \neq 1$, and r is any real number, then:

1. $\log_b(xy) = \log_b x + \log_b y$

2. $\log_b \left(\frac{x}{y} \right) = \log_b x - \log_b y$

3. $\log_b x^r = r \log_b x$

Common Logarithms [8.5]

7. $\log_{10} 10{,}000 = \log 10{,}000$
$= \log 10^4$
$= 4$

Common logarithms are logarithms with a base of 10. To save time in writing, we omit the base when working with common logarithms; that is,

$$\log x = \log_{10} x$$

Natural Logarithms [8.5]

8. $\ln e = 1$
$\ln 1 = 0$

Natural logarithms, written *ln x*, are logarithms with a base of e, where the number e is an irrational number (like the number π). A decimal approximation for e is 2.7183. All the properties of exponents and logarithms hold when the base is e.

Change of Base [8.6]

9. $\log_6 475 = \dfrac{\log 475}{\log 6}$
$\approx \dfrac{2.6767}{0.7782}$
≈ 3.44

If x, a, and b are positive real numbers, $a \neq 1$ and $b \neq 1$, then

$$\log_a x = \frac{\log_b x}{\log_b a}$$

COMMON MISTAKES

The most common mistakes that occur with logarithms come from trying to apply the three properties of logarithms to situations in which they don't apply. For example, a very common mistake looks like this:

$$\frac{\log 3}{\log 2} = \log 3 - \log 2 \quad \text{Mistake}$$

This is not a property of logarithms. To write the equation $\log 3 - \log 2$, we would have to start with

$$\log \frac{3}{2} \quad NOT \quad \frac{\log 3}{\log 2}$$

There is a difference.

$$\log \frac{3}{2} = 0.17609 \qquad \frac{\log 3}{\log 2} = 1.58496$$

Let $f(x) = 4^x$ and $g(x) = \left(\frac{1}{3}\right)^x$. Evaluate each of the following. [8.1]

1. $f(3) - g(2)$

2. $\dfrac{g(3) - g(2)}{3 - 2}$

Graph each exponential function. [1.1]

3. $f(x) = -5^x$

4. $g(x) = 3^{-x}$

For each of the following one-to-one functions, find the equation of the inverse. Write the inverse using the notation $f^{-1}(x)$. [8.2]

5. $f(x) = 2x - 5$

6. $f(x) = x^3 - 2$

7. $f(x) = -\dfrac{1}{2}x + 4$

8. $f(x) = \dfrac{3x + 2}{5x + 1}$

Sketch the graph of each function and its inverse. Find $f^{-1}(x)$ for Problem 3. [8.2]

9. $f(x) = \dfrac{1}{3}x + \dfrac{4}{3}$

10. $f(x) = x^2 - 2$

Write each of the following equations in logarithmic form. [8.3]

11. $4^{-2} = \dfrac{1}{16}$

12. $81 = 3^4$

Write each of the following equations in exponential form. [8.3]

13. $\log_3 \dfrac{1}{81} = -4$

Solve for x. [8.3]

14. $\log_4 x = 3$

15. $\log_x 16 = 4$

16. $\log_3 27 = x$

17. $\log_{25} 5 = x$

Graph each of the following. [8.3]

18. $y = \log_4 x$

19. $y = \log_{1/4} x$

Simplify each of the following. [8.3]

20. $\log_4 4$

21. $\log_4 8$

22. $\log_{10} 0.0001$

23. $\log_5 (\log_{32} 2)$

24. $\log_9 (\log_8 2)$

25. $\log_5 (\log_3 3)$

Evaluate each of the following. Round to the nearest hundredth if necessary. [8.3, 8.4, 8.5]

26. $\log_{16} 4$

27. $\log_8 24$

28. $\log 57{,}300$

29. $\log 0.0507$

30. $\ln 27.4$

31. $\ln 0.024$

Use the properties of logarithms to expand each expression. [8.4]

32. $\log_7 y^3$

33. $\log_8 \sqrt{z}$

34. $\log_{10} x^2 y^4$

35. $\log_{10} \dfrac{5}{4y}$

36. $\log_3 \dfrac{5x^3}{y}$

37. $\log \dfrac{\sqrt{x}}{y^5 \sqrt[3]{z}}$

Write each expression as a single logarithm. [8.4]

38. $\dfrac{1}{2} \log_4 x + 3 \log_4 y$

39. $3 \log x - 4\log y + \dfrac{1}{2} \log z$

40. $\dfrac{1}{3} \log_{10} x - \dfrac{1}{4} \log_{10} y$

41. $2 \log_3 x + 3 \log_3 y - \log_3 z$

Find the following logarithms. [8.5]

42. log 42,600

43. log 0.4260

44. log 4,900

45. log 900

Use a calculator to find x. Round to the nearest thousandth. [8.5]

46. $\log x = 2.6532$

47. $\log x = -1.2518$

Solve for x. Round to the nearest thousandth if necessary. [8.4, 8.5, 8.6]

48. $5 = 2^x$

49. $4^{2x-3} = 64$

50. $\log_{27} x + \log_{27}(x + 8) = \dfrac{2}{3}$

51. $\ln x = 4$

52. $\log_4 x + \log_4 2 = 3$

53. $\log_2 x + \log_2(x - 2) = 3$

54. $\log x = \log_3 9$

55. $\log_2(x + 3) - \log_2(x - 3) = 2$

Simplify each of the following expressions. [8.5]

56. $\ln e^{-3}$

57. $\ln \sqrt{e}$

Solve each exponential equation. Use a calculator to write the answer in decimal form. Round answers to four decimal places where appropriate. [8.6]

58. $2^{5x} = \dfrac{1}{16}$

59. $81^x = 27^{x-4}$

Use the change-of-base property to calculate each of the following logarithms. Round answers to four decimal places, where appropriate. [8.6]

60. $\log_9 27$

61. $\log_{12} 3$

62. Chemistry Use the formula $pH = -\log[H^+]$ to find the pH of a solution in which $[H^+] = 9.3 \times 10^{-6}$. [8.5]

63. Compound Interest If $400 is deposited in an account that earns 6% annual interest compounded quarterly, how much money will be in the account after 7 years? [8.1]

64. Compound Interest How long will it take $600 to become $2,400 if the $600 is deposited in an account that earns 9% annual interest compounded quarterly? [8.6]

Determine if the function is one-to-one. [8.2]

65. $f(x) = \{(3, 6), (4, 2), (5, -3), (6, 2)\}$

66. $g(x) = \{(-1, 2), (0, 4), (2, 8), (3, 10)\}$

67.

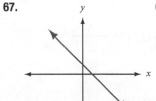

68.

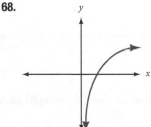

69.

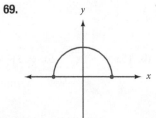

70.

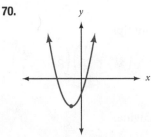

Simplify.

1. $-3 - 2[4 + 8(-1 - 5)]$

2. $4(3x + 1) - 7(2x - 2)$

3. $\dfrac{8}{3} \div \dfrac{1}{6} \cdot \dfrac{3}{2}$

4. $\dfrac{8}{3} - \dfrac{1}{6} + \dfrac{3}{2}$

5. $4\sqrt{32} - 3\sqrt{8} + 2\sqrt{18}$

6. $(\sqrt{5} + 3\sqrt{3})(2\sqrt{5} + \sqrt{3})$

7. $[(5 + i) - (2 - 3i)] - (2 - 5i)$

8. $\dfrac{(1 + i)^2}{2}$

9. $\log_{13}(\log_{17}17)$ **10.** $\log_9[\log_3(\log_2 8)]$

11. $\dfrac{1 - \dfrac{1}{x - 2}}{1 + \dfrac{1}{x - 2}}$ **12.** $1 - \dfrac{x}{1 - \dfrac{1}{x}}$

Reduce the following to lowest terms.

13. $\dfrac{137}{274}$ **14.** $\dfrac{216}{168}$

Multiply.

15. $\left(4t^2 + \dfrac{1}{5}\right)\left(5t^2 - \dfrac{1}{4}\right)$ **16.** $(5x + 3)(x^2 - 5x + 3)$

Divide.

17. $\dfrac{x^2 - 6x + 3}{x - 1}$ **18.** $\dfrac{12x^2 + 23x - 3}{3x + 2}$

Add or subtract.

19. $\dfrac{6}{(x + 3)(x - 3)} - \dfrac{5}{(x + 3)(x - 2)}$

20. $\dfrac{3}{6x^2 - 5x - 4} + \dfrac{1}{6x^2 - 11x + 4}$

Solve.

21. $8 - 5(3x + 2) = 4$

22. $\dfrac{3}{4}(12x - 7) + \dfrac{1}{4} = 13$

23. $32x^2 = 4x + 3$

24. $4x - 4 = x^2$

25. $\dfrac{1}{x + 2} + \dfrac{1}{x - 5} = 1$

26. $\dfrac{1}{(x + 3)(x - 2)} + \dfrac{4}{(x + 2)(x - 2)} = \dfrac{-2}{(x + 2)(x + 3)}$

27. $x - 4\sqrt{x} - 5 = 0$

28. $2(3y + 1)^2 + (3y + 1) - 10 = 0$

29. $\log_4 x = 4$ **30.** $\log_x 0.2 = -1$

31. $\log_6(x + 2) - \log_6(x + 5) = 1$

32. $\log_3(x - 1) + \log_3(x - 6) = 2$

Solve each system.

33. $-x + 5y = 19$
$2x - 9y = -34$

34. $4x + 7y = 24$
$-8x - 9y = -48$

35. $x + 3y = -12$
$x - 3z = -4$
$y + 3z = 0$

36. $6x - y + 9z = 0$
$6x + 6y + 9z = 0$
$8x + y + z = -11$

Graph each line.

37. $5x + 4y = 4$

38. $x = 2$

Write in symbols.

39. The difference of 8 and y is $-3x$.

40. The difference of $2a$ and $3b$ is less than their sum.

Write in scientific notation.

41. 0.00000129

42. 83,000,000

Factor completely.

43. $64x^3 + 27$

44. $48a^4 + 36a^2 + 6$

Specify the domain and range for the relation and state whether the relation is a function.

45. $\{(3, -2), (3, -4), (5, -1)\}$

46. $\{(1, 2), (2, 1), (3, 2)\}$

The illustration shows the yearly sales of different brands of cookies. Use the information to answer Questions 47 and 48.

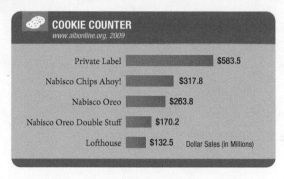

47. Write the combined total for Nabisco Oreo and Oreo Double Stuff cookies in scientific notation.

48. What is the difference in sales between Private Label cookies and Chips Ahoy! cookies?

49. Unit Analysis Convert 844.8 ft/min to miles per hour.

50. Translations Sketch the basic graph and the graph of the given function by hand on the same axes. Label the coordinates of two or three points on the graph of the given function.

$$f(x) = (x + 4)^2 + 1$$

Graph each exponential function. [8.1]

1. $f(x) = 2^x$

2. $g(x) = 3^{-x}$

Sketch the graph of each function and its inverse. Find $f^{-1}(x)$ for Problem 3. [8.2]

3. $f(x) = 2x - 3$

4. $f(x) = x^2 - 4$

Solve for x. [8.3]

5. $\log_4 x = 3$

6. $\log_x 5 = 2$

Graph each of the following [8.3]

7. $y = \log_2 x$

8. $y = \log_{1/2} x$

Evaluate each of the following. [8.3, 8.4, 8.5]

9. $\log_8 4$

10. $\log_7 21$

11. $\log 23{,}400$

12. $\log 0.0123$

13. $\ln 46.2$

14. $\ln 0.0462$

Use the properties of logarithms to expand each expression. [8.4]

15. $\log_2 \dfrac{8x^2}{y}$

16. $\log \dfrac{\sqrt{x}}{(y^4)\sqrt[5]{z}}$

Write each expression as a single logarithm. [8.4]

17. $2 \log_3 x - \dfrac{1}{2} \log_3 y$

18. $\dfrac{1}{3} \log x - \log y - 2 \log z$

Use a calculator to find x. [8.5]

19. $\log x = 4.8476$

20. $\log x = -2.6478$

Solve for x. [8.4, 8.6]

21. $5 = 3^x$

22. $4^{2x-1} = 8$

23. $\log_5 x - \log_5 3 = 1$

24. $\log_2 x + \log_2(x - 7) = 3$

25. pH Find the pH of a solution in which $[H^+] = 6.6 \times 10^{-7}$. [8.5]

26. Compound Interest If $400 is deposited in an account that earns 10% annual interest compounded twice a year, how much money will be in the account after 5 years? [8.1]

27. Compound Interest How long will it take $600 to become $1,800 if the $600 is deposited in an account that earns 8% annual interest compounded 4 times a year? [8.6]

28. Depreciation If a car depreciates in value 20% per year for the first 5 years after it is purchased for P_0 dollars, then its value after t years will be $V(t) = P_0(1 - r)^t$ for $0 \le t \le 5$. To the nearest dollar, find the value of a car 4 years after it is purchased for $18,000. [8.1]

Determine if the function is one-to-one. [8.2]

29. $f(x) = \{(4, 7), (5, 8), (7, 4), (8, 5)\}$

30. $g(x) = \{(1, 7), (2, 0), (3, 1), (4, 0)\}$

31.

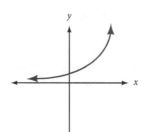

32.

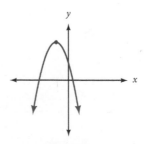

9

Sequences and Series

© istockphoto.com/zhudifeng

Suppose you run up a balance of $1,000 on a credit card that charges 1.65% interest each month (i.e., an annual rate of 19.8%). If you stop using the card and make the minimum payment of $20 each month, how long will it take you to pay off the balance on the card? The answer can be found by using the formula

$$U_n = (1.0165)U_{n-1} - 20$$

where U_n stands for the current unpaid balance on the card, and U_{n-1} is the previous month's balance. The table and figure below were created from this formula and a graphing calculator. As you can see from the table, the balance on the credit card decreases very little each month.

Monthly Credit Card Balances

Previous Balance $U_{(n-1)}$	Monthly Interest Rate	Payment Number n	Monthly Payment	New Balance U_n
$1,000.00	1.65%	1	$20	$996.50
$996.50	1.65%	2	$20	$992.94
$992.94	1.65%	3	$20	$989.32
$989.32	1.65%	4	$20	$985.64
$985.64	1.65%	5	$20	$981.90

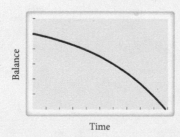

Balance

Time

719

Don't complain about anything, ever.

Do you complain to your classmates about your teacher? If you do, it could be getting in the way of your success in the class.

I have students who tell me that they like the way I teach and that they are enjoying my class. I have other students, in the same class, who complain to each other about me. They say I don't explain things well enough. Are the complaining students giving themselves a reason for not doing well in the class? I think so. They are shifting the responsibility for their success from themselves to me. It's not their fault they are not doing well, it's mine. When these students are alone, trying to do homework, they start thinking about how unfair everything is and they lose their motivation to study. Without intending to, they have set themselves up to fail by making their complaints more important than their progress in the class.

What happens when you stop complaining? You put yourself back in charge of your success. When there is no one to blame if things don't go well, you are more likely to do well. I have had students tell me that, once they stopped complaining about a class, the teacher became a better teacher and they started to actually enjoy going to class.

If you find yourself complaining to your friends about a class or a teacher, make a decision to stop. When other people start complaining to each other about the class or the teacher, walk away; don't participate in the complaining session. Try it for a day, or a week, or for the rest of the term. It may be difficult to do at first, but I'm sure you will like the results, and if you don't, you can always go back to complaining.

9.1

OBJECTIVES

A   Find the terms of a sequence given a general term.

B   Find the terms of a sequence using a recursion formula.

C   Find a general term for a sequence.

©iStockphoto.com/Yurkakul

Imagine you work as a production assistant for a long-running television series. Your first-year salary totaled $27,000. Each year thereafter, you received a 3% increase. With this information, we can calculate each of your annual salaries. Your first five salaries are as follows:

$27,000 $27,810 $28,644.30 $29,503.63 $30,388.74

This list is called a *sequence* and we can represent it with the following formula:

$$a_n = 27{,}000(1.03)^{n-1}$$

You'll understand more about how we arrived at this formula after reading this section.

A   Sequences

Many of the sequences in this chapter will be familiar to you on an intuitive level because you have worked with them for some time now. Here are some of those sequences:

The sequence of odd numbers

$$1, 3, 5, 7, \ldots$$

The sequence of even numbers

$$2, 4, 6, 8, \ldots$$

The sequence of squares

$$1^2, 2^2, 3^2, 4^2, \ldots = 1, 4, 9, 16, \ldots$$

The numbers in each of these sequences can be found from the formulas that define functions. For example, the sequence of even numbers can be found from the function

$$f(x) = 2x$$

by finding $f(1), f(2), f(3), f(4)$, and so forth. This gives us justification for the formal definition of a sequence.

> **DEFINITION** *sequence*
>
> A *sequence* is a function whose domain is the set of positive integers 1, 2, 3, 4, ...

As you can see, sequences are simply functions with a specific domain. If we want to form a sequence from the function $f(x) = 3x + 5$, we simply find $f(1), f(2), f(3)$, and so on. Doing so gives us the sequence

$$8, 11, 14, 17, \ldots$$

because $f(1) = 3(1) + 5 = 8$, $f(2) = 3(2) + 5 = 11$, $f(3) = 3(3) + 5 = 14$, and $f(4) = 3(4) + 5 = 17$.

Notation   Because the domain for a sequence is always the set $\{1, 2, 3, \ldots\}$, we can simplify the notation we use to represent the terms of a sequence. Using the letter a instead of f, and subscripts instead of numbers enclosed by parentheses, we can represent the sequence from the previous discussion as follows:

$$a_n = 3n + 5$$

Instead of $f(1)$, we write a_1 for the *first term* of the sequence.
Instead of $f(2)$, we write a_2 for the *second term* of the sequence.
Instead of $f(3)$, we write a_3 for the *third term* of the sequence.
Instead of $f(4)$, we write a_4 for the *fourth term* of the sequence.
Instead of $f(n)$, we write a_n for the *nth term* of the sequence.

The nth term is also called the *general term* of the sequence. The general term is used to define the other terms of the sequence. That is, if we are given the formula for the general term a_n, we can find any other term in the sequence. The following examples illustrate.

EXAMPLE 1

Find the first four terms of the sequence whose general term is given by $a_n = 2n - 1$.

SOLUTION   The subscript notation a_n works the same way function notation works. To find the first, second, third, and fourth terms of this sequence, we simply substitute 1, 2, 3, and 4 for n in the formula $2n - 1$:

$$\text{If} \qquad \text{the general term is } a_n = 2n - 1$$
$$\text{then} \qquad \text{the first term is } a_1 = 2(1) - 1 = 1$$
$$\text{the second term is } a_2 = 2(2) - 1 = 3$$
$$\text{the third term is } a_3 = 2(3) - 1 = 5$$
$$\text{the fourth term is } a_4 = 2(4) - 1 = 7$$

The first four terms of this sequence are the odd numbers 1, 3, 5, and 7. The whole sequence can be written as

$$1, 3, 5, 7, \ldots, 2n - 1, \ldots$$

Because each term in this sequence is larger than the preceding term, we say the sequence is an *increasing sequence*.

EXAMPLE 2

Write the first four terms of the sequence defined by

$$a_n = \frac{1}{n+1}$$

SOLUTION   Replacing n with 1, 2, 3, and 4, we have, respectively, the first four terms:

$$\text{First term} = a_1 = \frac{1}{1+1} = \frac{1}{2}$$

$$\text{Second term} = a_2 = \frac{1}{2+1} = \frac{1}{3}$$

$$\text{Third term} = a_3 = \frac{1}{3+1} = \frac{1}{4}$$

$$\text{Fourth term} = a_4 = \frac{1}{4+1} = \frac{1}{5}$$

The sequence defined by

$$a_n = \frac{1}{n+1}$$

can be written as

$$\frac{1}{2}, \frac{1}{3}, \frac{1}{4}, \cdots, \frac{1}{n+1}, \cdots$$

Because each term in the sequence is smaller than the term preceding it, the sequence is said to be a *decreasing sequence*.

EXAMPLE 3

Find the fifth and sixth terms of the sequence whose general term is given by

$$a_n = \frac{(-1)^n}{n^2}$$

SOLUTION   For the fifth term, we replace n with 5. For the sixth term, we replace n with 6:

$$\text{Fifth term} = a_5 = \frac{(-1)^5}{5^2} = \frac{-1}{25}$$

$$\text{Sixth term} = a_6 = \frac{(-1)^6}{6^2} = \frac{1}{36}$$

The sequence in Example 3 can be written as

$$-1, \frac{1}{4}, -\frac{1}{9}, \frac{1}{16}, \cdots, \frac{(-1)^n}{n^2}, \cdots$$

Because the terms alternate in sign — if one term is positive, then the next term is negative — we call this an *alternating sequence*. The first three examples all illustrate how we work with a sequence in which we are given a formula for the general term.

USING TECHNOLOGY *Finding Sequences on a Graphing Calculator*

Method 1: Using a Table

We can use the table function on a graphing calculator to view the terms of a sequence. To view the terms of the sequence $a_n = 3n + 5$, we set $Y_1 = 3X + 5$. Then we use the table setup feature on the calculator to set the table minimum to 1, and the table increment to 1 also. Here is the setup and result for a TI-83.

Table Setup	Y Variables Setup	Resulting Table	
		X	Y
Table minimum $=1$	$Y_1 = 3X + 5$		
Table increment $=1$		1	8
Independent variable: Auto		2	11
Dependent variable: Auto		3	14
		4	17
		5	20

To find any particular term of a sequence, we change the independent variable setting to Ask, and then input the number of the term of the sequence we want to find. For example, if we want term a_{100}, then we input 100 for the independent variable, and the table gives us the value of 305 for that term.

Method 2: Using the Built-in seq(Function

Using this method, first find the seq(function. On a TI-83 it is found in the LIST OPS menu. To find terms a_1 through a_7 for $a_n = 3n + 5$, we first bring up the seq(function on our calculator, then we input the following four items, in order, separated by commas: 3X+5, X, 1, 7. Then we close the parentheses. Our screen will look like this:

$$\text{seq}(3X+5, X, 1, 7)$$

Pressing ENTER displays the first five terms of the sequence. Pressing the right arrow key repeatedly brings the remaining members of the sequence into view.

Method 3: Using the Built-in Seq Mode

Press the MODE key on your TI-83 and then select Seq (it's next to Func Par and Pol). Go to the Y variables list and set $n\text{Min} = 1$ and $u(n) = 3n+5$. Then go to the TBLSET key to set up your table like the one shown in Method 1. Pressing TABLE will display the sequence you have defined.

B Recursion Formulas

Let's go back to one of the first sequences we looked at in this section:

$$8, 11, 14, 17, \ldots$$

Each term in the sequence can be found by simply substituting positive integers for n in the formula $a_n = 3n + 5$. Another way to look at this sequence, however, is to notice that each term can be found by adding 3 to the preceding term; so, we could give all the terms of this sequence by simply saying

Start with 8, and then add 3 to each term to get the next term.

The same idea, expressed in symbols, looks like this:

$$a_1 = 8 \quad \text{and} \quad a_n = a_{n-1} + 3 \quad \text{for } n > 1$$

This formula is called a *recursion formula* because each term is written recursively in terms of the term or terms that precede it.

EXAMPLE 4

Write the first four terms of the sequence given recursively by

$$a_1 = 4 \quad \text{and} \quad a_n = 5a_{n-1} \quad \text{for } n > 1$$

SOLUTION   The formula tells us to start the sequence with the number 4, and then multiply each term by 5 to get the next term. Therefore,

$$a_1 = 4$$
$$a_2 = 5a_1 = 5(4) = 20$$
$$a_3 = 5a_2 = 5(20) = 100$$
$$a_4 = 5a_3 = 5(100) = 500$$

The sequence is 4, 20, 100, 500, ...

USING TECHNOLOGY   *Recursion Formulas on a Graphing Calculator*

· ·

We can use a TI-83 graphing calculator to view the sequence defined recursively as

$$a_1 = 8, a_n = a_{n-1} + 3$$

First, put your TI-83 calculator in sequence mode by pressing the $\boxed{\text{MODE}}$ key, and then selecting Seq (it's next to Func Par and Pol). Go to the Y variables list and set $n\text{Min} = 1$ and $u(n) = u(n-1)+3$. (The u is above the 7, and the n is on the $\boxed{\text{X, T, } \theta, n}$ and is automatically displayed if that key is pressed when the calculator is in the Seq mode.) Pressing $\boxed{\text{TABLE}}$ will display the sequence you have defined.

🅒 Finding the General Term

In the first four examples, we found some terms of a sequence after being given the general term. In the next two examples, we will do the reverse. That is, given some terms of a sequence, we will find the formula for the general term.

EXAMPLE 5

Find a formula for the nth term of the sequence 2, 8, 18, 32, ...

SOLUTION Solving a problem like this involves some guessing. Looking over the first four terms, we see each is twice a perfect square:

$$2 = 2(1)$$
$$8 = 2(4)$$
$$18 = 2(9)$$
$$32 = 2(16)$$

If we write each square with an exponent of 2, the formula for the nth term becomes obvious:

$$a_1 = 2 \ = 2(1)^2$$
$$a_2 = 8 \ = 2(2)^2$$
$$a_3 = 18 = 2(3)^2$$
$$a_4 = 32 = 2(4)^2$$

$$\cdot$$
$$\cdot$$
$$\cdot$$

$$a_n = 2(n)^2 = 2n^2$$

The general term of the sequence 2, 8, 18, 32, . . . is $a_n = 2n^2$.

EXAMPLE 6

Find the general term for the sequence $2, \dfrac{3}{8}, \dfrac{4}{27}, \dfrac{5}{64}, \ldots$

SOLUTION The first term can be written as $\frac{2}{1}$. The denominators are all perfect cubes. The numerators are all 1 more than the base of the cubes in the denominators:

$$a_1 = \frac{2}{1} = \frac{1+1}{1^3}$$

$$a_2 = \frac{3}{8} = \frac{2+1}{2^3}$$

$$a_3 = \frac{4}{27} = \frac{3+1}{3^3}$$

$$a_4 = \frac{5}{64} = \frac{4+1}{4^3}$$

Observing this pattern, we recognize the general term to be

$$a_n = \frac{n+1}{n^3}$$

> **NOTE** Finding the nth term of a sequence from the first few terms is not always automatic. That is, it sometimes takes a while to recognize the pattern. Don't be afraid to guess at the formula for the general term. Many times, an incorrect guess leads to the correct formula.

GETTING READY FOR CLASS

After reading through the preceding section, respond in your own words and in complete sentences.

A. How are subscripts used to denote the terms of a sequence?
B. What is the relationship between the subscripts used to denote the terms of a sequence and function notation?
C. What is a decreasing sequence?
D. What is meant by a recursion formula for a sequence?

Problem Set 9.1

Write the first five terms of the sequences with the following general terms.

1. $a_n = 3n + 1$ **2.** $a_n = 2n + 3$ **3.** $a_n = 4n - 1$

4. $a_n = n + 4$ **5.** $a_n = n$ **6.** $a_n = -n$

7. $a_n = n^2 + 3$ **8.** $a_n = n^3 + 1$ **9.** $a_n = \dfrac{n}{n + 3}$

10. $a_n = \dfrac{n + 1}{n + 2}$ **11.** $a_n = \dfrac{1}{n^2}$ **12.** $a_n = \dfrac{1}{n^3}$

13. $a_n = 2^n$ **14.** $a_n = 3^{-n}$ **15.** $a_n = 1 + \dfrac{1}{n}$

16. $a_n = n - \dfrac{1}{n}$ **17.** $a_n = (-2)^n$ **18.** $a_n = (-3)^n$

19. $a_n = 4 + (-1)^n$ **20.** $a_n = 10 + (-2)^n$

21. $a_n = (-1)^{n+1} \cdot \dfrac{n}{2n - 1}$ **22.** $a_n = (-1)^n \cdot \dfrac{2n + 1}{2n - 1}$

23. $a_n = n^2 \cdot 2^{-n}$ **24.** $a_n = n^n$

Write the first five terms of the sequences defined by the following recursion formulas.

25. $a_1 = 3$ $a_n = -3a_{n-1}$ $n > 1$

26. $a_1 = 3$ $a_n = a_{n-1} - 3$ $n > 1$

27. $a_1 = 1$ $a_n = 2a_{n-1} + 3$ $n > 1$

28. $a_1 = 1$ $a_n = a_{n-1} + n$ $n > 1$

29. $a_1 = 2$ $a_n = 2a_{n-1} - 1$ $n > 1$

30. $a_1 = -4$ $a_n = -2a_{n-1}$ $n > 1$

31. $a_1 = 5$ $a_n = 3a_{n-1} - 4$ $n > 1$

32. $a_1 = -3$ $a_n = -2a_{n-1} + 5$ $n > 1$

33. $a_1 = 4$ $a_n = 2a_{n-1} - a_1$ $n > 1$

34. $a_1 = -3$ $a_n = -2a_{n-1} - n$ $n > 1$

Determine the general term for each of the following sequences.

35. $4, 8, 12, 16, 20, \ldots$ **36.** $7, 10, 13, 16, \ldots$

37. $1, 4, 9, 16, \ldots$ **38.** $3, 12, 27, 48, \ldots$

39. $4, 8, 16, 32, \ldots$ **40.** $-2, 4, -8, 16, \ldots$

41. $\dfrac{1}{4}, \dfrac{1}{8}, \dfrac{1}{16}, \dfrac{1}{32}, \ldots$ **42.** $\dfrac{1}{4}, \dfrac{2}{9}, \dfrac{3}{16}, \dfrac{4}{25}, \ldots$

43. $5, 8, 11, 14, \ldots$ **44.** $7, 5, 3, 1, \ldots$

45. $-2, -6, -10, -14, \ldots$ **46.** $-2, 2, -2, 2, \ldots$

47. $1, -2, 4, -8, \ldots$ **48.** $-1, 3, -9, 27, \ldots$

49. $\log_2 3, \log_3 4, \log_4 5, \log_5 6, \ldots$ **50.** $0, \dfrac{3}{5}, \dfrac{8}{10}, \dfrac{15}{17}, \ldots$

APPLYING THE CONCEPTS

51. **Salary Increase** The entry level salary for a teacher is $28,000 with 4% increases after every year of service.

   a. Write a sequence for this teacher's salary for the first 5 years.

   b. Find the general term of the sequence in part **a**.

52. **Holiday Account** To save money for holiday presents, a person deposits $5 in a savings account on January 1, and then deposits an additional $5 every week thereafter until Christmas.

   a. Write a sequence for the money in that savings account for the first 10 weeks of the year.

   b. Write the general term of the sequence in part **a**.

   c. If there are 50 weeks from January 1 to Christmas, how much money will be available for spending on Christmas presents?

53. **Akaka Falls** If a boulder fell from the top of Akaka Falls in Hawaii, the distance, in feet, the boulder would fall in each consecutive second would be modeled by a sequence whose general term is $a_n = 32n - 16$, where n represents the number of seconds.

   a. Write a sequence for the first 5 seconds the boulder falls.

   b. What is the total distance the boulder fell in 5 seconds?

   c. If Akaka Falls is approximately 420 feet high, will the boulder hit the ground within 5 seconds?

54. **Polygons** The formula for the sum of the interior angles of a polygon with n sides is $a_n = 180°(n - 2)$.

   a. Write a sequence to represent the sum of the interior angles of a polygon with 3, 4, 5, and 6 sides.

   b. What would be the sum of the interior angles of a polygon with 20 sides?

   c. What happens when $n = 2$ to indicate that a polygon cannot be formed with only two sides?

55. **Pendulum** A pendulum swings 10 feet left to right on its first swing. On each swing following the first, the pendulum swings $\frac{4}{5}$ of the previous swing.

   a. Write a sequence for the distance traveled by the pendulum on the first, second, and third swing.

   b. Write a general term for the sequence, where n represents the number of the swing.

   c. How far will the pendulum swing on its tenth swing? (Round to the nearest hundredth.)

56. **Row Seating** If row A (in front) has 8 seats across, row B has 9 seats across, and row C has 10 seats across, and the pattern continues for all rows, then how many seats will be in row W?

GETTING READY FOR THE NEXT SECTION

Simplify.

57. $-2 + 6 + 4 + 22$

58. $9 - 27 + 81 - 243$

59. $-8 + 16 - 32 + 64$

60. $-4 + 8 - 16 + 32 - 64$

61. $(1 - 3) + (4 - 3) + (9 - 3) + (16 - 3)$

62. $(1 - 3) + (9 + 1) + (16 + 1) + (25 + 1) + (36 + 1)$

63. $-\dfrac{1}{3} + \dfrac{1}{9} - \dfrac{1}{27} + \dfrac{1}{81}$

64. $\dfrac{1}{2} + \dfrac{2}{3} + \dfrac{3}{4} + \dfrac{4}{5}$

65. $\dfrac{1}{3} + \dfrac{1}{2} + \dfrac{3}{5} + \dfrac{2}{3}$

66. $\dfrac{1}{16} + \dfrac{1}{32} + \dfrac{1}{64}$

EXTENDING THE CONCEPTS

67. As n increases, the terms in the sequence

$$a_n = \left(1 + \frac{1}{n}\right)^n$$

get closer and closer to the number e (the same e we used in defining natural logarithms). It takes some fairly large values of n, however, before we can see this happening. Use a calculator to find a_{100}, $a_{1,000}$, $a_{10,000}$, and $a_{100,000}$, and compare them to the decimal approximation we gave for the number e.

68. The sequence

$$a_n = \left(1 + \frac{1}{n}\right)^{-n}$$

gets close to the number $\frac{1}{e}$ as n becomes large. Use a calculator to find approximations for a_{100} and $a_{1,000}$, and then compare them to $\frac{1}{2.7183}$.

69. Write the first ten terms of the sequence defined by the recursion formula

$$a_1 = 1, \ a_2 = 1, a_n = a_{n-1} + a_{n-2} \quad n > 2$$

70. Write the first ten terms of the sequence defined by the recursion formula

$$a_1 = 2, \ a_2 = 2, a_n = a_{n-1} + a_{n-2} \quad n > 2$$

71. Simplify each complex fraction in the following sequence, and then compare this sequence with the sequence you wrote in Problem 69.

$$1 + \frac{1}{1 + 1}, 1 + \frac{1}{1 + \dfrac{1}{1 + 1}}$$

$$1 + \frac{1}{1 + \dfrac{1}{1 + \dfrac{1}{1 + 1}}}, \ldots$$

72. Write the first five terms of the sequence given by the following recursion formula.

$$a_1 = \frac{3}{2}, a_n = 1 + \frac{1}{a_{n-1}} \quad n > 1$$

9.2

OBJECTIVES

A Expand and simplify an expression in summation notation.

B Write an expression in expanded form as an expression in summation notation.

In February 2006, 400 skydivers set a record for the world's largest formation in free fall. The group jumped from five Royal Thai Air Force planes that flew 23,000 feet above Udon Thani, Thailand. As they fell toward Earth, the divers linked hands to create a shape that resembled a 10-point pinwheel. Suppose one of these skydivers fell 16 feet during the first second of his descent. During the second second, he fell 48 feet, and 80 feet during the third second. Using the topics we'll present in this section, we can calculate the diver's falling distance during his fifth second as 144 feet. You'll see this problem later in the problem set, but for now, let's discuss a few new terms you'll need to work the problem successfully.

There is an interesting relationship between the sequence of odd numbers and the sequence of squares that is found by adding the terms in the sequence of odd numbers.

$$
\begin{aligned}
1 &= 1 \\
1 + 3 &= 4 \\
1 + 3 + 5 &= 9 \\
1 + 3 + 5 + 7 &= 16
\end{aligned}
$$

When we add the terms of a sequence, the result is called a series.

DEFINITION *series*

The sum of a number of terms in a sequence is called a *series*.

A sequence can be finite or infinite, depending on whether the sequence ends at the nth term. For example,

$$1, 3, 5, 7, 9$$

is a finite sequence, but

$$1, 3, 5, \ldots$$

is an infinite sequence. Associated with each of the preceding sequences is a series found by adding the terms of the sequence:

$$1 + 3 + 5 + 7 + 9 \qquad \text{Finite series}$$

$$1 + 3 + 5 + \ldots \qquad \text{Infinite series}$$

9.2 VIDEOS

ⓐ Summation Notation for a Series

In this section, we will consider only finite series. We can introduce a new kind of notation here that is a compact way of indicating a finite series. The notation is called *summation notation*, or *sigma notation* because it is written using the Greek letter sigma. The expression

$$\sum_{i=1}^{4} (8i - 10)$$

is an example of an expression that uses summation notation. The summation notation in this expression is used to indicate the sum of all the expressions $8i - 10$ from $i = 1$ up to and including $i = 4$. That is,

$$\sum_{i=1}^{4} (8i - 10) = (8 \cdot 1 - 10) + (8 \cdot 2 - 10) + (8 \cdot 3 - 10) + (8 \cdot 4 - 10)$$
$$= -2 + 6 + 14 + 22$$
$$= 40$$

The letter i as used here is called the *index of summation*, or just *index* for short.
Here are some examples illustrating the use of summation notation.

EXAMPLE 1

Expand and simplify $\displaystyle\sum_{i=1}^{5} (i^2 - 1)$.

SOLUTION   We replace i in the expression $i^2 - 1$ with all consecutive integers from 1 up to 5, including 1 and 5:

$$\sum_{i=1}^{5} (i^2 - 1) = (1^2 - 1) + (2^2 - 1) + (3^2 - 1) + (4^2 - 1) + (5^2 - 1)$$
$$= 0 + 3 + 8 + 15 + 24$$
$$= 50$$

EXAMPLE 2

Expand and simplify $\displaystyle\sum_{i=3}^{6} (-2)^i$.

SOLUTION   We replace i in the expression $(-2)^i$ with the consecutive integers beginning at 3 and ending at 6:

$$\sum_{i=3}^{6} (-2)^i = (-2)^3 + (-2)^4 + (-2)^5 + (-2)^6$$
$$= -8 + 16 + (-32) + 64$$
$$= 40$$

USING TECHNOLOGY *Summing Series on a Graphing Calculator*

A TI-83 graphing calculator has a built-in sum(function that, when used with the seq(function, allows us to add the terms of a series. Let's repeat Example 1 using our graphing calculator. First, we go to LIST and select MATH. The fifth option in that list is sum(, which we select. Then we go to LIST again and select OPS. From that list we select seq(. Next we enter X^2−1, X, 1, 5, and then we close both sets of parentheses. Our screen shows the following:

$$\text{sum(seq(X^2}-1, \text{X}, 1, 5)) \qquad \text{which will give us } \sum_{i=1}^{5} (i^2 - 1)$$

When we press $\boxed{\text{ENTER}}$ the calculator displays 50, which is the same result we obtained in Example 1.

EXAMPLE 3

Expand $\sum_{i=2}^{5} (x^i - 3)$.

SOLUTION We must be careful not to confuse the letter x with i. The index i is the quantity we replace by the consecutive integers from 2 to 5, not x:

$$\sum_{i=2}^{5} (x^i - 3) = (x^2 - 3) + (x^3 - 3) + (x^4 - 3) + (x^5 - 3).$$

B Expanded Form

In the first three examples, we were given an expression with summation notation and asked to expand it. The next examples in this section illustrate how we can write an expression in expanded form as an expression involving summation notation.

EXAMPLE 4

Write with summation notation $1 + 3 + 5 + 7 + 9$.

SOLUTION A formula that gives us the terms of this sum is

$$a_i = 2i - 1$$

where i ranges from 1 up to and including 5. Notice we are using the subscript i in exactly the same way we used the subscript n in the previous section — to indicate the general term. Writing the sum

$$1 + 3 + 5 + 7 + 9$$

with summation notation looks like this:

$$\sum_{i=1}^{5} (2i - 1)$$

EXAMPLE 5

Write with summation notation $3 + 12 + 27 + 48$.

SOLUTION   We need a formula, in terms of i, that will give each term in the sum. Writing the sum as

$$3 \cdot 1^2 + 3 \cdot 2^2 + 3 \cdot 3^2 + 3 \cdot 4^2$$

we see the formula

$$a_i = 3 \cdot i^2$$

where i ranges from 1 up to and including 4. Using this formula and summation notation, we can represent the sum

$$3 + 12 + 27 + 48$$

as

$$\sum_{i=1}^{4} 3i^2$$

EXAMPLE 6

Write with summation notation

$$\frac{x+3}{x^3} + \frac{x+4}{x^4} + \frac{x+5}{x^5} + \frac{x+6}{x^6}$$

SOLUTION   A formula that gives each of these terms is

$$a_i = \frac{x+i}{x^i}$$

where i assumes all integer values between 3 and 6, including 3 and 6. The sum can be written as

$$\sum_{i=3}^{6} \frac{x+i}{x^i}$$

GETTING READY FOR CLASS

After reading through the preceding section, respond in your own words and in complete sentences.

A. What is the difference between a sequence and a series?

B. Explain the summation notation $\sum_{i=1}^{4}$ in the series $\sum_{i=1}^{4}(2i + 1)$.

C. When will a finite series result in a numerical value versus an algebraic expression?

D. Determine for what values of n the series $\sum_{i=1}^{n}(-1)^i$ will be equal to 0. Explain your answer.

Problem Set 9.2

Expand and simplify each of the following.

1. $\displaystyle\sum_{i=1}^{4}(2i+4)$ **2.** $\displaystyle\sum_{i=1}^{5}(2i+4)$ **3.** $\displaystyle\sum_{i=2}^{3}i^2-1$ **4.** $\displaystyle\sum_{i=3}^{6}(i^2+1)$

5. $\displaystyle\sum_{i=1}^{4}(i^2-3)$ **6.** $\displaystyle\sum_{i=2}^{6}(2i^2+1)$ **7.** $\displaystyle\sum_{i=1}^{4}\frac{i}{1+i}$ **8.** $\displaystyle\sum_{i=1}^{3}\frac{i^2}{2i-1}$

9. $\displaystyle\sum_{i=1}^{4}(-3)^i$ **10.** $\displaystyle\sum_{i=1}^{4}\left(-\frac{1}{3}\right)^i$ **11.** $\displaystyle\sum_{i=3}^{6}(-2)^i$ **12.** $\displaystyle\sum_{i=4}^{6}\left(-\frac{1}{2}\right)^i$

13. $\displaystyle\sum_{i=2}^{6}(-2)^i$ **14.** $\displaystyle\sum_{i=2}^{5}(-3)^i$ **15.** $\displaystyle\sum_{i=1}^{5}\left(-\frac{1}{2}\right)^i$ **16.** $\displaystyle\sum_{i=3}^{6}\left(-\frac{1}{3}\right)^i$

17. $\displaystyle\sum_{i=2}^{5}\frac{i-1}{i+1}$ **18.** $\displaystyle\sum_{i=2}^{4}\frac{i^2-1}{i^2+1}$

Expand the following.

19. $\displaystyle\sum_{i=1}^{5}(x+i)$ **20.** $\displaystyle\sum_{i=2}^{7}(x+1)^i$ **21.** $\displaystyle\sum_{i=1}^{4}(x-2)^i$ **22.** $\displaystyle\sum_{i=2}^{5}\left(x+\frac{1}{i}\right)^2$

23. $\displaystyle\sum_{i=1}^{5}\frac{x+i}{x-1}$ **24.** $\displaystyle\sum_{i=1}^{6}\frac{x-3i}{x+3i}$ **25.** $\displaystyle\sum_{i=3}^{8}(x+i)^i$ **26.** $\displaystyle\sum_{i=1}^{5}(x+i)^{i+1}$

27. $\displaystyle\sum_{i=3}^{6}(x-2i)^{i+3}$ **28.** $\displaystyle\sum_{i=5}^{8}\left(\frac{x-i}{x+i}\right)^{2i}$

Write each of the following sums with summation notation. Do not calculate the sum.

29. $2+4+8+16$

30. $3+5+7+9+11$

31. $4+8+16+32+64$

32. $3+8+15+24$

Write each of the following sums with summation notation. Do not calculate the sum. *Note:* More than one answer is possible.

33. $5+9+13+17+21$

34. $3-6+12-24+48$

35. $-4+8-16+32$

36. $15+24+35+48+63$

37. $\dfrac{3}{4}+\dfrac{4}{5}+\dfrac{5}{6}+\dfrac{6}{7}+\dfrac{7}{8}$

38. $\dfrac{1}{2}+\dfrac{2}{3}+\dfrac{3}{4}+\dfrac{4}{5}$

39. $\dfrac{1}{3}+\dfrac{2}{5}+\dfrac{3}{7}+\dfrac{4}{9}$

40. $\dfrac{3}{1}+\dfrac{5}{3}+\dfrac{7}{5}+\dfrac{9}{7}$

41. $(x-2)^6+(x-2)^7+(x-2)^8+(x-2)^9$

42. $(x+1)^3+(x+2)^4+(x+3)^5+(x+4)^6+(x+5)^7$

43. $\left(1+\dfrac{1}{x}\right)^2+\left(1+\dfrac{2}{x}\right)^3+\left(1+\dfrac{3}{x}\right)^4+\left(1+\dfrac{4}{x}\right)^5$

44. $\dfrac{x-1}{x+2}+\dfrac{x-2}{x+4}+\dfrac{x-3}{x+6}+\dfrac{x-4}{x+8}+\dfrac{x-5}{x+10}$

45. $\dfrac{x}{x+3}+\dfrac{x}{x+4}+\dfrac{x}{x+5}$ **46.** $\dfrac{x-3}{x^3}+\dfrac{x-4}{x^4}+\dfrac{x-5}{x^5}+\dfrac{x-6}{x^6}$

47. $x^2(x+2)+x^3(x+3)+x^4(x+4)$ **48.** $x(x+2)^2+x(x+3)^3+x(x+4)^4$

49. Repeating Decimals Any repeating, nonterminating decimal may be viewed as a series. For instance, $\frac{2}{3} = 0.6 + 0.06 + 0.006 + 0.0006 + \cdots$. Write the following fractions as series.

a. $\frac{1}{3}$ **b.** $\frac{2}{9}$ **c.** $\frac{3}{11}$

50. Repeating Decimals Refer to the previous exercise, and express the following repeating decimals as fractions.

a. $0.55555\cdots$ **b.** $1.33333\cdots$ **c.** $0.29292929\cdots$

APPLYING THE CONCEPTS

51. Skydiving A skydiver jumps from a plane and falls 16 feet the first second, 48 feet the second second, and 80 feet the third second. If he continues to fall in the same manner, how far will he fall the seventh second? What is the total distance he falls in 7 seconds?

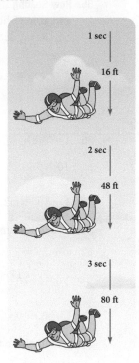

52. Bacterial Growth After 1 day, a colony of 50 bacteria reproduces to become 200 bacteria. After 2 days, they reproduce to become 800 bacteria. If they continue to reproduce at this rate, how many bacteria will be present after 4 days?

Start   1 day   2 days

53. **Akaka Falls** In Section 9.1, when a boulder fell from the top of Akaka Falls in Hawaii, the sequence generated during the first 5 seconds the boulder fell was 16, 48, 80, 112, and 144.

a. Write a finite series that represents the sum of this sequence.

b. The general term of this sequence was given as $a_n = 32n - 16$. Write the series produced in part **a** in summation notation.

54. **Pendulum** A pendulum swings 12 feet left to right on its first swing, and on each swing following the first, swings $\frac{3}{4}$ of the previous swing. The distance the pendulum travels in 5 seconds can be expressed in summation notation

$$\sum_{i=1}^{5} 12\left(\frac{3}{4}\right)^{i-1}$$

Expand the summation notation and simplify. Round your final answer to the nearest tenth.

55. **Row Seating** If row A (in front) has 8 seats across, row B has 9 seats across, and row C has 10 seats across and the pattern continues for all rows, then how many seats will be total (inclusive) from rows A to W?

GETTING READY FOR THE NEXT SECTION

Simplify.

56. $2 + 9(8)$

57. $\frac{1}{2} + 9\left(\frac{1}{2}\right)$

58. $\frac{10}{2}\left(\frac{1}{2} + 5\right)$

59. $\frac{10}{2}(2 + 74)$

60. $3 + (n - 1)2$

61. $7 + (n - 1)3$

Solve each system of equations.

62. $x + 2y = 7$
$x + 7y = 17$

63. $x + 3y = 14$
$x + 9y = 32$

EXTENDING THE CONCEPTS

Solve each of the following equations for x.

64. $\sum_{i=1}^{4}(x - i) = 16$

65. $\sum_{i=3}^{8}(2x + i) = 30$

66. $\sum_{i=2}^{5}\left(x + \frac{1}{i}\right) = \frac{1}{60}$

67. $\sum_{i=1}^{4}\left(x - \frac{1}{i^2}\right) = -2$

68. Use the formula $\sum_{i=1}^{n} i = \frac{n(n + 1)}{2}$ to find the sum of the first 100 natural numbers.

69. Use the formula $\sum_{i=1}^{n} i^3 = \left[\frac{n(n + 1)}{2}\right]^2$ to find the sum of the first 40 positive cubes.

ARITHMETIC SEQUENCES

©iStockphoto.com/Indysystem

OBJECTIVES

A   Give the common difference for an arithmetic sequence.

B   Find the general term for an arithmetic sequence.

C   Find the *n*th partial sum of an arithmetic sequence.

9.3

Suppose the following table represents the balance of your savings account for the first four months of the year:

Savings Account Balance	
January	$730
February	$850
March	$970
April	$1,090

The sequence

$$730, 850, 970, 1090$$

is called an *arithmetic sequence*, because we obtained each term by adding 120 to each preceding term. In this section, we will review and extend this type of sequence. Let's begin with a formal definition.

> **DEFINITION**   *arithmetic sequence*
>
> An *arithmetic sequence* is a sequence of numbers in which each term is obtained from the preceding term by adding the same amount each time. An arithmetic sequence is also called an *arithmetic progression.*

Common Difference for an Arithmetic Sequence

The sequence

$$2, 6, 10, 14, \ldots$$

is an example of an arithmetic sequence, because each term is obtained from the preceding term by adding 4 each time. The amount we add each time — in this case, 4 — is called the *common difference*, because it can be obtained by subtracting any two consecutive terms. (The term with the larger subscript must be written first.) The common difference is denoted by d.

9.3   VIDEOS

737

EXAMPLE 1

Give the common difference d for the arithmetic sequence 4, 10, 16, 22, ...

SOLUTION   Because each term can be obtained from the preceding term by adding 6, the common difference is 6. That is, $d = 6$.

EXAMPLE 2

Give the common difference for 100, 93, 86, 79, ...

SOLUTION   The common difference in this case is $d = -7$, since adding -7 to any term always produces the next consecutive term.

EXAMPLE 3

Give the common difference for $\frac{1}{2}$, 1, $\frac{3}{2}$, 2, ...

SOLUTION   The common difference is $d = \frac{1}{2}$.

B The General Term

The general term a_n of an arithmetic progression can always be written in terms of the first term a_1 and the common difference d. Consider the sequence from Example 1:

$$4, 10, 16, 22, \ . \ . \ .$$

We can write each term in terms of the first term 4 and the common difference 6:

$$4, \qquad 4 + (1 \cdot 6), \qquad 4 + (2 \cdot 6), \qquad 4 + (3 \cdot 6), \ . \ . \ .$$
$$a_1, \qquad a_2, \qquad a_3, \qquad a_4, \ . \ . \ .$$

Observing the relationship between the subscript on the terms in the second line and the coefficients of the 6's in the first line, we write the general term for the sequence as

$$a_n = 4 + (n - 1)6$$

We generalize this result to include the general term of any arithmetic sequence.

Arithmetic Sequences
. . . .

The ***general term*** of an arithmetic progression with first term a_1 and common difference d is given by

$$a_n = a_1 + (n - 1)d$$

EXAMPLE 4

Find the general term for the sequence

$$7, 10, 13, 16, \ . \ . \ .$$

SOLUTION The first term is $a_1 = 7$, and the common difference is $d = 3$. Substituting these numbers into the formula given earlier, we have

$$a_n = 7 + (n - 1)3$$

which we can simplify, if we choose, to

$$a_n = 7 + 3n - 3$$
$$= 3n + 4$$

EXAMPLE 5

Find the general term of the arithmetic progression whose third term a_3 is 7 and whose eighth term a_8 is 17.

SOLUTION According to the formula for the general term, the third term can be written as $a_3 = a_1 + 2d$, and the eighth term can be written as $a_8 = a_1 + 7d$. Because these terms are also equal to 7 and 17, respectively, we can write

$$a_3 = a_1 + 2d = 7$$
$$a_8 = a_1 + 7d = 17$$

To find a_1 and d, we simply solve the system:

$$a_1 + 2d = 7$$
$$a_1 + 7d = 17$$

We add the opposite of the top equation to the bottom equation. The result is

$$5d = 10$$
$$d = 2$$

To find a_1, we simply substitute 2 for d in either of the original equations and get

$$a_1 = 3$$

The general term for this progression is

$$a_n = 3 + (n - 1)2$$

which we can simplify to

$$a_n = 2n + 1$$

C nth Partial Sum

The sum of the first n terms of an arithmetic sequence is denoted by S_n. The following theorem gives the formula for finding S_n, which is sometimes called the *nth partial sum*.

THEOREM

The sum of the first n terms of an arithmetic sequence whose first term is a_1 and whose nth term is a_n is given by

$$S_n = \frac{n}{2} (a_1 + a_n)$$

Proof We can write S_n in expanded form as

$$S_n = a_1 + [a_1 + d] + [a_1 + 2d] + \cdots + [a_1 + (n-1)d]$$

We can arrive at this same series by starting with the last term a_n and subtracting d each time. Writing S_n this way, we have

$$S_n = a_n + [a_n - d] + [a_n - 2d] + \cdots + [a_n - (n-1)d]$$

If we add the preceding two expressions term by term, we have

$$2S_n = (a_1 + a_n) + (a_1 + a_n) + (a_1 + a_n) + \cdots + (a_1 + a_n)$$

$$2S_n = n(a_1 + a_n)$$

$$S_n = \frac{n}{2}(a_1 + a_n)$$

EXAMPLE 6

Find the sum of the first 10 terms of the arithmetic progression $2, 10, 18, 26, \ldots$.

SOLUTION The first term is 2, and the common difference is 8. The tenth term is

$$a_{10} = 2 + 9(8)$$

$$= 2 + 72$$

$$= 74$$

Substituting $n = 10$, $a_1 = 2$, and $a_{10} = 74$ into the formula

$$S_n = \frac{n}{2}(a_1 + a_n)$$

we have

$$S_{10} = \frac{10}{2}(2 + 74)$$

$$= 5(76)$$

$$= 380$$

The sum of the first 10 terms is 380.

GETTING READY FOR CLASS

After reading through the preceding section, respond in your own words and in complete sentences.

A. Explain how to determine if a sequence is arithmetic.

B. What is a common difference?

C. Suppose the value of a_5 is given. What other possible pieces of information could be given to have enough information to obtain the first 10 terms of the sequence?

D. Explain the formula $a_n = a_1 + (n-1)d$ in words so that someone who wanted to find the nth term of an arithmetic sequence could do so from your description.

Problem Set 9.3

Determine which of the following sequences are arithmetic progressions. For those that are arithmetic progressions, identify the common difference d.

1. $1, 2, 3, 4, \ldots$ **2.** $4, 6, 8, 10, \ldots$ **3.** $1, 2, 4, 7, \ldots$ **4.** $1, 2, 4, 8, \ldots$

5. $50, 45, 40, \ldots$ **6.** $1, \dfrac{1}{2}, \dfrac{1}{4}, \dfrac{1}{8}, \ldots$ **7.** $1, 4, 9, 16, \ldots$ **8.** $5, 7, 9, 11, \ldots$

9. $\dfrac{1}{3}, 1, \dfrac{5}{3}, \dfrac{7}{3}, \ldots$ **10.** $5, 11, 17, \ldots$

Each of the following problems refers to arithmetic sequences.

11. If $a_1 = 3$ and $d = 4$, find a_n and a_{24}.

12. If $a_1 = 5$ and $d = 10$, find a_n and a_{100}.

13. If $a_1 = 6$ and $d = -2$, find a_{10} and S_{10}.

14. If $a_1 = 7$ and $d = -1$, find a_{24} and S_{24}.

15. If $a_6 = 17$ and $a_{12} = 29$, find the term a_1, the common difference d, and then find a_{30}.

16. If $a_5 = 23$ and $a_{10} = 48$, find the first term a_1, the common difference d, and then find a_{40}.

17. If the third term is 16 and the eighth term is 26, find the first term, the common difference, and then find a_{20} and S_{20}.

18. If the third term is 16 and the eighth term is 51, find the first term, the common difference, and then find a_{50} and S_{50}.

19. If $a_1 = 3$ and $d = 4$, find a_{20} and S_{20}.

20. If $a_1 = 40$ and $d = -5$, find a_{25} and S_{25}.

21. If $a_4 = 14$ and $a_{10} = 32$, find a_{40} and S_{40}.

22. If $a_7 = 0$ and $a_{11} = -\dfrac{8}{3}$, find a_{61} and S_{61}.

23. If $a_6 = -17$ and $S_6 = -12$, find a_1 and d.

24. If $a_{10} = 12$ and $S_{10} = 40$, find a_1 and d.

25. Find a_{85} for the sequence $14, 11, 8, 5, \ldots$

26. Find S_{100} for the sequence $-32, -25, -18, -11, \ldots$

27. If $S_{20} = 80$ and $a_1 = -4$, find d and a_{39}.

28. If $S_{24} = 60$ and $a_1 = 4$, find d and a_{116}.

29. Find the sum of the first 100 terms of the sequence $5, 9, 13, 17, \ldots$

30. Find the sum of the first 50 terms of the sequence $8, 11, 14, 17, \ldots$

31. Find a_{35} for the sequence $12, 7, 2, -3, \ldots$

32. Find a_{45} for the sequence $25, 20, 15, 10, \ldots$

33. Find the tenth term and the sum of the first 10 terms of the sequence $\dfrac{1}{2}, 1, \dfrac{3}{2}, 2, \ldots$

34. Find the 15th term and the sum of the first 15 terms of the sequence $-\dfrac{1}{3}, 0, \dfrac{1}{3}, \dfrac{2}{3}, \ldots$

APPLYING THE CONCEPTS

Straight-Line Depreciation Recall from a previous section that straight-line depreciation is an accounting method used to help spread the cost of new equipment over a number of years. The value at any time during the life of the machine can be found with a linear equation in two variables. For income tax purposes, however, it is the value at the end of the year that is most important, and for this reason sequences can be used.

35. Value of a Copy Machine A large copy machine sells for $18,000 when it is new. Its value decreases $3,300 each year after that. We can use an arithmetic sequence to find the value of the machine at the end of each year. If we let a_0 represent the value when it is purchased, then a_1 is the value after 1 year, a_2 is the value after 2 years, and so on.

  a. Write the first 5 terms of the sequence.

  b. What is the common difference?

  c. Construct a line graph for the first 5 terms of the sequence.

  d. Use the line graph to estimate the value of the copy machine 2.5 years after it is purchased.

  e. Write the sequence from part **a** using a recursive formula.

36. Value of a Forklift An electric forklift sells for $125,000 when new. Each year after that, it decreases $16,500 in value.

  a. Write an arithmetic sequence that gives the value of the forklift at the end of each of the first 5 years after it is purchased.

  b. What is the common difference for this sequence?

  c. Construct a line graph for this sequence using the template that follows.

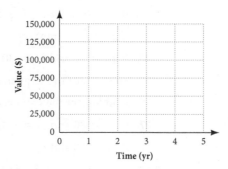

  d. Use the line graph to estimate the value of the forklift 3.5 years after it is purchased.

  e. Write the sequence from part **a** using a recursive formula.

37. Distance A rocket travels vertically 1,500 feet in its first second of flight, and then about 40 feet less each succeeding second. Use these estimates to answer the following questions.

  a. Write a sequence of the vertical distance traveled by a rocket in each of its first 6 seconds.

  b. Is the sequence in part **a** an arithmetic sequence? Explain why or why not.

  c. What is the general term of the sequence in part **a**?

38. Depreciation Suppose an automobile sells for N dollars new, and then depreciates 40% each year.

    a. Write a sequence for the value of this automobile (in terms of N) for each year.

    b. What is the general term of the sequence in part **a**?

    c. Is the sequence in part **a** an arithmetic sequence? Explain why it is or is not.

39. Triangular Numbers The first four triangular numbers are 1, 3, 6, 10, . . ., and are illustrated in the following diagram.

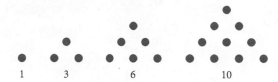

    a. Write a sequence of the first 15 triangular numbers.

    b. Write the recursive general term for the sequence of triangular numbers.

    c. Is the sequence of triangular numbers an arithmetic sequence? Explain why it is or is not.

40. Arithmetic Means Three (or more) arithmetic means between two numbers may be found by forming an arithmetic sequence using the original two numbers and the arithmetic means. For example, three arithmetic means between 10 and 34 may be found by examining the sequence 10, a, b, c, 34. For the sequence to be arithmetic, the common difference must be 6; therefore, $a = 16$, $b = 22$, and $c = 28$. Use this idea to answer the following questions.

    a. Find four arithmetic means between 10 and 35.

    b. Find three arithmetic means between 2 and 62.

    c. Find five arithmetic means between 4 and 28.

41. Paratroopers At the Fort Campbell Army Base, soldiers in the 101st Airborne Division are trained to be paratroopers. A paratrooper's free fall drop per second can be modeled by an arithmetic sequence whose first term is 16 feet and whose common difference is 32 feet.

    a. Write the general term for this arithmetic progression.

    b. How far would a paratrooper fall during the tenth second of free fall?

    c. Using the sum formula for an arithmetic sequence, what would be the total distance a paratrooper fell during 10 seconds of free fall?

42. College Fund When Jack's first grandchild, Kayla, was born, he decided to establish a college fund for her. On Kayla's first birthday, Jack deposited $1,000 into an account and decided that each year on her birthday he would deposit an amount $500 more than the previous year, through her 18th birthday.

    a. Write a sequence to represent the amount Jack deposited on her first through fifth birthdays. Does this sequence represent an arithmetic progression? If so, what is the common difference?

    b. How much will Jack deposit on Kayla's 18th birthday?

    c. What would be the total amount (excluding any interest earned) Jack has saved for Kayla's college fund by her 18th birthday?

GETTING READY FOR THE NEXT SECTION

Simplify.

43. $\dfrac{1}{8}\left(\dfrac{1}{2}\right)$

44. $\dfrac{1}{4}\left(\dfrac{1}{2}\right)$

45. $\dfrac{3\sqrt{3}}{3}$

46. $\dfrac{3}{\sqrt{3}}$

47. $2 \cdot 2^{n-1}$

48. $3 \cdot 3^{n-1}$

49. $\dfrac{ar^6}{ar^3}$

50. $\dfrac{ar^7}{ar^4}$

51. $\dfrac{\dfrac{1}{5}}{1-\dfrac{1}{2}}$

52. $\dfrac{\dfrac{9}{10}}{1-\dfrac{1}{10}}$

53. $\dfrac{3[(-2)^8 - 1]}{-2 - 1}$

54. $\dfrac{4\left[\left(\dfrac{1}{2}\right)^6 - 1\right]}{\dfrac{1}{2} - 1}$

EXTENDING THE CONCEPTS

55. Find a_1 and d for an arithmetic sequence with $S_7 = 147$ and $S_{13} = 429$.

56. Find a_1 and d for an arithmetic sequence with $S_{12} = 282$ and $S_{19} = 646$.

57. Find d for an arithmetic sequence in which $a_{15} - a_7 = -24$.

58. Find d for an arithmetic sequence in which $a_{25} - a_{15} = 120$.

59. Find the sum of the first 50 terms of the sequence $1, 1 + \sqrt{2}, 1 + 2\sqrt{2}, \ldots$.

60. Given $a_1 = -3$ and $d = 5$ in an arithmetic sequence, and $S_n = 890$, find n.

GEOMETRIC SEQUENCES

©iStockphoto.com/kkgas

9.4

OBJECTIVES

A Find the common ratio for a geometric sequence.

B Find the general term of a geometric sequence.

C Find the nth partial sum of a geometric sequence.

D Find the sum of an infinite geometric sequence.

Suppose you own a café that hosts a musical act every Thursday evening. After the first month, you notice that the number of cups of coffee you sold on Thursdays doubled from each of the previous Thursdays. The following sequence represents the numbers of cups sold on Thursdays for that first month:

$$8, 16, 32, 64$$

This type of sequence is called a *geometric sequence*, which is the focus of this section. Here's the definition:

> **DEFINITION** *geometric sequence*
>
> A sequence of numbers in which each term is obtained from the previous term by multiplying by the same amount each time is called a *geometric sequence*. Geometric sequences are also called *geometric progressions*.

A Geometric Sequences

The sequence

$$3, 6, 12, 24, \ldots$$

is an example of a geometric progression. Each term is obtained from the previous term by multiplying by 2. The amount by which we multiply each time — in this case, 2 — is called the *common ratio*. The common ratio is denoted by r and can be found by taking the ratio of any two consecutive terms. (The term with the larger subscript must be in the numerator.)

EXAMPLE 1

Find the common ratio for the geometric progression

$$\frac{1}{2}, \frac{1}{4}, \frac{1}{8}, \frac{1}{16}, \ldots$$

SOLUTION Because each term can be obtained from the term before it by multiplying by $\frac{1}{2}$, the common ratio is $\frac{1}{2}$. That is, $r = \frac{1}{2}$.

9.4 VIDEOS

EXAMPLE 2

Find the common ratio for $\sqrt{3}, 3, 3\sqrt{3}, 9, \ldots$

SOLUTION   If we take the ratio of the third term to the second term, we have

$$\frac{3\sqrt{3}}{3} = \sqrt{3}$$

The common ratio is $r = \sqrt{3}$.

B The General Term of a Geometric Sequence

Geometric Sequences

The **general term** a_n of a geometric sequence with first term a_1 and common ratio r is given by

$$a_n = a_1 r^{n-1}$$

To see how we arrive at this formula, consider the following geometric progression whose common ratio is 3:

$$2, 6, 18, 54, \ldots$$

We can write each term of the sequence in terms of the first term 2 and the common ratio 3:

$$2 \cdot 3^0, \qquad 2 \cdot 3^1, \qquad 2 \cdot 3^2, \qquad 2 \cdot 3^3, \ldots$$
$$a_1, \qquad\qquad a_2, \qquad\qquad a_3, \qquad\qquad a_4, \ldots$$

Observing the relationship between the two preceding lines, we find we can write the general term of this progression as

$$a_n = 2 \cdot 3^{n-1}$$

Because the first term can be designated by a_1 and the common ratio by r, the formula

$$a_n = 2 \cdot 3^{n-1}$$

coincides with the formula

$$a_n = a_1 r^{n-1}$$

EXAMPLE 3

Find the general term for the geometric progression

$$5, 10, 20, \ldots$$

SOLUTION   The first term is $a_1 = 5$, and the common ratio is $r = 2$. Using these values in the formula

$$a_n = a_1 r^{n-1}$$

we have

$$a_n = 5 \cdot 2^{n-1}$$

EXAMPLE 4

Find the tenth term of the sequence $3, \dfrac{3}{2}, \dfrac{3}{4}, \dfrac{3}{8}, \ldots$

SOLUTION The sequence is a geometric progression with first term $a_1 = 3$ and common ratio $r = \frac{1}{2}$. The tenth term is

$$a_{10} = 3\left(\frac{1}{2}\right)^9 = \frac{3}{512}$$

EXAMPLE 5

Find the general term for the geometric progression whose fourth term is 16 and whose seventh term is 128.

SOLUTION The fourth term can be written as $a_4 = a_1 r^3$, and the seventh term can be written as $a_7 = a_1 r^6$.

$$a_4 = a_1 r^3 = 16$$
$$a_7 = a_1 r^6 = 128$$

We can solve for r by using the ratio $\frac{a_7}{a_4}$.

$$\frac{a_7}{a_4} = \frac{a_1 r^6}{a_1 r^3} = \frac{128}{16}$$
$$r^3 = 8$$
$$r = 2$$

The common ratio is 2. To find the first term, we substitute $r = 2$ into either of the original two equations. The result is

$$a_1 = 2$$

The general term for this progression is

$$a_n = 2 \cdot 2^{n-1}$$

which we can simplify by adding exponents, because the bases are equal:

$$a_n = 2^n$$

C *n*th Partial Sum of a Geometric Sequence

As was the case in the preceding section, the sum of the first n terms of a geometric progression is denoted by S_n, which is called the *nth partial sum* of the progression.

THEOREM

The sum of the first n terms of a geometric progression with first term a_1 and common ratio r is given by the formula

$$S_n = \frac{a_1(r^n - 1)}{r - 1}$$

Proof We can write the sum of the first n terms in expanded form:

$$S_n = a_1 + a_1 r + a_1 r^2 + \cdots + a_1 r^{n-1} \tag{1}$$

Then multiplying both sides by r, we have

$$r S_n = a_1 r + a_1 r^2 + a_1 r^3 + \cdots + a_1 r^n \tag{2}$$

If we subtract the left side of equation (1) from the left side of equation (2) and do the same for the right sides, we end up with

$$rS_n - S_n = a_1 r^n - a_1$$

We factor S_n from both terms on the left side and a_1 from both terms on the right side of this equation:

$$S_n(r - 1) = a_1(r^n - 1)$$

Dividing both sides by $r - 1$ gives the desired result:

$$S_n = \frac{a_1(r^n - 1)}{r - 1}$$

EXAMPLE 6

Find the sum of the first 10 terms of the geometric progression $5, 15, 45, 135, \ldots$

SOLUTION   The first term is $a_1 = 5$, and the common ratio is $r = 3$. Substituting these values into the formula for S_{10}, we have the sum of the first 10 terms of the sequence:

$$S_{10} = \frac{5(3^{10} - 1)}{3 - 1}$$

$$= \frac{5(3^{10} - 1)}{2}$$

The answer can be left in this form. A calculator will give the result as 147,620.

D Infinite Geometric Series

Suppose the common ratio for a geometric sequence is a number whose absolute value is less than 1—for instance, $\frac{1}{2}$. The sum of the first n terms is given by the formula

$$S_n = \frac{a_1\left[\left(\frac{1}{2}\right)^n - 1 \right]}{\frac{1}{2} - 1}$$

As n becomes larger and larger, the term $\left(\frac{1}{2}\right)^n$ will become closer and closer to 0. That is, for $n = 10, 20,$ and 30, we have the following approximations:

$$\left(\frac{1}{2}\right)^{10} \approx 0.001$$

$$\left(\frac{1}{2}\right)^{20} \approx 0.000001$$

$$\left(\frac{1}{2}\right)^{30} \approx 0.000000001$$

so that for large values of n, there is little difference between the expression

$$\frac{a_1(r^n - 1)}{r - 1}$$

and the expression

$$\frac{a_1(0 - 1)}{r - 1} = \frac{-a_1}{r - 1} = \frac{a_1}{1 - r} \qquad \text{if} \qquad |r| < 1$$

In fact, the sum of the terms of a geometric sequence in which $|r| < 1$ actually becomes the expression

$$\frac{a_1}{1 - r}$$

as n approaches infinity. To summarize, we have the following:

> ### *The Sum of an Infinite Geometric Sequence*
>
> If a geometric sequence has first term a_1 and common ratio r such that $|r| < 1$, then the following is called an *infinite geometric series*:
>
> $$S = \sum_{i=0}^{\infty} a_1 r^i = a_1 + a_1 r + a_1 r^2 + a_1 r^3 + \cdots$$
>
> Its sum is given by the formula
>
> $$S = \frac{a_1}{1 - r}$$

EXAMPLE 7

Find the sum of the infinite geometric series

$$\frac{1}{5} + \frac{1}{10} + \frac{1}{20} + \frac{1}{40} + \cdots$$

SOLUTION The first term is $a_1 = \frac{1}{5}$, and the common ratio is $r = \frac{1}{2}$, which has an absolute value less than 1. Therefore, the sum of this series is

$$S = \frac{a_1}{1 - r} = \frac{\frac{1}{5}}{1 - \frac{1}{2}} = \frac{\frac{1}{5}}{\frac{1}{2}} = \frac{2}{5}$$

EXAMPLE 8

Show that $0.999 \ldots$ is equal to 1.

SOLUTION We begin by writing $0.999 \ldots$ as an infinite geometric series:

$$0.999 \ldots = 0.9 + 0.09 + 0.009 + 0.0009 + \cdots$$

$$= \frac{9}{10} + \frac{9}{100} + \frac{9}{1,000} + \frac{9}{10,000} + \cdots$$

$$= \frac{9}{10} + \frac{9}{10}\left(\frac{1}{10}\right) + \frac{9}{10}\left(\frac{1}{10}\right)^2 + \frac{9}{10}\left(\frac{1}{10}\right)^3 + \cdots$$

As the last line indicates, we have an infinite geometric series with $a_1 = \frac{9}{10}$ and $r = \frac{1}{10}$. The sum of this series is given by

$$S = \frac{a_1}{1 - r} = \frac{\frac{9}{10}}{1 - \frac{1}{10}} = \frac{\frac{9}{10}}{\frac{9}{10}} = 1$$

GETTING READY FOR CLASS

After reading through the preceding section, respond in your own words and in complete sentences.

A. What is a common ratio?

B. Explain the formula $a_n = a_1 r^{n-1}$ in words so that someone who wanted to find the nth term of a geometric sequence could do so from your description.

C. When is the sum of an infinite geometric series a finite number?

D. Explain how a repeating decimal can be represented as an infinite geometric series.

Problem Set 9.4

Identify those sequences that are geometric progressions. For those that are geometric, give the common ratio r.

1. $1, 5, 25, 125, \ldots$ **2.** $6, 12, 24, 48, \ldots$ **3.** $\dfrac{1}{2}, \dfrac{1}{6}, \dfrac{1}{18}, \dfrac{1}{54}, \ldots$

4. $5, 10, 15, 20, \ldots$ **5.** $4, 9, 16, 25, \ldots$ **6.** $-1, \dfrac{1}{3}, -\dfrac{1}{9}, \dfrac{1}{27}, \ldots$

7. $-2, 4, -8, 16, \ldots$ **8.** $1, 8, 27, 64, \ldots$ **9.** $4, 6, 8, 10, \ldots$

10. $1, -3, 9, -27, \ldots$

Each of the following problems gives some information about a specific geometric progression.

11. If $a_1 = 4$ and $r = 3$, find a_n. **12.** If $a_1 = 5$ and $r = 2$, find a_n.

13. If $a_1 = -2$ and $r = -\dfrac{1}{2}$, find a_6. **14.** If $a_1 = 25$ and $r = -\dfrac{1}{5}$, find a_6.

15. If $a_1 = 3$ and $r = -1$, find a_{20}. **16.** If $a_1 = -3$ and $r = -1$, find a_{20}.

17. If $a_1 = 10$ and $r = 2$, find S_{10}. **18.** If $a_1 = 8$ and $r = 3$, find S_5.

19. If $a_1 = 1$ and $r = -1$, find S_{20}. **20.** If $a_1 = 1$ and $r = -1$, find S_{21}.

21. Find a_8 for $\dfrac{1}{5}, \dfrac{1}{10}, \dfrac{1}{20}, \ldots$ **22.** Find a_8 for $\dfrac{1}{2}, \dfrac{1}{10}, \dfrac{1}{50}, \ldots$

23. Find S_5 for $-\dfrac{1}{2}, -\dfrac{1}{4}, -\dfrac{1}{8}, \ldots$ **24.** Find S_6 for $-\dfrac{1}{2}, 1, -2, \ldots$

25. Find a_{10} and S_{10} for $\sqrt{2}, 2, 2\sqrt{2}, \ldots$

26. Find a_8 and S_8 for $\sqrt{3}, 3, 3\sqrt{3}, \ldots$

27. Find a_6 and S_6 for $100, 10, 1, \ldots$

28. Find a_6 and S_6 for $100, -10, 1, \ldots$

29. If $a_4 = 40$ and $a_6 = 160$, find r. **30.** If $a_5 = \dfrac{1}{8}$ and $a_8 = \dfrac{1}{64}$, find r.

31. Given the sequence $-3, 6, -12, 24, \ldots$, find a_8 and S_8.

32. Given the sequence $4, 2, 1, \frac{1}{2}, \ldots$, find a_9 and S_9.

33. Given $a_7 = 13$ and $a_{10} = 104$, find r.

34. Given $a_5 = -12$ and $a_8 = 324$, find r.

Find the sum of each infinite geometric series.

35. $\dfrac{1}{2} + \dfrac{1}{4} + \dfrac{1}{8} + \cdots$ **36.** $\dfrac{1}{3} + \dfrac{1}{9} + \dfrac{1}{27} + \cdots$

37. $4 + 2 + 1 + \cdots$ **38.** $8 + 4 + 2 + \cdots$

39. $2 + 1 + \dfrac{1}{2} + \cdots$ **40.** $3 + 1 + \dfrac{1}{3} + \cdots$

41. $\dfrac{4}{3} - \dfrac{2}{3} + \dfrac{1}{3} + \cdots$ **42.** $6 - 4 + \dfrac{8}{3} + \cdots$

43. $\dfrac{2}{5} + \dfrac{4}{25} + \dfrac{8}{125} + \cdots$ **44.** $\dfrac{3}{4} + \dfrac{9}{16} + \dfrac{27}{64} + \cdots$

45. $\dfrac{3}{4} + \dfrac{1}{4} + \dfrac{1}{12} + \cdots$ **46.** $\dfrac{5}{3} + \dfrac{1}{3} + \dfrac{1}{15} + \cdots$

47. Show that 0.444 . . . is the same as $\frac{4}{9}$.

48. Show that 0.333 . . . is the same as $\frac{1}{3}$.

49. Show that 0.272727 . . . is the same as $\frac{3}{11}$.

50. Show that 0.545454 . . . is the same as $\frac{6}{11}$.

APPLYING THE CONCEPTS

Declining-Balance Depreciation The declining-balance method of depreciation is an accounting method businesses use to deduct most of the cost of new equipment during the first few years of purchase. The value at any time during the life of the machine can be found with a linear equation in two variables. For income tax purposes, however, it is the value at the end of the year that is most important, and for this reason sequences can be used.

51. Value of a Crane A construction crane sells for $450,000 if purchased new. After that, the value decreases by 30% each year. We can use a geometric sequence to find the value of the crane at the end of each year. If we let a_0 represent the value when it is purchased, then a_1 is the value after 1 year, a_2 is the value after 2 years, and so on.

   a. Write the first five terms of the sequence.

   b. What is the common ratio?

   c. Construct a line graph for the first five terms of the sequence.

   d. Use the line graph to estimate the value of the crane 4.5 years after it is purchased.

   e. Write the sequence from part **a** using a recursive formula.

52. Value of a Printing Press A large printing press sells for $375,000 when it is new. After that, its value decreases 25% each year.

   a. Write a geometric sequence that gives the value of the press at the end of each of the first 5 years after it is purchased.

   b. What is the common ratio for this sequence?

   c. Construct a line graph for this sequence.

   d. Use the line graph to estimate the value of the printing press 1.5 years after it is purchased.

   e. Write the sequence from part **a** using a recursive formula.

53. Adding Terms Given the geometric series

$$\frac{1}{3} + \frac{1}{9} + \frac{1}{27} + \cdots$$

   a. Find the sum of all the terms.

   b. Find the sum of the first six terms.

   c. Find the sum of all but the first six terms.

54. Bouncing Ball A ball is dropped from a height of 20 feet. Each time it bounces, it returns to $\frac{7}{8}$ of the height it fell from. If the ball is allowed to bounce an infinite number of times, find the total vertical distance that the ball travels.

55. Stacking Paper Assume that a thin sheet of paper is 0.002 inch thick. The paper is torn in half, and the two halves placed together.

    a. How thick is the pile of torn paper?

    b. The pile of paper is torn in half again, and then the two halves placed together and torn in half again. The paper is large enough so this process can be performed a total of 5 times. How thick is the pile of torn paper?

    c. Refer to the tearing and piling process described in part **b.** Assuming that somehow the original paper is large enough, how thick is the pile of torn paper if 25 tears are made?

56. Pendulum A pendulum swings 15 feet left to right on its first swing. On each swing following the first, the pendulum swings $\frac{4}{5}$ of the previous swing.

    a. Write the general term for this geometric sequence.

    b. If the pendulum is allowed to swing an infinite number of times, what is the total distance the pendulum will travel?

57. Salary Increases After completing her MBA degree, an accounting firm offered Jane a job with a starting salary of $60,000, with a guaranteed annual raise of 4% of her previous year's salary.

    a. Write a finite sequence to represent Jane's first 5 years of income with this company. (Round each calculation to the nearest dollar.)

    b. Write the general term for this geometric sequence.

    c. Find the sum of Jane's income for the first 10 years with the company.

GETTING READY FOR THE NEXT SECTION

Simplify.

58. $(x + y)^0$

59. $(x + y)^1$

Expand and multiply.

60. $(x + y)^2$

61. $(x + y)^3$

Simplify.

62. $\dfrac{6 \cdot 5 \cdot 4 \cdot 3 \cdot 2 \cdot 1}{(2 \cdot 1)(4 \cdot 3 \cdot 2 \cdot 1)}$

63. $\dfrac{7 \cdot 6 \cdot 5 \cdot 4 \cdot 3 \cdot 2 \cdot 1}{(5 \cdot 4 \cdot 3 \cdot 2 \cdot 1)(2 \cdot 1)}$

EXTENDING THE CONCEPTS

64. Find the fraction form for 0.636363 . . . .

65. Find the fraction form for 0.123123123 . . . .

66. The sum of an infinite geometric series is 6 and the first term is 4. Find the common ratio.

67. The sum of an infinite geometric series is $\frac{5}{2}$ and the first term is 3. Find the common ratio.

68. Find the sum of the first 16 terms of the sequence $1, \sqrt{2}, 2, 2\sqrt{2}, \ldots$.

69. Find the sum of the first 12 terms of the sequence $2, 2\sqrt{3}, 6, 6\sqrt{3}, \ldots$.

70. Express the sum $1 + x + x^2 + x^3 + \ldots$ as a fraction.

71. Express the sum $1 - x^2 + x^4 - x^6 + \ldots$ as a fraction.

72. Find the sum: $\displaystyle\sum_{i=1}^{6} 12\left(\frac{2}{3}\right)^{i}$

73. Find the sum: $\displaystyle\sum_{i=1}^{5} 8\left(-\frac{1}{2}\right)^{i+2}$

74. Sierpinski Triangle In the sequence that follows, the figures are moving toward what is known as the Sierpinski triangle. To construct the figure in Stage 2, we remove the triangle formed from the midpoints of the sides of the shaded region in Stage 1. Likewise, the figure in Stage 3 is found by removing the triangles formed by connecting the midpoints of the sides of the shaded regions in Stage 2. If we repeat this process infinitely many times, we arrive at the Sierpinski triangle.

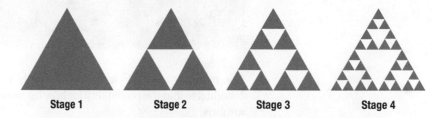

Stage 1   Stage 2   Stage 3   Stage 4

a. If the shaded region in Stage 1 has an area of 1, find the area of the shaded regions in Stages 2 through 4.

b. Do the areas you found in part **a.** form an arithmetic sequence or a geometric sequence?

c. The Sierpinski triangle is the triangle that is formed after the process of forming the stages shown in the figure is repeated infinitely many times. What do you think the area of the shaded region of the Sierpinski triangle will be?

d. Suppose the perimeter of the shaded region of the triangle in Stage 1 is 1. If we were to find the perimeters of the shaded regions in the other stages, would we have an increasing sequence or a decreasing sequence?

THE BINOMIAL EXPANSION

OBJECTIVES

A Calculate binomial coefficients using factorial notation.

B Expand a binomial using the binomial formula.

C Find the general term of a binomial expansion.

©iStockphoto.com/cristiani

The binomial $(\frac{1}{2} + \frac{1}{2})^7$ represents a family with 7 children. We can expand this binomial to find the probability for how many of the children are boys and how many are girls.

The purpose of this section is to write and apply the formula for the expansion of expressions of the form $(x + y)^n$, where n is any positive integer. To write the formula, we must generalize the information in the following chart:

$$(x + y)^0 = 1$$
$$(x + y)^1 = x + y$$
$$(x + y)^2 = x^2 + 2xy + y^2$$
$$(x + y)^3 = x^3 + 3x^2y + 3xy^2 + y^3$$
$$(x + y)^4 = x^4 + 4x^3y + 6x^2y^2 + 4xy^3 + y^4$$
$$(x + y)^5 = x^5 + 5x^4y + 10x^3y^2 + 10x^2y^3 + 5xy^4 + y^5$$

NOTE The polynomials on the right have been found by expanding the binomials on the left — we just haven't shown the work.

There are a number of similarities to notice among the polynomials on the right:

1. In each polynomial, the sequence of exponents on the variable x decreases to 0 from the exponent on the binomial at the left. (The exponent 0 is not shown, since $x^0 = 1$.)

2. In each polynomial, the exponents on the variable y increase from 0 to the exponent on the binomial at the left. (Because $y^0 = 1$, it is not shown in the first term.)

3. The sum of the exponents on the variables in any single term is equal to the exponent on the binomial at the left.

9.5 VIDEOS

The pattern in the coefficients of the polynomials on the right can best be seen by writing the right side again without the variables. It looks like this:

row 0   1

row 1   1   1

row 2   1   2   1

row 3   1   3   3   1

row 4   1   4   6   4   1

row 5   1   5   10   10   5   1

This triangle-shaped array of coefficients is called *Pascal's triangle*. Each entry in the triangular array is obtained by adding the two numbers above it. Each row begins and ends with the number 1. If we were to continue Pascal's triangle, the next two rows would be

row 6   1   6   15   20   15   6   1

row 7   1   7   21   35   35   21   7   1

The coefficients for the terms in the expansion of $(x + y)^n$ are given in the nth row of Pascal's triangle.

Ⓐ Factorial Notation and the Binomial Coefficient

There is an alternative method of finding these coefficients that does not involve Pascal's triangle. The alternative method involves *factorial notation*.

DEFINITION

The expression *n!* is read "*n* factorial" and is the product of all the consecutive integers from *n* down to 1. For example,

$$1! = 1$$
$$2! = 2 \cdot 1 = 2$$
$$3! = 3 \cdot 2 \cdot 1 = 6$$
$$4! = 4 \cdot 3 \cdot 2 \cdot 1 = 24$$
$$5! = 5 \cdot 4 \cdot 3 \cdot 2 \cdot 1 = 120$$

The expression 0! is defined to be 1. We use factorial notation to define binomial coefficients as follows.

DEFINITION   *binomial coefficient*

The expression $\binom{n}{r}$ is called a *binomial coefficient* and is defined by

$$\binom{n}{r} = \frac{n!}{r!(n-r)!}$$

EXAMPLE 1

Calculate the following binomial coefficients:

$$\binom{7}{5}, \binom{6}{2}, \binom{3}{0}$$

SOLUTION   We simply apply the definition for binomial coefficients:

$$\binom{7}{5} = \frac{7!}{5!(7-5)!}$$

$$= \frac{7!}{5! \cdot 2!}$$

$$= \frac{7 \cdot 6 \cdot 5 \cdot 4 \cdot 3 \cdot 2 \cdot 1}{(5 \cdot 4 \cdot 3 \cdot 2 \cdot 1)(2 \cdot 1)}$$

$$= \frac{42}{2}$$

$$= 21$$

$$\binom{6}{2} = \frac{6!}{2!(6-2)!}$$

$$= \frac{6!}{2! \cdot 4!}$$

$$= \frac{6 \cdot 5 \cdot 4 \cdot 3 \cdot 2 \cdot 1}{(2 \cdot 1)(4 \cdot 3 \cdot 2 \cdot 1)}$$

$$= \frac{30}{2}$$

$$= 15$$

$$\binom{3}{0} = \frac{3!}{0!(3-0)!}$$

$$= \frac{3!}{0! \cdot 3!}$$

$$= \frac{3 \cdot 2 \cdot 1}{(1)(3 \cdot 2 \cdot 1)}$$

$$= 1$$

If we were to calculate all the binomial coefficients in the following array, we would find they match exactly with the numbers in Pascal's triangle. That is why they are called binomial coefficients — because they are the coefficients of the expansion of $(x + y)^n$.

$$\binom{0}{0}$$

$$\binom{1}{0} \quad \binom{1}{1}$$

$$\binom{2}{0} \quad \binom{2}{1} \quad \binom{2}{2}$$

$$\binom{3}{0} \quad \binom{3}{1} \quad \binom{3}{2} \quad \binom{3}{3}$$

$$\binom{4}{0} \quad \binom{4}{1} \quad \binom{4}{2} \quad \binom{4}{3} \quad \binom{4}{4}$$

$$\binom{5}{0} \quad \binom{5}{1} \quad \binom{5}{2} \quad \binom{5}{3} \quad \binom{5}{4} \quad \binom{5}{5}$$

Using the new notation to represent the entries in Pascal's triangle, we can summarize everything we have noticed about the expansion of binomial powers of the form $(x + y)^n$.

🅑 The Binomial Expansion

If x and y represent real numbers and n is a positive integer, then the following formula is known as the *binomial expansion* or *binomial formula*:

$$(x + y)^n = \binom{n}{0} x^n y^0 + \binom{n}{1} x^{n-1} y^1 + \binom{n}{2} x^{n-2} y^2 + \cdots + \binom{n}{n} x^0 y^n$$

It does not make any difference, when expanding binomial powers of the form $(x + y)^n$, whether we use Pascal's triangle or the formula

$$\binom{n}{r} = \frac{n!}{r!(n-r)!}$$

to calculate the coefficients. We will show examples of both methods.

EXAMPLE 2

Expand $(x - 2)^3$.

SOLUTION   Applying the binomial formula, we have

$$(x - 2)^3 = \binom{3}{0} x^3 (-2)^0 + \binom{3}{1} x^2 (-2)^1 + \binom{3}{2} x^1 (-2)^2 + \binom{3}{3} x^0 (-2)^3$$

The coefficients

$$\binom{3}{0}, \binom{3}{1}, \binom{3}{2}, \text{and} \binom{3}{3}$$

can be found in the third row of Pascal's triangle. They are 1, 3, 3, and 1:

$$(x - 2)^3 = 1x^3 (-2)^0 + 3x^2 (-2)^1 + 3x^1 (-2)^2 + 1x^0 (-2)^3$$
$$= x^3 - 6x^2 + 12x - 8$$

EXAMPLE 3

Expand $(3x + 2y)^4$.

SOLUTION   The coefficients can be found in the fourth row of Pascal's triangle.

1, 4, 6, 4, 1

Here is the expansion of $(3x + 2y)^4$:

$$(3x + 2y)^4 = 1(3x)^4 + 4(3x)^3 (2y) + 6(3x)^2 (2y)^2 + 4(3x)(2y)^3 + 1(2y)^4$$
$$= 81x^4 + 216x^3 y + 216x^2 y^2 + 96xy^3 + 16y^4$$

EXAMPLE 4

Write the first three terms in the expansion of $(x + 5)^9$.

SOLUTION   The coefficients of the first three terms are

$$\binom{9}{0}, \binom{9}{1}, \text{and} \binom{9}{2}$$

which we calculate as follows:

$$\binom{9}{0} = \frac{9!}{0! \cdot 9!} = \frac{9 \cdot 8 \cdot 7 \cdot 6 \cdot 5 \cdot 4 \cdot 3 \cdot 2 \cdot 1}{(1)(9 \cdot 8 \cdot 7 \cdot 6 \cdot 5 \cdot 4 \cdot 3 \cdot 2 \cdot 1)} = \frac{1}{1} = 1$$

$$\binom{9}{1} = \frac{9!}{1! \cdot 8!} = \frac{9 \cdot 8 \cdot 7 \cdot 6 \cdot 5 \cdot 4 \cdot 3 \cdot 2 \cdot 1}{(1)(8 \cdot 7 \cdot 6 \cdot 5 \cdot 4 \cdot 3 \cdot 2 \cdot 1)} = \frac{9}{1} = 9$$

$$\binom{9}{2} = \frac{9!}{2! \cdot 7!} = \frac{9 \cdot 8 \cdot 7 \cdot 6 \cdot 5 \cdot 4 \cdot 3 \cdot 2 \cdot 1}{(2 \cdot 1)(7 \cdot 6 \cdot 5 \cdot 4 \cdot 3 \cdot 2 \cdot 1)} = \frac{72}{2} = 36$$

From the binomial formula, we write the first three terms:

$$(x + 5)^9 = 1 \cdot x^9 + 9 \cdot x^8(5) + 36x^7(5)^2 + \cdots$$
$$= x^9 + 45x^8 + 900x^7 + \cdots$$

ⓒ The kth Term of a Binomial Expansion

If we look at each term in the expansion of $(x + y)^n$ as a term in a sequence, $a_1, a_2, a_3, \ldots$, we can write

$$a_1 = \binom{n}{0} x^n y^0$$

$$a_2 = \binom{n}{1} x^{n-1} y^1$$

$$a_3 = \binom{n}{2} x^{n-2} y^2$$

$$a_4 = \binom{n}{3} x^{n-3} y^3 \quad \text{and so on}$$

To write the formula for the general term, we simply notice that the exponent on y and the number below n in the coefficient are both 1 less than the term number. This observation allows us to write the following:

> **Operation Symbols**
> •
>
> The kth term in the expansion of $(x + y)^n$ is
>
> $$a_k = \binom{n}{k-1} x^{n-(k-1)} y^{k-1}$$

EXAMPLE 5

Find the fifth term in the expansion of $(2x + 3y)^{12}$.

SOLUTION   Applying the preceding formula, we have

$$a_5 = \binom{12}{4} (2x)^8 (3y)^4$$

$$= \frac{12!}{4! \cdot 8!} (2x)^8 (3y)^4$$

Notice that once we have one of the exponents, the other exponent and the denominator of the coefficient are determined: The two exponents add to 12 and match the numbers in the denominator of the coefficient.

Making the calculations from the preceding formula, we have

$$a_5 = 495(256x^8)(81y^4)$$

$$= 10,264,320x^8 y^4$$

GETTING READY FOR CLASS

After reading through the preceding section, respond in your own words and in complete sentences.

A. What is Pascal's triangle?

B. Why is $\binom{n}{0} = 1$ for any natural number?

C. State the binomial formula.

D. When is the binomial formula more efficient than multiplying to expand a binomial raised to a whole-number exponent?

SPOTLIGHT ON SUCCESS

Buddy | instructor

What does math mean to me? Well, from a young age, I always knew math was my favorite subject in school. I never knew that math was something of which you could make a career. I like math because it is beautiful and mechanical, as well as confusing, thought-provoking, and hard to understand. Math made me comfortable asking my teachers the simple questions, "Why?" or "Why is that true?" which I think are very important questions with most things in life. Math became very philosophical to me and changed a lot of my world views.

Math made me a critical thinker. Every statement in math requires a very detailed and rigorous proof, but in the end, there is always a moment of immense coherence where hard evidence and logic made a difficult idea become clear. That moment is a tremendous experience. In math, you can make arguments about whether something is true or false and prove to someone it is one way or the other. I liked the idea of critically analyzing the truth value of any mathematical statement. A lot of things in math were hard to accept as true, and some seemingly obvious things were hard to accept as being not true. This is where the beauty lies in math. It helped me make sense of the world. I found some objective truths, which seemed rare in such a scary and chaotic world, and that was very satisfying for me.

I love being a part of math education. I want to share these wonderful moments and experiences that math has brought into my life. When I see one of my students have that moment of clarity, where an idea finally sticks, and the student begins to see the big picture and truly starts to understand an abstract idea, it is an incredibly rewarding feeling. Math means the world to me, and I want to share the wealth of endless knowledge that it can provide with as many students as I can.

Use the binomial formula to expand each of the following.

1. $(x + 2)^4$ **2.** $(x - 2)^5$ **3.** $(x + y)^6$ **4.** $(x - 1)^6$

5. $(2x + 1)^5$ **6.** $(2x - 1)^4$ **7.** $(x - 2y)^5$ **8.** $(2x + y)^5$

9. $(3x - 2)^4$ **10.** $(2x - 3)^4$ **11.** $(4x - 3y)^3$ **12.** $(3x - 4y)^3$

13. $(x^2 + 2)^4$ **14.** $(x^2 - 3)^3$ **15.** $(x^2 + y^2)^3$ **16.** $(x^2 - 3y)^4$

17. $(2x + 3y)^4$ **18.** $(2x - 1)^5$ **19.** $\left(\dfrac{x}{2} + \dfrac{y}{3}\right)^3$ **20.** $\left(\dfrac{x}{3} - \dfrac{y}{2}\right)^4$

21. $\left(\dfrac{x}{2} - 4\right)^3$ **22.** $\left(\dfrac{x}{3} + 6\right)^3$ **23.** $\left(\dfrac{x}{3} + \dfrac{y}{2}\right)^4$ **24.** $\left(\dfrac{x}{2} - \dfrac{y}{3}\right)^4$

Write the first four terms in the expansion of the following.

25. $(x + 2)^9$ **26.** $(x - 2)^9$ **27.** $(x - y)^{10}$ **28.** $(x + 2y)^{10}$

29. $(x + 3)^{25}$ **30.** $(x - 1)^{40}$ **31.** $(x - 2)^{60}$ **32.** $\left(x + \dfrac{1}{2}\right)^{30}$

33. $(x - y)^{18}$ **34.** $(x - 2y)^{65}$

Write the first three terms in the expansion of each of the following.

35. $(x + 1)^{15}$ **36.** $(x - 1)^{15}$ **37.** $(x - y)^{12}$ **38.** $(x + y)^{12}$

39. $(x + 2)^{20}$ **40.** $(x - 2)^{20}$

Write the first two terms in the expansion of each of the following.

41. $(x + 2)^{100}$ **42.** $(x - 2)^{50}$ **43.** $(x + y)^{50}$ **44.** $(x - y)^{100}$

45. Find the ninth term in the expansion of $(2x + 3y)^{12}$.

46. Find the sixth term in the expansion of $(2x + 3y)^{12}$.

47. Find the fifth term of $(x - 2)^{10}$.   **48.** Find the fifth term of $(2x - 1)^{10}$.

49. Find the sixth term in the expansion of $(x - 2)^{12}$.

50. Find the ninth term in the expansion of $(7x - 1)^{10}$.

51. Find the third term in the expansion of $(x - 3y)^{25}$.

52. Find the 24th term in the expansion of $(2x - y)^{26}$.

53. Write the formula for the 12th term of $(2x + 5y)^{20}$. Do not simplify.

54. Write the formula for the eighth term of $(2x + 5y)^{20}$. Do not simplify.

55. Write the first three terms of the expansion of $(x^2 y - 3)^{10}$.

56. Write the first three terms of the expansion of $(x - \frac{1}{x})^{50}$.

APPLYING THE CONCEPTS

57. Probability The third term in the expansion of $\left(\frac{1}{2} + \frac{1}{2}\right)^7$ will give the probability that in a family with 7 children, 5 will be boys and 2 will be girls. Find the third term.

58. Probability The fourth term in the expansion of $\left(\frac{1}{2} + \frac{1}{2}\right)^8$ will give the probability that in a family with 8 children, 3 will be boys and 5 will be girls. Find the fourth term.

MAINTAINING YOUR SKILLS

Solve each equation. Write your answers to the nearest hundredth.

59. $5^x = 7$ **60.** $10^x = 15$ **61.** $8^{2x-1} = 16$ **62.** $9^{3x-1} = 27$

63. Compound Interest How long will it take $400 to double if it is invested in an account with an annual interest rate of 1% compounded four times a year?

64. Compound Interest How long will it take $200 to become $800 if it is invested in an account with an annual interest rate of 1% compounded four times a year?

Find each of the following to the nearest hundredth.

65. $\log_4 20$ **66.** $\log_7 21$ **67.** $\ln 576$ **68.** $\ln 5,760$

69. Solve the formula $A = 10e^{5t}$ for t.   **70.** Solve the formula $A = Pe^{-5t}$ for t.

EXTENDING THE CONCEPTS

71. Calculate both $\binom{8}{5}$ and $\binom{8}{3}$ to show that they are equal.

72. Calculate both $\binom{10}{8}$ and $\binom{10}{2}$ to show that they are equal.

73. Simplify $\binom{20}{12}$ and $\binom{20}{8}$.

74. Simplify $\binom{15}{10}$ and $\binom{15}{5}$.

75. Show that $\binom{n}{r}$ and $\binom{n}{n-r}$ are equal.

76. Use the results of Problem 75 to find an easy way to compute $\binom{24}{23}$.

77. Find the term containing x^7 in the expansion of $(x + 3y)^{10}$.

78. Find the term containing y^8 in the expansion of $(x - 2y)^{10}$.

79. Compute: $\binom{200}{197}$

80. Compute: $\binom{6}{0} + \binom{6}{1} + \binom{6}{2} + \binom{6}{3} + \binom{6}{4} + \binom{6}{5} + \binom{6}{6}$

81. What is the value of n if the coefficients of the 13th and 15th terms of the expansion of $(x + 1)^n$ are the same?

82. Compute: $\dfrac{10!}{5!} \cdot \dfrac{6!}{12!}$

83. Pascal's Triangle Copy the first eight rows of Pascal's triangle into the eight rows of the following triangular array. (Each number in Pascal's triangle will go into one of the hexagons in the array.) Next, color in each hexagon that contains an odd number. What pattern begins to emerge from this coloring process?

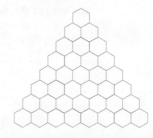

EXAMPLES

1. In the sequence
$1, 3, 5, \ldots, 2n - 1, \ldots,$
$a_1 = 1, a_2 = 3, a_3 = 5,$ and
$a_n = 2n - 1.$

Sequences [9.1]

A *sequence* is a function whose domain is the set of positive integers. The terms of a sequence are denoted by

$$a_1, a_2, a_3, \ldots, a_n, \ldots$$

where a_1 (read "a sub 1") is the first term, a_2 the second term, and a_n the nth or *general term*.

2. $\displaystyle\sum_{i=3}^{6}(-2)^i$
$= (-2)^3 + (-2)^4 + (-2)^5 + (-2)^6$
$= -8 + 16 + (-32) + 64$
$= 40$

Summation Notation [9.2]

The notation

$$\sum_{i=1}^{n} a_i = a_1 + a_2 + a_3 + \cdots + a_n$$

is called *summation notation* or *sigma notation*. The letter i as used here is called the *index of summation* or just *index*.

3. For the sequence $3, 7, 11, 15, \ldots,$
$a_1 = 3$ and $d = 4.$ The general term is
$$a_n = 3 + (n - 1)4$$
$$= 4n - 1$$
Using this formula to find the tenth term, we have
$$a_{10} = 4(10) - 1 = 39$$
The sum of the first 10 terms is
$$S_{10} = \frac{10}{2}(3 + 39) = 210$$

Arithmetic Sequences [9.3]

An *arithmetic sequence* is a sequence in which each term comes from the preceding term by adding a constant amount each time. If the first term of an arithmetic sequence is a_1 and the amount we add each time (called the *common difference*) is d, then the nth term of the progression is given by

$$a_n = a_1 + (n - 1)d$$

The sum of the first n terms of an arithmetic sequence is

$$S_n = \frac{n}{2}(a_1 + a_n)$$

S_n is called the *nth partial sum*.

4. For the geometric progression
$3, 6, 12, 24, \ldots, a_1 = 3$ and $r = 2.$
The general term is
$$a_n = 3 \cdot 2^{n-1}$$
The sum of the first 10 terms is
$$S_{10} = \frac{3(2^{10} - 1)}{2 - 1} = 3{,}069$$

Geometric Sequences [9.4]

A *geometric sequence* is a sequence of numbers in which each term comes from the previous term by multiplying by a constant amount each time. The constant by which we multiply each term to get the next term is called the *common ratio*. If the first term of a geometric sequence is a_1 and the common ratio is r, then the formula that gives the general term a_n is

$$a_n = a_1 r^{n-1}$$

The sum of the first n terms of a geometric sequence is given by the formula

$$S_n = \frac{a_1(r^n - 1)}{r - 1}$$

The Sum of an Infinite Geometric Series [9.4]

5. The sum of the series

$$\frac{1}{3} + \frac{1}{6} + \frac{1}{12} + \cdots$$

is

$$S = \frac{\frac{1}{3}}{1 - \frac{1}{2}} = \frac{\frac{1}{3}}{\frac{1}{2}} = \frac{2}{3}$$

If a geometric sequence has first term a_1 and common ratio r such that $|r| < 1$, then the following is called an *infinite geometric series*:

$$S = \sum_{i=0}^{\infty} a_1 r^i = a_1 + a_1 r + a_1 r^2 + a_1 r^3 + \cdots$$

Its sum is given by the formula

$$S = \frac{a_1}{1 - r}$$

Factorials [9.5]

The notation $n!$ is called *n factorial* and is defined to be the product of each consecutive integer from n down to 1. That is,

$$0! = 1 \qquad \text{(By definition)}$$
$$1! = 1$$
$$2! = 2 \cdot 1$$
$$3! = 3 \cdot 2 \cdot 1$$
$$4! = 4 \cdot 3 \cdot 2 \cdot 1$$

and so on.

Binomial Coefficients [9.5]

6. $\dbinom{7}{3} = \dfrac{7!}{3!(7-3)!}$

$= \dfrac{7!}{3! \cdot 4!}$

$= \dfrac{7 \cdot 6 \cdot 5 \cdot \cancel{4 \cdot 3 \cdot 2 \cdot 1}}{(3 \cdot 2 \cdot 1)(\cancel{4 \cdot 3 \cdot 2 \cdot 1})}$

$= 35$

The notation $\dbinom{n}{r}$ is called a *binomial coefficient* and is defined by

$$\binom{n}{r} = \frac{n!}{r!(n-r)!}$$

Binomial coefficients can be found by using the formula above or by *Pascal's triangle*, which is

```
                1
             1     1
          1     2     1
       1     3     3     1
    1     4     6     4     1
 1     5    10    10     5     1
```

and so on.

Binomial Expansion [9.5]

7. $(x + 2)^4$

$= x^4 + 4x^3 \cdot 2 + 6x^2 \cdot 2^2 + 4x \cdot 2^3 + 2^4$

$= x^4 + 8x^3 + 24x^2 + 32x + 16$

If n is a positive integer, then the formula for expanding $(x + y)^n$ is given by

$$(x + y)^n = \binom{n}{0} x^n y^0 + \binom{n}{1} x^{n-1} y^1 + \binom{n}{2} x^{n-2} y^2 + \cdots + \binom{n}{n} x^0 y^n$$

Chapter 9 Review

Write the first five terms of the sequences with the following general terms. [9.1]

1. $a_n = 6n + 2$

2. $a_n = n^2 - 4$

3. $a_n = n^3 - 6$

4. $a_n = \dfrac{n + 2}{n + 5}$

5. $a_n = \dfrac{1}{n^3}$

6. $a_n = n - \dfrac{1}{n}$

Write the first five terms of the sequences defined by the following recursion formulas. [9.1]

7. $a_1 = 5, a_n = -3a_{n-1} - 1, n > 1$

8. $a_1 = 5, a_n = 2a_{n-1} + n, n > 1$

9. $a_1 = 3 \qquad a_n = a_{n-1} - 3 \qquad n > 1$

10. $a_1 = -4 \qquad a_n = -2a_{n-1} \qquad n > 1$

Give the general term for each sequence. [9.1]

11. $3, 5, 7, 9, \ldots$

12. $1, -3, 9, -27, \ldots$

13. $\dfrac{1}{6}, -\dfrac{1}{12}, \dfrac{1}{24}, -\dfrac{1}{48}, \ldots$

14. $3, 12, 27, 48, \ldots$

15. $-2, 2, -2, 2, \ldots$

16. $0, \dfrac{3}{5}, \dfrac{8}{10}, \dfrac{15}{17}, \ldots$

Expand and simplify each of the following. [9.2]

17. $\displaystyle\sum_{i=1}^{4}(4i - 2)$

18. $\displaystyle\sum_{i=2}^{3}(i^2 + 4)$

19. $\displaystyle\sum_{i=4}^{6}\left(-\dfrac{1}{2}\right)^i$

20. $\displaystyle\sum_{i=2}^{4}\left(\dfrac{i^2 + 1}{i^2 - 1}\right)$

21. $\displaystyle\sum_{i=1}^{5}(x + i)$

22. $\displaystyle\sum_{i=2}^{7}(x + 1)^i$

23. $\displaystyle\sum_{i=2}^{5}\left(x + \dfrac{1}{i}\right)^2$

24. $\displaystyle\sum_{i=1}^{5}(x + i)^{i+1}$

Write each sum using summation notation. [9.2]

25. $9 + 15 + 21 + 27$

26. $\dfrac{4}{3} + \dfrac{5}{4} + \dfrac{6}{5} + \dfrac{7}{6} + \dfrac{8}{7} + \dfrac{9}{8}$

27. $(x + 3)^6 + (x + 6)^7 + (x + 9)^8 + (x + 12)^9$

28. $\dfrac{x - 3}{x + 4} + \dfrac{x - 4}{x + 6} + \dfrac{x - 5}{x + 8} + \dfrac{x - 6}{x + 10}$

29. If $a_1 = 5$ and $d = 10$, find a_n and a_{100}. [9.3]

30. If $a_1 = 7$ and $d = -1$, find a_{24} and S_{24}. [9.3]

31. Find the first term of an arithmetic progression if $a_7 = 35$ and $a_{12} = 60$. [9.3]

Find the sum of the first 10 terms of the following arithmetic progressions. [9.3]

32. $\dfrac{2}{5}, 1, \dfrac{8}{5}, \dfrac{11}{5}, \ldots$

33. $\dfrac{3}{5}, 1, \dfrac{7}{5}, \dfrac{9}{5}, \ldots$

34. If $a_5 = 23$ and $a_{10} = 48$, find the first term a_1, the common difference d, and then find a_{40}. [9.3]

35. If the third term is 16 and the eighth term is 51, find the first term, the common difference, and then find a_{50} and S_{50}. [9.3]

36. If $S_{24} = 60$ and $a_1 = 4$, find d and a_{116}.

37. Find a_{45} for the sequence 25, 20, 15, 10, ...

38. Find the second term of a geometric progression for which $a_1 = 4$ and $a_4 = -108$. [9.4]

39. Write a formula for the sum of the first 50 terms of the geometric progression 3, 15, 75, ... [9.4]

Find the sum of each infinite geometric series. [9.4]

40. $\dfrac{1}{4} + \dfrac{1}{16} + \dfrac{1}{64} + \ldots$

41. $8 + 4 + 2 + \ldots$

42. $3 + 1 + \dfrac{1}{3} + \ldots$

43. $6 - 4 + \dfrac{8}{3} + \ldots$

44. $\dfrac{5}{3} + \dfrac{1}{3} + \dfrac{1}{15} + \ldots$

Determine whether each sequence is arithmetic, geometric, or neither. If the sequence is arithmetic, give the common difference d. If the sequence is geometric, give the common ratio r. [9.3, 9.4]

45. 1, 4, 7, 10, ...

46. $1, \dfrac{1}{2}, \dfrac{1}{4}, \dfrac{1}{8}, \ldots$

47. 5, 11, 17, ...

48. 3, -15, 75, -375, ...

49. 5, 10, 15, 20, ...

50. 8, 14, 33, 75, ...

51. $-1, \dfrac{1}{3}, -\dfrac{1}{9}, \dfrac{1}{27}, \ldots$

52. 95, 85, 75, 65, ...

Use the binomial formula to expand each of the following. [9.5]

53. $(x + 2)^4$

54. $(2x - y)^4$

55. $(x^2 - 3y)^4$

56. $\left(\dfrac{x}{3} - \dfrac{y}{2} \right)^4$

57. Find the first 3 terms in the expansion of $(x - 3)^{11}$. [9.5]

Write the first four terms in the expansion of the following.

58. $(x - 2)^9$

59. $(x + 2y)^{10}$

60. Find the seventh term in $(x + 2y)^{11}$. [9.5]

Simplify.

1. $\dfrac{5(-6) + 3(-2)}{4(-3) + 3}$

2. $9 + 5(4y + 8) + 10y$

3. $18a^7b^{-4}/36a^2b^{-8}$

4. $\dfrac{y^2 - y - 6}{y^2 - 4}$

5. $8^{-2/3}$

6. $\log_3 27$

Factor completely.

7. $ab^3 + b^3 + 6a + 6$ **8.** $8x^2 - 5x - 3$

Solve.

9. $6 - 2(5x - 1) + 4x = 20$

10. $|4x - 3| + 2 = 3$

11. $(x - 1)(x - 2) = 12$ **12.** $1 - \dfrac{2}{x} = \dfrac{8}{x^2}$

13. $t - 6 = \sqrt{t - 4}$ **14.** $(4x - 3)^2 = -50$

15. $8t^3 - 27 = 0$ **16.** $16x^4 - 13x^2 = 5$

Solve and graph the solution on the number line.

17. $-3y - 2 < 7$ **18.** $|2x + 5| - 2 < 9$

Graph on a rectangular coordinate system.

19. $2x - 3y = 6$ **20.** $y < \dfrac{1}{2}x + 3$

21. $y = x^2 - 2x - 3$ **22.** $y = \left(\dfrac{1}{2}\right)^x$

Solve each system.

23. $5x - 3y = -4$ **24.** $x + 2y = 0$
       $x + 2y = 7$ $3y + z = -3$
                                   $2x - z = 5$

Multiply.

25. $3y^2 - 3y/3y - 12 \cdot y^2 - 2y - 8/y^2 + 3y + 2$

26. $(x^{3/5} + 2)(x^{3/5} - 2)$

27. $(2 + 3i)(1 - 4i)$

28. Add: $-3/x^2 - 2x - 8 + 4/x^2 - 16$

29. Combine: $4\sqrt{50} + 3\sqrt{8}$

30. Rationalize the denominator: $\dfrac{3}{\sqrt{7} - \sqrt{3}}$

31. If $f(x) = 4x + 1$, find $f^{-1}(x)$.

32. Find x if $\log x = 3.9786$.

33. Find $\log_6 14$ to the nearest hundredth.

Find the general term for each sequence.

34. $3, 14, 25, \dots$ **35.** $16, 8, 4, \dots$

36. Solve $S = 2x^2 + 4xy$ for y.

37. Expand and simplify.

$$\sum_{i=1}^{5} (2i - 1)$$

38. Find the slope of the line through $(2, -3)$ and $(-1, -3)$.

39. Find the equation of the line through $(2, 5)$ and $(6, -3)$. Write your answer in slope-intercept form.

40. If $C(t) = 80\left(\dfrac{1}{2}\right)^{t/5}$, find $C(5)$ and $C(10)$.

41. Find an equation with solutions $t = -\dfrac{3}{4}$ and $t = \dfrac{1}{5}$.

42. If $f(x) = 3x - 7$ and $g(x) = 4 - x$, find $(f \infty g)(x)$.

43. Solve the system.
$$3x - 5y = 2$$
$$2x + 4y = 1$$

44. Find the first term in the expansion of $(2x - y)^5$.

45. Add -8 to the difference of -7 and 4.

46. Factor $x^3 - \dfrac{1}{8}$.

47. Geometry Two supplementary angles are such that one is 5 times as large as the other. Find the two angles.

48. Variation y varies directly with x. If y is 24 when x is 8, find y when x is 2.

49. Number Problem One number is 3 times another. The sum of their reciprocals is $\dfrac{4}{3}$. Find the numbers.

50. Mixture How much 30% alcohol solution and 70% alcohol solution must be mixed to get 16 gallons of 60% alcohol solution?

51. Unit Analysis Convert 1,689.6 ft/min to miles per hour.

52. Piecewise Functions Graph the piecewise function.

$$f(x) = \begin{cases} x + 1 & \text{if } x < -2 \\ 1 & \text{if } -2 \le x \le 2 \\ x + 3 & \text{if } x > 2 \end{cases}$$

Write the first five terms of the sequences with the following general terms. [9.1]

1. $a_n = 3n - 5$

2. $a_1 = 3, a_n = a_{n-1} + 4, n > 1$

3. $a_n = n^2 + 1$

4. $a_n = 2n^3$

5. $a_n = \dfrac{n+1}{n^2}$

6. $a_1 = 4, a_n = -2a_{n-1}, n > 1$

Give the general term for each sequence. [9.1]

7. 6, 10, 14, 18, . . .

8. 1, 2, 4, 8, . . .

9. $\dfrac{1}{2}, \dfrac{1}{4}, \dfrac{1}{8}, \dfrac{1}{16}, \ldots$

10. $-3, 9, -27, 81, \ldots$

11. Expand and simplify each of the following. [9.2]

a. $\displaystyle\sum_{i=1}^{5}(5i + 3)$

b. $\displaystyle\sum_{i=3}^{5}(2^i - 1)$

c. $\displaystyle\sum_{i=2}^{6}(i^2 + 2i)$

12. Find the first term of an arithmetic progression if $a_5 = 11$ and $a_9 = 19$ [9.3]

13. Find the second term of a geometric progression for which $a_3 = 18$ and $a_5 = 162$. [9.4]

Find the sum of the first 10 terms of the following arithmetic progressions. [9.3]

14. 5, 11, 17, . . .

15. 25, 20, 15, . . .

16. Write a formula for the sum of the first 50 terms of the geometric progression 3, 6, 12, ... [9.4]

17. Find the sum of $\dfrac{1}{2} + \dfrac{1}{6} + \dfrac{1}{18} + \dfrac{1}{54} + \ldots$ [9.4]

Use the binomial formula to expand each of the following. [9.5]

18. $(x - 3)^4$

19. $(2x - 1)^5$

20. Find the first 3 terms in the expansion of $(x - 1)^{20}$. [9.5]

21. Find the sixth term in $(2x - 3y)^8$. [9.5]

10

Conic Sections

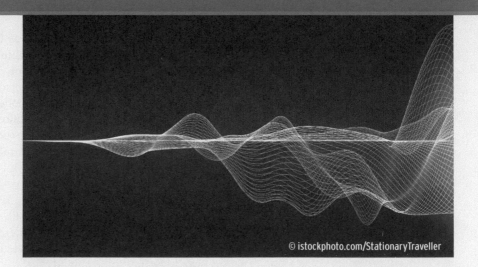

© istockphoto.com/StationaryTraveller

One of the curves we will study in this chapter has interesting reflective properties. Figure 1(A) shows how you can draw one of these curves (an ellipse) using thumbtacks, string, pencil, and paper. Elliptical surfaces will reflect sound waves that originate at one focus through the other focus. This property of ellipses allows doctors to treat patients with kidney stones using a procedure called lithotripsy. A lithotripter is an elliptical device that creates sound waves that crush the kidney stone into small pieces, without surgery. The sound wave originates at one focus of the lithotripter. The energy from it reflects off the surface of the lithotripter and converges at the other focus, the location of the kidney stone. Below (Figure 1(B)) is a cross-section of a lithotripter, with a patient positioned so the kidney stone is at the other focus.

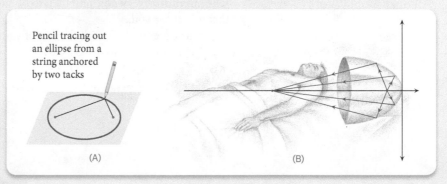

Pencil tracing out an ellipse from a string anchored by two tacks

(A)

(B)

FIGURE 1

By studying the conic sections in this chapter, you will be better equipped to understand some of the more technical equipment that exists in the world outside of class.

Dear Student,

Now that you are close to finishing this course, I want to pass on a couple of things that have helped me a great deal with my career. I'll introduce each one with a quote:

Do something for the person you will be 5 years from now.

I have always made sure that I arranged my life so that I was doing something for the person I would be 5 years later. For example, when I was 20 years old, I was in college. I imagined that the person I would be as a 25-year-old would want to have a college degree, so I made sure I stayed in school. That's all there is to this. It is not a hard, rigid philosophy. It is a soft, behind-the-scenes foundation. It does not include ideas such as "Five years from now I'm going to graduate at the top of my class from the best college in the country." Instead, you think, "Five years from now I will have a college degree, or I will still be in school working toward it."

This philosophy led to a community college teaching job, writing textbooks, creating videos with the textbooks, then to MathTV and the book you are reading right now. Along the way there were many other options and directions that I didn't take, but all the choices I made were due to keeping the person I would be in 5 years in mind.

It's easier to ride a horse in the direction it is going.

I started my college career thinking that I would become a dentist. I enrolled in all the courses that were required for dental school. When I completed the courses, I applied to a number of dental schools, but wasn't accepted. I kept going to school, and applied again the next year, again, without success. My life was not going in the direction of dental school, even though I had worked hard to put it in that direction. So I did a little inventory of the classes I had taken and the grades I earned, and realized that I was doing well in mathematics. My life was actually going in that direction, so I decided to see where that would take me. It was a good decision.

It is a good idea to work hard toward your goals, but it is also a good idea to take inventory every now and then to be sure you are headed in the direction that is best for you.

I wish you good luck with the rest of your college years, and with whatever you decide you want to do as a career.

Pat McKeague

10.1

OBJECTIVES

A Find the distance between two points using the distance formula.

B Write the equation of a circle, given its center and radius.

C Find the center and radius of a circle from its equation, and then sketch the graph.

© sxc.hu, peaces, 2009

Because of their perfect symmetry, circles have been used for thousands of years in many disciplines, including art, science, and religion. The photograph above is of Stonehenge, a 4,500-year-old site in England. The arrangement of the stones is based on a circular plan that is thought to have both religious and astronomical significance.

Conic sections include ellipses, circles, hyperbolas, and parabolas. They are called conic sections because each can be found by slicing a cone with a plane as shown in Figure 1. We begin our work with conic sections by studying circles. Before we find the general equation of a circle, we must first derive what is known as the *distance formula*.

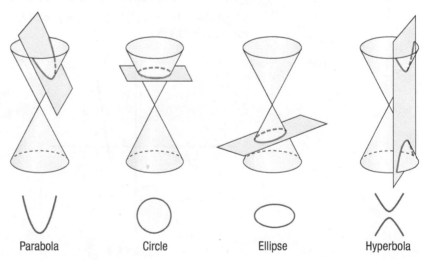

Parabola   Circle   Ellipse   Hyperbola

FIGURE 1

A The Distance Formula

Suppose (x_1, y_1) and (x_2, y_2) are any two points in the first quadrant. (Actually, we could choose the two points to be anywhere on the coordinate plane. It is just more convenient to have them in the first quadrant.) We can name the points P_1 and P_2, respectively, and draw the diagram shown in Figure 2.

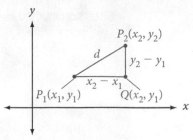

FIGURE 2

Notice the coordinates of point Q. The x-coordinate is x_2 because Q is directly below point P_2. The y-coordinate of Q is y_1 because Q is directly across from point P_1. It is evident from the diagram that the length of P_2Q is $y_2 - y_1$ and the length of P_1Q is $x_2 - x_1$. Using the Pythagorean theorem, we have

$$(P_1P_2)^2 = (P_1Q)^2 + (P_2Q)^2$$

or

$$d^2 = (x_2 - x_1)^2 + (y_2 - y_1)^2$$

Taking the square root of both sides, we have

$$d = \sqrt{(x_2 - x_1)^2 + (y_2 - y_1)^2}$$

We know this is the positive square root, because d is the distance from P_1 to P_2 and must therefore be positive. This formula is called the *distance formula*.

EXAMPLE 1

Find the distance between $(3, 5)$ and $(2, -1)$.

SOLUTION If we let $(3, 5)$ be (x_1, y_1) and $(2, -1)$ be (x_2, y_2) and apply the distance formula, we have

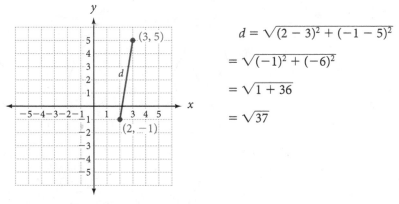

$$d = \sqrt{(2 - 3)^2 + (-1 - 5)^2}$$
$$= \sqrt{(-1)^2 + (-6)^2}$$
$$= \sqrt{1 + 36}$$
$$= \sqrt{37}$$

FIGURE 3

EXAMPLE 2

Find x if the distance from $(x, 5)$ to $(3, 4)$ is $\sqrt{2}$.

SOLUTION   Using the distance formula, we have

$$\sqrt{2} = \sqrt{(x-3)^2 + (5-4)^2} \qquad \text{Distance formula}$$

$$2 = (x-3)^2 + 1^2 \qquad \text{Square each side.}$$

$$2 = x^2 - 6x + 9 + 1 \qquad \text{Expand } (x-3)^2.$$

$$0 = x^2 - 6x + 8 \qquad \text{Simplify.}$$

$$0 = (x-4)(x-2) \qquad \text{Factor.}$$

$$x = 4 \quad \text{or} \quad x = 2 \qquad \text{Set factors equal to 0.}$$

The two solutions are 4 and 2, which indicates that two points, (4, 5) and (2, 5), are $\sqrt{2}$ units from (3, 4).

ⓑ Circles

The photograph on the left is another of Stonehenge, the 4,500-year-old site in England mentioned in this section's opening. More recently, the design shown in the photo on the right began appearing in agricultural fields in England in the 1990s. Whoever made these designs chose the circle as their basic shape.

Masterfile

Masterfile

We can model circles very easily in algebra by using equations that are based on the distance formula.

> **THEOREM** *Circle Theorem*
>
> The *equation of the circle* with center at (a, b) and radius r is given by
>
> $$(x-a)^2 + (y-b)^2 = r^2$$

Proof   By definition, all points on the circle are a distance r from the center (a, b). If we let (x, y) represent any point on the circle, then (x, y) is r units from (a, b). Applying the distance formula, we have

$$r = \sqrt{(x-a)^2 + (y-b)^2}$$

Squaring both sides of this equation gives the equation of the circle:

$$(x-a)^2 + (y-b)^2 = r^2$$

We can use the circle theorem to find the equation of a circle, given its center and radius, or to find its center and radius, given the equation.

EXAMPLE 3

Find the equation of the circle with center at $(-3, 2)$ having a radius of 5.

SOLUTION We have $(a, b) = (-3, 2)$ and $r = 5$. Applying our theorem for the equation of a circle yields

$$[x - (-3)]^2 + (y - 2)^2 = 5^2$$
$$(x + 3)^2 + (y - 2)^2 = 25$$

EXAMPLE 4

Give the equation of the circle with radius 3 whose center is at the origin.

SOLUTION The coordinates of the center are $(0, 0)$, and the radius is 3. The equation must be

$$(x - 0)^2 + (y - 0)^2 = 3^2$$
$$x^2 + y^2 = 9$$

We can see from Example 4 that the equation of any circle with its center at the origin and radius r will be

$$x^2 + y^2 = r^2$$

C Graphing Circles

EXAMPLE 5

Find the center and radius, and sketch the graph of the circle whose equation is
$$(x - 1)^2 + (y + 3)^2 = 4$$

SOLUTION Writing the equation in the form
$$(x - a)^2 + (y - b)^2 = r^2$$
we have
$$(x - 1)^2 + [y - (-3)]^2 = 2^2$$
The center is at $(1, -3)$, and the radius is 2. The graph is shown in Figure 4.

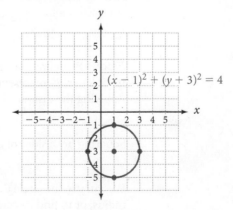

FIGURE 4

EXAMPLE 6

Sketch the graph of $x^2 + y^2 = 9$.

SOLUTION Because the equation can be written in the form

$$(x - 0)^2 + (y - 0)^2 = 3^2$$

it must have its center at $(0, 0)$ and a radius of 3. The graph is shown in Figure 5.

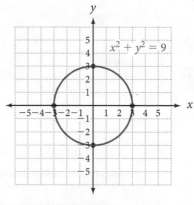

FIGURE 5

EXAMPLE 7

Sketch the graph of $x^2 + y^2 + 6x - 4y - 12 = 0$.

SOLUTION To sketch the graph, we must find the center and radius of our circle. We can do so easily if the equation is in standard form. That is, if it has the form

$$(x - a)^2 + (y - b)^2 = r^2$$

To put our equation in standard form, we start by using the addition property of equality to group all the constant terms together on the right side of the equation. In this case, we add 12 to each side of the equation. We do this because we are going to add our own constants later to complete the square.

$$x^2 + y^2 + 6x - 4y = 12$$

Next, we group all the terms containing x together and all terms containing y together, and we leave some space at the end of each group for the numbers we will add when we complete the square on each group.

$$x^2 + 6x \qquad + y^2 - 4y \qquad = 12$$

To complete the square on x, we add 9 to each side of the equation. To complete the square on y, we add 4 to each side of the equation.

$$x^2 + 6x + 9 + y^2 - 4y + 4 = 12 + 9 + 4$$

The first three terms on the left side can be written as $(x + 3)^2$. Likewise, the last three terms on the left side simplify to $(y - 2)^2$. The right side simplifies to 25.

$$(x + 3)^2 + (y - 2)^2 = 25$$

Writing 25 as 5^2, we have our equation in standard form.

$$(x + 3)^2 + (y - 2)^2 = 5^2$$

From this last line, it is apparent that the center is at $(-3, 2)$ and the radius is 5.

Using this information, we create the graph shown in Figure 6.

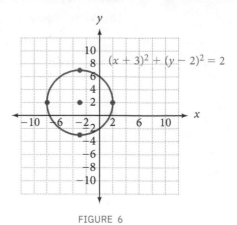

FIGURE 6

After reading through the preceding section, respond in your own words and in complete sentences.

A. Describe the distance formula in words, as if you were explaining to someone how they should go about finding the distance between two points.

B. What is the mathematical definition of a circle?

C. How are the distance formula and the equation of a circle related?

D. When graphing a circle from its equation, why is completing the square sometimes useful?

Problem Set 10.1

Find the distance between the following points.

1. $(3, 7)$ and $(6, 3)$ **2.** $(4, 7)$ and $(8, 1)$

3. $(0, 9)$ and $(5, 0)$ **4.** $(-3, 0)$ and $(0, 4)$

5. $(3, -5)$ and $(-2, 1)$ **6.** $(-8, 9)$ and $(-3, -2)$

7. $(-1, -2)$ and $(-10, 5)$ **8.** $(-3, -8)$ and $(-1, 6)$

9. Find x so the distance between $(x, 2)$ and $(1, 5)$ is $\sqrt{13}$.

10. Find x so the distance between $(-2, 3)$ and $(x, 1)$ is 3.

11. Find x so the distance between $(x, 5)$ and $(3, 9)$ is 5.

12. Find y so the distance between $(-4, y)$ and $(2, 1)$ is 8.

13. Find x so the distance between $(x, 4)$ and $(2x + 1, 6)$ is 6.

14. Find y so the distance between $(3, y)$ and $(7, 3y - 1)$ is 6.

Write the equation of the circle with the given center and radius.

15. Center $(3, -2)$; $r = 3$ **16.** Center $(-2, 4)$; $r = 1$

17. Center $(-5, -1)$; $r = \sqrt{5}$ **18.** Center $(-7, -6)$; $r = \sqrt{3}$

19. Center $(0, -5)$; $r = 1$ **20.** Center $(0, -1)$; $r = 7$

21. Center $(0, 0)$; $r = 2$ **22.** Center $(0, 0)$; $r = 5$

Give the center and radius, and sketch the graph of each of the following circles.

23. $x^2 + y^2 = 4$ **24.** $x^2 + y^2 = 16$

25. $(x - 1)^2 + (y - 3)^2 = 25$ **26.** $(x - 4)^2 + (y - 1)^2 = 36$

27. $(x + 2)^2 + (y - 4)^2 = 8$ **28.** $(x - 3)^2 + (y + 1)^2 = 12$

29. $(x + 2)^2 + (y - 4)^2 = 17$ **30.** $x^2 + (y + 2)^2 = 11$

31. $x^2 + y^2 + 2x - 4y = 4$ **32.** $x^2 + y^2 - 4x + 2y = 11$

33. $x^2 + y^2 - 6y = 7$ **34.** $x^2 + y^2 - 4y = 5$

35. $x^2 + y^2 + 2x = 1$ **36.** $x^2 + y^2 + 10x = 0$

37. $x^2 + y^2 - 4x - 6y = -4$ **38.** $x^2 + y^2 - 4x + 2y = 4$

39. $x^2 + y^2 + 2x + y = \dfrac{11}{4}$ **40.** $x^2 + y^2 - 6x - y = -\dfrac{1}{4}$

41. $4x^2 + 4y^2 - 4x + 8y = 11$ **42.** $36x^2 + 36y^2 - 24x - 12y = 31$

Each of the following circles passes through the origin. In each case, find the equation.

43.

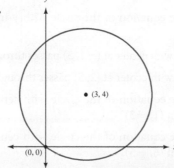

44.

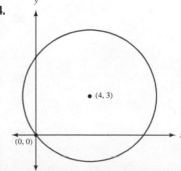

777

45. Find the equations of circles A, B, and C in the following diagram. The three points are the centers of the three circles.

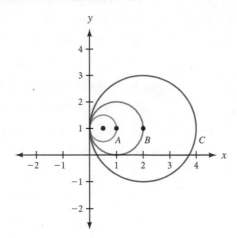

46. Each of the following circles passes through the origin. The centers are as shown. Find the equation of each circle.

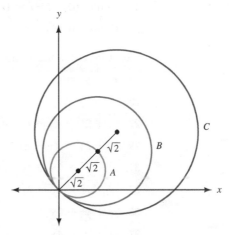

47. Find the equation of the circle with center at the origin that contains the point $(3, 4)$.

48. Find the equation of the circle with center at the origin that contains the point $(-5, 12)$.

49. Find the equation of the circle with center at the origin and x-intercepts 3 and -3.

50. Find the equation of the circle with y-intercepts 4 and -4 and center at the origin.

51. A circle with center at $(-1, 3)$ passes through the point $(4, 3)$. Find the equation.

52. A circle with center at $(2, 5)$ passes through the point $(-1, 4)$. Find the equation.

53. Find the equation of the circle with center at $(-2, 5)$, which passes through the point $(1, -3)$.

54. Find the equation of the circle with center at $(4, -1)$, which passes through the point $(6, -5)$.

55. Find the equation of the circle with center on the y-axis and y-intercepts at -2 and 6.

56. Find the equation of the circle with center on the x-axis and x-intercepts at -8 and 2.

57. Find the circumference and area of the circle $x^2 + (y - 3)^2 = 18$. Leave your answer in terms of π.

58. Find the circumference and area of the circle $(x + 2)^2 + (y + 6)^2 = 12$. Leave your answer in terms of π.

59. Find the circumference and area of the circle $x^2 + y^2 + 4x + 2y = 20$. Leave your answer in terms of π.

60. Find the circumference and area of the circle $x^2 + y^2 - 6x + 2y = 6$. Leave your answer in terms of π.

APPLYING THE CONCEPTS

61. Search Area A 3-year-old child has wandered away from home. The police have decided to search a circular area with a radius of 6 blocks. The child turns up at his grandmother's house, 5 blocks east and 3 blocks north of home. Was he found within the search area?

62. Placing a Bubble Fountain A circular garden pond with a diameter of 12 feet is to have a bubble fountain. The water from the bubble fountain falls in a circular pattern with a radius of 1.5 feet. If the center of the bubble fountain is placed 4 feet west and 3 feet north of the center of the pond, will all the water from the fountain fall inside the pond? What is the farthest distance from the center of the pond that water from the fountain will fall?

63. Ferris Wheel A giant Ferris wheel has a diameter of 240 feet and sits 12 feet above the ground. As shown in the diagram below, the wheel is 500 feet from the entrance to the amusement park. The xy-coordinate system containing the wheel has its origin on the ground at the center of the entrance. Write an equation that models the shape of the wheel.

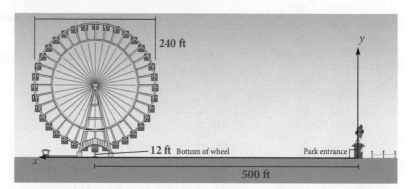

64. Magic Rings A magician is holding two rings that seem to lie in the same plane and intersect in two points. Each ring is 10 inches in diameter.

a. Find the equation of each ring if a coordinate system is placed with its origin at the center of the first ring and the x-axis contains the center of the second ring.

b. Find the equation of each ring if a coordinate system is placed with its origin at the center of the second ring and the x-axis contains the center of the first ring.

GETTING READY FOR THE NEXT SECTION

Solve.

65. $y^2 = 9$ **66.** $x^2 = 25$ **67.** $-y^2 = 4$ **68.** $-x^2 = 16$

69. $-x^2 = 9$ **70.** $y^2 = 100$

71. Divide $4x^2 + 9y^2$ by 36   **72.** Divide $25x^2 + 4y^2$ by 100

Find the x-intercepts and the y-intercepts

73. $3x - 4y = 12$ **74.** $y = 3x^2 + 5x - 2$

75. If $\dfrac{x^2}{25} + \dfrac{y^2}{9} = 1$, find y when x is 3.   **76.** If $\dfrac{x^2}{25} + \dfrac{y^2}{9} = 1$, find y when x is -4.

EXTENDING THE CONCEPTS

A circle is *tangent* to a line if it touches, but does not cross, the line.

77. Find the equation of the circle with center at $(2, 3)$ if the circle is tangent to the y-axis.

78. Find the equation of the circle with center at $(3, 2)$ if the circle is tangent to the x-axis.

79. Find the equation of the circle with center at $(2, 3)$ if the circle is tangent to the vertical line $x = 4$.

80. Find the equation of the circle with center at $(3, 2)$ if the circle is tangent to the horizontal line $y = 6$.

Find the distance from the origin to the center of each of the following circles.

81. $x^2 + y^2 - 6x + 8y = 144$

82. $x^2 + y^2 - 8x + 6y = 144$

83. $x^2 + y^2 - 6x - 8y = 144$

84. $x^2 + y^2 + 8x + 6y = 144$

10.2

 OBJECTIVES

A Graph an ellipse from its standard form.

B Graph a hyperbola from its standard form.

C Convert ellipses and hyperbolas to standard form by completing the square to graph the curve.

Like the planets in our solar system, Halley's comet orbits the Sun in an elliptical path. While it takes Earth 1 year to complete one orbit around the Sun, it takes Halley's comet 76 years. The first known sighting of Halley's comet was in 239 B.C. Its most famous appearance occurred in 1066 A.D., when it was seen at the Battle of Hastings.

The Orbit of Halley's Comet

 Ellipses

This section is concerned with the graphs of ellipses and hyperbolas. To begin, we will consider only those graphs that are centered about the origin.

Suppose we want to graph the equation

$$\frac{x^2}{25} + \frac{y^2}{9} = 1$$

We can find the y-intercepts by letting $x = 0$, and we can find the x-intercepts by letting $y = 0$:

When	$x = 0$	When	$y = 0$
	$\frac{0^2}{25} + \frac{y^2}{9} = 1$		$\frac{x^2}{25} + \frac{0^2}{9} = 1$
	$y^2 = 9$		$x^2 = 25$
	$y = \pm 3$		$x = \pm 5$

The graph crosses the y-axis at $(0, 3)$ and $(0, -3)$ and the x-axis at $(5, 0)$ and $(-5, 0)$. Graphing these points and then connecting them with a smooth curve gives the graph shown in Figure 1.

781

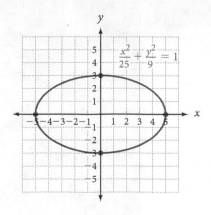

FIGURE 1

We can find other ordered pairs on the graph by substituting in values for x (or y) and then solving for y (or x). For example, if we let $x = 3$, then

$$\frac{3^2}{25} + \frac{y^2}{9} = 1$$

$$\frac{9}{25} + \frac{y^2}{9} = 1$$

$$0.36 + \frac{y^2}{9} = 1$$

$$\frac{y^2}{9} = 0.64 \qquad \text{Add } -0.36 \text{ to each side.}$$

$$y^2 = 5.76 \qquad \text{Multiply each side by 9.}$$

$$y = \pm 2.4 \qquad \text{Square root of each side}$$

This would give us the two ordered pairs $(3, -2.4)$ and $(3, 2.4)$.

A graph of the type shown in Figure 1 is called an *ellipse*. If we were to find some other ordered pairs that satisfy our original equation, we would find that their graphs lie on the ellipse. Also, the coordinates of any point on the ellipse will satisfy the equation. We can generalize these results as follows.

The Ellipse

The graph of any equation of the form

$$\frac{x^2}{a^2} + \frac{y^2}{b^2} = 1 \qquad \text{Standard form}$$

will be an *ellipse* centered at the origin. The ellipse will cross the x-axis at $(a, 0)$ and $(-a, 0)$. It will cross the y-axis at $(0, b)$ and $(0, -b)$. When a and b are equal, the ellipse will be a circle. Each of the points $(a, 0)$, $(-a, 0)$, $(0, b)$, and $(0, -b)$ is a *vertex* (intercept) of the graph.

The most convenient way to graph an ellipse is to locate the intercepts (vertices).

EXAMPLE 1

Sketch the graph of $4x^2 + 9y^2 = 36$.

SOLUTION   To write the equation in the form

$$\frac{x^2}{a^2} + \frac{y^2}{b^2} = 1$$

we must divide both sides by 36:

$$\frac{4x^2}{36} + \frac{9y^2}{36} = \frac{36}{36}$$

$$\frac{x^2}{9} + \frac{y^2}{4} = 1$$

The graph crosses the *x*-axis at (3, 0), (−3, 0) and the *y*-axis at (0, 2), (0, −2). (See Figure 2.)

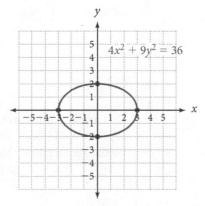

FIGURE 2

B Hyperbolas

The photo below shows Europa, one of Jupiter's moons, as it was photographed by the Galileo space probe in the late 1990s. To speed up the trip from Earth to Jupiter—nearly a billion miles—Galileo made use of the *slingshot effect*. This involves flying a hyperbolic path very close to a planet so that gravity can be used to gain velocity as the space probe hooks around the planet (Figure 3).

NASA

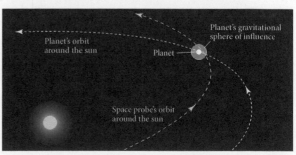

FIGURE 3

We use the rest of this section to consider equations that produce hyperbolas. Consider the equation

$$\frac{x^2}{9} - \frac{y^2}{4} = 1$$

If we were to find a number of ordered pairs that are solutions to the equation and connect their graphs with a smooth curve, we would have Figure 4.

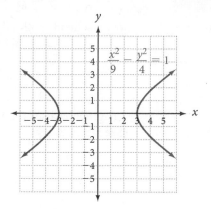

FIGURE 4

This graph is an example of a *hyperbola*. Notice that the graph has x-intercepts at $(3, 0)$ and $(-3, 0)$. The graph has no y-intercepts and hence does not cross the y-axis. We can show this by substituting $x = 0$ into the equation

$$\frac{0^2}{9} - \frac{y^2}{4} = 1 \qquad \text{Substitute 0 for } x.$$

$$-\frac{y^2}{4} = 1 \qquad \text{Simplify left side.}$$

$$y^2 = -4 \qquad \text{Multiply each side by } -4.$$

for which there is no real solution.

We want to produce reasonable sketches of hyperbolas without having to build extensive tables. We can produce the graphs we are after by using what are called *asymptotes* for our graphs. The discussion that follows is intended to give you some insight as to why these asymptotes exist. However, even if you don't understand this discussion completely, you will still be able to graph hyperbolas.

Asymptotes for hyperbolas   Let's solve the equation we graphed above for y:

$$\frac{x^2}{9} - \frac{y^2}{4} = 1 \qquad \text{Original equation}$$

$$-\frac{y^2}{4} = -\frac{x^2}{9} + 1 \qquad \text{Add } -\frac{x^2}{9} \text{ to each side.}$$

$$y^2 = \frac{4x^2}{9} - 4 \qquad \text{Multiply each side by } -4.$$

$$y = \pm\sqrt{\frac{4x^2}{9} - 4} \qquad \text{Square root property of equality}$$

To understand what comes next, you need to see that for very large values of x, the following expressions are almost the same:

$$\sqrt{\frac{4x^2}{9} - 4} \qquad \text{and} \qquad \sqrt{\frac{4x^2}{9}} = \frac{2}{3}x$$

This is because the 4 becomes insignificant compared with $\frac{4x^2}{9}$ for very large values of x. In fact, the larger x becomes, the closer these two expressions are to being equal. The table on the next page is intended to help you see this fact.

x	$\sqrt{\dfrac{4x^2}{9} - 4}$	$\sqrt{\dfrac{4x^2}{9}} = \dfrac{2}{3}x$
1	undefined	0.67
10	6.35959	6.66667
100	66.63666	66.66667
1000	666.66367	666.66667
10000	6666.66637	6666.66667

Extending the idea presented above, we can say that, for very large values of x, the graphs of the equations

$$y = \pm\sqrt{\frac{4x^2}{9} - 4} \qquad \text{and} \qquad y = \pm\frac{2}{3}x$$

will be close to each other. Further, the larger x becomes, the closer the graphs are to one another. (Using a similar line of reasoning, we can draw the same conclusion for values of x on the other side of the origin, -10, -100, $-1,000$, and $-10,000$.) Believe it or not, this helps us find the shape of our hyperbola. We simply note that the graph of

$$\frac{x^2}{9} - \frac{y^2}{4} = 1$$

crosses the x-axis at -3 and 3, and that as x gets further and further from the origin, the graph looks more like the graph of $y = \frac{2}{3}x$ and $y = -\frac{2}{3}x$.

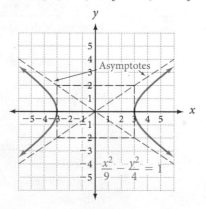

FIGURE 5

The lines $y = \frac{2}{3}x$ and $y = -\frac{2}{3}x$ are asymptotes for the graph of the hyperbola $\frac{x^2}{9} - \frac{y^2}{4} = 1$. The further we get from the origin, the closer the hyperbola is to these lines.

Asymptotes from a Rectangle In Figure 5, note the rectangle that has its sides parallel to the x- and y-axes and that passes through the x-intercepts and the points on the y-axis corresponding to the square roots of the number below y^2, $+2$, and -2. The lines that connect opposite corners of the rectangle are the *asymptotes* for the graph of the hyperbola

$$\frac{x^2}{9} - \frac{y^2}{4} = 1$$

EXAMPLE 2

Graph the equation $\frac{y^2}{9} - \frac{x^2}{16} = 1$.

SOLUTION In this case the y-intercepts are 3 and -3, and the x-intercepts do not exist. We can use the square roots of the number below x^2, however, to find the asymptotes associated with the graph. The sides of the rectangle used to draw the asymptotes must pass through 3 and -3 on the y-axis, and 4 and -4 on the x-axis. Figure 6 shows the rectangle, the asymptotes, and the hyperbola.

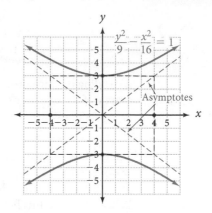

FIGURE 6

Here is a summary of what we have for hyperbolas.

> ### Hyperbolas Centered at the Origin
> •
>
> The graph of the equation
> $$\frac{x^2}{a^2} - \frac{y^2}{b^2} = 1$$
> will be a *hyperbola centered at the origin.* The graph will have *x-intercepts (vertices)* at $-a$ and a.
>
> The graph of the equation
> $$\frac{y^2}{b^2} - \frac{x^2}{a^2} = 1$$
> will be a *hyperbola centered at the origin.* The graph will have *y-intercepts (vertices)* at $-b$ and b.
>
> As an aid in sketching either of these equations, the asymptotes can be found by graphing the lines $y = \frac{b}{a}x$ and $y = -\frac{b}{a}x$, or by drawing lines through opposite corners of the rectangle whose sides pass through $-a$, a, $-b$, and b on the axes.

Ⓒ Ellipses and Hyperbolas Not Centered at the Origin

The following equation is that of an ellipse with its center at the point $(4, 1)$:

$$\frac{(x-4)^2}{9} + \frac{(y-1)^2}{4} = 1$$

To see why the center is at $(4, 1)$, we substitute x' (read "x prime") for $x - 4$ and y' for $y - 1$ in the equation. That is,

If $\qquad\qquad\qquad\qquad\qquad x' = x - 4$

and $\qquad\qquad\qquad\qquad\quad y' = y - 1$

the equation $\qquad \dfrac{(x-4)^2}{9} + \dfrac{(y-1)^2}{4} = 1$

becomes $\qquad\qquad \dfrac{(x')^2}{9} + \dfrac{(y')^2}{4} = 1$

This is the equation of an ellipse in a coordinate system with an x'-axis and a y'-axis. We call this new coordinate system the *$x'y'$-coordinate system*. The center of our ellipse is at the origin in the $x'y'$-coordinate system. The question is this: What are the coordinates of the center of this ellipse in the original xy-coordinate system? To answer this question, we go back to our original substitutions:

$$x' = x - 4$$
$$y' = y - 1$$

In the $x'y'$-coordinate system, the center of our ellipse is at $x' = 0$, $y' = 0$ (the origin of the $x'y'$ system). Substituting these numbers for x' and y', we have

$$0 = x - 4$$
$$0 = y - 1$$

Solving these equations for x and y will give us the coordinates of the center of our ellipse in the xy-coordinate system. As you can see, the solutions are $x = 4$ and $y = 1$. Therefore, in the xy-coordinate system, the center of our ellipse is at the point $(4, 1)$. Figure 7 shows the graph.

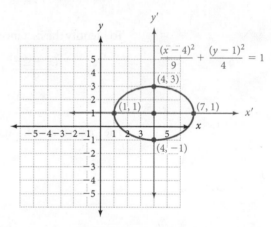

FIGURE 7

The coordinates of all points labeled in Figure 7 are given with respect to the xy-coordinate system. The x'- and y'-axes are shown simply for reference in our discussion. Note that the horizontal distance from the center to the vertices is 3, the square root of the denominator of the $(x - 4)^2$ term. Likewise, the vertical distance from the center to the other vertices is 2, the square root of the denominator of the $(y - 1)^2$ term.

We summarize the information above with the following:

An Ellipse with Center at (h, k)

The graph of the equation

$$\frac{(x - h)^2}{a^2} + \frac{(y - k)^2}{b^2} = 1$$

will be an *ellipse with center at (h, k).* The vertices of the ellipse will be at the points $(h + a, k)$, $(h - a, k)$, $(h, k + b)$, and $(h, k - b)$.

EXAMPLE 3

Graph the ellipse: $x^2 + 9y^2 + 4x - 54y + 76 = 0$

SOLUTION   To identify the coordinates of the center, we must complete the square on x and also on y. To begin, we rearrange the terms so that those containing x are together, those containing y are together, and the constant term is on the other side of the equal sign. Doing so gives us the following equation:

$$x^2 + 4x \qquad + 9y^2 - 54y \qquad = -76$$

Before we can complete the square on y, we must factor 9 from each term containing y:

$$x^2 + 4x \qquad + 9(y^2 - 6y \quad) = -76$$

To complete the square on x, we add 4 to each side of the equation. To complete the square on y, we add 9 inside the parentheses. This increases the left side of the equation by 81 since each term within the parentheses is multiplied by 9. Therefore, we must add 81 to the right side of the equation also.

$$x^2 + 4x + 4 + 9(y^2 - 6y + 9) = -76 + 4 + 81$$
$$(x + 2)^2 + 9(y - 3)^2 = 9$$

To identify the distances to the vertices, we divide each term on both sides by 9:

$$\frac{(x + 2)^2}{9} + \frac{9(y - 3)^2}{9} = \frac{9}{9}$$
$$\frac{(x + 2)^2}{9} + \frac{(y - 3)^2}{1} = 1$$

The graph is an ellipse with center at $(-2, 3)$, as shown in Figure 8.

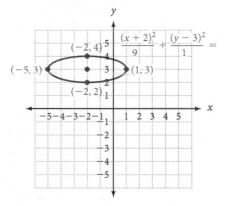

FIGURE 8

The ideas associated with graphing hyperbolas whose centers are not at the origin parallel the ideas just presented about graphing ellipses whose centers have been moved off the origin. Without showing the justification for doing so, we state the following guidelines for graphing hyperbolas:

Hyperbolas with Centers at (h, k)

The graphs of the equations

$$\frac{(x-h)^2}{a^2} - \frac{(y-k)^2}{b^2} = 1 \quad \text{and} \quad \frac{(y-k)^2}{b^2} - \frac{(x-h)^2}{a^2} = 1$$

will be *hyperbolas with their centers at (h, k)*. The vertices of the graph of the first equation will be at the points $(h + a, k)$ and $(h - a, k)$, and the vertices for the graph of the second equation will be at $(h, k + b)$ and $(h, k - b)$. In either case, the asymptotes can be found by connecting opposite corners of the rectangle that contains the four points $(h + a, k)$, $(h - a, k)$, $(h, k + b)$, and $(h, k - b)$.

EXAMPLE 4

Graph the hyperbola: $4x^2 - y^2 + 4y - 20 = 0$

SOLUTION   To identify the coordinates of the center of the hyperbola, we need to complete the square on y. (Because there is no linear term in x, we do not need to complete the square on x. The x-coordinate of the center will be $x = 0$.)

$$4x^2 - y^2 + 4y - 20 = 0$$

$$4x^2 - y^2 + 4y = 20 \qquad \text{Add 20 to each side.}$$

$$4x^2 - 1(y^2 - 4y) = 20 \qquad \text{Factor } -1 \text{ from each term containing } y.$$

To complete the square on y, we add 4 to the terms inside the parentheses. Doing so adds -4 to the left side of the equation because everything inside the parentheses is multiplied by -1. To keep from changing the equation, we must add -4 to the right side also.

$$4x^2 - 1(y^2 - 4y + 4) = 20 - 4 \qquad \text{Add } -4 \text{ to each side.}$$

$$4x^2 - 1(y - 2)^2 = 16 \qquad y^2 - 4y + 4 = (y-2)^2$$

$$\frac{4x^2}{16} - \frac{(y-2)^2}{16} = \frac{16}{16} \qquad \text{Divide each side by 16.}$$

$$\frac{x^2}{4} - \frac{(y-2)^2}{16} = 1 \qquad \text{Simplify each term.}$$

This is the equation of a hyperbola with center at (0, 2). The graph opens to the right and left as shown in Figure 9.

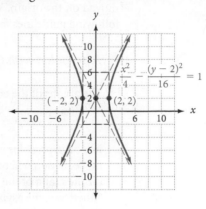

FIGURE 9

After reading through the preceding section, respond in your own words and in complete sentences.

A. How do we find the *x*-intercepts of a graph from the equation?

B. What is an ellipse?

C. How can you tell by looking at an equation if its graph will be an ellipse or a hyperbola?

D. Are the points on the asymptotes of a hyperbola in the solution set of the equation of the hyperbola? (That is, are the asymptotes actually part of the graph?) Explain.

Graph each of the following. Be sure to label both the x- and y-intercepts.

1. $\dfrac{x^2}{9} + \dfrac{y^2}{16} = 1$ **2.** $\dfrac{x^2}{25} + \dfrac{y^2}{4} = 1$ **3.** $\dfrac{x^2}{16} + \dfrac{y^2}{9} = 1$

4. $\dfrac{x^2}{4} + \dfrac{y^2}{25} = 1$ **5.** $\dfrac{x^2}{3} + \dfrac{y^2}{4} = 1$ **6.** $\dfrac{x^2}{4} + \dfrac{y^2}{3} = 1$

7. $4x^2 + 25y^2 = 100$ **8.** $4x^2 + 9y^2 = 36$ **9.** $x^2 + 8y^2 = 16$

10. $12x^2 + y^2 = 36$

Graph each of the following. Show the intercepts and the asymptotes in each case.

11. $\dfrac{x^2}{9} - \dfrac{y^2}{16} = 1$ **12.** $\dfrac{x^2}{25} - \dfrac{y^2}{4} = 1$ **13.** $\dfrac{x^2}{16} - \dfrac{y^2}{9} = 1$

14. $\dfrac{x^2}{4} - \dfrac{y^2}{25} = 1$ **15.** $\dfrac{y^2}{9} - \dfrac{x^2}{16} = 1$ **16.** $\dfrac{y^2}{25} - \dfrac{x^2}{4} = 1$

17. $\dfrac{y^2}{36} - \dfrac{x^2}{4} = 1$ **18.** $\dfrac{y^2}{4} - \dfrac{x^2}{36} = 1$ **19.** $x^2 - 4y^2 = 4$

20. $y^2 - 4x^2 = 4$ **21.** $16y^2 - 9x^2 = 144$ **22.** $4y^2 - 25x^2 = 100$

Find the x- and y-intercepts, if they exist, for each of the following. Do not graph.

23. $0.4x^2 + 0.9y^2 = 3.6$ **24.** $1.6x^2 + 0.9y^2 = 14.4$ **25.** $\dfrac{x^2}{0.04} - \dfrac{y^2}{0.09} = 1$

26. $\dfrac{y^2}{0.16} - \dfrac{x^2}{0.25} = 1$ **27.** $\dfrac{25x^2}{9} + \dfrac{25y^2}{4} = 1$ **28.** $\dfrac{16x^2}{9} + \dfrac{16y^2}{25} = 1$

Graph each of the following ellipses. In each case, label the coordinates of the center and the vertices.

29. $\dfrac{(x-4)^2}{4} + \dfrac{(y-2)^2}{9} = 1$ **30.** $\dfrac{(x-2)^2}{4} + \dfrac{(y-4)^2}{9} = 1$

31. $4x^2 + y^2 - 4y - 12 = 0$ **32.** $4x^2 + y^2 - 24x - 4y + 36 = 0$

33. $x^2 + 9y^2 + 4x - 54y + 76 = 0$ **34.** $4x^2 + y^2 - 16x + 2y + 13 = 0$

Graph each of the following hyperbolas. In each case, label the coordinates of the center and the vertices and show the asymptotes.

35. $\dfrac{(x-2)^2}{16} - \dfrac{y^2}{4} = 1$ **36.** $\dfrac{(y-2)^2}{16} - \dfrac{x^2}{4} = 1$

37. $9y^2 - x^2 - 4x + 54y + 68 = 0$ **38.** $4x^2 - y^2 - 24x + 4y + 28 = 0$

39. $4y^2 - 9x^2 - 16y + 72x - 164 = 0$ **40.** $4x^2 - y^2 - 16x - 2y + 11 = 0$

41. Find x when $y = 4$ in the equation $\dfrac{x^2}{25} + \dfrac{y^2}{16} = 1$.

42. Find x when $y = 3$ in the equation $\dfrac{x^2}{25} + \dfrac{y^2}{16} = 1$.

43. Find y when $x = -3$ in the equation $\dfrac{x^2}{9} + \dfrac{y^2}{16} = 1$.

44. Find y when $x = -2$ in the equation $\dfrac{x^2}{9} + \dfrac{y^2}{16} = 1$.

45. The longer line segment connecting opposite vertices of an ellipse is called the *major axis* of the ellipse. Give the length of the major axis of the ellipse you graphed in Problem 3.

46. The shorter line segment connecting opposite vertices of an ellipse is called the *minor axis* of the ellipse. Give the length of the minor axis of the ellipse you graphed in Problem 3.

APPLYING THE CONCEPTS

Some of the problems that follow use the major and minor axes mentioned in Problems 45 and 46. The diagram below shows the minor axis and the major axis for an ellipse.

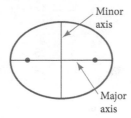

In any ellipse, the length of the major axis is 2a, and the length of the minor axis is 2b (these are the same a and b that appear in the general equations of an ellipse). Each of the two points shown on the major axis is a focus of the ellipse. If the distance from the center of the ellipse to each focus is c, then it is always true that $a^2 = b^2 + c^2$. You will need this information for some of the problems that follow.

47. The Colosseum The Colosseum in Rome seated 50,000 spectators around a central elliptical arena. The base of the Colosseum measured 615 feet long and 510 feet wide. Write an equation for the elliptical shape of the Colosseum.

48. Archway A new theme park is planning an archway at its main entrance. The arch is to be in the form of a semi-ellipse with the major axis as the span. If the span is to be 40 feet and the height at the center is to be 10 feet, what is the equation of the ellipse? How far left and right of center could a 6-foot man walk upright under the arch?

49. The Ellipse President's Park, located between the White House and the Washington Monument in Washington, DC, is also called The Ellipse. The park is enclosed by an elliptical path with major axis 458 meters and minor axis 390 meters. What is the equation for the path around The Ellipse?

50. **Garden Trellis** John is planning to build an arched trellis for the entrance to his botanical garden. If the arch is to be in the shape of the upper half of an ellipse that is 6 feet wide at the base and 9 feet high, what is the equation for the ellipse?

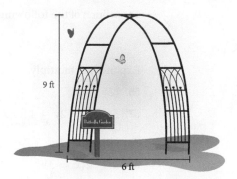

51. **Elliptical Pool Table** A children's science museum plans to build an elliptical pool table to demonstrate that a ball rolled from a particular point (focus) will always go into a hole located at another particular point (the other focus). The focus needs to be 1 foot from the vertex of the ellipse. If the table is to be 8 feet long, how wide should it be? *Hint:* The distance from the center to each focus point is represented by c and is found by using the equation $a^2 = b^2 + c^2$.

52. **Entering the Zoo** A zoo is planning a new entrance. Visitors are to be "funneled" into the zoo between two tall brick fences. The bases of the fences will be in the shape of a hyperbola. The narrowest passage east and west between the fences will be 24 feet. The total north–south distance of the fences is to be 50 feet. Write an equation for the hyperbolic shape of the fences if the center of the hyperbola is to be placed at the origin of the coordinate system.

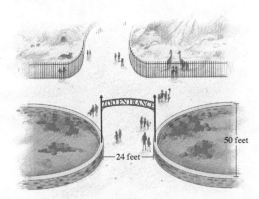

GETTING READY FOR THE NEXT SECTION

53. Which of the following are solutions to $x^2 + y^2 < 16$?

$$(0, 0) \quad (4, 0) \quad (0, 5)$$

54. Which of the following are solutions to $y \geq x^2 - 16$?

$$(0, 0) \quad (-2, 0) \quad (0, -2)$$

Expand and multiply.

55. $(2y + 4)^2$

56. $(y + 3)^2$

57. Solve $x - 2y = 4$ for x.

58. Solve $2x + 3y = 6$ for y.

Simplify.

59. $x^2 - 2(x^2 - 3)$

60. $x^2 + (x^2 - 4)$

Factor.

61. $5y^2 + 16y + 12$

62. $3x^2 + 17x - 28$

Solve.

63. $y^2 = 4$

64. $x^2 = 25$

65. $-x^2 + 6 = 2$

66. $5y^2 + 16y + 12 = 0$

EXTENDING THE CONCEPTS

The equation of the tangent line to the ellipse $\frac{x^2}{a^2} + \frac{y^2}{b^2} = 1$ at the point $P(x_1, y_1)$ is given by the equation $\frac{x_1 x}{a^2} + \frac{y_1 y}{b^2} = 1$. Use this fact to find the tangent line to each ellipse at the given point.

67. $x^2 + 3y^2 = 12$, point $= (-3, 1)$

68. $x^2 + 3y^2 = 84$, point $= (3, -5)$

The equation of the tangent line to the hyperbola $\frac{x^2}{a^2} - \frac{y^2}{b^2} = 1$ at the point $P(x_1, y_1)$ is given by the equation $\frac{x_1 x}{a^2} - \frac{y_1 y}{b^2} = 1$. Use this fact to find the tangent line to each hyperbola at the given point.

69. $5x^2 - 4y^2 = 80$, point $= (-6, 5)$

70. $3x^2 - y^2 = 12$, point $= (4, -6)$

SECOND-DEGREE INEQUALITIES AND NONLINEAR SYSTEMS

OBJECTIVES

A Graph second-degree inequalities.

B Solve systems of nonlinear equations.

C Graph the solution sets to systems of inequalities.

© sxc.hu/profile/somadjinn, Nicolas Raymond, 2011

10.3

The tarantula is a very large arachnid often kept as a pet. Venom vents through his two curved fangs, but is rarely hazardous to humans. As a defense mechanism, the tarantula's abdomen is covered with a layer of barbed hairs called urticating hairs. The tarantula uses his back pair of legs to flick these hairs off his abdomen towards a threatening target. Tarantula owners are advised to wear eye protection when handling the spider. One such owner learned this lesson the difficult way. The man was cleaning a stain on the glass wall of the tarantula's tank. He turned toward the spider, who hit him in the face with the hairs. A few of the hairs lodged in the man's cornea and began to irritate his eye. Doctors treated the man's condition with antibiotics after being unable to extract the hairs.

We can use this scenario to learn about systems of nonlinear equations. Suppose we can picture the man's eye as a circle given by the equation $x^2 + y^2 = 9$, and the projection of the hairs given by the equation $y = -\frac{1}{3}x + 1$. The solution for this system can be used to picture where the hairs lodged in the man's eye. In this section, we will learn how to solve systems of nonlinear equations, such as this one. But first, let's explore second-degree inequalities.

 Second-Degree Inequalities

A *second-degree inequality* is an inequality that contains at least one squared variable.

10.3 VIDEOS

EXAMPLE 1

Graph $x^2 + y^2 < 16$.

SOLUTION The boundary is $x^2 + y^2 = 16$, which is a circle with center at the origin and a radius of 4. Because the inequality sign is $<$, the boundary is not included in the solution set and must therefore be represented with a broken line. The graph of the boundary is shown in Figure 1.

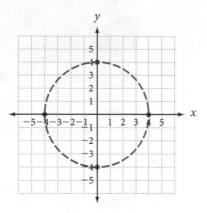

FIGURE 1

The solution set for $x^2 + y^2 < 16$ is either the region inside the circle or the region outside the circle. To see which region represents the solution set, we choose a convenient point not on the boundary and test it in the original inequality. The origin $(0, 0)$ is a convenient point. Because the origin satisfies the inequality $x^2 + y^2 < 16$, all points in the same region will also satisfy the inequality. The graph of the solution set is shown in Figure 2.

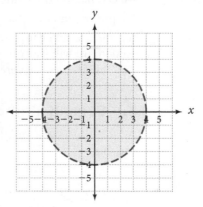

FIGURE 2

EXAMPLE 2

Graph the inequality $y \leq x^2 - 2$.

SOLUTION   The parabola $y = x^2 - 2$ is the boundary and is included in the solution set. Using $(0, 0)$ as the test point, we see that $0 \leq 0^2 - 2$ is a false statement, which means that the region containing $(0, 0)$ is not in the solution set. (See Figure 3.)

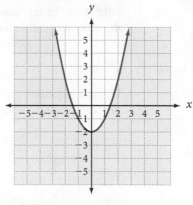

FIGURE 3

EXAMPLE 3

Graph $4y^2 - 9x^2 < 36$.

SOLUTION   The boundary is the hyperbola $4y^2 - 9x^2 = 36$ and is not included in the solution set. Testing $(0, 0)$ in the original inequality yields a true statement, which means that the region containing the origin is the solution set. (See Figure 4.)

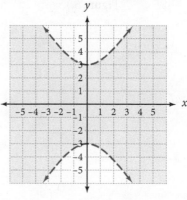

FIGURE 4

B Nonlinear Systems

In an earlier chapter, we worked with equations of linear systems. Now we will practice solving nonlinear systems, which include at least one nonlinear equation. A *nonlinear equation* is an equation whose graph is something other than a line, such as a circle, ellipse, hyperbola, or a parabola. In most cases, we can use the substitution method to solve nonlinear systems. However, we can use the elimination method if like variables are raised to the same power in both equations.

EXAMPLE 4

Solve the system.

$$x^2 + y^2 = 4$$

$$x - 2y = 4$$

SOLUTION In this case, the substitution method is the most convenient. Solving the second equation for x in terms of y, we have

$$x - 2y = 4$$

$$x = 2y + 4$$

We now substitute $2y + 4$ for x in the first equation in our original system and proceed to solve for y:

$(2y + 4)^2 + y^2 = 4$	
$4y^2 + 16y + 16 + y^2 = 4$	Expand $(2y + 4)^2$.
$5y^2 + 16y + 16 = 4$	Simplify left side.
$5y^2 + 16y + 12 = 0$	Add -4 to each side.
$(5y + 6)(y + 2) = 0$	Factor.
$5y + 6 = 0$ or $y + 2 = 0$	Set factors equal to 0.
$y = -\dfrac{6}{5}$ $y = -2$	Solve.

These are the y-coordinates of the two solutions to the system. Substituting $y = -\frac{6}{5}$ into $x - 2y = 4$ and solving for x gives us $x = \frac{8}{5}$. Using $y = -2$ in the same equation yields $x = 0$. The two solutions to our system are $\left(\frac{8}{5}, -\frac{6}{5}\right)$ and $(0, -2)$. Although graphing the system is not necessary, it does help us visualize the situation. (See Figure 5.)

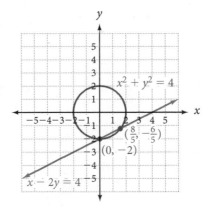

FIGURE 5

EXAMPLE 5

Solve the system.

$$16x^2 - 4y^2 = 64$$

$$x^2 + y^2 = 9$$

SOLUTION Because each equation is of the second degree in both x and y, it is easier to solve this system by eliminating one of the variables by addition. To eliminate y, we multiply the bottom equation by 4 and add the result to the top equation:

$$\begin{array}{r} 16x^2 - 4y^2 = 64 \\ 4x^2 + 4y^2 = 36 \\ \hline 20x^2 = 100 \end{array}$$

$$x^2 = 5$$

$$x = \pm\sqrt{5}$$

The x-coordinates of the points of intersection are $\sqrt{5}$ and $-\sqrt{5}$. We substitute each back into the second equation in the original system and solve for y:

When

$$x = \sqrt{5}$$

$$(\sqrt{5})^2 + y^2 = 9$$

$$5 + y^2 = 9$$

$$y^2 = 4$$

$$y = \pm 2$$

When

$$x = -\sqrt{5}$$

$$(-\sqrt{5})^2 + y^2 = 9$$

$$5 + y^2 = 9$$

$$y^2 = 4$$

$$y = \pm 2$$

The four points of intersection are $(-\sqrt{5}, 2)$, $(-\sqrt{5}, -2)$, $(\sqrt{5}, 2)$, and $(\sqrt{5}, -2)$. Graphically the situation is as shown in Figure 6.

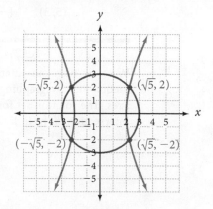

FIGURE 6

EXAMPLE 6

Solve the system.

$$x^2 - 2y = 2$$

$$y = x^2 - 3$$

SOLUTION   We can solve this system using the substitution method. Replacing y in the first equation with $x^2 - 3$ from the second equation, we have

$$x^2 - 2(x^2 - 3) = 2$$

$$-x^2 + 6 = 2$$

$$x^2 = 4$$

$$x = \pm 2$$

Using either $+2$ or -2 in the equation $y = x^2 - 3$ gives us $y = 1$. The system has two solutions: $(2, 1)$ and $(-2, 1)$.

EXAMPLE 7

The sum of the squares of two numbers is 34. The difference of their squares is 16. Find the two numbers.

SOLUTION   Let x and y be the two numbers. The sum of their squares is $x^2 + y^2$, and the difference of their squares is $x^2 - y^2$. (We can assume here that x^2 is the larger number.) The system of equations that describes the situation is

$$x^2 + y^2 = 34$$

$$x^2 - y^2 = 16$$

We can eliminate y by simply adding the two equations. The result of doing so is

$$2x^2 = 50$$

$$x^2 = 25$$

$$x = \pm 5$$

Substituting $x = 5$ into either equation in the system gives $y = \pm 3$. Using $x = -5$ gives the same results, $y = \pm 3$. The four pairs of numbers that are solutions to the original problem are

$$(5, 3) \qquad (-5, 3) \qquad (5, -3) \qquad (-5, -3)$$

We now turn our attention to systems of inequalities. To solve a system of inequalities by graphing, we simply graph each inequality on the same set of axes. The solution set for the system is the region common to both graphs — the intersection of the individual solution sets.

Graphing Systems of Inequalities

EXAMPLE 8

Graph the solution set for the system

$$x^2 + y^2 \leq 9$$

$$\frac{x^2}{4} + \frac{y^2}{25} \geq 1$$

SOLUTION   The boundary for the top inequality is a circle with center at the origin and a radius of 3. The solution set lies inside the boundary. The boundary for the second inequality is an ellipse. In this case, the solution set lies outside the boundary. (See Figure 7.)

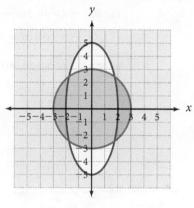

FIGURE 7

The solution set is the intersection of the two individual solution sets.

GETTING READY FOR CLASS

After reading through the preceding section, respond in your own words and in complete sentences.

A. What is the significance of a broken line when graphing inequalities?

B. Describe, in words, the set of points described by $(x - 3)^2 + (y - 2)^2 < 9$.

C. When solving the nonlinear systems whose graphs are a line and a circle, how many possible solutions can you expect?

D. When solving the nonlinear systems whose graphs are both circles, how many possible solutions can you expect?

Problem Set 10.3

Graph each of the following inequalities.

1. $x^2 + y^2 \leq 49$

2. $x^2 + y^2 < 49$

3. $(x - 2)^2 + (y + 3)^2 < 16$

4. $(x + 3)^2 + (y - 2)^2 \geq 25$

5. $y < x^2 - 6x + 7$

6. $y \geq x^2 + 2x - 8$

7. $\dfrac{x^2}{25} - \dfrac{y^2}{9} \geq 1$

8. $\dfrac{x^2}{25} - \dfrac{y^2}{9} \leq 1$

9. $4x^2 + 25y^2 \leq 100$

10. $25x^2 - 4y^2 > 100$

11. $\dfrac{(x + 2)^2}{25} + \dfrac{(y - 1)^2}{9} \leq 1$

12. $\dfrac{(x - 1)^2}{16} - \dfrac{(y + 1)^2}{16} < 1$

13. $16x^2 - 9y^2 \geq 144$

14. $16y^2 - 9x^2 < 144$

15. $9x^2 + 4y^2 + 36x - 8y + 4 < 0$

16. $9x^2 - 4y^2 + 36x + 8y \geq 4$

17. $9y^2 - x^2 + 18y + 2x > 1$

18. $x^2 + y^2 - 6x - 4y \leq 12$

Graph the solution sets to the following systems.

19. $x^2 + y^2 < 9$
  $y \geq x^2 - 1$

20. $x^2 + y^2 \leq 16$
  $y < x^2 + 2$

21. $\dfrac{x^2}{9} + \dfrac{y^2}{25} \leq 1$
  $\dfrac{x^2}{4} - \dfrac{y^2}{9} > 1$

22. $\dfrac{x^2}{4} + \dfrac{y^2}{16} \geq 1$
  $\dfrac{x^2}{9} - \dfrac{y^2}{25} < 1$

23. $4x^2 + 9y^2 \leq 36$
  $y > x^2 + 2$

24. $9x^2 + 4y^2 \geq 36$
  $y < x^2 + 1$

25. $x + y \leq 3$
  $x - 3y \leq 3$

26. $x - y \leq 4$
  $x + 2y \leq 4$

27. $x + y \leq 2$
  $-x + y \leq 2$

28. $x - y \leq 3$
  $-x - y \leq 3$

29. $x + y \leq 4$
  $x \geq 0$

30. $x - y \leq 2$
  $x \geq 0$

31. $2x + 3y \leq 6$
  $x \geq 0$

32. $x + 2y \leq 10$
  $3x + 2y \leq 12$

33. $x^2 + y^2 \leq 25$
  $\dfrac{x^2}{9} - \dfrac{y^2}{16} > 1$

34. $\dfrac{x^2}{16} + \dfrac{y^2}{25} < 1$
  $\dfrac{y^2}{4} - \dfrac{x^2}{1} \geq 1$

35. $x + y \leq 2$
  $y > x^2$

36. $\dfrac{x^2}{9} + \dfrac{y^2}{16} \leq 1$
  $\dfrac{x^2}{16} + \dfrac{y^2}{9} > 1$

Solve each of the following systems of equations.

37. $x^2 + y^2 = 9$
  $2x + y = 3$

38. $x^2 + y^2 = 9$
  $x + 2y = 3$

39. $x^2 + y^2 = 16$
  $x + 2y = 8$

40. $x^2 + y^2 = 16$
  $x - 2y = 8$

41. $x^2 + y^2 = 25$
  $x^2 - y^2 = 25$

42. $x^2 + y^2 = 4$
  $2x^2 - y^2 = 5$

43. $x^2 + y^2 = 9$
  $y = x^2 - 3$

44. $x^2 + y^2 = 4$
  $y = x^2 - 2$

45. $x^2 + y^2 = 16$
  $y = x^2 - 4$

46. $x^2 + y^2 = 1$
  $y = x^2 - 1$

47. $3x + 2y = 10$
  $y = x^2 - 5$

48. $4x + 2y = 10$
  $y = x^2 - 10$

49. $y = x^2 + 2x - 3$
  $y = \quad -x + 1$

50. $y = -x^2 - 2x + 3$
  $y = \quad\quad x - 1$

51. $y = x^2 - 6x + 5$
  $y = \quad\quad x - 5$

52. $y = x^2 - 2x - 4$
  $y = \quad\quad x - 4$

53. $4x^2 - 9y^2 = 36$
  $4x^2 + 9y^2 = 36$

54. $4x^2 + 25y^2 = 100$
  $4x^2 - 25y^2 = 100$

55. $x - y = 4$
  $x^2 + y^2 = 16$

56. $x + y = 2$
  $x^2 - y^2 = 4$

57. $2x^2 - y = 1$
  $x^2 + y = 7$

58. $x^2 + y^2 = 52$
  $y = x + 2$

59. $y = x^2 - 3$
  $y = x^2 - 2x - 1$

60. $y = 8 - 2x^2$
  $y = x^2 - 1$

61. $4x^2 + 5y^2 = 40$
  $4x^2 - 5y^2 = 40$

62. $x + 2y^2 = -4$
  $3x - 4y^2 = -3$

APPLYING THE CONCEPTS

63. Number Problem The sum of the squares of two numbers is 89. The difference of their squares is 39. Find the numbers.

64. Number Problem The difference of the squares of two numbers is 35. The sum of their squares is 37. Find the numbers.

65. Consider the equations for the three circles below. They are

Circle A	Circle B	Circle C
$(x + 8)^2 + y^2 = 64$	$x^2 + y^2 = 64$	$(x - 8)^2 + y^2 = 64$

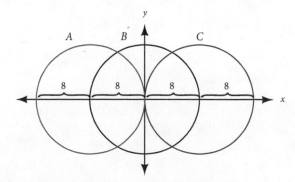

a. Find the points of intersection of circles A and B.

b. Find the points of intersection of circles B and C.

66. A magician is holding two rings that seem to lie in the same plane and intersect in two points. Each ring is 10 inches in diameter. If a coordinate system is placed with its origin at the center of the first ring and the x-axis contains the center of the second ring, then the equations are as follows:

$$\text{First Ring} \qquad \text{Second Ring}$$
$$x^2 + y^2 = 25 \qquad (x - 5)^2 + y^2 = 25$$

Find the points of intersection of the two rings.

MAINTAINING YOUR SKILLS

Expand and simplify.

67. $(x + 2)^4$ **68.** $(x - 2)^4$ **69.** $(2x + y)^3$ **70.** $(x - 2y)^3$

71. Find the first two terms in the expansion of $(x + 3)^{50}$.

72. Find the first two terms in the expansion of $(x - y)^{75}$.

EXTENDING THE CONCEPTS

Solve each of the following systems of equations.

73. $xy = 9$ **74.** $xy = 9$

      $x - y = 8$ $x - 2y = 7$

75. $\dfrac{3}{x} - \dfrac{4}{y} = 12$ **76.** $\dfrac{2}{x} - \dfrac{1}{y} = 4$

      $\dfrac{2}{x} + \dfrac{3}{y} = 6$ $\dfrac{3}{x} + \dfrac{2}{y} = 2$

77. Find where the line $y = 9x$ intersects the parabola $y = x^2$.

78. Find where the line $y = mx$ intersects the parabola $y = m^2x^2$. Assume m $\neq$ 0.

SPOTLIGHT ON SUCCESS

Nathan | student instructor

*Keep steadily before you the fact that
all true success depends at last upon yourself.*

—Theodore T. Hunger

Math has always come fairly easily for me and is the academic subject I have enjoyed most. I knew I wanted to attend Cal Poly San Luis Obispo for its high job placement and prestige, but I had no idea what I wanted to study. I decided to major in Mathematics because it is so universal but not so specialized or concentrated that I would get stuck in a field that I did not enjoy. I felt that if I kept studying math and its related fields, I would set myself up to be successful later in life, as math is the foundation for engineering, physics, and other science-related fields. I have not looked back on my decision. I know it will be a degree that I am proud to have achieved.

I appreciate the consistency that math offers in its problems and in its solutions. I like that math can be simplified into smaller easier-to-understand parts, and its answers are almost always definite. It provides challenges that I enjoy solving, like completing a puzzle piece by piece. In the end, I am able to enjoy the success I have put together for myself.

1. The distance between $(5, 2)$ and $(-1, 1)$ is
$$d = \sqrt{(5 + 1)^2 + (2 - 1)^2} = \sqrt{37}$$

Distance Formula [10.1]

The distance between the two points (x_1, y_1) and (x_2, y_2) is given by the formula
$$d = \sqrt{(x_2 - x_1)^2 + (y_2 - y_1)^2}$$

The Circle [10.1]

2. The graph of the circle $(x - 3)^2 + (y + 2)^2 = 25$ will have its center at $(3, -2)$ and the radius will be 5.

The graph of any equation of the form
$$(x - a)^2 + (y - b)^2 = r^2$$
will be a circle having its center at (a, b) and a radius of r.

The Ellipse [10.2]

3. The ellipse $\frac{x^2}{9} + \frac{y^2}{4} = 1$ will cross the x-axis at 3 and -3 and will cross the y-axis at 2 and -2.

Any equation that can be put in the form

$$\frac{x^2}{a^2} + \frac{y^2}{b^2} = 1$$

will have an ellipse for its graph. The x-intercepts will be at a and $-a$, and the y-intercepts will be at b and $-b$.

The Hyperbola [10.2]

4. The hyperbola $\frac{x^2}{4} - \frac{y^2}{9} = 1$ will cross the x-axis at 2 and -2. It will not cross the y-axis.

The graph of an equation that can be put in either of the forms

$$\frac{x^2}{a^2} - \frac{y^2}{b^2} = 1 \quad \text{or} \quad \frac{y^2}{a^2} - \frac{x^2}{b^2} = 1$$

will be a hyperbola. The x-intercepts for the first equation will be at a and $-a$. The y-intercepts for the second equation will be at a and $-a$. Two straight lines, called *asymptotes*, are associated with the graph of every hyperbola. Although the asymptotes are not part of the hyperbola, they are useful in sketching the graph.

Second-Degree Inequalities in Two Variables [10.3]

5. The graph of the inequality
$$x^2 + y^2 < 9$$
is all points inside the circle with center at the origin and radius 3. The circle itself is not part of the solution and therefore is shown with a broken curve.

We graph second-degree inequalities in two variables in much the same way that we graphed linear inequalities; that is, we begin by graphing the boundary, using a solid curve if the boundary is included in the solution (this happens when the inequality symbol is $\geq$ or $\leq$) or a broken curve if the boundary is not included in the solution (when the inequality symbol is $>$ or $<$). After we have graphed the boundary, we choose a test point that is not on the boundary and try it in the original inequality. A true statement indicates we are in the region of the solution. A false statement indicates we are not in the region of the solution.

Systems of Nonlinear Equations [10.3]

6. We can solve the system

$$x^2 + y^2 = 4$$
$$x = 2y + 4$$

by substituting $2y + 4$ from the second equation for x in the first equation:

$$(2y + 4)^2 + y^2 = 4$$
$$4y^2 + 16y + 16 + y^2 = 4$$
$$5y^2 + 16y + 12 = 0$$
$$(5y + 6)(y + 2) = 0$$

$$y = -\frac{6}{5} \quad \text{or} \quad y = -2$$

Substituting these values of y into the second equation in our system gives $x = \frac{8}{5}$ and $x = 0$. The solutions are $\left(\frac{8}{5}, -\frac{6}{5}\right)$ and $(0, -2)$.

A system of nonlinear equations is two equations, at least one of which is not linear, considered at the same time. The solution set for the system consists of all ordered pairs that satisfy both equations. In most cases we use the substitution method to solve these systems; however, the addition method can be used if like variables are raised to the same power in both equations. It is sometimes helpful to graph each equation in the system on the same set of axes to anticipate the number and approximate positions of the solutions.

Chapter 10 Review

Find the distance between the points. [10.1]

1. $(4, -6)$ and $(-1, 6)$ **2.** $(-4, 3)$ and $(2, -5)$

3. $(4, 7)$ and $(8, 1)$ **4.** $(-8, 9)$ and $(-3, -2)$

5. Find y so that $(-5, y)$ is $5\sqrt{2}$ units from $(2, 4)$. [10.1]

6. Find x so the distance between $(-2, 3)$ and $(x, 1)$ is 3. [10.1]

7. Give the equation of the circle with center at $(-4, 2)$ and radius 6. [10.1]

8. Give the equation of the circle with center at $(0, -1)$ and radius 7. [10.1]

9. Give the equation of the circle with center at the origin that contains the point $(9, 12)$. [10.1]

10. A circle with center at $(2, 5)$ passes through the point $(-1, 4)$. Find the equation. [10.1]

11. Find the equation of the circle with center at $(4, -1)$, which passes through the point $(6, -5)$. [10.1]

12. Find the equation of the circle with center on the x-axis and x-intercepts at -8 and 2. [10.1]

13. Find the circumference and area of the circle
$$(x + 2)^2 + (y + 6)^2 = 12$$
Leave your answer in terms of π. [10.1]

14. Find the circumference and area of the circle
$$x^2 + y^2 - 6x + 2y = 6$$
Leave your answer in terms of π. [10.1]

Find the center and radius of the circle. [10.1]

15. $x^2 + (y + 5)^2 = 9$ **16.** $x^2 + y^2 - 6x + 10y = 2$

Graph each of the following. [10.1, 10.2, 10.3, 10.4]

17. $(x - 3)^2 + (y + 1)^2 = 16$ **18.** $\dfrac{x^2}{81} + \dfrac{y^2}{36} = 1$

19. $x^2 + (y + 2)^2 = 11$ **20.** $x^2 + y^2 - 4y = 5$

21. $4x^2 + 9y^2 = 36$ **22.** $12x^2 + y^2 = 36$

23. $\dfrac{y^2}{4} - \dfrac{x^2}{36} = 1$ **24.** $4y^2 - 25x^2 = 100$

25. $9x^2 - y^2 = 81$

26. $36x^2 + 144x + 4y^2 - 24y + 36 = 0$

27. $(x - 3)^2 + (y + 1)^2 \geq 16$

28. $y > x^2 - 3$

Find the x- and y-coordinates for the x- and y-intercepts, if they exist, for each of the following equations. Do not graph. [10.3]

29. $\dfrac{y^2}{0.16} + \dfrac{x^2}{0.25} = 1$ **30.** $\dfrac{16x^2}{9} + \dfrac{16y^2}{25} = 1$

Graph each of the following ellipses. In each case, label the coordinates of the center and the vertices. [10.3]

31. $\dfrac{(x - 2)^2}{4} + \dfrac{(y - 4)^2}{9} = 1$

32. $4x^2 + y^2 - 24x - 4y + 36 = 0$

33. Find y when $x = -5$ in the equation $\dfrac{x^2}{25} + \dfrac{y^2}{16} = 1$.

34. Find y when $x = 4$ in the equation $\dfrac{x^2}{25} + \dfrac{y^2}{16} = 1$.

Find the x- and y-coordinates for the x- and y-intercepts, if they exist, for each of the following. Do not graph. [10.4]

35. $1.6x^2 - 0.9y^2 = 14.4$

36. $\dfrac{16x^2}{9} - \dfrac{16y^2}{25} = 1$

37. Find x when $y = -2$ in the equation $\dfrac{x^2}{25} - \dfrac{y^2}{16} = 1$.

38. Find the equations of the asymptotes of the hyperbola.
$$\frac{y^2}{25} - \frac{x^2}{16} = 1$$

Solve the following systems. [10.5]

39. $x^2 + y^2 = 25$
$15x + 3y = 15$

40. $x^2 + y^2 = \dfrac{25}{4}$
$y = x^2 - \dfrac{5}{2}$

Graph each of the following inequalities. [10.5]

41. $(x + 3)^2 + (y - 2)^2 \geq 25$

42. $y \geq x^2 + 2x - 8$

43. $25x^2 - 4y^2 > 100$

44. $16y^2 - 9x^2 < 144$

45. Graph the solution set to the system. [10.5]
$$x^2 + y^2 > 25$$
$$y \leq -x^2 + 6$$

Solve each of the following systems of equations. [10.5]

46. $x^2 + y^2 = 9$
$x + 2y = 3$

47. $x^2 + y^2 = 4$
$2x^2 - y^2 = 5$

48. $4x + 2y = 10$
$y = x^2 - 10$

49. $4x^2 + 25y^2 = 100$
$4x^2 - 25y^2 = 100$

Graph the solution sets to the following systems. [10.5]

50. $\dfrac{x^2}{4} + \dfrac{y^2}{16} \geq 1$
$\dfrac{x^2}{9} - \dfrac{y^2}{25} < 1$

51. $9x^2 + 4y^2 \geq 36$
$y < x^2 + 1$

52. $x - y \leq 4$
$x + 2y \leq 4$

53. $x - y \leq 3$
$-x - y \leq 3$

54. $x + 2y \leq 10$
$3x + 2y \leq 12$

55. $\dfrac{x^2}{9} + \dfrac{y^2}{16} \leq 1$
$\dfrac{x^2}{16} + \dfrac{y^2}{9} > 1$

Match each equation with its graph. [10.1, 10.2]

56. $x^2 + y^2 = 81$

57. $9x^2 + y^2 = 81$

58. $x^2 - y^2 = 9$

59. $x^2 + 9y^2 = 81$

A.
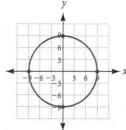

B.
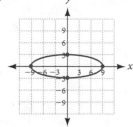

C.
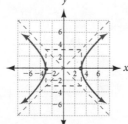

D.
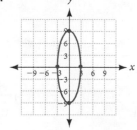

Find an equation for each graph. [10.1, 10.2]

60.

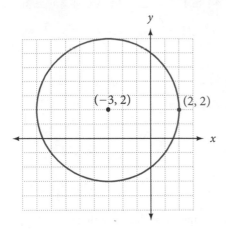

61.

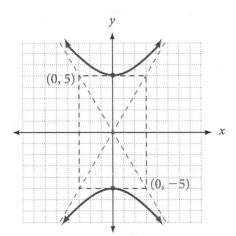

62.

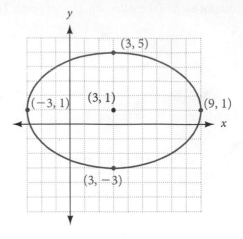

Simplify.

1. $\dfrac{-3 - 5}{6 - 7}$

2. $\dfrac{3(-2) + 4(-5)}{-8 - 5}$

3. $\dfrac{19a^4b^{-7}}{38a^{-3}b^4}$

4. $\dfrac{x^{5/3}y^2}{x^{1/6}y^{5/6}}$

5. $\dfrac{36x^3y^2 - 12x^2y^3 + 18x^4y}{6x^2y}$

6. $\dfrac{y^2 + 7y + 12}{y^2 - 9}$

7. $\log_6 36$

8. $\log_2 8$

Factor completely.

9. $ab^3 - b^3 + 4a - 4$

10. $7x^2 - 20x - 3$

Solve.

11. $4 + 2(3x - 8) - 7x = 3$

12. $6 - 7(2x + 1) = -8$

13. $(x - 2)(x + 5) = 8$

14. $1 - \dfrac{3}{x} = \dfrac{10}{x^2}$

15. $t - 6 = \sqrt{t + 14}$

16. $\sqrt{6x - 3} = -3$

17. $(7x + 2)^2 = -48$

18. $(x - 6)^2 = -5$

Solve and graph the solution on the number line.

19. $|3x + 6| + 1 > 7$

20. $|4x - 7| - 3 < 10$

Graph on a rectangular coordinate system.

21. $y < -\dfrac{1}{2}x + 1$

22. $2x - y < -2$

23. $y = -(x + 1)^2 + 3$

24. $y = (x - 2)^2 - 3$

Multiply.

25. $\dfrac{y^2 - 3y}{2y^2 + 12y + 10} \cdot \dfrac{2y + 10}{y^2 - 5y + 6}$

26. $\dfrac{x^2 - 9}{x^2 - 2x - 15} \cdot \dfrac{x^2 - 3x - 10}{x^3 + x^2 - 2x}$

27. $(3 - 2i)(1 + 5i)$

28. $(5 + i)(6 - 8i)$

Rationalize the denominator.

29. $\dfrac{3}{\sqrt{5} - \sqrt{2}}$

30. $\dfrac{3\sqrt{10}}{2\sqrt{10} + 1}$

Find the inverse, $f^{-1}(x)$.

31. $f(x) = \dfrac{1 - x}{3}$

32. $f(x) = 5x + 2$

33. Find x to the nearest hundredth if $\log x = 3.2164$.

34. Find $\log_{17} 13$ to the nearest hundredth.

35. Solve $2mn - 1 = nx + 4$ for x.

36. Solve $S = 3x^2 + 5xy - 2y$ for y.

37. Find the slope of the line through $(5, 4)$ and $(-6, 2)$.

38. Find the slope and y-intercept for $5x - 6y = 18$.

39. If $f(x) = -\dfrac{5}{4}x + 2$, find $f(16)$.

40. If $C(t) = 60\left(\dfrac{1}{2}\right)^{t/15}$, find $C(15)$ and $C(30)$.

41. Find an equation with solutions $x = \dfrac{1}{2}$ and $x = 4$.

42. Find an equation with solutions $t = -\dfrac{4}{3}$ and $t = \dfrac{1}{8}$.

43. If $f(x) = 2x + 5$ and $g(x) = 3 - x$, find $(f \circ g)(x)$.

44. If $f(x) = 9 - x^2$ and $g(x) = 3x + 2$, find $(g \circ f)(x)$.

45. Add -7 to the difference of -3 and 2.

46. Subtract 8 from the sum of -7 and 4.

47. Direct Variation y varies directly with x. If y is 36 when x is 12, find y when x is 7.

48. Inverse Variation y varies inversely with the square of x. If y is 2 when x is 4, find y when x is 5.

49. Mixture How much 40% alcohol solution and 90% alcohol solution must be mixed to get 50 gallons of 85% alcohol solution?

50. Mixture How many gallons of 10% alcohol solution and 70% alcohol solution must be mixed to get 8 gallons of 55% alcohol solution?

The chart shows the extent to which Americans say they know a foreign language. Use the information to answer Questions 51 and 52. Round to the nearest person.

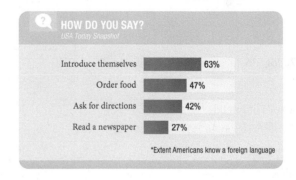

51. If 425,382 people were surveyed, how many say they can ask for directions in another language?

52. If 5,347,265 people were surveyed, how many say they cannot introduce themselves?

53. Unit Analysis Convert 3379.2 feet per minute to miles per hour.

54. Piecewise Functions Sketch the basic graph and the graph of the given function by hand on the same axes. Label the coordinates of two or three points on the graph of the given function.

$$f(x) = 2 + (x - 3)^2$$

1. Find x so that $(x, 2)$ is $2\sqrt{5}$ units from $(-1, 4)$. [10.1]

2. Give the equation of the circle with center at $(-2, 4)$ and radius 3. [10.1]

3. Give the equation of the circle with center at the origin that contains the point $(-3, -4)$. [10.1]

4. Find the center and radius of the circle. [10.1]

$$x^2 + y^2 - 10x + 6y = 5$$

Graph each of the following. [10.2, 10.3]

5. $4x^2 - y^2 = 16$

6. $\dfrac{x^2}{25} + \dfrac{y^2}{4} = 1$

7. $(x - 2)^2 + (y + 1)^2 \le 9$

8. $9x^2 + 4y^2 - 72x - 16y + 124 = 0$

Solve the following systems. [10.3]

9. $x^2 + y^2 = 25$
   $x^2 + y^2 = 16$

10. $2x + y = 5$
    $y = x^2 - 4$

Answers to Odd-Numbered Problems

Chapter 1

PROBLEM SET 1.1

1. $x + 5 = 2$ **3.** $6 - x = y$ **5.** $2t < y$ **7.** $x + y < x - y$ **9.** $\frac{1}{2}(t - s) < 2(t + s)$ **11. a.** 19 **b.** 27 **c.** 27
13. a. 16 **b.** 12 **c.** 18 **15. a.** 33 **b.** 33 **17. a.** 144 **b.** 74 **c.** 144 **19. a.** 23 **b.** 41 **c.** 65
21. a. 39 **b.** 7 **c.** 5 **23. a.** 32 **b.** 24 **25.** 11 **27.** 87 **29.** 4 **31.** 22 **33.** 16
35. 625 **37.** 1 **39.** $\frac{3}{5} = 0.6$ **41.** 8,700 **43.** 5.046 **45.** 5 **47.** 800 **49.** 625 **51.** 31 **53.** 1,000
55. 750 **57.** 0.5 **59.** 1.6 **61.** 320 **63.** 1.7917 **65.** 2.0793 **67.** 785 **69.** 185.12 **71.** 12.3106
73. -0.1587 **75.** 196 **77.** 650 **79. a.** 7 **b.** 10 **c.** 25 **d.** 32
81. a. 121 **b.** 121 **c.** 101 **d.** 81 **83. a.** 1 **b.** 2,400 **c.** 0.52 **85.** $A \cup B = \{0, 1, 2, 3, 4, 5, 6\}$
87. $A \cap B = \{2, 4\}$ **89.** $B \cap C = \{1, 3, 5\}$ **91.** $A \cup (B \cap C) = \{0, 1, 2, 3, 4, 5, 6\}$ **93.** $\{0, 2\}$ **95.** $\{0, 6\}$
97. $\{0, 1, 2, 3, 4, 5, 6, 7\}$ **99.** 1, 2 **101.** $-6, -5.2, 0, 1, 2, 2.3, \frac{9}{2}$ **103.** $-\sqrt{7}, -\pi, \sqrt{17}$ **105.** False **107.** True
109. True **111.** True **113.** $2 \cdot 7 \cdot 19$ **115.** $3 \cdot 37$ **117.** $3^2 \cdot 41$ **119.** $\frac{3}{7}$ **121.** $\frac{11}{21}$ **123.** $\frac{3}{5}$ **125.** $\frac{5}{9}$
127. $\frac{3}{4}$ **129.** 40 **131.** $\frac{5}{11}$ **133. a.** 2.9 seconds **b.** 7.5 seconds **135. a.** $5^1 + 8^1 = 5 + 8 = 13 = 13^1$
b. $2^1 + 3^1 = 2 + 3 = 5 = 5^1$ **c.** $3^2 + 4^2 = 9 + 16 = 25 = 5^2$ **d.** $7^2 + 24^2 = 49 + 576 = 625 = 25^2$ **137.** 720
139. The left side is 20. The right side is 8.

PROBLEM SET 1.2

	Number	Opposite	Reciprocal
1.	4	-4	$\frac{1}{4}$
3.	$-\frac{1}{2}$	$\frac{1}{2}$	-2
5.	5	-5	$\frac{1}{5}$
7.	$\frac{3}{8}$	$-\frac{3}{8}$	$\frac{8}{3}$
9.	$-\frac{1}{6}$	$\frac{1}{6}$	-6
11.	3	-3	$\frac{1}{3}$

13. Answers will vary **15.** 0 **17.** 2 **19.** $\frac{3}{4}$ **21.** π
23. -4 **25.** -2 **27.** $-\frac{3}{4}$ **29.** $\frac{21}{40}$ **31.** 2 **33.** $\frac{8}{27}$
35. $\frac{1}{10,000}$ **37.** $\frac{72}{385}$ **39.** 1 **41.** $6 + x$ **43.** $a + 8$
45. $15y$ **47.** x **49.** a **51.** x **53.** $3x + 18$ **55.** $12x + 8$
57. $15a + 10b$ **59.** $\frac{4}{3}x + 2$ **61.** $2 + y$ **63.** $40t + 8$
65. $9x + 3y - 6z$ **67.** $3x + 7y$ **69.** $6x + 7y$ **71.** $3x + 1$
73. $2x - 1$ **75.** $x + 2$ **77.** $a - 3$ **79.** $x + 24$ **81.** $3x - 2y$
83. $3x + 8y$ **85.** $8x + 5y$ **87.** $15x + 10$ **89.** $8y + 32$

91. $15t + 9$ **93.** $28x + 11$ **95.** $\frac{7}{15}$ **97.** $\frac{29}{35}$ **99.** $\frac{35}{144}$ **101.** $\frac{949}{1260}$ **103.** $\frac{47}{105}$ **105.** 15 **107.** $14a + 7$
109. $12x + 2$ **111.** $24a + 15$ **113.** $8x + 13$ **115.** $17x + 14y$ **117.** $17b + 9a$
119. Commutative property of addition **121.** Commutative property of multiplication **123.** Additive inverse
125. Commutative property of addition **127.** Associative and commutative properties of multiplication
129. Commutative and associative properties of addition **131.** Distributive property
133. a. Women **b.** Positive **c.** 2013 **135.** $x^2 + 8x$ **137.** False **139.** False

PROBLEM SET 1.3

1. 4 **3.** -4 **5.** -10 **7.** -4 **9.** $\frac{19}{12}$ **11.** $-\frac{32}{105}$ **13.** -8 **15.** -12 **17.** $-7x$ **19.** 13 **21.** -14
23. $6a$ **25.** -15 **27.** 15 **29.** -24 **31.** $-10x$ **33.** x **35.** y **37.** $-8x + 6$ **39.** $-3a + 4$ **41.** 39
43. 11 **45.** -5 **47.** 11 **49.** 7 **51.** -44 **53.** 2 **55.** $\frac{4}{3}$ **57.** 0 **59.** Undefined **61.** $-\frac{2}{3}$ **63.** $x - 5$
65. $x - 7$ **67.** $-6x + 9y$ **69.** $3x - 18$ **71.** $3a + 6$ **73.** $3x + 60$ **75.** $11y$ **77.** $14x + 12$ **79.** $7m - 15$
81. $-2x + 9$ **83.** $7y + 10$ **85.** $-20x + 5$ **87.** $-11x + 10$ **89.** $4x + 13$ **91.** Undefined **93.** 0 **95.** $-\frac{2}{3}$
97. 32 **99.** 64 **101.** $-\frac{1}{18}$ **103.** $\frac{5}{3}$ **105.** 11 **107.** 12

109.

		Sum	Difference	Product	Quotient
a	b	$a + b$	$a - b$	ab	$\dfrac{a}{b}$
3	12	15	-9	36	$\frac{1}{4}$
-3	12	9	-15	-36	$-\frac{1}{4}$
3	-12	-9	15	-36	$-\frac{1}{4}$
-3	-12	-15	9	36	$\frac{1}{4}$

111.

x	$3(5x - 2)$	$15x - 6$	$15x - 2$
-2	-36	-36	-32
-1	-21	-21	-17
0	-6	-6	-2
1	9	9	13
2	24	24	28

113. a. 1   **b.** $\frac{3}{2}$ **c.** -1 **d.** 135   **115.** 1.5282   **117.** -0.0794 **119.** -0.0714 **121.** 3.4   **123.** 1.6

125. 1,200   **127.** 190   **129.** Sante Fe 10:10 p.m.; Detroit 3:30 a.m.   **131. a.** 7:57:37 for Patrick and 8:55:16 for Lucy

b. 0:57:39   **c.** 0:29:26   **d.** 0:28:10

PROBLEM SET 1.4

1. 16   **3.** -16 **5.** -0.027 **7.** 32   **9.** $\frac{1}{8}$ **11.** $\frac{25}{36}$ **13.** x^9 **15.** 64   **17.** $-6a^6$ **19.** $\frac{1}{9}$ **21.** $-\frac{1}{32}$
23. $\frac{16}{9}$ **25.** 17   **27.** $80x^3y^6$ **29.** $72x^3y^3$ **31.** $15xy$ **33.** $12x^7y^6$ **35.** $54a^6b^2c^4$ **37.** $2x^3$ **39.** $4x$ **41.** $-4xy^4$
43. $-2x^2y^2$ **45.** $\frac{1}{x^{10}}$ **47.** a^{10} **49.** $\frac{1}{t^6}$ **51.** x^{12} **53.** x^{18} **55.** $\frac{1}{x^{22}}$ **57.** $\frac{a^3b^7}{4}$ **59.** $\frac{y^{38}}{x^{16}}$ **61.** $\frac{16y^{16}}{x^8}$ **63.** x^4y^6
65. 1   **67.** 3.78×10^5 **69.** 4.9×10^3 **71.** 3.7×10^{-4} **73.** 4.95×10^{-3} **75.** 5,340   **77.** 7,800,000   **79.** 0.00344
81. 0.49   **83.** 8×10^4 **85.** 2×10^9 **87.** 2.5×10^{-6} **89.** 1.8×10^{-7} **91.** 50

93.

Expanded Form	Scientific Notation $n \times 10^r$
0.000357	3.57×10^{-4}
0.00357	3.57×10^{-3}
0.0357	3.57×10^{-2}
0.357	3.57×10^{-1}
3.57	3.57×10^0
35.7	3.57×10^1
357	3.57×10^2
3,570	3.57×10^3
35,700	3.57×10^4

95.

Jupiter's Moon	Period (seconds)	
Io	153,000	1.53×10^5
Europa	307,000	3.07×10^5
Ganymede	618,000	6.18×10^5
Callisto	1,440,000	1.44×10^6

97. 6.3×10^8 **99. a.** 3.5×10^8 **b.** 2.0×10^8 **c.** 7.5×10^7
101. 4.22×10^{11} **103.** $\frac{1}{x^3}$ **105.** y^3 **107.** x^5
109. $x^{-6a + 5}$ or $\frac{1}{x^{6a - 5}}$ **111.** $x^{8m - 9}$ **113.** $x^{-2a + 12}$ or $\frac{1}{x^{2a - 12}}$
115. $x^{-4m + 9}$ or $\frac{1}{x^{4m - 9}}$ **117.** $x^{12a - 8}$

PROBLEM SET 1.5

1. Trinomial, degree 2, leading coefficient 5   **3.** Binomial, degree 1, leading coefficient 3
5. Trinomial, degree 2, leading coefficient 8   **7.** Polynomial, degree 3, leading coefficient 4
9. Monomial, degree 0, leading coefficient $-\frac{3}{4}$ **11.** Trinomial, degree 3, leading coefficient 6
13. $2x^2 + 7x - 15$ **15.** $x^2 + 9x + 4$ **17.** $a^2 + 5a + 4$ **19.** $x^2 - 13x + 3$ **21.** $-3x$ **23.** $x^2 - 6x - 5$
25. $10x - 5$ **27.** $9x - 35$ **29.** $15x - 64$ **31.** -2 **33.** 226   **35.** -15 **37.** $12x^3 - 10x^2 + 8x$
39. $a^3 - b^3$ **41.** $8x^3 + y^3$ **43.** $6a^2 + 13a + 6$ **45.** $20 - 2t - 6t^2$ **47.** $x^6 - 2x^3 - 15$ **49.** $18t^2 - \frac{2}{9}$
51. $4a^2 - 12a + 9$ **53.** $25x^2 + 20xy + 4y^2$ **55.** $25 - 30t^3 + 9t^6$ **57.** $4a^2 - 9b^2$ **59.** $9r^4 - 49s^2$
61. $x^3 - 6x^2 + 12x - 8$ **63.** $8x^3 - 12x^2 + 6x - 1$ **65.** $a^2b^2 + b^2 + 8a^2 + 8$ **67.** $3xy^2 + 4x - 6y^2 - 8$
69. $3x^2 + 12x + 14$ **71.** $24x$ **73.** $5x$ **75.** $8x^2$ **77.** $2x^3$ **79.** -5 **81.** $-2x + 3$ **83.** $9x + 6$
85. a. $3x - 3a$ **b.** $2x - 2a$ **87.** $2x^3 + 7x - 15$ **89.** $6x^3 - 11x^2y + 11xy^2 - 12y^3$ **91.** $-12x^4$ **93.** $20x^5$
95. a^6 **97.** $x - 5$ **99.** $x - 7$ **101.** $5x - 1$ **103.** $x - 1$ **105.** $3x + 8y$ **107.** $x^2 + 4x - 5$ **109.** $4a^2 - 30a + 56$
111. $32a^2 + 20a - 18$ **113.** 48 feet; 48 feet   **115.** $P + Pr + \frac{Pr^2}{4}$ **117.** $x^{4n} - 9$ **119.** $28x^{2n} - 13x^n - 6$
121. $x^{3n} + 6x^{2n} + 12x^n + 8$ **123.** $2x^{3n} + 7x^{2n} - 19x^n - 20$

PROBLEM SET 1.6

1. $5x^2(2x - 3)$ **3.** $9y^3(y^3 + 2)$ **5.** $3ab(3a - 2b)$ **7.** $7xy^2(3y^2 + x)$ **9.** $3(a - 2)(a - 5)$ **11.** $4x(x + 1)(x - 5)$
13. $(a - 2b)(5x - 3y)$ **15.** $3(x + y)^2(x^2 - 2y^2)$ **17.** $(x + 5)(2x + 3)(x + 2)$ **19.** $(x + 1)(3y + 2a)$

21. $(xy + 1)(x + 3)$ **23.** $(x - a)(x - b)$ **25.** $(b + 5)(a - 1)$ **27.** $(x + 3)(x + 4)$ **29.** $(x + 3)(x - 4)$

31. $(y + 3)(y - 2)$ **33.** $(2 - x)(8 + x)$ **35.** $(2 + x)(6 + x)$ **37.** $(x + 2y)(x + y)$ **39.** $(a + 6b)(a - 3b)$

41. $(x - 8a)(x + 6a)$ **43.** $(x - 6b)^2$ **45.** $(2x - 3)(x + 5)$ **47.** $(2x - 5)(x + 3)$ **49.** $(2x - 3)(x - 5)$

51. $(2x - 5)(x - 3)$ **53.** Prime   **55.** $(2 + 3a)(1 + 2a)$ **57.** $15(4y + 3)(y - 1)$ **59.** $x^2(3x - 2)(2x + 1)$

61. $10r(2r - 3)^2$ **63.** $(4x + y)(x - 3y)$ **65.** $(2x - 3a)(5x + 6a)$ **67.** $(3a + 4b)(6a - 7b)$ **69.** $200(1 + 2t)(3 - 2t)$

71. $y^2(3y - 2)(3y + 5)$ **73.** $2a^2(3 + 2a)(4 - 3a)$ **75.** $2x^2y^2(4x + 3y)(x - y)$ **77.** $100(3x^2 + 1)(x^2 + 3)$

79. $(x - 3)^2$ **81.** $(a - 6)^2$ **83.** $(5 - t)^2$ **85.** $\left(\frac{1}{2}x + 3\right)^2$ **87.** $(4a + 5b)^2$ **89.** $\left(\frac{1}{5} + \frac{1}{4}t^2\right)^2$ **91.** $(x + 5)^2$

93. $(7x + 8y)(7x - 8y)$ **95.** $\left(2a + \frac{1}{2}\right)\left(2a - \frac{1}{2}\right)$ **97.** $\left(x + \frac{3}{5}\right)\left(x - \frac{3}{5}\right)$ **99.** $\left(5 - \frac{t}{10}\right)\left(5 + \frac{t}{10}\right)$

101. $(4a^2 + 9)(2a + 3)(2a - 3)$ **103.** $(x - 5 + y)(x - 5 - y)$ **105.** $(a + 4 + 5b)(a + 4 - 5b)$ **107.** $(x + 2)(x + 5)(x - 5)$

109. $(2x + 3)(x + 2)(x - 2)$ **111.** $(x + 3)(2x + 3)(2x - 3)$ **113.** $(x - y)(x^2 + xy + y^2)$ **115.** $(a + 2)(a^2 - 2a + 4)$

117. $\left(\frac{1}{2}y - 4\right)\left(\frac{1}{4}y^2 + 2y + 16\right)$ **119.** $10(r - 5)(r^2 + 5r + 25)$ **121.** $(4 + 3a)(16 - 12a + 9a^2)$

123. $\left(5t + \frac{1}{3}\right)\left(25t^2 - \frac{5t}{3} + \frac{1}{9}\right)$ **125.** $h(t) = 16(6 - t)(1 + t)$; $h(6)$ ft $= 0$; $h(3)$ ft $= 192$

127. $(P + Pr) + (P + Pr)r = (P + Pr)(1 + r) = P(1 + r)(1 + r) = P(1 + r)^2$ **129.** $(2x^3 + 5y^2)(4x^3 + 3y^2)$

131. $(3x - 5)(x + 100)$ **133.** $(\frac{1}{4}x + 1)(\frac{1}{2}x + 2)$ **135.** $(2x + 0.5)(x + 0.5)$

PROBLEM SET 1.7

1. $(x - 6)(x + 4)$ **3.** $(x - 6)(x + 1)$ **5.** $(x - 3)(x - 2)$ **7.** $(x - 5)^2$ **9.** $(2x + 1)(x - 3)$

11. $(7x - 3)(3x - 2)$ **13.** $(x + 4)(x - 4)$ **15.** $(a + 1)(a - 1)$ **17.** $(a + 4b)(a - 4b)$ **19.** $(3x + 7)(3x - 7)$

21. $(4x^2 + 7)(4x^2 - 7)$ **23.** $(t + 3)(t - 3)(t^2 + 9)$ **25.** $(a + b)(a^2 - ab + b^2)$ **27.** $(x - 2)(x^2 + 2x + 4)$

29. $(x + 1)(x^2 - x + 1)$ **31.** $(2x + 1)(4x^2 - 2x + 1)$ **33.** $10(3x - 2)(2x - 3)$ **35.** $x(x + 3)(x + 2)$

37. $x(2x + 1)(x - 3)$ **39.** $x(x - 6)(x + 4)$ **41.** $6(x + 4)$ **43.** $100x(x - 3)$ **45.** $5(2a + 3)(2a - 3)$

47. $a(3a + 4)(3a - 4)$ **49.** $-2y(x + 3)(x - 2)$ **51.** $(a + 2)(x + 3)$ **53.** $(x - 3a)(x - 2)$ **55.** $(x + 2)(x + 3)(x - 3)$

57. $(x + 3)(x + 2)(x - 2)$ **59.** $(x + 3)(2x + 3)(2x - 3)$ **61.** $(2x + 1)(x + 3)(x - 3)$ **63.** $(4x + 1)(x - 8)$

65. prime   **67.** $5x(6x - 7)(5x + 8)$ **69.** $(12x - 5)(2x + 1)$ **71.** $(x - 1)(x^2 + x + 1)(x + 1)(x^2 - x + 1)$

73. $(a + 5)(x + 3)^2$ **75.** $3(x - 7)(2a + 5)(2a - 5)$ **77.** $\left(r + \frac{1}{3}\right)\left(r - \frac{1}{3}\right)$ **79.** $(t + 2 + y)(t + 2 - y)$

81. $\left(5t + \frac{1}{3}\right)\left(25t^2 - \frac{5}{3}t + \frac{1}{9}\right)$ **83.** $(15t + 16)(t - 1)$ **85.** $100(x - 3)(x + 2)$ **87.** $4x(x^2 + 4y^2)$

89. $(5 - t + x)(5 - t - x)$ **91.** $(5x + 7)(6x + 11)$ **93.** $A = P(1 + 0.08)$; \$540

95. $2(4 - t)(1 + 8t)$

97. $(x^{2n} + 1)(x^n + 1)(x^n - 1)$ **99.** $(x^n - 3)(x^{2n} + 3x^n + 9)$

101. $(x^{2n} + y^{2m})(x^n + y^m)(x^n - y^m)$

103. $(x^n - 1)(x^{2n} + x^n + 1)(x^{6n} + x^{3n} + 1)$

105. $(x^{2n} + 1)(x^{4n} - x^{2n} + 1)$

Time t (seconds)	0	1	2	3	4
Height h (feet)	8	54	68	50	0

PROBLEM SET 1.8

1. 2.7 miles   **3.** 742 miles per hour   **5.** 0.45 miles per hour   **7.** 19.5 miles per hour   **9.** 4.7 miles $= 10{,}000$ steps; Yes

11. Smallest: 1.1 acres, Rose Bowl: 1.7 acres, Largest: 1.9 acres   **13.** 2.4g   **15.** 120g   **17.** 1.0×10^{19} miles

CHAPTER 1 REVIEW

1. $\frac{24}{36}$ **2.** $\frac{28}{36}$ **3.** $\frac{54}{60}$ **4.** $-\frac{35}{60}$ **5.** $-\frac{16}{3}, \frac{3}{16}, \frac{16}{3}$ **6.** $5, -\frac{1}{5}, 5$ **7.** $\frac{3}{8}, -\frac{8}{3}, \frac{3}{8}$ **8.** $-a, \frac{1}{a}, |a|$

9. 7   **10.** $\frac{5}{6}$ **11.** $\sqrt{2}$ **12.** -5 **13.** $<$ **14.** $<$ **15.** $>$ **16.** $<$ **17.** 3   **18.** -5

19. -8 **20.** 9   **21.** 9   **22.** 2   **23.** $5 \cdot 7 \cdot 11$ **24.** $3 \cdot 5 \cdot 7^2$ **25.** $\frac{7}{13}$ **26.** $\frac{13}{31}$

27. $\frac{19}{15}$ **28.** 40   **29.** 12 cm; 6.75 cm^2 **30.** 5 in.; 1 in.^{2} **31.** Commutative property of multiplication

32. Associative and commutative properties of addition   **33.** Multiplicative inverse property

34. Distributive property   **35.** $2x + 15$ **36.** $7x + 2$ **37.** $9x + 8y$ **38.** $4a + 6b + 4$

39. $4x + 2y - 8$ **40.** $10x + 2$ **41.** $\frac{157}{168}$ **42.** $\frac{43}{48}$ **43.** $\frac{31}{15} = 2\frac{1}{15}$ **44.** $-\frac{4}{15}$ **45.** 4

46. $11a$ **47.** $-2 + 4x$ **48.** $-5x + 10$ **49.** $-5x$ **50.** -11 **51.** Undefined   **52.** 0

53. -59 **54.** 15   **55.**

x	$(x+1)^2$	x^2+1	x^2+2x+1
-2	1	5	1
-1	0	2	0
0	1	1	1
1	4	2	4
2	9	5	9

56. a. -44 **b.** 121 **c.** 0 **d.** 49   **57.** 16   **58.** -16 **59.** $\frac{9}{16}$ **60.** $-\frac{64}{125}$ **61.** $\frac{1}{4x^4}$ **62.** $27x^6y^6$

63. $\frac{27}{8}$ **64.** $-4x^5y^5$ **65.** $-21x^7y^3$ **66.** $\frac{6a^3}{b^3}$ **67.** $-\frac{4a^5}{5b}$ **68.** $4a^4b^8$ **69.** $\frac{6}{5}x^3 - \frac{6}{5}x^2 - \frac{8}{5}x - \frac{6}{5}$

70. $-14x - 58$ **71.** $-5x^2 - 31x + 28$ **72.** $6x^3 + 14x^2 - 27x + 10$ **73.** $9a^8 - 42a^4 + 49$

74. $4x^2 - 9$ **75.** $3x^3 - 17x^2 - 28x$ **76.** $14x^2 - \frac{1}{14}$ **77.** $(x-1)(x-5)$ **78.** $3(5x^2 - 4)(x^2 + 3)$

79. $(3x + 2y)(3x - 2y)(9x^2 + 4y^2)$ **80.** $(6x - y)(a + 3b^2)$ **81.** $(y - \frac{1}{3})(y^2 + \frac{1}{3}y + \frac{1}{9})$

82. $3x^2y^4(x - 3y)(x + 8y)$ **83.** 4.82×10^5 **84.** 7.28×10^6 **85.** 4.21×10^{-3} **86.** 5.26×10^{-2}

87. 0.00629   **88.** 3,290   **89.** 0.0631   **90.** 48,200,000   **91.** 3.0×10^{-3} **92.** 2.0×10^{-4}

CHAPTER 1 TEST

1. $2a - 3b < 2a + 3b$ **2.** $\{1, 2, 3, 4, 6\}$ **3.** $2 \cdot 5 \cdot 7 \cdot 11$ **4.** Commutative property of addition

5. Multiplicative identity property   **6.** -19 **7.** -149 **8.** 2   **9.** $\frac{59}{72}$ **10.** $-4x$ **11.** $-5x - 8$ **12.** $-\frac{5}{2}$

13. x^8 **14.** $\frac{1}{32}$ **15.** a^2 **16.** $\frac{2a^{12}}{b^{15}}$ **17.** 6.53×10^6 **18.** 3×10^8 **19.** $\frac{3}{4}x^3 - \frac{5}{4}x^2 - 2x - 1$ **20.** $4x + 75$

21. $6y^2 + y - 35$ **22.** $2x^3 + 3x^2 - 26x + 15$ **23.** $64 - 48t^3 + 9t^6$ **24.** $1 - 36y^2$ **25.** $2(3x^2 - 1)(2x^2 + 5)$

26. $(4a^2 + 9y^2)(2a + 3y)(2a - 3y)$ **27.** $(7a - b^2)(x^2 - 2y)$ **28.** $\left(t + \frac{1}{2}\right)\left(t^2 - \frac{1}{2}t + \frac{1}{4}\right)$

29. $(x - 5 + b)(x - 5 - b)$ **30.** $(9 + x^2)(3 + x)(3 - x)$ **31.** 742 mph   **32.** 1,012 mph

Chapter 2

PROBLEM SET 2.1

1. 3   **3.** $-\frac{4}{3}$ **5.** 7,000   **7.** -3 **9.** $-1, 6$ **11.** $-\frac{4}{3}, 0, \frac{4}{3}$ **13.** $-5, 1$ **15.** $0, -3, \frac{3}{2}$ **17.** $-3, -\frac{3}{2}, \frac{3}{2}$

19. a. $\frac{5}{8}$ **b.** $10x - 8$ **c.** $16x^2 - 34x + 15$ **d.** $\frac{3}{2}, \frac{5}{8}$ **21. a.** $\frac{25}{9}$ **b.** $-\frac{5}{3}, \frac{5}{3}$ **c.** $-3, 3$ **d.** $\frac{5}{3}$ **23.** $-\frac{9}{2}$ **25.** $0, 2, 3$

27. $-\frac{7}{640}$ **29.** $\frac{7}{10}$ **31.** 24   **33.** $\frac{46}{15}$ **35.** $-2, \frac{5}{3}$ **37.** $-3, 0$ **39.** 5   **41.** 4   **43.** 6,000   **45.** $-4, -2$ **47.** No solution

49. No solution   **51.** All real numbers   **53.** 30,000   **55.** 3,750 lbs   **57.** $\frac{1}{2}$ **59.** 62.5   **61.** 0   **63.** $\frac{5}{4}$ **65.** 1, 3   **67.** 13   **69.** 2

71. $-\frac{3}{2}$ **73.** $-\frac{21}{2}$ **75.** No solution   **77.** $-3, -2, 2, 3$

PROBLEM SET 2.2

1. -3 **3.** 0   **5.** $\frac{3}{2}$ **7.** 4   **9.** $\frac{8}{5}$ **11.** $-\frac{7}{640}$ **13.** 675   **15. a.** 3,400 **b.** 3,400   **17. a.** 23 **b.** 23   **19.** $c = 2$ **21.** 3

23. 2,400   **25.** $\frac{25}{3}$ **27.** $\frac{14}{3}$ **29.** $r = \frac{d}{t}$ **31.** $t = \frac{d}{r + c}$ **33.** $\ell = \frac{A}{w}$ **35.** $t = \frac{I}{pr}$ **37.** $T = \frac{PV}{nR}$ **39.** $x = \frac{y - b}{m}$ **41.** $F = \frac{9}{5}C + 32$

43. $v = \frac{h - 16t^2}{t}$ **45.** $d = \frac{A - a}{n - 1}$ **47.** $y = -\frac{2}{3}x + 2$ **49.** $y = \frac{3}{5}x + 3$ **51.** $y = \frac{1}{3}x + 2$ **53.** $x = \frac{5}{a - b}$ **55.** $h = \frac{S - \pi r^2}{2\pi r}$

57. $x = \frac{4}{3}y - 4$ **59.** $x = -\frac{10}{a - c}$ **61.** $y = \frac{1}{2}x + \frac{3}{2}$ **63.** $y = -2x - 5$ **65.** $y = -\frac{2}{3}x + 1$ **67.** $y = -\frac{1}{2}x + \frac{7}{2}$

69. a. $y = 4x - 1$ **b.** $y = -\frac{1}{2}x$ **c.** $y = -3$ **71.** $y = -\frac{1}{4}x + 2$ **73.** $y = \frac{3}{5}x - 3$

75. a. $-\frac{15}{4} = -3.75$ **b.** -7 **c.** $y = \frac{4}{5}x + 4$ **d.** $x = \frac{5}{4}y - 5$ **77.** $\frac{9}{5}$ tons   **79.** 2, 3 seconds   **81.** 6 miles per hour

83. 42 miles per hour   **85.** 6.8 feet per second   **87.** 3.1 mm^3 **89.** \$7 or \$10   **91.** 13,330 KB   **93.** 8

95. Shar: 128.4 beats per min, Sara: 140.4 beats per min   **97.** $2x - 3$ **99.** $x + y = 180$ **101.** 30   **103.** 8.5   **105.** 6,000

107. $x = -\frac{a}{b}y + a$ **109.** $a = \frac{bc}{b - c}$

PROBLEM SET 2.3

1. 10 feet by 20 feet   **3.** 7 feet   **5.** 5 inches   **7.** 4 meters   **9.** $92.00   **11.** $200.00   **13.** $86.47   **15.** $468 million
17. 20°, 160°   **19.** **a.** 20.4°, 69.6°   **b.** 38.4°, 141.6°   **21.** 27°, 72°, 81°   **23.** 102°, 44°, 34°   **25.** 43°, 43°, 94°   **27.** 24 ft
29. 6, 8, 10   **31.** $w = 2, \ell = 8$ **33.** $h = 4, b = 18$ **35.** $6,000 at 8%; $3,000 at 9%   **37.** $5,000 at 12%; $10,000 at 10%
39. $4,000 at 8%; $2,000 at 9%   **41.** 3 hours

43.

Speed (miles per hour)	Distance (miles)
20	10
30	15
40	20
50	25
60	30
70	35

45.

Time (hours)	Distance Upstream (miles)	Distance Downstream (miles)
1	6	14
2	12	28
3	18	42
4	24	56
5	30	70
6	36	84

47. 30 fathers, 45 sons

49. $54

51. 44 minutes

53.

t	0	$\frac{1}{4}$	1	$\frac{7}{4}$	2
h	0	7	16	7	0

55.

Hot Coffee Sales	
Year	Sales (billions of dollars)
2005	7
2006	7.5
2007	8
2008	8.6
2009	9.2

57.

w (ft)	l (ft)	A (ft²)
2	22	44
4	20	80
6	18	108
8	16	128
10	14	140
12	12	144

59.

Age (years)	Maximum Heart Rate (beats per minute)
18	202
19	201
20	200
21	199
22	198
23	197

61.

Resting Heart Rate (beats per minute)	Training Heart Rate (beats per minute)
60	144
62	144.8
64	145.6
68	147.2
70	148
72	148.8

63. -5 **65.** 6

PROBLEM SET 2.4

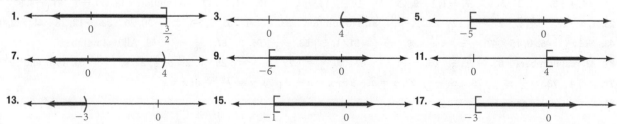

19.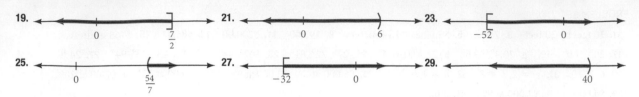

21.

23.

25.

27.

29.

31. $(-\infty, -2]$ **33.** $[1, \infty)$ **35.** $(-\infty, 3)$ **37.** $(-\infty, -1]$ **39.** $[-17, \infty)$ **41.** $(-\infty, 10]$ **43.** $(1, \infty)$ **45.** $(435, \infty)$

47. $y < -\frac{3}{2}x + 3$ **49.** $y < \frac{2}{5}x - 2$ **51.** $y \le \frac{3}{7}x + 3$ **53.** $y \le \frac{1}{2}x + 1$

55. **a.** 3   **b.** 2   **c.** No   **d.** $x > 2$ **57.** At least 291   **59.** $x < 2$ **61.** $x > -\frac{8}{3}$

63. $x \ge 6$; the width is at least 6 meters.   **65.** $x > 6$; the shortest side is even and greater than 6 inches.

67.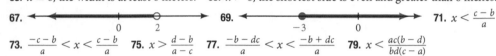

69.

71. $x < \frac{c - b}{a}$

73. $\frac{-c - b}{a} < x < \frac{c - b}{a}$ **75.** $x > \frac{d - b}{a - c}$ **77.** $\frac{-b - dc}{a} < x < \frac{-b + dc}{a}$ **79.** $x < \frac{ac(b - d)}{bd(c - a)}$

PROBLEM SET 2.5

1. a. $(-\infty, -3) \cup (1, \infty)$

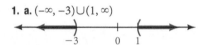

b. $(-3, \infty)$

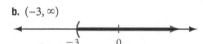

c. All real numbers, $(-\infty, \infty)$

3. a. $(-\infty, 1) \cup [4, \infty)$

b. $(-\infty, 4]$

c. All real numbers, $(-\infty, \infty)$

5. a. $(-6, 2]$

b. $(-\infty, -6)$

c. $[2, \infty]$

d. No solution, $\varnothing$

7. $[3, 7]$

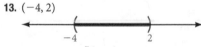

9. $(-4, 2)$

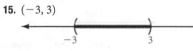

11. $[4, 6]$

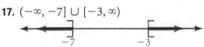

13. $(-4, 2)$

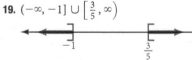

15. $(-3, 3)$

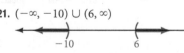

17. $(-\infty, -7] \cup [-3, \infty)$

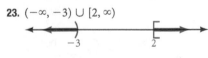

19. $(-\infty, -1] \cup \left[\frac{3}{5}, \infty\right)$

21. $(-\infty, -10) \cup (6, \infty)$

23. $(-\infty, -3) \cup [2, \infty)$

25. a. 1   **b.** 16   **c.** no   **d.** $x > 16$ **27.** $-2 < x \le 4$ **29.** $x < -4$ or $x \ge 1$ **31. a.** $p \le \$2.00$ **b.** $p < \$1.00$ **c.** $p > \$1.25$

d. $p \ge \$1.75$ **33. a.** 1983 and earlier   **b.** 1991 and later   **35.** $x \le 0.08\left(\frac{24,000}{12}\right)$; \$160 or less

37. Eggs to hatching: $0.7 \le r \le 0.8$; Hatching to fledgling: $0.5 \le r \le 0.7$; Fledglings to first breeding: $r < 0.5$

39. $3,000 + 100x \le 7,000$; $x \le 40$ **41.** 5   **43.** -6 **45.** No solution   **47.** $\frac{5}{2}$

PROBLEM SET 2.6

1. $-4, 4$ **3.** $-2, 2$ **5.** $\varnothing$ **7.** $-1, 1$ **9.** $\varnothing$ **11.** $\frac{17}{3}, \frac{7}{3}$ **13.** $-\frac{5}{2}, \frac{5}{6}$ **15.** $-1, 5$ **17.** No solution   **19.** $20, -4$ **21.** $-4, 8$

23. $1, 4$ **25.** $-\frac{1}{7}, \frac{9}{7}$ **27.** $-3, 12$ **29.** $\frac{2}{3}, -\frac{10}{3}$ **31.** No solution   **33.** $\frac{3}{2}, -1$ **35.** $5, 25$ **37.** $-30, 26$ **39.** $-12, 28$ **41.** $-2, 0$

43. $-\frac{1}{2}, \frac{7}{6}$ **45.** $0, 15$ **47.** $-\frac{23}{7}, -\frac{11}{7}$ **49.** $-5, \frac{3}{5}$ **51.** $1, \frac{1}{9}$ **53.** $-\frac{1}{2}$ **55.** 0   **57.** $-\frac{1}{6}, -\frac{7}{4}$ **59.** All real numbers

61. All real numbers   **63.** $-\frac{3}{10}, \frac{3}{2}$ **65.** $-\frac{1}{10}, -\frac{3}{5}$ **67. a.** $\frac{5}{4} = 1.25$ **b.** $\frac{5}{4} = 1.25$ **c.** 2   **d.** $\frac{1}{2}, 2$ **e.** $\frac{1}{3}, 4$ **69.** 1987 and 1995

71. $x < 4$ **73.** $-\frac{11}{3} \le a$ **75.** $t \le -\frac{3}{2}$ **77.** $x = a - b$ or $x = a + b$ **79.** $x = \frac{-b - c}{a}$ or $x = \frac{-b + c}{a}$

81. $x = -\frac{a}{b}y - a$ or $x = -\frac{a}{b}y + a$

PROBLEM SET 2.7

1. $-3 < x < 3$

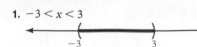

3. $x \le -2$ or $x \ge 2$

5. $-3 < x < 3$

7. $t < -7$ or $t > 7$

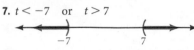

9. $\varnothing$

11. All real numbers

13. $-4 < x < 10$

15. $a \le -9$ or $a \ge -1$

17. $\varnothing$

19. $-1 < x < 5$

21. $y \le -5$ or $y \ge -1$

23. $k \le -5$ or $k \ge 2$

25. $-1 < x < 7$

27. $a \le -2$ or $a \ge 1$

29. $-6 < x < \frac{8}{3}$

31. $[-2, 8]$ **33.** $\left(-2, \frac{4}{3}\right)$ **35.** $(-\infty, -5] \cup [-3, \infty)$ **37.** $\left(-\infty, -\frac{7}{2}\right) \cup \left(-\frac{3}{2}, \infty\right)$ **39.** $\left[-1, \frac{11}{5}\right]$ **41.** $\left(\frac{5}{3}, 3\right)$

43. $x < 2$ or $x > 8$

45. $x \le -3$ or $x \ge 12$

47. $x < 2$ or $x > 6$

49. $0.99 < x < 1.01$ **51.** $x \le -\frac{3}{5}$ or $x \ge -\frac{2}{5}$ **53.** $\frac{5}{9} \le x \le \frac{7}{9}$ **55.** $x < -\frac{2}{3}$ or $x > 0$ **57.** $x \le \frac{2}{3}$ or $x \ge 2$
59. $-\frac{1}{6} \le x \le \frac{3}{2}$ **61.** $-0.05 < x < 0.25$ **63.** $|x| \le 4$ **65.** $|x - 5| \le 1$ **67.** **a.** 3   **b.** $-2, \frac{4}{5}$ **c.** no   **d.** $x < -2$ or $x > \frac{4}{5}$
69. $55 \le x \le 75$ **71.** $12.5 \le p \le 13.5$ **73.** x^{10} **75.** $\frac{4b^8}{a^6}$ **77.** $1125x^{17}y^{14}$ **79.** 3.59×10^{-2} **81.** 0.025 **83.** 4×10^{13}
85. $x < a - b$ or $x > a + b$ **87.** $\frac{b - c}{a} < x < \frac{b + c}{a}$ **89.** $x \le \frac{-c - b}{a}$ or $x \ge \frac{c - b}{a}$ **91.** $\frac{a(-1 - bc)}{b} < x < \frac{a(-1 + bc)}{b}$
93. $-\frac{1}{2}a < x < \frac{3}{2}a$

CHAPTER 2 REVIEW

1. Not a solution   **2.** A solution   **3.** Not a solution   **4.** A solution   **5.** 11   **6.** $-\frac{5}{3}$ **7.** $\frac{4}{5}$ **8.** 6   **9.** $-2, 3$ **10.** $-\frac{3}{2}, 4$
11. $-\frac{1}{2}, \frac{4}{5}$ **12.** $-\frac{5}{3}, \frac{5}{3}$ **13.** $-\frac{1}{9}, \frac{1}{9}$ **14.** 400   **15.** $1, 4$ **16.** $-4, -3, 3$ **17.** $-2, -\frac{2}{3}, \frac{2}{3}$ **18.** $\frac{10}{7}$ **19.** 5   **20.** $y = \frac{1}{3}$
21. $y = -3$ **22.** $x = -7$ **23.** $x = -14$ **24.** $y = \frac{1}{2}x - 2$ **25.** $x = \frac{dy - by}{a - c}$ **26.** $r = \frac{d}{t} - c$ **27.** $b = \frac{2A}{h}$ **28.** $C = \frac{5}{9}(F - 32)$
29. $v = \frac{h + 16t^2}{t}$ **30.** $y = -\frac{2}{7}x - 2$ **31.** $P = \frac{A}{1 + rt}$ **32.** $y = 5x - 5$ **33.** $y = 2x - 5$ **34.** $y = \frac{2}{3}x + 6$ **35.** $y = 3x + 7$
36. 34 inches by 17 inches   **37.** 5 meters   **38.** $25°, 155°$ **39.** $36°, 54°$ **40.** $138.24   **41.** 14.0 g   **42.** $71.72
43. $500 at 9%, $2,500 at 11%

44.

45.

46.

47.

48.

49.

50. $\left[-\frac{5}{9}, \infty\right)$ **51.** $(4, \infty)$ **52.** $\left(-\infty, -\frac{12}{7}\right)$ **53.** $[500, \infty)$ **54.** $(-4, \infty)$

55. $(-\infty, -6]$

56. $[4, \infty)$

57. $(-16, \infty)$

58. $\{3\}$ **59.** $\{3, 4\}$ **60.** $\{1, 2, 3, 4, 5, 7\}$ **61.** $\{1, 3, 7\}$ **62.** $\{0, 1, 2, 3, 4, 5, 6, 7\}$ **63.** $\{4, 5\}$

64.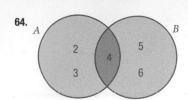

65. $(-\infty, \infty)$ **66.** $3 \le x \le 18$ **67.** $x > 8$ or $x < 5$

68. $\left[0, \frac{1}{2}\right]$

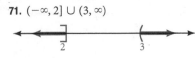

69. $[-4, 0]$

70. $(-\infty, 3) \cup (5, \infty)$

71. $(-\infty, 2] \cup (3, \infty)$

72. $28, 36$ **73.** $-2, -\frac{22}{3}$ **74.** $\frac{9}{2}, -2$ **75.** $\varnothing$ **76.** $2 \le y$ or $y \le \frac{2}{3}$

77. $-\frac{16}{5} \le x \le 2$ **78.** All real numbers   **79.** $\varnothing$

CHAPTER 2 CUMULATIVE REVIEW

1. 36 **2.** 5 **3.** -5 **4.** -3 **5.** 2 **6.** 54 **7.** 18 **8.** 45 **9.** -2 **10.** -6 **11.** 2 **12.** $\frac{5}{3}$ **13.** $\frac{3}{4}$ **14.** $\frac{1}{3}$

15. $\frac{2}{3}$ **16.** $\frac{271}{168}$ **17.** 9 **18.** -25 **19.** $\{1, 3, 5, 6, 7, 8\}$ **20.** $\{3\}$ **21.** $\{9, 12\}$ **22.** $\{5, 6, 7\}$ **23.** $2x + 14$

24. $12x - 1$ **25.** $x + 6 + \frac{9}{x}$ **26.** $30x + 10y$ **27.** $-x + 3$ **28.** $7x^3 + x^2 + 15$ **29.** $32x^{10}y^5$ **30.** $\frac{3x}{y^8}$

31. $8x^2 + 6x - 5$ **32.** $x^2 - \frac{3}{2}x + \frac{9}{16}$ **33.** 1.6×10^8 **34.** $(x + a)(x - b)$ **35.** $(x + 3)(x - 5)$ **36.** 2 **37.** $\frac{2}{15}$

38. 6 **39.** $\frac{1}{4}$ **40.** $-2, 2$ **41.** $-\frac{21}{5}, 3$ **42.** $t = 100$ **43.** $n = 3$ **44.** $m = \frac{y - b}{x}$ **45.** $r = \frac{s - a}{S}$ **46.** $F = \frac{9}{5}C + 32$

47. $x = \frac{5}{a - b}$ **48.** $17°, 73°$ **49.** 8 feet by 24 feet   **50.** $(-1, \infty)$ **51.** $[-2, 4]$ **52.** $x < -2$ or $x > \frac{8}{3}$

53. $\frac{1}{4} < t < \frac{5}{4}$ **54.** 2.8 hours per day   **55.** 0.9 hours per day

CHAPTER 2 TEST

1. 28 **2.** -3 **3.** $-\frac{1}{3}, 2$ **4.** $0, 5$ **5.** $-\frac{7}{4}$ **6.** 2 **7.** $-5, 2$ **8.** $-4, -2, 4$ **9.** $w = \frac{P - 2l}{2}$ **10.** $B = \frac{2A - hb}{h}$

11. $y = \frac{5}{2}x - 5$ **12.** $y = 3x - 7$ **13.** 6 inches by 12 inches   **14.** $55°, 125°$ **15.** 6 inches, 8 inches, 10 inches

16. 0 seconds, 2 seconds

17. $x > 10$

18. $y \ge -4$

19. $x > -4$

20. $n \le -2$

21. $[-6, \infty)$

22. $(-\infty, 4)$

23. $(-\infty, 6)$

24. $[-52, \infty)$

25. $2, 6$ **26.** $\varnothing$ **27.** $x < -1$ or $x > \frac{4}{3}$

28. $-\frac{2}{3} \le x \le 4$

29. All real numbers

30. $\varnothing$

Chapter 3

PROBLEM SET 3.1

1.

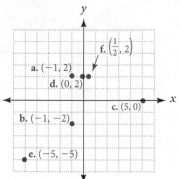

3. **a.** $(4, 1)$ **b.** $(-4, 3)$ **c.** $(-2, -5)$ **d.** $(2, -2)$ **e.** $(0, 5)$ **f.** $(-4, 0)$ **g.** $(1, 0)$

5. b **7.** b **9.** $y = x + 3$ **11.** $y = |x| - 3$

13. **a.**

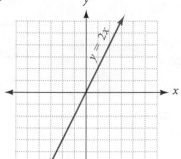

b.

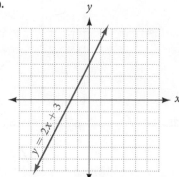

c.

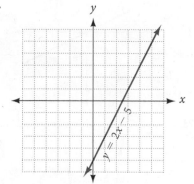

15. **a.**

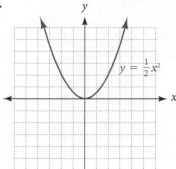

b.

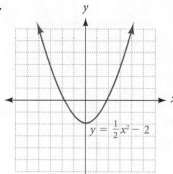

c.

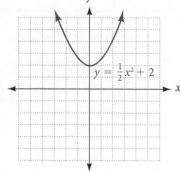

17.

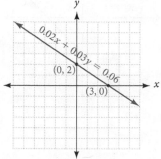

19. a. 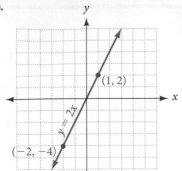 **b.** 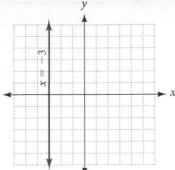 **c.**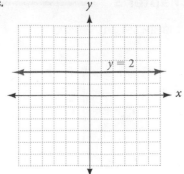

21. a. 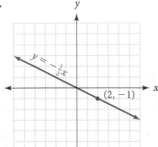 **b.** 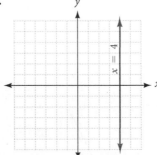 **c.**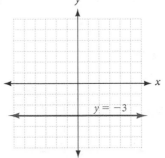

23. 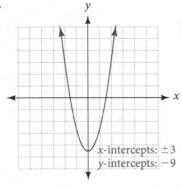 **25.** 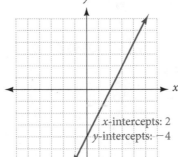 **27.**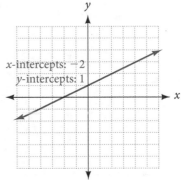

29. 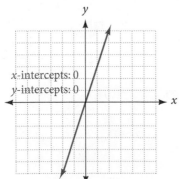 **31.** 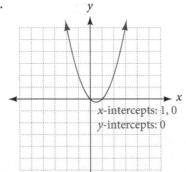 **33.**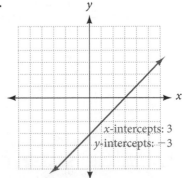

35. a. -7 **b.** -4 **c.** $-\frac{4}{3}$ **d.**

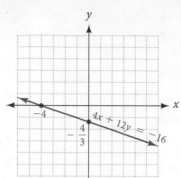

e. $y = -\frac{1}{3}x - \frac{4}{3}$

37. a. Yes   **b.** No   **c.** Yes   **39.**

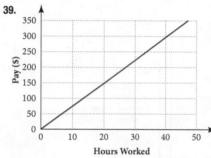

41.

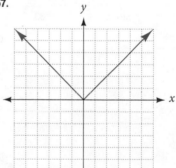

43. a. 60   **b.** 70   **c.** 10   **d.** 6:30 & 7:00   **e.** about 22 minutes   **45.** $-\frac{6}{100}$ **47.** 1   **49.** $-\frac{4}{3}$ **51.** undefined   **53. a.** $\frac{3}{2}$ **b.** $-\frac{3}{2}$

55.

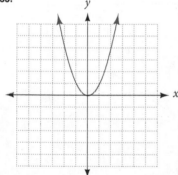

57.

59.

61. x-intercept $= \frac{c}{a}$; y-intercept $= \frac{c}{b}$ **63.** x-intercept $= a$; y-intercept $= b$

PROBLEM SET 3.2

1. $\frac{3}{2}$ **3.** Undefined slope   **5.** $\frac{2}{3}$ **7.**

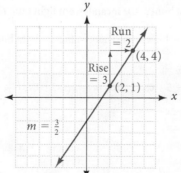

9.

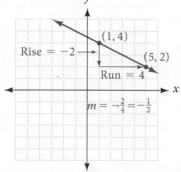

11.

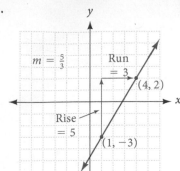

13.

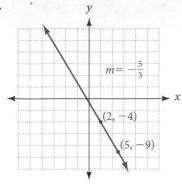

15.

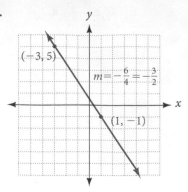

17.

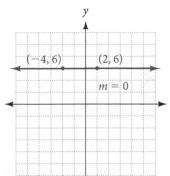

19.

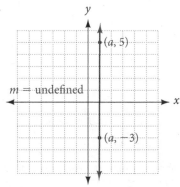

21. $a = 5$ **23.** $b = 2$
25. $x = 3$ **27.** $x = -4$

29.

x	y
0	2
3	0

Slope $= -\frac{2}{3}$

31.

x	y
0	-5
3	-3

Slope $= \frac{2}{3}$

33.

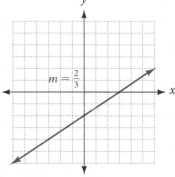

35. $\frac{1}{5}$ **37.** 0 **39.** -1 **41.** $-\frac{3}{2}$ **43. a.** Yes **b.** No **45.** 17.5 mph **47.** 120 ft/sec

49. a. 10 minutes **b.** 20 minutes **c.** 20°C per minute **d.** 10°C per minute **e.** 1st minute

51. 1,333 solar thermal collector shipments/year. Between Year 1 and Year 10, the number of solar thermal collector shipments increased an average of 1,333 shipments per year.

53. a. .05 watts/lumens gained. For every additional lumen, the incandescent light bulb needs to use on average an extra .05 watts.
b. .014 watts/lumens gained. For every additional lumen, the energy-efficient light bulb needs to use on average an extra .014 watts.
c. Energy-efficient light bulb. Answers may vary.

55. -1 **57.** -2 **59.** $y = mx + b$ **61.** $y = -2x - 5$ **63.** 5 **65.** $m = x + 2$ **67.** $\left(\frac{5}{4}, \frac{7}{4}\right)$ **69.** $(0, 2)$ and $(2, 6)$

71.  **73.** 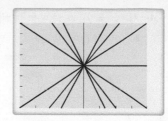 **75.**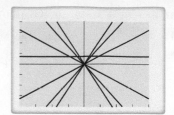

PROBLEM SET 3.3

1. $y = -4x - 3$ **3.** $y = -\frac{2}{3}x$ **5.** $y = -\frac{2}{3}x + \frac{1}{4}$ **7. a.** 3 **b.** $-\frac{1}{3}$ **9. a.** -3 **b.** $\frac{1}{3}$ **11. a.** $-\frac{2}{5}$ **b.** $\frac{5}{2}$

13. Slope $= 3$, y-intercept $= -2$, perpendicular slope $= -\frac{1}{3}$

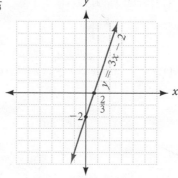

15. Slope $= \frac{2}{3}$, y-intercept $= -4$, perpendicular slope $= -\frac{3}{2}$

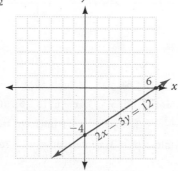

17. Slope $= -\frac{4}{5}$, y-intercept $= 4$, perpendicular slope $= \frac{5}{4}$

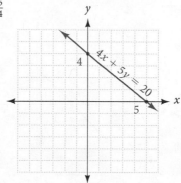

19. Slope $= \frac{1}{2}$, y-intercept $= -4$, $y = \frac{1}{2}x - 4$ **21.** Slope $= -\frac{2}{3}$, y-intercept $= 3$, $y = -\frac{2}{3}x + 3$ **23.** $y = 2x - 1$

25. $y = -\frac{1}{2}x - 1$ **27.** $y = -3x + 1$ **29.** $y = \frac{2}{3}x + \frac{14}{3}$ **31.** $y = -\frac{1}{4}x - \frac{13}{4}$ **33.** $3x + 5y = -1$ **35.** $x - 12y = -8$

37. $6x - 5y = 3$ **39.** $(0, -4), (2, 0); y = 2x - 4$ **41.** $(-2, 0), (0, 4); y = 2x + 4$

43. **a.** $x: \frac{10}{3}, y: -5$ **b.** $(4, 1)$, answers may vary **c.** $y = \frac{3}{2}x - 5$ **d.** no

45. **a.** 2 **b.** $\frac{3}{2}$ **c.** -3 **d.** **e.** $y = 2x - 3$

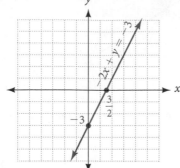

47. **a.** Slope $= \frac{1}{2}$;
x-intercept $= 0$; y-intercept $= 0$

b. Undefined;
x-intercept $= 3$; y-intercept $=$ none

c. Slope $= 0$;
x-intercept $=$ none; y-intercept $= -2$

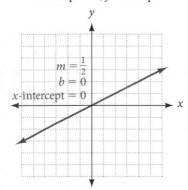

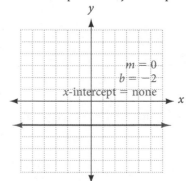

49. $y = 3x + 7$ **51.** $y = -\frac{5}{2}x - 13$ **53.** $y = \frac{1}{4}x + \frac{1}{4}$ **55.** $y = -\frac{2}{3}x + 2$ **57.** **a.** $32 = \frac{9}{5}(0) + 32$; answers will vary **b.** $86°$

59. **a.** $190{,}000 **b.** $19 **c.** $6.50 **61.** $y = 3.4x - 6713$ **63.** $(0, 0), (4, 0)$ **65.** $(0, 0), (2, 0)$ **67.** $y = -\frac{3}{2}x + 12$

69. $y = -\frac{4}{3}x - \frac{3}{2}$ **71.** $y = -\frac{3}{2}x + 3$; slope $= -\frac{3}{2}$; y-intercept $= 3$; x-intercept $= 2$

73. $y = \frac{3}{2}x + 3$; slope $= \frac{3}{2}$; y-intercept $= 3$; x-intercept $= -2$ **75.** x-intercept $= a$; y-intercept $= b$; slope $= -\frac{b}{a}$

PROBLEM SET 3.4

1.

3.

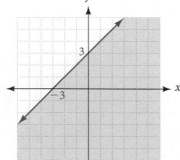

5.

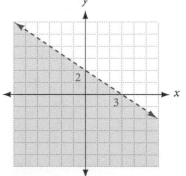

7.

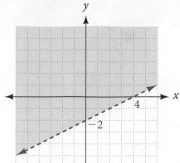

9.

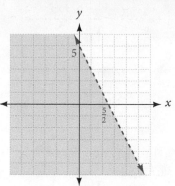

11.

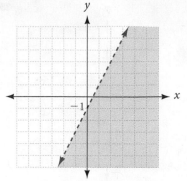

13.

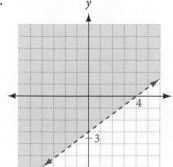

15.

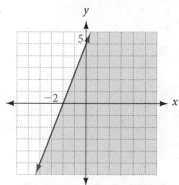

17. $y > -x + 4$ **19.** $y \le \frac{1}{2}x + 2$

21.

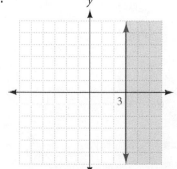

23.

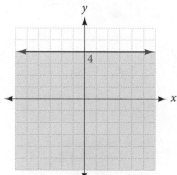

25.

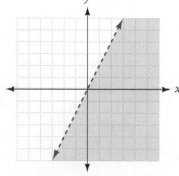

27.

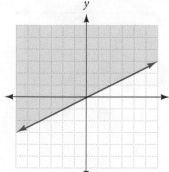

29.

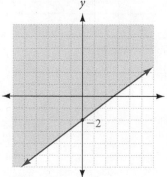

31.

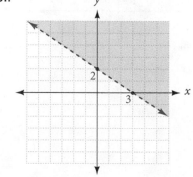

33.

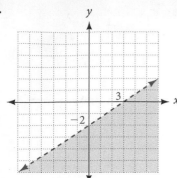

35.

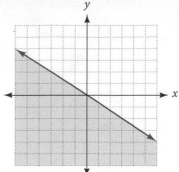

37.

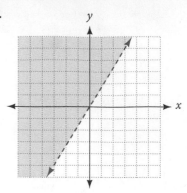

39.

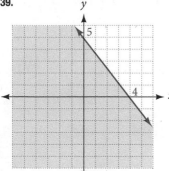

41.

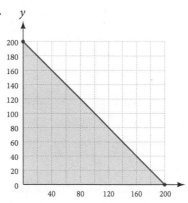

4⁁.

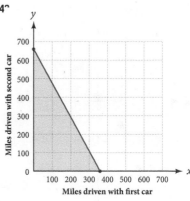

45.

x	y
0	0
10	75
20	150

47.

x	y
0	0
$\frac{1}{2}$	3.75
1	7.5

49.

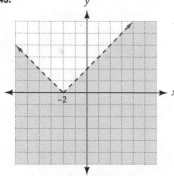

51.

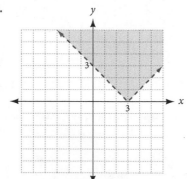

53.

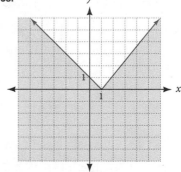

55.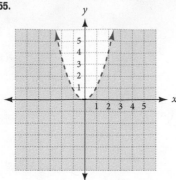

57.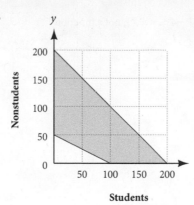

PROBLEM SET 3.5

1. Domain = {1, 3, 5, 7}; Range = {2, 4, 6, 8}; a function   **3.** Domain = {0, 1, 2, 3}; Range = {4, 5, 6}; a function
5. Domain = {a, b, c, d}; Range = {3, 4, 5}; a function   **7.** Domain = {a}; Range = {1, 2, 3, 4}; not a function   **9.** Yes   **11.** No
13. No   **15.** Yes   **17.** Yes   **19.** Domain = {$x \mid -5 \leq x \leq 5$}, Range = {$y \mid 0 \leq y \leq 5$}
21. Domain = {$x \mid -5 \leq x \leq 3$}; Range = {$y \mid y = 3$}   **23.** Domain = All real numbers; Range = {$y \mid y \geq -1$}; a function
25. Domain = All real numbers; Range = {$y \mid y \geq 4$}; a function
27. Domain = {$x \mid x \geq -1$}; Range = All real numbers; not a function
29. Domain = All real numbers, Range = {$y \mid y \geq 0$}; a function
31. Domain = {$x \mid x \geq 0$}, Range = All real numbers; not a function
33. **a.** $y = 8.5x$ for $10 \leq x \leq 40$ **b.**

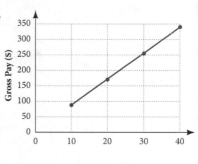

TABLE 4 Weekly Wages		
Hours Worked	Function Rule	Gross Pay ($)
x	$y = 8.5x$	y
10	$y = 8.5(10)$	85
20	$y = 8.5(20)$	170
30	$y = 8.5(30)$	255
40	$y = 8.5(40)$	340

c.

d. Domain = {$x \mid 10 \leq x \leq 40$}; Range = {$y \mid 85 \leq y \leq 340$}   **e.** Minimum = $85; Maximum = $340

35. Domain = {2010, 2011, 2012, 2013, 2014, 2015}, Range = {184, 207, 208, 214, 220, 226}

37. a. III **b.** I **c.** II **d.** IV **39.** 113   **41.** -9 **43. a.** 6 **b.** 7.5   **45. a.** 27 **b.** 6 **47.** 1   **49.** -3 **51.** $-\frac{6}{5}$ **53.** $-\frac{35}{32}$

55. Domain = All real numbers; Range = {$y \mid y \leq 5$}; function   **57.** Domain = {$x \mid x \geq 3$}; Range = All real numbers; not a function

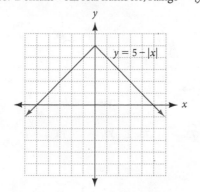

59. Domain = $\{x \mid -4 \le x \le 4\}$; Range = $\{y \mid -4 \le y \le 4\}$; not a function

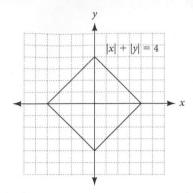

PROBLEM SET 3.6

1. -1 **3.** -11 **5.** 2 **7.** 4 **9.** $a^2 + 3a + 4$ **11.** $2a + 7$ **13.** 1 **15.** -9 **17.** 8 **19.** 0 **21.** $3a^2 - 4a + 1$

23. $3a^2 + 8a + 5$ **25.** 4 **27.** 0 **29.** 2 **31.** 24 **33.** -1 **35.** $2x^2 - 19x + 12$ **37.** 99 **39.** $\frac{3}{10}$ **41.** $\frac{2}{5}$ **43.** undefined

45. **a.** $a^2 - 7$ **b.** $a^2 - 6a + 5$ **c.** $x^2 - 2$ **d.** $x^2 + 4x$ **e.** $a^2 + 2ab + b^2 - 4$ **f.** $x^2 + 2xh + h^2 - 4$

47.

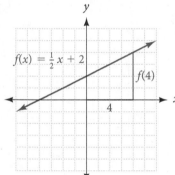

49. $x = 4$ **51.**

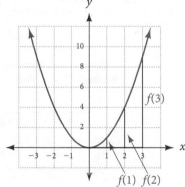

53. $V(3) = 300$, the painting is worth \$300 in 3 years; $V(6) = 600$, the painting is worth \$600 in 6 years.

55. **a.** True **b.** False **c.** True **d.** False **e.** True

57. **a.** \$5,625 **b.** \$1,500 **c.** $\{t \mid 0 \le t \le 5\}$ **d.**

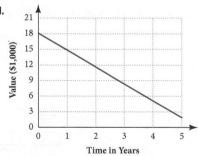

e. $\{V(t) \mid 1,500 \le V(t) \le 18,000\}$

f. About 2.42 years

59. 196 **61.** 4 **63.** 1.6 **65.** 3 **67.** 2,400 **69.** **a.** 2 **b.** 0 **c.** 1 **d.** 4 **71.** 0, 1 **73.** -6 **75.** 2 **77.** $-\frac{1}{3}$ **79.** $4 + h$

81. $\frac{-1}{4 + 2h}$

PROBLEM SET 3.7

1. 30   **3.** −6   **5.** 40   **7.** $\frac{81}{5}$ **9.** 64   **11.** 108   **13.** 300   **15.** ±2   **17.** 1600   **19.** ±8   **21.** $\frac{50}{3}$ pounds

23. **a.** $T = 4P$ **b.**   **c.** 70 pounds per square inch

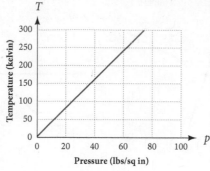

25. 12 pounds per square inch   **27.** **a.** $f = \frac{80}{d}$ **b.**   **c.** An f-stop of 8

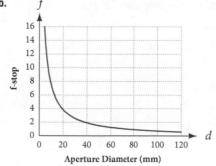

29. $\frac{1504}{15}$ square inches   **31.** 1.5 ohms   **33.** **a.** $P = 0.21\sqrt{L}$ **b.**   **c.** 3.15

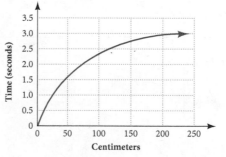

35. $.6M - 42$ **37.** $16x^3 - 40x^2 + 33x - 9$ **39.** $4x^2 - 3x$ **41.** $6x^2 - 2x - 4$ **43.** 11

PROBLEM SET 3.8

1. $6x + 2$ **3.** $-2x + 8$ **5.** $8x^2 + 14x - 15$ **7.** $\frac{2x + 5}{4x - 3}$ **9.** $4x - 7$ **11.** $3x^2 - 10x + 8$ **13.** $-2x + 3$ **15.** $3x^2 - 11x + 10$

17. $9x^3 - 48x^2 + 85x - 50$ **19.** $x - 2$ **21.** $\frac{1}{x - 2}$ **23.** $3x^2 - 7x + 3$ **25.** $6x^2 - 22x + 20$ **27.** 15   **29.** 98   **31.** $\frac{3}{2}$ **33.** 1

35. 40   **37.** 147   **39.** **a.** 81   **b.** 29   **c.** $(x + 4)^2$ **d.** $x^2 + 4$ **41.** **a.** −2   **b.** −1   **c.** $16x^2 + 4x - 2$ **d.** $4x^2 + 12x - 1$

43. $(f \circ g)(x) = 5\left[\frac{x + 4}{5}\right] - 4 = x + 4 - 4 = x,\ (g \circ f)(x) = \frac{(5x - 4) + 4}{5} = \frac{5x}{5} = x$

45.

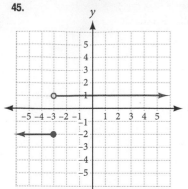

47.

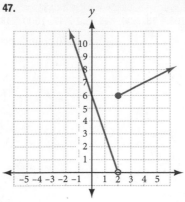

49.

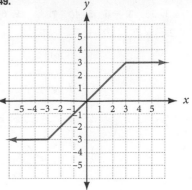

51.   **a.** $R(x) = 11.5x - 0.05x^2$ **b.** $C(x) = 2x + 200$ **c.** $P(x) = -0.05x^2 + 9.5x - 200$ **d.** $\overline{C}(x) = 2 + \frac{200}{x}$

53.   **a.** $M(x) = 220 - x$ **b.** $M(24) = 196$ **c.** 142   **d.** 135   **e.** 128

55. y is $\frac{1}{9}$ of its original value.   **57.** z is 18 times its original value.

PROBLEM SET 3.9

1. $y = x + 3$ **3.** $y = |x| - 3$

5.   **a.**

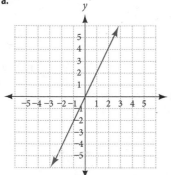

b.

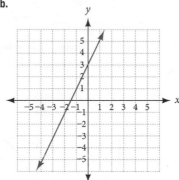

c.

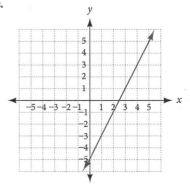

7.   **a.**

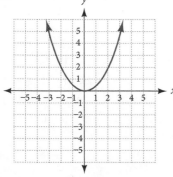

b.

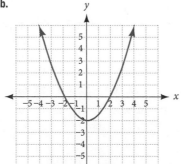

c.

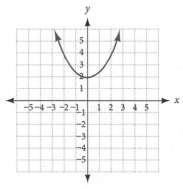

9. $y = \dfrac{1}{x-4}$

11.

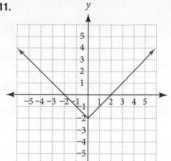

13.

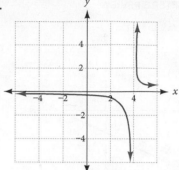

15.

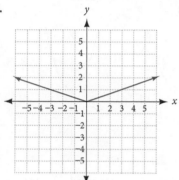

17.

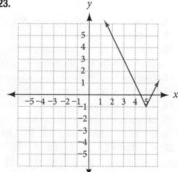

19.

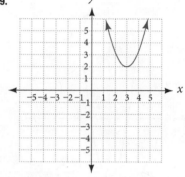

21.

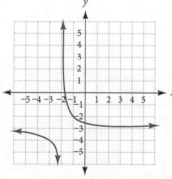

23.

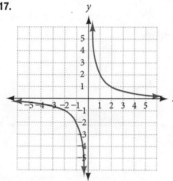

25. $y = |x+1| - 2$ **27.** 12

29. 28   **31.** $-\dfrac{7}{4}$ **33.** $w = \dfrac{P - 2l}{2}$

35. $[-6, \infty)$

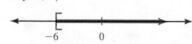

37. $(-\infty, 6)$

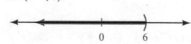

39. 6, 2   **41.** $\varnothing$

CHAPTER 3 REVIEW

1.

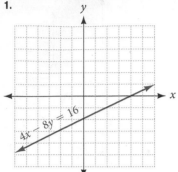

2.

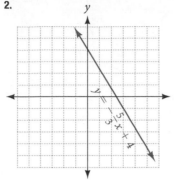

3.

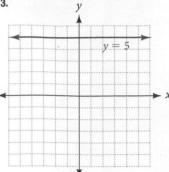

4.

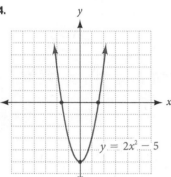

5.

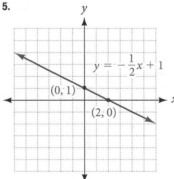

6.

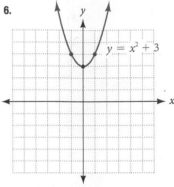

7.

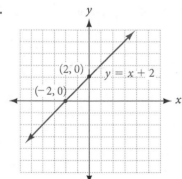

8. Domain = {2, 3}; Range = {0, −4, 1}; not a function
9. Domain = {3, 4, 5}; Range = {0, 2, −3}; a function
10. Domain = {a, b, c, d}; Range = {4, 5}; a function
11. Domain = {a, b, c, d}; Range = {1}; a function
12. Domain = All real numbers; Range = $\{y \mid y \geq 9\}$; a function
13. Domain = $\{x \mid x \geq 1\}$; Range = All real numbers ; not a function
14. Domain = $\{x \mid x \leq 3\}$; Range = All real numbers ; not a function
15. Domain = $\{x \mid x \geq -3\}$; Range = All real numbers ; not a function
16. 3   **17.** 0   **18.** −13   **19.** 14   **20.** $2a - 5$ **21.** $a^2 + 15a + 58$
22. $\frac{1}{2}$ **23.** −2   **24.** 1   **25.** 9   **26.** 0   **27.** 40   **28.** $-4x^2 - 2x + 4$ **29.** 11
30. Undefined   **31.** $\frac{1}{2}$ **32.** 13   **33.** 1   **34.** $\frac{8}{5}$ **35.** 29   **36.** $-\frac{1}{2}$ **37.** $-\frac{5}{2}$
38. 3   **39.** 5   **40.** $-\frac{7}{12}$ **41.** −1   **42.** −2   **43.** $\frac{4}{5}$ **44.** −4   **45.** 6
46. $y = -5x + 4$ **47.** $y = -7x - 3$ **48.** $m = 4, b = -3$

49. $m = \frac{4}{5}, b = -2$ **50.** $y = 3x - 4$ **51.** $y = \frac{3}{2}x + 8$ **52.** $f(x) = -\frac{1}{3}x - 3$ **53.** $f(x) = 2$
54. $y = -\frac{4}{5}x + \frac{12}{5}$ **55.** $y = -x - 2$ **56.** $3x - 5y = 1$ **57.** $x + y = 0$ **58.** $y = -\frac{1}{3}x + \frac{7}{3}$ **59.** $y = -3x + 4$
60. $y = \frac{3}{2}x + 6$ **61.**

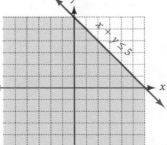

62.

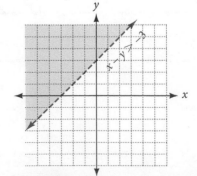

63.

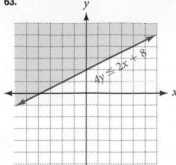

64.

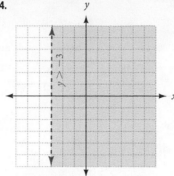

65.

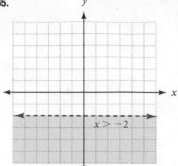

66.

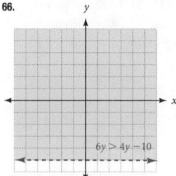

67.

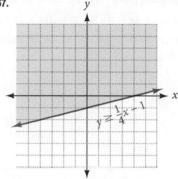

68.

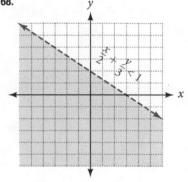

69.

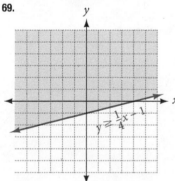

70.

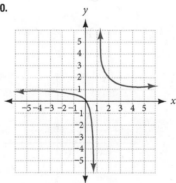

71. $\frac{3}{2}$ **72.** 112

CHAPTER 3 CUMULATIVE REVIEW

1. -36 **2.** -4 **3.** 43   **4.** 169   **5.** 18   **6.** 54   **7.** 34   **8.** 33   **9.** 33   **10.** -11 **11.** $\frac{6}{7}$ **12.** $\frac{7}{13}$ **13.** $\frac{49}{72}$ **14.** $\frac{1}{96}$
15. $\frac{1}{2}$ **16.** $\frac{53}{9}$ **17.** $8x - 10y$ **18.** $2x - 9y$ **19.** $2a - 6$ **20.** $7x^3 + x^2 + 15$ **21.** $-20a^3b^5$ **22.** $\frac{125b}{a^2}$ **23.** $18x^3 + 8x^2 + 14x$
24. $x^3 + \frac{9}{4}x^2 + \frac{27}{16}x + \frac{27}{64}$ **25.** 3×10^6 **26.** $(4a + 7)(2a - 1)$ **27.** $\left(t + \frac{1}{3}\right)\left(t2 - \frac{1}{3}t + \frac{1}{9}\right)$ **28.** -4 **29.** -18 **30.** $\frac{7}{6}$
31. 4   **32.** -8 **33.** $3, \frac{1}{3}$ **34.** $-\frac{1}{2}$ **35.** $y = -4x + 11$ **36.** $y = 6x - 13$ **37.** $-2 \leq x \leq 7$ **38.** $x > \frac{5}{3}$ or $x < -5$

39.

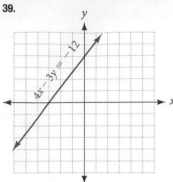

40.

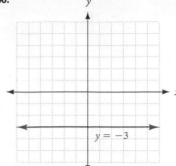

41.

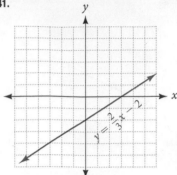

42.

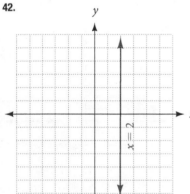

43. -1 **44.** 8   **45.** Undefined slope   **46.** $y = -\frac{3}{5}x + 7$ **47.** $m = \frac{1}{2}$; $b = -2$

48. $y = 2x + 6$ **49.** $y = \frac{1}{2}x + 4$ **50.** $y = \frac{5}{7}x - 13$ **51.** 3   **52.** 1   **53.** 104

54. $x^2 + 3x - 3$ **55.** Domain $= \{-2, -5, 1\}$; Range $= \{3, 4\}$; a function

56. 15%   **57.** $16.40   **58.** 58.5% more   **59.** 13.6% more

CHAPTER 3 TEST

1. x-intercept $= 3$, y-intercept $= 6$, slope $= -2$ **2.** x-intercept $= -\frac{3}{2}$, y-intercept $= -3$, slope $= -2$

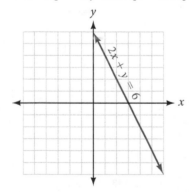

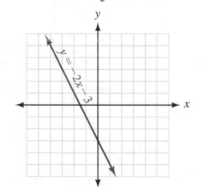

3. x-intercept $= -\frac{8}{3}$, y-intercept $= 4$, slope $= \frac{3}{2}$ **4.** x-intercept $= -2$, no y-intercept, undefined slope

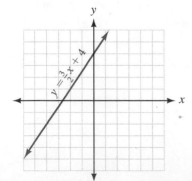

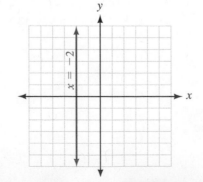

5. $y = 2x + 5$ **6.** $y = -\frac{3}{7}x + \frac{5}{7}$ **7.** $y = \frac{2}{5}x - 5$ **8.** $y = -\frac{1}{3}x - \frac{7}{3}$ **9.** $x = 4$

10.

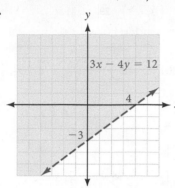

11.

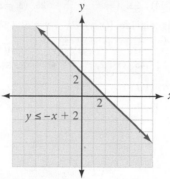

12. Domain $= \{-3, -2\}$; Range $= \{0, 1\}$; not a function   **13.** Domain $=$ All real numbers; Range $= \{y \mid y \geq -9\}$; is a function

14. 11   **15.** -4 **16.** 8   **17.** 4   **18.** 18   **19.** $\frac{81}{4}$ **20.** $\frac{2000}{3}$ pounds

21.

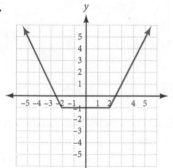

22.

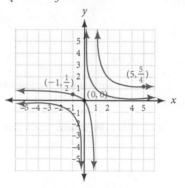

Chapter 4

PROBLEM SET 4.1

1.

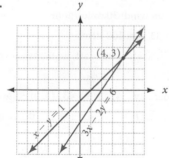

3.

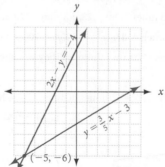

5.

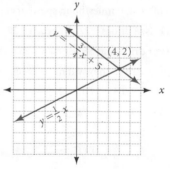

7. Lines are parallel; no solution

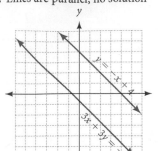

9. $\left(\frac{4}{3}, 1\right)$ **11.** $\left(1, -\frac{1}{2}\right)$ **13.** $\left(\frac{1}{2}, -3\right)$ **15.** $\left(-\frac{8}{3}, 5\right)$ **17.** $(2, 2)$ **19.** $\varnothing$

21. $(10, 24)$ **23.** $\left(-\frac{32}{7}, -\frac{50}{21}\right)$ **25.** $(3, -3)$ **27.** $\left(\frac{4}{3}, -2\right)$ **29.** $(2, 4)$

31. Lines coincide: $\{(x, y) \mid 2x - y = 5\}$ **33.** Lines coincide; $\left\{(x, y) \mid x = \frac{3}{2}y\right\}$

35. $\left(-\frac{15}{43}, -\frac{27}{43}\right)$ **37.** $\left(\frac{60}{43}, \frac{46}{43}\right)$ **39.** $\left(\frac{9}{41}, -\frac{11}{41}\right)$ **41.** $\left(-\frac{11}{7}, -\frac{20}{7}\right)$

43. $\left(2, \frac{4}{3}\right)$ **45.** $(-12, -12)$ **47.** Lines are parallel; $\varnothing$ **49.** $y = 5, z = 2$ **51.** $\left(\frac{3}{2}, \frac{3}{8}\right)$

53. $\left(-4, -\frac{8}{3}\right)$

55. a. $-y$ **b.** -2 **c.** -2 **d.** **e.** $(0, -2)$

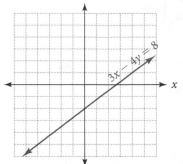

57. $(6{,}000, 4{,}000)$ **59.** $(4, 0)$ **61.** -10 **63.** $3y + 2z$

65. 1 **67.** 3 **69.** $10x - 2z$ **71.** $9x + 3y - 6z$

73. $a = -\frac{8}{3}, b = 16$ **75.** 24 square units **77.** $\left(\frac{1}{2}, \frac{1}{2}\right)$

PROBLEM SET 4.2

1. $(1, 2, 1)$ **3.** $(2, 1, 3)$ **5.** $(2, 0, 1)$ **7.** $\left(\frac{1}{2}, \frac{2}{3}, -\frac{1}{2}\right)$ **9.** No solution, inconsistent system **11.** $(4, -3, -5)$

13. No unique solution **15.** $(4, -5, -3)$ **17.** No unique solution **19.** $\left(\frac{1}{2}, 1, 2\right)$ **21.** $\left(\frac{1}{2}, \frac{1}{3}, \frac{1}{4}\right)$ **23.** $\left(\frac{10}{3}, -\frac{5}{3}, -\frac{1}{3}\right)$

25. $\left(\frac{1}{4}, -\frac{1}{3}, \frac{1}{8}\right)$ **27.** $(6, 8, 12)$ **29.** $(-141, -210, -104)$ **31.** 4 amp, 3 amp, 1 amp **33.** \$100 per G, \$250 per PG, \$300 per R

35. $2 + 3x$ **37.** $-\frac{160}{9}$ **39.** 320 **41.** $2x + 5y$ **43.** 6 **45.** $(-1, 5)$ **47.** $\left(\frac{26}{27}, -\frac{13}{18}, -\frac{26}{47}\right)$ **49.** $a = -16, b = 96, c = 56$

51. $(1, 2, 3, 4)$

PROBLEM SET 4.3

1. 5, 13 **3.** 10, 16 **5.** 1, 3, 4 **7.** 225 adult and 700 children's tickets **9.** \$12,000 at 6%, \$8,000 at 7%

11. \$4,000 at 6%, \$8,000 at 7.5% **13.** \$200 at 6%, \$1,400 at 8%, \$600 at 9% **15.** 6 gallons of 20%, 3 gallons of 50%

17. 5 gallons of 20%, 10 gallons of 14% **19.** 12.5 lbs of oats, 12.5 lbs of nuts

21. speed of boat: 9 miles/hour, speed of current: 3 miles/hour **23.** airplane: 270 miles/hour, wind: 30 miles/hour

25. 12 nickels, 8 dimes **27.** 3 of each **29.** 110 nickels **31.** $x = -200p + 700$; when $p = \$3, x = 100$ items

33. \$4.50 per drink and \$6 per taco **35.** The system has no solutions. They will always be half an hour apart, no matter how long they drive. They will never meet.

37. $h = -16t^2 + 64t + 80$ **39.** $B = 5$ hours per day; $C = 2$ hours per day; $N = 4$ hours per day **41.** No **43.** $(4, 0)$

45. $x > 435$ **47.**

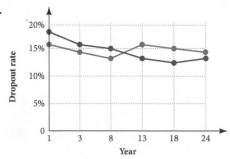

49. a. $M = 0.6x + 8.2$ **b.** $F = -0.4x + 8.2$ **c.** Year 10

PROBLEM SET 4.4

1.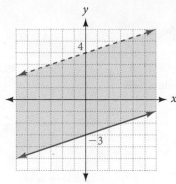

3.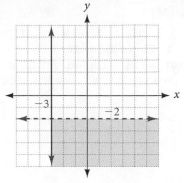

5.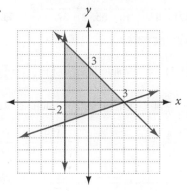

7.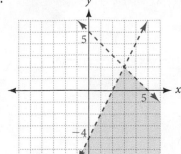

9.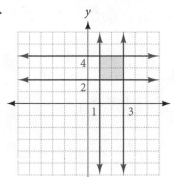

11.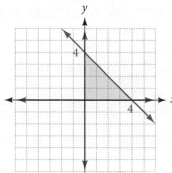

13.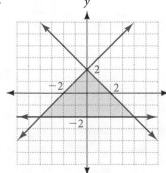

15.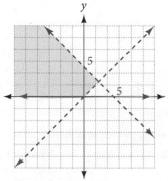

17.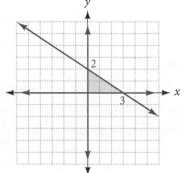

19. $x + y \leq 4$ $-x + y < 4$ **21.** $x + y \geq 4$ $-x + y < 4$

23. a. $0.55x + 0.65y \leq 40$
$x \geq 2y$
$x > 15$
$y \geq 0$

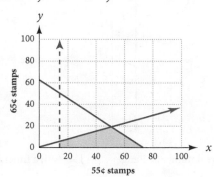

b. 10 65-cent stamps

25. a. x = 50 hours   **b.** No, y = 275 is not in the shaded region.   **27.** −2   **29.** 16   **31.** 11   **33.** $x = 4, x = -2$

PROBLEM SET 4.5

1. 3   **3.** 5   **5.** −1   **7.** 0   **9.** 10   **11.** 2   **13.** −3   **15.** −2   **17.** −2, 5   **19.** 4, −8
21. −3, 3   **23.** 3   **25.** 0   **27.** 3
29. 8   **31.** 6   **33.** −228   **35.** 27   **37.** 0   **39.** −57   **41.** −3, 2, 4   **43.** 2, 3   **45.** $y(1) - mx = b; y = mx + b$
47. a. $y = 0.3x + 3.4$ **b.** $y = 0.3(2) + 3.4$; 4 billion   **49.**

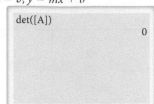

51. 171   **53.** −10   **55.** 3   **57.** $-\dfrac{5}{11}$ **59.** 13   **61.** 22   **63.** 13   **65.** 4   **67.** 4   **69.** 171   **71.** −3, 2, 4   **73.** 3, 2

PROBLEM SET 4.6

1. (3, 1)   **3.** No solution   **5.** $\left(-\dfrac{15}{43}, -\dfrac{27}{43}\right)$ **7.** $\left(\dfrac{60}{43}, \dfrac{46}{43}\right)$ **9.** (2, 0)   **11.** $\left(\dfrac{474}{323}, \dfrac{40}{323}\right)$ **13.** (0, 0, −2)

15. $\left(\dfrac{1}{2}, \dfrac{5}{2}, 1\right)$ **17.** No unique solution   **19.** $\left(-\dfrac{10}{91}, -\dfrac{9}{13}, \dfrac{107}{91}\right)$ **21.** $\left(\dfrac{83}{18}, -\dfrac{7}{9}, -\dfrac{17}{18}\right)$ **23.** $\left(\dfrac{111}{53}, \dfrac{57}{53}, \dfrac{80}{53}\right)$

25. $\left(\dfrac{114}{5}, -\dfrac{16}{5}, \dfrac{28}{5}\right)$ **27.** $\left(-\dfrac{139}{13}, \dfrac{18}{13}, \dfrac{94}{13}\right)$ **29.** (3, 1, 2)   **31.** x = 50 items   **33.** 11   **35.** 0   **37.** (3, −2)

39. $x = -\dfrac{1}{a-b} = \dfrac{1}{b-a}; y = \dfrac{1}{a-b}$ **41.** $x = \dfrac{1}{a^2 + ab + b^2}; y = \dfrac{a+b}{a^2 + ab + b^2}$ **43.** $x + 2y = 1$ **45.** $x - 3y + 4z = 5$
 $3x + 4y = 0$ $3x + 4y + z = 0$
 $2x - 2y - 6z = -5$

47. $\left(\dfrac{2}{7}, \dfrac{1}{7}, \dfrac{69}{14}, \dfrac{39}{14}\right)$ **49.** $\left(\dfrac{1}{a+2}, \dfrac{1}{a+2}, \dfrac{1}{a+2}\right)$

PROBLEM SET 4.7

1. (2, 3)   **3.** (−1, −2)   **5.** (7, 1)   **7.** (1, 2, 1)   **9.** (2, 0, 1)   **11.** (1, 1, 2)   **13.** (4, 1, 5)   **15.** $\left(4, \dfrac{10}{3}\right)$

17. When you convert your matrix back to equations, you will end up with an equation that is a false statement.

19. When you convert your matrix back to equations, you will end up with an equation that is a true statement.

21. x-intercept = 3; y-intercept = 6, slope = −2   **23.** x-intercept = −2; no y-intercept, undefined slope

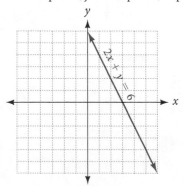

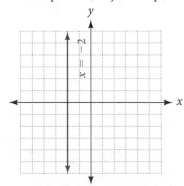

25. $y = -\dfrac{3}{7}x + \dfrac{5}{7}$ **27.** x = 4   **29.** Domain = All real numbers; Range = $\{y \mid y \geq -9\}$; a function   **31.** −4   **33.** 4   **35.** $\dfrac{81}{4}$

CHAPTER 4 REVIEW

1.

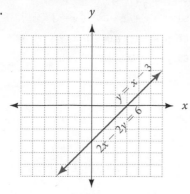

2.

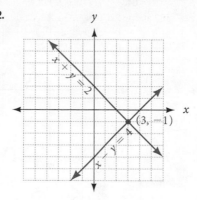

3. $(-3, 5)$ **4.** $(0, -2)$ **5.** $(5, 4)$ **6.** Inconsistent (no solution)   **7.** Dependent (lines coincide)   **8.** $(8, 4)$ **9.** $(2, 5)$

10. $(4, 2)$ **11.** Inconsistent (no solution)   **12.** $(1, -1)$ **13.** $(2, -1)$ **14.** $\left(-\frac{5}{11}, \frac{46}{11}\right)$ **15.** $\left(3, \frac{2}{5}\right)$ **16.** Dependent (lines coincide)

17. $(18, 25)$ **18.** Dependent (lines coincide)   **19.** $\left(-\frac{5}{2}, -3\right)$ **20.** $\left(\frac{14}{11}, \frac{20}{11}\right)$ **21.** $\left(\frac{1}{2}, 1, 1\right)$ **22.** $\left(\frac{3}{4}, -2, \frac{1}{2}\right)$

23. Inconsistent (no solution)   **24.** $\left(\frac{1}{3}, -\frac{1}{5}, -\frac{1}{6}\right)$ **25.** $(2, -6, 3)$ **26.** $(1, 0)$ **27.** $(-3, -5)$ **28.** $(10, 24)$ **29.** $(18, -12)$

30. $(1, -2, 3)$ **31.** 2 **32.** 27 **33.** $\frac{3}{2}$ **34.** $-6, 12$ **35.** $-\frac{2}{3}, 4$ **36.** 34 **37.** -39 **38.** 374 **39.** $(2, -1)$ **40.** $\left(-\frac{63}{5}, -\frac{64}{5}\right)$

41. $\left(-\frac{523}{2}, -\frac{451}{2}\right)$ **42.** Inconsistent solution; $\varnothing$ **43.** $\left(\frac{7}{17}, \frac{21}{17}, \frac{2}{17}\right)$ **44.** $\left(\frac{3}{5}, \frac{7}{5}, \frac{1}{5}\right)$ **45.** $(1, -1, -1)$

46. Left as an exercise for the reader   **47.** 197,454 women   **48.** 3 and 13   **49.** 16 and 33   **50.** $1,750 at 12%; $3,500 at 10%

51. 543 adult tickets, 362 children's tickets   **52.** 6 ounces of 30%, 4 ounces of 80%   **53.** Boat: 7 mph; Current: 3 mph

54. 2 quarters, 5 dimes, 15 nickels   **55.** $x = -300p + 900$; when $p = \$3, x = 0$ bracelets   **56.** Length = 11 in.; width = 7 in.

57.

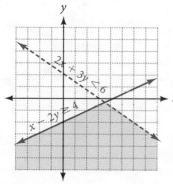

58.

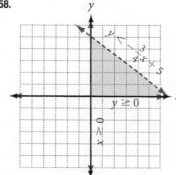

59.

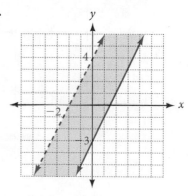

60.

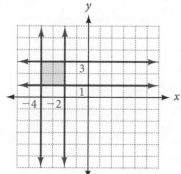

61.

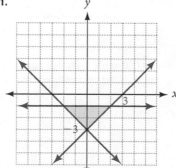

62.

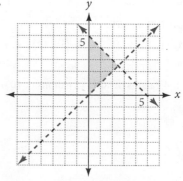

63.

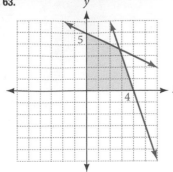

64.

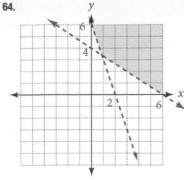

65. $y \le \frac{1}{3}x + 2, y < -\frac{3}{2}x + 3$

66. $y > x + 2, y \ge -\frac{3}{2}x - 3$

67. $p = \$500$; 4,000 units

68. $p = \$6,000$; 18,000 units

CHAPTER 4 CUMULATIVE REVIEW

1. -40 **2.** 16 **3.** 1 **4.** -49 **5.** -4 **6.** -35 **7.** -46 **8.** -16 **9.** $12y^3 - 9y^2 + 12y + 10$ **10.** $4x^3 - 2x^2 - 14x - 3$

11. $4x^2 - 25$ **12.** $15x^2 - \frac{1}{15}$ **13.** $2(x^2 + 4)(3x^2 - 5)$ **14.** $(3a - 4x)(2b + y)$ **15.** $\left(y - \frac{1}{4}\right)\left(y^2 + \frac{1}{4}y + \frac{1}{16}\right)$ **16.** $-5, 4$

17. $-4, -3, 4$ **18.** $21x - 17$ **19.** $2x + 37$ **20.** $-4x - 15y$ **21.** $2x - 3y$ **22.** 1 **23.** 0 **24.** $\varnothing$ **25.** $\varnothing$ **26.** $y = -2x + 2$

27. $y = \frac{3}{4}x + 4$ **28.** $x = \frac{-4}{a - b}$ **29.** $x = \frac{-1}{a - c}$ **30.** $[-3, \infty)$ **31.** $\left(-\infty, -\frac{4}{3}\right] \cup [2, \infty)$ **32.** $\{0, 1, 3, 5, 7, 10, 15\}$ **33.** $\{5\}$

34. $(-4, -5)$ **35.** Lines are parallel; $\varnothing$ **36.** Lines coincide **37.** $(0, -2, 4)$

38.

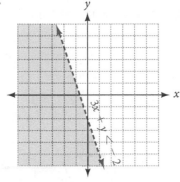

39. -9

40.

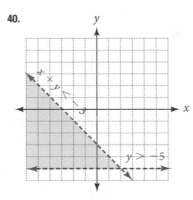

41.

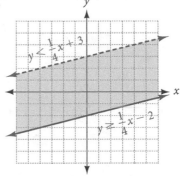

42. -1 **43.** $m = \frac{2}{7}, b = -2$ **44.** $y = -\frac{1}{2}x + 13$ **45.** $y = -\frac{3}{4}x - 3$ **46.** $y = 2x - 16$ **47.** -4 **48.** 4 **49.** $x^2 + x - 4$

50. $-x^2 + 2x + 2$ **51.** 1 **52.** 2 **53.** -1 **54.** 8

CHAPTER 4 TEST

1. $(1, 2)$ **2.** $(3, 2)$ **3.** $(15, 12)$ **4.** $\left(-\frac{54}{13}, -\frac{58}{13}\right)$ **5.** $(1, 2)$ **6.** $(3, -2, 1)$ **7.** lines coincide: $\{(x, y) \mid 2x + 4y = 3\}$

8. $\left(\frac{5}{11}, -\frac{15}{11}, -\frac{1}{11}\right)$ **9.** $5, 9$ **10.** $\$4,000$ at 5%; $\$8,000$ at 6% **11.** 340 adult, 410 children

12. 4 gallons of 30%, 12 gallons of 70% **13.** boat: 8 mph; current: 2 mph **14.** 11 nickels, 3 dimes, 1 quarter

15.

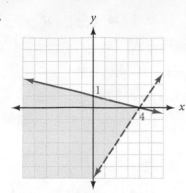

16.

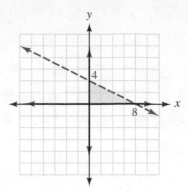

17. $\dfrac{4}{7}$ **18.** -26 **19.** lines coincide; $\{(x, y) \mid 2x + 4y = 3\}$ **20.** $\begin{bmatrix} 2 & -1 & 3 & | & 1 \\ 1 & -4 & -1 & | & 6 \\ 3 & -2 & 1 & | & 4 \end{bmatrix}$ **21.** $(6, -2)$

Chapter 5

PROBLEM SET 5.1

1. $\dfrac{2}{3}$ **3.** $\dfrac{2x}{3}$ **5.** $\dfrac{2x}{y}$ **7.** $\dfrac{3ab}{2}$ **9.** $g(0) = -3, g(-3) = 0, g(3) = 3, g(-1) = -1, g(1) =$ undefined

11. $h(0) = -3, h(-3) = 3, h(3) = 0, h(-1)$ is undefined, $h(1) = -1$ **13.** $\{x \mid x \neq 1\}$ **15.** $\{x \mid x \neq 2\}$ **17.** $\{t \mid t \neq 4, t \neq -4\}$

19. $\dfrac{x - 4}{6}$ **21.** $(a^2 + 9)(a + 3)$ **23.** $\dfrac{2y + 3}{y + 1}$ **25.** $\dfrac{x - 2}{x - 1}$ **27.** $\dfrac{x - 3}{x + 2}$ **29.** $\dfrac{x^2 - x + 1}{x - 1}$ **31.** $-\dfrac{4a}{3}$ **33.** $\dfrac{b - 1}{b + 1}$ **35.** $\dfrac{7x - 3}{7x + 5}$

37. $\dfrac{4x + 3}{4x - 3}$ **39.** $\dfrac{x + 5}{2x - 7}$ **41.** $\dfrac{a^2 - ab + b^2}{a - b}$ **43.** $\dfrac{2x - 2}{x}$ **45.** $\dfrac{x + 3}{y - 4}$ **47.** $x + 2$ **49.** $\dfrac{x^2 + 2x + 4}{x + 2}$ **51.** $\dfrac{4x^2 + 6x + 9}{2x + 3}$ **53.** -1

55. $-(y + 6)$ **57.** $-\dfrac{3a + 1}{3a - 1}$ **59.** 3 **61.** $x + a$ **63.** a. 4 b. 4 **65.** a. 5 b. 5 **67.** a. $x + a$ b. $2x + h$

69. a. $x + a$ b. $2x + h$ **71.** a. $x + a - 3$ b. $2x + h - 3$ **73.** a. 2 b. -4 c. Undefined   d. 2

75.

Weeks	Weight (lb)
x	$W(x)$
0	20.0
1	194.3
4	184
12	173.3
24	168

77. $G(2) = 90\%$ **79.** 1.66 g/mL   **81.** $\dfrac{2}{3}$ **83.** $20x^2y^2$ **85.** $72x^4y^5$ **87.** $(x + 2)(x - 2)$

89. $x^2(x - y)$ **91.** $2(y + 1)(y - 1)$ **93.** $x^2 + 2xy + y^2 + 2xz + 2yz + 4z^2$

95. $a^2 - ab + ac + b^2 - 2bc + c^2$ **97.** $x - y - 1$ **99.** $\dfrac{(x + 1)^2}{(x - 1)^3}$

PROBLEM SET 5.2

1. $\dfrac{1}{6}$ **3.** $\dfrac{9}{4}$ **5.** $\dfrac{1}{2}$ **7.** $\dfrac{15y}{x^2}$ **9.** $\dfrac{b}{a}$ **11.** $\dfrac{2y^5}{z^3}$ **13.** $\dfrac{x + 3}{x + 2}$ **15.** $y + 1$ **17.** $\dfrac{3(x + 4)}{x - 2}$ **19.** $\dfrac{y^2}{xy + 1}$ **21.** $\dfrac{x^2 + 9}{x^2 - 9}$ **23.** $\dfrac{1}{4}$ **25.** 1

27. $\dfrac{(a - 2)(a + 2)}{a - 5}$ **29.** $\dfrac{2}{x + 1}$ **31.** 1 **33.** $x - 3$ **35.** $\dfrac{9t^2 - 6t + 4}{4t^2 - 2t + 1}$ **37.** $\dfrac{x + 3}{x + 4}$ **39.** $\dfrac{a - b}{5}$ **41.** $\dfrac{5c - 1}{3c - 2}$ **43.** $\dfrac{5a - b}{9a^2 + 15ab + 25b^2}$

45. 2 **47.** $x(x - 1)(x + 1)$ **49.** $\dfrac{(a + 4b)(a - 3b)}{(a - 4b)(a + 5b)}$ **51.** $\dfrac{2y - 1}{2y - 3}$ **53.** $\dfrac{(y - 2)(y + 1)}{(y + 2)(y - 1)}$ **55.** $\dfrac{x - 1}{x + 1}$ **57.** $\dfrac{x - 2}{x + 3}$ **59.** $\dfrac{w(y - 1)}{w - x}$

61. $\dfrac{(m + 2)(x + y)}{(2x + y)^2}$ **63.** $3x$ **65.** $2(x + 5)$ **67.** $x - 2$ **69.** $-(y - 4)$ or $4 - y$ **71.** $(a - 5)(a + 1)$

73. a. $\dfrac{5}{21}$ b. $\dfrac{5x + 3}{25x^2 + 15x + 9}$ c. $\dfrac{5x - 3}{25x^2 + 15x + 9}$ d. $\dfrac{5x + 3}{5x - 3}$ **75.** $\dfrac{2}{3}$ **77.** $\dfrac{47}{105}$ **79.** $x - 7$ **81.** $(x + 1)(x - 1)$

83. $2(x + 5)$ **85.** $(a - b)(a^2 + ab + b^2)$ **87.** $\dfrac{x^4 - x^2y^2 + y^4}{x^2 + y^2}$ **89.** $\dfrac{(a + 5)(a - 1)}{3a^2 - 2a + 1}$ **91.** $\dfrac{a(c - 1)}{a - b}$

PROBLEM SET 5.3

1. $\dfrac{5}{4}$ **3.** $\dfrac{1}{3}$ **5.** $\dfrac{41}{24}$ **7.** $\dfrac{19}{144}$ **9.** $\dfrac{31}{24}$ **11.** 1 **13.** -1 **15.** $\dfrac{1}{x + y}$ **17.** 1 **19.** $\dfrac{a^2 + 2a - 3}{a^3}$ **21.** $\dfrac{3x + 4}{x^2}$ **23.** $\dfrac{5a^2 + 6b^2}{2ab}$

25. $\dfrac{2b + 3a}{a^2 b^2}$ **27.** 1   **29. a.** $\dfrac{1}{16}$ **b.** $\dfrac{9}{4}$ **c.** $\dfrac{13}{24}$ **d.** $\dfrac{5x + 15}{(x - 3)^2}$ **e.** $\dfrac{x + 3}{5}$ **f.** $\dfrac{x - 2}{x - 3}$ **31.** $\dfrac{1}{2}$ **33.** $\dfrac{1}{5}$ **35.** $\dfrac{x + 3}{2(x + 1)}$ **37.** $\dfrac{a - b}{a^2 + ab + b^2}$

39. $\dfrac{2y - 3}{4y^2 + 6y + 9}$ **41.** $\dfrac{2(2x - 3)}{(x - 3)(x - 2)}$ **43.** $\dfrac{1}{2t - 7}$ **45.** $\dfrac{4}{(a - 3)(a + 1)}$ **47.** $\dfrac{-4x^2}{(2x + 1)(2x - 1)(4x^2 + 2x + 1)}$ **49.** $\dfrac{2}{(2x + 3)(4x + 3)}$

51. $\dfrac{a}{(a + 4)(a + 5)}$ **53.** $\dfrac{x + 1}{(x - 2)(x + 3)}$ **55.** $\dfrac{x - 1}{(x + 1)(x + 2)}$ **57.** $\dfrac{1}{(x + 2)(x + 1)}$ **59.** $\dfrac{1}{(x + 2)(x + 3)}$ **61.** $\dfrac{4x + 5}{2x + 1}$ **63.** $\dfrac{22 - 5t}{4 - t}$

65. $\dfrac{2x^2 + 3x - 4}{2x + 3}$ **67.** $\dfrac{2x - 3}{2x}$ **69.** $\dfrac{1}{2}$ **71.** $\dfrac{3}{x + 4}$ **73.** $\dfrac{(2x + 1)(x + 5)}{(x - 2)(x + 1)(x + 3)}$ **75.** $\dfrac{51}{10} = 5.1$ **77.** $x + \dfrac{4}{x} = \dfrac{x^2 + 4}{x}$

79. $\dfrac{1}{x} + \dfrac{1}{x + 1} = \dfrac{2x + 1}{x(x + 1)}$ **81.** $\dfrac{6}{5}$ **83.** $x + 2$ **85.** $3 - x$ **87.** $(x + 2)(x - 2)$ **89.** $\dfrac{x - 1}{x + 3}$ **91.** $\dfrac{a^2 + ab + a - v}{u + v}$

93. $\dfrac{6x + 5}{(4x - 1)(3x - 4)}$ **95.** $\dfrac{y^3 + y}{(y + 1)^2(y - 1)^2}$

PROBLEM SET 5.4

1. $\dfrac{9}{8}$ **3.** $\dfrac{2}{15}$ **5.** $\dfrac{119}{20}$ **7.** $\dfrac{1}{x + 1}$ **9.** $\dfrac{a + 1}{a - 1}$ **11.** $\dfrac{y - x}{y + x}$ **13.** $\dfrac{1}{(x + 5)(x - 2)}$ **15.** $\dfrac{1}{a^2 - a + 1}$ **17.** $\dfrac{x + 3}{x + 2}$ **19.** $\dfrac{a + 3}{a - 2}$ **21.** $\dfrac{x - 3}{x}$

23. $\dfrac{x + 4}{x + 2}$ **25.** $\dfrac{x - 3}{x + 3}$ **27.** $\dfrac{a - 1}{a + 1}$ **29.** $-\dfrac{x}{3}$ **31.** $\dfrac{y^2 + 1}{2y}$ **33.** $\dfrac{-x^2 + x - 1}{x - 1}$ **35.** $\dfrac{5}{3}$ **37.** $\dfrac{2x - 1}{2x + 3}$ **39.** $-\dfrac{1}{x(x + h)}$ **41.** $\dfrac{3c + 4a - 2b}{5}$

43. $\dfrac{(t - 4)(t + 1)}{(t + 6)(t - 3)}$ **45.** $\dfrac{(5b - 1)(b + 5)}{2(2b - 11)}$ **47.** $-\dfrac{3}{2x + 14}$ **49.** $2m - 9$ **51. a.** $\dfrac{-4}{ax}$ **b.** $\dfrac{-1}{(x + 1)(a + 1)}$ **c.** $-\dfrac{a + x}{a^2 x^2}$

53. a. As v approaches 0, the denominator approaches 1   **b.** $v = \dfrac{fs}{h} - s$ **55. a.** $v = \dfrac{2v_1 v_2}{v_1 + v_2}$ **b.** $v = 39.4$ mph

57. $xy - x$ **59.** $2x + 6$ **61.** $xa + xb$ **63.** $(x - 2)(x - 1)$ **65.** $y(x - 1)$ **67.** $-4, 5$ **69.** $\dfrac{1}{2}$ **71.** $\dfrac{5}{16}$ **73.** $\dfrac{a + b}{a + 8b}$ **75.** $\dfrac{r^3 s^3}{s - r}$

77. $\dfrac{qt(q + t)}{q^2 + t^2}$

PROBLEM SET 5.5

1. $-\dfrac{35}{3}$ **3.** $-\dfrac{18}{5}$ **5.** $\dfrac{36}{11}$ **7.** 2   **9.** 5   **11.** 2   **13.** $-3, 4$ **15.** $1, -\dfrac{4}{3}$ **17.** Possible solution -1, which does not check; $\varnothing$

19. 5   **21.** $-\dfrac{1}{2}, \dfrac{5}{3}$ **23.** $\dfrac{2}{3}$ **25.** 18   **27.** Possible solution 4, which does not check; $\varnothing$

29. Possible solutions 3 and -4; only -4 checks; -4 **31.** -6 **33.** -5 **35.** $\dfrac{53}{17}$

37. Possible solutions 1 and 2; only 2 checks; 2   **39.** Possible solution 3, which does not check; $\varnothing$ **41.** $\dfrac{22}{3}$

43. a. $-\dfrac{9}{5}, 5$ **b.** $\dfrac{9}{2}$ **c.** no solution   **45. a.** $\dfrac{1}{3}$ **b.** 3   **c.** 9   **d.** 4   **e.** $\dfrac{1}{3}, 3$ **47. a.** $\dfrac{6}{(x - 4)(x + 3)}$ **b.** $\dfrac{x - 3}{x - 4}$ **c.** 5

49. $x = \dfrac{ab}{a - b}$ **51.** $y = \dfrac{x - 3}{x - 1}$ **53.** $y = \dfrac{1 - x}{3x - 2}$

55.

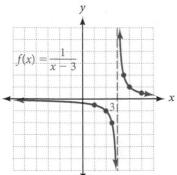

57.

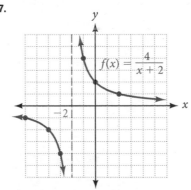

59.

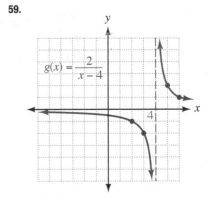

61.

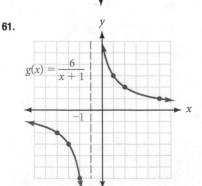

63. $\dfrac{24}{5}$ feet   **65.** $R = 4$ ohms   **67.** 2,358   **69.** 12.3   **71.** 3   **73.** $9, -1$ **75.** 60

77. $-\dfrac{5}{2}, -\dfrac{2}{3}, \dfrac{5}{2}$ **79.** $-2, 2, 3$ **81.** $x = -\dfrac{a}{5}$ **83.** $v = \dfrac{16t^2 + s}{t}$ **85.** $f = \dfrac{pg}{g - p}$

PROBLEM SET 5.6

1. $\frac{1}{x} + \frac{1}{3x} = \frac{20}{3}$; $\frac{1}{5}$ and $\frac{3}{5}$ **3.** $x + \frac{1}{x} = \frac{10}{3}$; 3 or $\frac{1}{3}$ **5.** $\frac{1}{x} + \frac{1}{x+1} = \frac{7}{12}$; 3, 4   **7.** $\frac{7+x}{9+x} = \frac{5}{6}$; 3

9. a./b.

	d	r	t
Upstream	1.5	$5 - x$	$\dfrac{1.5}{5 - x}$
Downstream	3	$5 + x$	$\dfrac{3}{5 + x}$

c. They are the same. $\frac{1.5}{5 - x} = \frac{3}{5 + x}$

d. The speed of the current is $\frac{5}{3}$ mph.

11. $\frac{8}{x + 2} + \frac{8}{x - 2} = 3$; 6 mph   **13. a./b.**

	d	r	t
Train A	150	$x + 15$	$\dfrac{150}{x + 15}$
Train B	120	x	$\dfrac{120}{x}$

c. They are the same, $\frac{150}{x + 5} = \frac{120}{x}$ **d.** The speed of the train A is 75 mph, train B is 60 mph

15. 540 mph   **17.** 54 mph   **19.** 20 mph   **21.** 16 hours   **23.** 15 hours   **25.** 5.25 minutes   **27.** 3.6 hours

29. $10 = \frac{1}{3}\left[\left(x + \frac{2}{3}x\right) + \frac{1}{3}\left(x + \frac{2}{3}x\right)\right]$; $x = \frac{27}{2}$ **31.** 3.6 ft   **33. a.** 30 grams   **b.** 3.25 moles

35.

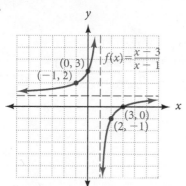

$f(x) = \dfrac{x - 3}{x - 1}$

$(0, 3)$, $(-1, 2)$, $(3, 0)$, $(2, -1)$

37.

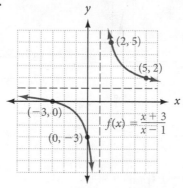

$f(x) = \dfrac{x + 3}{x - 1}$

$(2, 5)$, $(5, 2)$, $(-3, 0)$, $(0, -3)$

39.

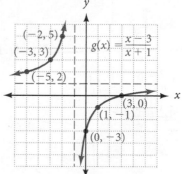

$g(x) = \dfrac{x - 3}{x + 1}$

$(-2, 5)$, $(-3, 3)$, $(-5, 2)$, $(3, 0)$, $(1, -1)$, $(0, -3)$

41. 2   **43.** $-2x^2y^2$ **45.** 185.12   **47.** $4x^3 - 8x^2$ **49.** $4x^3 - 6x - 20$ **51.** $-3x + 9$ **53.** $(x + a)(x - a)$ **55.** $(x - 7y)(x + y)$

PROBLEM SET 5.7

1. $2x^2 - 4x + 3$ **3.** $-2x^2 - 3x + 4$ **5.** $2y^2 + \frac{5}{2} - \frac{3}{2y^2}$ **7.** $-\frac{5}{2}x + 4 + \frac{3}{x}$ **9.** $4ab^3 + 6a^2b$ **11.** $-xy + 2y^2 + 3xy^2$

13. $x + 2$ **15.** $a - 3$ **17.** $5x + 6y$ **19.** $x^2 + xy + y^2$ **21.** $(y^2 + 4)(y + 2)$ **23.** $(x + 2)(x + 5)$ **25.** $(2x + 3)(2x - 3)$

27. $x - 7 + \frac{7}{x + 2}$ **29.** $2x + 5 + \frac{2}{3x - 4}$ **31.** $2x^2 - 5x + 1 + \frac{4}{x + 1}$ **33.** $y^2 - 3y - 13$ **35.** $x - 3$ **37.** $3y^2 + 6y + 8 + \frac{37}{2y - 4}$

39. $a^3 + 2a^2 + 4a + 6 + \frac{17}{a - 2}$ **41.** $y^3 + 2y^2 + 4y + 8$ **43.** $x^2 - 2x + 1$ **45.** $(x + 3)(x + 2)(x + 1)$

47. $(x + 3)(x + 4)(x - 2)$ **49.** Yes   **51.** Same   **53. a.** $(x - 2)(x^2 - x + 3)$ **b.** $(x - 5)(x^3 - x + 1)$

55. a.

x	1	5	10	15	20
C(x)	2.15	2.75	3.50	4.25	5.00

b. $\overline{C}(x) = \frac{2}{x} + 0.15$ **c.**

x	1	5	10	15	20
$\overline{C}(x)$	2.15	0.55	0.35	0.28	0.25

d. It decreases.

e. $y = C(x)$: domain = $\{x \mid 1 \le x \le 20\}$; range = $\{y \mid 2.15 \le y \le 5.00\}$

$y = \overline{C}(x)$: domain = $\{x \mid 1 \le x \le 20\}$; range = $\{y \mid 0.25 \le y \le 2.15\}$

57. a. $T(100) = 11.95$, $T(400) = 32.95$, $T(500) = 39.95$ **b.** $\overline{T}(m) = \frac{4.95 + .07}{m}$ **c.** $\overline{T}(100) = 0.1195$, $\overline{T}(400) = 0.082$, $\overline{T}(500) = 0.0799$

59. $\frac{2}{3a}$ **61.** $(x-3)(x+2)$ **63.** 1   **65.** $\frac{3-x}{x+3}$ **67.** No solution   **69.** $4x^3 - x^2 + 3$ **71.** $0.5x^2 - 0.4x + 0.3$ **73.** $\frac{3}{2}x - \frac{5}{2}$
75. $\frac{2}{3}x + \frac{1}{3} + \frac{2}{3x-1}$

CHAPTER 5 REVIEW

1. $\frac{6x+5}{2x+7}$ **2.** $-\frac{2x+3y}{3x+4y}$ **3.** $\frac{y-3}{y^2-3y+9}$ **4.** $\frac{1}{x^2-3x+9}$ **5.** $x-y$ **6.** $\frac{x-2}{x-1}$ **7.** 2   **8.** $x+a-2$ **9.** $\{x \mid x \neq 3\}$
10. $\{t \mid t \neq 5, t \neq -5\}$ **11.** 201.9 feet/min   **12. a.** 3 **b.** 3   **13. a.** $2x+2+h$ **b.** $x+a+2$ **14.** $x+3$ **15.** $\frac{xy-1}{x^2}$
16. $\frac{x-3}{x+3}$ **17.** $5(x+3)$ **18.** 1   **19.** $\frac{5}{x}$ **20.** $\frac{2x+4}{x^2+2x+4}$ **21.** $\frac{k-1}{k^2-k+1}$ **22.** $-\frac{(r^2+rs+s^2)(r+s)^2}{r^2+s^2}$ **23.** $\frac{93}{140}$ **24.** $\frac{25}{24}$
25. -1 **26.** 1   **27.** $\frac{1}{a-9}$ **28.** $\frac{3(4x-3)}{(x-1)(2x-1)}$ **29.** $\frac{2x}{(x+1)(x+3)}$ **30.** $\frac{x+1}{(x-1)(x+3)}$ **31.** $f(x) - g(x) = \frac{1}{5}$
32. $\frac{4t^2-6t+9}{9t^2-3t+1}$ **33.** $\frac{a}{(a+5)(a+4)}$ **34.** $\frac{x-1}{(x+1)(x+2)}$ **35.** $\frac{3x-2}{3x}$ **36.** $\frac{1}{3}$ **37.** $x-1$ **38.** $\frac{2x+y}{x+2y}$ **39.** $\frac{2x-1}{2x+1}$
40. $-\frac{x}{a}$ **41.** $\frac{4a+7}{4a+9}$ **42.** $\frac{x+4}{x+3}$ **43.** $\frac{2x^2-x-2}{2x-1}$ **44.** $\frac{6x+1}{6x-5}$ **45.** $-\frac{1}{x^2+7x}$ **46.** $-\frac{14x+49}{x^2(x+7)^2}$ **47.** $-8, 1$ **48.** No solution
49. $\frac{5}{8}$ **50.** $-\frac{5}{3}, 2$ **51.** $-\frac{5}{3}, 3$ **52.** $5, 6$ **53.** 3   **54.** No solution   **55. a.** $-4, 4$ **b.** $\frac{3}{2}$ **c.** 0

56.

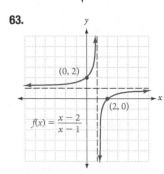

57.

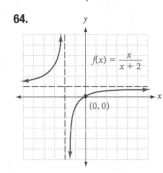

58. Vertical asymptote: $x = 1, 4$; Domain: $\{x \mid x \neq 1, 4\}$

59. Vertical asymptote: $x = -2, \frac{1}{3}$; Domain: $\left\{ x \mid x \neq -2, \frac{1}{3} \right\}$

60. 6   **61.** 11 mph   **62.** 30 mins

63.

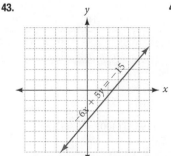

64.

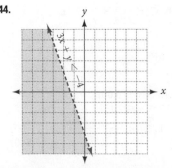

65. $(y^2+9)(y+3)$ **66.** $(x+2)(x-5)$

67. $2x^2y - 5x^3 + 7y^2$ **68.** $3x^2 + 2x + 4 + \frac{8}{3x-5}$

CHAPTER 5 CUMULATIVE REVIEW

1. $-\frac{125}{27}$ **2.** $\frac{49}{64}$ **3.** -40 **4.** 93   **5.** 9   **6.** 12   **7.** $\frac{1}{11}$ **8.** 1   **9.** $-\frac{11}{5}$ **10.** $2a - 9b > 2a + 9b$ **11.** 139   **12.** -141 **13.** 94
14. 90   **15.** $-27x - 20$ **16.** $20x$ **17.** $\frac{1}{(y+3)(y+2)}$ **18.** $\frac{1}{y-x}$ **19.** $20t^4 - \frac{1}{20}$ **20.** $\frac{x-9}{x+3}$ **21.** $3x^{5n} - 2x^{2n}$
22. $2a^4 + 2a^3 + 2a^2 + 3a + 3 - \frac{5}{a-1}$ **23.** x^5 **24.** x^7y^2 **25.** $x+5$ **26.** $\frac{1}{a^2+a+1}$ **27.** $y = -\frac{3}{5}x + 7$ **28.** $y = -1$
29. $2^2 \cdot 3^2 \cdot 11$ **30.** $(x-5)(x+8)$ **31.** $(x-6+y)(x+6+y)$ **32.** $\left(y - \frac{4}{5}x\right)\left(y^2 + \frac{4}{5}xy + \frac{16}{25}x^2\right)$
33. 14   **34.** -5 **35.** -1 **36.** $\frac{3}{2}$ **37.** $\frac{31}{3}$ **38.** $-\frac{2}{3}, 6$ **39.** $(1, 3)$ **40.** $\left(\frac{57}{31}, \frac{8}{31}\right)$ **41.** $x \geq -6$ **42.** $2 < x < 6$

43.

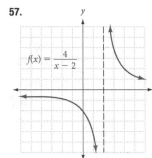

44.

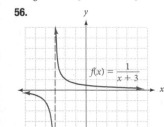

45. 4   **46.** 10   **47.** 7   **48.** 3   **49.** 1.2 miles/hour

CHAPTER 5 TEST

1. $x + y$ **2.** $\dfrac{x-1}{x+1}$ **3.** $2(a+4)$ **4.** $4(a+3)$ **5.** $x+3$ **6.** $\dfrac{38}{105}$ **7.** $\dfrac{7}{8}$ **8.** $\dfrac{1}{a-3}$ **9.** $\dfrac{3(x-1)}{x(x-3)}$ **10.** $\dfrac{x}{(x+4)(x+5)}$

11. $\dfrac{x+4}{(x+1)(x+2)}$ **12.** $\dfrac{3a+8}{3a+10}$ **13.** $\dfrac{x-3}{x-2}$ **14.** $-\dfrac{3}{5}$ **15.** no solution (3 does not check)   **16.** $\dfrac{3}{13}$ **17.** $-2, 3$

18.

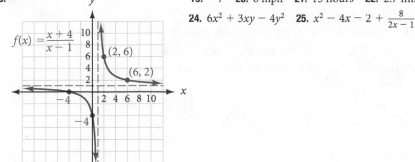

19. -7 **20.** 6 mph   **21.** 15 hours   **22.** 2.7 miles   **23.** 1,012 mph

24. $6x^2 + 3xy - 4y^2$ **25.** $x^2 - 4x - 2 + \dfrac{8}{2x-1}$

Chapter 6

PROBLEM SET 6.1

1. 12   **3.** Not a real number   **5.** -7 **7.** -3 **9.** 2   **11.** Not a real number   **13.** 0.2   **15.** 0.2   **17.** $6a^4$ **19.** $3a^4$ **21.** xy^2

23. $2x^2y$ **25.** $2a^3b^5$ **27.** 6   **29.** -3 **31.** 2   **33.** -2 **35.** 2   **37.** $\dfrac{9}{5}$ **39.** $\dfrac{4}{5}$ **41.** 9   **43.** 125   **45.** 8   **47.** $\dfrac{1}{3}$ **49.** $\dfrac{1}{27}$

51. $\dfrac{6}{5}$ **53.** $\dfrac{8}{27}$ **55.** 7   **57.** $\dfrac{3}{4}$ **59.** $x^{4/5}$ **61.** a **63.** $\dfrac{1}{x^{2/5}}$ **65.** $x^{1/6}$ **67.** $x^{9/25}y^{1/2}z^{1/5}$ **69.** $\dfrac{b^{7/4}}{a^{1/8}}$ **71.** $y^{3/10}$ **73.** $\dfrac{1}{a^2b^4}$

75. a. 5   **b.** 0.5   **c.** 50   **d.** 0.05   **77. a.** $4a^2b^4$ **b.** $2ab^2\sqrt[3]{2ab^2}$ **c.** $2ab^2$ **79.** $(9^{1/2} + 4^{1/2})^2 = (3+2)^2 = 25 \neq 13 = 9 + 4$

81. $(a^{1/2})^{1/2} = a^{1/4}$ **83.** 25 mph   **85.** 12 inches   **87.** 15,000 units   **89.** 587 millibars   **91.** 1.618   **93.** $\dfrac{13}{8}$. The denominator is

the sum of the 2 previous denominators, and the numerator is the sum of the 2 previous numerators.   **95. a.** 420 picometers

b. 594.0 picometers   **c.** 5.94×10^{-10} meters   **97.** 5   **99.** 6   **101.** $4x^2y$ **103.** $5y$ **105.** 3   **107.** 2   **109.** $2ab$ **111.** 25

113. $48x^4y^2$ **115.** $4x^6y^6$ **117.** When $x = 2$, $y = 1.7$ **119.** When $x = 10$, $y = 5.6$ **121.** $x = 1$, $y = 1$, and $x = 0$, $y = 0$

123. a. 1.62 micrograms   **b.** 0.87 micrograms   **c.** 0.00293 micrograms   **d.** 0.0000029 micrograms

PROBLEM SET 6.2

1. $x + x^2$ **3.** $a^2 - a$ **5.** $6x^3 - 8x^2 + 10x$ **7.** $12x^2 - 36y^2$ **9.** $x^{4/3} - 2x^{2/3} - 8$ **11.** $a - 10a^{1/2} + 21$

13. $20y^{2/3} - 7y^{1/3} - 6$ **15.** $10x^{4/3} + 21x^{2/3}y^{1/2} + 9y$ **17.** $t + 10t^{1/2} + 25$ **19.** $x^3 + 8x^{3/2} + 16$ **21.** $a - 2a^{1/2}b^{1/2} + b$

23. $4x - 12x^{1/2}y^{1/2} + 9y$ **25.** $a - 3$ **27.** $x^3 - y^3$ **29.** $t - 8$ **31.** $4x^3 - 3$ **33.** $x + y$ **35.** $a - 8$ **37.** $8x + 1$

39. $t - 1$ **41.** $2x^{1/2} + 3$ **43.** $3x^{1/3} - 4y^{1/3}$ **45.** $3a - 2b$ **47.** $3(x-2)^{1/2}(4x - 11)$ **49.** $5(x-3)^{7/5}(x-6)$

51. $3(x+1)^{1/2}(3x^2 + 3x + 2)$ **53.** $(x^{1/3} - 2)(x^{1/3} - 3)$ **55.** $(a^{1/5} - 4)(a^{1/5} + 2)$

57. $(2y^{1/3} + 1)(y^{1/3} - 3)$ **59.** $(3t^{1/5} + 5)(3t^{1/5} - 5)$ **61.** $(2x^{1/7} + 5)^2$ **63.** $\dfrac{3+x}{x^{1/2}}$ **65.** $\dfrac{x+5}{x^{1/3}}$ **67.** $\dfrac{x^3 + 3x^2 + 1}{(x^3 + 1)^{1/2}}$

69. $\dfrac{-4}{(x^2+4)^{1/2}}$ **71.** 2   **73.** 27   **75.** 0.871   **77.** 15.8%   **79.** 5.9%   **81.** ≈ 3.54 seconds   **83.** ≈ 1.87 years

85. $\dfrac{1}{x^2 + 9}$ **87.** $3x - 4x^3y$ **89.** $5x - 4$ **91.** $x^2 + 5x + 25$ **95.** xy **97.** $x^{17/20}$ **99.** $m^{1/12}$

PROBLEM SET 6.3

1. $2\sqrt{2}$ **3.** $7\sqrt{2}$ **5.** $12\sqrt{2}$ **7.** $4\sqrt{5}$ **9.** $4\sqrt{3}$ **11.** $15\sqrt{3}$ **13.** $3\sqrt[3]{2}$ **15.** $4\sqrt[3]{2}$ **17.** $6\sqrt[3]{2}$ **19.** $2\sqrt[5]{2}$ **21.** $3x\sqrt{2x}$

23. $2y\sqrt[4]{2y^3}$ **25.** $2xy^2\sqrt[3]{5xy}$ **27.** $4abc^2\sqrt[3]{3b}$ **29.** $2bc\sqrt[3]{6a^2c}$ **31.** $2xy^2\sqrt[3]{2x^3y^2}$ **33.** $3xy^2z\sqrt[5]{x^2}$ **35.** $2\sqrt{3}$

37. $\sqrt{-20}$; not real number   **39.** $\dfrac{\sqrt{11}}{2}$ **41. a.** $\dfrac{\sqrt{5}}{2}$ **b.** $\dfrac{2\sqrt{5}}{5}$ **c.** $2 + \sqrt{3}$ **d.** 1

43. a. $2 + \sqrt{3}$ **b.** $-2 + \sqrt{5}$ **c.** $\dfrac{-2 - 3\sqrt{3}}{6}$ **45.** $\dfrac{2\sqrt{3}}{3}$ **47.** $\dfrac{5\sqrt{6}}{6}$ **49.** $\dfrac{\sqrt{2}}{2}$ **51.** $\dfrac{\sqrt{5}}{5}$ **53.** $2\sqrt[3]{4}$ **55.** $\dfrac{2\sqrt[3]{3}}{3}$ **57.** $\dfrac{\sqrt[4]{24x^2}}{2x}$

59. $\frac{\sqrt[4]{8y^3}}{y}$ **61.** $\frac{\sqrt[3]{36xy^2}}{3y}$ **63.** $\frac{\sqrt[3]{6xy^2}}{3y}$ **65.** $\frac{3x\sqrt{15xy}}{5y}$ **67.** $\frac{5xy\sqrt{6xz}}{2z}$ **69. a.** $\frac{\sqrt{2}}{2}$ **b.** $\frac{\sqrt[3]{4}}{2}$ **c.** $\frac{\sqrt[4]{8}}{2}$ **71.** $5\,|x|$

73. $3\,|xy|\,\sqrt{3x}$ **75.** $|x - 5|$ **77.** $|2x + 3|$ **79.** $2\,|a(a + 2)|$ **81.** $2\,|x|\,\sqrt{x - 2}$

83. $\sqrt{9 + 16} \overset{?}{=} \sqrt{9} + \sqrt{16}$
$\qquad \sqrt{25} \overset{?}{=} 3 + 4$
$\qquad\quad 5 \neq 7$

85. $5\sqrt{13}$ feet **87.** **a.** ≈ 89.4 miles **b.** ≈ 126.5 miles **c.** ≈ 154.9 miles **89.** 536.7 ft/s **91.** $2.09\ m^2$

95. $\sqrt{2}, \sqrt{3}, 2, \sqrt{5}, \sqrt{6}, \sqrt{7}; a_{10} = \sqrt{11}; a_{100} = \sqrt{101}$ **97.** $7x$ **99.** $27xy^2$ **101.** $\frac{5}{6}x$ **103.** $3\sqrt{2}$ **105.** $5y\sqrt{3xy}$

107. $2a\sqrt[3]{ab^2}$ **109.** $12\sqrt[3]{5}$ **111.** $6\sqrt[3]{49}$ **113.** $\frac{\sqrt[10]{a^7}}{a}$ **115.** $\frac{\sqrt[20]{a^9}}{a}$ **117.**

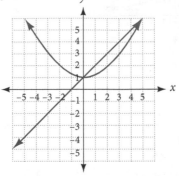

119. ≈ 0.76 apart **121.** 0

PROBLEM SET 6.4

1. $7\sqrt{5}$ **3.** $-x\sqrt{7}$ **5.** $\sqrt[3]{10}$ **7.** $9\sqrt[5]{6}$ **9.** $0 \cdot$ **11.** $\sqrt{5}$ **13.** $-32\sqrt{2}$ **15.** $-3x\sqrt{2}$ **17.** $-2\sqrt[3]{2}$ **19.** $8x\sqrt[3]{xy^2}$

21. $3a^2b\sqrt{3ab}$ **23.** $11ab\sqrt[3]{3a^2b}$ **25.** $10xy\sqrt[4]{3y}$ **27.** $\sqrt{2}$ **29.** $\frac{8\sqrt{5}}{15}$ **31.** $\frac{(x-1)\sqrt{x}}{x}$ **33.** $\frac{3\sqrt{2}}{2}$ **35.** $\frac{5\sqrt{6}}{6}$ **37.** $\frac{8\sqrt[3]{25}}{5}$

39. $\sqrt{12} \approx 3.464; 2\sqrt{3} \approx 2(1.732) = 3.464$ **41.** $\sqrt{8} + \sqrt{18} \approx 2.828 + 4.243 = 7.071; \sqrt{50} \approx 7.071; \sqrt{26} \approx 5.099$

43. $8\sqrt{2x}$ **45.** 5 **53.** $\sqrt{2}$ **55. a.** $\sqrt{2}:1 \approx 1.414:1$ **b.** $5:\sqrt{2}$ **c.** $5:4$ **57.** 6 **59.** $4x^2 + 3xy - y^2$ **61.** $x^2 + 6x + 9$

63. $x^2 - 4$ **65.** $6\sqrt{2}$ **67.** 6 **69.** $9x$ **71.** $\frac{\sqrt{6}}{2}$ **73.** $\frac{5ab^2\sqrt{c}}{3}$ **75.** $-\frac{b}{a}$ **77.** $\sqrt{7}$ **79.** $17ab\sqrt{3b} - 2a\sqrt{2b}$

PROBLEM SET 6.5

1. $3\sqrt{2}$ **3.** $10\sqrt{21}$ **5.** 720 **7.** 54 **9.** $\sqrt{6} - 9$ **11.** $24 + 6\sqrt[3]{4}$ **13.** $7 + 2\sqrt{6}$ **15.** $x + 2\sqrt{x} - 15$ **17.** $34 + 20\sqrt{3}$

19. $19 + 8\sqrt{3}$ **21.** $x - 6\sqrt{x} + 9$ **23.** $4a - 12\sqrt{ab} + 9b$ **25.** $x + 4\sqrt{x - 4}$ **27.** $x - 6\sqrt{x - 5} + 4$ **29.** 1 **31.** $a - 49$

33. $25 - x$ **35.** $x - 8$ **37.** $10 + 6\sqrt{3}$ **39.** $\frac{\sqrt{3} + 1}{2}$ **41.** $\frac{5 - \sqrt{5}}{4}$ **43.** $\frac{x + 3\sqrt{x}}{x - 9}$ **45.** $\frac{10 + 3\sqrt{5}}{11}$ **47.** $\frac{3\sqrt{x} + 3\sqrt{y}}{x - y}$

49. $2 + \sqrt{3}$ **51.** $\frac{11 - 4\sqrt{7}}{3}$ **53. a.** $2\sqrt{x}$ **b.** $x - 4$ **c.** $x + 4\sqrt{x} + 4$ **d.** $\frac{x + 4\sqrt{x} + 4}{x - 4}$

55. a. 10 **b.** 23 **c.** $27 + 10\sqrt{2}$ **d.** $\frac{27 + 10\sqrt{2}}{23}$ **57. a.** $\sqrt{6} + 2\sqrt{2}$ **b.** $2 + 2\sqrt{3}$ **c.** $1 + \sqrt{3}$ **d.** $\frac{-1 + \sqrt{3}}{2}$

59. a. 1 **b.** -1 **61.** $(\sqrt[3]{2} + \sqrt[3]{3})(\sqrt[3]{4} - \sqrt[3]{6} + \sqrt[3]{9}) = \sqrt[3]{8} - \sqrt[3]{12} + \sqrt[3]{18} + \sqrt[3]{12} - \sqrt[3]{18} + \sqrt[3]{27} = 2 + 3 = 5$

63. $10\sqrt{3}$ **65.** $x + 6\sqrt{x} + 9$ **67.** 75 **69.** $\frac{5\sqrt{2}}{4}$ second; $\frac{5}{2}$ second **71.** Answers will vary **73.** Answers will vary

75. $t^2 + 10t + 25$ **77.** x **79.** 7 **81.** $-4, -3$ **83.** $-6, -3$ **85.** $-5, -2$ **87.** Yes **89.** No **91.** $\frac{\sqrt{xy}}{2}$ **93.** $\frac{2a^2}{b}$

95. $\frac{y^2\sqrt[5]{z}}{xz}$ **97.** $\frac{3\sqrt{x} + 3\sqrt{xy}}{1 - y}$ **99.** $\frac{\sqrt{b} + 1}{b - 1}$ **101.** $\frac{x - 4 - 2\sqrt{x - 4}}{x - 8}$ **103.** $\frac{x + \sqrt{x^2 - 9}}{3}$ **105.** $\frac{\sqrt[3]{x^2} - 2\sqrt[3]{x} + 4}{x + 8}$ **107.** $\frac{\sqrt[3]{9} - \sqrt[3]{6} + \sqrt[3]{4}}{5}$

PROBLEM SET 6.6

1. 4 **3.** $\emptyset$ **5.** 5 **7.** $\emptyset$ **9.** $\frac{39}{2}$ **11.** $\emptyset$ **13.** 5 **15.** 3 **17.** $-\frac{32}{3}$ **19.** $3, 4$ **21.** $-1, -2$ **23.** $1, -2$; only 1 checks **25.** -3

27. $1, -1$ **29.** $\emptyset$ **31.** 7 **33.** $0, 3$ **35.** -4 **37.** 8 **39.** 0 **41.** 9 **43.** 0 **45.** 8

47. Possible solution 9, which does not check; $\emptyset$ **49. a.** 100 **b.** 40 **c.** $\emptyset$ **d.** Possible solutions 5, 8; only 8 checks

51. a. 3 **b.** 9 **c.** 3 **d.** $\emptyset$ **e.** 4. **f.** $\emptyset$ **g.** Possible solutions 1, 4; only 4 checks **53.** $h = 100 - 16t^2$ **55.** $\frac{392}{121} \approx 3.24$ feet

57. 40 watts **59.** 216 cubic feet **61.** 17 feet

63.

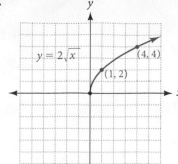

65.

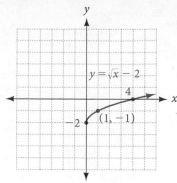

67.

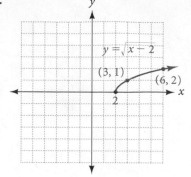

69.

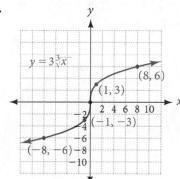

71.

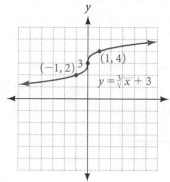

73.

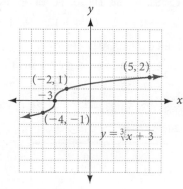

75. 5   **77.** $2\sqrt{3}$ **79.** -1 **81.** 1   **83.** 4   **85.** 2   **87.** $10 - 2x$ **89.** $2 - 3x$ **91.** $6 + 7x - 20x^2$ **93.** $8x - 12x^2$

95. $4 + 12x + 9x^2$ **97.** $4 - 9x^2$ **99.** Possible solutions 2, 6; only 6 checks   **101.** $1, -1, -3$ **103.** $y = \frac{1}{8}x^2$

105. Left as an exercise for the reader

PROBLEM SET 6.7

1. $6i$ **3.** $-5i$ **5.** $6i\sqrt{2}$ **7.** $-2i\sqrt{3}$ **9.** 1   **11.** -1 **13.** $-i$ **15.** $x = 3, y = -1$ **17.** $x = -2, y = -\frac{1}{2}$

19. $x = -8, y = -5$ **21.** $x = 7, y = \frac{1}{2}$ **23.** $x = \frac{3}{7}, y = \frac{2}{5}$ **25.** $5 + 9i$ **27.** $5 - i$ **29.** $2 - 4i$ **31.** $1 - 6i$ **33.** $2 + 2i$

35. $-1 - 7i$ **37.** $6 + 8i$ **39.** $2 - 24i$ **41.** $-15 + 12i$ **43.** $18 + 24i$ **45.** $10 + 11i$ **47.** $21 + 23i$ **49.** $-2 + 2i$ **51.** $2 - 11i$

53. $-21 + 20i$ **55.** $-2i$ **57.** $-7 - 24i$ **59.** 5   **61.** 40   **63.** 13   **65.** 164   **67.** $-3 - 2i$ **69.** $-2 + 5i$ **71.** $\frac{8}{13} + \frac{12}{13}i$

73. $-\frac{18}{13} - \frac{12}{13}i$ **75.** $-\frac{5}{13} + \frac{12}{13}i$ **77.** $\frac{13}{15} - \frac{2}{5}i$ **79.** $R = -11 - 7i$ ohms   **81.** $-\frac{3}{2}$ **83.** $-3, \frac{1}{2}$ **85.** $\frac{5}{4}$ or $\frac{4}{5}$

87.–91. Left as an exercise for the reader

CHAPTER 6 REVIEW

1. 7   **2.** -3 **3.** Not a real number   **4.** 0.9   **5.** $\frac{1}{9}$ **6.** $\frac{7}{12}$ **7.** $x^{13/15}$ **8.** ab^4 **9.** $7x^4y^6$ **10.** $3x^2y^4$ **11.** $\frac{7b^4}{3a}$ **12.** $x^{1-n}y^{2-n^3}$

13. $y^{2/10} = y^{1/5}$ **14.** $b^{3/12} = b^{1/4}$ **15.** $2ab\sqrt[3]{9ac^2}$ **16.** $2xz^2\sqrt[5]{2x^3y^4z}$ **17.** $-4x^5y^6\sqrt[3]{2y^2}$ **18.** $2x^3y^5\sqrt[4]{3xy^3}$

19. $5xy^3\sqrt{2xy}$ **20.** $3xy^2\sqrt[3]{5x^2y}$ **21.** $\frac{\sqrt{35}}{7}$ **22.** $\frac{2ab^2\sqrt{15ab}}{5c}$ **23.** $\frac{2x^2\sqrt{21xy}}{7y}$ **24.** $\frac{5xy\sqrt{6yz}}{3z}$ **25.** $\frac{\sqrt{6}}{6}$

26. $\frac{5\sqrt[3]{9}}{3}$ **27.** $3\sqrt{5} + 6$ **28.** $\frac{x - 2\sqrt{5x} + 5}{x - 5}$ **29.** $\frac{\sqrt[3]{10xy^2}}{2y}$ **30.** $\frac{\sqrt[4]{15y^3}}{3y}$ **31.** $\frac{3 + x}{x^{1/3}}$ **32.** $\frac{6\sqrt{x^4 - 6}}{x^4 - 6}$ **33.** $10\sqrt[6]{7}$ **34.** 0

35. $-4\sqrt{3}$ **36.** $5ab\sqrt[3]{5b^2}$ **37.** $\frac{(x + 1)\sqrt{x}}{x}$ **38.** $\frac{17\sqrt{15}}{15}$ **39.** $\frac{11\sqrt[3]{49}}{7}$ **40.** $6xy^2\sqrt[4]{5x}$ **41.** $6a^4 - 3a^6$ **42.** $9a^{8/5} - 12a^{4/5} + 4$

43. $x + 2\sqrt{x} - 15$ **44.** $53 - 8\sqrt{15}$ **45.** $\frac{21 + 2\sqrt{7}}{59}$ **46.** $4 - \sqrt{15}$ **47.** 9   **48.** 12   **49.** 17   **50.** No solution   **51.** -1

52.

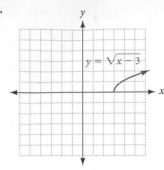

$y = \sqrt{x - 3}$

53.

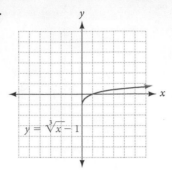

$y = \sqrt[3]{x} - 1$

54. C   **55.** A   **56.** D   **57.** B   **58.** E   **59.** F   **60.** $-9i$ **61.** $-5i\sqrt{3}$ **62.** $x = -\frac{1}{2}, y = -\frac{5}{3}$ **63.** $x = \frac{11}{4}, y = -2$
64. $x = -1; y = 3$ **65.** $4 + 12i$ **66.** $4 - i$ **67.** $18 + i$ **68.** $16 - 30i$ **69.** $\frac{15 + 8i}{17}$ **70.** $i^{39} = (i^4)^9 \cdot i^3 = 1 \cdot i^3 = -i$

CHAPTER 6 CUMULATIVE REVIEW

1. 5   **2.** 14   **3.** -4 **4.** $2\sqrt[4]{8}$ **5.** 14   **6.** $\frac{1}{6}$ **7.** $-\frac{5}{2}$ **8.** 45   **9.** $\frac{8}{9}$ **10.** $3x - 5y$ **11.** $\frac{x - 1}{x + 3}$ **12.** $\frac{a^5 b^6}{4}$ **13.** $3x - 2\sqrt{x} - 5$
14. $50 + 4i$ **15.** $\frac{11}{13} + \frac{3i}{13}$ **16.** $-5 - 2\sqrt{5}$ **17.** $28x^2 + 10xy - 2y^2$ **18.** $(4x + 3)(x - 7)$ **19.** $\frac{20x^6}{y^8}$ **20.** $\frac{x - 2\sqrt{xy} - y}{x - y}$
21. -5 **22.** 5   **23.** $\frac{-4 \pm \sqrt{7}}{3}$ **24.** $\frac{1}{2}, -4$ **25.** 16   **26.** $-2, 0, \frac{2}{3}$ **27.** $-5, 1$ **28.** 3   **29.** $\frac{1}{4}, 2$ **30.** 4
31. $20 \le x \le 30$

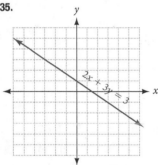

32. $x < -\frac{5}{7}$ or $x > \frac{1}{7}$

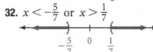

33. $(0, 5)$ **34.** $(1, -5)$

35.

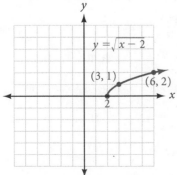

$2x + 3y = 3$

36.

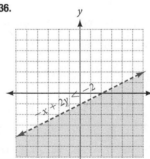

$x - 2y < -2$

37. 4   **38.** 11   **39.** $2\sqrt[4]{3}$ **40.** $\frac{5\sqrt{7} + 15}{-2}$
41. $(3a - 4b)(3a + 4b)(9a^2 + 16b^2)$
42. $4(3a^2 - 7)(2a^2 + 1)$ **43.** $y = -\frac{3}{4}$
44. $z = -54$ **46.** 4   **47.** 20   **48.** 2
49. $-\$80,557,200$ **50.** $\$438,632,000$ **51.** 2.4 mph

CHAPTER 6 TEST

1. $\frac{1}{9}$ **2.** $\frac{7}{5}$ **3.** $a^{5/12}$ **4.** $\frac{x^{13/12}}{y}$ **5.** $7x^4 y^5$ **6.** $2x^2 y^4$ **7.** $2a$ **8.** $x^{n^2 - n} y^{1 - n^3}$ **9.** $6a^2 - 10a$ **10.** $16a^3 - 40a^{3/2} + 25$
11. $(3x^{1/3} - 1)(x^{1/3} + 2)$ **12.** $(3x^{1/3} - 7)(3x^{1/3} + 7)$ **13.** $\frac{x + 4}{x^{1/2}}$ **14.** $\frac{3}{(x^2 - 3)^{1/2}}$ **15.** $5xy^2 \sqrt{5xy}$ **16.** $2x^2 y^2 \sqrt[3]{5xy^2}$ **17.** $\frac{\sqrt{6}}{3}$
18. $\frac{2a^2 b \sqrt{15bc}}{5c}$ **19.** $-6\sqrt{3}$ **20.** $-3ab \sqrt[3]{3}$ **21.** $x + 3\sqrt{x} - 28$ **22.** $21 - 6\sqrt{6}$ **23.** $\frac{5 + 5\sqrt{3}}{2}$ **24.** $\frac{x - 2\sqrt{2x} + 2}{x - 2}$
25. 8 (1 does not check)   **26.** -4 **27.** -3

28.

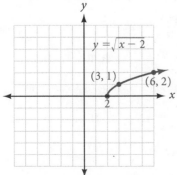

$y = \sqrt{x - 2}$

$(3, 1)$ $(6, 2)$

29.

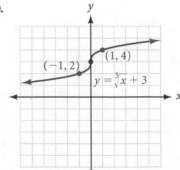

$(-1, 2)$ $(1, 4)$

$y = \sqrt[3]{x} + 3$

30. $x = \frac{1}{2}, y = 7$ **31.** $6i$ **32.** $17 - 6i$
33. $9 - 40i$ **34.** $-\frac{5}{13} - \frac{12}{13}i$
35. $i^{38} = (i^2)^{19} = (-1)^{19} = -1$

Chapter 7

PROBLEM SET 7.1

1. ± 5 **3.** $\pm 3i$ **5.** $\pm \dfrac{\sqrt{3}}{2}$ **7.** $\pm 2i\sqrt{3}$ **9.** $\pm \dfrac{3\sqrt{5}}{2}$ **11.** $-2, 3$ **13.** $\dfrac{-3 \pm 3i}{2}$ **15.** $\dfrac{-2 \pm 2i\sqrt{2}}{5}$ **17.** $-4 \pm 3i\sqrt{3}$ **19.** $\dfrac{3 \pm 2i}{2}$

21. $36, 6$ **23.** $4, 2$ **25.** $25, 5$ **27.** $\dfrac{25}{4}, \dfrac{5}{2}$ **29.** $\dfrac{49}{4}, \dfrac{7}{2}$ **31.** $\dfrac{1}{16}, \dfrac{1}{4}$ **33.** $\dfrac{1}{9}, \dfrac{1}{3}$ **35.** $-6, 2$ **37.** $-3, -9$ **39.** $1 \pm 2i$ **41.** $4 \pm \sqrt{15}$

43. $\dfrac{5 \pm \sqrt{37}}{2}$ **45.** $1 \pm \sqrt{5}$ **47.** $\dfrac{4 \pm \sqrt{13}}{3}$ **49.** $\dfrac{3 \pm i\sqrt{71}}{8}$ **51.** $\dfrac{-2 \pm \sqrt{7}}{3}$ **53.** $\dfrac{5 \pm \sqrt{47}}{2}$ **55.** $\dfrac{5 \pm i\sqrt{19}}{4}$ **57. a.** No   **b.** $\pm 3i$

59. a. $0, 6$ **b.** $0, 6$ **61. a.** $-7, 5$ **b.** $-7, 5$ **63.** No   **65. a.** $\dfrac{7}{5}$ **b.** 3 **c.** $\dfrac{7 \pm 2\sqrt{2}}{5}$ **d.** $\dfrac{71}{5}$ **e.** 3 **67.** $\dfrac{\sqrt{3}}{2}$ inch, 1 inch

69. $\sqrt{2}$ inches   **71.** 781 feet   **73.** 7.3% to the nearest tenth   **75.** 2%   **77.** 1.56 AU   **79.** $d_{\text{grand}} = \sqrt{9 + 5x^2}$

81. $20\sqrt{2} \approx 28$ feet   **83.** 169   **85.** 49   **87.** $\dfrac{85}{12}$ **89.** $(3t - 2)(9t^2 + 6t + 4)$ **91.** $x = \pm 2a$ **93.** $x = -a$ **95.** $x = 0, -2a$

97. $x = \dfrac{-p \pm \sqrt{p^2 - 4q}}{2}$ **99.** $x = \dfrac{-p \pm \sqrt{p^2 - 12q}}{6}$

PROBLEM SET 7.2

1. $-3, -2$ **3.** $2 \pm \sqrt{3}$ **5.** $1, 2$ **7.** $\dfrac{2 \pm i\sqrt{14}}{3}$ **9.** $0, 5$ **11.** $0, -\dfrac{4}{3}$ **13.** $\dfrac{3 \pm \sqrt{5}}{4}$ **15.** $-3 \pm \sqrt{17}$ **17.** $\dfrac{-1 \pm i\sqrt{5}}{2}$ **19.** 1

21. $\dfrac{1 \pm i\sqrt{47}}{6}$ **23.** $4 \pm \sqrt{2}$ **25.** $\dfrac{1}{2}, 1$ **27.** $-\dfrac{1}{2}, 3$ **29.** $\dfrac{-1 \pm i\sqrt{7}}{2}$ **31.** $1 \pm \sqrt{2}$ **33.** $\dfrac{-3 \pm \sqrt{5}}{2}$ **35.** $3, -5$

37. $2, -1 \pm i\sqrt{3}$ **39.** $-\dfrac{3}{2}, \dfrac{3 \pm 3i\sqrt{3}}{4}$ **41.** $\dfrac{1}{5}, \dfrac{-1 \pm i\sqrt{3}}{10}$ **43.** $0, \dfrac{-1 \pm i\sqrt{5}}{2}$ **45.** $0, 1 \pm i$ **47.** $0, \dfrac{-1 \pm i\sqrt{2}}{3}$ **49.** a and b

51. a. $\dfrac{5}{3}, 0$ **b.** $\dfrac{5}{3}, 0$ **53.** No, $2 \pm i\sqrt{3}$ **55.** Yes   **57.** 2 seconds   **59.** 20 or 60 items   **61.** 169   **63.** 0   **65.** ± 12

67. $x^2 - x - 6$ **69.** $x^3 - 4x^2 - 3x + 18$ **71.** $-2\sqrt{3}, \sqrt{3}$ **73.** $\dfrac{-1 \pm \sqrt{3}}{\sqrt{2}} = \dfrac{-\sqrt{2} \pm \sqrt{6}}{2}$ **75.** $-2i, i$ **77.** $-i, 4i$

PROBLEM SET 7.3

1. $D = 16$, two rational   **3.** $D = 0$, one rational   **5.** $D = 5$, two irrational   **7.** $D = 17$, two irrational   **9.** $D = 36$, two rational

11. $D = 116$, two irrational   **13.** ± 10 **15.** ± 12 **17.** 9   **19.** -16 **21.** $\pm 2\sqrt{6}$ **23.** $x^2 - 7x + 10 = 0$ **25.** $t^2 - 3t - 18 = 0$

27. $y^3 - 4y^2 - 4y + 16 = 0$ **29.** $2x^2 - 7x + 3 = 0$ **31.** $4t^2 - 9t - 9 = 0$ **33.** $6x^3 - 5x^2 - 54x + 45 = 0$

35. $10a^2 - a - 3 = 0$ **37.** $9x^3 - 9x^2 - 4x + 4 = 0$ **39.** $x^4 - 13x^2 + 36 = 0$ **41.** $x^2 - 7 = 0$ **43.** $x^2 + 25 = 0$

45. $x^2 - 2x + 2 = 0$ **47.** $x^2 + 4x + 13 = 0$ **49.** $x^3 + 7x^2 - 5x - 75 = 0$ **51.** $x^4 - 18x^2 + 81 = 0$ **53.** $-3, -2, -1$

55. $-4, -3, 2$ **57.** $1 \pm i$ **59.** $5, 4 \pm 3i$ **61.** $x^2 + 4x - 5$ **63.** $4a^2 - 30a + 56$ **65.** $32a^2 + 20a - 18$ **67.** $\pm \dfrac{1}{2}$

69. No solution   **71.** 1   **73.** $-2, 4$ **75.** $-2, \dfrac{1}{4}$ **77.** $-2, 1, i, -i$ **79.** $x = -a$ is a solution of multiplicity 3   **81.** $-1, 5$

83. $1 + \sqrt{2} \approx 2.41, 1 - \sqrt{2} \approx -0.41$ **85.** $-1, \dfrac{1}{2}, 1$ **87.** $\dfrac{1}{2}, \sqrt{2} \approx 1.41, -\sqrt{2} \approx -1.41$ **89.** $\dfrac{2}{3}, 1 + i, 1 - i$

PROBLEM SET 7.4

1. $1, 2$ **3.** $-8, -\dfrac{5}{2}$ **5.** $\pm 3, \pm i\sqrt{3}$ **7.** $\pm 2i, \pm i\sqrt{5}$ **9.** $\dfrac{7}{2}, 4$ **11.** $-\dfrac{9}{8}, \dfrac{1}{2}$ **13.** $\pm \dfrac{\sqrt{30}}{6}, \pm i$ **15.** $\pm \dfrac{\sqrt{21}}{3}, \pm \dfrac{i\sqrt{21}}{3}$ **17.** $4, 25$

19. Only 25 checks   **21.** Only $\dfrac{25}{9}$ checks   **23.** $27, 38$ **25.** $4, 12$ **27.** $t = \dfrac{v \pm \sqrt{v^2 + 64h}}{32}$ **29.** $x = \dfrac{-4 \pm 2\sqrt{4 - k}}{k}$ **31.** $x = -y$

33. $t = \dfrac{1 \pm \sqrt{1 + h}}{4}$ **35. a.**   *y*   **b.** 630 ft.

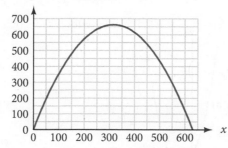

37. a. $l + 2w = 160$ **b.** $A = -2w^2 + 160w$ **c.** **d.** 3,200 square yards

w	l	A
50	60	3,000
45	70	3,150
40	80	3,200
35	90	3,150
30	100	3,000

39. -2 **41.** 1,322.5 **43.** $-\dfrac{7}{640}$ **45.** 1, 5 **47.** $-3, 1$ **49.** $\dfrac{3}{2} \pm \dfrac{1}{2}i$ **51.** 9, 3 **53.** 1, 1 **55.** $1, 2, -1 \pm i\sqrt{3}, \dfrac{-1 + i\sqrt{3}}{2}$
57. $\pm 1, \pm 2, \pm 2i, i$ **59.** $\dfrac{11}{3}, \dfrac{7}{3}$ **61.** $-2, 3, \dfrac{1 \pm i\sqrt{7}}{2}$ **63.** $\pm 1, \dfrac{\pm i\sqrt{5}}{2}$ **65.** $\pm i\sqrt{2}, \pm \dfrac{i\sqrt{10}}{2}$ **67.** $\dfrac{1}{2}, 1$
69. x-intercepts $= -2, 0, 2$; y-intercept $= 0$ **71.** x-intercepts $= -3, -1/3, 3$; y-intercept $= -9$

PROBLEM SET 7.5

1. x-intercepts $= -3, 1$; vertex $= (-1, -4)$ **3.** x-intercepts $= -5, 1$; vertex $= (-2, 9)$

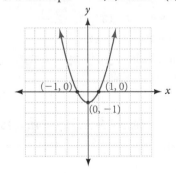

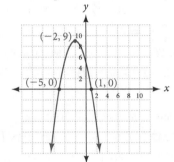

5. x-intercepts $= -1, 1$; vertex $= (0, -1)$ **7.** x-intercepts $= -3, 3$; vertex $= (0, 9)$

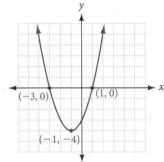

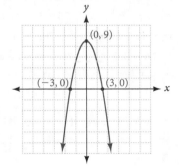

9. x-intercepts $= -1, 3$; vertex $= (1, -8)$ **11.** x-intercepts $= 1 - \sqrt{5}, 1 + \sqrt{5}$; vertex $= (1, -5)$

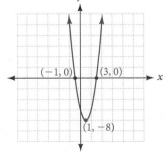

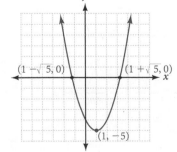

13.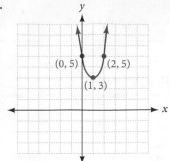

15.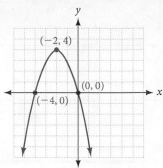

17.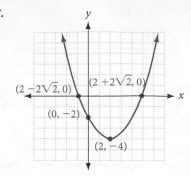

19.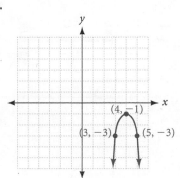

21. vertex $= (2, -8)$

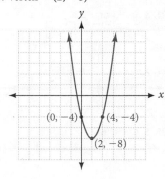

23. vertex $= (1, -4)$

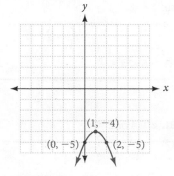

25. vertex $= (0, 1)$

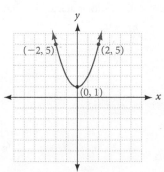

27. vertex $= (0, -3)$

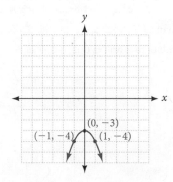

29. vertex $= \left(-\frac{2}{3}, -\frac{1}{3}\right)$

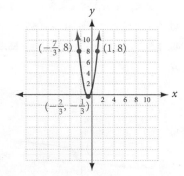

31. $(3, -4)$ lowest   **33.** $(1, 9)$ highest   **35.** $(2, 16)$ highest   **37.** $(-4, 16)$ highest   **39.** 875 patterns; maximum profit \$731.25
41. The ball is in her hand when $h(t) = 0$, which means $t = 0$ or $t = 2$ seconds. Maximum height is $h(1) = 16$ feet.
43. Maximum $R = \$3,600$ when $p = \$6.00$ **45.** Maximum $R = \$7,225$ when $p = \$8.50$ **47.** $y = -\frac{1}{135}(x - 90)^2 + 60$

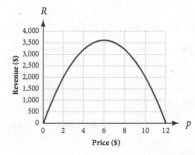

49. $-2, 4$ **51.** $-\frac{1}{2}, \frac{2}{3}$ **53.** 3   **55.** $y = (x - 2)^2 - 4$ **57. a.** $(x - 2)(x + 2)$ **b.** $(x - 2)(x + 2) = 0$
  c. See Figure 8 in problem set   **d.** $0 = x^2 - 4; (x - 2)(x + 2) = 0; x = -2, 2$

PROBLEM SET 7.6

1. $x < 3$ or $x > 2$

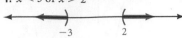

3. $-3 \leq x \leq 4$

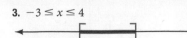

5. $x \leq -3$ or $x \geq -2$

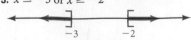

7. $\frac{1}{3} < x < \frac{1}{2}$

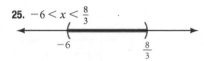

9. $-3 < x < 3$

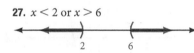

11. $x \leq -\frac{3}{2}$ or $x \geq \frac{3}{2}$

13. $-1 < x < \frac{3}{2}$

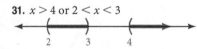

15. All real numbers

17. $\varnothing$

19. $2 < x < 3$ or $x > 4$

21. $x \leq -3$ or $-2 \leq x \leq -1$

23. $-4 < x \leq 1$

25. $-6 < x < \frac{8}{3}$

27. $x < 2$ or $x > 6$

29. $x < -3$ or $2 < x < 4$

31. $x > 4$ or $2 < x < 3$

33. $5 \leq x < 6$

35. a. $-2 < x < 2$ **b.** $x < -2$ or $x > 2$ **c.** $x = -2$ or $x = 2$ **37. a.** $-2 < x < 5$ **b.** $x < -2$ or $x > 5$ **c.** $x = -2$ or $x = 5$

39. a. $x < -1$ or $1 < x < 3$ **b.** $-1 < x < 1$ or $x > 3$ **c.** $x = -1$ or $x = 1$ or $x = 3$ **41.** $x \geq 4$; the width is at least 4 inches

43. $5 \leq p \leq 8$; charge at least $5 but no more than $8 for each speaker **45.** $300, $1,800,000 **47.** $30

49. 1.5625 **51.** 0.6549 **53.** $\frac{2}{3}$ **55.** Possible solutions 1 and 6; only 6 checks; 6

57.

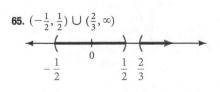

59. $1 - \sqrt{2} < x < 1 + \sqrt{2}$

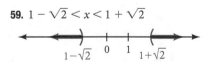

61. $x < 4 - \sqrt{3}$ or $x > 4 + \sqrt{3}$

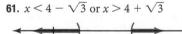

63. $(-\infty, -3) \cup (-2, 3)$

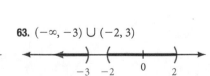

65. $\left(-\frac{1}{2}, \frac{1}{2}\right) \cup \left(\frac{2}{3}, \infty\right)$

67. $(-\infty, -6) \cup \left(\frac{5}{4}, \infty\right)$

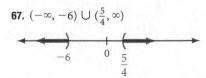

69. $\left(-\frac{8}{3}, \frac{5}{12}\right)$

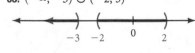

CHAPTER 7 REVIEW

1. $\pm 7i$ **2.** $\pm 2i\sqrt{2}$ **3.** $-3, \frac{1}{3}$ **4.** $\frac{3 \pm 3i\sqrt{2}}{2}$ **5.** $-3, 9$ **6.** $-6, 3$ **7.** $\frac{4}{5}, \frac{-2 \pm 2i\sqrt{3}}{5}$ **8.** $1 \pm i\sqrt{5}$ **9.** $\frac{3 \pm 3i\sqrt{7}}{4}$

10. $\frac{3 \pm i\sqrt{31}}{4}$ **11.** $-1 \pm 2i$ **12.** 4 **13.** $16, 4$ **14.** $\frac{9}{4}, \frac{3}{2}$ **15.** $r = 4 \pm \frac{\sqrt{A}}{6}$ **16.** $4 \pm \sqrt{3}$ **17.** $-1, 4$ **18.** $\frac{5 \pm i\sqrt{31}}{14}$

19. $-6, 1$ **20.** $\frac{5 \pm \sqrt{21}}{2}$ **21.** $\frac{-3 \pm \sqrt{5}}{2}$ **22.** None **23.** $6\,\text{cm}, 6\sqrt{3}\,\text{cm}$ **24.** $\frac{1}{2}$ second, $\frac{5}{2}$ seconds **25.** 2 or 300 watches

26. 27 **27.** 25 **28.** Two rational solutions **29.** $D = -8$, two complex **30.** $3, \frac{-3 \pm 3i\sqrt{3}}{2}$ **31.** $-\frac{1}{4}, \frac{1 \pm i\sqrt{3}}{8}$ **32.** $0, 2 \pm i$

33. $0, \frac{-1 \pm i\sqrt{79}}{8}$ **34.** $5x^2 - 12x - 9 = 0$ **35.** $x^3 - 3x^2 - 16x + 48 = 0$ **36.** $36x^2 + 36x + 13 = 0$ **37.** $x^2 - 4x - 1 = 0$

38. $x^3 + x^2 - 21x - 45 = 0$ **39.** $2, 3 \pm 2i$ **40.** $-\frac{9}{4}, -\frac{5}{3}$ **41.** $\pm i\sqrt{2}, \pm \frac{2\sqrt{3}}{3}$ **42.** $\pm \frac{3i}{2}, \pm \sqrt{5}$ **43.** $-1, \frac{10}{9}$ **44.** 9 **45.** $\frac{25}{4}$

46. $6, \frac{19}{2}$ **47.** $t = \frac{3 \pm \sqrt{9 + 4h}}{8}$

48. $(-1, -8)$

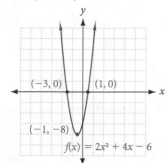

49. $(5, -2)$

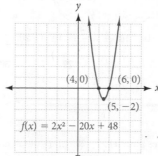

50. $(3, 1)$

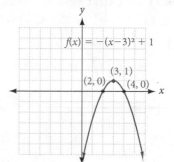

51. $(1, 4)$

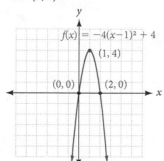

52. $(0, -2)$

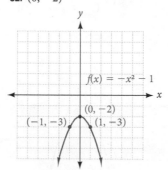

53. $(1, 3)$

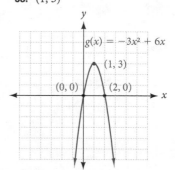

54. $(2, -10)$

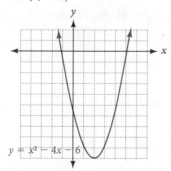

55. $(-1, 5)$

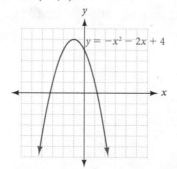

56. $y = 3(x - 4)^2 - 3$

57. $102.50

58. $-5 \le x \le 2$

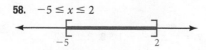

59. $x < -\frac{1}{3}$ or $x > 3$

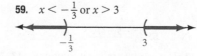

60. $x \le -4$ or $x \ge 4$

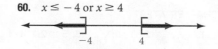

61. $-\frac{2}{3} < x < \frac{2}{3}$

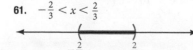

62. $x < 2$ or $3 < x < 4$

63. $-3 \le x \le -2$ or $x \ge -1$

64. $x \leq 1$ or $x > 3$

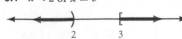

65. $x < -2$ or $1 < x < 5$

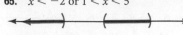

66. $-3 \leq x < -1$

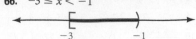

67. $x < 2$ or $x \geq 3$

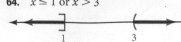

68. a. $x < -3$ or $-1 < x < 2$
 b. $-3 < x < -1$ or $x > 2$
 c. $x = -3$ or $x = -1$ or $x = 2$

69. $\frac{5}{3} < x < 3$; the width must be between $\frac{5}{3}$ and 3 inches

70. $7 \leq p \leq 10$; charge at least \$7 but no more than \$10 for each calculator

CHAPTER 7 CUMULATIVE REVIEW

1. 4 **2.** 78 **3.** $-\frac{125}{8}$ **4.** $\frac{9}{16}$ **5.** $-13x + 13$ **6.** $-2x + 13$ **7.** xy^7 **8.** $\frac{a^{21}}{b^3}$ **9.** $3\sqrt[3]{2}$ **10.** $4x^2\sqrt{2x}$ **11.** $-\frac{3}{16}$ **12.** $\frac{4}{25}$
13. $-\frac{3}{11}$ **14.** $-\frac{25}{4}$ **15.** $x - 7y$ **16.** $x - 5$ **17.** $1 + 4i$ **18.** $\frac{9}{13} + \frac{7}{13}i$ **19.** 4 **20.** $\frac{51}{4}$ **21.** $-3, 3$ **22.** $\varnothing$ **23.** $-3, -5$
24. $a = 4$ **25.** $\frac{3 \pm 2\sqrt{2}}{2}$ **26.** $0, \frac{9 \pm \sqrt{65}}{8}$ **27.** $1, 4$ **28.** $\frac{-1 \pm i\sqrt{35}}{9}$ **29.** $-4, -3$ **30.** 9 **31.** 9 **32.** $-125, 216$

33. $-36 \leq x \leq -18$

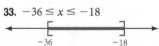

34. $x \leq -1, x \geq -\frac{1}{3}$

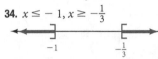

35. $-3 < x < 6$

36. $x < -2$ or $x \geq 4$

37. Lines parallel $\varnothing$ **38.** Lines parallel $\varnothing$

39.

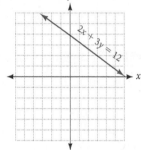

40.

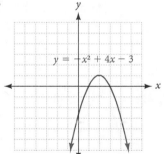
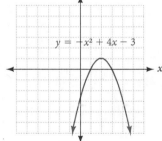

41. $y = \frac{3}{2}x + \frac{5}{2}$ **42.** $y = 4x - 2$ **43.** $(x - 3 - y)(x - 3 + y)$ **44.** $(x + 3)(x - 4)$ **45.** $3\sqrt[3]{2}$ **46.** $\frac{-3 + \sqrt{15}}{2}$ **47.** $\{1, 4\}$
48. $\{2, 8\}$ **49.** $30°, 50°, 100°$ **50.** \$3,500 at 2%; \$2,500 at 3% **51.** $-\frac{25}{9}$ **52.** 4 **53.** 328,290 people **54.** 38,372 people
55. 4.8 mph **56.**

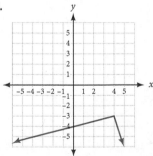

CHAPTER 7 TEST

1. $-\frac{9}{2}, \frac{1}{2}$ **2.** $3 \pm i\sqrt{2}$ **3.** $5 \pm 2i$ **4.** $1 \pm i\sqrt{2}$ **5.** $\frac{5}{2}, \frac{-5 \pm 5i\sqrt{3}}{4}$ **6.** $-1 \pm i\sqrt{5}$ **7.** $r = \pm\frac{\sqrt{A}}{8} - 1$ **8.** $2 \pm \sqrt{2}$
9. $\frac{1}{2}$ or $\frac{3}{2}$ sec **10.** 15 or 100 cups **11.** 9 **12.** $D = 81$; two rational solutions **13.** $3x^2 - 13x - 10 = 0$
14. $x^3 - 7x^2 - 4x + 28 = 0$ **15.** $\pm\sqrt{2}, \pm\frac{1}{2}i$ **16.** $\frac{1}{2}, 1$ **17.** $\frac{1}{4}, 9$ **18.** $t = \frac{7 + \sqrt{49 + 16h}}{16}$

19. vertex: $(1, -4)$

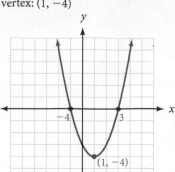

20. vertex: $(1, 9)$

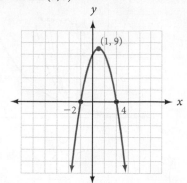

21. $-2 \leq x \leq 3$

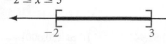

22. $x < -3$ or $x > \frac{1}{2}$

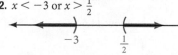

23. profit = $900

Chapter 8

PROBLEM SET 8.1

1. 1   **3.** 2   **5.** $\frac{1}{27}$ **7.** 13   **9.** $\frac{7}{12}$ **11.** $\frac{3}{16}$

13.

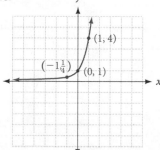

15.

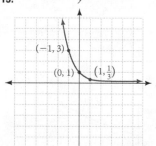

17.

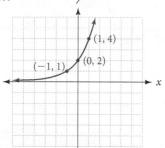

19.

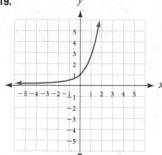

21.

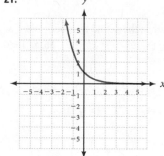

23.

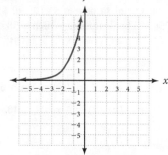

25.

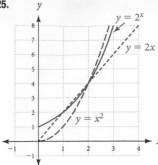

27.

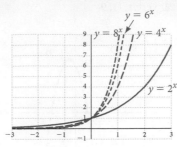

29. $h = 6 \cdot \left(\frac{2}{3}\right)^n$; 5th bounce: $6\left(\frac{2}{3}\right)^5 \approx 0.79$ feet   **31.** 4.27 days

33. **a.** $A(t) = 1,200\left(1 + \frac{.06}{4}\right)^{4t}$ **b.** \$1,932.39   **c.** About 11.64 years   **d.** \$1,939.29

35. **a.** \$129,138.48   **b.** $\{t \mid 0 \le t \le 6\}$ **c.**

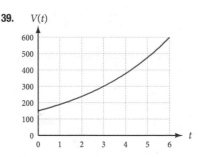

d. $\{V(t) \mid 52,942.05 \le V(t) \le 450,000\}$
e. After approximately 4 years and 8 months

37. $f(1) = 200, f(2) = 800, f(3) = 3,200$ **39.**

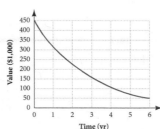

41. **a.** \$0.42   **b.** \$1.00   **c.** \$3.17   **d.** \$17.84

43. 1,258,525 bankruptcies, which is 58,474 less than the actual number.   **45.** **a.** 251,437 cells   **b.** 12,644 cells   **c.** 32 cells

47. $y = \frac{x+3}{2}$ **49.** $y = \pm\sqrt{x+3}$ **51.** $y = \frac{2x-4}{x-1}$ **53.** $y = x^2 + 3$ **55.** **a.** 1   **b.** $\frac{1}{3}$ **c.** 3   **d.** 1   **e.** $\frac{1}{2}$ **f.** 2   **g.** 3   **h.** $\frac{1}{2}$

PROBLEM SET 8.2

1. $f^{-1}(x) = \frac{x+1}{3}$ **3.** $f^{-1}(x) = \sqrt[3]{x}$ **5.** $f^{-1}(x) = \frac{x-3}{x-1}$ **7.** $f^{-1}(x) = 4x + 3$ **9.** $f^{-1}(x) = 2(x+3) = 2x + 6$

11. $f^{-1}(x) = \frac{3}{2}(x+3) = \frac{3}{2}x + \frac{9}{2}$ **13.** $f^{-1}(x) = \sqrt[3]{x+4}$ **15.** $f^{-1}(x) = \frac{x+3}{4-2x}$ **17.** $f^{-1}(x) = \frac{1-x}{3x-2}$

19.

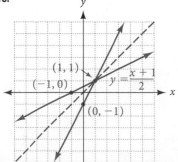

21.

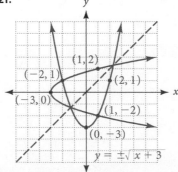

23.

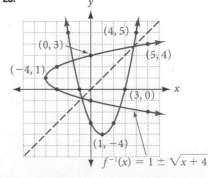

25.

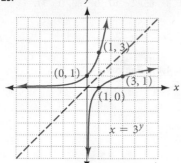

27.

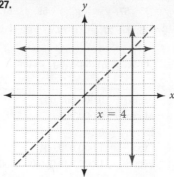

29.

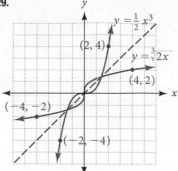

31.

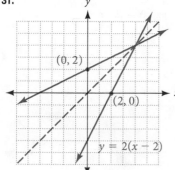

33.

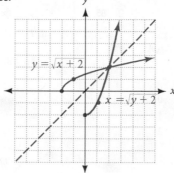

35. **a.** Yes   **b.** No   **c.** Yes

37. **a.** 4   **b.** $\frac{4}{3}$ **c.** 2   **d.** 2   **39.** $f^{-1}(x) = \frac{1}{x}$ **41.** $f^{-1}(x) = 7(x + 2) = 7x + 14$

43. **a.** -3 **b.** -6 **c.** 2   **d.** 3   **e.** -2 **f.** 3   **g.** inverses   **45.** **a.** 729.4   **b.** $s^{-1}(t) = \frac{t - 249.4}{16}$ **c.** 2006

47. **a.** 6629.33 ft/s   **b.** $f^{-1}(m) = \frac{15m}{22}$ **c.** 1.36 mph   **49.** $\frac{1}{9}$ **51.** $\frac{2}{3}$ **53.** $\sqrt[3]{4}$ **55.** 3   **57.** 4   **59.** 4   **61.** 1

63. $f^{-1}(x) = \frac{x - 5}{3}$ **65.** $f^{-1}(x) = \sqrt[3]{x - 1}$ **67.** $f^{-1}(x) = \frac{2x - 4}{x - 1}$ **69.** **a.** 1   **b.** 2   **c.** 5   **d.** 0   **e.** 1   **f.** 2   **g.** 2   **h.** 5

PROBLEM SET 8.3

1. $\log_2 16 = 4$ **3.** $\log_5 125 = 3$ **5.** $\log_{10} 0.01 = -2$ **7.** $\log_2 \frac{1}{32} = -5$ **9.** $\log_{1/2} 8 = -3$ **11.** $\log_3 27 = 3$ **13.** $10^2 = 100$

15. $2^6 = 64$ **17.** $8^0 = 1$ **19.** $10^{-3} = 0.001$ **21.** $6^2 = 36$ **23.** $5^{-2} = \frac{1}{25}$ **25.** 9   **27.** $\frac{1}{125}$ **29.** 4   **31.** $\frac{1}{3}$ **33.** 2   **35.** $\sqrt[3]{5}$

37. 2   **39.** 6   **41.** $\frac{2}{3}$ **43.** $-\frac{1}{2}$ **45.** $\frac{1}{64}$

47.

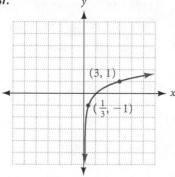

49.

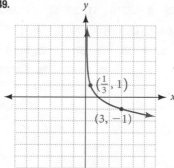

51.

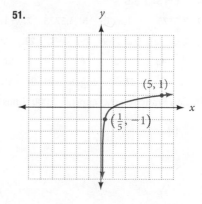

53.

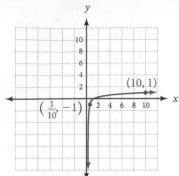

55. $y = 3^x$ **57.** $y = \log_{1/3} x$ **59.** 4 **61.** $\frac{3}{2}$ **63.** 3 **65.** 1 **67.** 0 **69.** 0 **71.** $\frac{1}{2}$

73. $\frac{3}{2}$ **75.** 1 **77.** -2 **79.** 0 **81.** $\frac{1}{2}$

83.

Prefix	Multiplying Factor	$\log_{10}$ (Multiplying Factor)
Nano	0.000 000 001	-9
Micro	0.000 001	-6
Deci	0.1	-1
Giga	1,000,000,000	9
Peta	1,000,000,000,000,000	15

85. 2 **87.** 10^8 times as large **89.** 120 **91.** 4 **93.** $-4, 2$ **95.** $-\frac{11}{8}$ **97.** $2^3 = (x + 2)(x)$ **99.** $3^4 = \frac{x-2}{x+1}$

101. a.

x	$f(x)$
-1	$\frac{1}{8}$
0	1
1	8
2	64

b.

x	$f^{-1}(x)$
$\frac{1}{8}$	-1
1	0
8	1
64	2

c. $f(x) = 8^x$ **d.** $f^{-1}(x) = \log_8 x$

PROBLEM SET 8.4

1. $\log_3 4 + \log_3 x$ **3.** $\log_6 5 - \log_6 x$ **5.** $5 \log_2 y$ **7.** $\frac{1}{3} \log_9 z$ **9.** $2 \log_6 x + 4 \log_6 y$ **11.** $\frac{1}{2} \log_5 x + 4 \log_5 y$

13. $\log_b x + \log_b y - \log_b z$ **15.** $\log_{10} 4 - \log_{10} x - \log_{10} y$ **17.** $2 \log_{10} x + \log_{10} y - \frac{1}{2} \log_{10} z$

19. $3 \log_{10} x + \frac{1}{2} \log_{10} y - 4 \log_{10} z$ **21.** $\frac{2}{3} \log_b x + \frac{1}{3} \log_b y - \frac{4}{3} \log_b z$ **23.** $\frac{2}{3} \log_3 x + \frac{1}{3} \log_3 y - 2 \log_3 z$

25. $2\log_a 2 + 5\log_a x - 2\log_a 3 - 2$ **27.** $\log_b xz$ **29.** $\log_3 \frac{x^2}{y^3}$ **31.** $\log_{10} \sqrt{x}\sqrt[3]{y}$ **33.** $\log_2 \frac{x^3\sqrt{y}}{z}$ **35.** $\log_2 \frac{\sqrt{x}}{y^3 z^4}$

37. $\log_{10} \frac{x^{3/2}}{y^{3/4} z^{4/5}}$ **39.** $\log_5 \frac{\sqrt{x} \cdot \sqrt[3]{y^2}}{z^4}$ **41.** $\log_3 \frac{x-4}{x+4}$ **43.** $\frac{2}{3}$ **45.** 18 **47.** 3 **49.** 3 **51.** 4 **53.** 4 **55.** 1 **57.** 0 **59.** $\frac{3}{2}$

61. 27 **63.** $\frac{5}{3}$ **67. a.** 1.602 **b.** 2.505 **c.** 3.204 **69.** pH $= 6.1 + \log_{10} x - \log_{10} y$ **71.** 2.52 **73.** 1 **75.** 1 **77.** 4

79. 2.5×10^{-6} **81.** 51 **83.** 2 **85.** $2u + \frac{1}{2}v$ **87.** $\frac{5}{2}u + \frac{7}{2}v$ **89.** $-\frac{1}{2}$

PROBLEM SET 8.5

1. 2.5775 **3.** 1.5775 **5.** 3.5775 **7.** -1.4225 **9.** 4.5775 **11.** 2.7782 **13.** 3.3032 **15.** -2.0128 **17.** -1.5031 **19.** -0.3990

21. 759 **23.** 0.00759 **25.** 1,430 **27.** 0.00000447 **29.** 0.0000000918 **31.** 10^{10} **33.** 10^{-10} **35.** 10^{20} **37.** $\frac{1}{100}$ **39.** 1,000

41. $\frac{1}{e}$ **43.** 25 **45.** $\frac{1}{8}$ **47.** 1 **49.** 5 **51.** x **53.** 4 **55.** -3 **57.** $\frac{3}{2}$ **59.** $\ln 10 + 3t$ **61.** $\ln A - 2t$ **63.** $2 + 3t\log 1.01$

65. $rt + \ln P$ **67.** $3 - \log 4.2$ **69.** 2.7080 **71.** -1.0986 **73.** 2.1972 **75.** 2.7724

77. San Francisco was approx. 2,000 times greater. **79.** e **81.** 2009 **83.** 3.19

x	$(1 + x)^{1/x}$
1	2
0.5	2.25
0.1	2.5937
0.01	2.7048
0.001	2.7169
0.0001	2.7181
0.00001	2.7183

85. 1.78×10^{-5} **87.** 3.16×10^{5} **89.** 2.00×10^{8} **91.**

Location	Date	Magnitude M	Shockwave T
Moresby Island	Jan. 23	4.0	1.00×10^{4}
Vancouver Island	Apr. 30	5.3	1.99×10^{5}
Quebec City	June 29	3.2	1.58×10^{3}
Mould Bay	Nov. 13	5.2	1.58×10^{5}
St. Lawrence	Dec. 14	3.7	5.01×10^{3}

SOURCE: *National Resources Canada, National Earthquake Hazards Program.*

93. 12.9%   **95.** 5.3%   **97.** $\frac{7}{10}$ **99.** 3.1250   **101.** 1.2575   **103.** $t\log 1.05$ **105.** $.05t$

PROBLEM SET 8.6

1. 1.4650   **3.** 0.6826   **5.** -1.5440 **7.** -0.6477 **9.** $-\frac{1}{3}$ **11.** $x = 2$ **13.** -0.1845 **15.** 0.1845   **17.** 1.6168   **19.** 2.1131

21. -1 **23.** 1.2275   **25.** 0.3054   **27.** 42.5528   **29.** 6.0147   **31.** 1.333   **33.** 0.7500   **35.** 1.3917   **37.** 0.7186   **39.** 2.6356

41. 4.1632   **43.** 5.8435   **45.** -1.0642 **47.** 2.3026   **49.** 10.7144   **51.** 11.72 years   **53.** 9.25 years   **55.** 8.75 years

57. 18.58 years   **59.** 11.55 years   **61.** 18.31 years   **63.** 11.45 years   **65.** October 2028   **67.** 2009   **69.** 1992   **71.** 10.07 years

73. 2000   **75. a.** 15.625 grams   **b.** 520 minutes   **77.** $(-2, -23)$, lowest   **79.** $\left(\frac{3}{2}, 9\right)$, highest   **81.** 2 seconds, 64 feet

83. $t = \frac{1}{r} \ln\frac{A}{P}$ **85.** $t = -\frac{1}{k} \frac{\log P - \log A}{\log 2}$ **87.** $t = \frac{\log A - \log P}{\log(1 - r)}$ **89.** 3   **91.** $\frac{1}{4}$ **93.** $\frac{8}{15}$

CHAPTER 8 REVIEW

1. $\frac{575}{9}$ **2.** $-\frac{2}{27}$ **3.**

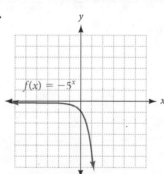

4.

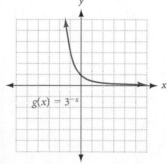

5. $f^{-1}(x) = \frac{x + 5}{2}$ **6.** $f^{-1}(x) = \sqrt[3]{x + 2}$ **7.** $f^{-1}(x) = -2x + 8$ **8.** $f^{-1}(x) = \frac{2 - x}{5x - 3}$

9. $f^{-1}(x) = 3x - 4$ **10.** $f^{-1}(x) = \pm\sqrt{x + 2}$

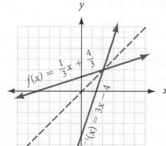

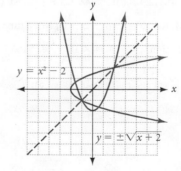

11. $\log_4 \frac{1}{16} = -2$ **12.** $\log_3 81 = 4$ **13.** $3^{-4} = \frac{1}{81}$ **14.** 64   **15.** 2   **16.** 3   **17.** $\frac{1}{2}$

18.

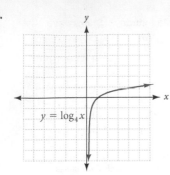

19.

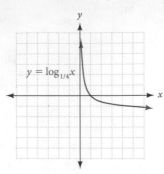

20. 1   **21.** $\frac{3}{2}$ **22.** -4 **23.** -1 **24.** $-\frac{1}{2}$ **25.** 0   **26.** $\frac{1}{2}$ **27.** 1.53   **28.** 4.76   **29.** -1.29 **30.** 3.31   **31.** -3.73

32. $3 \log_7 y$ **33.** $\frac{1}{2}\log_8 z$ **34.** $2 \log_{10} x + 4 \log_{10} y$ **35.** $\log_{10} 5 - \log_{10} 4 - \log_{10} y$ **36.** $\log_3 5 + 3 \log_3 x - \log_3 y$

37. $\frac{1}{2}\log x - 5 \log y - \frac{1}{3}\log z$ **38.** $\log_4 y^3 \sqrt{x}$ **39.** $\log \frac{x^3 \sqrt{z}}{y^4}$ **40.** $\log_{10}\left(\frac{\sqrt[3]{x}}{\sqrt[4]{y}}\right)$ **41.** $\log_3\left(\frac{x^2 y^3}{z}\right)$ **42.** 4.6294   **43.** -0.3706

44. 3.6902   **45.** 2.9542   **46.** 449.987   **47.** 0.056   **48.** 2.322   **49.** 3   **50.** 1   **51.** e^4 **52.** 32   **53.** 4, -2 does not check

54. 100   **55.** 5   **56.** -3 **57.** $\frac{1}{2}$ **58.** $-\frac{4}{5}$ **59.** -12 **60.** $\frac{3}{2}$ **61.** 0.4421   **62.** 5.03   **63.** \$606.89   **64.** 15.6 years   **65.** No

66. Yes   **67.** Yes   **68.** Yes   **69.** No   **70.** No

CHAPTER 8 CUMULATIVE REVIEW

1. 85   **2.** $-2x + 18$ **3.** 24   **4.** 4   **5.** $16\sqrt{2}$ **6.** $19 + 7\sqrt{15}$ **7.** $1 + 9i$ **8.** i **9.** 0   **10.** 0   **11.** $\frac{x - 3}{x - 1}$ **12.** $\frac{x^2 - x + 1}{1 - x}$

13. $\frac{1}{2}$ **14.** $\frac{9}{7}$ **15.** $20t^4 - \frac{1}{20}$ **16.** $5x^3 - 22x^2 + 9$ **17.** $x - 5 - \frac{2}{x - 1}$ **18.** $4x + 5 - \frac{13}{3x + 2}$ **19.** $\frac{1}{(x - 3)(x - 2)}$

20. $\frac{8x - 2}{(3x - 4)(2x + 1)(2x - 1)}$ **21.** $-\frac{2}{5}$ **22.** 2   **23.** $-\frac{1}{4}, \frac{3}{8}$ **24.** 2   **25.** $\frac{5 \pm \sqrt{53}}{2}$ **26.** $-\frac{10}{7}$ **27.** 25   **28.** $-\frac{7}{6}, \frac{1}{3}$ **29.** 256   **30.** 5

31. $\varnothing$ **32.** $\frac{7 + \sqrt{61}}{2}$ **33.** $(1, 4)$ **34.** $(6, 0)$ **35.** $\left(0, -4, \frac{4}{3}\right)$ **36.** $\left(-\frac{3}{2}, 0, 1\right)$

37.

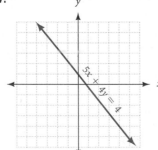

38.

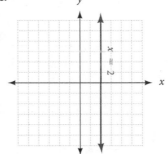

39. $8 - y = -3x$ **40.** $2a - 3b < 2a + 3b$ **41.** 1.29×10^{-6} **42.** 8.3×10^7 **43.** $(4x + 3)(16x^2 - 12x + 9)$

44. $6(4a^2 + 1)(2a^2 + 1)$ **45.** Domain $= \{3, 5\}$; Range $= \{-4, -2, -1\}$; Not a function

46. Domain $= \{1, 2, 3\}$; Range $= \{1, 2\}$; A function   **47.** $\$4.34 \times 10^8$ **48.** \$265.7 million   **49.** 9.6 mph

CHAPTER 8 TEST

1.

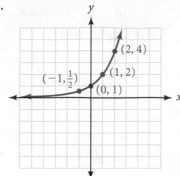

2.

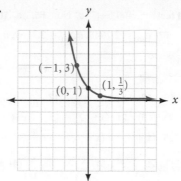

3.

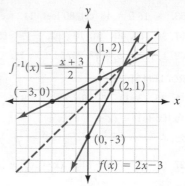

4.

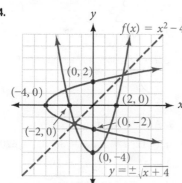

5. 64   **6.** $\sqrt{5}$

7.

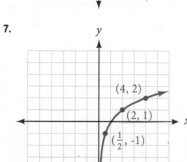

8.

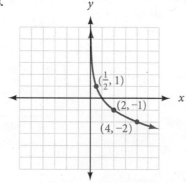

9. $\frac{2}{3}$ **10.** 1.5646   **11.** 4.3692   **12.** −1.9101   **13.** 3.8330   **14.** −3.0748   **15.** $3 + 2\log_2 x - \log_2 y$

16. $\frac{1}{2}\log x - 4\log y - \frac{1}{5}\log z$ **17.** $\log_3 \frac{x^2}{\sqrt{y}}$ **18.** $\log_3 \frac{\sqrt[3]{x}}{yz^2}$ **19.** 70,404   **20.** 0.00225   **21.** 1.4650   **22.** $\frac{5}{4}$ **23.** 15

24. 8 (−1 does not check)   **25.** 6.18   **26.** $651.56   **27.** 13.87 years   **28.** $7,373   **29.** Yes   **30.** No   **31.** Yes   **32.** No

Chapter 9

PROBLEM SET 9.1

1. 4, 7, 10, 13, 16   **3.** 3, 7, 11, 15, 19   **5.** 1, 2, 3, 4, 5   **7.** 4, 7, 12, 19, 28   **9.** $\frac{1}{4}, \frac{2}{5}, \frac{3}{6}, \frac{4}{7}, \frac{5}{8}$ **11.** $1, \frac{1}{4}, \frac{1}{9}, \frac{1}{16}, \frac{1}{25}$

13. 2, 4, 8, 16, 32   **15.** $2, \frac{3}{2}, \frac{4}{3}, \frac{5}{4}, \frac{6}{5}$ **17.** −2, 4, −8, 16, −32   **19.** 3, 5, 3, 5, 3   **21.** $1, -\frac{2}{3}, \frac{3}{5}, -\frac{4}{7}, \frac{5}{9}$ **23.** $\frac{1}{2}, 1, \frac{9}{8}, 1, \frac{25}{32}$

25. 3, −9, 27, −81, 243   **27.** 1, 5, 13, 29, 61   **29.** 2, 3, 5, 9, 17   **31.** 5, 11, 29, 83, 245   **33.** 4, 4, 4, 4, 4   **35.** $a_n = 4n$

37. $a_n = n^2$ **39.** $a_n = 2^{n+1}$ **41.** $a_n = \frac{1}{2^{n+1}}$ **43.** $a_n = 3n + 2$ **45.** $a_n = -4n + 2$ **47.** $a_n = (-2)^{n-1}$

49. $a_n = \log_{n+1}(n+2)$ **51.** **a.** \$28,000, \$29,120, \$30,284.80, \$31,496.19, \$32,756.04   **b.** $a_n = 28,000(1.04)^{n-1}$

53. **a.** 16 feet, 48 feet, 80 feet, 112 feet, 144 feet   **b.** 400 feet   **c.** No   **55.** **a.** 10, 8, $\frac{32}{5}$ **b.** $a_n = 10\left(\frac{4}{5}\right)^{n-1}$ **c.** 1.34 feet

57. 30   **59.** 40   **61.** 18   **63.** $-\frac{20}{81}$ **65.** $\frac{21}{10}$ **67.** $a_{100} = 2.7048,\ a_{1000} = 2.7169,\ a_{10,000} = 2.7181,\ a_{100,000} = 2.7183$

69. 1, 1, 2, 3, 5, 8, 13, 21, 34, 55   **71.** $\frac{3}{2}, \frac{5}{3}, \frac{8}{5}$

PROBLEM SET 9.2

1. 36   **3.** 11   **5.** 18   **7.** $\frac{163}{60}$ **9.** 60   **11.** 40   **13.** 44   **15.** $-\frac{11}{32}$ **17.** $\frac{21}{10}$

19. $(x + 1) + (x + 2) + (x + 3) + (x + 4) + (x + 5)$ **21.** $(x - 2) + (x - 2)^2 + (x - 2)^3 + (x - 2)^4$

23. $\frac{x+1}{x-1} + \frac{x+2}{x-1} + \frac{x+3}{x-1} + \frac{x+4}{x-1} + \frac{x+5}{x-1}$ **25.** $(x + 3)^3 + (x + 4)^4 + (x + 5)^5 + (x + 6)^6 + (x + 7)^7 + (x + 8)^8$

27. $(x - 6)^6 + (x - 8)^7 + (x - 10)^8 + (x - 12)^9$ **29.** $\sum\limits_{i=1}^{4} 2^i$ **31.** $\sum\limits_{i=2}^{6} 2^i$ **33.** $\sum\limits_{i=1}^{5}(4i + 1)$ **35.** $\sum\limits_{i=2}^{5} -(-2)^i$ **37.** $\sum\limits_{i=3}^{7} \frac{1}{i+1}$

39. $\sum\limits_{i=1}^{4} \frac{i}{2i+1}$ **41.** $\sum\limits_{i=6}^{9}(x - 2)^i$ **43.** $\sum\limits_{i=1}^{4}\left(1 + \frac{i}{x}\right)^{i+1}$ **45.** $\sum\limits_{i=3}^{5} \frac{x}{x+i}$ **47.** $\sum\limits_{i=2}^{4} x^i(x + i)$

49. a. $0.3 + 0.03 + 0.003 + 0.0003 + \ldots$ **b.** $0.2 + 0.02 + 0.002 + 0.0002 + \ldots$ **c.** $0.27 + 0.0027 + 0.000027 + \ldots$

51. seventh second: 208 feet; total: 784 feet   **53.** **a.** $16 + 48 + 80 + 112 + 144$ **b.** $\sum\limits_{i=1}^{5}(32i - 16)$ **55.** 437 seats   **57.** 5

59. 380   **61.** $3n + 4$ **63.** $(5, 3)$ **65.** $x = -\frac{1}{4}$ **67.** $-\frac{83}{576}$ **69.** 672,400

PROBLEM SET 9.3

1. Arithmetic, $d = 1$ **3.** Not arithmetic   **5.** Arithmetic, $d = -5$ **7.** Not arithmetic   **9.** Arithmetic, $d = \frac{2}{3}$

11. $a_n = 4n - 1; a_{24} = 95$ **13.** $a_{10} = -12; S_{10} = -30$ **15.** $a_1 = 7; d = 2; a_{30} = 65$ **17.** $a_1 = 12; d = 2; a_{20} = 50; S_{20} = 620$

19. $a_{20} = 79, S_{20} = 820$ **21.** $a_{40} = 122, S_{40} = 2540$ **23.** $a_1 = 13, d = -6$ **25.** $a_{85} = -238$ **27.** $d = \frac{16}{19}, a_{39} = 28$

29. 20,300   **31.** -158 **33.** $a_{10} = 5 ; S_{10} = \frac{55}{2}$

35. a. \$18,000, \$14,700, \$11,400, \$8,100, \$4,800   **b.** $-\$3,300$ **c.**
d. \$9,750   **e.** $a_0 = 18,000, a_n = a_{n-1} - 3,300$ for $n \geq 1$

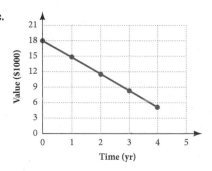

37. a. 1,500, 1,460, 1,420, 1,380, 1,340, 1,300
  b. It is arithmetic because the same amount is subtracted from each succeeding term.
  c. $a_n = 1,500 - (n - 1)40 = 1,540 - 40n$

39. a. 1, 3, 6, 10, 15, 21, 28, 36, 45, 55, 66, 78, 91, 105, 120   **b.** $a_1 = 1; a_n = n + a_{n-1}$ for $n \geq 2$
  c. No, it is not arithmetic because the same amount is not added to each term.

41. a. $a_n = 32n - 16$ **b.** $a_{10} = 304$ feet   **c.** 1600 feet   **43.** $\frac{1}{16}$ **45.** $\sqrt{3}$ **47.** 2^n **49.** r^3 **51.** $\frac{2}{5}$ **53.** -255

55. $a_1 = 9, d = 4$ **57.** $d = -3$ **59.** $50 + 1225\sqrt{2}$

PROBLEM SET 9.4

1. 5   **3.** $\frac{1}{3}$ **5.** Not geometric   **7.** -2 **9.** Not geometric   **11.** $a_n = 4 \cdot 3^{n-1}$ **13.** $a_6 = -2\left(-\frac{1}{2}\right)^5 = \frac{1}{16}$

15. $a_{20} = 3(-1)^{19} = -3$ **17.** 10,230   **19.** $S_{20} = \frac{1((-1)^{20} - 1)}{-1 - 1} = 0$ **21.** $a_8 = \frac{1}{5}\left(\frac{1}{2}\right)^7 = \frac{1}{640}$ **23.** $-\frac{31}{32}$ **25.** 32, 62 + 31\sqrt{2}$

27. $\frac{1}{1000}$, 111.111   **29.** $r = \pm 2$ **31.** $a_8 = 384, S_8 = 255$ **33.** $r = 2$ **35.** $S = \frac{-\frac{1}{2}}{1 - \frac{1}{2}} = 1$ **37.** 8   **39.** 4   **41.** $\frac{8}{9}$

43. $S = \frac{\frac{2}{5}}{1 - \frac{2}{5}} = \frac{2}{3}$ **45.** $S = \frac{\frac{3}{4}}{1 - \frac{1}{3}} = \frac{9}{8}$ **47.–49.** Left as an exercise for the reader

51. a. \$450,000, \$315,000, \$220,500, \$154,350, \$108,045   **b.** 0.7   **c.**
d. \$90,000   **e.** $a_0 = 450,000$, $a_n = 0.7a_{n-1}$

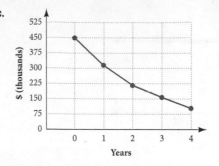

53. a. $\frac{1}{2}$ **b.** $\frac{364}{729}$ **c.** $\frac{1}{1,458}$

55. a. 0.004 inches   **b.** 0.064 inches   **c.** 67,108.864 inches

57. a. \$60,000; \$64,200; \$64,896; \$67,492; \$70,192;
b. $a_n = 60,000(1.04)^{n-1}$ **c.** \$720,367

59. $x + y$ **61.** $x^3 + 3x^2y + 3xy^2 + y^3$ **63.** 21   **65.** $\frac{41}{333}$

67. $r = -\frac{1}{5}$ **69.** $S_{12} = 728(1 + \sqrt{3})$ **71.** $\frac{1}{1 + x^2}$ **73.** $-\frac{11}{16}$

PROBLEM SET 9.5

1. $x^4 + 8x^3 + 24x^2 + 32x + 16$ **3.** $x^6 + 6x^5y + 15x^4y^2 + 20x^3y^3 + 15x^2y^4 + 6xy^5 + y^6$

5. $32x^5 + 80x^4 + 80x^3 + 40x^2 + 10x + 1$ **7.** $x^5 - 10x^4y + 40x^3y^2 - 80x^2y^3 + 80xy^4 - 32y^5$

9. $81x^4 - 216x^3 + 216x^2 - 96x + 16$ **11.** $64x^3 - 144x^2y + 108xy^2 - 27y^3$ **13.** $x^8 + 8x^6 + 24x^4 + 32x^2 + 16$

15. $x^6 + 3x^4y^2 + 3x^2y^4 + y^6$ **17.** $16x^4 + 96x^3y + 216x^2y^2 + 216xy^3 + 81y^4$ **19.** $\frac{x^3}{8} + \frac{x^2y}{4} + \frac{xy^2}{6} + \frac{y^3}{27}$

21. $\frac{x^3}{8} - 3x^2 + 24x - 64$ **23.** $\frac{x^4}{81} + \frac{2x^3y}{27} + \frac{x^2y^2}{6} + \frac{xy^3}{6} + \frac{y^4}{16}$ **25.** $x^9 + 18x^8 + 144x^7 + 672x^6$

27. $x^{10} - 10x^9y + 45x^8y^2 - 120x^7y^3$ **29.** $x^{25} + 75x^{24} + 2,700x^{23} + 62,100x^{22}$ **31.** $x^{60} - 120x^{59} + 7,080x^{58} - 273,760x^{57}$

33. $x^{18} - 18x^{17}y + 153x^{16}y^2 - 816x^{15}y^3$ **35.** $x^{15} + 15x^{14} + 105x^{13}$ **37.** $x^{12} - 12x^{11}y + 66x^{10}y^2$ **39.** $x^{20} + 40x^{19} + 760x^{18}$

41. $x^{100} + 200x^{99}$ **43.** $x^{50} + 50x^{49}y$ **45.** $51,963,120x^4y^8$ **47.** $3,360x^6$ **49.** $-25,344x^7$ **51.** $2,700x^{23}y^2$

53. $\binom{20}{11}(2x)^9(5y)^{11} = \frac{20!}{11!9!}(2x)^9(5y)^{11}$ **55.** $x^{20}y^{10} - 30x^{18}y^9 + 405x^{16}y^8$ **57.** $\frac{21}{128}$ **59.** $x \approx 1.21$ **61.** $\frac{7}{6}$ **63.** ≈ 69.4 years

65. 2.16   **67.** 6.36   **69.** $t = \frac{1}{5}\ln\left(\frac{A}{10}\right)$ **71.** 56   **73.** 125, 970   **75.** Left as an exercise for the reader   **77.** $3240x^7y^3$

79. 1,313,400   **81.** $n = 26$

CHAPTER 9 REVIEW

1. 8, 14, 20, 26, 32   **2.** −3, 0, 5, 12, 21   **3.** −5, 2, 21, 58, 119   **4.** $\frac{1}{2}, \frac{4}{7}, \frac{5}{8}, \frac{2}{3}, \frac{7}{10}$ **5.** $1, \frac{1}{8}, \frac{1}{27}, \frac{1}{64}, \frac{1}{125}$ **6.** $0, \frac{3}{2}, \frac{8}{3}, \frac{15}{4}, \frac{24}{5}$

7. 5, −16, 47, −142, 425   **8.** 5, 12, 27, 58, 121   **9.** 3, 0, −3, −6, −9   **10.** −4, 8, −16, 32, −64   **11.** $a_n = 2n + 1$

12. $a_n = (-3)^{n-1}$ **13.** $a_n = \frac{1}{(-3)(-2)^n}$ **14.** $a_n = 3n^2$ **15.** $a_n = -2 \cdot (-1)^{n+1}$ **16.** $a_n = \frac{n^2 - 1}{n^2 + 1}$ **17.** 32   **18.** 21

19. $\frac{3}{64}$ **20.** $\frac{81}{20}$ **21.** $5x + 15$ **22.** $(x + 1)^2 + (x + 1)^3 + (x + 1)^4 + (x + 1)^5 + (x + 1)^6 + (x + 1)^7$

23. $\left(x + \frac{1}{2}\right)^2 + \left(x + \frac{1}{3}\right)^2 + \left(x + \frac{1}{4}\right)^2 + \left(x + \frac{1}{5}\right)^2$ **24.** $(x + 1)^2 + (x + 2)^3 + (x + 3)^4 + (x + 4)^5 + (x + 5)^6$

25. $\sum_{i=1}^{4}(6i + 3)$ **26.** $\sum_{i=1}^{6}\frac{i+3}{i+2}$ **27.** $\sum_{i=1}^{4}(x + 3i)^{i+5}$ **28.** $\sum_{i=1}^{4}\frac{x-i-2}{x+2i+2}$ **29.** $a_n = 10n - 5; a_{100} = 995$

30. $a_{24} = -16; S_{24} = -108$ **31.** $a_1 = 5$ **32.** 31   **33.** 24   **34.** $a_1 = 3; d = 5; a_{40} = 198$

35. $a_1 = 2, d = 7, a_{50} = 345, S_{50} = 8,675$ **36.** $d = -\frac{3}{23}, a_{116} = -11$ **37.** −195   **38.** −12   **39.** $S_{50} = \frac{3(5^{50} - 1)}{4}$ **40.** $\frac{1}{3}$

41. 16   **42.** $\frac{9}{2}$ **43.** $\frac{18}{5}$ **44.** $\frac{25}{12}$ **45.** Arithmetic, $d = 3$ **46.** Geometric, $r = \frac{1}{2}$ **47.** Arithmetic, $d = 6$

48. Geometric, $r = -5$ **49.** Arithmetic, $d = 5$ **50.** Neither   **51.** Geometric, $r = -\frac{1}{3}$ **52.** Arithmetic, $d = -10$

53. $x^4 + 8x^3 + 24x^2 + 32x + 16$ **54.** $16x^4 - 32x^3y + 24x^2y^2 - 8xy^3 + y^4$ **55.** $x^8 - 12x^6y + 54x^4y^2 - 108x^2y^3 + 81y^4$

56. $\frac{x^4}{81} - \frac{2x^3y}{27} + \frac{x^2y^2}{6} - \frac{xy^3}{6} + \frac{y^4}{16}$ **57.** $x^{11} - 33x^{10} + 495x^9$ **58.** $x^9 - 18x^8 + 144x^7 - 672x^6$

59. $x^{10} + 20x^9y + 180x^8y^2 + 960x^7y^3$ **60.** $29,568x^5y^6$

CHAPTER 9 CUMULATIVE REVIEW

1. 4   **2.** $30y + 49$ **3.** $\frac{a^5b^4}{2}$ **4.** $\frac{y-3}{y-2}$ **5.** $\frac{1}{4}$ **6.** 3   **7.** $(a + 1)(b^3 + 6)$ **8.** $(x - 1)(8x + 3)$ **9.** −2   **10.** $1, \frac{1}{2}$

11. −2, 5   **12.** −2, 4   **13.** 8   **14.** $\frac{3 \pm 5i\sqrt{2}}{4}$ **15.** $\frac{3}{2}, \frac{-3 \pm 3i\sqrt{3}}{4}$ **16.** $\pm\frac{\sqrt{10}}{2}$ and $\pm\frac{\sqrt{3}}{3i}$

17. $y > 3$

18. $-8 < x < 3$

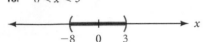

19.

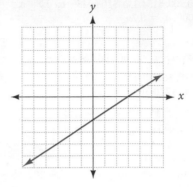

20.

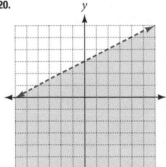

21.

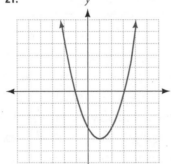

22.

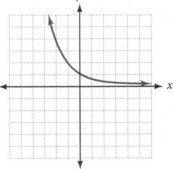

23. $(1, 3)$ **24.** $(4, -2, 3)$ **25.** $\dfrac{y(y - 1)}{y + 1}$ **26.** $x^{6/5} - 4$ **27.** $14 - 5i$ **28.** $\dfrac{1}{(x + 4)(x + 2)}$ **29.** $26\sqrt{2}$

30. $\dfrac{3\sqrt{7} + 3\sqrt{3}}{4}$ **31.** $f^{-1}(x) = \dfrac{x - 1}{4}$ **32.** $9{,}519$ **33.** 1.47 **34.** $a_n = 11_n - 8$ **35.** $a_n = 16\left(\dfrac{1}{2}\right)^{n-1}$ **36.** $y = \dfrac{S - 2x^2}{4x}$

37. 25 **38.** 0 **39.** $y = -2x + 9$ **40.** $C(5) = 40, C(10) = 20$ **41.** $20t^2 + 11t - 3 = 0$ **42.** $(f \circ g)(x) = 5 - 3x$

43. $\dfrac{13}{22}, -\dfrac{1}{22}$ **44.** $32x^5$ **45.** -19 **46.** $\left(x - \dfrac{1}{2}\right)\left(x^2 + \dfrac{1}{2}x + \dfrac{1}{4}\right)$ **47.** $30°, 150°$ **48.** 6 **49.** $1, 3$

50. 4 gallons of 30%, 12 gallons of 70%   **51.** 19.2 mph   **52.**

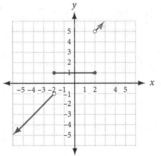

CHAPTER 9 TEST

1. $-2, 1, 4, 7, 10$ **2.** $3, 7, 11, 15, 19$ **3.** $2, 5, 10, 17, 26$ **4.** $2, 16, 54, 128, 250$ **5.** $2, \dfrac{3}{4}, \dfrac{4}{9}, \dfrac{5}{16}, \dfrac{6}{25}$ **6.** $-8, 16, -32, 64, -128$

7. $a_n = a_{n-1} + 4, a_1 = 6$ **8.** $a_n = 2a_{n-1}, a_1 = 1$ **9.** $a_n = \dfrac{1}{2}a_{n-1}, a_1 = \dfrac{1}{2}$ **10.** $a_n = -3a_{n-1}, a_1 = -3$

11. a. 90 **b.** 53 **c.** 130 **12.** 3 **13.** 6 **14.** 320 **15.** 25 **16.** $S_{50} = \dfrac{3(2^{50} - 1)}{2 - 1}$ **17.** $\dfrac{3}{4}$ **18.** $x^4 - 12x^3 + 54x^2 - 108x + 81$

19. $32x^5 - 80x^4 + 80x^3 - 40x^2 + 10x - 1$ **20.** $x^{20} - 20x^{19} + 190x^{18}$ **21.** $-108{,}864x^3y^5$

Chapter 10

PROBLEM SET 10.1

1. 5 **3.** $\sqrt{106}$ **5.** $\sqrt{61}$ **7.** $\sqrt{130}$ **9.** 3 or -1 **11.** 0 or 6   **13.** $x = -1 \pm 4\sqrt{2}$ **15.** $(x - 3)^2 + (y + 2)^2 = 9$

17. $(x + 5)^2 + (y + 1)^2 = 5$ **19.** $x^2 + (y + 5)^2 = 1$ **21.** $x^2 + y^2 = 4$

23. center = $(0, 0)$; radius = 2

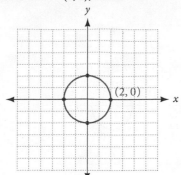

25. center = $(1, 3)$; radius = 5

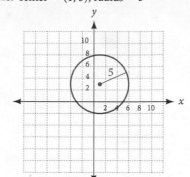

27. center = $(-2, 4)$; radius = $2\sqrt{2}$

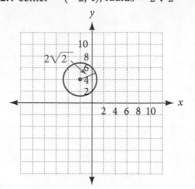

29. center = $(-2, 4)$; radius = $\sqrt{17}$

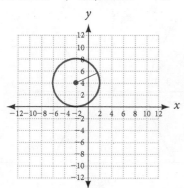

31. center = $(-1, 2)$; radius = 3

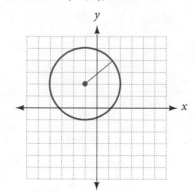

33. center = $(0, 3)$; radius = 4

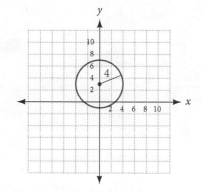

35. center = $(-1, 0)$; radius = $\sqrt{2}$

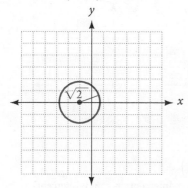

37. center = $(2, 3)$; radius = 3

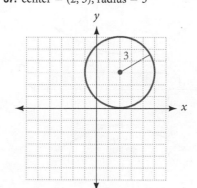

39. center = $\left(-1, -\frac{1}{2}\right)$; radius = 2

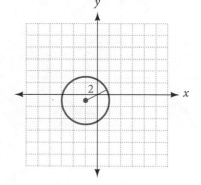

41. center = $\left(\frac{1}{2}, -1\right)$; radius = 2

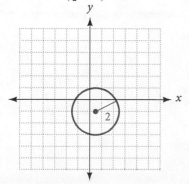

43. $(x - 3)^2 + (y - 4)^2 = 25$

45. a. $\left(x - \frac{1}{2}\right)^2 + (y - 1)^2 = \frac{1}{4}$ b. $(x - 1)^2 + (y - 1)^2 = 1$
  c. $(x - 2)^2 + (y - 1)^2 = 4$

47. $x^2 + y^2 = 25$ **49.** $x^2 + y^2 = 9$ **51.** $(x + 1)^2 + (y - 3)^2 = 25$

53. $(x + 2)^2 + (y - 5)^2 = 73$ **55.** $x^2 + (y - 2)^2 = 16$ **57.** $C = 6\pi\sqrt{2}, A = 18\pi$

59. $C = 10\pi, A = 25\pi$ **61.** Yes **63.** $(x - 500)^2 + (y - 132)^2 = 120^2$ **65.** $y = \pm 3$

67. $y = \pm 2i$ **69.** $x = \pm 3i$ **71.** $\frac{x^2}{9} + \frac{y^2}{4}$ **73.** x-intercept 4, y-intercept -3 **75.** $\pm\frac{12}{5}$

77. $(x - 2)^2 + (y - 3)^2 = 4$ **79.** $(x - 2)^2 + (y - 3)^2 = 4$ **81.** 5 **83.** 5

PROBLEM SET 10.2

1.

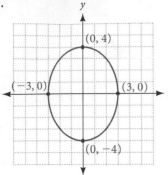

3.

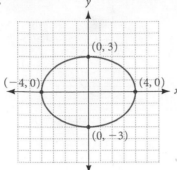

5.

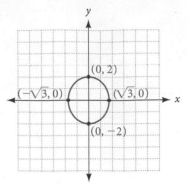

7.

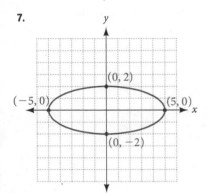

9.

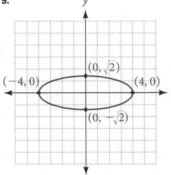

11.

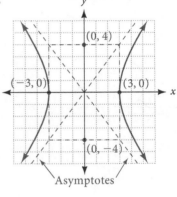

Asymptotes

13.

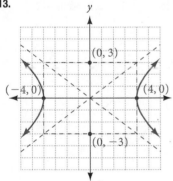

15.

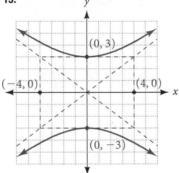

17.

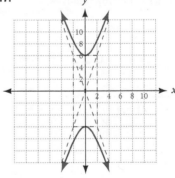

19.

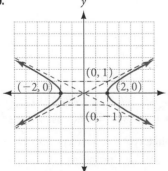

21.

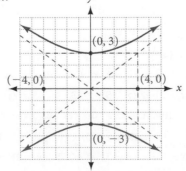

23. x-intercepts $= \pm 3$, y-intercepts $= \pm 2$

25. x-intercepts $= \pm 0.2$, no y-intercepts

27. x-intercepts $= \pm \frac{3}{5}$, y-intercepts $= \pm \frac{2}{5}$

29.

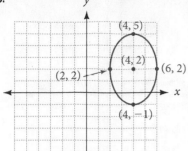

31.

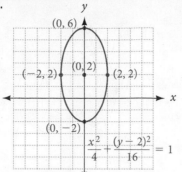

$$\frac{x^2}{4} + \frac{(y-2)^2}{16} = 1$$

33.

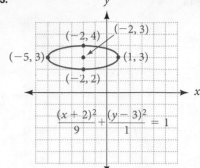

$$\frac{(x+2)^2}{9} + \frac{(y-3)^2}{1} = 1$$

35.

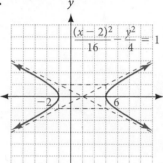

$$\frac{(x-2)^2}{16} - \frac{y^2}{4} = 1$$

37.

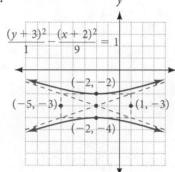

$$\frac{(y+3)^2}{1} - \frac{(x+2)^2}{9} = 1$$

39.

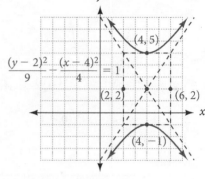

$$\frac{(y-2)^2}{9} - \frac{(x-4)^2}{4} = 1$$

41. $x = 0$ **43.** $y = 0$ **45.** 8 **47.** $\frac{x^2}{307.5^2} + \frac{y^2}{255^2} = 1$ **49.** $\frac{x^2}{229^2} + \frac{y^2}{195^2} = 1$ **51.** 5.3 feet **53.** $(0, 0)$ **55.** $4y^2 + 16y + 16$

57. $x = 2y + 4$ **59.** $-x^2 + 6$ **61.** $(5y + 6)(y + 2)$ **63.** $y = \pm 2$ **65.** $x = \pm 2$ **67.** $y = x + 4$ **69.** $y = -\frac{3}{2}x - 4$

PROBLEM SET 10.3

1.

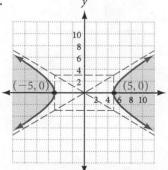

3.

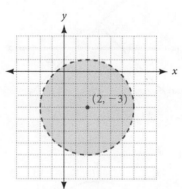

5.

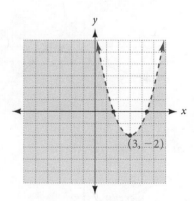

7.

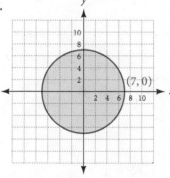

9.

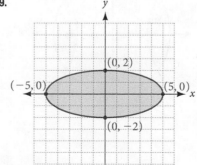

11.

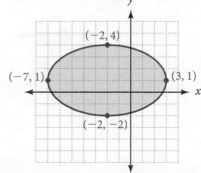

13.

15.

17.

19.

21.

23.

25.

27.

29.

31.

33.

35.

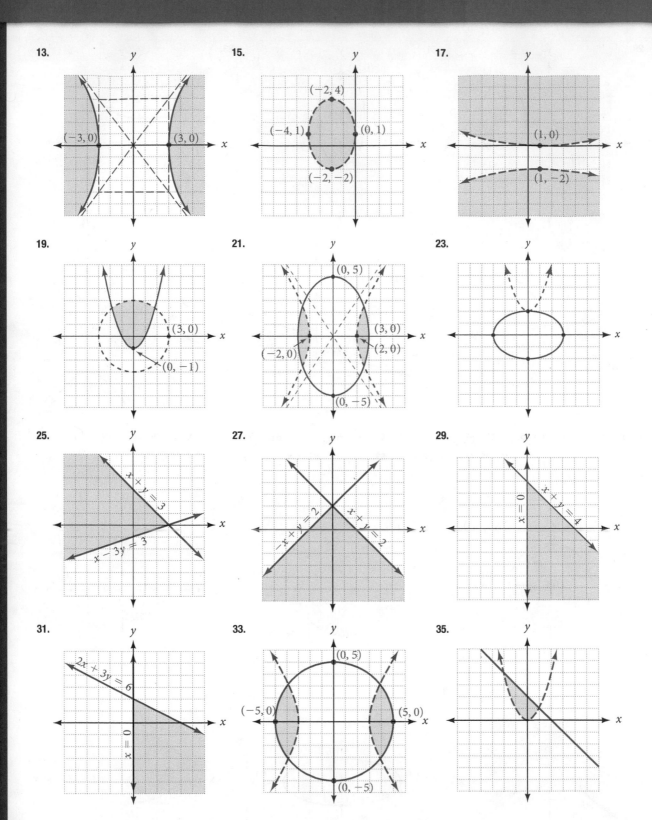

37. $(0, 3), \left(\frac{12}{5}, -\frac{9}{5}\right)$ **39.** $(0, 4), \left(\frac{16}{5}, \frac{12}{5}\right)$ **41.** $(5, 0), (-5, 0)$ **43.** $(0, -3), (\sqrt{5}, 2), (-\sqrt{5}, 2)$

45. $(0, -4), (\sqrt{7}, 3), (-\sqrt{7}, 3)$ **47.** $(-4, 11), \left(\frac{5}{2}, \frac{5}{4}\right)$ **49.** $(-4, 5), (1, 0)$ **51.** $(2, -3), (5, 0)$ **53.** $(3, 0), (-3, 0)$

55. $(4, 0), (0, -4)$ **57.** $\left(\frac{2\sqrt{6}}{3}, \frac{13}{3}\right), \left(-\frac{2\sqrt{6}}{3}, \frac{13}{3}\right)$ **59.** $(1, -2)$ **61.** $(\sqrt{10}, 0), (-\sqrt{10}, 0)$

63. $8, 5$ or $-8, -5$ or $8, -5$ or $-8, 5$ **65. a.** $(-4, 4\sqrt{3})$ and $(-4, -4\sqrt{3})$ **b.** $(4, 4\sqrt{3})$ and $(4, -4\sqrt{3})$

67. $x^4 + 8x^3 + 24x^2 + 32x + 16$ **69.** $8x^3 + 12x^2y + 6xy^2 + y^3$ **71.** $x^{50} + 150x^{49}$ **73.** $(-1, -9), (9, 1)$ **75.** $\left(\frac{17}{60}, -\frac{17}{6}\right)$

77. $(0, 0), (9, 81)$

CHAPTER 10 REVIEW

1. 13 **2.** 10 **3.** $2\sqrt{13}$ **4.** $\sqrt{146}$ **5.** $3, 5$ **6.** $-2 \pm \sqrt{5}$ **7.** $(x + 4)^2 + (y - 2)^2 = 36$ **8.** $x^2 + (y + 1)^2 = 49$

9. $x^2 + y^2 = 15^2$ **10.** $(x - 2)^2 + (y - 5)^2 = 10$ **11.** $(x - 4)^2 + (y + 1)^2 = 20$ **12.** $(x + 3)^2 + y^2 = 25$

13. $C = 4\pi\sqrt{3}, A = 12\pi$ **14.** $C = 8\pi, A = 16\pi$ **15.** Center $= (0, -5)$; Radius $= 3$ **16.** Center $= (3, -5)$; Radius $= 6$

17.
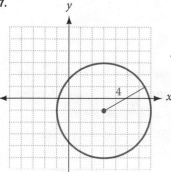

18.
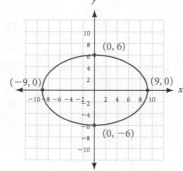

19.
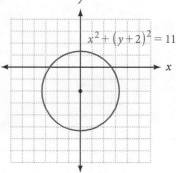

20.
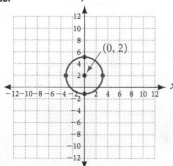

21.
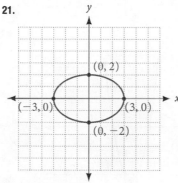

22.
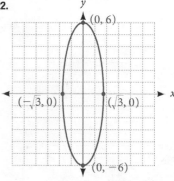

23.
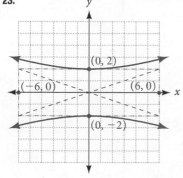

24.
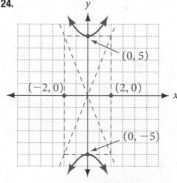

25.
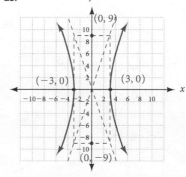

26.

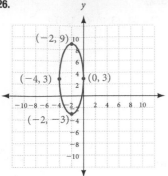

27.

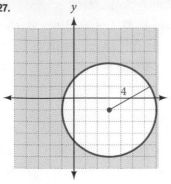

28.

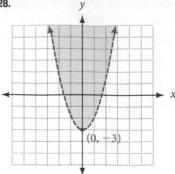

29. x-intercepts $= \pm 0.5$; y-intercepts $= \pm 0.4$;   **30.** x-intercepts $= \pm\frac{3}{4}$; y-intercepts $= \pm\frac{5}{4}$

31.

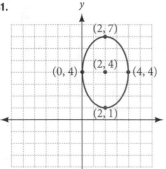

32.

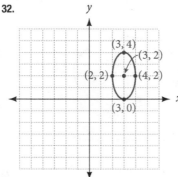

33. $y = 0$

34. $y = \pm\frac{12}{5}$

35. x-intercepts $= \pm 3$; no y-intercepts

36. x-intercepts $= \pm\frac{3}{4}$; no y-intercepts

37. $x = \pm\frac{5\sqrt{5}}{2}$ **38.** $y = \pm\frac{5}{4}x$

39. $(0, 5)$, $\left(\frac{25}{13}, -\frac{60}{13}\right)$

40. $\left(0, -\frac{5}{2}\right)$, $\left(2, \frac{3}{2}\right)$, $\left(-2, \frac{3}{2}\right)$

41.

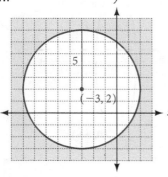

42.

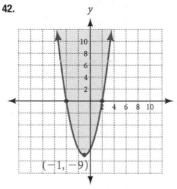

43.

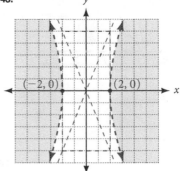

44.

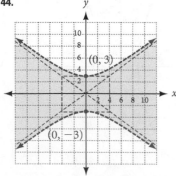

45.

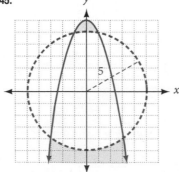

46. $(3, 0), \left(-\frac{9}{5}, \frac{12}{5}\right)$ **47.** $(\sqrt{3}, 1), (\sqrt{3}, -1), (-\sqrt{3}, 1), (-\sqrt{3}, -1)$ **48.** $(-5, 15), (3, -1)$ **49.** $(5, 0), (-5, 0)$

50.

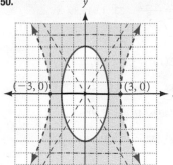

51.

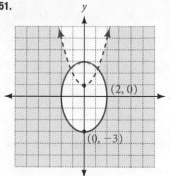

52.

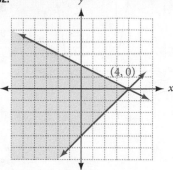

53.

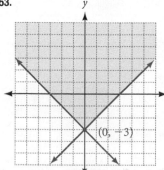

54.

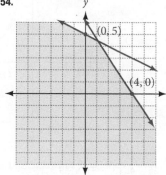

55.

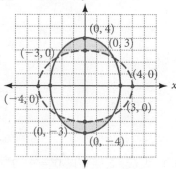

56. A   **57.** D   **58.** C   **59.** B   **60.** $(x + 3)^2 + (y - 2)^2 = 25$ **61.** $\frac{y^2}{25} - \frac{x^2}{9} = 1$ **62.** $\frac{(x - 3)^2}{36} + \frac{(y - 1)^2}{16} = 1$

CHAPTER 10 CUMULATIVE REVIEW

1. 8   **2.** 2   **3.** $\frac{a^7}{2b^{11}}$ **4.** $x^{3/2}y^{7/6}$ **5.** $6xy - 2y^2 + 3x^2$ **6.** $\frac{y + 4}{y - 3}$ **7.** 2   **8.** 3   **9.** $(b^3 + 4)(a - 1)$ **10.** $(7x + 1)(x - 3)$ **11.** -15

12. $\frac{1}{2}$ **13.** $-6, 3$ **14.** $-2, 5$ **15.** 11, 2 does not check   **16.** $\varnothing$ **17.** $\frac{-2 \pm 4i\sqrt{3}}{7}$ **18.** $6 \pm i\sqrt{5}$

19. $x < -4$ or $x > 0$

20. $-\frac{3}{2} < x < 5$

21.

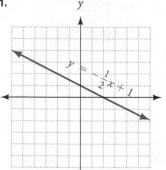

22.

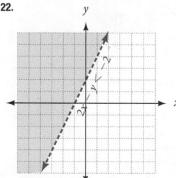

23.

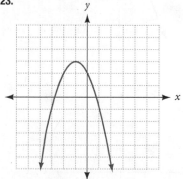

24.

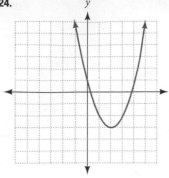

25. $\dfrac{y}{(y+1)(y-2)}$ **26.** $\dfrac{x-3}{x(x-1)}$ **27.** $13 + 13i$ **28.** $38 - 34i$ **29.** $\sqrt{5} + \sqrt{2}$

30. $\dfrac{20 - \sqrt{10}}{13}$ **31.** $f^{-1}(x) = 1 - 3x$ **32.** $f^{-1}(x) = \dfrac{x-2}{5}$ **33.** $1{,}645.89$

34. 0.91 **35.** $x = \dfrac{2mn - 5}{n}$ **36.** $y = \dfrac{S - 3x^2}{5x - 2}$ **37.** $\dfrac{2}{11}$

38. Slope $= \dfrac{5}{6}$; y-intercept $= -3$ **39.** -18 **40.** $C(15) = 30$; $C(30) = 15$

41. $2x^2 - 9x + 4 = 0$ **42.** $24t^2 + 29t - 4 = 0$ **43.** $-2x + 11$ **44.** $-3x^2 + 29$

45. -12 **46.** -11 **47.** 21 **48.** $\dfrac{32}{25}$ **49.** 5 gal of 40%; 45 gal of 90%

50. 2 gal of 10%; 6 gal of 70%   **51.** 178,660 people   **52.** 1,978,488 people

53. 38.4 mph   **54.**

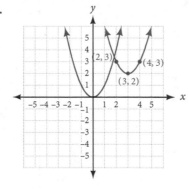

CHAPTER 10 TEST

1. $-5, 3$ **2.** $(x + 2)^2 + (y - 4)^2 = 9$ **3.** $x^2 + y^2 = 25$ **4.** $(5, -3), \sqrt{39}$

5.

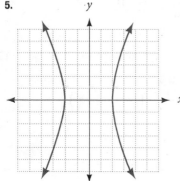

6.

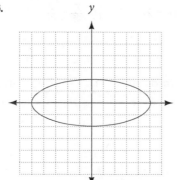

7.

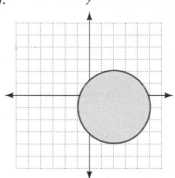

8.

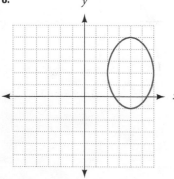

9. $(0, 5), (4, -3)$ **10.** $(0, -4), (-\sqrt{7}, 3), (\sqrt{7}, 3)$

Index